Environmental Ethics

Environmental Ethics
Concepts, Policy, and Theory

Joseph DesJardins
College of Saint Benedict
Saint Joseph, Minnesota

Mayfield Publishing Company
Mountain View, California
London • Toronto

Library of Congress Cataloging-in-Publication Data

DesJardins, Joseph.
    Environmental ethics : concepts, policy, and theory / Joseph
DesJardins.
        p.   cm.
    Includes index.
    ISBN 1–55934–986–7
    1. Environmental ethics.   I. Title.
GE42.D48   1998
079′.1—dc21   98-5946
                                                     CIP

Manufactured in the United States of America

10   9   8   7   6   5   4   3   2   1

Mayfield Publishing Company
1280 Villa Street
Mountain View, California 94041

Sponsoring editor, Ken King; *production editor,* Carla White Kirschenbaum; *manuscript editor,* Jennifer Gordon; *text and cover designer,* Linda Robertson; *design manager,* Jeanne Schreiber; *art manager,* Amy Folden; *manufacturing manager,* Randy Hurst. The text was set in 10/12 Berkeley Book by G&S Typesetters and printed on 45# Baycoat Velvet by Banta Book Group, Harrisonburg.

Cover image: © Carr Clifton/Minden Pictures

To Linda, Michael, and Matthew

Contents

Preface

This book is modeled on the structure of my courses in environmental ethics. After many years of experimenting, I have settled on an approach that works for me. I hope that it will work for you too. In any case, this book is flexible enough to support a variety of teaching styles and teaching goals.

Philosophical ethics is constantly challenged to balance theoretical issues with applied issues. Ethics, even philosophical ethics, should be engaged with the pressing value issues of day-to-day living. But if we remain exclusively at the level of practice, we risk failing to do philosophy, for philosophical ethics is also essentially abstract. With Socrates as a guide, philosophers have always recognized the critical importance of stepping back from everyday life—of abstracting themselves from it—to reflect on the nature and values of that life. If we fail to abstract ourselves from the practical issues of life, we cease to do philosophy; if we don't apply our philosophy to practice, we cease to do ethics.

So, too, in teaching environmental ethics we are challenged to balance theory and practice. We want students to think for themselves and to be capable of articulating and defending their own views on important matters of environmental policy. As philosophers, our role here is to ensure that students understand the conceptual and value implications of these policies. But we also want them to understand the need, and perhaps experience the joy, of abstract reasoning.

So how does this book balance these goals? I structure my own classes and this book around four general themes: context, concepts, policy, and theory. In class, I tell students that the goal of the course is to contribute to their development as educated citizens. This book provides some resources to help this development. Part I (Chapters 1 to 3) establishes a context for beginning the intellectual journey. Part II (Chapters 4 to 7) introduces students to the basic concepts and categories essential for understanding environmental policy issues. Part III (Chapters 8 to 14) focuses on the ethical and conceptual aspects of specific environmental policy issues. Part IV (Chapters 15 to 18) steps back from these policy issues to reflect more systematically and generally on environmental ethics.

The goal of Part I is to prepare students by providing a framework for understanding and analysis. Students need to know the fundamental language and concepts of ethics, so Chapter 1 provides an introduction to philosophical ethics, with an emphasis on public policy. I believe it is also important that students recognize the importance of historical antecedents to present thinking. Chapter 2 contributes this context by reviewing a variety of historical views on environmental concerns. Finally, Chapter 3 offers some general and differing interpretations of the present situation. Many students assume that we are currently in an environmental crisis. I have found it helpful, pedagogically and philosophically, to have students step back and examine that assumption at the very beginning. Chapter 3 aims to provide just this opportunity.

With this context, we turn to some concepts and categories that will prove indispensable for developing a mature understanding of any environmental issue. The assumption is that students will not be capable of truly understanding environmental controversies until they have some familiarity with these basic ideas. Economics plays an indisputable and dominant role in public policy making. Chapter 4 addresses a wide range of ethical and conceptual issues involved in economic analysis. But economic value is not the only value issue at stake in environmental matters. Chapter 5 introduces some important considerations of aesthetic and spiritual value. Chapters 6 and 7 address two of the best-known topics of environmental ethics: ethical extensionism and ethical holism. Chapter 6 examines the question of whether things other than human beings deserve ethical standing. The focus of this chapter is on the ethical status of individual living things. Chapter 7 expands the scope of this question and considers the implications of ecological thinking, particularly its emphasis on ecological systems and wholes, for environmental ethics.

At this point students should be equipped for an analysis of specific topics of environmental policy. Although the following seven chapters address policy issues, they do so from a philosophical and ethical perspective. The readings do not provide detailed empirical and scientific analyses. Rather, the emphasis remains on the ethical aspects of policy issues. In turn, we examine pollution (Chapter 8), ethics and animals (Chapter 9), land use (Chapter 10), wilderness preservation (Chapter 11), population and economic growth (Chapter 12), environmental justice (Chapter 13), and international relations (Chapter 14).

Only after having worked through specific issues and controversies do we step back again and consider environmental ethics in a more general and theoretical way. In my own teaching there have been times when I looked at more theoretical perspectives earlier in the semester. Almost always, this approach proved unsatisfactory. Students either had difficulty understanding how such theories as deep ecology or social ecology were relevant to their more practical environmental concerns, or they assumed that these theories could function as major premises from which specific practical conclusions and prescriptions could be drawn. I believe that students get a much better appreciation of the relevance of abstract theorizing after having worked through specific policy controversies and that they get a better understanding of these policy debates if they approach them without prior theoretical assumptions.

Chapter 15 looks at what are called radical environmental philosophies. Deep ecology, social ecology, and ecofeminism all suggest that environmental problems will not be resolved without also addressing more fundamental questions and values. They are radical in the sense of recommending broad social and cultural changes as solutions to the environmental crisis. Chapter 16 continues the recognition that ethical issues are not only philosophical, they are also political in the original sense of that term. The readings in this chapter suggest to students that environmental concerns ought to have no political or cultural boundaries. Chapter 17 introduces students to broader cultural perspectives on environmental topics. Finally, Chapter 18 reflects on widespread policy disagreements and agreements, and the struggle for theoretical and ideological unity. These readings are representative of a more pragmatic approach that has been evident among some environmental philosphers in recent years. Pragmatism, perhaps, is not a bad place to end a textbook that began with tension between theory and practice.

In choosing the readings, I have also attempted to provide a balance of philosophers with nonphilosophers, classical sources with the most recent writings, theorists with policy experts. I have included selections from philosophers, historians, scientists, theologians, feminists, journalists, economists of both neoclassical and sustainable perspectives, non-Western authors, and environmental activists. Some selections are classics and appear in most other texts, others are peculiar to this book. In choosing selections I sought essays that presented important ideas in an accessible style for nonspecialist undergraduate students.

As another tool for balancing theory and practice, each of the middle eleven chapters—the application and policy chapters—include several discussion cases. I use these short cases as entry points into discussion of the chapter's readings or to highlight a particular issue that arises from within a reading. I have used these in class and have found them to be a helpful way to facilitate student discussion and to connect abstract philosophical questions to practical policy concerns. They are not intended as detailed case studies but rather as discussion points to encourage further thinking and to motivate further questioning. By challenging students to simply identify the ethical issues involved in these cases, I have found that it is a helpful place to begin discussion.

Acknowledgments

I am pleased to take this opportunity to thank the many people who contributed to this book in a variety of ways. The original idea for the book developed in conversations with Ken King, philosophy editor at Mayfield. Ken provided all the essential editorial ingredients: advice, criticism, direction, encouragement, pressure, support, and late-night phone calls. It is good to have him back in the philosophical community.

I began work on this book while on a sabbatical from The College of St. Benedict. I am grateful for the continuing support of this collegial academic community. The students in my Senior Seminar class during the fall semester of 1997 willingly helped with a trial run of many of these selections and cases. Their advice and editorial comments helped make this book more "user-friendly."

The following individuals reviewed early versions of the proposal and outline of this book: Daniel Holbrook, Washington State University; Dr. Louisa Moon, Mira Costa College; Kristin Shrader-Frechette, University of South Florida; Homes Rolston III, Colorado State University; Arthur Skidmore, University of Kansas; Charles Blatz, University of Toledo; and H. Phillips Hamlin, University of Tennessee-Knoxville. Their advice did much to shape the final product, particularly helping to overcome my tendency to slip into abstractions and overlook the pressing practical issues of policy. In particular, I would also like to thank Eric Katz of the New Jersey Institute of Technology and Russell Swanson of Florida State University who provided very helpful advice on later versions of the book.

It has been a pleasure working with everyone associated with Mayfield Publishing. I especially would like to acknowledge the skillful work and support of Carla White Kirschenbaum, production editor, and Jennifer Gordon, copyeditor.

Finally, time spent on a project such as this is time not spent with one's family. My greatest debt is owed to Linda, Michael, and Matthew. As a small repayment of that debt, I dedicate this book to them.

PART I   *A Context for Environmental Ethics*

The ultimate goal of this textbook is to help you develop into a thoughtful and informed citizen who is capable of articulating and defending ethically responsible positions on matters of environmental policy. The intent is not to *tell* you what to think. As a democratic citizen and autonomous adult, you have the freedom to form your own opinions on such matters. But with that freedom comes a responsibility to ensure that your opinions are well founded. A fundamental philosophical assumption is that all opinions are not equally valid or equally reasonable. This textbook aims to provide the resources to form well-founded and reasonable opinions.

To further this aim, the book is divided into four parts. Part I situates environmental issues within philosophical, historical, and contemporary contexts. Part II introduces several fundamental topics that are important for analyzing any environmental policy issue. Part III contains a variety of specific environmental policy controversies. Finally, Part IV provides the opportunity to step back from these controversies and consider more systematic and comprehensive perspectives on environmental matters.

Part I consists of three chapters that provide a context for understanding environmental issues. Chapter 1 surveys some influential ethical theories and connects those theories to environmental topics. These ethical theories will help you understand and analyze contemporary environmental policy issues. Chapter 2 provides readings that give a historical context for contemporary Western thinking about the natural world. Understanding how people in the past thought about the natural world proves useful for understanding and analyzing how they think about it today. Finally, Chapter 3 offers some widely diverse opinions on the present state of environmental policy. One often hears claims that present generations face a significant environmental crisis, but not everyone agrees with such pessimistic assessments. Chapter 3 provides three general perspectives on the present state of the environment. One perspective is optimistic about the present environmental situation. From this perspective, if properly understood, natural resources are infinite. A second perspective acknowledges that significant progress has been made in recent years but recognizes that much work remains to be done. A final perspective forecasts significant environmental havoc if present population and lifestyle trends continue.

My hope is that Part I will serve as an introduction to environmental ethics and policy by broadening your initial perspective on these issues. If this book is a journey through environmental ethics and policy, this initial part is a large-scale map that indicates in broad outline where we have been and where, in general terms, we are at present.

CHAPTER 1

Philosophical Ethics and Environmental Public Policy

"We are dealing with no small thing," Socrates once said, "but with how we ought to live." There is perhaps no question more fundamental for human beings than this: How ought we to live?

Within the last century, and particularly within the last few decades, it has become clear that human activities are capable of causing widespread disruption within the natural environment. Strong evidence suggests that humans are responsible for the greatest mass extinctions since the end of the dinosaur age 65 million years ago. Human activity threatens the very atmosphere and climate of our planet. Escalating population and economic growth in the next few decades will intensify pressures on the natural environment. Throughout the world, human beings are degrading natural systems, depleting resources, accumulating toxic wastes, polluting the air, water, and soil. Is this how we ought to live?

PHILOSOPHICAL ETHICS

When Western philosophy began twenty-five hundred years ago in Greece, environmental issues as we know them simply did not exist. Worldwide human population was insignificant, and technology that could transform natural systems, other than in a minor way, did not exist. The earth's ecosystem sustained both the quantity and quality of human life. At the approach of the twenty-first century, the situation has changed. But what does philosophy have to do with any of this? At first glance, philosophy and environmental policy seem to be at opposite ends of the spectrum. Philosophy seems abstract whereas environmental issues like pollution and species extinction seem practical. Nonetheless, as we will see throughout this book, philosophy may play an important role in helping us to understand and address environmental issues.

We can start examining philosophical ethics by returning to Socrates's question about how we ought to live. Like all cultures, the Greeks had a set of beliefs, attitudes, and values that guided their lives. In fact, the word *ethics* is derived from the Greek word *ethos,* meaning customs and conventions. Most Greeks would have answered Socrates's question by claiming that we ought to live an ethical life. Like people everywhere, an ethical life for the Greeks would have been a life lived according to the customary beliefs, attitudes, and values of their own culture. Most often, these customary beliefs are connected to a culture's religious worldview. To be ethical is to conform to what is typically done, to obey the conventions and rules of one's society and religion. In this sense, ethics is identical to ethos.

Philosophical ethics is not satisfied with this answer to the question of how we ought to live and denies that conformity and obedience are the best guides to how we should live. From the very beginning, philosophy rejected authority as the source of ethics and has instead defended the use of reason as the foundation of ethics. Philosophical ethics seeks a reasoned analysis of custom and a reasoned defense of how we ought to live. Distinguishing between what people *do* value and what people *should* value, philosophical ethics asks us to step back and rationally evaluate our customary beliefs and values. This branch of philosophy requires us to abstract ourselves from what is normally or typically done and to reflect on whether what is done should be done, and whether what is valued should be valued. The difference between what *is* valued and what *should* be valued is the difference between ethos and ethics.

Much of what we do, as individuals and as a society, has an impact on the natural environment. Many of our actions—from having children to driving a car—have environmental implications. Environmental ethics is the practice of rationally reflecting upon how we should live in relation to our natural environment. As a branch of philosophical ethics, environmental ethics asks us to step back from our daily decisions, to step back from our own ethos, and to reflect upon how our decisions affect the natural environment. We should reflect on daily events and echo Socrates's question: How ought we to live in relation to our natural surroundings?

MORALITY AND SOCIAL ETHICS

How ought we to live? This fundamental question of ethics can be interpreted in two ways. "We" can mean each one of us individually, or it might mean all of us collectively. In the first sense, this is a question about how I should live my life, how I should act, what I should do, what kind of person I should be. This meaning of ethics is sometimes referred to as *morality*. In the second sense, this is a question about how a society ought to be structured, about how we ought to live together. This area is sometimes referred to as *social ethics,* and it raises questions of public policy, law, civic virtues, and political philosophy.

Environmental ethics certainly raises both kinds of questions. For example, many individuals choose to live their lives as vegetarians because they believe, upon reflection, that eating animal flesh is unethical. These individuals believe that a moral life should not include eating meat. Environmental aphorisms like "tread softly upon the earth" and "live simply so others may simply live" offer moral advice for the type of life each of us should live.

Social ethics addresses issues concerning the structure and institutions of society. Most environmental issues also raise questions of social ethics. Passing a law that protects endangered species is a matter of social ethics, as is giving tax incentives to encourage mass transit, regulating development, creating national wilderness areas, and prohibiting air and water pollution.

Although this book will examine both types of ethical questions (as well as philosophical issues in areas other than ethics), our emphasis will be on social ethics and issues of public policy. This emphasis may be justified simply by the size of environmental problems. Such issues as population growth, global warming, nuclear waste disposal, species extinction, and wilderness preservation cannot be resolved adequately by individual action alone. Most environmental problems exist on a scale that only can be addressed by public policy. But there are other reasons for emphasizing social ethics. It will be helpful to think a bit about the connection between individual morality and social ethics.

Many people assume that moral questions are all that matter. From this perspective, ethics is simply concerned with personal morality, and if we can only get individuals to live moral lives then the correct environmental policies will follow from individual action. But this view misrepresents the connection between individual action and public policy in two ways.

First, there are important value questions about how we should live that make little sense from the individual perspective. Consider how an individual might think about global warming. Strong evidence suggests that burning fossil fuels increases the amount of greenhouse gases in the atmosphere and thus can lead to increased global temperatures. Under such circumstances, deciding to drive a car can raise an ethical question: Do I have a responsibility to stop driving as a means to reduce greenhouse gases? Consider how a person might answer this question solely from an individual point of view.

It would be reasonable for an individual to weigh the benefits of driving her own car against the costs of increased greenhouse gases. Given the amount of gas burned in any single car (on average 2 to 3 gallons per hour) and given the fact that an average person might drive her car only a few hours a day, this gas alone will make no detectable difference in the amount of atmospheric carbon dioxide. Given the convenience and freedom associated with driving one's own car, it is unlikely that an individual will judge her own action as making enough of a difference to matter. Strictly from an individual point of view, it seems reasonable enough to choose to drive.

Now, recognize that across the world hundreds of millions of individuals face the same decision each day. At this rate, we are talking about hundreds and hundreds of millions of gallons of gasoline a day. This amount of fossil fuel use does make a difference that matters. Thus, when extended across a population of a hundred million people, a decision that seems minor to an individual can be monumental to a society.

We are now faced with questions that would never arise from an individual point of view: Should we create policies to regulate the use of fossil fuels? For example, we might increase taxes on gasoline, require automobile manufacturers to improve milage efficiency, subsidize mass transit, provide tax incentives for alternative fuel transportation, or even prohibit private automobiles in urban areas. Of course, we might choose to do nothing and leave this issue up to individuals. But note, each of these policies raises questions that would never arise from the individual perspective. Are there public policies that could influence individual choice in a way that can solve an environmental problem? If we recognize that these are important options, and if we see that they would not arise from an individual perspective, then we have seen the justification for treating issues of public policy as separate from individual moral questions.

The second misrepresentation in relying on personal morality alone to set public policy is that it is sometimes easy for us to underestimate the influence that social practices have on individual choices. Individuals choose to do things based on their individual beliefs, their wants, their interests, their desires. But we need to question the source of these beliefs, wants, interests, and desires. The factors that influence individual choice arise out of the culture and society in which the individual lives. Individuals are shaped, to various degrees, by their culture and society. In its social dimension, ethics must address the legitimacy of these cultural and social factors.

For example, consider what a typical middle-class North American believes is desirable for living a comfortable life. At the risk of overgeneralizing, there is strong market evidence for what the typical U.S. citizen desires: a single-family home on a large building lot, several televisions, several cars, brand-name shoes and clothing, a variety of electronic devices, convenient local shopping centers, low taxes, and significant government services like schools, police, sewer system, water, good roads, and so on. It would not be too much of an exaggeration to say that the American ethos defines this as the good life, as the way we ought to live.

But where do these expectations and desires come from? It is hard to imagine that any of us, if placed in a different culture at a different time, would independently come up with a desire for Nike shoes, gourmet lattés, McDonald's, portable CD players, and the like. Social ethics, unlike individual morality, steps away from the wants and expectations of individuals to reflect on the social influences shaping individual choice. Is the consumer lifestyle that many North Americans take for granted the way we ought to live? Are there social arrangements that would encourage us

to adopt environmentally benign lifestyles? These are questions that may be raised from the perspective of social ethics and would unlikely be raised from the perspective of individual morality.

Social Ethics and Public Policy

Our focus, then, will be on social ethics. How should we live in community with other human beings and with the natural world? We will be concerned, then, with issues of ethics and public policy. Before turning to a survey of various influential theories of social ethics, let us think first about the relation between ethics and public policy.

Three questions characterize our focus on ethics and environmental policy: What should we do? Why should we do it? How should we attain this goal? The first two questions concern our ends and the justification of those ends. The final question, the crucial question of public policy, concerns the means chosen to achieve our ends.

An example can help develop this framework. Over the last several years my hometown has been considering a local ordinance that would create new zoning rules to regulate development of sensitive natural areas. Urban growth over the past few decades has destroyed all but a small portion of the wetlands, prairies, and oak forests that were native to this area. In the face of this, local citizens and government officials asked these three questions: What should be done? Why should it be done? How should it be done?

In this case, some people argued that we, as a community, should preserve the remaining natural areas. Some argued that we should try to restore and reclaim some of the lost areas. Others, fortunately a minority, argued that we should do nothing to the status quo and allow continued development of natural areas. Over time, a public consensus was reached to try to preserve natural areas in a way that does not prevent development.

During the public debates surrounding this issue, a variety of justifications emerged, even among those who agreed about the goal of preservation. Some argued that we should preserve these natural areas because they were important to a stable local economy. Others argued that the plants and wildlife themselves deserved to be treated with respect. Still others argued for the beauty of natural areas. Some people argued on historical grounds that these areas connected us to our ancestors who first settled this area in the nineteenth century. Some argued on spiritual grounds that natural areas are our closest connection to God's creation and provide occasions for spiritual reflection and meditation. Even though there was wide agreement about the ends of public policy, there was widespread disagreement about the justification for these policies.

Finally, much of the political disagreement arose over the proper means for attaining these ends. Some argued that new laws were required to prohibit any building in sensitive natural areas. Some thought that local government should purchase these areas and preserve them as parklands. Some thought that scientific experts in ecology and wildlife management should work with builders and developers to manage these areas. Some thought that tax incentives to private industry would provide an efficient means for preservation. Still others argued that these policies would be best left to the working of a competitive market. As is often the case in public policy, the final result was a compromise: New zoning laws require builders in sensitive natural areas to work with a team of scientific experts to design a development plan that preserves a portion of the natural resources in that area.

This example is typical of many controversies surrounding environmental policy. Disagreements abound. One important role for philosophical thinking is to identify exactly the issues in dispute. This is sometimes more difficult than it may seem. As we enter into the policy debates in the readings that follow, it will be helpful to return to these questions regularly. We need to ask: What is the goal or end sought by this author? Why does this author believe that this goal is important? How should we pursue this goal? Is this the most effective and efficient means to the end? Do these means treat all people fairly and with respect? Do these means encourage or discourage beneficial social practices?

Controversies raised by environmental policy questions represent some of the most significant challenges that we will confront in the next generation. In the face of such challenges, it would be easy to dismiss philosophical reasoning as too abstract or irrelevant. But this would be a mistake, for philosophical questions are often at the heart of environmental controversies. Taking a stand on any of the three fundamental policy questions—What? Why? How?—presupposes answers to fundamental questions of ethics. Responsible solutions require that these answers be adopted explicitly, rather than implicitly and without careful consideration.

We can think of philosophical ethics as offering a context for examining environmental policy by providing the concepts and categories necessary for understanding the ethical dimensions of environmental policy. The history of philosophical ethics is a history of careful and systematic thinking about the basic concepts and categories of ethics. To better prepare us for a philosophical examination of environmental policy and ethics, we turn now to a survey of various philosophical traditions in ethics and social policy.

GREEK ETHICS

Ancient Greek philosophy is primarily identified with three major figures: Socrates, Plato, and Aristotle. Socrates (ca. 470–399 B.C.E.) wrote nothing that is known to exist. What we know of his thinking comes to us through the extensive writing of his student, Plato (ca. 428–348 B.C.E.). Plato, and his student Aristotle (384–322 B.C.E.), wrote extensive and monumental works in virtually all branches of philosophy.

It might seem strange to begin a textbook in environmental ethics with a discussion of ancient Greek philosophers. But there are some important lessons that we can learn from each of these great thinkers.

Philosophical ethics truly begins with Socrates. The portrait that emerges from Plato's writings is of a man who spent much of his time questioning and challenging the beliefs and values of the Athenian people. Socrates seemed particularly interested in questioning those with power and authority: religious leaders, politicians, business leaders. Needless to say, this didn't go over very well with these authorities, and Socrates was eventually put on trial, convicted, and executed by the Athenian citizens.

As mentioned earlier, the word *ethics* comes from the Greek word *ethos,* which refers to what is customary or conventional. In this sense, living an ethical life is living according to the customary and conventional values of one's culture. Socrates spent his life encouraging the citizens of Athens to step back from these customary beliefs and examine them rationally. This stepping back and rationally examining the values, beliefs, and attitudes of one's own culture symbolizes the beginning of philosophical ethics. Throughout this book you will be asked to follow Socrates's example. Challenge yourself to step back from what is considered ordinary, customary, normal, conventional. Ask yourself: Why do I believe this? What assumptions am I making? Am I satisfied with the reasons that support these beliefs and values? Have I really considered the alternatives? In doing this, you will be following Socrates and entering one of the oldest traditions of Western culture: the tradition of philosophical reflection. By following Socrates, you will be taking responsibility for your own choices, making up your own mind about these important issues, and embracing the real freedom that thinking for yourself offers.

Although Socrates truly represents the beginning of philosophical ethics, Plato offers us the first full-fledged ethical theory. Plato is the first who offers Western culture a vision of the good life that is defended by and integrated within a metaphysics and epistemology. Metaphysics is the branch of philosophy that deals with the nature of reality. Epistemology is the branch that investigates standards of knowledge and rationality. Like many philosophers who follow him, Plato believed that by studying *human nature* (metaphysics), we can come to *know* (epistemology) how we *should live* (ethics).

In connecting ethics to an underlying human nature, Plato begins the oldest and most influential tradition in Western ethics. In this tradition, we are asked to distinguish *accidental* characteristics of human beings from *essential* characteristics. Human beings vary according to numerous characteristics: height, weight, color, age, shape, sex, abilities, and so on. But underlying these accidental characteristics is the reality that we are all humans. Thus, there must be some human nature, some essential characteristic, that makes all humans human. In other words, all of the specific differences that distinguish people from one another exist only at the surface and mask a deeper underlying reality. All humans have, essentially, the same nature. This human nature or essence was called the human psyche by the Greeks and was later translated as *soul*.

Plato concluded that human nature, the human psyche or soul, is best understood in terms of three components: the appetites, the will, and reason. Each component has its own function or activity and can best be understood in terms of this characteristic or natural activity. The appetitive part of the soul, closely identified with our bodies, is the section of the soul that wants and desires. The will is the active part of the human soul. We seek, we do, we act, and the will is that part of the soul that performs these actions. Finally, the human soul is also home to reason; we think, we calculate, we know. It follows, according to Plato, that the good life is a life in which each part of the soul performs its proper function. Ideally, we should desire (the appetites) and pursue (the will) only those things that reason judges to be good. Thus, the good life for humans is a life spent in pursuit of the fulfillment of human nature.

This general approach to ethics is captured in the slogan (adapted by the U.S. Army's recruiting effort) to "be all that you can be." This suggests that you have a potential—what you "can be"—that you haven't yet attained. What you *should* do, given this capacity, is strive to fulfill your potential.[1] .

But this tradition recognizes that, for a variety of reasons, humans do not always fulfill this nature. Ethical norms, rules, and practices are what guide us from the way we are to the way we should be. However, because not all people attain knowledge of what is good, society should be arranged as an aristocracy. Those few people who do come to know what is good (Plato called them philosopher-kings) should govern. The goal of government and public policy should be to help all people attain, as much as possible, their full potential.

Two aspects of Plato's thinking are worth considering in an environmental philosophy textbook. First, Plato begins a long tradition of thought that makes a sharp distinction between human nature and the natural world. Part of humanity, of course, is connected to the physical world through the body. But this is the part of humanity that is constantly changing and hence not what is most real and enduring about humans. Thus, not only is there a distinction between the physical and the nonphysical worlds, it is a hierarchical distinction in which the physical world is less real and less good than the world of ideas. Only the nonphysical world of the *forms* (Plato's word for the unchanging essences of things) is real; the rest, the changing world of accidental characteristics, is mere appearance.

To use one of Plato's own images, life in the physical world is like living in a cave in which the objects of our experience are mere shadows of what is truly real. Fulfilling our potential as human beings, attaining the good life, requires that we transcend this physical world and escape out of the cave into the world of ideas. In short, humans are essentially separate from the natural world. Our bodies keep us tied to the physical world, but it is a tie that we should strive to transcend. Plato himself believed that the philosopher, the person who was able to transcend the physical world, had a responsibility to return to the cave to serve, teach, and lead. But Plato and many thinkers in the tradition that followed also recognized a tension that exists between the intellectual life and the physical world.

Plato's political views are the second aspect of his thought that relate to environmental issues. Plato was a strong critic of democracy. He believed that, lacking knowledge of the good, most people fail to live according to reason. Therefore, the majority of people are governed by their appetites and desires rather than by reason. Democracy, therefore, is rule by the appetites. In a democracy, people use government to get what they want and, for the majority of people, what they want is not necessarily what is good.

This critique of democracy can be extended from politics to economics. Although he lived more than two thousand years before its invention, it is safe to say that Plato would be an equally strong critic of free market capitalism. In this economic system, goods and services are produced to meet consumer demand. The law of supply and demand tells us that supply (what gets produced) is a function of demand (what people want). Because, again, most people do not know what is good, we can assume that consumer demand will not result in the production of what is good.

Extrapolating from his critique of democracy, we can infer that Plato would have very little tolerance for consumerist lifestyles and consumption patterns. (At several places in his writing, Plato criticized members of his own society who were overly concerned with acquiring wealth and status.) Because many environmentalists believe that this lifestyle is responsible for such environmental predicaments as global warming, ozone depletion, rain forest destruction, pollution, resource depletion, and wilderness destruction, environmentalists could find support for their position in Plato's ethics.

This general approach to ethics would find a sympathetic audience among some public policy officials today. Some people believe, as did Plato, that there are right and wrong ways to live and that certain experts can come to know such answers. On this view, policy should be set by those experts—perhaps scientific experts in ecology or economics—who know what is the best environmental policy. In addition, environmental decisions should not be left to the will of the people because the majority of people are motivated by their own selfish desires. Especially in a culture dominated by consumerism and marketing, important decisions should not be left in the hands of the people. Policy should be guided by experts who are knowledgeable of the common good.

Of course, this approach faces major challenges, not the least of which is justifying the claim that one expert knows what is good for others. (Plato would have had little faith in the experts trained in economics

and ecology.) This approach also denies that individual freedom is a core social value. To the degree that you value democratic decision procedures and individual freedom of choice, you will be skeptical of a Platonic approach to ethics and politics.

On the other hand, Plato's ethics can curtail a troubling tendency of democratic societies. Too often those who value democracy and freedom fail to distinguish their views from an "anything goes" relativism. Relativism is a philosophy that denies any objective legitimacy to value questions. A central tenet of Plato's philosophy is that there are common social goods that transcend the personal opinions of individuals. Personal judgments and preferences are not the best measure of what is good, of what is valuable, of what is beautiful. We can use Socrates and Plato as reminders that not everything is merely a matter of opinion, not all values are equal, and not everything we want is good.

Aristotle's ethical views have also had a significant influence on contemporary thinking. Something very much like Aristotle's thinking underlies, usually implicitly, the policies of some environmentalists. Like Plato, Aristotle believed that ethical judgments are derived from an account of human nature. Once we know something about the nature of human beings, we can know something about what humans should be and about how they should live. Unlike Plato, Aristotle believed that individual objects, not abstract ideas, are what is most real. For Aristotle, scientific reasoning, not abstract philosophical reasoning, can lead us to an adequate understanding of human nature. Aristotle saw human beings as very much a part of the natural world, rather than separate from it. (But note that Aristotle's science was in important respects quite different from modern science.)

It would be difficult to overestimate the grandeur of Aristotle's work. He was, without question, one of the greatest minds in human history. He wrote extensively and influentially on astronomy, physics, biology, art, theater, logic, psychology, ethics, politics, metaphysics. But the connection between his study of biology and his ethics is most helpful for our purposes.

The heart of Aristotle's biological studies, as it is for all great biologists, was his careful, descriptive taxonomy. Aristotle did extensive work cataloguing, classifying, and categorizing various life forms based on their similarities and differences. He was the first to give the familiar concepts of genus and species their central role in the biological sciences. Scientific understanding locates an object in a class of similar objects (genus) and then distinguishes it from other members of that class in terms of its specific differences (species).

According to Aristotle, scientific understanding requires that we be able to explain four facets—what he called the "four causes"—of every object: the *material, formal, efficient,* and *final* causes. To understand something scientifically, we must know what it is made of (matter); the structure that accounts for this material being what it is rather than something else made of the same stuff (form or genus); how it came to be what it is (efficient cause); and what it does that distinguishes it from other things (final cause).

This emphasis on the final cause separates Aristotelean science from modern science. No doubt influenced by his study of biology, Aristotle concluded that all things have a characteristic or natural activity. Ultimately, the specific difference that distinguishes members of the same genus is a unique, characteristic activity. The goal, function, end, or purpose of this activity is what the Greeks identified as an object's *telos.* Accordingly, we don't fully understand an object until we have grasped its natural function, purpose, or telos. Given the central role that telos plays in both Aristotle's science and ethics, his approach is identified as *teleological.*

A crucial aspect of teleological theories is the connection that they establish between scientific fact and ethical value. Once we know an object's telos, we have also come to understand the good for that object. Consider a teleological explanation of a heart. To understand the heart, according to Aristotle, we need to understand what it is made out of (its matter: muscle tissue), how this matter is formed (an internal organ connected to the circulatory system), how it came to be (perhaps the developmental history of organ formation), and its function (through rhythmic contractions, it pumps blood throughout the organic system).

Once we know the heart's telos—its function in maintaining blood circulation—we have a way of distinguishing good hearts from bad hearts. A good heart is one that pumps blood adequately and efficiently over a long period; a bad heart is one that fails to do this.

Given this relationship between an object's telos and its good, we have strong and objective grounds for making value judgments. We can know an object's good as scientifically as we can know how it functions. The judgment that this is a good heart is, given a teleo-

logical orientation, as objective as the judgment that a heart pumps blood.

Let's follow Aristotle's general taxonomy just a bit further. According to Aristotle, all things that exist exist either by nature or are brought into existence by some other cause. Rocks, trees, and humans are natural objects; tables, houses, and clothes are nonnatural artifacts. Natural objects have an internal principle of change; that is, their characteristic activity comes from their own natures, their telos is inherent. The telos of artifacts (for example, the purpose of a table) comes from some external source. So, we have natural things and nonnatural things.

Of natural things, some are alive and some are not. That is, some have a psyche (or soul) and some do not. Being the careful biologist that he was, Aristotle noticed that there are a variety of activities that characterize living things. Some living things have all these activities; some have only a few. These activities include nutrition, growth, decay, reproduction, appetite, sensation, locomotion, and thinking. Plants are the simplest forms of life; they take in nutrition, grow, reproduce, decay. Thus, plants have one type of psyche or soul. Animals have this plus other powers; they also have the capacities of sensation and movement. Thus, animals have another type of soul. Humans have all of this as well as the capacities of thought and reason. Humans, then, have yet another type of soul.

Therefore, unlike Plato, Aristotle is more likely to consider human beings as a part of nature. They are more complex than other living things (they possess all the activities of animals and plants plus some additional ones), but they remain a part of the natural world. Further, just as humans have a distinctive and natural telos, so, too, do other natural objects. Therefore, these other objects—besides being "ensouled"—have their own natural good. Thus, in Aristotelean terms, it makes perfect scientific sense to talk about the good of any living thing. Every natural object has its own intrinsic good, a good that is independent of how humans value or use it. As we shall see later in this book, many contemporary environmental controversies debate this very question. Do natural objects such as plants, animals, and ecosystems have their own intrinsic good? Or are they good only to the degree that humans value them as instruments for human purposes?

Before evaluating these views, we will consider later developments of Aristotelean thinking. During the thirteenth century, Aristotle's philosophy was appropriated by Saint Thomas Aquinas (1224–1274) and integrated with Christian theology. This Christian version of Aristoteleanism has also had a profound impact on Western thought.

NATURAL LAW ETHICS

Aquinas interpreted the teleological aspect of Aristotle as evidence of a divine plan. There are characteristic activities—purposes—in nature because nature was created by an all-knowing and all-powerful God. Thus, the purposefulness of nature is God's purpose, and the laws of nature are God's divine law. The final cause in Aristotle's science becomes the divine plan for natural objects. Natural objects should pursue and fulfill their natural goals because those are the goals of a benevolent creator.

Aquinas's philosophy is the prime example of another long philosophical traditional called *natural law*. The natural law tradition in ethics holds that laws of nature discoverable by human reason are part of a plan for how nature ought to be. Phrases like the "law of nature" and "natural law" often are ambiguous. In one sense, "law" refers to the descriptive regularities discovered by science. Thus, it seems a law of nature that gases expand when heated. This law *describes* how the world is in fact. But "law" is also used to *prescribe* how things ought to be. Thus, the law tells me that I ought to stop at red lights, pay my taxes, refrain from drinking and driving.

The difference between these two senses of law is the difference between what *is* and what *ought* to be, a distinction sometimes explained in terms of facts and values. Law in the first sense deals with what is, with facts; law in the second sense deals with what ought to be, with values. The unique element of natural law ethics is that it combines these two senses of law. Aristotle bridges the gap between descriptive factual laws and prescriptive evaluative laws by relying on the validity of teleological explanations. Aquinas offers further explanation of this connection by reference to a divine plan. The descriptive laws of nature prescribe how nature ought to be because these laws reflect the plans of a benevolent, omniscient creator. This natural law tradition encourages us to fulfill our natural potentials because in doing so we are living in accord with a divine plan.

The philosophies of Aristotle and Aquinas are relevant to an environmental ethics text because these views, and some others very similar to them, have been used on both sides of environmental debates. Being able to recognize these patterns will prepare us for a full rational assessment of a variety of environmental philosophies.

Some aspects of the natural law tradition, and certain particular passages from Aristotle's and Aquinas's writings, are thought by some critics to have encouraged environmental abuse. We spoke earlier of Aristotle's distinctions between living and nonliving things and between plant, animal, and human life (souls). Implicit within these distinctions was a rank ordering. At one point Aristotle concludes that

> Plants exist for the sake of animals . . . all other animals exist for the sake of man, tame animals for the use he can make of them as well as for the food they provide. . . . If we are right in believing that nature makes nothing without some end in view, nothing to no purpose, it must be that nature has made all things specifically for the sake of man.[2]

Aquinas further developed this point. This hierarchy (think of it as a rank ordering) of beings, sometimes referred to as the Great Chain of Being, begins with mere bodies (for example, rocks), moves on through plants, animals, humans, angels, and ends in the most perfect being, God. God is the most perfect being; angels, being pure spiritual beings, are next. Humans are less perfect than angels because they have bodies and are therefore limited physically. But, humans are superior to animals because they can think and reason; animals are superior beings to plants because they have sensation and locomotion. Plants are superior to mere bodies because they are alive.

According to Aquinas's theology, this hierarchy of being reinforced biblical passages in which God gave to humankind dominion over the rest of creation. The God of Genesis creates humans in his own image and likeness, blesses them, and says, "Be fruitful and multiply, fill the earth and subdue it, rule over the fish in the sea, the birds in the sky and every living thing that moves upon the earth" (Gen. 1: 28–29). According to Aquinas, the rationale for this domination was the hierarchy of being that could be traced to the philosophy of Aristotle.

But themes from Aristotle and Aquinas can also be found in defenses of modern environmentalism. Environmentalists who defend moral standing for animals and other living things sometimes rely on Aristotelean language to explain their view. Some biocentric ethicists argue that all living things deserve respect because, by virtue of the life principle (modern thinkers seldom call this a soul), each living thing has its own good, or its own sake. Having a good of their own establishes other living things as deserving moral consideration. It means that they possess a good or worth that is independent of human interests.

More generally, the teleological approach of Aristotle and Aquinas has a sympathetic audience among environmentalists who do find an inherent order and harmony in nature. Often influenced by ecology, some environmentalists argue that natural ecosystems are well ordered and working toward their own good. Individual elements of an ecosystem contribute to the overall balance and harmony by filling a unique niche. According to this view, undisturbed natural ecosystems are good; human interference with natural ecosystems frustrates a natural harmony. Paralleling Aristotelean thought, many preservationists believe that science (ecology) can discover a purpose, harmony, or direction within natural ecosystems. Knowing these facts, we can infer evaluative conclusions about what is good for ecosystems. Typically, this approach is used by preservationists to defend policies of noninterference and conservation.

However, there are several significant challenges to applying any form of teleological thinking to environmental ethics. The first challenge denies that we can infer from the fact that something is natural to the conclusion that it is good. Often associated with the eighteenth-century philosopher David Hume, this challenge claims that there is an unbridgeable logical gap between facts and values or between "is" and "ought." Reasoning to a value conclusion simply on the basis of some natural fact is, according to a twentieth-century view, committing the naturalistic fallacy. At best, further reasons need to be given to explain why something that is natural is also good. The fact that it is natural, alone, proves nothing. But, if we can give this further reason, then the fact that it is natural becomes irrelevant.

Defenders of this challenge cite numerous natural facts—ranging from death and disease to natural disasters—as counterexamples to naturalistic ethics. Smallpox, the HIV virus, Alzheimer's disease, floods, tornadoes, and droughts are all natural events that cause great (and undeserved) suffering and death. Those who would claim that what is natural is good have their work cut out for them in the face of such facts.

Perhaps more fundamentally, the teleological approach that is implicit in this philosophy is rejected by contemporary science and most modern thinkers. In so far as there is an order or balance in nature (and many scientists deny that there is), it is not a purposive, forward-looking order. Ordered or not, the natural world as seen through the eyes of modern science is simply the result of prior causes.

Put in Aristotelean terms, modern science (beginning with Copernicus and Galileo) rejects the notion of a final cause. The categories of scientific explanation—starting with the scientific revolution in the sixteenth and seventeenth centuries—include three, not four, causes. There is matter (material cause), structure (formal cause), and forces that work upon matter (efficient cause). These forces work according to mechanistic laws. Whatever is can be explained fully in terms of what it is made from, how it is put together, and what forces are at work upon it. Reference to purposes, ends, telos is irrelevant if not misguided.

It is reasonable to understand the scientific revolution in physics as being led by Galileo's rejection of final causality. The scientific revolution in astronomy was led by Copernicus's and eventually Newton's rejection of final causality. More relevant for our purposes, the scientific revolution in biology can be understood as Darwin's rejection of final causality. (It is testimony to the intellect of Aristotle that it was his science that gets rejected by each of these great thinkers.)

Consider the Darwinian revolution in biology and its ethical significance. (This may help explain why evolution is so feared and resented, especially among some religious people, even today.) Evolutionary theory tells us that the apparent order found in the natural world, where plant and animal species seem so well designed for survival, can be fully explained without reference to any purpose. Just as in other sciences, we explain evolutionary phenomena by reference to prior causes. Forward-looking references (such as purposes, goals, intents, designs, ends) are unnecessary to explain scientifically the behavior of plants and animals. The order discovered in nature comes from a process of natural selection working on a normal range of genetic diversity. Individuals who are better adapted to their environment tend to survive to reproduce. Those who are not, do not.

To use a common example, we often speak in terms of species evolving *in order to* survive. So we might say that giraffes evolved their long necks in order to reach food high off the ground. Thus, the giraffe has found its niche, doesn't have to compete with other animals, and survives. This suggests a design or telos for the giraffe's long neck. And this suggests a natural plan, perhaps a divine plan, that accounts for the harmony of ecosystems. But this is a mistaken view of how evolution works.

Natural selection would claim that the giraffe did not evolve a long neck *in order to* reach food but that, having longer necks, some animals were better able to reach food, and therefore they had greater reproductive success than those that did not. Over time, this adaptive difference led to the evolution of giraffes. Longer necks occurred originally not because of some design but simply as a result of random genetic deviation among species.

The significance of this, according to critics of teleology, is that the purposive language of teleology, and the connection between facts and values that it implies, can be abandoned in favor of the more mechanical language of the physical sciences. The nature that we observe today, and any apparent order therein, is the result of hundreds of millions of years of random change and natural selection. From these facts alone we can conclude nothing about the value of the particular biosphere in which we live.

This is not to say that the Aristotelean and natural law traditions are without merit. The structure and complexity that are found in the natural world, especially those discovered by ecologists, can and are used in environmental policy debates. The natural world is a fascinating place, and its value is defended in many of the readings that follow. The challenges to Aristotelean and natural law ethics should only caution us not to jump too quickly from natural facts to ethical values.

MODERN SOCIAL ETHICS

In many ways the scientific revolutions in physics, astronomy, and biology paralleled revolutions in social and political thought. The great democratic revolutions of the seventeenth and eighteenth centuries also represented a rejection of Greek and medieval thought. The political theory that developed out of Plato, Aristotle, and Aquinas held that society should be ruled by an elite—whether an aristocracy, a king, or a pope—who knew what was best and who would rule for the common good. Modern social and political theory can be understood as beginning with a rejec-

tion of this view in favor of democratic political structures and wide-ranging civil liberties.

Remember Plato's critique of democracy. Plato mistrusted democracy because it placed political authority in the hands of people who lacked knowledge of the good. Instead, he believed that political decisions should be made by those people, typically a small elite group, who know what is good. Plato also mistrusted freedom, believing that free individuals would likely choose to pursue their individual desires rather than what is good for them.

Because all people had the same nature, *the* good was identical for all people. Therefore, those with knowledge of the good are justified in telling others what to do, even if the others do not want to do it. Indeed, Plato thought that too much freedom would ultimately undermine democracy and lead to tyranny as democratic freedoms degenerate into anarchy and chaos.

By the seventeenth century, many philosophers were challenging this social and political philosophy. They recognized that those who claimed to be ruling for the common good were often corrupt and self-interested people ruling for their own good. There were also serious misgivings, especially once the teleological framework was cast into doubt, about ever truly knowing the good. Widespread skepticism about the objectivity of values arose. Finally, this traditional political philosophy could not be reconciled with the growing commitment to the value of individual freedom.

In light of this, social philosophers were challenged to develop an ethical theory that accomplished several things. Essentially, the theory needed to protect individual freedom of choice (against dogmatic and authoritarian theories) while not succumbing to a complete ethical relativism (which would claim that right or wrong is relative to individuals or to cultures). The two ethical theories discussed in the following sections—utilitarian ethics and deontological ethics—can be understood as attempting to meet these challenges.

UTILITARIAN ETHICS

Utilitarianism is the next tradition in ethics that we shall consider. Utilitarianism has had a significant impact on the modern world and has been especially influential in creating environmental policy.

It will be helpful to start our consideration by locating utilitarianism within its historical context. Roots of utilitarian thinking can be found in the works of Thomas Hobbes (1588–1679), David Hume (1711–1776), and Adam Smith (1723–1790), but the classic formulations are found in the works of Jeremy Bentham (1748–1832) and John Stuart Mill (1806–1873). Each of these social philosophers was writing with the background of the great democratic revolutions of the seventeenth and eighteenth centuries.

Utilitarianism tells us that we can determine the ethical significance of any action by looking to the consequences of that act. Utilitarianism is typically identified with the policy of maximizing the overall good, or, in a slightly different version, producing the greatest good for the greatest number. Acts that accomplish this are good; those that do not are bad.

This emphasis on the *overall* good and on producing the *greatest number* directly opposed authoritarian policies of an elite aristocracy. Thus, utilitarianism strongly supported democratic institutions and policies. Government exists for the well-being of all, not to further the interests of an aristocracy. Utilitarianism is also different from egoism, which would encourage individuals to seek their own good. Utilitarian acts are judged by their consequences for the general good. However, consistent with the utilitarian commitment to democratic equality, the general good includes the well-being of every individual affected by the action.

Thus, utilitarianism is a *consequentialist* ethics. Good and bad acts are determined by their consequences. (How else might we judge acts? Well, sometimes we determine that we should or should not do something *as a matter of principle,* regardless of consequences. We'll look at this approach in more detail in the next section.) In this way, utilitarians tend to be practical thinkers. Acts are never right or wrong in all cases in every situation. It will all depend on the consequences. For example, lying is neither right nor wrong in itself. There might be situations in which lying will produce greater overall good than telling the truth. In such a situation, it would be ethically right to tell a lie.

In this way, utilitarianism acknowledges two fundamentally different types of value. Most value judgments are judgments of *instrumental value;* some act (for example, telling the truth) is valuable as a means (an instrument) to the end of producing some good. But there must be this good, something valued for its own sake, which is the end for which other acts aim.

There must be something of *intrinsic value* by which we can judge the consequences of our acts. All utilitarians agree that good and bad is a function of maximizing the overall good. They disagree, and therefore we find different versions of utilitarianism, over what constitutes this good that is valued not for its consequences but in itself.

In general, the utilitarian position is that *happiness* is the ultimate good. The only thing that is and can be valued for its own sake is happiness. (Does it sound absurd to you to claim that unhappiness is good and happiness is bad?) The goal of ethics, both individually and as a matter of public policy, should be to maximize the overall happiness. But, what exactly is happiness?

Jeremy Bentham argued that only pleasure, or at least the absence of pain, was intrinsically valuable. Happiness, according to Bentham, must be understood in terms of pleasure and the absence of pain; unhappiness is understood as pain, or the deprivation of pleasure. On Bentham's view, pleasure and pain are the two fundamental motivational factors of human nature.

> Nature has placed mankind under the governance of two sovereign masters, *pain and pleasure.* It is for them alone to point out what we ought to do, as well as to determine what we shall do. . . . They govern us in all we do, in all we say, in all we think.[3]

Consider, then, Bentham's utilitarian reasoning. Only pleasure and the absence of pain is valued for its own sake. Only pleasure and the absence of pain, therefore, are objectively and indisputably good. If pleasure and the absence of pain are good, more pleasure (or less pain) is better and maximum pleasure (or minimum pain) is best. Therefore, maximizing pleasure (the utilitarian principle) is the fundamental, objective, and indisputable ethical principle.

As we will see in later readings, this perspective has major implications for environmental issues. If pleasure is good and pain bad—especially if they are the most fundamental forms of good and bad—and if we have an ethical responsibility to maximize pleasure and minimize pain, then our ethical responsibilities must extend, logically, to any and all creatures capable of sensing pleasure and pain. In Bentham's words,

> The day *may come,* when the rest of the animal creation may acquire those rights which never

could have been withholden from them but by the hand of tyranny. The French have already discovered that the blackness of the skin is no reason why a human being should be abandoned without redress to the caprice of a tormentor. It may come one day to be recognized that the number of the legs, the villosity of the skin, or the termination of the *os sacrum*, are reasons equally insufficient for abandoning a sensitive being to the same fate. . . . The question is not, Can they *reason?* nor Can they *talk?* but Can they *suffer?*[4]

Here in Bentham's writings we find one of the very first explicit suggestions by a Western philosopher that animals deserve moral consideration. We will see this view developed by contemporary philosophers in later chapters.

Although agreeing with the general framework of Bentham's utilitarianism, John Stuart Mill defended a different understanding of happiness. Mill believed that there is a qualitative dimension to happiness that is missed by Bentham's focus on pleasure. Human happiness is not mere hedonism. According to Mill, humans are capable of enjoying a variety of experiences that produce happiness. Besides the pleasures of sensation that Bentham mentions, humans also experience social and intellectual pleasures that are qualitatively different from, and superior to, mere feelings. In a famous passage, Mill claims that "it is better to be a human being dissatisfied than a pig satisfied; better to be a Socrates dissatisfied than a fool satisfied."[5]

But the claim that there is a form of happiness that is qualitatively better than sensations of pleasure is controversial. How do we know or how can we prove that it is better to be Socrates dissatisfied than a fool satisfied? Mill's answer has significant political and social implications.

To decide which pleasures and which type of happiness are better, according to Mill, we should consult with someone with the experience of both. Such experienced and competent judges are the best test for determining the highest happiness: "Of two pleasures, if there be one to which all or almost all who have experience of both give a decided preference, . . . that is the more desirable pleasure."[6] And if disagreement continues beyond this, Mill suggests

From the verdict of the only competent judges, I apprehend there can be no appeal. On a question which is the best worth having of two plea-

sures . . . the judgment of those who are qualified by knowledge of both, or if they differ, that of the majority among them, must be admitted as final.[7]

Thus, Mill acknowledges that not all opinions are equal. Some people are more competent and more qualified than others in judging what is good. Here Mill might agree with Plato that inexperienced and uneducated citizens frustrate democratic politics. But where Plato's response was to sanction a political class system in which citizens were segregated according to natural abilities, Mill remained committed to the political equality of all citizens. Like Plato, Mill did not support an uncritical majority rule in which every opinion of what is good is treated as equally valid. However, we can remain faithful to democratic principles by developing citizens as competent judges through experience and education. People need to be educated and experienced in a variety of pleasures before they are competent to judge. Once they are experienced, then majority-rule democracy is the best way to make decisions. (But because there are never any guarantees that democratic decisions will be best, Mill was a strong defender of minority rights to hold and express unpopular opinions.)

Thus, in John Stuart Mill we find one of the classic defenses of liberal democracy and liberal education. The most fundamental ethical principle commits us to arranging society in such a way that we maximize the happiness for the greatest number of people. The best means for attaining this goal is an educated citizenry making decisions through a majority-rule democracy. The best method for securing an educated citizenry is to allow individuals the freedom of choice to pursue their own ends. Even when those choices are unwise, individuals are gaining the experience needed to distinguish between good and bad, higher and lower, pleasures.

Perhaps utilitarianism's greatest contribution to social and political thought has come through its influence in economics. With roots in Adam Smith as well as in John Stuart Mill, the ethics of twentieth-century neoclassical economics—essentially what we think of as free market capitalism—is decidedly utilitarian. It is in this way that utilitarianism has had an overwhelming impact on environmental issues.

Under free market economies, economic activity aims to satisfy consumer demand. The law of supply and demand tells us that economies should, and healthy economies do, produce (supply) those goods and services that consumers want (demand). Because scarcity and competition prevent everyone from getting all that they want, the end or goal of free market economics is to optimally satisfy wants. Free markets accomplish this goal by allowing individuals to decide for themselves what they most want and then bargain for these goods in a free and competitive marketplace. This process of allowing individuals to set their own preferences and bid for them in the marketplace will, over time and under the right conditions, guarantee the optimal satisfaction of wants.

This brief description suggests how free market economics fits the utilitarian framework. The end or goal of economic activity, what economists often refer to as utility or welfare, is the maximum satisfaction of consumer demand. We do the most good for the greatest number when we get as many people as possible as much of what they want as possible. The good is defined in terms of satisfying one's wants. But because scarcity and competition prevent us from getting all that we want, individuals are left to rank their wants or, in other terms, to establish their own preferences. Thus, free market economics can be thought of as a version of preference utilitarianism where the utilitarian goal is the maximum satisfaction of preferences.

Given this goal, free market economics advises us that the most efficient means to attain that goal is to structure our economy according to the principles of free market capitalism. We should allow individuals the freedom to bargain for themselves in an open, free, and competitive marketplace. Self-interested individuals will always be seeking ways to improve their own position. Agreements (contracts) will occur only in those situations where both parties agree that a transaction will improve their own position. In such a situation, the competition among rational and self-interested individuals will continuously work to promote the greatest overall good. Whenever a situation occurs in which one or more individuals can attain an improvement in their own happiness without a net loss in others' happiness, market forces will guarantee that this occurs. Thus, the market is seen as the most efficient means to the utilitarian end of maximizing happiness.

We will examine the debates surrounding such claims at length in Chapter 4. For now, let us consider some general challenges to the ethics of utilitarianism. We can classify these challenges into two groups: problems raised from within utilitarian perspectives that involve finding a defensible version of utilitarianism

and problems from outside utilitarianism that challenge the plausibility of the entire utilitarian project.

We will mention two challenges that are debated from within utilitarian perspectives. First, utilitarians must find a defensible way to measure happiness. Phrases like "maximize the overall good" and "the greatest good for the greatest number" require some form of measurement and comparison. (How else would you know that this situation rather than another has maximized the good?) Bentham went to great lengths to develop a "hedonistic calculus" to help quantify pleasures. Mill left it to the judgment of a majority of well-informed, competent judges. Economists use such measures as the gross national product or average family income for determining overall happiness. Bentham, Mill, and neoclassical economics all seek a scientific, measurable ethics. But there simply is no consensus among utilitarians on how to measure and determine the overall good.

This problem is only compounded by the fact that utilitarians are committed to considering all the consequences to all affected parties. Many environmental problems highlight how difficult this can be. Consider the consequences of nuclear power and storage of nuclear waste. Some nuclear waste remains highly toxic for tens and hundreds of thousands of years. It is hard to see how a utilitarian could ever hope to calculate the consequences of burying nuclear waste. Yet this is exactly what is required by the utilitarian principle. (Attempts to shift focus, as economists often do, to the "expected" utility of an act is to abandon utilitarianism. At that point we have adopted an ethics not of consequences but of intentions, and that is no longer utilitarianism. We'll see this view developed in the following section.)

The second problem with the utilitarian perspective deals with differing versions of the good and the implications for human freedom. Historically, utilitarians have been social and political liberals. That is, they all place a very high value on individual freedom of choice. But, there is a tension between objective accounts of the good and individual freedom. Simply put, free individuals do not always choose to do what is good for them. The more utilitarians emphasize freedom, the more likely they hold more subjective accounts of the good. On this view, good is simply a matter of opinion, or individual desires, preferences, and wants. However, this seems to abandon the entire project of ethics because, after all, people often desire what is trivial, immoral, bad. On the other hand, the more utilitarians are willing to specify a content for the good life, the more they need to abandon the commitment to individual freedom. If we know what is truly good, then individuals ought to act in certain ways (to maximize the good) even if they don't want to. Finding a balance between individual freedom and the overall good is a challenge that confronts most versions of utilitarianism.

The final challenge goes directly to the core of utilitarianism. The essence of utilitarianism is its consequentialism. Good and bad acts are judged by their consequences. In short, the end justifies the means. But this seems to deny one of the earliest and most fundamental ethical principles: The ends *don't* justify the means.

Consider the following examples. Suppose you had good reason to believe that the overall happiness of a society could be increased by enslaving a minority of its citizens. Or suppose you could receive great personal benefits by betraying a close friend. These challenges would point out that slavery and betrayal of friends are wrong, no matter what the consequences.

These challenges can be explained in terms of rules or principles. When we say that the ends don't justify the means what we are typically saying is that there are certain rules or principles we should follow no matter what the consequences. Put another way, we have certain *duties* or *obligations* that we ought to obey even when doing so does not produce a net increase in overall happiness. Examples of such duties are those required by such principles as justice, loyalty, and respect, as well as the duties that flow from our roles as parent, spouse, friend, citizen. The approach to ethics that emphasizes duties and obligations is called *deontological ethics,* from the Greek word for duty. (We will examine that ethical tradition in a later section.)

Utilitarian Ethics and Public Policy

Before moving on to deontological ethics, it will be helpful to connect utilitarian ethics to the basic questions of public policy that we discussed in earlier sections. We said there that social ethics focuses on three questions: What should we do? Why should we do it? How should we do it? Utilitarians have answers to each question.

What should we do? What are our ends? Utilitarians begin by identifying an end that they believe is indisputably good: maximum happiness for all concerned. From the utilitarian perspective, public policy ought to be aimed at making as many people as happy as possible. Most commonly, this happiness is interpreted as getting as many people as much of what they want as possible.

Why should this be the aim of public policy? How does utilitarianism justify this goal? Utilitarians would begin their justification by getting skeptics to admit that their own happiness is good. (If happiness is equated with the satisfaction of wants, not valuing one's own happiness would be the same as not wanting what one wants!) If any individual happiness is good, more happiness is better and maximum happiness is best. The only way for a skeptic to deny this utilitarian conclusion is by claiming that his own happiness is more important or more valuable than the happiness of any other individual. But this, so the utilitarian would argue, would violate the requirement that ethics be impartial; it would be claiming that one individual is ethically more valuable than other similarly situated individuals. Because logic and fairness require us to treat similar individuals similarly, utilitarians conclude that maximizing the overall happiness is the justified goal of public policy.

There is more disagreement among utilitarians concerning the third question: How should we attain our ends? In general, utilitarians are committed to any means that attain the ends of maximum happiness. On this view, policy questions are really pragmatic questions that depend on the specific circumstances of time and place. Nevertheless, some general policy patterns can be identified.

One version of utilitarianism holds that there are experts who can predict policy outcomes and carry out those policies that will attain our ends. These experts, often trained in the social sciences, are familiar with the specifics of how society works and can therefore determine which policy will maximize the overall good.

This approach to public policy underlies one theory of the entire administrative and bureaucratic side of government. On this view, the legislative body (from Congress to local city councils) establishes the public goals and the administrative side (presidents, governors, mayors) executes (administers) polices to fulfill these goals. The people working within the administration (government bureaucrats) should know how the social and political system works and use this knowledge to carry out the mandate of the legislature. The government is filled with such people, typically trained in such fields as economics, law, social science, public policy, political science.

A second influential version of utilitarian policy invokes the tradition of Adam Smith and claims that competitive markets are the best means for attaining utilitarian goals. This version promotes policies that deregulate private industry, protect property rights, allow for free exchanges, and encourage competition. In such situations, the self-interest of rational individuals results, as if led by "an invisible hand" in Smith's terms, in the maximum satisfaction of individual happiness.

The dispute between these two versions of utilitarian policy—what we might call the "expert" and the "market" versions—characterize many contemporary environmental disputes. One clear example is the ongoing controversy concerning management of federal forestlands. The U.S. Forest Service was established in the late nineteenth century with a primary mission of furnishing a continuous supply of timber to U.S. industry. In 1960 Congress expanded the goals of the Forest Service to include "outdoor recreation, range, timber, watershed, and wildlife and fish purposes." This multiple-use philosophy remains the governing goal of present Forest Service practice.

The original director of the Forest Service, and major influence in its creation, was the well-known conservationist Gifford Pinchot. Pinchot was very explicit in explaining the Forest Service's mission in utilitarian terms. Pinchot argued that experts trained in the science of forestry are best able to manage the forests to "serve the greatest good for the greatest number for the longest time."[8] Pinchot believed that scientific forestry could manage these natural resources and direct their use in such a way that they provided maximum benefits to all citizens.

Some recent critics argue that this approach is too bureaucratic and leads to great inefficiencies in the allocation of scarce resources. The market version argues that government experts cannot know what citizens want and have little incentive to do anything more than protect their own turf.[9] This approach argues that "Americans can have all the wilderness, timber, wildlife, fish, and other forest resources they want" by selling these resources to the highest bidder.[10] Competitive markets would allocate timber to those who most value it, wilderness to those who most value it, recreation areas to those who most value it. We would achieve the same goals advocated by Pinchot and his experts, but we would achieve them in a much more efficient manner by leaving these decisions to the workings of a competitive marketplace. This economic version of utilitarianism, and the economic approach to public policy in general, will be examined in Chapter 4.

There is no question that utilitarian reasoning dominates among policymakers and policy administrators. Policy experts at all levels are focused on results and on getting things done, which makes the utilitarian emphasis on consequences particularly attractive to them. It seems obvious that policy should be judged by its results and consequences.

The utilitarian emphasis on measuring, comparing, and quantifying also reinforces the view that policymakers should be neutral administrators. The standard view is that policy goals should be left to the democratic decisions of the people. The people decide what they want and what makes them happy; the job of policy is simply to help them attain those goals as efficiently as possible. Efficiency is simply another word for maximizing happiness.

Finally, like utilitarians, policy experts are concerned with the well-being of the whole community. Their focus is on the collective or aggregate good. By their very nature, policy matters involve a broad social perspective. This, too, is consistent with the utilitarian emphasis on the overall good.

Despite these close connections between utilitarianism and public policy, serious ethical challenges remain. We turn now to a major alternative to utilitarian ethics: deontological ethics.

DEONTOLOGICAL ETHICS

Deontology is not a familiar word to most people. Don't be put off by it. "Deontology" comes from a Greek phrase that would translate as the study of duty. This approach to ethics, both in matters of individual morality and public policy, emphasizes that sometimes the correct path is determined not by its consequences but by certain duties that we must fulfill regardless of consequences. Synonyms for *duty* include obligations, commitments, and responsibility. Thus, deontological ethics studies duties, obligations, commitments, and the connected concept of rights.

The deontological approach is most often associated with the eighteenth-century German philosopher Immanuel Kant (1724–1804). Kant faults utilitarianism for its assertion that our acts should always be judged by their consequences to the overall good. Deontology denies the utilitarian belief that the ends do justify the means. It holds that there are some things that we should or should not do, regardless of the consequences.

To understand why the ends don't justify the means, we need to emphasize that utilitarian ends are focused on the collective or aggregate good: Utilitarianism is concerned with the well-being of the whole. (This is one reason utilitarianism is so attractive to policymakers.) But many of us have a deep commitment to the dignity of individuals. We believe that individuals should not be used as a mere means to the greater overall good. A prominent way of explaining this is to say that individuals have *rights* that should not be sacrificed simply to produce a net increase in the collective good.

Consider debates concerning child labor in the developing world. Some policymakers in impoverished countries believe the best means for raising the standard of living within their country is to increase exports. This thereby brings in hard currency with which the country can pay for food, medicine, and education (and repay debts!). Increasing exports will raise the standard of living for all citizens and thereby meet the utilitarian goal of improving the collective good. However, to increase exports a country must be capable of selling their goods at costs below that of competing countries. Because labor is a major production cost, keeping labor costs low helps the country as a whole. Unfortunately, one means for maintaining low labor costs is to employ young children. (Child labor in the manufacture of sneakers and clothing are only the most well-publicized instances of an all too common phenomenon.)

Is it ethical to use young children in such circumstances? Defenders of this practice argue, typically on good utilitarian grounds, that the children are better off with the jobs than without them, that they contribute to their own family's income, and that they contribute to the overall welfare of their society. Critics claim, on deontological grounds, that it is unethical to treat young children this way even if there are beneficial results. On this view, child labor is ethically equivalent to child abuse and slavery. It is wrong on principle.

Within the Kantian tradition of deontology, our ethical duty is explained in terms of a principle that Kant called the *categorical imperative* (an imperative is a command or duty; "categorical" means that it is without exception). Our primary duty, according to Kant, is to act only in those ways in which the maxim of our acts could be made a universal law. This is a pretty abstract way of saying something that is fairly intuitive. The "maxim" of our acts can be thought of as the intention behind our acts. The maxim answers the question: Why am I doing this?

Kant tells us that we should act only according to those maxims that could be universally accepted and acted on. For example, Kant believed that truthtelling, but not lying, could be made a universal law. If everyone lied whenever it suited them, rational communication would be impossible. Thus, lying is unethical. This condition of universality, not unlike the Golden Rule, prohibits us from giving our personal point of view privileged status over others' points of view. It is a strong requirement of impartiality and equality for ethics.

Kant also provided two other versions of this categorical imperative that are less abstract. He claimed that ethics requires us to treat all people as ends and never only as means. In yet another formulation, we are required to treat people as subjects, not as objects. These formulations restate the commitment to treat people as capable of thinking and choosing for themselves. Humans are subjects; they perform the act rather than being acted upon, to use the familiar subject–object categories from grammar. They have their own ends and purposes and therefore should not be treated simply as a means to the ends of others.

Thus, on this Kantian theory, our fundamental ethical duty is to treat people with respect, to treat them as equally capable of living an autonomous life. But, because each person has this same fundamental duty toward others, each of us can be said to have the right to be treated with respect, the right to be treated as an end and never only as a means. I have the right to pursue my own autonomously chosen ends as long as I do not in turn treat other people as means to my ends.

From this we see that the Kantian tradition in ethics is committed to two basic values: equality and freedom, or what is more commonly called liberty. Equality flows from the requirement of impartiality; I am rationally required to treat each individual as equally deserving respect. Freedom follows from our natures as autonomous agents; to respect an individual is to allow them the freedom to form their own ends and make their own choices. A common way of expressing this Kantian insight is to say that each individual has a fundamental right to as much freedom as is compatible with equal freedom for all.

As we have seen, some utilitarians are also committed to equality and freedom. But where utilitarians view these as instrumentally valuable, deontologists take them as fundamental ways of respecting the dignity of individuals. Utilitarians value freedom and equality as means to the end of overall happiness; deontologists value them as ends in themselves, as part of what it means to treat people as subjects.

Philosophers will sometimes claim that rights and duties are *correlative*. This is to say that my rights establish your duties, and my duties correspond to the rights of others. The deontological tradition focuses on duties, which can be thought of as establishing the ethical limits of my behavior. From my perspective, duties are what I owe to others. Other people in turn have certain claims on my behavior. They have, in other words, certain rights against me.

To return to the earlier example, the Kantian would object to child labor because such practices violate our duty to treat children with respect. We violate the rights of children when we treat them as mere means to the ends of production and economic growth. We are treating them *merely* as means because, as children, they are incapable of rationally and freely choosing their own ends.

From this beginning, the deontological or rights-based approach to ethics gets more complex. A complete theory must specify the rights we have and how they are justified, the range and scope of rights, and some process for prioritizing rights and resolving conflicts between different rights. As preparation for evaluating many of the debates to follow, we will pursue these questions briefly.

One way to understand rights is to think of them as protecting interests. We often make a distinction between a person's wants and interests. When we claim that each individual has a right to freedom, surely we can't mean that they have a right to do whatever they want. Wants (or desires) are psychological states of an individual. They are what people will pursue. Wants are subjectively known, in the sense that individuals enjoy a privileged status for knowing what they want. (Imagine disagreeing with a person's claim that she wants something.) Interests work for a person's benefit and are objectively connected to what is good for that person. People don't always want what it is in their interest to have.

For example, if given the choice my children would want to eat sugar-coated breakfast cereal each morning. Their mother and I deny them this on the grounds that it is not in their interests to eat such food. In this case, wants and interests conflict. Likewise, many college students want to skip class, but it is not in their interests to do so. On the other hand, wants and interests may coincide. You want a good education and good health, both of which are in your interests to have.

As we have seen, some versions of utilitarianism take happiness, understood as the satisfaction of wants, as the final goal of ethics. This version would either deny the distinction between wants and interests (interests being simply strong wants) or argue that the best way to decide what is in people's interest is to let them decide for themselves—that is, let them pursue their own wants. Either way, utilitarians believe that all wants/interests equally deserve to be satisfied to the degree that they equally produce happiness. If your desire for hiking in a wilderness area produces as much happiness as my desire to drive an all-terrain vehicle in that area, each equally deserves satisfaction. Given this equality, the utilitarian commitment to maximally satisfy as many wants as possible seems a reasonable strategy.

But deontologists argue that wants and interests are not equal. They argue that at least some interests are so important to the well-being of an individual that they should not be sacrificed simply for a net increase in the overall happiness. Rights serve to protect these interests from being sacrificed.

Suppose that you own some wooded land on a scenic hillside outside of town. The citizens of your community decide that your property would make a great location for an amusement park. Imagine that you are the only person who disagrees. On utilitarian grounds, it would seem that your land would best serve the overall good by being used for a park. However, your property *rights* prevent the community from taking your land (at least without just compensation) to serve the public. Rights are sometimes described as trumps that override the collective will. They function this way because they protect certain central interests that are ethically more important than the mere wants, or happiness, of others.

The connection between rights and interests is important for two reasons. First, it provides a way of determining what rights we have. By identifying central interests and distinguishing them from mere wants, we can determine the range of human rights. Second, some philosophers argue that this provides a means of determining what things *have* rights. As we will see in later readings, some argue that individual animals and perhaps trees or even ecosystems have interests. If that is true, then we could argue that these objects have rights that we ethically ought to respect.

So what rights do we have? The challenge is to develop an account that creates neither too many nor too few rights. Here's another example from my local community. City planners have a blueprint for road construction throughout the area. One of the planned roads would cut through and destroy a rare oak woodland within the city. When the plan was announced, local residents objected to the road on a variety of environmental grounds. The director of the regional planning group answered protesters by claiming that local citizens "have a right to uncongested roads." Surely this theory of rights is too extensive. The connection between rights and duties that we mentioned previously is a good test for this. If rights imply duties, and if people have a right to uncongested roads, then it would seem that someone (local government?) has the duty to provide enough roads to prevent people from ever having to sit in a traffic jam. It is difficult to see how this could be done without wreaking havoc on the well-being of many people by raising taxes, destroying neighborhoods, taking away property, and so forth.

This suggests that we do not get rights simply by wanting something very badly. Critics charge that this is a problem with rights-based ethics; it encourages self-centered individualism through people trying to privilege their own selfish wants by calling them rights. Anything that anybody wants eventually gets called a right, and thereby people come to expect society to provide this for them. But we also don't want to have too narrow a view of rights. Too weak an account, or too few rights, collapses the entire theory toward utilitarianism.

We can at least sketch a general account of rights by returning to the original idea of respect and the elements of autonomy and dignity on which it is based. We have already seen that equal freedom is a basic right that follows from respect. But what else is necessary for individuals to live a reasonable life as autonomous individuals? Being alive certainly is, so we could identify life as a central interest that ought to be acknowledged as a right. Freedom from threats of physical harm and a right to the necessities for life (food, clothing, shelter), or at least the freedom to pursue these goods, are other basic rights. Conditions necessary to exercise our autonomous choices are also candidates. Examples might include the right to participate in collective social and political decisions, the right to express one's opinions, the right to information (press, speech), the freedom to associate with others, the right to pursue one's own religious commitments, and so on. In short, basic rights might well include the sorts of rights identified in fundamental political documents such as the U.N. Declaration of Human Rights and the U.S. Bill of Rights.

Protecting the safety and security of individuals in this way, and granting them these types of freedoms, goes a long way toward giving people control over their own lives. It offers substantial protection for individual autonomy. The interests that these rights protect seem central to living a decent and meaningful human life. (This is, admittedly, a fairly narrow conception of rights. Many philosophers argue that beside protection from harms, people also have rights to such positive goods as health care and education. They would argue that without such goods, individuals would have little chance of either real freedom or meaningful happiness. I have adopted this narrow view simply as a baseline for beginning our study of ethical theory.)

We can now turn to the scope of rights. What, exactly, is involved in respecting someone's rights? Some philosophers argue that the correlativity of rights and duties can help identify the scope of our rights. If rights imply duties, then our rights extend only to those things that others can reasonably be said to owe to us. For example, what is involved in respecting someone's right to life? What duties do we owe, for example, to starving children throughout the world?

There are two common ways for describing rights and duties. Some argue that we have duties of only *not* causing harm to others. These are identified as "negative" rights and duties. On this view, the right to life means only the right not to be killed. Thus, as long as I am not the cause of the starvation, I have no duty to starving children. (Although it would be charitable, decent, and kind to aid them, it is not something that I have an obligation to do.) Others argue that the right to life is empty if it does not include a "positive" right to those goods necessary to secure my life. This would imply that others have a positive duty to provide food to starving children, even if they were not the cause of this starvation. But this, claim critics, would require too much of others. It suggests that we all have an obligation, individually or collectively through our governments, to go around reducing suffering and starvation at all times. This, critics charge, represents an unreasonable restriction on individual freedom. Yet, if rights are based on the fundamental ethical notion of respect for the dignity of each individual, it is difficult to see how we wouldn't have some positive duties to others. Even when we are not the cause of the harm, it seems only reasonable that we have a duty to help others, especially in cases where doing so would not put our own safety and well-being in jeopardy.

Finally, we need to think about situations where rights conflict. For example, consider a variety of cases concerning property rights. A farmer has a right to use his property to grow crops in order to make a living. But this right might conflict with the rights of neighbors who live downstream from the farmer. Do the farmer's rights include the right to use pesticides and chemical fertilizers in a way that pollutes the stream and harms his neighbors? Do the farmer's property rights include the right to fill in a swamp or cut down all the trees on his property and convert the land for crops? Does he have a right to convert his land into a housing development? A shopping center? An industrial park?

There are various strategies for dealing with conflicting right claims. One strategy is to admit that real conflicts can occur and that there is no way to resolve every possible conflict ahead of time. This approach suggests that all rights are simply *prima facie* rights that establish an initial claim that can be overridden in particular contexts and situations. This answer is troubling to many philosophers because it hints at an ethical relativism—all values are relative, therefore there are no ultimate ethical answers—that seems to deny the validity of ethics. Even philosophers who accept the possibility that ethical conflicts are ultimately irresolvable offer some means for settling conflicts.

An important first step in resolving conflict is to recognize that my rights are limited by the requirement of equal rights for others. Thus the landowner who sends toxic pollution to his neighbors is not exercising his property rights because his rights do not include the right to harm others. My right to freedom is not violated when I am prevented from stealing from you.

It is also important to distinguish between ethical and legal rights. Some of the goods that we claim as rights are granted by a legal system: the right to drive a car, the right to deduct home mortgage interest payments from one's taxes, the right to receive Social Security payments, the right to be paid a minimum wage. In general, rights granted by the legal system can be taken away by the legal system. In this sense, legal rights really are simply policies (that is, utilitarian matters) that are formally recognized in law. Whenever the legal system decides that any particular legal right is no longer desirable or no longer serves the public interest, it can be taken away. Thus, legal rights do not provide categorical protection against the will of the majority.

Ethical rights can be claimed by individuals independently of any legal system. When laws enforce ethical rights, as they do in many cases, it is more a matter of laws *recognizing* the right rather than creating it. In general, we can say that ethical rights are connected to the fundamental value of respect for individuals. It can be helpful to distinguish between *basic* and *derivative* rights. Some rights—for example, equal treatment and liberty—are essentially tied to respect. These are basic rights in the sense that to deny them is to deny the basic respect that each individual deserves. Other rights are connected to respect less directly. Derivative rights are derived from important social values, like respect, but may not be essential for protecting respect and dignity. The rights of privacy, property, free press, and to bear arms serve to protect and promote respect. However, we could imagine situations in which people enjoy respect but do not claim these rights.

This survey suggests that it is possible to rank rights according to their underlying justification. We can use this ordering as a general guide for resolving conflicts. As in all ethical matters, however, there are few hard and fast rules. This method for categorizing and ranking rights should serve only as a guide for our more detailed analysis in specific cases.

Deontological Ethics and Public Policy

As we did with utilitarianism, it will be helpful to connect deontological ethics to the basic questions of public policy. We said earlier that social ethics focuses on three questions: What should we do? Why should we do it? How should we do it? Deontologists have answers to each question.

What should we do? What are our ends? In general, deontologists accept any social end that respects individuals as free and equal citizens. Social goals should be left to the decisions of individuals whenever possible. The justification for this conclusion lies in the fundamental value of equal freedom, or liberty, for all.

Liberalism is the common term for social and political philosophies that take individual liberty as the fundamental value. Deontologists in the Kantian tradition are classic liberals.[11] (So, too, are many utilitarians. Utilitarians who define the good in terms of the individual happiness associated with freely chosen interests and preferences are liberals in this sense. Both Bentham and Mill are commonly seen as social and political liberals.) Liberal social policy is reluctant to specify any particular answer to the question of what we should do. Because social decisions should be left to individuals to decide, there is no one final vision of the good life.

This point is sometimes explained in terms of the distinction between substantive and procedural accounts of justice, or between the good and the right. Liberals are skeptical of any detailed, or substantive, version of the good life. Because this should derive from individual decisions, it is difficult to know beforehand the content of social policy. But this is not to say that anything goes. There are limits to ethical social policies, but these limits are best understood in terms of the procedures followed in establishing them. As long as the correct procedures are followed—procedures that respect individual rights—whatever policies that result are taken to be just. In setting social policy, liberals give a priority to individual rights over any version of what is good. In this, liberals contrast most clearly with the classical social theories of Plato, Aristotle, and Aquinas but also with any social theory that claims to specify how we should live the good life.

What are the procedures that a liberal social policy should follow? Liberals believe that the role of government is to remain neutral among differing versions of the good life. Government's role is modeled on the judge, or neutral arbitrator, who ensures that the rules are followed and that everyone is given a fair and equal opportunity to participate. Once the rules are followed, or once we guarantee that individual rights are protected, liberals are receptive to utilitarian policies of maximizing happiness.

What are the rules of liberal social policy? In general, we can say that they derive from the rights of individuals. Democratic participation is a major requirement. If we are to respect individuals as free and equal, then anyone whose interests are affected by a public decision should have a right to participate in making that decision. But there are ethical limits on even those policies that are democratically decided. Democratic policies must still respect the rights of individuals.

Some environmental issues pose serious challenges for liberal social policy. Issues like pollution are easily enough handled within liberal policy. Actions such as polluting the air or water or dumping toxic wastes pose clear threats to the well-being of individuals and can be justifiably restricted by liberal policy. But other issues, like wilderness preservation and the protection of endangered species, are more problematic.

Consider recent debates concerning the Boundary Waters Canoe Area, a federally protected wilderness area in northern Minnesota. A few hundred lakes within the Boundary Waters, with a couple of exceptions, are designated as "canoe only," and motorized vehicles of any type are prohibited within the area. The exceptions are lakes that were widely used by motorized boats when the wilderness area was created. Since 1976, access to the area has been limited through a permit system administered by the U.S. Forest Service. The motorized exceptions are due to expire in the late 1990s, but controversy surrounds this policy. Should the region be accessible to only a limited number of canoeists or should it be more widely accessible to the general public?

We can see how these restrictions might seem unfair. Government policy appears to favor the preferences of one group of citizens over those of others. Such a policy seems to deny the political neutrality required by the liberal commitment to equal freedom for all. Shouldn't government policy provide equal opportunity for all? This same question can be raised for any government policy that restricts use of public lands. Shouldn't the backcountry at Yellowstone and other national parks be accessible to all? Shouldn't roads be built into all wilderness areas to provide access to the disabled and elderly or to those who would prefer to drive their RVs and campers into the wilderness? Critics of restrictive policies, including those associated with the "wise-use" movement, argue that public lands should be available to all. They claim that public land should be open not only to backpackers and canoeists, but to hunters, ranchers, the elderly, loggers, and enthusiasts of all-terrain vehicles, motorboats, and dirt bikes. Any other policy violates the fundamental liberal commitment to equal treatment for all. But, defenders argue that opening up wilderness areas will deny the opportunity for a wilderness experience for all. The goods associated with canoeing or backpacking in a wilderness area will be impossible to obtain with motorboats roaring by. Defenders argue that equal access for all will deny to them the opportunity to pursue their own interests.

On closer look, we can recognize that the liberal commitment to neutrality does not require that we ignore differences among people. Liberal equality does not require that each individual be treated identically, only that the interests of each individual be given equal consideration. Thus, to the degree that government provides resources for wilderness experiences, government should provide resources for those who favor other forms of recreation.

Liberal public policy is also stretched by environmental policies that take into consideration the interests of the nonhuman natural world. Traditional liberals locate the respect due to each individual in the human capacity for autonomous (free and rational) choice. This suggests that only beings with this capacity deserve ethical consideration. In turn, this means that policies such as preserving biodiversity and protecting endangered species must be connected to the interests of humans. Liberal public policy is strongly anthropocentric, or human-centered. But this runs counter to many environmentalists who believe that other living things deserve to be treated with respect in their own right, not solely as means to our ends. Versions of this debate can be traced through many of the following readings.

ENVIRONMENTAL ETHICS: AN OVERVIEW

This final section provides a general overview of the philosophical landscape of environmental ethics. These categories can help you understand many of the controversies discussed in this book. They can also help you in developing your own environmental philosophy as you answer the question: How should I live?

Many environmental issues easily fit within the categories of traditional ethics. Allocation of scarce resources and pollution are examples of problems that can be analyzed quite well using traditional concepts of responsibility, harm, rights, and duties. In this way, environmental ethics fits a model, common to fields such as medical ethics and business ethics, of applying standard ethical theory to practical problems of everyday life. *Applied ethics* helps us to understand and analyze issues, evaluate alternatives, and defend policies. On this model, we apply well-developed ethical theories to environmental issues as a means of resolving controversies and setting policies.

Some environmental issues do not easily fit within the categories of traditional ethics. As issues concerning the depletion of natural resources developed, it became clear that environmentalism was raising new ethical questions. Some of the initial crucial questions concerned the scope of morality: Who or what is in-

cluded within the domain of ethical consideration? Some of the first attention was focused on future generations: What responsibilities, if any, do present generations have to people living in the distant future? This question simply had not been addressed by traditional ethical theories; however, with some work these theories could be extended to apply to such new issues. *Ethical extensionism* represents this step beyond the more standard applied ethics model.

The initial extensions of ethics remained anthropocentric, with ethics being extended to include the well-being of future human beings. Other developments within environmental ethics challenged this limitation. *Nonanthropocentric ethics* defends ethical standing for nonhuman living beings. Animals, plants, or both deserve direct ethical consideration. Although many philosophers had previously argued that we have duties *regarding* animals, it has only been within the last few decades that philosophers have considered that we have direct duties *to* animals. The distinction between anthropocentric and nonanthropocentric ethics is a fundamental divide within environmental ethics.

Another important distinction developed out of the growing influence of the science of ecology. Ecology, and its fundamental category of ecosystems, emphasizes the interconnectedness of the natural world. In the eyes of many ethicists and ecologists, the ecosystem and not its individual members should be the primary holder of ethical value. Ecological "wholes" such as an ecosystem, a species, or a population are more valuable than any particular member of that whole. *Holism* and *individualism* represent another fundamental schism within environmental ethics.

Some environmental philosophers believe that challenges such as nonanthropocentrism and holism stretch traditional ethical theories beyond the breaking point. These people argue that environmental issues require a more radical rethinking of ethics and philosophy. *Environmental philosophy* addresses a wide range of philosophical issues including metaphysics (Are ecosystems more "real" than individuals?), epistemology (Are teleological explanations appropriate in ecology?), aesthetics (Are judgments of natural beauty defensible?), as well as ethics and social philosophy. A variety of more comprehensive environmental philosophies have also been developed in recent years.

Readings in this textbook will follow this general outline: Part I provides a context for understanding environmental issues. Chapter 1 has introduced the discipline of philosophical ethics. Chapter 2 provides an overview of how environmental issues have been understood throughout Western history. Chapter 3 considers the reality of an environmental crisis from a contemporary perspective. Part II (Chapters 4 to 7), examines a variety of issues that will provide a basis for understanding all environmental topics. These chapters examine the issues of economics, aesthetic and spiritual values, extensionism and anthropocentrism, and holism. Part III (Chapters 8 to 14) examines a range of specific environmental controversies and public policy issues. Finally, the readings in Part IV (Chapters 15 to 18) introduce several more comprehensive environmental philosophies.

Notes

1. Advertising slogans have also captured this aspect of Plato's thought. Anheuser-Busch cautions beer drinkers to "know when to say when." This suggests that reason (knowing when) should guide the will (saying when) in controlling the appetites (the desire for another drink).
2. Aristotle, *The Politics,* book I, chap. 8, p. 1256b, in *Basic Works of Aristotle,* ed. Richard McKeon (New York: Random House, 1941).
3. Jeremy Bentham, *An Introduction to the Principles of Morals and Legislation* (Oxford: Clarendon, 1907), 1.
4. The italics are Bentham's. This passage appears in *Principles of Morals and Legislation,* chap. 17, sec. 1, footnote to par. 4.
5. John Stuart Mill, *Utilitarianism,* ed. George Sher (Indianapolis: Hackett, 1979), 10.
6. Ibid., 8.
7. Ibid., 11.
8. Gifford Pinchot, *The Training of a Forester* (Philadelphia: Lippincott, 1914), 13.
9. For an extended version of this critique, see Randal O'Toole, *Reforming the Forest Service* (Washington, DC: Island Press, 1988).

10. Ibid., xii.
11. Recent partisan politics have skewed the classic meaning of the term *liberalism.* In this partisan sense, a liberal is someone who favors big government, is fiscally irresponsible, and supports increased welfare and decreased military spending. These liberals are usually associated with the left wing of the Democratic party. A classic liberal is someone who takes individual freedom as a fundamental social value. This might include partisan liberals, but it also identifies as liberals defenders of the free market and those who seek less government. Therefore, many Republicans are liberals in this sense.

DISCUSSION AND STUDY QUESTIONS FOR CHAPTER 1

1. The natural law tradition in ethics holds that every natural object has its own purpose or function. Similarly, some contemporary environmentalists hold that ecosystems have a natural purpose or function. Do you believe that these environmentalists assume an ethics like that of natural law?
2. Plato's political theory mistrusted democracy because of its commitment to individual freedom. According to Plato, this freedom encourages people to pursue only short-term desires and ignore objects of more lasting value. Is this critique relevant to contemporary society?
3. If a utilitarian were in charge of a public agency such as the Environmental Protection Agency, how would public policy decisions concerning such things as air and water pollution be decided?
4. Kantians argue that the human capacity for free and rational choice is the foundation for moral rights. Are animals capable of making such choices? Might there be another criterion to qualify someone for moral rights?

CHAPTER 2

The Environment in Western Thought

This chapter and Chapter 3 provide a basis from which we can begin our analysis and evaluation of environmental policy and philosophy. In order to make an informed and reasoned analysis, it will be helpful to have an understanding of both the past and present contexts in which environmental controversies are situated.

This chapter provides a cultural context by presenting selections from a variety of influential Western sources. As the old saying goes, "Those who don't study history are doomed to repeat it," suggesting that we can and must learn from the past. It would be a mistake to think that our environmental beliefs, attitudes, and values spring forth anew each generation. How we think about and value the environment is, in part, a product of a long history. Understanding our own attitudes on environmental issues, as well as the attitudes of those who disagree with us, requires stepping back and reflecting on the origins and history of these issues.

We may think that beliefs, attitudes, and values are simply chosen or abandoned at will. Sometimes this is true and appropriate. I used to believe in Santa Claus—I no longer do. At one time I avoided spicy foods—now I enjoy them. I was once a Yankee fan—now I dislike them. On this view, beliefs, attitudes, and values are like a suit of clothes that I can choose to put on or not. But many are not so casually related to us. Many beliefs, attitudes, and values go to the core of our own identity. They define who we are, and aban-doning them, even if we could, would take a great toll. I could no more stop valuing my children than stop breathing. If I woke up some morning hating students and no longer valuing philosophy, I would no longer be Joe DesJardins.

The beliefs and values that we are raised with, that we learn growing up, are often those that form an important core of our being. Often, these are the beliefs, attitudes, and values that are central to our social and cultural circumstances. Thus, understanding one's own culture and history is crucial to understanding one's own self.

This observation is important for understanding many public policy controversies. It is easy to think that people (usually the "other side") can and should change their beliefs when shown alternatives or given reasons. When they don't, it is also easy to dismiss others as stubborn or ignorant. But if the beliefs are central to their own understanding—and this is particularly true of beliefs and values that have a religious, political, and ethical pedigree—those beliefs are not something that can or should be easily abandoned.

Contemporary attitudes toward the natural world have a long history. Judging those attitudes without knowing their history will likely prove inadequate on rational, and perhaps ethical, grounds. Dismissing deeply held beliefs and values can easily deny people the respect that is their due. Thus, we turn to a survey of some historically important sources of Western attitudes toward the natural world.

From *The Book of Genesis*
The Origin of the World and of Mankind

There is perhaps no more influential source than the Bible in the development of Western thinking. This first selection presents the story of creation as it is described in Genesis. This story has been referenced by those on both sides of environmental debates to provide support for their positions. On one hand, we are told that God reflected on each phase of creation and "saw that it was good." This suggests that the earth and all the creatures that dwell upon it are good in themselves. The natural world has, in terms that we will see later in this book, an intrinsic value that is independent of the uses that humans make of it. On the other hand, only humans are made "in the image and likeness" of God, and this suggests that humans enjoy a privileged status in creation. Like other creatures, humans are commanded to be "fruitful and multiply" but unlike the other creatures, humans are told to "conquer" the earth and are made "masters" of the fish, birds, and animals.

In reading this selection (Genesis 1: 1–3.24), it is important to keep two questions distinct. First, there is a textual question concerning what these passages mean: How should these passages be understood? The second question is more historical: How have these passages been understood in the Judeo-Christian tradition? How have they, in fact, influenced people's thinking?

1. THE CREATION AND THE FALL

THE FIRST ACCOUNT
OF THE CREATION

1. ¹In the beginning God created the heavens and the earth. ²Now the earth was a formless void, there was darkness over the deep, and God's spirit hovered over the water.

³God said, "Let there be light," and there was light. ⁴God saw that light was good, and God divided light from darkness. ⁵God called light "day," and darkness he called "night." Evening came and morning came: the first day.

⁶God said, "Let there be a vault in the waters to divide the waters in two." And so it was. ⁷God made the vault, and it divided the waters above the vault from the waters under the vault. ⁸God called the vault "heaven." Evening came and morning came: the second day.

⁹God said, "Let the waters under heaven come together into a single mass, and let dry land appear." And so it was. ¹⁰God called the dry land "earth" and the mass of waters "seas," and God saw that it was good.

¹¹God said, "Let the earth produce vegetation: seed-bearing plants, and fruit trees bearing fruit with their seed inside, on the earth." And so it was. ¹²The earth produced vegetation: plants bearing seed in their several kinds, and trees bearing fruit with their seed inside in their several kinds. God saw that it was good. ¹³Evening came and morning came: the third day.

¹⁴God said, "Let there be lights in the vault of heaven to divide day from night, and let them indicate festivals, days and years. ¹⁵Let them be lights in the vault of heaven to shine on the earth." And so it was. ¹⁶God made the two great lights: the greater light to govern the day, the smaller light to govern the night, and the stars. ¹⁷God set them in the vault of heaven to shine on the earth, ¹⁸to govern the day and the night and to divide light from darkness. God saw that it was good. ¹⁹Evening came and morning came: the fourth day.

²⁰God said, "Let the waters teem with living creatures, and let birds fly above the earth within the vault of heaven." And so it was. ²¹God created great sea-serpents and every kind of living creature with which the waters teem, and every kind of winged creature. God saw that it was good. ²²God blessed them, saying, "Be fruitful and multiply, and fill the waters of the seas; and let the birds multiply upon the earth." ²³Evening came and morning came: the fifth day.

²⁴God said, "Let the earth produce every kind of living creature: cattle, reptiles, and every kind of wild beast." And so it was. ²⁵God made every kind of wild beast, every kind of cattle, and every kind of land reptile. God saw that it was good.

²⁶God said, "Let us make man in our own image, in the likeness of ourselves, and let them be masters of the fish of the sea, the birds of heaven, the cattle, all

the wild beasts and all the reptiles that crawl upon the earth."

²⁷ God created man in the image of himself, in the image of God he created him; male and female he created them.

²⁸ God blessed them, saying to them, "Be fruitful, multiply, fill the earth and conquer it. Be masters of the fish of the sea, the birds of heaven and all living animals on the earth." ²⁹ God said, "See, I give you all the seed-bearing plants that are upon the whole earth, and all the trees with seed-bearing fruit; this shall be your food. ³⁰ To all wild beasts, all birds of heaven and all living reptiles on the earth I give all the foliage of plants for food." And so it was. ³¹ God saw all he had made, and indeed it was very good. Evening came and morning came: the sixth day.

2. ¹ Thus heaven and earth were completed with all their array. ² On the seventh day God completed the work he had been doing. He rested on the seventh day after all the work he had been doing. ³ God blessed the seventh day and made it holy, because on that day he had rested after all his work of creating.

⁴ Such were the origins of heavens and earth when they were created.

THE SECOND ACCOUNT OF THE CREATION. PARADISE

⁵ At the time when Yahweh God made earth and heaven there was as yet no wild bush on the earth nor had any wild plant yet sprung up, for Yahweh God had not sent rain on the earth, nor was there any man to till the soil. ⁶ However, a flood was rising from the earth and watering all the surface of the soil. ⁷ Yahweh God fashioned man of dust from the soil. Then he breathed into his nostrils a breath of life, and thus man became a living being.

⁸ Yahweh God planted a garden in Eden which is in the east, and there he put the man he had fashioned. ⁹ Yahweh God caused to spring up from the soil every kind of tree, enticing to look at and good to eat, with the tree of life and the tree of the knowledge of good and evil in the middle of the garden. ¹⁰ A river flowed from Eden to water the garden, and from there it divided to make four streams. ¹¹ The first is named the Pishon, and this encircles the whole land of Havilah where there is gold. ¹² The gold of this land is pure; bdellium and onyx stone are found there. ¹³ The second river is named the Gihon, and this encircles the whole land of Cush. ¹⁴ The third river is named the Tigris, and this flows to the east of Ashur. The fourth river is the Euphrates. ¹⁵ Yahweh God took the man and settled him in the garden of Eden to cultivate and take care of it. ¹⁶ Then Yahweh God gave the man this admonition, "You may eat indeed of all the trees in the garden. ¹⁷ Nevertheless of the tree of the knowledge of good and evil you are not to eat, for on the day you eat of it you shall most surely die."

¹⁸ Yahweh God said, "It is not good that the man should be alone. I will make him a helpmate." ¹⁹ So from the soil Yahweh God fashioned all the wild beasts and all the birds of heaven. These he brought to the man to see what he would call them; each one was to bear the name the man would give it. ²⁰ The man gave names to all the cattle, all the birds of heaven and all the wild beasts. But no helpmate suitable for man was found for him. ²¹ So Yahweh God made the man fall into a deep sleep. And while he slept, he took one of his ribs and enclosed it in flesh. ²² Yahweh God built the rib he had taken from the man into a woman, and brought her to the man. ²³ The man exclaimed: "This at last is bone from my bones, and flesh from my flesh! This is to be called woman, for this was taken from man."

²⁴ This is why a man leaves his father and mother and joins himself to his wife, and they become one body.

²⁵ Now both of them were naked, the man and his wife, but they felt no shame in front of each other.

THE FALL

3. ¹ The serpent was the most subtle of all the wild beasts that Yahweh God had made. It asked the woman, "Did God really say you were not to eat from any of the trees in the garden?" ² The woman answered the serpent, "We may eat the fruit of the trees in the garden. ³ But of the fruit of the tree in the middle of the garden God said, 'You must not eat it, nor touch it, under pain of death.'" ⁴ Then the serpent said to the woman, "No! you will not die! ⁵ God knows in fact that on the day you eat it your eyes will be opened and you will be like gods, knowing good and evil." ⁶ The woman saw that the tree was good to eat and pleasing to the eye, and that it was desirable for the knowledge that it could give. So she took some of its fruit and ate it. She gave some also to her husband who was with her, and he ate it. ⁷ Then the eyes of both of them were

opened and they realized that they were naked. So they sewed fig leaves together to make themselves loincloths.

⁸ The man and his wife heard the sound of Yahweh God walking in the garden in the cool of the day, and they hid from Yahweh God among the trees of the garden. ⁹ But Yahweh God called to the man, "Where are you?" he asked. ¹⁰ "I heard the sound of you in the garden," he replied. "I was afraid because I was naked, so I hid." ¹¹ "Who told you that you were naked?" he asked. "Have you been eating of the tree I forbade you to eat?" ¹² The man replied, "It was the woman you put with me; she gave me the fruit, and I ate it." ¹³ Then Yahweh God asked the woman. "What is this you have done?" The woman replied, "The serpent tempted me and I ate."

¹⁴ Then Yahweh God said to the serpent, because you have done this, "be accursed beyond all cattle, all wild beasts. You shall crawl on your belly and eat dust every day of your life. ¹⁵ I will make you enemies of each other: you and the woman, your offspring and her offspring. It will crush your head and you will strike its heel."

¹⁶ To the woman he said: "I will multiply your pains in childbearing, you shall give birth to your children in pain. Your yearning shall be for your husband, yet he will lord it over you."

¹⁷ To the man he said, "Because you listened to the voice of your wife and ate from the tree of which I had forbidden you to eat, accursed be the soil because of you. With suffering shall you get your food from it every day of your life. ¹⁸ It shall yield you brambles and thistles, and you shall eat wild plants. ¹⁹ With sweat on your brow shall you eat your bread, until you return to the soil, as you were taken from it. For dust you are and to dust you shall return."

²⁰ The man named his wife "Eve" because she was the mother of all those who live. ²¹ Yahweh God made clothes out of skins for the man and his wife, and they put them on. ²² Then Yahweh God said, "See, the man has become like one of us, with his knowledge of good and evil. He must not be allowed to stretch his hand out next and pick from the tree of life also, and eat some and live for ever." ²³ So Yahweh God expelled him from the garden of Eden, to till the soil from which he had been taken. ²⁴ He banished the man, and in front of the garden of Eden he posted the cherubs, and the flame of a flashing sword, to guard the way to the tree of life.

For Further Discussion

1. In this translation of Genesis, after creating the earth, God "saw that it was good." Does this suggest to you that the goodness of natural objects is independent of human choices and human uses? How might this natural goodness be explained?

2. In what ways might humans be created in the image of God? How, if at all, might this relate to the recognition of the goodness of other parts of creation?

3. In the second account of creation given at Chapter 2, verse 15, God commands humans to "cultivate and take care of" Eden. Can this command be reconciled with the earlier statement at Chapter 1, verse 28, to "fill the earth and conquer it"?

From **The Book of Job**

This second selection is taken from the Job (38: 1–42.6). After suffering untold calamities, Job curses God, and God replies by reminding Job, and all humans, of the relative place of humankind in creation. Our selection is God's answer to Job. This is an angry speech that criticizes human arrogance concerning the natural world ("Did you proclaim the rules that govern the heavens, or determine the laws of nature on earth?"). It is a response that celebrates the diversity and complexity of animal life while reminding humans that they are not the lords of creation.

38 Then the Lord answered Job out of the
     tempest:

² 　Who is this whose ignorant words
　　cloud my design in darkness?
³ 　Brace yourself and stand up like a man;
　　I will ask questions, and you shall answer.
⁴ 　Where were you when I laid the earth's
　　 foundations?
　　Tell me, if you know and understand.

5 Who settled its dimensions? Surely you should
   know.
   Who stretched his measuring-line over it?
6 On what do its supporting pillars rest?
   Who set its corner-stone in place,
7 when the morning stars sang together
   and all the sons of God shouted aloud?
8 Who watched over the birth of the sea,
   when it burst in flood from the womb?—
9 when I wrapped it in a blanket of cloud
   and cradled it in fog,
10 when I established its bounds,
   fixing its doors and bars in place,
11 and said, "Thus far shall you come and no
   farther,
   and here your surging waves shall halt."
12 In all your life have you ever called up the dawn
   or shown the morning its place?
13 Have you taught it to grasp the fringes of the
   earth
   and shake the Dog-star from its place;
14 to bring up the horizon in relief as clay under a
   seal,
   until all things stand out like the folds of a
   cloak,
15 when the light of the Dog-star is dimmed
   and the stars of the Navigator's Line go out one
   by one?
16 Have you descended to the springs of the sea
   or walked in the unfathomable deep?
17 Have the gates of death been revealed to you?
   Have you ever seen the door-keepers of the place
   of darkness?
18 Have you comprehended the vast expanse of the
   world?
   Come, tell me all this, if you know.
19 Which is the way to the home of light
   and where does darkness dwell?
20 And can you then take each to its appointed
   bound
   and escort it on its homeward path?
21 Doubtless you know all this; for you were born
   already,
   so long is the span of your life!
22 Have you visited the storehouse of the snow
   or seen the arsenal where hail is stored,
23 which I have kept ready for the day of calamity,
   for war and for the hour of battle?
24 By what paths is the heat spread abroad
   or the east wind carried far and wide over the
   earth?

25 Who has cut channels for the downpour
   and cleared a passage for the thunderstorm,
26 for rain to fall on land where no man lives
   and on the deserted wilderness,
27 clothing lands waste and derelict with green
   and making grass grow on thirsty ground?
28 Has the rain a father?
   Who sired the drops of dew?
29 Whose womb gave birth to the ice,
   and who was the mother of the frost from
   heaven,
30 which lays a stony cover over the waters
   and freezes the expanse of ocean?
31 Can you bind the cluster of the Pleiades
   or loose Orion's belt?
32 Can you bring out the signs of the zodiac in
   their season
   or guide Aldebaran and its train?
33 Did you proclaim the rules that govern the
   heavens,
   or determine the laws of nature on earth?
34 Can you command the dense clouds
   to cover you with their weight of waters?
35 If you bid lightning speed on its way,
   will it say to you, "I am ready"?
36 Who put wisdom in depths of darkness
   and veiled understanding in secrecy?
37 Who is wise enough to marshal the rain-clouds
   and empty the cisterns of heaven,
38 when the dusty soil sets hard as iron,
   and the clods of earth cling together?
39 Do you hunt her prey for the lioness
   and satisfy the hunger of young lions,
40 as they crouch in the lair
   or lie in wait in the covert?
41 Who provides the raven with its quarry
   when its fledglings croak for lack of food?

39 Do you know when the mountain-goats are
born or attend the wild doe when she is in
labour?

2 Do you count the months that they carry their
   young
   or know the time of their delivery,
3 when they crouch down to open their wombs
   and bring their offspring to the birth,
4 when the fawns grow and thrive in the open
   forest,
   and go forth and do not return?
5 Who has let the wild ass of Syria range at will
   and given the wild ass of Arabia its freedom?—

6 whose home I have made in the wilderness
and its lair in the saltings;

7 it disdains the noise of the city
and is deaf to the driver's shouting;

8 it roams the hills as its pasture
and searches for anything green.

9 Does the wild ox consent to serve you,
does it spend the night in your stall?

10 Can you harness its strength with ropes,
or will it harrow the furrows after you?

11 Can you depend on it, strong as it is,
or leave your labour to it?

12 Do you trust it to come back
and bring home your grain to the threshing-
floor?

13 The wings of the ostrich are stunted;
her pinions and plumage are so scanty

14 that she abandons her eggs to the ground,
letting them be kept warm by the sand.

15 She forgets that a foot may crush them,
or a wild beast trample on them;

16 she treats her chicks heartlessly as if they were
not hers,
not caring if her labour is wasted

17 (for God has denied her wisdom
and left her without sense),

18 while like a cock she struts over the uplands,
scorning both horse and rider.

19 Did you give the horse his strength?
Did you clothe his neck with a mane?

20 Do you make him quiver like a locust's wings,
when his shrill neighing strikes terror?

21 He shows his mettle as he paws and prances;
he charges the armoured line with all his might.

22 He scorns alarms and knows no dismay;
he does not flinch before the sword.

23 The quiver rattles at his side,
the spear and sabre flash.

24 Trembling with eagerness, he devours the
ground
and cannot be held in when he hears the horn;

25 at the blast of the horn he cries "Aha!"
and from afar he scents the battle:

26 Does your skill teach the hawk to use its pinions
and spread its wings towards the south?

27 Do you instruct the vulture to fly high
and build its nest aloft?

28 It dwells among the rocks and there it lodges;
its station is a crevice in the rock;

29 from there it searches for food,
keenly scanning the distance,

30 that its brood may be gorged with blood;
and where the slain are, there the vulture is.

41 Can you pull out the whale with a gaff
or can you slip a noose round its tongue?

2 Can you pass a cord through its nose
or put a hook through its jaw?

3 Will it plead with you for mercy
or beg its life with soft words?

4 Will it enter into an agreement with you
to become your slave for life?

5 Will you toy with it as with a bird
or keep it on a string like a song-bird for your
maidens?

6 Do trading-partners haggle over it
or merchants share it out?

40 Then the Lord said to Job:

2 Is it for a man who disputes with the Almighty
to be stubborn?
Should he that argues with God answer back?

3 And Job answered the Lord:

4 What reply can I give thee, I who carry no
weight?
I put my finger to my lips.

5 I have spoken once and now will not answer
again;
twice have I spoken, and I will do so no more.

6 Then the Lord answered Job out of the tempest:

7 Brace yourself and stand up like a man;
I will ask questions, and you shall answer.

8 Dare you deny that I am just
or put me in the wrong that you may be right?

9 Have you an arm like God's arm,
can you thunder with a voice like his?

10 Deck yourself out, if you can, in pride and dignity,
array yourself in pomp and splendour;

11 unleash the fury of your wrath,
look upon the proud man and humble him;

12 look upon every proud man and bring him low,
throw down the wicked where they stand;

13 hide them in the dust together,
and shroud them in an unknown grave.

14 Then I in my turn will acknowledge
that your own right hand can save you.

15 Consider the chief of the beasts, the crocodile,
who devours cattle as if they were grass:

16 what strength is in his loins!
what power in the muscles of his belly!

17 His tail is rigid as a cedar,
the sinews of his flanks are closely knit,

18   his bones are tubes of bronze,
    and his limbs like bars of iron.

19   He is the chief of God's works,
    made to be a tyrant over his peers;

20   for he takes the cattle of the hills for his prey
    and in his jaws he crunches all wild beasts.

21   There under the thorny lotus he lies,
    hidden in the reeds and the marsh;

22   the lotus conceals him in its shadow,
    the poplars of the stream surround him.

23   If the river is in spate, he is not scared,
    he sprawls at his ease though the stream is in
        flood.

24   Can a man blind his eyes and take him
    or pierce his nose with the teeth of a trap?

41 Can you fill his skin with harpoons

7   or his head with fish-hooks?

8   If ever you lift your hand against him,
    think of the struggle that awaits you, and let be.

9   No, such a man is in desperate case,
    hurled headlong at the very sight of him.

10 How fierce he is when he is roused!
    Who is there to stand up to him?

11 Who has ever attacked him unscathed?
    Not a man under the wide heaven.

12 I will not pass over in silence his limbs,
    his prowess and the grace of his proportions.

13 Who has ever undone his outer garment
    or penetrated his doublet of hide?

14 Who has ever opened the portals of his face?
    for there is terror in his arching teeth.

15 His back is row upon row of shields,
    enclosed in a wall of flints;

16 one presses so close on the other
    that air cannot pass between them,

17 each so firmly clamped to its neighbour
    that they hold and cannot spring apart.

18 His sneezing sends out sprays of light,
    and his eyes gleam like the shimmer of dawn.

19 Firebrands shoot from his mouth,
    and sparks come streaming out;

20 his nostrils pour forth smoke
    like a cauldron on a fire blown to full heat.

21 His breath sets burning coals ablaze,
    and flames flash from his mouth.

22 Strength is lodged in his neck,
    and untiring energy dances ahead of him.

23 Close knit is his underbelly,
    no pressure will make it yield.

24 His heart is firm as a rock,
    firm as the nether millstone.

25 When he raises himself, strong men take fright,
    bewildered at the lashings of his tail.

26 Sword or spear, dagger or javelin,
    if they touch him, they have no effect.

27 Iron he counts as straw,
    and bronze as rotting wood.

28 No arrow can pierce him,
    and for him sling-stones are turned into chaff;

29 to him a club is a mere reed,
    and he laughs at the swish of the sabre.

30 Armoured beneath with jagged sherds,
    he sprawls on the mud like a threshing-sledge.

31 He makes the deep water boil like a cauldron,
    he whips up the lake like ointment in a mixing-
        bowl.

32 He laves a shining trail behind him,
    and the great river is like white hair in his wake.

33 He has no equal on earth;
    for he is made quite without fear.

34 He looks down on all creatures, even the highest;
    he is king over all proud beasts.

42 Then Job answered the Lord:

2   I know that thou canst do all things
    and that no purpose is beyond thee.

3   But I have spoken of great things which I have
        not understood,
    things too wonderful for me to know.

5   I knew of thee then only by report,
    but now I see thee with my own eyes.

6   Therefore I melt away;
    I repent in dust and ashes.

EPILOGUE

7   When the Lord had finished speaking to Job, he said to Eliphaz the Temanite, "I am angry with you and your two friends, because you have not spoken as you ought about me, as my servant Job has done. So now take seven bulls and seven rams, go to my servant Job and offer a whole-offering for yourselves, and he will intercede for you; I will surely show him favour by not being harsh with you because you have not spoken as you ought about me, as he has done." Then Eliphaz the Temanite and Bildad the Shuhite and Zophar the Naamathite went and carried out the Lord's command, and the Lord showed favour to Job when he had interceded for his friends. So the Lord restored Job's fortunes and doubled all his possessions.

11 Then all Job's brothers and sisters and his former acquaintance came and feasted with him in his home, and they consoled and comforted him for all the misfortunes which the Lord had brought on him; and each of them gave him a
12 sheep and a gold ring. Furthermore, the Lord blessed the end of Job's life more than the beginning; and he had fourteen thousand head of small cattle and six thousand camels, a thousand yoke of oxen and as many she-asses. He had
13 14 seven sons and three daughters; and he named his eldest daughter Jemimah, the second Keziah
15 and the third Kerenhappuch. There were no women in all the world so beautiful as Job's daughters; and their father gave them an inheritance with their brothers.
16 Thereafter Job lived another hundred and forty
17 years, he saw his sons and his grandsons to four generations, and died at a very great age.

For Further Discussion

1. This lecture to Job is a strong criticism of human arrogance. The God of this speech has a clear notion of the appropriate place of humanity in creation. How do you compare this speech with the command from Genesis to conquer the earth?

2. What lesson do you think is intended by this speech? What follows if humans are ignorant of God's purposes?

3. Throughout this lecture, God points out many situations in which humans are ignorant of the workings of nature. Since this was written, humans have learned much that they did not know in previous times. How much of this speech would still be relevant today?

4. Besides pointing out human ignorance about the natural world, God also argues the very limited ability of humans to control that world. Again, much has changed since this was written. Do humans have a right to be more confident in their ability to understand and control nature today?

The Canticle of Brother Sun

Saint Francis of Assisi

Despite the reprimand to Job, much of the Western tradition has been more influenced by the command to subdue and conquer the world. One notable exception within the Christian tradition was Saint Francis of Assisi (1182–1226). Founder of the Franciscan order of friars in the Catholic Church, Saint Francis treated nonhuman creatures with a love and respect that he believed was due all of God's creations. Our third selection, "The Canticle of Brother Sun," is a poem that acclaims the blessedness of the entire natural world.

1. Most High, all-powerful, good Lord,
   Yours are the praises, the glory, the honor, and all blessing.

2. To You alone, Most High, do they belong,
   and no man is worthy to mention Your name.

3. Praised be You, my Lord, with all your creatures,
   especially Sir Brother Sun,
   Who is the day and through whom You give us light.

4. And he is beautiful and radiant with great splendor;
   and bears a likeness of You, Most High One.

5. Praised be You, my Lord, through Sister Moon and the stars,
   in heaven You formed them clear and precious and beautiful.

6. Praised be You, my Lord, through Brother Wind,
   and through the air, cloudy and serene, and every kind of weather
   through which You give sustenance to Your creatures.

7. Praised be You, my Lord, through Sister Water,
   which is very useful and humble and precious and chaste.

8. Praised be You, my Lord, through Brother Fire,
through whom You light the night
and he is beautiful and playful and robust and
   strong.

9. Praised be You, my Lord, through our Sister
   Mother Earth,
who sustains and governs us,
and who produces varied fruits with colored
   flowers and herbs.

10. Praised be You, my Lord, through those who give
   pardon for Your love
and bear infirmity and tribulation.

11. Blessed are those who endure in peace
for by You, Most High, they shall be crowned.

12. Praised be You, my Lord, through our Sister
   Bodily Death,
from whom no living man can escape.

13. Woe to those who die in mortal sin.
Blessed are those whom death will find in Your
   most holy will,
for the second death shall do them no harm.

14. Praise and bless my Lord and give Him thanks
and serve Him with great humility.

For Further Discussion

1. Saint Francis calls for praise of God through the
sun, moon, wind, water, and "Sister Mother
Earth." How could humans praise their God
through their attitudes toward natural objects?

2. Saint Francis speaks of the sun, moon, and earth
as God's creatures. How do you interpret this
phrase? Do you think of the earth as a creature?

3. This poem suggests that the natural world is the
vehicle through which God gives sustenance to
humans. Natural objects therefore are gifts from
God to humans. How ought we to treat such gifts?

4. It could be said that Saint Francis attributes only
instrumental value to natural objects. The sun,
moon, wind, water, and earth are praised for what
they provide to humans. Do you agree with that
view?

Differences Between Rational and Other Creatures

Saint Thomas Aquinas

Saint Thomas Aquinas (1225–1274) was a more influential medieval thinker than Saint Francis. Aquinas, a philosopher and theologian, was a major force in integrating Aristotle's philosophy with Christian theology. This synthesis, described in Chapter 1 as natural law philosophy, influenced Christian and Catholic thinking well into the twentieth century. In this selection, Aquinas presents a more hierarchical view of creation than that offered by Saint Francis. Humans are distinct from the rest of creation in that, being created in the image and likeness of God, they possess intellect and reason. Reason gives humans the capacity for free and responsible choice, they have dominion over their own actions. Creatures lacking this capacity—all creatures except humans as it turns out—cannot be the cause of their own actions and therefore are necessarily subject to the acts of others. Like slaves, creatures lacking an intellect can rightfully be used for the sake of creatures with an intellect.

In the first place then, the very condition of the rational creature, in that it has dominion over its actions, requires that the care of providence should be bestowed on it for its own sake: whereas the condition of other things that have not dominion over their actions shows that they are cared for, not for their own sake, but as being directed to other things. Because that which acts only when moved by another, is like an instrument; whereas that which acts by itself, is like a principal agent. Now an instrument is required, not for its own sake, but that the principal agent may use it. Hence whatever is done for the care of the instruments must be referred to the principal agent as its end: whereas any such action directed to the principal agent as such, either by the agent itself or by another, is for the sake of the same principal agent. Accordingly intellectual creatures are ruled by God, as though He cared for them for their own sake, while other creatures are ruled as being directed to rational creatures.

Again. That which has dominion over its own act, is free in its action, because *he is free who is cause of himself:* whereas that which by some kind of necessity is moved by another to act, is subject to slavery. Therefore every other creature is naturally under slavery; the intellectual nature alone is free. Now, in every government provision is made for the free for their own sake; but for slaves that they may be useful to the free. Accordingly divine providence makes provision for the intellectual creature for its own sake, but for other creatures for the sake of the intellectual creature.

Moreover. Whenever certain things are directed to a certain end, if any of them are unable of themselves to attain to the end, they must needs be directed to those that attain to the end, which are directed to the end for their own sake. Thus the end of the army is victory, which the soldiers obtain by their own action in fighting, and they alone in the army are required for their own sake; whereas all others, to whom other duties are assigned, such as the care of horses, the preparing of arms, are requisite for the sake of the soldiers of the army. Now, it is clear from what has been said, that God is the last end of the universe, whom the intellectual nature alone obtains in Himself, namely by knowing and loving Him, as was proved above. Therefore the intellectual nature alone is requisite for its own sake in the universe, and all others for its sake.

Further. In every whole, the principal parts are requisite on their own account for the completion of the whole, while others are required for the preservation or betterment of the former. Now, of all the parts of the universe, intellectual creatures hold the highest place, because they approach nearest to the divine likeness. Therefore divine providence provides for the intellectual nature for its own sake, and for all others for its sake.

Besides. It is clear that all the parts are directed to the perfection of the whole: since the whole is not on account of the parts, but the parts on account of the whole. Now, intellectual natures are more akin to the whole than other natures: because, in a sense, the intellectual substance is all things, inasmuch as by its intellect it is able to comprehend all things; whereas every other substance has only a particular participation of being. Consequently God cares for other things for the sake of intellectual substances.

Besides. Whatever happens to a thing in the course of nature happens to it naturally. Now, we see that in the course of nature the intellectual substance uses all others for its own sake; either for the perfection of the intellect, which sees the truth in them as in a mirror;

or for the execution of its power and development of its knowledge, in the same way as a craftsman develops the conception of his art in corporeal matter; or again to sustain the body that is united to an intellectual soul, as is the case in man. It is clear, therefore, that God cares for all things for the sake of intellectual substances.

Moreover. If a man seek something for its own sake, he seeks it always, because *what is per se, is always:* whereas if he seek a thing on account of something else, he does not of necessity seek it always but only in reference to that for the sake of which he seeks it. Now, as we proved above, things derive their being from the divine will. Therefore whatever is always is willed by God for its own sake; and what is not always is willed by God, not for its own sake, but for another's. Now, intellectual substances approach nearest to being always, since they are incorruptible. They are, moreover, unchangeable, except in their choice. Therefore intellectual substances are governed for their own sake, as it were; and others for the sake of intellectual substances.

The fact that all the parts of the universe are directed to the perfection of the whole is not in contradiction with the foregoing conclusion: since all the parts are directed to the perfection of the whole, in so far as one part serves another. Thus in the human body it is clear that the lungs belong to the body's perfection, in that they serve the heart: wherefore there is no contradiction in the lungs being for the sake of the heart, and for the sake of the whole animal. In like manner that other natures are on account of the intellectual is not contrary to their being for the perfection of the universe: for without the things required for the perfection of the intellectual substance, the universe would not be complete.

Nor again does the fact that individuals are for the sake of the species militate against what has been said. Because through being directed to their species, they are directed also to the intellectual nature. For a corruptible thing is directed to man, not on account of only one individual man, but on account of the whole human species. Yet a corruptible thing could not serve the whole human species, except as regards its own entire species. Hence the order whereby corruptible things are directed to man, requires that individuals be directed to the species.

When we assert that intellectual substances are directed by divine providence for their own sake, we do not mean that they are not also referred by God and for the perfection of the universe. Accordingly they are said to be provided for on their own account, and others on account of them, because the goods bestowed on them by divine providence are not given them for another's profit: whereas those bestowed on others are in the divine plan intended for the use of intellectual substances. Hence it is said (Deut. iv. 19): *Lest thou see the sun and the moon and the other stars, and being deceived by error, thou adore and serve them, which the Lord thy God created for the service of all the nations that are under heaven:* and (Ps. viii. 8): *Thou hast subjected all things under his feet, all sheep and oxen: moreover, the beasts also of the field:* and (Wis. xii. 18): *Thou, being master of power, judgest with tranquillity, and with great favour disposest of us.*

Hereby is refuted the error of those who said it is sinful for a man to kill dumb animals: for by divine providence they are intended for man's use in the natural order. Hence it is no wrong for man to make use of them, either by killing or in any other way whatever. For this reason the Lord said to Noe (Gen. ix. 3): *As the green herbs I have deliverd all flesh to you.*

And if any passages of Holy Writ seem to forbid us to be cruel to dumb animals, for instance to kill a bird with its young: this is either to remove man's thoughts from being cruel to other men, and lest through being cruel to animals one become cruel to human beings: or because injury to an animal leads to the temporal hurt of man, either of the doer of the deed, or of another: or on account of some signification: thus the Apostle expounds the prohibition against *muzzling the ox that treadeth the corn.*

For Further Discussion

1. Aquinas begins with a fundamental distinction between creatures that have dominion over their own actions and those that do not. What do you think he means by this? Do you think any nonhuman animals might have dominion over their actions in the sense that Aquinas means?

2. Aquinas believes that creatures without the freedom to direct their own actions therefore must not have their own natural ends. Thus, they can rightfully be treated as "slaves" that can be used for the ends of intellectual creatures. What reasons might he have to believe this?

3. Aquinas quotes a passage from the Book of Deuteronomy that warns against the error of treating the sun and moon as objects of adoration. How does this view compare to the poem by Saint Francis in the previous selection?

4. Aquinas concludes with an observation that there are reasons not to be cruel to animals, but these reasons are all connected to potential harms to other humans. Do you agree with that view?

Animals as Automata

René Descartes

René Descartes (1596–1650) is perhaps the best-known philosophical defender of a categorical distinction between humans and the rest of nature. Descartes was writing at a time when science, particularly the mechanistic physics of Galileo, was challenging the view that nature was purposive or directional (see the discussion of teleology in Chapter 1). This scientific revolution claimed that the natural world operated according to strict deterministic causal laws. Descartes accepted the truth of this for the physical world but argued that humans were both physical and mental beings. Cartesian dualism, the claim that humans have a body and a mind, allowed for both the truth of mechanistic science and the distinctiveness of human beings. Humans have a mind in virtue of which they think, reason, and are conscious and free. In this selection, Descartes explains how animals, despite appearances, are really no more than automata or moving machines. Descartes does not deny life (the "heat of the heart") nor sensation to animals; he only denies that they have the mental capacity of being conscious of their own life and sensations. Thus, humans are absolved from crime "when they eat or kill animals."

I

I had explained all these matters in some detail in the Treatise which I formerly intended to publish. And afterwards I had shown there, what must be the fabric of the nerves and muscles of the human body in order that the animal spirits therein contained should have the power to move the members, just as the heads of animals, a little while after decapitation, are still observed to move and bite the earth, notwithstanding that they are no longer animate; what changes are necessary in the brain to cause wakefulness, sleep and dreams; how light, sounds, smells, tastes, heat and all other qualities pertaining to external objects are able to imprint on it various ideas by the intervention of the senses; how hunger, thirst and other internal affections can also convey their impressions upon it; what should be regarded as the "common sense" by which these ideas are received, and what is meant by the memory which retains them, by the fancy which can change them in diverse ways and out of them constitute new ideas, and which, by the same means, distributing the animal spirits through the muscles, can cause the members of such a body to move in as many diverse ways, and in a manner as suitable to the objects which present themselves to its senses and to its internal passions, as can happen in our own case apart from the direction of our free will. And this will not seem strange to those, who, knowing how many different *automata* or moving machines can be made by the industry of man, without employing in so doing more than a very few parts in comparison with the great multitude of bones, muscles, nerves, arteries, veins, or other parts that are found in the body of each animal. From this aspect the body is regarded as a machine which, having been made by the hands of God, is incomparably better arranged, and possesses in itself movements which are much more admirable, than any of those which can be invented by man. Here I specially stopped to show that if there had been such machines, possessing the organs and outward form of a monkey or some other animal without reason, we should not have had any means of ascertaining that they were not of the same nature as those animals. On the other hand, if there were machines which bore a resemblance to our body and imitated our actions as far as it was morally possible to do so, we should always have two very certain tests by which to recognise that, for all that, they were not real men. The first is, that they could never use speech or other signs as we do when placing our thoughts on record for the benefit of others. For we can easily understand a machine's being constituted so that it can utter words, and even emit some responses to action on it of a corporeal kind, which brings about a change in its organs; for instance, if it is touched in a particular part it may ask what we wish to say to it; if in another part it may exclaim that it is being hurt, and so on. But it never happens that it arranges its speech in various ways, in order to reply appropriately to everything that may be said in its presence, as even the lowest type of man can do. And the second difference is, that although machines can perform certain things as well as or perhaps better than any of us can do, they infallibly fall short in others, by the which means we may discover that they did not act from knowledge, but only from the disposition of their organs. For while reason is a universal instrument which can serve for all contingencies, these organs have need of some special adaptation for every particular action. From this it follows that it is morally impossible that there should be sufficient diversity in any machine to allow it to act in all the events of life in the same way as our reason causes us to act.

By these two methods we may also recognise the difference that exists between men and brutes. For it is a very remarkable fact that there are none so depraved and stupid, without even excepting idiots, that they cannot arrange different words together, forming of them a statement by which they make known their thoughts; while, on the other hand, there is no other animal, however perfect and fortunately circumstanced it may be, which can do the same. It is not the want of organs that brings this to pass, for it is evident that magpies and parrots are able to utter words just like ourselves, and yet they cannot speak as we do, that is, so as to give evidence that they think of what they say. On the other hand, men who, being born deaf and dumb, are in the same degree, or even more than the brutes, destitute of the organs which serve the others for talking, are in the habit of themselves inventing certain signs by which they make themselves understood by those who, being usually in their company, have leisure to learn their language. And this does not merely show that the brutes have less reason than men, but that they have none at all, since it is clear that very little is required in order to be able to talk. And when we notice the inequality that exists between animals of the same species, as well as between men, and observe that some are more capable of receiving instruction than others, it is not credible that a monkey or a parrot, selected as the most perfect of its species, should not in these matters equal the stupidest child to be found, or at least a child whose mind is clouded, unless in the case of the brute the soul were of an entirely different nature from ours. And we ought not to confound speech with natural movements which betray passions and may be imitated by machines as well as be manifested by animals; nor must we think, as did some of the ancients, that brutes talk, although we do not understand their language. For if this were true, since they have many organs which are allied to our own, they could communicate their thoughts to us just as easily as to those of their own race. It is also a very remarkable fact that although there are many animals which exhibit more dexterity than we do in some of their actions, we at the same time observe that they do not manifest any dexterity at all in many others. Hence the fact that they do better than we do, does not prove that they are endowed with mind, for in this case they would have more reason than any of us, and would surpass us in all other things. It rather shows that they have no reason at all, and that it is nature which acts in them according to the disposition of their organs, just as a clock, which is only composed of wheels and weights is able to tell the hours and measure the time more correctly than we can do with all our wisdom.

I had described after this the rational soul and shown that it could not be in any way derived from the power of matter, like the other things of which I had spoken, but that it must be expressly created. I showed, too, that it is not sufficient that it should be lodged in the human body like a pilot in his ship, unless perhaps for the moving of its members, but that it is necessary that it should also be joined and united more closely to the body in order to have sensations and appetites similar to our own, and thus to form a true man. In conclusion, I have here enlarged a little on the subject of the soul, because it is one of the greatest importance. For next to the error of those who deny God, which I think I have already sufficiently refuted, there is none which is more effectual in leading feeble spirits from the straight path of virtue, than to imagine that the soul of the brute is of the same nature as our own, and that in consequence, after this life we have nothing to fear or to hope for, any more than the flies and ants. As a matter of fact, when one comes to know how greatly they differ, we understand much better the reasons which go to prove that our soul is in its nature entirely independent of body, and in consequence that it is not liable to die with it. And then, inasmuch as we observe no other causes capable of

destroying it, we are naturally inclined to judge that it is immortal.

II

I cannot share the opinion of Montaigne and others who attribute understanding or thought to animals. I am not worried that people say that men have an absolute empire over all the other animals; because I agree that some of them are stronger than us, and believe that there may also be some who have an instinctive cunning capable of deceiving the shrewdest human beings. But I observe that they only imitate or surpass us in those of our actions which are not guided by our thoughts. It often happens that we walk or eat without thinking at all about what we are doing; and similarly, without using our reason, we reject things which are harmful for us, and parry the blows aimed at us. Indeed, even if we expressly willed not to put our hands in front of our head when we fall, we could not prevent ourselves. I think also that if we had no thought we would eat, as the animals do, without having to learn to; and it is said that those who walk in their sleep sometimes swim across streams in which they would drown if they were awake. As for the movements of our passions, even though in us they are accompanied with thought because we have the faculty of thinking, it is none the less very clear that they do not depend on thought, because they often occur in spite of us. Consequently they can also occur in animals, even more violently than they do in human beings, without our being able to conclude from that that they have thoughts.

In fact, none of our external actions can show anyone who examines them that our body is not just a self-moving machine but contains a soul with thoughts, with the exception of words, or other signs that are relevant to particular topics without expressing any passion. I say words or other signs, because deaf-mutes use signs as we use spoken words; and I say that these signs must be relevant, to exclude the speech of parrots, without excluding the speech of madmen, which is relevant to particular topics even though it does not follow reason. I add also that these words or signs must not express any passion, to rule out not only cries of joy or sadness and the like, but also whatever can be taught by training to animals. If you teach a magpie to say good-day to its mistress, when it sees her approach, this can only be by making the utterance of this word the expression of one of its passions. For instance it will be an expression of the hope of eating, if it has always been given a titbit when it says it. Similarly, all the things which dogs, horses, and monkeys are taught to perform are only expressions of their fear, their hope, or their joy; and consequently they can be performed without any thought. Now it seems to me very striking that the use of words, so defined, is something peculiar to human beings. Montaigne and Charron may have said that there is more difference between one human being and another than between a human being and an animal; but there has never been known an animal so perfect as to use a sign to make other animals understand something which expressed no passion; and there is no human being so imperfect as not to do so, since even deaf-mutes invent special signs to express their thoughts. This seems to me a very strong argument to prove that the reason why animals do not speak as we do is not that they lack the organs but that they have no thoughts. It cannot be said that they speak to each other and that we cannot understand them; because since dogs and some other animals express their passions to us, they would express their thoughts also if they had any.

I know that animals do many things better than we do, but this does not surprise me. It can even be used to prove they act naturally and mechanically, like a clock which tells the time better than our judgement does. Doubtless when the swallows come in spring, they operate like clocks. The actions of honeybees are of the same nature, and the discipline of cranes in flight, and of apes in fighting, if it is true that they keep discipline. Their instinct to bury their dead is no stranger than that of dogs and cats who scratch the earth for the purpose of burying their excrement; they hardly ever actually bury it, which shows that they act only by instinct and without thinking. The most that one can say is that though the animals do not perform any action which shows us that they think, still, since the organs of their body are not very different from ours, it may be conjectured that there is attached to those organs some thoughts such as we experience in ourselves, but of a very much less perfect kind. To which I have nothing to reply except that if they thought as we do, they would have an immortal soul like us. This is unlikely, because there is no reason to believe it of some animals without believing it of all, and many of them such as oysters and sponges are too imperfect for this to be credible. But I am afraid of boring you with this discussion, and my only desire is to show you that I am, etc.

III

But there is no prejudice to which we are all more accustomed from our earliest years than the belief that dumb animals think. Our only reason for this belief is the fact that we see that many of the organs of animals are not very different from ours in shape and movement. Since we believe that there is a single principle within us which causes these motions—namely the soul, which both moves the body and thinks—we do not doubt that some such soul is to be found in animals also. I came to realize, however, that there are two different principles causing our motions: one is pure mechanical and corporeal and depends solely on the force of the spirits and the construction of our organs, and can be called the corporeal soul; the other is the incorporeal mind, the soul which I have defined as a thinking substance. Thereupon I investigated more carefully whether the motions of animals originated from both these principles or from one only. I soon saw clearly that they could all originate from the corporeal and mechanical principle, and I thenceforward regarded it as certain and established that we cannot at all prove the presence of a thinking soul in animals. I am not disturbed by the astuteness and cunning of dogs and foxes, or all the things which animals do for the sake of food, sex, and fear; I claim that I can easily explain the origin of all of them from the constitution of their organs.

But though I regard it as established that we cannot prove there is any thought in animals, I do not think it is thereby proved that there is not, since the human mind does not reach into their hearts. But when I investigate what is most probable in this matter, I see no argument for animals having thoughts except the fact that since they have eyes, ears, tongues, and other sense-organs like ours, it seems likely that they have sensation like us; and since thought is included in our mode of sensation, similar thought seems to be attributable to them. This argument, which is very obvious, has taken possession of the minds of all men from their earliest age. But there are other arguments, stronger and more numerous, but not so obvious to everyone, which strongly urge the opposite. One is that it is more probable that worms and flies and caterpillars move mechanically than that they all have immortal souls.

It is certain that in the bodies of animals, as in ours, there are bones, nerves, muscles, animal spirits, and other organs so disposed that they can by themselves, without any thought, give rise to all animals the motions we observe. This is very clear in convulsive movements when the machine of the body moves despite the soul, and sometimes more violently and in a more varied manner than when it is moved by the will.

Second, it seems reasonable, since art copies nature, and men can make various automata which move without thought, that nature should produce its own automata, much more splendid than artificial ones. These natural automata are the animals. This is especially likely since we have no reason to believe that thought always accompanies the disposition of organs which we find in animals. It is much more wonderful that a mind should be found in every human body than that one should be lacking in every animal.

But in my opinion the main reason which suggests that the beasts lack thought is the following. Within a single species some of them are more perfect than others, as men are too. This can be seen in horses and dogs, some of whom learn what they are taught much better than others. Yet, although all animals easily communicate to us, by voice or bodily movement, their natural impulses of anger, fear, hunger and so on, it has never yet been observed that any brute animal reached the stage of using real speech, that is to say, of indicating by word or sign something pertaining to pure thought and not to natural impulse. Such speech is the only certain sign of thought hidden in a body. All men use it, however stupid and insane they may be, and though they may lack tongue and organs of voice; but no animals do. Consequently it can be taken as a real specific difference between men and dumb animals.

For brevity's sake I here omit the other reasons for denying thought to animals. Please note that I am speaking of thought, and not of life or sensation. I do not deny life to animals, since I regard it as consisting simply in the heat of the heart; and I do not deny sensation, in so far as it depends on a bodily organ. Thus my opinion is not so much cruel to animals as indulgent to men—at least to those who are not given to the superstitions of Pythagoras—since it absolves them from the suspicion of crime when they eat or kill animals.

Perhaps I have written at too great length for the sharpness of your intelligence; but I wished to show you that very few people have yet sent me objections which were as agreeable as yours. Your kindness and candour has made you a friend of that most respectful admirer of all who seek true wisdom, etc.

For Further Discussion

1. Descartes describes animals as automata or "moving machines." Why does he believe that animals are so different from humans?

2. What, precisely, do you think is the difference between humans and other animals? How does your view compare to Descartes's?

3. Descartes believes that the ability to speak, in words or other signs, is a crucial difference between humans and animals. Do you agree? Do animals speak in any meaningful sense?

4. Descartes speaks of the human soul as an internal cause of motion. In this sense, he concludes that "some such soul is to be found in animals also." How does he distinguish the human soul from the soul of animals?

Duties to Animals

Immanuel Kant

Descartes was not alone in considering rationality to be the distinctive human characteristic; Immanuel Kant (1724–1804) also held this view. Kant offered a strong defense of categorical duties to human beings (see Chapter 1) because, as rational beings, humans are ends in themselves. Because animals lack this capacity, humans can justifiably treat animals as means or objects rather than as ends or subjects. Humans do, however, have indirect duties to animals. These are duties that concern animals but are truly duties to other human beings. Our next selection is Kant's reflections on the ethical status of animals.

Baumgarten speaks of duties towards beings which are beneath us and beings which are above us. But so far as animals are concerned, we have no direct duties. Animals are not self-conscious and are there merely as a means to an end. That end is man. We can ask, "Why do animals exist?" But to ask, "Why does man exist?" is a meaningless question. Our duties towards animals are merely indirect duties towards humanity. Animal nature has analogies to human nature, and by doing our duties to animals in respect of manifestations of human nature, we indirectly do our duty towards humanity. Thus, if a dog has served his master long and faithfully, his service, on the analogy of human service, deserves reward, and when the dog has grown too old to serve, his master ought to keep him until he dies. Such action helps to support us in our duties towards human beings, where they are bounden duties. If then any acts of animals are analogous to human acts and spring from the same principles, we have duties towards the animals because thus we cultivate the corresponding duties towards human beings. If a man shoots his dog because the animal is no longer capable of service, he does not fail in his duty to the dog, for the dog cannot judge, but his act is inhuman and damages in himself that humanity which it is his duty to show towards mankind. If he is not to stifle his human feelings, he must practise kindness towards animals, for he who is cruel to animals becomes hard also in his dealing with men. We can judge the heart of a man by his treatment of animals. Hogarth depicts this in his engravings. He shows how cruelty grows and develops. He shows the child's cruelty to animals, pinch the tail of a dog or a cat; he then depicts the grown man in his cart running over a child; and lastly, the culmination of cruelty in murder. He thus brings home to us in a terrible fashion the rewards of cruelty, and this should be an impressive lesson to children. The more we come in contact with animals and observe their behaviour, the more we love them, for we see how great is their care for their young. It is then difficult for us to be cruel in thought even to a wolf. Leibnitz used a tiny worm for purposes of observation, and then carefully replaced it with its leaf on the tree so that it should not come to harm through any act of his. He would have been sorry—a natural feeling for a humane man—to destroy such a creature for no reason. Tender feelings towards dumb animals develop humane feelings towards mankind. In England butchers and doctors do not sit on a jury because they are accustomed to the sight of death and hardened. Vivisectionists, who use living animals for their experiments, certainly act cru-

elly, although their aim is praiseworthy, and they can justify their cruelty, since animals must be regarded as man's instruments; but any such cruelty for sport cannot be justified. A master who turns out his ass or his dog because the animal can no longer earn its keep manifests a small mind. The Greeks' ideas in this respect were highminded, as can be seen from the fable of the ass and the bell of ingratitude. Our duties towards animals, then, are indirect duties towards mankind.

Our duties towards immaterial beings are purely negative. Any course of conduct which involves dealings with spirits is wrong. Conduct of this kind makes men visionaries and fanatics, renders them superstitious, and is not in keeping with the dignity of mankind; for human dignity cannot subsist without a healthy use of reason, which is impossible for those who have commerce with spirits. Spirits may exist or they may not; all that is said of them may be true; but we know them not and can have no intercourse with them. This applies to good and to evil spirits alike. Our Ideas of good and evil are coordinate, and as we refer all evil to hell so we refer all good to heaven. If we personify the perfection of evil, we have the Idea of the devil. If we believe that evil spirits can have an influence upon us, can appear and haunt us at night, we become a prey to phantoms and incapable of using our powers in a reasonable way. Our duties towards such beings must, therefore, be negative.

DUTIES TOWARDS INANIMATE OBJECTS

Baumgarten speaks of duties towards inanimate objects. These duties are also indirectly duties towards mankind. Destructiveness is immoral; we ought not to destroy things which can still be put to some use. No man ought to mar the beauty of nature; for what he has no use for may still be of use to some one else. He need, of course, pay no heed to the thing itself, but he ought to consider his neighbour. Thus we see that all duties towards animals, towards immaterial beings and towards inanimate objects are aimed indirectly at our duties towards mankind.

For Further Discussion

1. Kant introduces an important distinction between direct and indirect duties. What is this distinction, and how does he apply it to our relationship to animals?

2. Kant describes cruelty toward animals as inhuman because it can make people more likely to act unethically toward humans. Do you believe that there is a connection between human attitudes toward animals and toward other humans?

3. Kant believes that we have indirect duties to inanimate objects. We ought not to destroy objects that might "still be put to use," nor should we "mar the beauty of nature." Do you agree with his reasoning?

Walking

Henry David Thoreau

Nineteenth-century American philosophy, represented in the writings of Ralph Waldo Emerson and Henry David Thoreau (1817–1862), reached conclusions about the transcendence of nature that differed from many previous Western thinkers. The American transcendentalists argued that humans become separated from God through the artifacts and creations of civilization. To approach God and become in tune with their own spiritual natures, humans need to retreat from the civilized world and return to a simpler lifestyle. The wilderness, the natural world, is the place where humans can come closest to experiencing their most authentic natures. Thoreau's essay is an eloquent expression of these views.

I wish to speak a word for Nature, for absolute freedom and wildness, as contrasted with a freedom and culture merely civil,—to regard man as an inhabitant, or a part and parcel of Nature, rather than a member of society. I wish to make an extreme statement, if so I may make an emphatic one, for there are enough champions of civilization: the minister and the school committee and every one of you will take care of that.

I have met with but one or two persons in the course of my life who understood the art of Walking, that is, of taking walks,—who had a genius, so to speak, for *sauntering*, which word is beautifully derived "from idle people who roved about the country, in the Middle Ages, and asked charity, under pretense of going *à la Sainte Terre,*" to the Holy Land, till the children exclaimed, "There goes a *Sainte-Terrer,*" a Saunterer, a Holy-Lander. They who never go to the Holy Land in their walks, as they pretend, are indeed mere idlers and vagabonds; but they who do go there are saunterers in the good sense, such as I mean.

Some, however, would derive the word from *sans terre,* without land or a home, which, therefore, in the good sense, will mean, having no particular home, but equally at home everywhere. For this is the secret of successful sauntering. He who sits still in a house all the time may be the greatest vagrant of all; but the saunterer, in the good sense, is no more vagrant than the meandering river, which is all the while sedulously seeking the shortest course to the sea. But I prefer the first, which, indeed, is the most probable derivation. For every walk is a sort of crusade, preached by some Peter the Hermit in us, to go forth and reconquer this Holy Land from the hands of the Infidels.

It is true, we are but faint-hearted crusaders, even the walkers, nowadays, who undertake no persevering, never-ending enterprises. Our expeditions are but tours, and come round again at evening to the old hearth-side from which we set out. Half the walk is but retracing our steps. We should go forth on the shortest walk, perchance, in the spirit of undying adventure, never to return,—prepared to send back our embalmed hearts only as relics to our desolate kingdoms. If you are ready to leave father and mother, and brother and sister, and wife and child and friends, and never see them again,—if you have paid your debts, and made your will, and settled all your affairs, and are a free man, then you are ready for a walk.

To come down to my own experience, my companion and I, for I sometimes have a companion, take pleasure in fancying ourselves knights of a new, or rather an old, order,—not Equestrians or Chevaliers, not Ritters or Riders, but Walkers, a still more ancient and honorable class, I trust. The chivalric and heroic spirit which once belonged to the Rider seems now to reside in, or perchance to have subsided into, the

Walker,—not the Knight, but Walker, Errant. He is a sort of fourth estate, outside of Church and State and People.

We have felt that we almost alone hereabouts practiced this noble art; though, to tell the truth, at least if their own assertions are to be received, most of my townsmen would fain walk sometimes, as I do, but they cannot. No wealth can buy the requisite leisure, freedom, and independence which are the capital in this profession. It comes only by the grace of God. It requires a direct dispensation from Heaven to become a walker. You must be born into the family of the Walkers. *Ambulator nascitur, non fit.* Some of my townsmen, it is true, can remember and have described to me some walks which they took ten years ago, in which they were so blessed as to lose themselves for half an hour in the woods; but I know very well that they have confined themselves to the highway ever since, whatever pretensions they may make to belong to this select class. No doubt they were elevated for a moment as by the reminiscence of a previous state of existence, when even they were foresters and outlaws.

> "When he came to grene wode,
>   In a mery mornynge,
> There he herde the notes small
>   Of byrdes mery syngynge.
> "It is ferre gone, sayd Robyn,
>   That I was last here;
> Me lyste a lytell for to shote
>   At the donne dere."

I think that I cannot preserve my health and spirits, unless I spend four hours a day at least—and it is commonly more than that—sauntering through the woods and over the hills and fields, absolutely free from all worldly engagements. You may safely say, A penny for your thoughts, or a thousand pounds. When sometimes I am reminded that the mechanics and shopkeepers stay in their shops not only all the forenoon, but all the afternoon too, sitting with crossed legs, so many of them,—as if the legs were made to sit upon, and not to stand or walk upon,—I think that they deserve some credit for not having all committed suicide long ago.

I, who cannot stay in my chamber for a single day without acquiring some rust, and when sometimes I have stolen forth for a walk at the eleventh hour, or four o'clock in the afternoon, too late to redeem the day, when the shades of night were already beginning to be mingled with the daylight, have felt as if I had committed some sin to be atoned for,—I confess that I am astonished at the power of endurance, to say nothing of the moral insensibility, of my neighbors who confine themselves to shops and offices the whole day for weeks and months, aye, and years almost together. I know not what manner of stuff they are of,— sitting there now at three o'clock in the afternoon, as if it were three o'clock in the morning. Bonaparte may talk of the three-o'clock-in-the-morning courage, but it is nothing to the courage which can sit down cheerfully at this hour in the afternoon over against one's self whom you have known all the morning, to starve out a garrison to whom you are bound by such strong ties of sympathy. I wonder that about this time, or say between four and five o'clock in the afternoon, too late for the morning papers and too early for the evening ones, there is not a general explosion heard up and down the street, scattering a legion of antiquated and house-bred notions and whims to the four winds for an airing,—and so the evil cure itself.

How womankind, who are confined to the house still more than men, stand it I do not know; but I have ground to suspect that most of them do not *stand* it at all. When, early in a summer afternoon, we have been shaking the dust of the village from the skirts of our garments, making haste past those houses with purely Doric or Gothic fronts, which have such an air of repose about them, my companion whispers that probably about these times their occupants are all gone to bed. Then it is that I appreciate the beauty and the glory of architecture, which itself never turns in, but forever stands out and erect, keeping watch over the slumberers.

No doubt temperament, and, above all, age, have a good deal to do with it. As a man grows older, his ability to sit still and follow indoor occupations increases. He grows vespertinal in his habits as the evening of life approaches, till at last he comes forth only just before sundown, and gets all the walk that he requires in half an hour.

But the walking of which I speak has nothing in it akin to taking exercise, as it is called, as the sick take medicine at stated hours,—as the swinging of dumbbells or chairs; but is itself the enterprise and adventure of the day. If you would get exercise, go in search of the springs of life. Think of a man's swinging dumbbells for his health, when those springs are bubbling up in far-off pastures unsought by him!

Moreover, you must walk like a camel, which is said to be the only beast which ruminates when walking. When a traveler asked Wordsworth's servant to show him her master's study, she answered, "Here is his library, but his study is out of doors."

Living much out of doors, in the sun and wind, will no doubt produce a certain roughness of character,—will cause a thicker cuticle to grow over some of the finer qualities of our nature, as on the face and hands, or as severe manual labor robs the hands of some of their delicacy of touch. So staying in the house, on the other hand, may produce a softness and smoothness, not to say thinness of skin, accompanied by an increased sensibility to certain impressions. Perhaps we should be more susceptible to some influences important to our intellectual and moral growth, if the sun had shone and the wind blown on us a little less; and no doubt it is a nice matter to proportion rightly the thick and thin skin. But methinks that is a scurf that will fall off fast enough,—that the natural remedy is to be found in the proportion which the night bears to the day, the winter to the summer, thought to experience. There will be so much the more air and sunshine in our thoughts. The callous palms of the laborer are conversant with finer tissues of self-respect and heroism, whose touch thrills the heart, than the languid fingers of idleness. That is mere sentimentality that lies abed by day and thinks itself white, far from the tan and callus of experience.

When we walk, we naturally go to the fields and woods: what would become of us, if we walked only in a garden or a mall? Even some sects of philosophers have felt the necessity of importing the woods to themselves, since they did not go to the woods. "They planted groves and walks of Platanes," where they took *subdiales ambulationes* in porticos open to the air. Of course it is of no use to direct our steps to the woods, if they do not carry us thither. I am alarmed when it happens that I have walked a mile into the woods bodily, without getting there in spirit. In my afternoon walk I would fain forget all my morning occupations and my obligations to society. But it sometimes happens that I cannot easily shake off the village. The thought of some work will run in my head and I am not where my body is,—I am out of my senses. In my walks I would fain return to my senses. What business have I in the woods, if I am thinking of something out of the woods? I suspect myself, and cannot help a shudder, when I find myself so implicated even in what are called good works,—for this may sometimes happen.

My vicinity affords many good walks; and though for so many years I have walked almost every day, and sometimes for several days together, I have not yet exhausted them. An absolutely new prospect is a great happiness, and I can still get this any afternoon. Two or three hours' walking will carry me to as strange a country as I expect ever to see. A single farmhouse which I had not seen before is sometimes as good as the dominions of the King of Dahomey. There is in fact a sort of harmony discoverable between the capabilities of the landscape within a circle of ten miles' radius, or the limits of an afternoon walk, and the three-score years and ten of human life. It will never become quite familiar to you.

Nowadays almost all man's improvements, so called, as the building of houses and the cutting down of the forest and of all large trees, simply deform the landscape, and make it more and more tame and cheap. A people who would begin by burning the fences and let the forest stand! I saw the fences half consumed, their ends lost in the middle of the prairie, and some worldly miser with a surveyor looking after his bounds, while heaven had taken place around him, and he did not see the angels going to and fro, but was looking for an old post-hole in the midst of paradise. I looked again, and saw him standing in the middle of a boggy Stygian fen, surrounded by devils, and he had found his bounds without a doubt, three little stones, where a stake had been driven, and looking nearer, I saw that the Prince of Darkness was his surveyor.

I can easily walk ten, fifteen, twenty, any number of miles, commencing at my own door, without going by any house, without crossing a road except where the fox and the mink do: first along by the river, and then the brook, and then the meadow and the woodside. There are square miles in my vicinity which have no inhabitant. From many a hill I can see civilization and the abodes of man afar. The farmers and their works are scarcely more obvious than woodchucks and their burrows. Man and his affairs, church and state and school, trade and commerce, and manufactures and agriculture, even politics, the most alarming of them all,—I am pleased to see how little space they occupy in the landscape. Politics is but a narrow field, and that still narrower highway yonder leads to it. I sometimes direct the traveler thither. If you would go to the political world, follow the great road,—follow that market-man, keep his dust in your eyes, and it will lead you straight to it; for it, too, has its place merely, and does not occupy all space. I pass from it as from a bean-field into the forest, and it is forgotten. In one half-hour I can walk off to some portion of the earth's surface where a man does not stand from one year's end to another, and there, consequently, politics are not, for they are but as the cigar-smoke of a man.

The village is the place to which the roads tend, a sort of expansion of the highway, as a lake of a river. It is the body of which roads are the arms and legs,—

a trivial or quadrivial place, the thoroughfare and ordinary of travelers. The word is from the Latin *villa,* which together with via, a way, or more anciently *ved* and *vella*. Varro derives from *veho,* to carry, because the villa is the place to and from which things are carried. They who got their living by teaming were said *vellaturam facere*. Hence, too, the Latin word *vilis* and our vile, also *villain*. This suggests what kind of degeneracy villagers are liable to. They are wayworn by the travel that goes by and over them, without traveling themselves.

Some do not walk at all; others walk in the highways; a few walk across lots. Roads are made for horses and men of business. I do not travel in them much, comparatively, because I am not in a hurry to get to any tavern or grocery or livery-stable or depot to which they lead. I am a good horse to travel, but not from choice a roadster. The landscape-painter uses the figures of men to mark a road. He would not make that use of my figure. I walk out into a nature such as the old prophets and poets, Menu, Moses, Homer, Chaucer, walked in. You may name it America, but it is not America; neither Americus Vespucius, nor Columbus, nor the rest were the discoverers of it. There is a truer account of it in mythology than in any history of America, so called, that I have seen.

. . .

At present, in this vicinity, the best part of the land is not private property; the landscape is not owned, and the walker enjoys comparative freedom. But possibly the day will come when it will be partitioned off into so-called pleasure-grounds, in which a few will take a narrow and exclusive pleasure only,—when fences shall be multiplied, and man-traps and other engines invented to confine men to the *public* road, and walking over the surface of God's earth shall be construed to mean trespassing on some gentleman's grounds. To enjoy a thing exclusively is commonly to exclude yourself from the true enjoyment of it. Let us improve our opportunities, then, before the evil days come.

What is it that makes it so hard sometimes to determine whither we will walk? I believe that there is a subtle magnetism in Nature, which, if we unconsciously yield to it, will direct us aright. It is not indifferent to us which way we walk. There is a right way; but we are very liable from heedlessness and stupidity to take the wrong one. We would fain take that walk, never yet taken by us through this actual world, which is perfectly symbolical of the path which we love to travel in the interior and ideal world; and sometimes,

no doubt, we find it difficult to choose our direction, because it does not yet exist distinctly in our idea.

When I go out of the house for a walk, uncertain as yet whither I will bend my steps, and submit myself to my instinct to decide for me, I find, strange and whimsical as it may seem, that I finally and inevitably settle southwest, toward some particular wood or meadow or deserted pasture or hill in that direction. My needle is slow to settle,—varies a few degrees, and does not always point due southwest, it is true, and it has good authority for this variation, but it always settles between west and south-southwest. The future lies that way to me, and the earth seems more unexhausted and richer on that side. The outline which would bound my walks would be, not a circle, but a parabola, or rather like one of those cometary orbits which have been thought to be non-returning curves, in this case opening westward, in which my house occupies the place of the sun. I turn round and round irresolute sometimes for a quarter of an hour, until I decide, for a thousandth time, that I will walk into the southwest or west. Eastward I go only by force; but westward I go free. Thither no business leads me. It is hard for me to believe that I shall find fair landscapes or sufficient wildness and freedom behind the eastern horizon. I am not excited by the prospect of a walk thither; but I believe that the forest which I see in the western horizon stretches uninterruptedly toward the setting sun, and there are no towns nor cities in it of enough consequence to disturb me. Let me live where I will, on this side is the city, on that the wilderness, and ever I am leaving the city more and more, and withdrawing into the wilderness. I should not lay so much stress on this fact, if I did not believe that something like this is the prevailing tendency of my countrymen. I must walk toward Oregon, and not toward Europe. And that way the nation is moving, and I may say that mankind progress from east to west. Within a few years we have witnessed the phenomenon of a southeastward migration, in the settlement of Australia; but this affects us as a retrograde movement, and, judging from the moral and physical character of the first generation of Australians, has not yet proved a successful experiment. The eastern Tartars think that there is nothing west beyond Thibet. "The world ends there," say they; "beyond there is nothing but a shoreless sea." It is unmitigated East where they live.

We go eastward to realize history and study the works of art and literature, retracing the steps of the race; we go westward as into the future, with a spirit of enterprise and adventure. The Atlantic is a Lethean

stream, in our passage over which we have had an opportunity to forget the Old World and its institutions.
If we do not succeed this time, there is perhaps one
more chance for the race left before it arrives on the
banks of the Styx; and that is in the Lethe of the Pacific, which is three times as wide.

. . .

The West of which I speak is but another name for
the Wild; and what I have been preparing to say is, that
in Wildness is the preservation of the World. Every tree
sends its fibres forth in search of the Wild. The cities
import it at any price. Men plow and sail for it. From
the forest and wilderness come the tonics and barks
which brace mankind. Our ancestors were savages.
The story of Romulus and Remus being suckled by a
wolf is not a meaningless fable. The founders of every
state which has risen to eminence have drawn their
nourishment and vigor from a similar wild source. It
was because the children of the Empire were not suckled by the wolf that they were conquered and displaced
by the children of the northern forests who were.

I believe in the forest, and in the meadow, and in
the night in which the corn grows. We require an in-
fusion of hemlock spruce or arbor-vitæ in our tea.
There is a difference between eating and drinking for
strength and from mere gluttony. The Hottentots eagerly devour the marrow of the koodoo and other
antelopes raw, as a matter of course. Some of our
northern Indians eat raw the marrow of the Arctic
reindeer, as well as various other parts, including the
summits of the antlers, as long as they are soft. And
herein, perchance, they have stolen a march on the
cooks of Paris. They get what usually goes to feed the
fire. This is probably better than stall-fed beef and
slaughter-house pork to make a man of. Give me a
wildness whose glance no civilization can endure,—
as if we lived on the marrow of koodoos devoured raw.

. . .

Life consists with wildness. The most alive is the
wildest. Not yet subdued to man, its presence re-
freshes him. One who pressed forward incessantly and
never rested from his labors, who grew fast and made
infinite demands on life, would always find himself in
a new country or wilderness, and surrounded by the
raw material of life. He would be climbing over the
prostrate stems of primitive forest-trees.

Hope and the future for me are not in lawns and
cultivated fields, not in towns and cities, but in the
impervious and quaking swamps. When, formerly, I
have analyzed my partiality for some farm which I had
contemplated purchasing, I have frequently found that

I was attracted solely by a few square rods of imper-
meable and unfathomable bog,—a natural sink in one
corner of it. That was the jewel which dazzled me. I
derive more of my subsistence from the swamps which
surround my native town than from the cultivated gardens in the village. There are no richer parterres to my
eyes than the dense beds of dwarf andromeda (*Cassan-
dra calyculata*) which cover these tender places on the
earth's surface. Botany cannot go farther than tell me
the names of the shrubs which grow there,—the high
blueberry, panicled andromeda, lambkill, azalea, and
rhodora,—all standing in the quaking sphagnum. I often think that I should like to have my house front on
this mass of dull red bushes, omitting other flower
plots and borders, transplanted spruce and trim box,
even graveled walks,—to have this fertile spot under
my windows, not a few imported barrowfuls of soil
only to cover the sand which was thrown out in digging the cellar. Why not put my house, my parlor, behind this plot, instead of behind that meagre assemblage of curiosities, that poor apology for a Nature and
Art, which I call my front yard? It is an effort to clear
up and make a decent appearance when the carpenter
and mason have departed, though done as much for
the passer-by as the dweller within. The most tasteful
front-yard fence was never an agreeable object of study
to me; the most elaborate ornaments, acorn tops, or
what not, soon wearied and disgusted me. Bring your
sills up to the very edge of the swamp, then (though it
may not be the best place for a dry cellar), so that there
be no access on that side to citizens. Front yards are
not made to walk in, but, at most, through, and you
could go in the back way.

Yes, though you may think me perverse, if it were
proposed to me to dwell in the neighborhood of the
most beautiful garden that ever human art contrived,
or else of a Dismal Swamp, I should certainly decide
for the swamp. How vain, then, have been all your labors, citizens, for me!

My spirits infallibly rise in proportion to the outward dreariness. Give me the ocean, the desert, or the
wilderness! In the desert, pure air and solitude compensate for want of moisture and fertility. The traveler
Burton says of it: "Your *morale* improves; you become
frank and cordial, hospitable and single-minded. . . .
In the desert, spirituous liquors excite only disgust.
There is a keen enjoyment in a mere animal existence."
They who have been traveling long on the steppes of
Tartary say, "On reentering cultivated lands, the agita-
tion, perplexity, and turmoil of civilization oppressed
and suffocated us; the air seemed to fail us, and we felt

every moment as if about to die of asphyxia." When I would recreate myself, I seek the darkest wood, the thickest and most interminable and, to the citizen, most dismal, swamp. I enter a swamp as a sacred place, a *sanctum sanctorum*. There is the strength, the marrow, of Nature. The wildwood covers the virgin mould, and the same soil is good for men and for trees. A man's health requires as many acres of meadow to his prospect as his farm does loads of muck. There are the strong meats on which he feeds. A town is saved, not more by the righteous men in it than by the woods and swamps that surround it. A township where one primitive forest waves above while another primitive forest rots below,—such a town is fitted to raise not only corn and potatoes, but poets and philosophers for the coming ages. In such a soil grew Homer and Confucius and the rest, and out of such a wilderness comes the Reformer eating locusts and wild honey.

To preserve wild animals implies generally the creation of a forest for them to dwell in or resort to. So it is with man. A hundred years ago they sold bark in our streets peeled from our own woods. In the very aspect of those primitive and rugged trees there was, methinks, a tanning principle which hardened and consolidated the fibres of men's thoughts. Ah! already I shudder for these comparatively degenerate days of my native village, when you cannot collect a load of bark of good thickness, and we no longer produce tar and turpentine.

The civilized nations—Greece, Rome, England—have been sustained by the primitive forests which anciently rotted where they stand. They survive as long as the soil is not exhausted. Alas for human culture! little is to be expected of a nation, when the vegetable mould is exhausted, and it is compelled to make manure of the bones of its fathers. There the poet sustains himself merely by his own superfluous fat, and the philosopher comes down on his marrow-bones.

. . .

For Further Discussion

1. In the opening line of this selection, Thoreau equates nature with "absolute freedom and wildness." He contrasts this with civil freedom and culture. What do you think Thoreau means by this?

2. Thoreau observes that "nowadays almost all man's improvements, so called, as the building of houses and the cutting down of the forest and of all large trees, simply deform the landscape, and make it more and more tame and cheap." Do you agree? Many people today would argue that such improvements do not cheapen the landscape but make it more valuable. What do you think Thoreau means by "cheap" in this passage?

3. One famous line from this selection claims that "in Wildness is the preservation of the World." How do you interpret this passage? Do you agree?

The Historical Roots of Our Ecological Crisis

Lynn White

As we step back from this survey, we see several clear themes that have environmental implications. Despite exceptions such as Saint Francis, much of the Western tradition seems committed to treating humans as superior to the rest of the natural world. In this classic essay written in 1967, historian Lynn White (1907–1987) argues that the roots of our present ecological crisis lie in the Western tradition's belief in the superiority of humans. White calls Christianity the "most anthropocentric religion the world has seen." This tradition understands humans as sharing God's transcendence, a transcendence that justifies human exploitation of the natural world.

A conversation wtih Aldous Huxley not infrequently put one at the receiving end of an unforgettable monologue. About a year before his lamented death he was discoursing on a favorite topic: Man's unnatural treatment of nature and its sad results. To illustrate his point he told how, during the previous summer, he had returned to a little valley in England where he had spent many happy months as a child. Once it had been composed of delightful grassy glades; now it was becoming overgrown with unsightly brush because the rabbits that formerly kept such growth under control had largely succumbed to a disease, myxomatosis, that was deliberately introduced by the local farmers to reduce the rabbits' destruction of crops. Being something of a Philistine, I could be silent no longer, even in the interests of great rhetoric. I interrupted to point out that the rabbit itself had been brought as a domestic animal to England in 1176, presumably to improve the protein diet of the peasantry.

All forms of life modify their contexts. The most spectacular and benign instance is doubtless the coral polyp. By serving its own ends, it has created a vast undersea world favorable to thousands of other kinds of animals and plants. Ever since man became a numerous species he has affected his environment notably. The hypothesis that his fire-drive method of hunting created the world's great grasslands and helped to exterminate the monster mammals of the Pleistocene from much of the globe is plausible, if not proved. For 6 millennia at least, the banks of the lower Nile have been a human artifact rather than the swampy African jungle which nature, apart from man, would have made it. The Aswan Dam, flooding 5000 square miles, is only the latest stage in a long process. In many regions terracing or irrigation, overgrazing, the cutting of forests by Romans to build ships to fight Carthaginians or by Crusaders to solve the logistics problems of their expeditions, have profoundly changed some ecologies. Observation that the French landscape falls into two basic types, the open fields of the north and the *bocage* of the south and west, inspired Marc Bloch to undertake his classic study of medieval agricultural methods. Quite unintentionally, changes in human ways often affect nonhuman nature. It has been noted, for example, that the advent of the automobile eliminated huge flocks of sparrows that once fed on the horse manure littering every street.

The history of ecologic change is still so rudimentary that we know little about what really happened, or what the results were. The extinction of the European aurochs as late as 1627 would seem to have been a simple case of overenthusiastic hunting. On more intricate matters it often is impossible to find solid information. For a thousand years or more the Frisians and Hollanders have been pushing back the North Sea, and the process is culminating in our own time in the reclamation of the Zuider Zee. What, if any, species of animals, birds, fish, shore life, or plants have died out in the process? In their epic combat with Neptune, have the Netherlanders overlooked ecological values in such a way that the quality of human life in the Netherlands has suffered? I cannot discover that the questions have ever been asked, much less answered.

People, then, have often been a dynamic element in their own environment, but in the present state of historical scholarship we usually do not know exactly when, where, or with what effects man-induced changes came. As we enter the last third of the 20th century, however, concern for the problem of ecologic backlash is mounting feverishly. Natural science, conceived as the effort to understand the nature of things, had flourished in several eras and among several peoples. Similarly there had been an age-old accumulation of technological skills, sometimes growing rapidly, sometimes slowly. But it was not until about four generations ago that Western Europe and North America arranged a marriage between science and technology, a union of the theoretical and the empirical approaches to our natural environment. The emergence in widespread practice of the Baconian creed that scientific knowledge means technological power over nature can scarcely be dated before about 1850, save in the chemical industries, where it is anticipated in the 18th century. Its acceptance as a normal pattern of action may mark the greatest event in human history since the invention of agriculture, and perhaps in nonhuman terrestrial history as well.

Almost at once the new situation forced the crystallization of the novel concept of ecology; indeed, the word *ecology* first appeared in the English language in 1873. Today, less than a century later, the impact of our race upon the environment has so increased in force that it has changed in essence. When the first cannons were fired, in the early 14th century, they affected ecology by sending workers scrambling to the forests and mountains for more potash, sulfur, iron ore, and charcoal, with some resulting erosion and deforestation. Hydrogen bombs are of a different order: a war fought with them might alter the genetics of all life on this planet. By 1285 London had a smog problem arising from the burning of soft coal, but our present combustion of fossil fuels threatens to change the chemistry of the globe's atmosphere as a whole, with consequences which we are only beginning to guess. With the population explosion, the carcinoma of planless urbanism, the now geological deposits of sewage and garbage, surely no creature other than man has ever managed to foul its nest in such short order.

There are many calls to action, but specific proposals, however worthy as individual items, seem too partial, palliative, negative: ban the bomb, tear down the billboards, give the Hindus contraceptives and tell them to eat their sacred cows. The simplest solution to any suspect change is, of course, to stop it, or, better yet, to revert to a romanticized past: make those ugly gasoline stations look like Anne Hathaway's cottage or (in the Far West) like ghost-town saloons. The "wilderness area" mentality invariably advocates deep-freezing an ecology, whether San Gimignano or the High Sierra, as it was before the first Kleenex was dropped. But neither atavism nor prettification will cope with the ecologic crisis of our time.

What shall we do? No one yet knows. Unless we think about fundamentals, our specific measures may produce new backlashes more serious than those they are designed to remedy.

As a beginning we should try to clarify our thinking by looking, in some historical depth, at the presuppositions that underlie modern technology and science. Science was traditionally aristocratic, speculative, intellectual in intent; technology was lower-class, empirical, action-oriented. The quite sudden fusion of these two, towards the middle of the 19th century, is surely related to the slightly prior and contemporary democratic revolutions which, by reducing social barriers, tended to assert a functional unity of brain and hand. Our ecologic crisis is the product of an emerging, entirely novel, democratic culture. The issue is whether a democratized world can survive its own implications. Presumably we cannot, unless we rethink our axioms.

THE WESTERN TRADITIONS OF TECHNOLOGY AND SCIENCE

One thing is so certain that it seems stupid to verbalize it: both modern technology and modern science are distinctively *Occidental*. Our technology has absorbed elements from all over the world, notably from China; yet everywhere today, whether in Japan or in Nigeria, successful technology is Western. Our science is the heir to all the sciences of the past, especially perhaps to the work of the great Islamic scientists of the Middle Ages, who so often outdid the ancient Greeks in skill and perspicacity: al-Rāzī in medicine, for example; or ibn-al-Haytham in optics; or Omar Khayyám in mathematics. Indeed, not a few works of such geniuses seem to have vanished in the original Arabic and to survive only in medieval Latin translations that helped to lay the foundations for later Western developments. Today, around the globe, all significant science is Western in style and method, whatever the pigmentation or language of the scientists.

A second pair of facts is less well recognized because they result from quite recent historical scholarship. The leadership of the West, both in technology and in science, is far older than the so-called Scientific Revolution of the 17th century or the so-called Industrial Revolution of the 18th century. These terms are in fact outmoded and obscure the true nature of what they try to describe—significant stages in two long and separate developments. By A.D. 1000 at the latest—and perhaps, feebly, as much as 200 years earlier—the West began to apply water power to industrial processes other than milling grain. This was followed in the late 12th century by the harnessing of wind power. From simple beginnings, but with remarkable consistency of style, the West rapidly expanded its skills in the development of power machinery, labor-saving devices, and automation. Those who doubt should contemplate that most monumental achievement in the history of automation: the weight-driven mechanical clock, which appeared in two forms in the early 14th century. Not in craftsmanship but in basic technological capacity, the Latin West of the later Middle Ages far outstripped its elaborate, sophisticated, and esthetically magnificent sister cultures, Byzantium and Islam. In 1444 a great Greek ecclesiastic, Bessarion, who had gone to Italy, wrote a letter to a prince in Greece. He is amazed by the superiority of Western ships, arms, textiles, glass. But above all he is astonished by the spectacle of water-wheels sawing timbers and pumping the bellows to blast furnaces. Clearly, he had seen nothing of the sort in the Near East.

By the end of the 15th century the technological superiority of Europe was such that its small, mutually hostile nations could spill out over all the rest of the world, conquering, looting, and colonizing. The symbol of this technological superiority is the fact that Portugal, one of the weakest states of the Occident, was able to become, and to remain for a century, mistress of the East Indies. And we must remember that the technology of Vasco da Gama and Albuquerque was built by pure empiricism, drawing remarkably little support or inspiration from science.

In the present-day vernacular of understanding, modern science is supposed to have begun in 1543, when both Copernicus and Vesalius published their great works. It is no derogation of their accomplishments, however, to point out that such structures as the *Fabrica* and the *De revolutionibus* do not appear overnight. The distinctive Western tradition of science, in fact, began in the late 11th century with a massive movement of translation of Arabic and Greek scientific works into Latin. A few notable books—Theophrastus, for example—escaped the West's avid new appetite for science, but within less than 200 years, effectively the entire corpus of Greek and Muslim science was available in Latin, and was being eagerly read and criticized in the new European universities. Out of criticism arose new observation, speculation, and increasing distrust of ancient authorities. By the late 13th century Europe had seized global scientific leadership from the faltering hands of Islam. It would be as absurd to deny the profound originality of Newton, Galileo, or Copernicus as to deny that of the 14th century scholastic scientists like Buridan or Oresme on whose work they built. Before the 11th century, science scarcely existed in the Latin West, even in Roman times. From the 11th century onward, the scientific sector of Occidental culture has increased in a steady crescendo.

Since both our technological and our scientific movements got their start, acquired their character, and achieved world dominance in the Middle Ages, it would seem that we cannot understand their nature or their present impact upon ecology without examining fundamental medieval assumptions and developments.

MEDIEVAL VIEW OF MAN AND NATURE

Until recently, agriculture has been the chief occupation even in "advanced" societies; hence, any change in methods of tillage has much importance. Early plows, drawn by two oxen, did not normally turn the sod but merely scratched it. Thus, cross-plowing was needed and fields tended to be squarish. In the fairly light soils and semi-arid climates of the Near East and Mediterranean, this worked well. But such a plow was inappropriate to the wet climate and often sticky soils of northern Europe. By the latter part of the 7th century after Christ, however, following obscure beginnings, certain northern peasants were using an entirely new kind of plow, equipped with a vertical knife to cut the line of the furrow, a horizontal share to slice under the sod, and a moldboard to turn it over. The friction of this plow with the soil was so great that it normally required not two but eight oxen. It attacked the land with such violence that cross-plowing was not needed, and fields tended to be shaped in long strips.

In the days of the scratch-plow, fields were distributed generally in units capable of supporting a single family. Subsistence farming was the presupposition. But no peasant owned eight oxen: to use the new and more efficient plow, peasants pooled their oxen to form large plow-teams, originally receiving (it would appear) plowed strips in proportion to their contribution. Thus, distribution of land was based no longer on the needs of a family but, rather, on the capacity of a power machine to till the earth. Man's relation to the soil was profoundly changed. Formerly man had been part of nature; now he was the exploiter of nature. Nowhere else in the world did farmers develop any analogous agricultural implement. Is it coincidence that modern technology, with its ruthlessness toward nature, has so largely been produced by descendants of these peasants of northern Europe?

This same exploitive attitude appears slightly before A.D. 830 in Western illustrated calendars. In older calendars the months were shown as passive personifications. The new Frankish calendars, which set the style for the Middle Ages, are very different: they show men coercing the world around them—plowing, harvesting, chopping trees, butchering pigs. Man and nature are two things, and man is master.

These novelties seem to be in harmony with larger intellectual patterns. What people do about their ecology depends on what they think about themselves in relation to things around them. Human ecology is deeply conditioned by beliefs about our nature and destiny—that is, by religion. To Western eyes this is very evident in, say, India or Ceylon. It is equally true of ourselves and of our medieval ancestors.

The victory of Christianity over paganism was the greatest psychic revolution in the history of our culture. It has become fashionable today to say that, for better or worse, we live in "the post-Christian age." Certainly the forms of our thinking and language have largely ceased to be Christian, but to my eye the substance often remains amazingly akin to that of the past. Our daily habits of action, for example, are dominated by an implicit faith in perpetual progress which was unknown either to Greco-Roman antiquity or to the Orient. It is rooted in, and is indefensible apart from,

Judeo-Christian teleology. The fact that Communists share it merely helps to show what can be demonstrated on many other grounds: that Marxism, like Islam, is a Judeo-Christian heresy. We continue today to live, as we have lived for about 1700 years, very largely in a context of Christian axioms.

What did Christianity tell people about their relations with the environment?

While many of the world's mythologies provide stories of creation, Greco-Roman mythology was singularly incoherent in this respect. Like Aristotle, the intellectuals of the ancient West denied that the visible world had had a beginning. Indeed, the idea of a beginning was impossible in the framework of their cyclical notion of time. In sharp contrast, Christianity inherited from Judaism not only a concept of time as nonrepetitive and linear but also a striking story of creation. By gradual stages a loving and all-powerful God had created light and darkness, the heavenly bodies, and earth and all its plants, animals, birds, and fishes. Finally, God had created Adam and, as an afterthought, Eve to keep man from being lonely. Man named all the animals, thus establishing his dominance over them. God planned all of this explicitly for man's benefit and rule: no item in the physical creation had any purpose save to serve man's purposes. And, although man's body is made of clay, he is not simply part of nature: he is made in God's image.

Especially in its Western form, Christianity is the most anthropocentric religion the world has seen. As early as the 2nd century both Tertullian and St. Irenaeus of Lyons were insisting that when God shaped Adam he was foreshadowing the image of the incarnate Christ, the Second Adam. Man shares, in great measure, God's transcendence of nature. Christianity, in absolute contrast to ancient paganism and Asia's religions (except, perhaps, Zoroastrianism), not only established a dualism of man and nature but also insisted that it is God's will that man exploit nature for his proper ends.

At the level of the common people this worked out in an interesting way. In Antiquity every tree, every spring, every stream, every hill had its own *genius loci,* its guardian spirit. These spirits were accessible to men, but were very unlike men; centaurs, fauns, and mermaids show their ambivalence. Before one cut a tree, mined a mountain, or dammed a brook, it was important to placate the spirit in charge of that particular situation, and to keep it placated. By destroying pagan animism, Christianity made it possible to exploit nature in a mood of indifference to the feelings of natural objects.

It is often said that for animism the Church substituted the cult of saints. True; but the cult of saints is functionally quite different from animism. The saint is not *in* natural objects; he may have special shrines, but his citizenship is in heaven. Moreover, a saint is entirely a man; he can be approached in human terms. In addition to saints, Christianity of course also had angels and demons inherited from Judaism and perhaps, at one remove, from Zoroastrianism. But these were all as mobile as the saints themselves. The spirits *in* natural objects, which formerly had protected nature from man, evaporated. Man's effective monopoly on spirit in this world was confirmed, and the old inhibitions to the exploitation of nature crumbled.

When one speaks in such sweeping terms, a note of caution is in order. Christianity is a complex faith, and its consequences differ in differing contexts. What I have said may well apply to the medieval West, where in fact technology made spectacular advances. But the Greek East, a highly civilized realm of equal Christian devotion, seems to have produced no marked technological innovation after the late 7th century, when Greek fire was invented. The key to the contrast may perhaps be found in a difference in the tonality of piety and thought which students of comparative theology find between the Greek and the Latin Churches. The Greeks believed that sin was intellectual blindness, and that salvation was found in illumination, orthodoxy—that is, clear thinking. The Latins, on the other hand, felt that sin was moral evil, and that salvation was to be found in right conduct. Eastern theology has been intellectualist. Western theology has been voluntarist. The Greek saint contemplates; the Western saint acts. The implications of Christianity for the conquest of nature would emerge more easily in the Western atmosphere.

The Christian dogma of creation, which is found in the first clause of all the Creeds, has another meaning for our comprehension of today's ecologic crisis. By revelation, God had given man the Bible, the Book of Scripture. But since God had made nature, nature also must reveal the divine mentality. The religious study of nature for the better understanding of God was known as natural theology. In the early Church, and always in the Greek East, nature was conceived primarily as a symbolic system through which God speaks to men: the ant is a sermon to sluggards; rising flames are the symbol of the soul's aspiration. This view of nature was essentially artistic rather than scientific. While Byzantium preserved and copied great numbers of ancient Greek scientific texts, science as we conceive it could scarcely flourish in such an ambience.

However, in the Latin West by the early 13th century natural theology was following a very different bent. It was ceasing to be the decoding of the physical symbols of God's communication with man and was becoming the effort to understand God's mind by discovering how his creation operates. The rainbow was no longer simply a symbol of hope first sent to Noah after the Deluge: Robert Grosseteste, Friar Roger Bacon, and Theodoric of Freiberg produced startlingly sophisticated work on the optics of the rainbow, but they did it as a venture in religious understanding. From the 13th century onward, up to and including Leibnitz and Newton, every major scientist, in effect, explained his motivations in religious terms. Indeed, if Galileo had not been so expert an amateur theologian he would have got into far less trouble: the professionals resented his intrusion. And Newton seems to have regarded himself more as a theologian than as a scientist. It was not until the late 18th century that the hypothesis of God became unnecessary to many scientists.

It is often hard for the historian to judge, when men explain why they are doing what they want to do, whether they are offering real reasons or merely culturally acceptable reasons. The consistency with which scientists during the long formative centuries of Western science said that the task and the reward of the scientist was "to think God's thoughts after him" leads one to believe that this was their real motivation. If so, then modern Western science was cast in a matrix of Christian theology. The dynamism of religious devotion, shaped by the Judeo-Christian dogma of creation, gave it impetus.

AN ALTERNATIVE CHRISTIAN VIEW

We would seem to be headed toward conclusions unpalatable to many Christians. Since both *science* and *technology* are blessed words in our contemporary vocabulary, some may be happy at the notions, first, that, viewed historically, modern science is an extrapolation of natural theology and, second, that modern technology is at least partly to be explained as an Occidental, voluntarist realization of the Christian dogma

of man's transcendence of, and rightful mastery over, nature. But, as we now recognize, somewhat over a century ago science and technology—hitherto quite separate activities—joined to give mankind powers which, to judge by many of the ecologic effects, are out of control. If so, Christianity bears a huge burden of guilt.

I personally doubt that disastrous ecologic backlash can be avoided simply by applying to our problems more science and more technology. Our science and technology have grown out of Christian attitudes toward man's relation to nature which are almost universally held not only by Christians and neo-Christians but also by those who fondly regard themselves as post-Christians. Despite Copernicus, all the cosmos rotates around our little globe. Despite Darwin, we are *not,* in our hearts, part of the natural process. We are superior to nature, contemptuous of it, willing to use it for our slightest whim. The newly elected Governor of California, like myself a churchman but less troubled than I, spoke for the Christian tradition when he said (as is alleged), "when you've seen one redwood tree, you've seen them all." To a Christian a tree can be no more than a physical fact. The whole concept of the sacred grove is alien to Christianity and to the ethos of the West. For nearly 2 millennia Christian missionaries have been chopping down sacred groves, which are idolatrous because they assume spirit in nature.

What we do about ecology depends on our ideas of the man–nature relationship. More science and more technology are not going to get us out of the present ecologic crisis until we find a new religion, or rethink our old one. The beatniks, who are the basic revolutionaries of our time, show a sound instinct in their affinity for Zen Buddhism, which conceives of the man–nature relationship as very nearly the mirror image of the Christian view. Zen, however, is as deeply conditioned by Asian history as Christianity is by the experience of the West, and I am dubious of its viability among us.

Possibly we should ponder the greatest radical in Christian history since Christ: St. Francis of Assisi. The prime miracle of St. Francis is the fact that he did not end at the stake, as many of his left-wing followers did. He was so clearly heretical that a General of the Franciscan Order, St. Bonaventura, a great and perceptive Christian, tried to suppress the early accounts of Franciscanism. The key to an understanding of Francis is his belief in the virtue of humility—not merely for the individual but for man as a species.

Francis tried to depose man from his monarchy over creation and set up a democracy of all God's creatures. With him the ant is no longer simply a homily for the lazy, flames a sign of the thrust of the soul toward union with God; now they are Brother Ant and Sister Fire, praising the Creator in their own ways as Brother Man does in his.

Later commentators have said that Francis preached to the birds as a rebuke to men who would not listen. The records do not read so: he urged the little birds to praise God, and in spiritual ecstasy they flapped their wings and chirped rejoicing. Legends of saints, especially the Irish saints, had long told of their dealings with animals but always, I believe, to show their human dominance over creatures. With Francis it is different. The land around Gubbio in the Apennines was being ravaged by a fierce wolf. St. Francis, says the legend, talked to the wolf and persuaded him of the error of his ways. The wolf repented, died in the odor of sanctity, and was buried in consecrated ground.

What Sir Steven Runciman calls "the Franciscan doctrine of the animal soul" was quickly stamped out. Quite possibly it was in part inspired, consciously or unconsciously, by the belief in reincarnation held by the Cathar heretics who at that time teemed in Italy and southern France, and who presumably had got it originally from India. It is significant that at just the same moment, about 1200, traces of metempsychosis are found also in western Judaism, in the Provençal *Cabbala*. But Francis held neither to transmigration of souls nor to pantheism. His view of nature and of man rested on a unique sort of pan-psychism of all things animate and inanimate, designed for the glorification of their transcendent Creator, who, in the ultimate gesture of cosmic humility, assumed flesh, lay helpless in a manger, and hung dying on a scaffold.

I am not suggesting that many contemporary Americans who are concerned about our ecologic crisis will be either able or willing to counsel with wolves or exhort birds. However, the present increasing disruption of the global environment is the product of a dynamic technology and science which were originating in the Western medieval world against which St. Francis was rebelling in so original a way. Their growth cannot be understood historically apart from distinctive attitudes toward nature which are deeply grounded in Christian dogma. The fact that most people do not think of these attitudes as Christian is irrelevant. No new set of basic values has been accepted in our society to displace those of Christianity. Hence we shall continue to have a worsening ecologic crisis until we reject the Christian axiom that nature has no reason for existence save to serve man.

The greatest spiritual revolutionary in Western history, St. Francis, proposed what he thought was an alternative Christian view of nature and man's relation to it: he tried to substitute the idea of the equality of all creatures, including man, for the idea of man's limitless rule of creation. He failed. Both our present science and our present technology are so tinctured with orthodox Christian arrogance toward nature that no solution for our ecologic crisis can be expected from them alone. Since the roots of our trouble are so largely religious, the remedy must also be essentially religious, whether we call it that or not. We must rethink and refeel our nature and destiny. The profoundly religious, but heretical, sense of the primitive Franciscans for the spiritual autonomy of all parts of nature may point a direction. I propose Francis as a patron saint for ecologists.

For Further Discussion

1. White believes that the present ecological crisis is rooted in a particular Western understanding of science and technology, an understanding that itself is rooted in medieval views of humans and nature. How does White connect these medieval views and Western science and technology?

2. White identifies a particular interpretation of Christianity as influencing a destructive attitude toward the natural world. How can you compare what he says about this destructive attitude to the earlier selections from the Bible and Saint Thomas Aquinas?

3. White doubts that the present crisis can be avoided by further reliance on science and technology. Instead, he seems to encourage a rethinking of our attitudes toward nature and suggests that the "radical" Christian, Saint Francis, might provide guidance here. What does White identify as the key to understanding Saint Francis?

A Wilderness Condition

Roderick Nash

A more common American perspective viewed the wilderness as a threat rather than as salvation. The early American settlers and the pioneers who led the western migration saw the wilds as something to be conquered and tamed. The final reading in this chapter is Chapter 2 from historian Roderick Nash's book Wilderness and the American Mind. *Nash surveys the intellectual and religious roots of American attitudes toward the wilderness. Without question, these attitudes continue to influence contemporary environmental policy.*

> Looking only a few years through the vista of futurity what a sublime spectacle presents itself! Wilderness, once the chosen residence of solitude and savageness, converted into populous cities, smiling villages, beautiful farms and plantations!
> —Chillicothe (Ohio) *Supporter,* 1817

Alexis de Tocqueville resolved to see wilderness during his 1831 trip to the United States, and in Michigan Territory in July the young Frenchman found himself at last on the fringe of civilization. But when he informed the frontiersmen of his desire to travel for *pleasure* into the primitive forest, they thought him mad. The Americans required considerable persuasion from Tocqueville to convince them that his interests lay in matters other than lumbering or land speculation. Afterwards he generalized in his journal that "living in the wilds, [the pioneer] only prizes the works of man" while Europeans, like himself, valued wilderness because of its novelty. Expanding the point in *Democracy in America,* Tocqueville concluded: "in Europe people talk a great deal of the wilds of America, but the Americans themselves never think about them; they are insensible to the wonders of inanimate nature and they may be said not to perceive the mighty forests that surround them till they fall beneath the hatchet. Their eyes are fixed upon another sight," he added, "the . . . march across these wilds, draining swamps, turning the course of rivers, peopling solitudes, and subduing nature."

The unfavorable attitude toward wilderness that Tocqueville observed in Michigan also existed on other American frontiers. When William Bradford stepped off the *Mayflower* into a "hideous and desolate wilderness" he started a tradition of repugnance. With few exceptions later pioneers continued to regard wilderness with defiant hatred and joined the Chillicothe *Supporter* in celebrating the advance of civilization as the greatest of blessings. Under any circumstances the necessity of living in close proximity to wild country—what one of Bradford's contemporaries called "a Wilderness condition"—engendered strong antipathy. Two centuries after Bradford, a fur trader named Alexander Ross recorded his despair in encountering a "gloomy," "dreary," and "unhallowed wilderness" near the Columbia River.

Two components figured in the American pioneer's bias against wilderness. On the direct, physical level, it constituted a formidable threat to his very survival. The transatlantic journey and subsequent western advances stripped away centuries. Successive waves of frontiersmen had to contend with wilderness as uncontrolled and terrifying as that which primitive man confronted. Safety and comfort, even necessities like food and shelter, depended on overcoming the wild environment. For the first Americans, as for medieval Europeans, the forest's darkness hid savage men, wild beasts, and still stranger creatures of the imagination. In addition civilized man faced the danger of succumbing to the wildness of his surroundings and reverting to savagery himself. The pioneer, in short, lived too close to wilderness for appreciation. Understandably, his attitude was hostile and his dominant criteria utilitarian. The *conquest* of wilderness was his major concern.

Wilderness not only frustrated the pioneers physically but also acquired significance as a dark and sinister symbol. They shared the long Western tradition of imagining wild country as a moral vacuum, a cursed and chaotic wasteland. As a consequence, frontiersmen acutely sensed that they battled wild country not only for personal survival but in the name of nation, race, and God. Civilizing the New World meant enlightening darkness, ordering chaos, and changing evil into good. In the morality play of westward expansion, wilderness was the villain, and the pioneer, as hero, relished its destruction. The transformation of a wilderness into civilization was the reward for his sacrifices, the definition of his achievement, and the source of his pride. He applauded his successes in terms suggestive of the high stakes he attached to the conflict.

The discovery of the New World rekindled the traditional European notion that an earthly paradise lay somewhere to the west. As the reports of the first explorers filtered back the Old World began to believe that America might be the place of which it had dreamed since antiquity. One theme in the paradise myth stressed the material and sensual attributes of the new land. It fed on reports of fabulous riches, a temperate climate, longevity, and garden-like natural beauty. Promoters of discovery and colonization embellished these rumors. One Londoner, who likely never set foot in the New World, wrote lyrically of the richness of Virginia's soil and the abundance of its game. He even added: "nor is the present wildernesse of it without a particular beauty, being all over a naturall Grove of Oakes, Pines, Cedars . . . all of so delectable an aspect, that the melanchollyest eye in the World cannot look upon it without contentment, nor content himselfe without admiration." Generally, however, European portrayers of a material paradise in the New World completely ignored the "wildernesse" aspect, as inconsistent with the idea of beneficent nature. Illogically, they exempted America from the adverse conditions of life in other uncivilized places.

Anticipations of a second Eden quickly shattered against the reality of North America. Soon after he arrived the seventeenth-century frontiersman realized that the New World was the antipode of paradise. Previous hopes intensified the disappointment. At Jamestown the colonists abandoned the search for gold and turned, shocked, to the necessity of survival in a hostile environment. A few years later William Bradford recorded his dismay at finding Cape Cod wild and desolate. He lamented the Pilgrims' inability to find a vantage point "to view from this wilderness a more goodly country to feed their hopes." In fact, there was none. The forest stretched farther than Bradford and his generation imagined. For Europeans wild country was a single peak or heath, an island of uninhabited land surrounded by settlement. They at least knew its character and extent. But the seemingly boundless wilderness of the New World was something else. In the face of this vast blankness, courage failed and imagination multiplied fears.

Commenting on the arrival of the Puritans some years after, Cotton Mather indicated the change in attitude that contact with the New World produced. "Lady Arabella," he wrote, left an "earthly *paradise*" in England to come to America and "encounter the sorrows of a wilderness." She then died and "left that *wilderness* for the Heavenly *paradise*." Clearly the American wilderness was not paradise. If men expected to enjoy an idyllic environment in America, they would have to *make* it by conquering wild country. Mather realized in 1693 that "Wilderness" was the stage "thro' which we are passing to the Promised Land." Yet optimistic Americans continued to be fooled. "Instead of a garden," declared one traveler in the Ohio Valley in 1820, "I found a wilderness."

How frontiersmen described the wilderness they found reflected the intensity of their antipathy. The same descriptive phrases appeared again and again. Wilderness was "howling," "dismal," "terrible." In the 1650s John Eliot wrote of going "into a wilderness where nothing appeareth but hard labour [and] wants," and Edward Johnson described "the penuries of a Wildernesse." Cotton Mather agreed in 1702 about the "difficulties of a rough and hard wilderness," and in 1839 John Plumbe, Jr. told about "the hardships and privations of the wilderness" in Iowa and Wisconsin. Invariably the pioneers singled out wilderness as the root cause of their difficulties. For one thing, the physical character of the primeval forest proved baffling and frustrating to settlers. One chronicler of the "Wildernesse-worke" of establishing the town of Concord, Massachusetts portrayed in graphic detail the struggle through "unknowne woods," swamps, and flesh-tearing thickets. The town founders wandered lost for days in the bewildering gloom of the dense forest. Finally came the back-breaking labor of carving fields from the wilderness. Later generations who settled forested regions reported similar hardships. On every frontier obtaining cleared land, the symbol of civilization, demanded tremendous effort.

The pioneers' situation and attitude prompted them to use military metaphors to discuss the coming of civilization. Countless diaries, addresses, and memorials of the frontier period represented wilderness as an "enemy" which had to be "conquered," "subdued," and "vanquished" by a "pioneer army." The same phraseology persisted into the present century; an old Michigan pioneer recalled how as a youth he had engaged in a "struggle with nature" for the purpose of "converting a wilderness into a rich and prosperous civilization." Historians of westward expansion chose the same figure: "they conquered the wilderness, they subdued the forests, they reduced the land to fruitful subjection." The image of man and wilderness locked in mortal combat was difficult to forget. Advocates of a giant dam on the Colorado River system spoke in the 1950s of "that eternal problem of subduing the earth" and of "conquering the wilderness" while a President urged us in his 1961 inaugural address to "conquer the deserts." Wilderness, declared a correspondent to the *Saturday Evening Post* in 1965, "is precisely what man has been fighting against since he began his painful, awkward climb to civilization. It is the dark, the formless, the terrible, the old chaos which our fathers pushed back. . . . It is held at bay by constant vigilance, and when the vigilance slackens it swoops down for a melodramatic revenge." Such language animated the wilderness, investing it with an almost conscious enmity toward men, who returned it in full measure.

Along with the obstacle it offered to settlement and civilization, wilderness also confronted the frontier mind with terrifying creatures, both known and imagined. Wild men headed the menagerie. Initially Indians were regarded with pity and instructed in the Gospel, but after the first massacres most of the compassion changed to contempt. Sweeping out of the forest to strike, and then melting back into it, savages were almost always associated with wilderness. When Mary Rowlandson was captured in the 1670s on the Massachusetts frontier, she wrote that she went "mourning and lamenting, leaving farther my own Country, and travelling into the vast and howling Wilderness." The remainder of her account revealed an hysterical horror of her captors and of what she called "this Wilderness-condition." A century later J. Hector St. John Crevecoeur discussed the imminency of Indian attack as one of the chief "distresses" of frontier life and described the agony of waiting, gun in hand, for the first arrows to strike his home. "The wilderness," he observed, "is a harbour where it is impossible to find [the Indians] . . . a door through which they can enter our country whenever they please." Imagination and the presence of wild country could multiply fears. Riding through "savage haunts" on the Santa Fe Trail in the 1830s, Josiah Gregg noticed how "each click of a pebble" seemed "the snap of a firelock" and

"in a very rebound of a twig [was] the whisk of an arrow."

Wild animals added to the danger of the American wilderness, and here too the element of the unknown intensified feelings. Reporting in 1630 on the "discommodities" of New England, Francis Higginson wrote that "this Countrey being verie full of Woods and Wildernesses, doth also much abound with Snakes and Serpents of strange colours and huge greatnesse." There were some, he added, "that haue [have] Rattles in their Tayles that will not flye from a Man . . . but will flye upon him and sting him so mortally, that he will dye within a quarter of an houre after." Clearly there was some truth here and in the stories that echo through frontier literature of men whom "the savage Beasts had devoured . . . in the Wilderness," but often fear led to exaggeration. Cotton Mather, for instance, warned in 1707 of "the *Evening Wolves,* the rabid and howling *Wolves of the Wilderness* [which] would make . . . Havock among you, *and not leave the Bones till the morning.*" Granted this was a jeremiad intended to shock Mather's contemporaries into godly behavior, but his choice of imagery still reflected a vivid conception of the physical danger of wild country. Elsewhere Mather wrote quite seriously about the "Dragons," "Droves of Devils," and "Fiery flying serpents" to be found in the primeval forest. Indeed, legends and folktales from first contact until well into the national period linked the New World wilderness with a host of monsters, witches, and similar supernatural beings.

A more subtle terror than Indians or animals was the opportunity the freedom of wilderness presented for men to behave in a savage or bestial manner. Immigrants to the New World certainly sought release from oppressive European laws and traditions, yet the complete license of the wilderness was an overdose. Morality and social order seemed to stop at the edge of the clearing. Given the absence of restraint, might not the pioneer succumb to what John Eliot called "wilderness-temptations?" Would not the proximity of wildness pull down the level of all American civilization? Many feared for the worst, and the concern with the struggle against barbarism was widespread in the colonies. Seventeenth-century town "planters" in New England, for instance, were painfully aware of the dangers wilderness posed for the individual. They attempted to settle the northern frontier through the well-organized movement of entire communities. Americans like these pointed out that while liberty and solitude might be desirable to the man in a crowd, it was the gregarious tendency and controlling institutions of society that took precedence in the wilderness.

Yale's president, Timothy Dwight, spoke for most of his generation in regretting that as the pioneer pushed further and further into the wilds he became "less and less a civilized man." J. Hector St. John Crevecoeur was still more specific. Those who lived near "the great woods," he wrote in 1782, tend to be "regulated by the wildness of their neighborhood." This amounted to no regulation at all; the frontiersmen were beyond "the power of example, and check of shame." According to Crevecoeur, they had "degenerated altogether into the hunting state" and became ultimately "no better than carnivorous animals of a superior rank." He concluded that if man wanted happiness, "he cannot live in solitude, he must belong to some community bound by some ties."

The behavior of pioneers frequently lent substance to these fears. In the struggle for survival many existed at a level close to savagery, and not a few joined Indian tribes. Even the ultimate horror of cannibalism was not unknown among the mountain men of the Rockies, as the case of Charles "Big Phil" Gardner proved. Wilderness could reduce men to such a condition unless society maintained constant vigilance. Under wilderness conditions the veneer civilization laid over the barbaric elements in man seemed much thinner than in the settled regions.

It followed from the pioneer's association of wilderness with hardship and danger in a variety of forms, that the rural, controlled, state of nature was the object of his affection and goal of his labor. The pastoral condition seemed closest to paradise and the life of ease and contentment. Americans hardly needed reminding that Eden had been a garden. The rural was also the fruitful and as such satisfied the frontiersman's utilitarian instincts. On both the idyllic and practical counts wilderness was anathema.

Transforming the wild into the rural had Scriptural precedents which the New England pioneers knew well. Genesis 1:28, the first commandment of God to man, stated that mankind should increase, conquer the earth, and have dominion over all living things. This made the fate of wilderness plain. In 1629 when John Winthrop listed reasons for departing "into . . . the wilderness," an important one was that "the whole earth is the lords Garden & he hath given it to the sonnes of men, and with a general Condision, Gen. 1.28: Increase & multiply, replenish the earth & subdue it." Why remain in England, Winthrop argued, and "suffer a whole Continent . . . to lie waste without

any improvement." Discussing the point a year later, John White also used the idea of man's God-appointed dominion to conclude that he did not see "how men should make benefit of [vacant land] . . . but by habitation and culture." Two centuries later advocates of expansion into the wilderness used the same rhetoric. "There can be no doubt," declared Lewis Cass, soldier and senator from Michigan, in 1830, "that the Creator intended the earth should be reclaimed from a state of nature and cultivated." In the same year Governor George R. Gilmer of Georgia noted that this was specifically "by virtue of that command of the Creator delivered to man upon his formation—be fruitful, multiply, and replenish the earth, and subdue it." Wilderness was waste; the proper behavior toward it, exploitation.

Without invoking the Bible, others involved in the pioneering process revealed a proclivity for the rural and useful. Wherever they encountered wild country they viewed it through utilitarian spectacles: trees became lumber, prairies farms, and canyons the sites of hydroelectric dams. The pioneers' self-conceived mission was to bring these things to pass. Writing about his experience settling northern New York in the late eighteenth century, William Cooper declared that his "great primary object" was "to cause the Wilderness to bloom and fructify." Another popular expression of the waste-to-garden imagery appeared in an account of how the Iowa farmer "makes the wilderness blossom like the rose." Rural, garden-like nature was invariably the criterion of goodness to this mentality. A seventeenth-century account of New England's history noted the way a "howling wilderness" had, through the labors of settlers, become "pleasant Land." Speaking of the Ohio country in 1751, Christopher Gist noted that "it wants Nothing but Cultivation to make it a most delightful Country." Wilderness alone could neither please nor delight the pioneer. "Uncultivated" land, as an early nineteenth-century report put it, was "absolutely useless."

Enthusiasm for "nature" in America during the pioneering period almost always had reference to the rural state. The frequent celebrations of country life, beginning with Richard Steele's *The Husbandman's Calling* of 1668 and continuing through the more familiar statements of Robert Beverley, Thomas Jefferson, and John Taylor of Caroline, reveal only a contempt for the wild, native landscape as "unimproved" land. When wilderness scenery did appeal, it was not for its wildness but because it resembled a "Garden or Orchard in England." The case of Samuel Sewall is instructive, since his 1697 encomium to Plum Island north of Boston has been cited as the earliest known manifestation of love for the New World landscape. What actually appealed to Sewall, however, was not the island's wild qualities but its resemblance to an English countryside. He mentioned cattle feeding in the fields, sheep on the hills, "fruitful marshes," and, as a final pastoral touch, the doves picking up left-over grain after a harvest. In Plum Island Sewall saw the rural idyll familiar since the Greeks, hardly the American wilderness. Indeed, in the same tract, he singled out "a dark Wilderness Cave" as the fearful location for pagan rites.

Samuel Sewall's association of wild country with the ungodly is a reminder that wilderness commonly signified other than a material obstacle or physical threat. As a concept it carried a heavy load of ethical connotations and lent itself to elaborate figurative usage. Indeed, by the seventeenth century "wilderness" had become a favorite metaphor for discussing the Christian situation. John Bunyan's *Pilgrim's Progress* summarized the prevailing viewpoint of wilderness as the symbol of anarchy and evil to which the Christian was unalterably opposed. The book's opening phrase, "As I walk'd through the Wilderness of this World," set the tone for the subsequent description of attempts to keep the faith in the chaotic and temptation-laden existence on earth. Even more pointed in the meaning it attached to wilderness was Benjamin Keach's *Tropologia, or a Key to Open Scripture Metaphor.* In a series of analogies, Keach instructed his readers that as wilderness is "barren" so the world is devoid of holiness; as men lose their way in the wilds so they stray from God in the secular sphere; and as travelers need protection from beasts in wild country, so the Christian needs the guidance and help of God. "A Wilderness," Keach concluded, "is a solitary and dolesom Place: so is this World to a godly Man."

The Puritans who settled New England shared the same tradition regarding wilderness that gave rise to the attitudes of Bunyan and Keach. In the middle of his 1664 dictionary of the Indian language Roger Williams moralized: "the Wildernesse is a cleer resemblance of the world, where greedie and furious men persecute and devour the harmlesse and innocent as the wilde beasts pursue and devour the Hinds and Roes." The Puritans, especially, understood the Christian conception of wilderness, since they conceived of themselves as the latest in a long line of dissenting groups who had braved the wild in order to advance God's cause. They found precedents for coming to the New World in the twelfth-century Waldensians and in still earlier Christian hermits and ascetics who had sought the freedom of deserts or mountains. As enthusiastic practitioners of the art of typology (according to which events in the Old Testament were thought to prefigure later occurrences), the first New Englanders associated their migration with the Exodus. As soon as William Bradford reached Massachusetts Bay, he looked for "Pisgah," the mountain from which Moses had allegedly seen the promised land. Edward Johnson specifically compared the Puritans to "the ancient Beloved of Christ, whom he of old led by the hand from Egypt to Canaan, through that great and terrible Wildernesse." For Samuel Danforth the experience of John the Baptist seemed the closest parallel to the New England situation, although he too likened their mission to that of the children of Israel.

While the Puritans and their predecessors in perfectionism often fled to the wilderness from a corrupt civilization, they never regarded the wilderness itself as their goal. The driving impulse was always to carve a garden from the wilds; to make an island of spiritual light in the surrounding darkness. The Puritan mission had no place for wild country. It was, after all, a *city* on a hill that John Winthrop called upon his colleagues to erect. The Puritans, and to a considerable extent their neighbors in the plantations to the south, went to the wilderness in order to begin the task of redeeming the world from its "wilderness" state. Paradoxically, their sanctuary and their enemy were one and the same.

. . .

For Further Discussion

1. Nash begins this selection by contrasting the attitude toward wilderness of Alexis de Tocqueville and American frontiersmen. What, exactly, is the difference and what do you think explains it?

2. How could you compare the attitude toward wilderness of the American puritan settlers to that expressed by Thoreau in the earlier selection?

3. Do you think that the idea of wilderness depends on a particular cultural and historical background? Could Native Americans have attitudes toward the wilderness similar to the puritans' attitudes?

4. Do you think that the pioneer and puritan attitudes toward wilderness continue to influence public policy today?

DISCUSSION AND STUDY QUESTIONS FOR CHAPTER 2

1. Some interpret the biblical story of humans being made in the image and likeness of God to imply that humans have a unique place in creation, specifically, that humans and only humans have moral standing. Do you agree?

2. Genesis also commands humans to be fruitful and multiply. Does this injunction have any implications for your views on population policy and population control?

3. Descartes believed that animals were mere automata (machines) that could mimic conscious behavior but that in fact were not conscious. Do you think that animals are conscious in the way that humans are? Does this make any difference ethically?

4. Thoreau suggests that in wilderness is the salvation of the world. This contrasts sharply with some biblical and pioneer views of wilderness. What do you understand about the idea of wilderness? Is it a harsh and dangerous place? Is it benign and peaceful? Do such words even make sense when applied to nature?

CHAPTER 3

Is There an Environmental Crisis?

Chapter 2 provided a historical context for our thinking about environmental issues. This chapter again asks us to step back from particular issues and consider, in general, the present environmental situation. Are we in the middle of a serious environmental crisis? Are there reasons for hope and optimism, or does the future look bleak?

It is worth considering how these questions could be answered. How would one know if we are in an environmental crisis? At first glance, we might think that such a dispute could be resolved by the facts, and no doubt, some environmental disputes are factual disputes. But consider the issue of global warming as an example that is not so easily settled by an appeal to the facts.

Does the threat of global warming constitute an environmental crisis? Some facts are clear and beyond dispute. A variety of gases in the atmosphere act to trap solar energy (heat) close to the surface of the earth. These gases, functioning much like the window glass of a greenhouse, allow solar energy in but prevent it from escaping into space. Thus, these are often referred to as greenhouse gases. There is no reasonable dispute about these facts. There is also little argument that carbon dioxide, methane, nitrous oxides, and chlorofluorocarbons are the primary greenhouse gases.

But there is some disagreement concerning the relative amount of these gases that is produced by human activities. Human activities during the past few centuries, especially burning carbon-based fossil fuels, have contributed to some increase in greenhouse gases. But so have many natural events such as solar energy output and volcanic eruptions. Scientists can determine, in approximate terms, the quantity of these gases produced by human activity, but there is not yet agreement on the significance of this increase.

There is less agreement concerning the temperature effects of this increase. There is some evidence that worldwide temperatures have increased over the last century or so. But, there is no way of knowing, at present, whether this change is within the normal range of variation or whether it represents something more ominous. Because global temperatures have been recorded for just a short period, scientists can make only informed guesses about this question.

There is far less agreement about the wider climatic effects that even a significant change in temperature would create. Given the complexity of weather processes and the interconnectedness of atmosphere, oceans, temperatures, and clouds, it is almost impossible to know what would result from an increase in global temperatures. One scenario envisions polar ice caps melting, ocean levels rising, and widespread destructive flooding. Other scenarios forecast greater water evaporation, more clouds, and, hence, an eventual equilibrium of temperatures.

Given all of this, do we face a crisis concerning global warming? People of goodwill can look at the same facts and disagree about their significance, implications, and meaning. The disagreement, in part, is factual. But it also involves the general beliefs, attitudes, and values discussed in Chapter 2. As you read through the selections in this chapter, work to distinguish disagreements concerning facts, concerning values, and concerning attitudes.

The readings in this chapter offer three general perspectives on the present environmental situation. The first perspective argues that we do not face any immediate danger of environmental catastrophe. In fact, we should be optimistic about the future, particularly if we continue to rely on advances in science and technology. The next two readings take a more cautious approach. Both readings claim that, although we still have room to improve, the environment is getting cleaner, and recent environmental policies can claim credit for this progress. The final selection is much less sanguine about the present situation and future prospects. From this perspective, unless we take significant action to change our direction, humans will face grave environmental harms in the near future.

Natural Resources Are Infinite

Julian Simon

The optimistic perspective denies that there is an impending environmental crisis. This perspective often bases its optimism on a particular understanding of the history of science. Too often, controversy surrounding environmental issues is based on a lack of knowledge rather than on informed scientific understanding. For example, scientists simply do not know enough about atmospheric and climatic mechanisms to offer any reliable predictions about the future. What we do know is that science and technology have a good track record in solving problems and contributing to human well-being.

A similar argument concerning the supply of natural resources is used by Julian Simon in this reading. Simon argues that fears concerning depletion of natural resources are misguided because they are based on a mistaken understanding of what resources really are. Simon points out that human beings are interested not in resources themselves but in the services that resources provide. Because human creativity and technology can improve efficiencies and find substitutes that provide the same service, concern about the finitude of resources is unfounded.

Natural resources are not finite. Yes, you read correctly. This [essay] shows that the supply of natural resources is not finite in any economic sense, which is why their cost can continue to fall in the future.

On the face of it, even to inquire whether natural resources are finite seems like nonsense. Everyone "knows" that resources are finite, from C. P. Snow to Isaac Asimov to as many other persons as you have time to read about in the newspaper. And this belief has led many persons to draw far-reaching conclusions about the future of our world economy and civilization. A prominent example is the *Limits to Growth* group, who open the preface to their 1974 book, a sequel to the *Limits*, as follows.

> Most people acknowledge that the earth is finite. . . . Policy makers generally assume that growth will provide them tomorrow with the resources required to deal with today's problems. . . . Recently, however, concern about the consequences of population growth, increased environmental pollution, and the depletion of fossil fuels has cast doubt upon the belief that continuous growth is either possible or a panacea.[1]

(Note the rhetorical device embedded in the term "acknowledge" in the first sentence of the quotation. That word suggests that the statement is a fact, and that anyone who does not "acknowledge" it is simply refusing to accept or admit it.)

The idea that resources are finite in supply is so pervasive and influential that the President's 1972 Commission on Population Growth and the American Future based its policy recommendations squarely upon this assumption. Right at the beginning of its report the commission asked, "What does this nation stand for and where is it going? At some point in the future, the finite earth will not satisfactorily accommodate more human beings—nor will the United States. . . . It is both proper and in our best interest to participate fully in the worldwide search for the good life, which must include the eventual stabilization of our numbers."[2]

The assumption of finiteness is responsible for misleading many scientific forecasters because their conclusions follow inexorably from that assumption. From the *Limits to Growth* team again, this time on food: "The world model is based on the fundamental assumption that there is an upper limit to the total amount of food that can be produced annually by the world's agricultural system."[3]

THE THEORY OF DECREASING NATURAL-RESOURCE SCARCITY

We shall begin with a far-out example to see what contrasting possibilities there are. (Such an analysis of far-out examples is a useful and favorite trick of economists and mathematicians.) If there is just one person, Alpha Crusoe, on an island, with a single copper mine on his island, it will be harder to get raw copper next year if Alpha makes a lot of copper pots and bronze tools this year. And if he continues to use his mine, his son Beta Crusoe will have a tougher time getting copper than did his daddy.

Recycling could change the outcome. If Alpha decides in the second year to make new tools to replace the old tools he made in the first year, it will be easier for him to get the necessary copper than it was the first year because he can reuse the copper from the old tools without much new mining. And if Alpha adds fewer new pots and tools from year to year, the proportion of copper that can come from recycling can rise year by year. This could mean a progressive decrease in the cost of obtaining copper with each successive year for this reason alone, even while the total amount of copper in pots and tools increases.

But let us be "conservative" for the moment and ignore the possibility of recycling. Another scenario: If there are two people on the island, Alpha Crusoe and Gamma Defoe, copper will be more scarce for each of them this year than if Alpha lived there alone, unless by cooperative efforts they can devise a more complex but more efficient mining operation—say, one man on the surface and one in the shaft. Or, if there are two fellows this year instead of one, and if copper is therefore harder to get and more scarce, both Alpha and Gamma may spend considerable time looking for new lodes of copper. And they are likely to be successful in their search. This discovery may lower the cost of copper to them somewhat, but on the average the cost will still be higher than if Alpha lived alone on the island.

Alpha and Gamma may follow still other courses of action. Perhaps they will invent better ways of obtaining copper from a given lode, say a better digging tool, or they may develop new materials to substitute for copper, perhaps iron.

The cause of these new discoveries, or the cause of applying ideas that were discovered earlier, is the "shortage" of copper—that is, the increased cost of getting copper. So a "shortage" of copper causes the creation of its own remedy. This has been the key process in the supply and use of natural resources throughout history.

Discovery of an improved mining method or of a substitute product differs, in a manner that affects future generations, from the discovery of a new lode. Even after the discovery of a new lode, on the average it will still be more costly to obtain copper, that is, more costly than if copper had never been used enough to lead to a "shortage." But discoveries of improved mining methods and of substitute products, caused by the shortage of copper, can lead to lower costs of the services people seek from copper. Let's see how.

The key point is that a discovery of a substitute process or product by Alpha or Gamma can benefit innumerable future generations. Alpha and Gamma cannot themselves extract nearly the full benefit from their discovery of iron. (You and I still benefit from the discoveries of the uses of iron and methods of processing it that our ancestors made thousands of years ago.) This benefit to later generations is an example of what economists call an "externality" due to Alpha and Gamma's activities, that is, a result of their discovery that does not affect them directly.

So, if the cost of copper to Alpha and Gamma does not increase, they may not be impelled to develop improved methods and substitutes. If the cost of getting copper does rise for them, however, they may then bestir themselves to make a new discovery. The discovery may not immediately lower the cost of copper dramatically, and Alpha and Gamma may still not be as well off as if the cost had never risen. But subsequent generations may be better off because their ancestors suffered from increasing cost and "scarcity."

This sequence of events explains how it can be that people have been using cooking pots for thousands of years, as well as using copper for many other purposes, and yet the cost of a pot today is vastly cheaper by any measure than it was 100 or 1,000 or 10,000 years ago.

It is all-important to recognize that discoveries of improved methods and of substitute products are not just luck. They happen in response to "scarcity"—an increase in cost. Even after a discovery is made, there is a good chance that it will not be put into operation until there is need for it due to rising cost. This point is important: Scarcity and technological advance are not two unrelated competitors in a race; rather, each influences the other.

The last major U.S. governmental inquiry into raw materials was the 1952 President's Materials Policy Commission (Paley Commission), organized in response to fears of raw-material shortages during and just after World War II. The Paley Commission's report is distinguished by having some of the right logic, but exactly the wrong predictions, for its twenty-five year forecast.

> There is no completely satisfactory way to measure the real costs of materials over the long sweep of our history. But clearly the man-hours required per unit of output declined heavily from 1900 to 1940, thanks especially to improvements in production technology and the heavier use of energy and capital equipment per worker. This long-term decline in real costs is reflected in the downward drift of prices of various groups of materials in relation to the general level of prices in the economy.
>
> [But since 1940 the trend has been] soaring demands, shrinking resources, the consequences pressure toward rising real costs, the risk of wartime shortages, the strong possibility of an arrest or decline in the standard of living we cherish and hope to share.[4]

For the quarter century for which the commission predicted, however, costs declined rather than rose.

The two reasons why the Paley Commission's cost predictions were topsy-turvy should help keep us from making the same mistakes. First, the commission reasoned from the notion of finiteness and from a static technological analysis.

> A hundred years ago resources seemed limitless and the struggle upward from meager conditions of life was the struggle to create the means and methods of getting those materials into use. In this struggle we have by now succeeded all too well. . . . The nature of the problem can perhaps be successfully oversimplified by saying that the consumption of almost all materials is expanding at compound rates and is thus pressing harder and harder against resources which whatever else they may be doing are not similarly expanding.[5]

The second reason the Paley Commission went wrong is that it looked at the wrong facts. Its report gave too

much emphasis to the trends of costs over the short period from 1940 to 1950, which included World War II and therefore was almost inevitably a period of rising costs, instead of examining the longer period from 1900 to 1940, during which the commission knew that "the man-hours required per unit of output declined heavily."[6]

We must not repeat the same mistakes. We should look at cost trends for the longest possible period, rather than focus on a historical blip; the OPEC-led price rise in all resources after 1973 is for us as the temporary 1940–50 wartime reversal was for the Paley Commission. And the long-run trends make it very clear that the costs of materials, and their scarcity, continuously decline with the growth of income and technology.

RESOURCES AS SERVICES

As economists or as consumers, we are interested in the particular services that resources yield, not in the resources themselves. Examples of such services are an ability to conduct electricity, an ability to support weight, energy to fuel autos, energy to fuel electrical generators, and food calories.

The supply of a service will depend upon (a) which raw materials can supply that service with the present technology; (b) the availabilities of these materials at various qualities; (c) the costs of extracting and processing them; (d) the amounts needed at the present level of technology to supply the services that we want; (e) the extent to which the previously extracted materials can be recycled; (f) the cost of recycling; (g) the cost of transporting the raw materials and services; and (h) the social and institutional arrangements in force. What is relevant to us is not whether we can find any lead in existing lead mines but whether we can have the services of lead batteries at a reasonable price; it does not matter to us whether this is accomplished by recycling lead, by making batteries last forever, or by replacing lead batteries with another contraption. Similarly, we want intercontinental telephone and television communication, and, as long as we get it, we do not care whether this requires 100,000 tons of copper for cables or just a single quarter-ton communications satellite in space that uses no copper at all.[7]

Let us see how this concept of services is crucial to our understanding of natural resources and the economy. To return to Crusoe's cooking pot, we are interested in a utensil that we can put over the fire and cook with. After iron and aluminum were discovered, quite satisfactory cooking pots, perhaps even better than pots of copper, could be made of these materials. The cost that interests us is the cost of providing the cooking service rather than the cost of copper. If we suppose that copper is used only for pots and that iron is quite satisfactory for the same purpose, as long as we have cheap iron it does not matter if the cost of copper rises sky high. (But in fact that has not happened. As we have seen, the prices of the minerals themselves, as well as the prices of the services they perform, have fallen over the years.)

ARE NATURAL RESOURCES FINITE?

Incredible as it may seem at first, the term "finite" is not only inappropriate but is downright misleading when applied to natural resources, from both the practical and philosophical points of view. As with many of the important arguments in this world, the one about "finiteness" is "just semantic." Yet the semantics of resource scarcity muddle public discussion and bring about wrong-headed policy decisions.

The word "finite" originates in mathematics, in which context we all learn it as schoolchildren. But even in mathematics the word's meaning is far from unambiguous. It can have two principal meanings, sometimes with an apparent contradiction between them.[8] For example, the length of a one-inch line is finite in the sense that it is bounded at both ends. But the line within the endpoints contains an infinite number of points; these points cannot be counted, because they have no defined size. Therefore the number of points in that one-inch segment is not finite. Similarly, the quantity of copper that will ever be available to us is not finite, because there is no method (even in principle) of making an appropriate count of it, given the problem of the economic definition of "copper," the possibility of creating copper or its economic equivalent from other materials, and thus the lack of boundaries to the sources from which copper might be drawn.

Consider this quote about potential oil and gas from Sheldon Lambert, an energy forecaster. He begins, "It's like trying to guess the number of beans in a jar without knowing how big the jar is." So far so good. But then he adds, "God is the only one who knows—and even He may not be sure."[9] Of course

Lambert is speaking lightly. But the notion that some mind might know the "actual" size of the jar is misleading, because it implies that there is a fixed quantity of standard-sized beans. The quantity of a natural resource that might be available to us—and even more important the quantity of the services that can eventually be rendered to us by that natural resource—can never be known even in principle, just as the number of points in a one-inch line can never be counted even in principle. Even if the "jar" were fixed in size, it might yield ever more "beans." Hence resources are not "finite" in any meaningful sense.

To restate: A satisfactory *operational* definition of the quantity of a natural resource, or of the services we now get from it, is the only sort of definition that is of any use in policy decisions. The definition must tell us about the quantities of a resource (or of a particular service) that we can expect to receive in any particular year to come, at each particular price, conditional on other events that we might reasonably expect to know (such as use of the resource in prior years). And there is no reason to believe that at any given moment in the future the available quantity of any natural resource or service at present prices will be much smaller than it is now, or non-existent. Only such one-of-a-kind resources as an Arthur Rubenstein concert or a Julius Erving basketball game, for which there are no close replacements, will disappear in the future and hence are finite in quantity.

Why do we become hypnotized by the word "finite"? That is an interesting question in psychology, education, and philosophy. A first likely reason is that the word "finite" seems to have a precise and unambiguous meaning in any context, even though it does not. Second, we learn the word in the context of mathematics, where all propositions are tautologous definitions and hence can be shown logically to be true or false (at least in principle). But scientific subjects are empirical rather than definitional, as twentieth-century philosophers have been at great pains to emphasize. Mathematics is not a science in the ordinary sense because it does not deal with facts other than the stuff of mathematics itself, and hence such terms as "finite" do not have the same meaning elsewhere that they do in mathematics.

Third, much of our daily life about which we need to make decisions is countable and finite—our weekly or monthly salaries, the number of gallons of gas in a full tank, the width of the backyard, the number of greeting cards you sent out last year, or those you will send out next year. Since these quantities are finite, why shouldn't the world's total possible salary in the future, or the gasoline in the possible tanks in the future, or the number of cards you ought to send out, also be finite? Though the analogy is appealing, it is not sound. And it is in making this incorrect analogy that we go astray in using the term "finite."

A fourth reason that the term "finite" is not meaningful is that we cannot say with any practical surety where the bounds of a relevant resource system lie, or even if there are any bounds. The bounds for the Crusoes are the shores of their island, and so it was for early man. But then the Crusoes found other islands. Mankind traveled farther and farther in search of resources—finally to the bounds of continents, and then to other continents. When America was opened up, the world, which for Europeans had been bounded by Europe and perhaps by Asia too, was suddenly expanded. Each epoch has seen a shift in the bounds of the relevant resource system. Each time, the old ideas about "limits," and the calculations of "finite resources" within those bounds, were thereby falsified. Now we have begun to explore the sea, which contains amounts of metallic and other resources that dwarf any deposits we know about on land. And we have begun to explore the moon. Why shouldn't the boundaries of the system from which we derive resources continue to expand in such directions, just as they have expanded in the past? This is one more reason not to regard resources as "finite" in principle.

You may wonder, however, whether "non-renewable" energy resources such as oil, coal, and natural gas differ from the recyclable minerals in such a fashion that the foregoing arguments do not apply. Energy is particularly important because it is the "master resource"; energy is the key constraint on the availability of all other resources. Even so, our energy supply is non-finite, and oil is an important example. (1) The oil potential of a particular well may be measured, and hence is limited (though it is interesting and relevant that as we develop new ways of extracting hard-to-get oil, the economic capacity of a well increases). But the number of wells that will eventually produce oil, and in what quantities, is not known or measurable at present and probably never will be, and hence is not meaningfully finite. (2) Even if we make the unrealistic assumption that the number of potential wells in the earth might be surveyed completely and that we could arrive at a reasonable estimate of the oil that might be obtained with present technology (or even with technology that will be developed in the next 100 years), we still would have to reckon the future possibilities of

shale oil and tar sands—a difficult task. (3) But let us assume that we could reckon the oil potential of shale and tar sands. We would then have to reckon the conversion of coal to oil. That, too, might be done; yet we still could not consider the resulting quantity to be "finite" and "limited." (4) Then there is the oil that we might produce not from fossils but from new crops—palm oil, soybean oil, and so on. Clearly, there is no meaningful limit to this source except the sun's energy. The notion of finiteness does not make sense here, either. (5) If we allow for the substitution of nuclear and solar power for oil, since what we really want are the services of oil, not necessarily oil itself, the notion of a limit makes even less sense. (6) Of course the sun may eventually run down. But even if our sun were not as vast as it is, there may well be other suns elsewhere.

About energy from the sun: The assertion that our resources are ultimately finite seems most relevant to energy but yet is actually more misleading with respect to energy than with respect to other resources. When people say that mineral resources are "finite" they are invariably referring to the earth as a boundary, the "spaceship earth," to which we are apparently confined just as astronauts are confined to their spaceship. But the main source of our energy even now is the sun, no matter how you think of the matter. This goes far beyond the fact that the sun was the prior source of the energy locked into the oil and coal we use. The sun is also the source of the energy in the food we eat, and in the trees that we use for many purposes. In coming years, solar energy may be used to heat homes and water in many parts of the world. (Much of Israel's hot water has been heated by solar devices for years, even when the price of oil was much lower than it is now.) And if the prices of conventional energy supplies were to rise considerably higher than they now are, solar energy could be called on for much more of our needs, though this price rise seems unlikely given present technology. And even if the earth were sometime to run out of sources of energy for nuclear processes—a prospect so distant that it is a waste of time to talk about it—there are energy sources on other planets. Hence the notion that the supply of energy is finite because the earth's fossil fuels or even its nuclear fuels are limited is sheer nonsense.

Whether there is an "ultimate" end to all this—that is, whether the energy supply really is "finite" after the sun and all the other planets have been exhausted—is a question so hypothetical that it should be compared with other metaphysical entertainments such as calculating the number of angels that can dance on the head of a pin. As long as we continue to draw energy from the sun, any conclusion about whether energy is "ultimately finite" or not has no bearing upon present policy decisions. . . .

SUMMARY

A conceptual quantity is not finite or infinite in itself. Rather, it is finite or infinite if you make it so—by your own definitions. If you define the subject of discussion suitably, and sufficiently closely so that it can be counted, then it is finite—for example, the money in your wallet or the socks in your top drawer. But without sufficient definition the subject is not finite— for example, the thoughts in your head, the strength of your wish to go to Turkey, your dog's love for you, the number of points in a one-inch line. You can, of course, develop definitions that will make these quantities finite; but that makes it clear that the finiteness inheres in you and in your definitions rather than in the money, love, or one-inch line themselves. There is no necessity either in logic or in historical trends to suggest that the supply of any given resource is "finite."

Notes

1. Meadows, Dennis L., William W. Behrens III, Donella H. Meadows, Roger F. Naill, Jorgen Randers, and

Erich K. O. Zahn. *Dynamics of Growth in a Finite World* (Cambridge, MA: Wright-Allen, 1974), p. vii.
2. U.S., The White House. *Population and the American Future* (New York: Signet, 1972), pp. 2–3.
3. Meadows, et al., p. 265.
4. U.S., The White House. *Resources for the Future.* Four volumes. The President's Materials Policy Commission (Washington, DC: GPO, June, 1952).
5. Ibid., p. 2.
6. Ibid., p. 1.
7. Fuller, Buckminster. *Utopia or Oblivion* (New York: Bantam Press, 1977), p. 45.
8. I appreciate a discussion of this point with Alvin Roth.
9. Sheldon Lambert, quoted in *Newsweek* (June 27, 1977), 71.

For Further Discussion

1. Simon obviously disagrees with those who think that natural resources are finite. What understanding of "resources" makes this conclusion valid?

2. Much of Simon's reasoning assumes that new technologies can act as substitutes for present resources. If this is so, do the users of present resources have any duty to contribute to the development of substitutes?

3. Would Simon's position be more likely to encourage savings or waste? Does it matter either way?

4. Can you think of any resources that have already been depleted? Is Simon making an empirical claim about resources or a conceptual claim about the meaning of words? What difference would this make?

From *A Moment on the Earth*

Gregg Easterbrook

A number of writers have defended a more moderate position in recent years. What might be called ecorealism or ecosanity acknowledges that humans face real environmental problems, many of which have been caused by the very technology that people such as Julian Simon praise. But ecorealists also disagree with doomsayers who despair about the future. Ecorealists believe that much progress has been made, that environmental conditions are improving,

and that the environmental movement itself should receive credit for much of this.

Journalist Gregg Easterbrook's book, A Moment on the Earth, *from which this selection is taken, offers a detailed defense of the ecorealist position. This reading, which comprises the Preface and Chapter 34, presents Easterbrook's call for a more realistic, and optimistic, environmentalism and concludes with his outline of an "ecorealist manifesto."*

In the autumn of 1992 I was struck by this headline in the *New York Times:* "Air Found Cleaner in U.S. Cities." The accompanying story said that in the past five years air quality had improved sufficiently that nearly half the cities once violating federal smog standards no longer did so.

I was also struck by how the *Times* treated the article—as a small box buried on page A24. I checked the nation's other important news organizations and learned that none had given the finding prominence. Surely any news that air quality was in decline would have received front-page attention. The treatment suggested that the world was somehow disappointed by an inappropriately encouraging discovery.

American air is getting cleaner. Can this be happening on the same planet from which most current environmental commentary emanates? Vice President Al Gore has described the U.S. environmental situation as "extremely grave—the worst crisis our country has ever faced." The *worst:* worse than the enslavement of African-Americans, worse than the persecution of Native Americans, worse than the Civil War, worse than the Depression, worse than World War II. George Mitchell, till 1994 the majority leader of the Senate, has declared that "we risk turning our world into a lifeless desert" through environmental abuse. Gaylord Nelson, who as a senator in 1970 originated Earth Day and who is now a lawyer for the Wilderness Society, said in 1990 that current environmental problems "are a greater threat to the Earth's life-sustaining systems than a nuclear war."

And can this be the same planet from which most contemporary environmental writing emanates? *Silent Spring,* published by Rachel Carson in 1962, foretold such a widespread biological wipeout that today robins should be extinct, no longer greeting the spring with song. Instead the robin is today one of the two or three most prolific birds in the United States. Paul Ehrlich's *The Population Bomb,* released in 1968, predicted that general crop failures would "certainly" result in mass starvation in the United States by the 1980s. Instead the leading American agricultural problem of that decade was oversupply. The same book found it "not inconceivable" that some ghastly plague triggered by pollution would flash-kill half a billion people. Instead life expectancies have steadily increased, even in the overcrowded Third World. *The Limits to Growth,* a saturnine 1972 volume acclaimed at the time by critics in the United States, projected that petroleum would be exhausted by the 1990s. Instead oil prices hover near postwar lows, reflecting ample supply. *The Sinking Ark,* published in 1979 by the biologist Norman Myers, portrayed the vessel of nature as riddled with breaches and going under before our eyes, with thousands of species to become extinct during the 1980s. Instead there were at worst a handful of confirmed extinctions globally in that decade.

A Blueprint for Survival, a 1972 anthology that was a bestseller in the United Kingdom, decreed that environmental trends mean "the breakdown of society, and the irreversible disruption of the life support systems on this planet . . . are inevitable." *Green Rage,* a 1990 volume by Christopher Manes on "deep" ecology, spoke of humanity as engaged in a "lemming-like march into environmental oblivion." Other recent books and public-interest campaigns have proclaimed a mass "poisoning of America," general radiation calamities, catastrophic climate change, deadly drinking water, and exhaustion of the basic processes of life. Affairs are thought so unswervingly bleak that the writer Bill McKibben, in his much-discussed 1989 work *The End of Nature,* declared there is no need to wait for the worst. Nature has already ended: ultimately, irrevocably, horrendously.

Yet I look out my window and observe that the sky above the populous Washington, D.C., region where I live each year grows more blue. The sun not only continues to rise; it does so above a horizon that is progressively cleaner. Is everybody talking about the same world?

Let's contemplate smog for a moment. Findings like those described in the first paragraph hardly mean the battle against smog is over. But despite the impression given to the public by fashionably pessimistic commentary, underlying trends in air pollution were positive throughout the 1980s. In that decade ambient smog in the United States declined a composite 16 percent, even as economic output expanded and the number of automobiles increased rapidly. In the beginning of the 1980s there were about 600 air-quality-alert days each year in major cities. By the end of the 1980s there were about 300 such days annually. Air pollution from lead, by far the worst atmospheric poison, declined 89 percent during the 1980s; from carbon monoxide, also poisonous, went down 31 percent; ambient levels of sulfur dioxide, the main precursor of acid rain, declined 27 percent; nitrogen dioxide, another smog cause, went down 12 percent; in no smog category did ambient levels rise. In sum, American air was much less dirty in 1990 than in 1980, not more dirty as commonly believed.

Environmental Protection Agency figures from the 1990s show the improvement trend accelerating. In 1992, the number of Americans living in counties that failed some aspect of air-quality standards was 54 million—too many, but down from the 86 million people who lived in dirty-air counties in 1991, and only half the 100 million who lived in dirty air in 1982. In 1992 13 major cities, including Detroit and Pittsburgh, met federal standards for smog reduction for the first time, while no new cities were added to the violations list.

In 1993 I wrote an article for *Newsweek* presenting in detail the argument that the air grows cleaner. Later Senator Frank Lautenberg of New Jersey, chair of an important environmental subcommittee, waved the article before EPA administrator Carol Browner during a Senate hearing and declared himself "outraged" that *Newsweek* had printed such words. Senator Lautenberg did not challenge any of the factual material in the article. He appeared upset simply that positive environmental information was being reported. The good news scared him.

The good news should not scare anyone, particularly lovers of nature. Consider that recent improvements in air quality came mainly during a decade of Republican presidents—prominently Ronald Reagan, who labored under the garbled impression that trees cause more air pollution than cars. If a significant aspect of the environment got better even under Reagan, it sounds like something important *is* going on.

Something important *is* going on here: a fundamental, far-reaching shift toward the positive in environmental events. That shift is the subject of this book.

. . . [This] is not an attack on environmentalism. Ecological consciousness is a leading force for good in world affairs. Without the imperatives of modern environmentalism—without its three decades of unstinting pressure on government and industry—the Western world today might actually be in the kind of ecological difficulty conventional wisdom assumes it to be in. Instead, the Western world today is on the verge of the greatest ecological renewal that humankind has known; perhaps the greatest that the Earth has known. Environmentalists deserve the credit for this remarkable turn of events.

Yet our political and cultural institutions continue to read from a script of instant doomsday. Environmentalists, who are surely on the right side of history, are increasingly on the wrong side of the present, risking their credibility by proclaiming emergencies that do not exist. What some doctrinaire environmentalists wish were true for reasons of ideology has begun to obscure the view of what is actually true in "the laboratory of nature." It's time we began reading from a new script, one that reconciles the ideals of environmentalism with the observed facts of the natural world. Toward that end this book will advance the following premises:

- That in the Western world pollution will end within our lifetimes, with society almost painlessly adapting a zero-emissions philosophy.

- That several categories of pollution have *already* ended.

- That the environments of Western countries have been growing cleaner during the very period the public has come to believe they are growing more polluted.

- That First World industrial countries, considered the scourge of the global environment, are by most measures much cleaner than developing nations.

- That most feared environmental catastrophes, such as runaway global warming, are almost certain to be avoided.

- That far from becoming a new source of global discord, environmentalism, which binds nations to a common concern, will be the best thing that's ever happened to international relations.

- That nearly all technical trends are toward new devices and modes of production that are more efficient, use fewer resources, produce less waste, and cause less ecological disruption than technology of the past.

- That there exists no fundamental conflict between the artificial and the natural.

- That artificial forces which today harm nature can be converted into allies of nature in an incredibly short time by natural standards.

- Most important, that humankind, even a growing human population of many billions, can take a constructive place in the natural order.

None of these notions are now common currency. It is possible to find yourself hooted down for proposing them at some public forums. A few years ago at a speech at a Harvard Divinity School conference on environmental affairs I was hissed merely for saying "People are more important than plants and animals." What better barometer is there of how nonsensical doomsday thinking can become?

But that is a passing situation. In the near future the propositions stated above will be widely embraced by society and even by the intelligentsia. Collectively I call these views *ecorealism*.

Ecorealism will be the next wave of environmental thinking. The core principles of ecorealism are these: that logic, not sentiment, is the best tool for safeguarding nature; that accurate understanding of the actual state of the environment will serve the Earth better than expressions of panic; that in order to form a constructive alliance with nature, men and women must learn to think like nature.

The coming wave of ecorealism will enable people and governments to make rational distinctions between those environmental alarms that are genuine and those that are merely this week's fad. Once rational decision-making becomes the rule in environmental affairs, the pace of progress will accelerate.

Essential to the ecorealist awakening will be the understanding that in almost every ecological category, nature has for millions of centuries been generating worse problems than any created by people.

Consider, for example, that today U.S. factories, power plants, and vehicles emit about 19 million tons per year of sulfur dioxide, the chief cause of acid rain. That level is far too high. Yet in 1991, the Mount Pinatubo eruption in the Philippines emitted an estimated 30 million tons of sulfur dioxide in just a few hours. Less spectacular, ongoing natural processes such as volcanic outgassing and ocean chemistry put about 100 million tons of sulfur dioxide into the atmosphere annually.

That nature makes pollutants in no way excuses the industrial variety. The comparison simply points to an important aspect of the environment, understanding of which is absent from current debate: that nature has spent vast spans of time learning to cope with acid rain, greenhouse gases, climate change, deforestation, radiation, species loss, waste, and other problems we humans so quaintly believe ourselves hurling at the environment for the first time. This knowledge suggests that environmental mischief by women and men will harm the Earth much less than popular culture now assumes. It further suggests that if people have the sense to stop the pollution they make today, and clean up that which they made in the past, the environment will regenerate in an amazingly short time by nature's standards.

Environmental commentary is so fogbound in woe that few people realize measurable improvements have already been made in almost every area. In the United States air pollution, water pollution, ocean pollution, toxic discharges, acid rain emissions, soil loss, radiation exposure, species protection, and recycling are areas where the trend lines have been consistently positive for many years. Yet polls show that people believe the environment is getting worse. Some of this can be explained by the new dynamic of fashionable doomsaying. Today many environmentalists and authors compete to see who can stage the most theatrical display of despair; public officials who once denied that environmental problems exist attempt to compensate by exaggerating in the other direction; celebrities whose lifestyles hardly reflect an ethic of modest consumption pause at limousine doors to demand that SOMEBODY ELSE conserve.

A peculiar intellectual inversion has occurred in which good news about the environment is treated as something that ought to be hushed over, while bad news is viewed with relief. Suppose a satellite produced evidence that ozone depletion was all a data error: some elements of the environmental movement would be heartbroken. Vice President Gore has written, in *Earth in the Balance,* that journalists should downplay scientific findings of ecological improvement because good news may dilute the public sense of anxiety. Gore has even said that scientists who disagree with the doomsday premise are "unethical" and must be ignored.

To the ecorealist, fashionable pessimism about the environment could not be more wrong, if only because it denies the good done already. In some vexing policy areas such as crime or public education it is difficult to imagine where solutions reside. On environmental affairs I can promise you—and will show you—that public investments yield significant benefits within the lifetimes of the people who make the investment. The first round of environmental investments did not fail; they worked, which is a great reason to have more.

I consider this glorious if only because as a political liberal I long for examples of government action that serves the common good. The extraordinary success of modern environmental protection is such an example: perhaps the best instance of government-led social progress in our age.

For this reason I have trouble fathoming why guarded optimism about the environment is politically incorrect. I have no trouble imagining that this situation will change. In the coming ecorealist ethic we will all be environmental optimists, citing conservation and pollution prevention as that rare area where government action and public concern lead promptly to results beneficial to all. Someday even Vice President Gore will smile when he talks about the ecology. Perhaps not tomorrow. But soon.

Let's note here three things that ecorealism is not.

First, it is not a philosophy of don't worry, be happy. The ecorealist must acknowledge there exists a wide range of human actions careless, selfish, or destructive to the environment. The point of ecorealism is that this equation can change, and it is much closer to that moment of transformation than all but a few people realize.

Second, ecorealism is not an endorsement of the technological lifestyle. In the past many foolish projections have been made about the course of technical events: from the thinkers of the Enlightenment, who believed that the perfectibility of humanity was at hand, to those daffy 1950s *Popular Mechanics* articles about how we'd all be flying personal helicopters by now. The epitome of this genre was a popular 1842 book by a writer named J. A. Eltizer called *The Paradise Within Reach of All Men, By Power of Machinery*. The title of the volume says everything you need to know about it. Ecorealism does not posit that technology is anyone's benefactor. It's just not necessarily bad, as is now fashionably assumed. Technology is a tool, and as a tool such as a knife can be used either to cause mayhem or carve a walking stick, technology may be used wisely or foolishly. It is up to us to decide which it will be.

Through the course of this book you will find many pages devoted to reasons for guarded optimism about the ecology juxtaposed against few detailing the evidence for ecological despair. There are passionate arguments for the latter position, the best expression of which can be found in McKibben's haunting *The End of Nature*. But the arguments for despair have received extensive explication. The arguments for optimism are rarely presented. So here I emphasize the story you haven't heard at the expense of the story you have heard. Through this book you will also encounter many passages in which the pessimistic aspects of environmental thought are generalized in ways that will fall short of reflecting the substantial range of opinions, some buoyant, that may be found within contemporary ecological mentation. By this I do not mean to suggest that the green movement is monolithic in its embrace of doomsday thinking; of course it is not. But pessimism is the main current in contemporary environmental thought, and its refutation is the main concern of this book. To those many nondoctrinaire environmentalists and environmental thinkers whose positions my paraphrases will not fully reflect, I apologize in advance.

Third and last in the inventory of what ecorealism is not, ecorealism has nothing to do with a minor fad called wise use. The phrase "wise use" once had a progressive meaning in environmental letters but in recent years has been expropriated by reactionary fundraisers. Today lovers of nature ought to have no use for wise use. The wise use crowd, for instance, is nearly psychasthenic in its opposition to the Endangered Species Act. The ecorealist ought to support strengthening of the act, for reasons we shall see.

One reason I propose ecorealism is to create a language in which environmental protection can be discussed without descending into the oratorical quicksand of instant doomsday on the left and bulldozer apologetics on the right. Ecorealism offers a guiding ideal for those who care about the integrity of nature yet hold no brief for the extreme positions on either side. People sharing those values—a group that I figure at about 90 percent of the American population—need a vocabulary and a platform for reasoned ecological debate. Ecorealism will provide it. Such debate will make environmental protection clearheaded and rational, and thus ultimately stronger still.

There was a time when to cry alarm regarding environmental affairs was the daring position. Now that's the safe position: People get upset when you say things may turn out fine.

This book describes a possible sequence of events in which people, machines, and nature learn to work together to each other's benefit. Many, many things could go wrong with that vision. But why not set our sights on such a goal? Nothing makes more sense for our moment on the Earth.

Chapter 34: The Ecorealist Manifesto

In the first section of this book the goal was to consider environmental problems from the perspective of nature: a long-term, almost mythic purview of existence, yet one with relevance to daily ecological challenges that confront humanity.

The second section considered environmental problems from the perspective of women and men alive today: a short-term perspective, yet one that can be adjusted to incorporate the wisdom embodied in eons of natural transitions.

The final section will suggest how the long-term purview of nature might be combined with the short-term insights of genus *Homo* in ways that allow people, machines, and nature to learn to work together for each other's mutual benefit.

Before broaching that final topic let's summarize the principles of a new view of ecological thought, the new view called ecorealism. The founding concept of ecorealism is this: Logic, not sentiment, best serves the interests of nature.

If the worthy inclinations of environmentalism are to be transformed from an ephemeral late-twentieth-century political fashion to a lasting component of human thought, the ecological impulse must become grounded in rationality. The straightforward, rational case for the environment will prove more durable than the fiercest doomsday emotion. Love nature? Learn science and speak logic. Many lesser creatures will thank you.

And now some principles of ecorealism:

Rationalism

- If ecological rationalism sometimes shows that environmental problems are not as bad as expected, that means warnings will be all the more persuasive when genuine problems are found.

- Graduation from overstatement will make the environmental movement stronger, not weaker.

- The worst thing the environmental movement could become is another absentminded interest group stumbling along toward preconceived ends regardless of what the evidence suggests.

- Skeptical debate is good for the environmental movement. The public need not to be brainwashed into believing in ecological protection, since a clean environment is in everyone's interest. Thus the environmental movement must learn to entertain skeptical debate in a reasoned manner or will discredit itself, as all close-minded political movements eventually discredit themselves.

- Market forces and cost-benefit thinking aren't perfect but generally will be good for the environment. This is so if only because society may be able to afford several cost-effective conservation initiatives for the price of one poorly conceived program.

- Optimism not only flows from a reasoned reading of natural history, it will be an effective political tool.

Pollution

- In the Western world the Age of Pollution is nearly over. Almost every pollution issue will be solved within the lifetimes of readers of this book.

- In the West many forms of pollution have begun to decline in the very period that environmental doctrine has declared them growing worse.

- Most recoveries from pollution will happen faster than even optimists project.

- Weapons aside, technology is not growing more dangerous and wasteful. It grows cleaner and more resource-efficient.

- Clean technology will be the successor to high technology. Most brute-force systems of material production will be supplanted by production based mainly on knowledge. Nature's creatures make extremely sophisticated "products" with hardly any input of energy or resources. People will learn to do the same.

- Sometimes approximated environmental rules are good ideas even if the result is a less than perfect cleanup. Better to realize 90 percent of an ecological restoration fast than to spend decades conducting lawsuits on how to achieve 100 percent.

- As positive as trends are in the First World, they are negative in the Third. One reason the West must shake off instant-doomsday thinking about the United States and Western Europe is so that resources can be diverted to ecological protection in the developing world.

Change

- It is pointless for men and women to debate what the "correct" reality for nature might have been before people arrived on the scene. There has never been and can never be any fixed, correct environmental reality. There are only moments on the Earth, moments that may be good or bad.

- Every environment and habitat comes into existence fated to end. This is not sad. It should inspire women and men to seek to prolong moments on the Earth, through conservation.

- All environmental errors are reversible save one: extinction. Therefore the prevention of extinctions is a priority.

- Though humanity may today be a cause of species extinction, in a very short time by nature's standards it can become an agent for species preservation.

People

- People may not sit above animals and plants in any metaphysical sense, but clearly are superior in their placement in the natural order. Decent material conditions must be provided for all of the former before there can be long-term assurance of protection for the latter.

- Either humanity was created by a higher power, in which case it is absurd for environmental dogma to consider the human role in nature to be bad; or humanity rose to its position through purely natural processes, in which case it is absurd for environmental dogma to consider the human role in nature to be bad.

- However the deed was done, once genus *Homo* was called forth into being, the wholly spontaneous ordering of the environment ended. And unless there is an extinction of intellect, wholly spontaneous nature will never return. Nature is not diminished by this. A fairly straightforward reading of natural history suggests that evolution spent 3.8 billion years working assiduously to bring about the demise of the wholly spontaneous order, via the creation of intellect.

- In principle the human population is no enemy of nature. Someday that population may be many times larger than at present, without ecological harm. But the world of the present knows more people than current social institutions and technical knowledge can support at an adequate material standard. Thus short-term global population stabilization is desperately required, though the prospect of dramatic long-term expansion of the human population should not be discounted.

Nature

- Nature is not ending, nor is human damage to the environment "unprecedented." Nature has repelled forces of a magnitude many times greater than the worst human malfeasance.

- Nature is not ponderously slow. It's just old. Old and slow are quite different concepts. That the living world can adjust with surprising alacrity is the reason nature has been able to get old. Most natural recoveries from ecological duress happen with amazing speed.

- Significant human tampering with the environment has been in progress for at least ten millennia and perhaps longer. If nature has been interacting with genus *Homo* for thousands of years, then the living things that made it to the present day may be ones whose genetic treasury renders them best suited to resist human mischief. This does not ensure any creature will continue to survive any clash with hu-

mankind. It does make survival more likely than doomsday orthodoxy asserts.

- If nature's adjustment to the human presence began thousands of years ago, perhaps it will soon be complete. Far from reeling helplessly before a human onslaught, nature may be on the verge of reasserting itself.

- Nature still rules much more of the Earth than does genus *Homo*. To the statistical majority of nature's creatures the arrival of men and women goes unnoticed.

- To people the distinction between artificial and natural means a great deal. To nature it means nothing at all.

- The fundamental force of nature is not amoral struggle between hunter and hunted. Most living things center their existence on cooperation and coexistence, the sort of behavior women and men should emulate. This is one reason nature will soon be viewed again in the way it was by the thinkers of the eighteenth-century Enlightenment—as a trove of wisdom and an exemplar for society.

Where Do We Fit In?

- Nature, limited by spontaneous interactions among elements randomly disturbed, may have an upper-bound limit on its potential to foster life and to evolve. Yet nature appears to enjoy fostering life and evolving. So perhaps nature hoped to acquire new sets of abilities, such as action by design.

- Therefore maybe nature needs us.

For Further Discussion

1. Can you think of any reasons why we should not be content with improvements in air and water quality? Is Easterbrook correct in saying that many environmentalists should be, but are not, happy with the good news?

2. How would you define ecorealism? How is it different from doomsday thinking? Do you think that many environmentalists are overly pessimistic?

3. Easterbrook claims that environmental regulation is "perhaps the best instance of government-led social progress in our age." Do you agree?

4. Do you disagree with any of the principles of ecorealism listed at the conclusion of this selection? On what grounds, if any, do you disagree?

Our World Is Getting Cleaner

Joseph Bast
Peter Hill
Richard Rue

A theme similar to Easterbrook's is developed by Joseph Bast, Peter Hill, and Richard Rue in the following reading. Their position, as identified in their book's title Eco-Sanity, *is that "the world is becoming cleaner, not dirtier, over time." This selection from Chapter 2 of their book outlines their realistic analysis of air and water quality. Like Easterbrook, these authors believe that much work needs to be done but that, on the whole, we have reason for cautious optimism.*

MESSENGERS OF DOOM

If you read *Time* or *Newsweek,* or watch television programs such as *Phil Donahue* or the network evening news, you have been told repeatedly that our world is getting dirtier, not cleaner. For example,

- A widely reported study titled *Global 2000 Report to the President,* produced in 1980 by agencies of the federal government, concluded "if current trends continue, the world in 2000 will be more crowded, more polluted, less stable ecologically, and more vulnerable to disruption than the world we live in now."

- Popular books by activists such as Ralph Nader, Paul Ehrlich, and Paul Brodeur have carried such titles as *Laying Waste, America the Poisoned, Currents of Death,* and *Who's Poisoning America?* These books, and the extensive publicity surrounding them, describe a "cancer epidemic" caused by industrial pollution and toxic waste.

- The Vice President of the United States, Al Gore, writes in *Earth in the Balance* that "the volume of garbage is now so high that we are running out of places to put it," and that automobiles are "posing a mortal threat to the security of every nation that is more deadly than that of any military enemy we are ever again likely to confront."

- British ecologist Norman Myers, writing in February 1994, warned that mankind is "set to eliminate between one-third and two-thirds of Earth's species," causing "a massive draining of the planetary gene pool." By the middle of the next century, he claimed, "we shall have lost virtually all our topsoil, and no substitute for soil exists."

If these accounts are true, then surely life today is less safe and less healthy than it was a generation ago. Widespread concern over air and water pollution, solid waste disposal, and depletion of natural resources would seem to be justified.

But there is a big problem with what these reporters and experts say: *They are wrong.* Dr. Elizabeth Whelan, executive director of the American Council on Science and Health, author of twelve books on health issues, and a graduate of the Harvard School of Public Health and the Yale School of Medicine, writes:

> I have reviewed literally thousands of popular and scientific articles on the topic of environmental factors and human health. . . . What I found in my literature review was an astounding gap between the consensus in the scientific and medical community on environmental issues, versus what is being presented in popular publications, on television and radio, and in books for the layman.

Dr. Whelan is not alone. Dr. Julian Simon, professor of economics at the University of Maryland, and Herman Kahn, director of the prestigious Hudson Institute prior to his death in 1983, wrote:

> The original *Global 2000* is totally wrong in its specific assertions and its general conclusion. . . . Many of its arguments are illogical or misleading.

It paints an overall picture of global trends that is fundamentally wrong, partly because it relies on non-facts and partly because it misinterprets the facts it does present.

At the 1992 Earth Summit in Rio de Janeiro, a group of 425 scientists and economists, including 27 Nobel laureates, decried the inaccuracy and exaggeration that has come to characterize the positions of leading environmental organizations. The group's statement, which now bears the signatures of more than 2,700 scientists and intellectual leaders, included these words:

> We are . . . worried, at the dawn of the 21st Century, at the emergence of an irrational ideology that is opposed to scientific and industrial progress and impedes economic and social development. . . . We intend to assert science's responsibility and duties toward society as a whole. We do, however, forewarn the authorities in charge of our planet's destiny against decisions which are supported by pseudoscientific arguments or false and irrelevant data. . . . The greatest evils which stalk our earth are ignorance and oppression, not science, technology, and industry.

The authors have conducted their own review of the evidence and have reached the same conclusion: The views of the alarmists have been proven to be wrong. Most trends—cancer rates, air and water quality, human health, acres of wooded lands, and a dozen other indicators of environmental quality in the U.S.— show improvement, not decline. Global trends also show improvement, though less dramatically than trends in the U.S. Although generalizations are difficult to make in the complex arena of environmental quality, available evidence supports the following statements:

> Most Americans today live in an environment that is cleaner than it was at any time in the past half-century.

> The environment in the U.S. today is safer than it has been at any time in recorded history.

On the following pages we document these claims. Before proceeding, however, it is necessary to clarify several points. First, when we say the environment in the U.S. is *cleaner* than at any time in the past half-century, we mean average human exposure to potentially harmful pollutants is as low as it was during the 1930s or 1940s. When we say the environment today is *safer* than at any time in recorded history, we mean the probability that a substance or natural process (such as global climate change) will cause human injury is the lowest it has ever been.

Next, the phrase *environmental quality* merits definition as well. We measure environmental quality in terms of "clean" and "safe," as well as in terms of the *preservation* of natural processes and life forms. The proposition that environmental quality today is better than it was fifty years ago is difficult to defend, because each person places a different value on the individual components of such a broad measure. Does safe drinking water, for example, offset the extinction of the passenger pigeon, ivory-billed woodpecker, and Carolina parakeet? Whereas *clean* and *safe* can be objectively measured and reported, *quality* (it seems to us) is subjective and hence controversial. The authors believe environmental quality in the U.S. has improved since the 1960s and possibly since earlier times, but we recognize that others may disagree even after reviewing the evidence presented on the following pages.

Third, the world is cleaner today partly because environmentalists called attention to the need to protect the environment. We say "partly" because other social and economic forces were leading to greater environmental protection in the U.S. even before the modern environmental movement emerged, and these forces would have continued to operate with or without an organized environmental movement. Nevertheless, government regulations and extensive private efforts to protect the environment arose in response to the prodding of the environmental movement, and it is not likely that as great an effort would have been made if environmentalists hadn't sounded the alarm.

Finally, our effort in the pages that follow to "set the record straight" is not meant as a criticism of the environmental movement as a whole or of its goals (as we understand them). Just the opposite is true: The environmental movement can take credit for the substantial reductions in pollution and the expanded protection of wildlife that we document. Our criticism is focused on the "messengers of doom" who never issue retractions when their warnings are found to be false; who continue to issue warnings even after problems have been solved; and who are increasingly inclined to issue warnings that are without scientific support. If the environmental movement is to be effective in the 1990s and beyond, it must abandon these tactics and adopt ones that are more responsible and effective.

The Air Is Getting Cleaner
(in millions of metric tons, except lead in thousands of metric tons)

| | | | | % change | |
Emission	1940	1970	1990	1940–1990	1970–1990
Particulate matter	23.1	18.5	7.5	−67.5	−60.5
Sulfur oxides	17.6	28.3	21.2	20.4	−25.1
Nitrogen oxides	6.9	18.5	19.6	184.0	5.9
VOCs	15.2	25.0	18.7	23.0	−25.2
Carbon monoxide	82.6	101.4	60.1	−27.2	−40.7
Lead	n.a.	203.8	7.1	n.a.	−96.5
Total emissions	145.4	191.9	127.1	−12.6	−33.8

Source: U.S. Environmental Protection Agency, *National Air Pollutant Emission Estimates, 1940–1990,* March 1992.

Our first premise, which we will attempt to prove in this chapter, is that the *world is becoming cleaner, not dirtier, over time.*

AIR QUALITY

Let's start with the air we breathe. Is it cleaner today than it was ten years ago? Fifty years ago?

Measurements of Air Quality

Reliable measurements of air quality in the U.S. are available only for the years since 1975. All six of the pollutants tracked by the EPA show dramatic improvement between 1975 and 1991: Total suspended particle concentrations (dust and airborne ash) fell by 24 percent; sulfur dioxide by 50 percent; carbon monoxide by 53 percent; ozone by 25 percent; nitrogen dioxide by 24 percent; and lead by 94 percent. There can be little debate that air quality in the U.S. improved considerably during this period.

We can estimate air quality prior to 1975 by examining the levels of pollutant *emissions,* rather than the concentration of pollutants in the air. Estimates of air pollutant emissions in the U.S. are available for every year since 1940. As the table . . . shows, emissions of most pollutants increased between 1940 and 1970, but then fell (except for nitrogen oxides) between 1970 and 1990.

The picture presented by these numbers is somewhat mixed, with emissions of three pollutants falling since 1940 (although we don't have the exact figures for lead) and three increasing. Significantly, though,

total emissions fell by 12.6 percent between 1940 and 1990, and an even more impressive 33.8 percent between 1970 and 1990. Compliance with the Clean Air Act Amendments of 1990 is expected to significantly reduce sulfur oxide and nitrogen oxide emissions, two of the three pollutants whose emissions in 1990 were still above their 1940 levels.

In summary, air *quality* has improved dramatically since 1975, the first year for which reliable measurements are available. And while *emissions* of some air pollutants remain higher today than they were in 1940, overall emissions are lower and continue to fall.

Other developed countries have come close to the U.S. record for reducing air pollutant emissions. The table . . . shows total air pollutant emissions for nine countries in 1970 and 1985. Significantly, every coun-

Air Pollutant Emissions Are Falling Around the World
(total emissions in thousands of metric tons)

Country	1970	1985	1985 as percent of 1970
Netherlands	3,918	2,697	68.8
USA	190,300	133,000	69.9
Germany	24,047	17,020	70.8
UK	15,830	13,210	83.4
Italy	12,142	10,185	83.9
Sweden	3,394	2,942	86.7
France	13,340	11,712	87.8
Canada	22,146	19,600	88.5

Source: OECD figures analyzed in Mikhail S. Bernstam, *The Wealth of Nations and the Environment,* p. 17.

try reduced its total emissions. The U.S. ranked second only to the Netherlands, and it is likely that more current data would place the U.S. first among the nations in this list.

Why Air Quality Is Improving

Why has air quality improved so dramatically during the past twenty years? One reason is the enactment of local, state, and national laws setting air quality standards and requiring factories, utilities, and auto manufacturers to take steps to reduce air pollution. The most important of these laws was the Clean Air Act, passed by Congress in 1963 and amended in 1970 and 1990. Environmentalists played a major role in the passage of these and other laws.

Thanks largely to the efforts of automobile manufacturers to comply with the Clean Air Act, a car built in 1993 emits 97 percent less hydrocarbons and carbon monoxide and 90 percent less nitrogen oxide than a car built twenty years earlier. Between 1987 and 2000, the gradual retirement of older vehicles from the domestic auto and truck fleet will reduce hydrocarbon emissions by 50 percent, carbon monoxide emissions by 52 percent, and nitrogen oxide emissions by 34 percent. The combined effect of fleet turnover and use of reformulated gasoline in many American cities will mean that by the year 2000, new cars will emit a remarkable *99 percent* less hydrocarbons and carbon monoxide than they did in 1973.

Another reason air is getting cleaner is the greater use of electricity over time. Electricity is gradually replacing other sources of energy in manufacturing, services, and household appliances. This frequently results in greater energy efficiency and less pollution, because large electric generation plants operate at higher efficiency levels and under stricter emission controls than do smaller gas and oil burning engines. Electrification also makes possible the use of new devices—such as microwave ovens and fax machines—that are substitutes for more energy-intensive products. In 1991, for the first time, the industrial, commercial, and residential sectors of the U.S. economy consumed over half of their energy in the form of electricity.

A third reason for rising air quality is that industry is using raw materials more efficiently. New technologies allow businesses to capture and recycle gases and particles that once simply escaped into the air. Some examples of how technology is leading to cleaner air include "clean coal" burning processes—which eliminate up to 99 percent of potential sulfur dioxide emis-

sions—and the use of infrared heat to dry ink in printing processes and paint in automobile finishing processes.

Effect of Air Pollution on Human Health

What effect, if any, does air pollution have on human health? Have reductions in air pollutant emissions led to less disease or a lower incidence of cancer?

It is common to hear environmentalists assert that air pollution is a leading cause of cancer, but there is in fact very little evidence to support such a claim. One of the most complete and authoritative studies ever done of the question, *what causes cancer?* was conducted in 1981 by Sir Richard Doll and Dr. Richard Peto, two leading epidemiologists. Their report was commissioned by the U.S. Office of Technology Assessment and has been endorsed by the National Cancer Institute.

Doll and Peto estimate that air pollution accounts for *just 1 percent* of all U.S. cancer each year. Even this is a "crude estimate," they say, since they found no sound evidence linking air pollution to cancer rates. Because their calculations were based on observed rates of cancer in urban versus rural populations over many years, their finding applies to a period of time when air pollution was much worse than it is today. It is reasonable, then, to assume that the risk, if it exists at all, is even smaller today.

The relationship between pollution and human health will be revisited many times in later chapters of this book. It is not our purpose to give the final word on this complex issue, nor are we qualified to do so. It is possible that air pollution has a greater effect on human health than Doll, Peto, and other experts suspect. But we *can* report to the reader that we have not found reliable scientific evidence in support of this possibility, and our view appears to be supported by the largest part of the scientific community.

Conclusion

Air quality in the U.S. today is indisputably better than it was twenty years ago. In some important ways, it is better than even fifty years ago, though in other ways it may not be as good. Other developed countries seem to be following the same path as the U.S., reducing their total air pollutant emissions since 1970.

The sources of improving air quality include compliance with federal regulations, for which the environmental movement can take partial credit. In later chapters we will comment further on the character of these laws and whether they were cost-effective solutions to the air pollution problem; here it is sufficient to note that they played a very important role in improving air quality during the 1970s and 1980s.

New research is constantly being conducted, but at the time of this writing, the leading experts in the field have found no link between air pollution and cancer. It would appear that current levels of air pollution are too low to pose a threat to human health.

Regarding the air we breathe, we can confidently say that our world is getting cleaner, not dirtier, over time.

WATER QUALITY

In 1969, the Cuyahoga River in Ohio caught fire and burned for several hours, a result of the river's heavy chemical contamination. Dangerously high levels of pollution were reported in the drinking water of New Orleans in 1974, and Duluth sometime later. Lake Erie was so heavily polluted during the 1960s that many environmentalists declared it "dead," predicting that Lake Michigan wasn't far behind.

But water quality in the U.S. was already beginning to improve in the 1960s. The first National Water Quality Inventory, conducted in 1973, found that water pollution levels had decreased considerably in most major waterways during the decade of the 1960s. According to Dr. A. Myrick Freeman III, a senior fellow at Resources for the Future, between 93 and 96 percent of the nation's waters were fishable in 1972.

The cleanup of America's rivers and lakes continued during the 1970s and 1980s. Federal law required factories to install specific pollution control technologies and gave municipalities federal grants equal to 75 percent of the cost of planning and building new sewage treatment facilities. During the 1980s, approximately $23 billion a year was spent by governments and private industry to comply with the Federal Water Pollution Control Act Amendments of 1972. More than $75 billion in tax funds have been spent since the early 1970s to build municipal waste treatment plants. Additional billions of dollars were spent on water pollution abatement efforts unrelated to the 1972 law.

Not all of this money was spent efficiently. The federal law's inflexible technology mandates often meant that less-expensive ways to abate water pollution were not pursued. The law's focus on factories and municipal sewage systems—called point sources—ignored the reality that between 57 and 98 percent of pollutants entering a river or lake come from nonpoint sources, such as run-off from fields, lawns, and construction sites. Generous federal subsidies led municipalities to design "gold-plated" treatment facilities that were often oversized or unnecessarily complex. And because the federal government subsidized sewage *treatment* but not sewage *reduction*, relatively inexpensive ways to prevent sewage from being produced in the first place were disregarded.

Despite these public policy shortcomings, water quality continued to improve during the 1970s and 1980s. By the 1990s, three-quarters of the U.S. population was served by wastewater treatment facilities. Freeman notes that "some improvement in water quality [has occurred] since 1972. In terms of aggregate measures or national averages, it has not been dramatic. But there are local success stories of substantial cleanup in what had been seriously polluted water bodies."

The Cuyahoga River, for example, is now fishable. Swimming has resumed in the Hudson River north of New York City. Salmon spawn in Maine's once-polluted Androscoggin River, and the Great Lakes support a growing sport fishing industry. The presence of toxins in rivers and lakes has consistently trended downward since the 1960s. For example, the Michigan Department of Natural Resources recently reported that in the Great Lakes, "levels of PCB, DDT, mirex and mercury in lake trout and herring gull eggs decreased dramatically in the mid- to late-1970s after extensive controls and restrictions on the use of these chemicals were implemented." According to the Council on Environmental Quality, concentration levels continued to fall during the 1980s and early 1990s. In its August 1993 report to the International Joint Commission, the Virtual Elimination Task Force noted that "considerable progress has been made to reduce inputs of persistent toxic substances to the Great Lakes Basin Ecosystem. As a result, ecosystem health today is improved from conditions 20 years ago."

The table . . . shows that water quality in the Mississippi River dramatically exceeds that of rivers in other industrialized nations. The Rhine River in Germany, for example, has 3.4 times the concentration of nitrates as does the Mississippi; 7.5 times as much ammonium; and nearly twice the level of biological oxygen demand (indicative of higher amounts of organic pollution).

The quality of water in the world's oceans also appears to be good, although long-standing pollution

U.S. Rivers Are Cleaner

In the table below, water quality in the Mississippi River is compared to water quality in the largest rivers in France, Germany, and Britain. The concentration of each kind of pollution in the Mississippi has been set equal to 1 and levels in other river systems are expressed as multiples of the Mississippi figures. The Seine River in France, for example, has about twice the level of biological oxygen demand, five times the level of nitrates, and 3.6 times the level of phosphates as the Mississippi. (Higher levels of biological oxygen demand are an indication of higher amounts of organic pollutants.)

	Mississippi (U.S.)	Seine (France)	Rhine (Germany)	Thames (U.K.)
Biological Oxygen Demand	1.00	2.13	1.87	1.67
Nitrates	1.00	5.42	3.45	6.98
Phosphates	1.00	3.65	1.09	n.a.
Ammonium	1.00	22.43	7.57	9.19
Lead	1.00	4.38	0.61	0.33

Source: Authors' calculations based on OECD data, typically mg/liter, average levels of 1980s; data reported in Global Climate Coalition, *The U.S. vs. European Community: Environmental Performance*, August 1993, p. 9.

problems exist in coastal waters. Fish, shellfish, and other marine life suffer from the effects of coastal sewage treatment and industrial discharges, as well as run-off from farms, construction sites, and other land uses. The Council on Environmental Quality reported in 1993 that "in contrast to coastal regions, the open sea remains relatively clean. . . . The oil slicks and litter common along sea lanes remain, for the most part, of minor consequence to communities of organisms living in the open-ocean areas." . . . Following the enactment of federal legislation in 1990, 29 of the 35 U.S. coastal states and territories produced plans to protect coastal waters from pollution. Together, these states and territories encompass 94 percent of the U.S. coastline.

Finally, the safety of drinking water in the U.S. testifies to the high quality of water. "[T]he animal evidence provides no good reason to expect that chlorination of water or current levels of man-made pollution of water pose a significant carcinogenic hazard," according to Dr. Bruce N. Ames and his colleagues. Doll and Peto agree, saying "we know of no established human carcinogen that is ever present in sufficient quantities in large U.S. public water supplies to account for any material percentage of the total risk of cancer. . . . [I]t is again not plausible that any material percentage of the total number of cancers in the whole United States derives from this source." . . .

There is plenty of room for further improvements to water quality in the U.S. and other countries. Public policy changes could encourage municipalities to reduce the volume of sewage that is produced, rather than merely treat it. Problems with nonpoint sources of water pollution can often be addressed by community education programs, land use compacts, and small investments in used-oil drop-off programs and similar initiatives.

An even-handed assessment finds that water quality has consistently improved since the 1950s and 1960s. Thanks to the spread of modern water and sewage treatment technologies, water today is cleaner and safer than it has been since before reliable measures of water quality were available. While some problems remain to be solved, there can be little debate that water quality in the U.S. is getting better, not worse. . . .

CONCLUSION

Our World Is Getting Cleaner, not Dirtier

Air quality in the U.S. has improved significantly since 1975, and total air pollutant emissions are below the levels that prevailed in 1940. Current levels of air pollution are believed to be too low to have much of an impact on the incidence of human cancer. Other developed nations also have made progress in cleaning their air since 1970.

Water quality in the U.S. has steadily improved since its low point in the 1950s. New technologies and regulations mean factories and municipal sewers no longer foul our rivers and lakes. Fish have returned to many rivers and lakes once thought to be dead.

The tiny traces of additives and pesticides present in our food, once thought by leading environmentalists to pose a major threat to human health, have been found to pose such tiny risks that they are overwhelmed by the hypothetical risks of naturally occurring chemicals in many common foods. Thanks to the expanded use of high-yield seeds, fertilizers, and pesticides, the people of the world have never been better fed. Indeed, an end to the threat of famine around the world is in sight, if only civil wars and corrupt governments did not stand in the way.

Forested areas in America are increasing in size, and logging is now conducted on a sustainable yield basis. The old days of clear cutting and abandonment in forestry are behind us. Gone, too, are the days of open town dumps with their potential for contaminated leachate, vermin, and odors. In their place is recycling, diverting one-fifth of our garbage; the modern landfill with its liner, cap, and other safety features; and clean and safe waste-to-energy incineration.

Our descendants won't blame us for using up all the Earth's energy resources because there is enough oil and coal in the Earth to last hundreds of years at current rates of consumption. The dynamics of supply and demand mean our distant descendants will have plenty of time to find safe alternatives to fossil fuels, should they ever need them.

The Work that Still Needs to Be Done

Environmentalists can celebrate the progress made since the early 1960s, without being blind to the serious environmental problems that remain to be addressed:

- Air pollution remains a problem in some of America's cities. We must find ways to identify and control the remaining sources of air pollution—such as cars that are not properly tuned and other nonpoint sources.

- The battle against water pollution must shift away from point sources, which are now controlled, to nonpoint sources such as farming, construction, and lawn care practices that cause fertilizer and other substances to run off into rivers, lakes, oceans, and groundwater.

- Older landfills should be closed or retrofitted with new and safer technology. Better ways to minimize the effects of landfill operations on local communities still need to be found.

- We must improve our understanding of when recycling makes sense and when it doesn't. Rather than rely on the old rhetoric that "recycling is always good," we need to study and honestly report the trade-offs that sometimes make recycling inappropriate.

- Government policies that subsidize logging in ecologically fragile areas need to be changed, and government management of our forests and public lands could be improved.

This list is only suggestive of the things that need to be done; . . . Despite all the work that remains ahead of the environmental movement, our review of available information confirms our initial assertion: Significant progress has been made in protecting the environment. *Our world truly is becoming cleaner over time, not dirtier.*

Americans have less reason today to be afraid of their environment than at any previous moment in human history. Their hard work and major investments of tax dollars have purchased a cleaner environment for them and their children. Elizabeth Whelan, one of the first scientists to speak out against the doomsday rhetoric that fills the popular press these days, said it best ten years ago when she wrote:

> Americans and citizens all over the Western World can be proud of the strides we've taken to improve our environment. The innovative technological breakthroughs of the twentieth century have resulted in the healthiest population ever to live on this planet. While the advances have not come without risks, the ultimate benefits have in fact vastly outweighed the costs, despite what the scaremonger lobby would have us believe.

Notes

". . . if current trends continue" is from *Global 2000 Report to the President.*

Gore quotes are from Albert Gore, *Earth in the Balance,* pp. 151, 325.

Myers quotes are from Norman Myers, "What Ails the Globe?" *International Wildlife,* February 1994.

Whelan quote is from Elizabeth Whelan, *Toxic Terror,* p. 15.

Simon and Kahn quote is from Julian L. Simon and Herman Kahn, *The Resourceful Earth,* p. 6.

Statement by 425 scientists is reprinted in Global Climate Coalition, *Climate Watch* 1 (7) (June 1993), p. 2.

EPA figures on air quality are from the Council on Environmental Quality, *Environmental Quality 1992,* Table 40, p. 337.

". . . a car built in 1993" is from EPA, *National Air Quality Emissions Trends Report 1992,* October 1993.

Figures on fleet turnover are from Motor Vehicle Manufacturers Association, *Cleaner Motor Vehicles: The Challenge Being Met,* September 1988, p. 10.

Re electrification and pollution, see Edison Electric Institute, *Powering a Cleaner Environment,* 1993.

Re Doll and Peto, see Sir Richard Doll and Richard Peto, *Journal of the National Cancer Institute,* January 1981, p. 1248.

Reference to National Water Quality Inventory, fishable lakes and rivers, and $23 billion a year are from A. Myrick Freeman III, in Paul R. Portney, *Public Policies for Environmental Protection,* pp. 114, 116, 125.

$75 billion is from Paul N. Tramontozzi, *Reforming Water Pollution Regulation,* Center for the Study of American Business, August 1985, p. 10. See also James Lis and Kenneth Chilton, *Clean Water—Murky Policy,* Center for the Study of American Business, January 1992.

"Between 57 and 98 percent" is from A. Myrick Freeman III, in Paul R. Portney, *Public Policies for Environmental Protection,* p. 109.

Freeman quote is from A. Myrick Freeman III, in Paul R. Portney, *Public Policies for Environmental Protection,* p. 120.

Information on Cuyahoga and Androscoggin Rivers is from "Across the U.S., Cleaner Water, but . . . ," *U.S. News & World Report,* February 28, 1983, p. 30.

Michigan Department of Natural Resources quote is from "Water Quality and Pollution Control in Michigan: 1992 Report," Vol. 12, p. 84.

Reference to the Council on Environmental Quality is to *Environmental Quality 1992,* pp. 389–391.

International Joint Commission quote is from International Joint Commission, *A Strategy for Virtual Elimination of Persistent Toxic Substances,* Vol. 1, p. 6.

Council on Environmental Quality quote is from *Environmental Quality 1992,* pp. 32–33.

Ames quote is from Bruce N. Ames, Renae Magaw, and Lois Swirsky Gold, "Ranking Possible Carcinogenic Hazards," *Science* 236 (April 17, 1987), p. 272.

Doll and Peto quote is from Sir Richard Doll and Richard Peto, *Journal of the National Cancer Institute,* p. 1249.

Whelan quote is from Elizabeth Whelan, *Toxic Terror,* p. 300.

For Further Discussion

1. Compare the analysis of the present state of the environment offered by Bast, Hill, and Rue to that offered by Easterbrook. Do they differ in any important ways?

2. The authors of this selection define clean air and water in terms of "average human exposure to potentially harmful pollutants." Can you think of any other meaning by which air and water are not cleaner? Is average exposure a good criterion to use to make this determination?

3. The authors interpret environmental quality as environmental cleanliness, concluding that other meanings of quality can be subjective and controversial. Do you agree?

From **Beyond the Limits**

Donella Meadows
Dennis Meadows
Jorgen Randers

Donella Meadows and three of her colleagues at MIT published the ground-breaking book, The Limits to Growth, *in 1972. Based on computer models using various sets of population, pollution, industrialization, and resource assumptions, they argued that present growth trends will mean that human society will exceed the earth's capacity to support human life within the next century.*

In recent years, Meadows and some of her colleagues have updated their original projections. They argue that, based on these updated figures, their original conclusions remain valid. In fact, in many cases the predictions are worse than the ones suggested in the early 1970s. In this selection from Beyond the Limits, *Meadows and her colleagues review some of this recent work and compare their present conclusions to those defended in their original book.*

Twenty years ago, after working with global data and with a computer model called World3, we came to the following conclusions in a book titled *The Limits to Growth:*

1. If present growth trends in world population, industrialization, pollution, food production, and resource depletion continue unchanged, the limits to growth on this planet will be reached sometime within the next hundred years. The most probable result will be a sudden and uncontrollable decline in both population and industrial capacity.

2. It is possible to alter these growth trends and to establish a condition of ecological and economic stability that is sustainable far into the future. The state of global equilibrium could be designed so that the basic material needs of each person on earth are satisfied and each person has an equal opportunity to realize his or her individual human potential.

3. If the world's people decide to strive for this second outcome rather than the first, the sooner they begin working to attain it, the greater will be their chances of success.

Now with an updated model, with more extensive data, and after twenty years of growth and change in the world, we believe that our original conclusions are still valid, but they need to be strengthened:

1. Human use of many essential resources and generation of many kinds of pollutants have already surpassed rates that are physically sustainable. Without significant reductions in material and energy flows, there will be in the coming decades an uncontrolled decline in per capita food output, energy use, and industrial production.

2. This decline is not inevitable. To avoid it two changes are necessary. The first is a comprehensive revision of policies and practices that perpetuate growth in material consumption and in population. The second is a rapid, drastic increase in the efficiency with which materials and energy are used.

3. A sustainable society is technically and economically possible. It could be much more desirable than a society that tries to solve its problems by constant expansion. The transition to a sustainable society requires a careful balance between long-term and short-term goals and an emphasis on sufficiency, equity, and quality of life rather than on quantity of output. It requires more than productivity and more than technology; it also requires maturity, compassion, and wisdom.

These conclusions constitute a conditional warning, not a dire prediction. They offer a living choice, not a death sentence. The choice isn't a gloomy one. It does not mean that the poor must be frozen in their poverty or that the rich must become poor. It could mean the solution of problems such as poverty and unemployment that humanity has been working at inefficiently or fruitlessly by trying to maintain physical growth.

We hope that the world will make the choice for sustainability. We think that a better world is possible and that the acceptance of physical limits is the first step toward getting there. We see "easing down" not as a sacrifice, but as an opportunity to stop battering against the earth's limits and to start transcending self-imposed and unnecessary limits within human institutions, minds, and hearts.

OVERSHOOT

The following are characteristics of a society that has grown *beyond its limits*—a society that is drawing upon the earth's resources faster than they can be restored, and that is releasing wastes and pollutants faster than the earth can absorb them or render them harmless.

• Falling stocks of groundwaters, forests, fish, soils.

Population and Capital in the Global Ecosystem

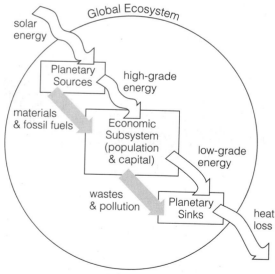

Population and capital are sustained by flows of fuels and nonrenewable resources from the planet, and they produce outflows of heat and waste, which contaminate the air, waters, and soils of the planet. (Source: R. Goodland et al.)

• Rising accumulations of wastes and pollutants.

• Capital, energy, materials, and labor devoted to exploitation of more distant, deeper, or more dilute resources.

• Capital, energy, materials, and labor compensating for what were once free natural services (sewage treatment, flood control, air purification, pest control, restoring soil nutrients, preserving species).

• Capital, energy, materials, and labor diverted to defend or gain access to resources that are concentrated in a few remaining places (such as oil in the Middle East).

• Deterioration in physical capital, especially in long-lived infrastructure.

• Reduced investment in human resources (education, health care, shelter) in order to meet consumption needs or to pay debts.

• Increasing conflict over resources or pollution emission rights. Less social solidarity, more hoarding, greater gaps between haves and have-nots.

A society like this is in a state of *overshoot*. To overshoot means to go too far, to grow so large so quickly that limits are exceeded. When an overshoot occurs, it induces stresses—in this case in both natural and social processes—that begin to work to slow and stop growth.

If humanity does not correct its condition of overshoot, problems like the ones listed above will worsen until human productive capacity, ingenuity, adaptability, and attention are overwhelmed. At that point overshoot will turn into collapse.

However, collapse is not the only possible outcome. The human society can ease down from beyond the limits. That need not mean reducing population or capital or living standards, though it certainly means reducing their growth. What must go down, and quickly, are *throughputs*—flows of material and energy from the supporting environment, through the economy, and back to the environment.

Fortunately, in a perverse way, the current global economy is so wasteful, inefficient, and inequitable that it has tremendous potential for reducing throughputs while raising the quality of life for everyone. While that is happening, other measures—nontechnical measures, evolutionary human measures—can restructure the social system so that overshoot never happens again.

LOOKING INTO THE FUTURE WITH WORLD3

To understand how the human economy and the environment may unfold in the future, we use a computer model called World3. World3 is, like all models, much, much simpler than the real world. It is, however, more dynamically sophisticated than many computer models. It is a nonlinear feedback model, one that tries to capture the forces behind population and capital growth, the layers of changing, interlinked environmental limits, and the delays in the physical and economic processes through which human society interacts with its environment.

World3 shows, in no uncertain terms, that if the world system continues to evolve with no significant changes, the most likely result is not only overshoot, but collapse, and within another few decades. One possible future, by no means the only one, is shown in Scenario 1.

In this scenario the world society proceeds along its historical path as long as possible without major policy change. Technology advances in agriculture, industry, and social services according to established patterns. The simulated world tries to bring all people into an industrial and then post-industrial economy.

The global population in this scenario rises from 1.6 billion in 1900 to over 5 billion in 1990 and over 6 billion in the year 2000. Total industrial output expands by a factor of 20 between 1900 and 1990, and it does so while using only 20 percent of the earth's total stock of nonrenewable resources. In 1990, 80 percent of these resources remain. Pollution in that year has just begun to rise significantly. Life expectancy is increasing, services and goods per capita are increasing, food production is increasing. But major changes are just ahead.

Just after the simulated year 2000 pollution rises high enough to begin to affect the fertility of the land. At the same time land erosion increases. Total food production begins to fall after 2015. That causes the economy to shift more investment into the agriculture sector. But agriculture has to compete for investment with a resource sector that is also beginning to sense some limits.

Between 1990 and 2020 in this scenario, population increases by 50 percent and industrial output by 85 percent. Therefore the nonrenewable resource use rate doubles. What was a 110-year supply of nonrenewable resources in 1990 is only a 30-year supply in 2020. So many resources have been used that much more capital and energy are required to find, extract, and refine what remains.

As both food and nonrenewable resources become harder to obtain in this simulated world, capital is diverted to producing more of them. That leaves less output to be invested in capital growth. Finally the

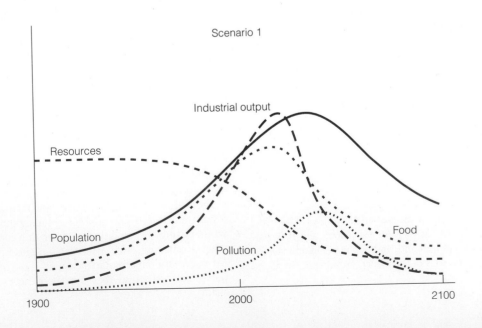

Scenario 1

Industrial output

Resources

Population

Pollution

Food

1900   2000   2100

industrial capital plant begins to decline, taking with it the service and agricultural sectors. For a short time the situation is especially serious, as population keeps rising, because of lags inherent in the age structure and in the process of social adjustment. Finally population too begins to decrease, as the death rate is driven upward by lack of food and health services.

This scenario is *not a prediction,* but we believe it is a possibility, one among many. Another very different possibility is shown in Scenario 10. To produce it we introduce technical, social, and economic measures quite different from those that are currently being pursued in the world. That is the purpose of a model, not to predict, but to test the "what if" possibilities.

In Scenario 10 people in the simulated world decide on an average family size of two starting in 1995, and they have available effective birth control technologies. They also set themselves a consumption limit. When every family attains roughly the material standard of living of present-day Europe, it says "enough" and turns its attention to achieving other, nonmaterial goals. Furthermore, starting in 1995, this world puts a high priority on developing and implementing technologies that increase the efficiency of resource use, decrease pollution emissions, control land erosion, and increase land yields.

We assume in Scenario 10 that these technologies come on only when needed and only after a development delay of twenty years, and that they have a capi-

tal cost. The capital is available for them, however, because in this restrained society, capital does not have to support rapid growth or to ameliorate a spiraling set of problems caused by growth. By the end of the twenty-first century in this scenario, the new technologies reduce nonrenewable resource use per unit of industrial output by 80 percent and pollution production per unit of output by 90 percent. Land yield declines slightly in the early twenty-first century as pollution rises (a delayed effect of pollution emissions around the end of the twentieth century), but by 2040 pollution begins to go down again. Land yield recovers and rises slowly for the rest of the century.

The population in Scenario 10 levels off at just under 8 billion and lives at its desired material standard of living for at least a century. Average life expectancy stays at just over eighty years, services per capita rise 210 percent above their 1990 levels, and there is sufficient food for everyone. Pollution peaks and falls before it causes irreversible damage. Nonrenewable resources deplete so slowly that half the original endowment is still present in the simulated year 2100.

We believe that the world could attain a sustainable state similar to that shown in Scenario 10. We think it is a picture not only of a feasible world, but of a desirable one, certainly more desirable than a world that keeps on growing until it is stopped by multiple crises.

Scenario 10 is not the only sustainable outcome the World3 model can produce. There are tradeoffs and

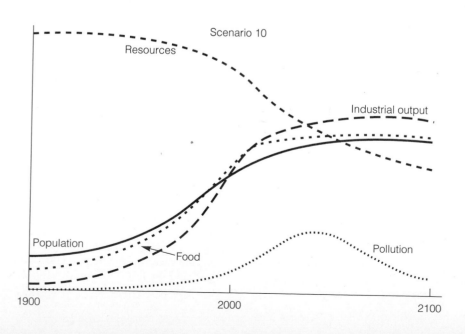

choices. There could be more food and less industrial output or vice versa, more people living with a smaller stream of industrial goods or fewer people living with more. The world society could take more time to make the transition to equilibrium—but it cannot delay forever, or even very long. When we postpone for twenty years the policies that brought about the sustainable world of Scenario 10, the population grows too large, pollution builds too high, resources are drained too much, and a collapse is no longer avoidable.

BEYOND THE LIMITS
AND STILL GROWING

Physical expansion is still the dominant behavior of human society, though the resource base is declining. In 1991 the human population was 5.4 billion. In that year the population grew by over 90 million people, a one-year addition equivalent to the total populations of Mexico plus Honduras, or to eight Calcuttas. World population is still growing exponentially. Under the most favorable circumstances, the World Bank projects that the population will not level off until late in the next century, at 12.5 billion people.

Industrial production is also growing, even more rapidly than the population. It has doubled over the past twenty years. Along with it have doubled, or more than doubled, the number of cars, the consumption of coal and natural gas, the electric generating capacity,

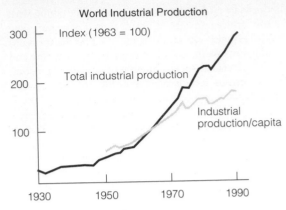

World Industrial Production

In 1991 world population growth rate was estimated to be 1.7%, corresponding to a doubling time of 40 years. (Sources: United Nations; D. J. Bogue)

World industrial production also shows clear exponential increase, despite fluctuations due to oil price shocks. The 1970 to 1990 growth rate in total production has averaged 3.3% per year. The per capita growth rate has been 1.5% per year. (Sources: United Nations; Population Reference Bureau)

the production of grain, the generation of garbage, the emissions of greenhouse gases. If the economy were to support 12.5 billion people, all living the way present North Americans live, it would have to expand at least twenty-fold—twenty more industrial worlds added to the existing one!

The industrial world that already exists is using the earth's resources unsustainably. It is not meeting the basic needs of all the world's people, and yet, given current knowledge and technology, all needs could be met without exceeding the earth's limits.

Fact: Of the more than 5 billion people on earth, over 1 billion at any time are eating less food than their bodies require. Every day an average of 35,000 people die of hunger-related causes, most of them children.

Possibility: If the food grown each year on earth were equitably distributed, and if less of it were lost to spoilage and waste, there would easily be enough to give all people an adequate, varied diet.

Fact: During the past twenty years deserts expanded by 288 million acres, nearly the size of France. Each year 16 to 17 million acres of cultivated lands are made unproductive because of erosion, and another 3.6 million acres because of salination and waterlogging. Soil erosion exceeds soil formation on a third of U.S. cropland and on 1,300 million acres in Asia, Africa, and Latin America. Fertilizers and pesticides acidify and al-

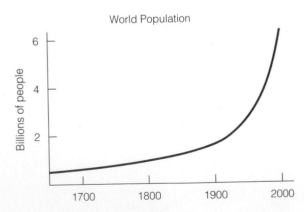

World Population

ter soils and run off to contaminate ground and surface waters.

Possibility: Farming methods that conserve and enhance soils are known and used by some farmers on every continent. In both temperate and tropic zones, some farms are obtaining high yields consistently without high rates of application of fertilizers and pesticides. More food could be grown, and it could be done in ways that are ecologically, economically, and socially sustainable.

Fact: Before the industrial revolution there were 14 billion acres of forest on the earth. Now there are 10 billion acres, only 3.6 billion of which are undisturbed primary forest. Half of the world's forest loss has occurred between 1950 and 1990. China has lost three-fourths of its forests. Europe has no primary forests left. The United States has lost one-third of its forested cover, and 85 percent of its primary forest. Half of the tropical forest is gone; half of what remains has been logged and degraded. At current logging rates the rest will be gone within fifty years and with it perhaps half the species of life on earth.

Possibility: Forest cutting could be greatly reduced while demand for wood products is still fulfilled. Half of U.S. wood consumption could be saved by increasing the efficiency of sawmills and construction, by doubling the rate of paper recycling, and by reducing the use of disposable paper products. Logging could be conducted so as to reduce its negative impact on soils, streams, and unharvested trees. Fast-growing forests could be replanted on already logged lands. High-yield

agriculture could reduce the need to clear forests for food, and more efficient stoves could reduce the need for firewood.

Fact: The average North American uses forty times as much energy as the average person in a developing country. The energy use of the human economy grew sixtyfold between 1860 and 1985, and it is projected to grow by another 75 percent by the year 2020. At present, 88 percent of the commercial energy used in the world comes from the fossil fuels—coal, oil, and gas.

Fact: Fossil fuels are limited both by their sources (the deposits in the earth) and by their sinks (the atmosphere, waters, and soils to which their combustion products and refinery by-products go). The primary product of fossil fuel combustion is carbon dioxide, a greenhouse gas, which is rapidly increasing in the atmosphere. Its atmospheric concentration is now higher than it has been for the past 160,000 years, and it is still growing exponentially.

Possibility: Through efficiency measures alone the world could maintain its total energy use at or below the current level with no reduction in productivity, comfort, or convenience in the rich countries and with steady economic improvement in the poor countries. Efficiencies of that magnitude would make it possible to supply most or all the world's energy needs from solar-based renewables, whose costs are dropping steadily.

Fact: As inputs of energy and materials to the human economy increase, outputs of waste and pollution also increase. Some forms of pollution, such as lead in gasoline and DDT, have been greatly reduced,

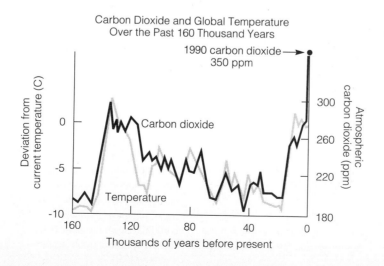

Carbon Dioxide and Global Temperature
Over the Past 160 Thousand Years

primarily by outright bans. In rich countries some widespread pollutants, such as nitrogen oxides in air and phosphates in streams, have been reduced or held constant, at considerable expense. Other kinds of pollutants, particularly nuclear wastes, hazardous wastes, and greenhouse gases, continue to grow unabated.

Fact: Knowledge about the environment and concern for it has grown enormously all over the world. In twenty years the number of environmental ministries in the world's governments has risen from ten to over one hundred. Global monitoring systems now exist, and global conferences, information networks, and agreements have been put into place.

Possibility: The reorganization of manufacturing and farming practices to reduce pollution outputs has barely begun. Increased efficiencies of fuel and material use and more complete material recycling will reduce both depletion of sources and pollution of sinks. Great reductions in pollution will be a natural result of pricing products to include their environmental costs, and of the adoption of the idea of sufficiency—simply reducing unnecessary, wasteful consumption.

All over the earth soils, forests, surface waters, groundwaters, wetlands, and the diversity of nature are being degraded. Even in places where renewable resources appear to be stable, such as the forests of North America or the soils of Europe, the quality or health of the resource is in question. Deposits of fossil fuels and high-grade ores are being drawn down. There is no plan and no sufficient investment program to power the industrial economy after nonrenewable resources are gone. Pollutants are accumulating; their sinks are overflowing. The chemical composition of the entire atmosphere is being changed.

If only one or a few resource stocks were falling while others were stable or rising, one might argue that growth could continue by the substitution of one resource for another. But when many stocks are eroding and many sinks are filling, there can be no doubt that human withdrawals of material and energy have grown too far. They have overshot their sustainable limits.

THE DYNAMICS OF OVERSHOOT AND COLLAPSE

A growing population and economy can approach the limits to its physical carrying capacity in one of four ways:

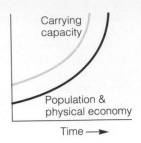

1. It can keep growing without interruption, as long as its limits are far away or growing faster than it is.

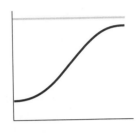

2. It can level off smoothly, slowing and then stopping in a smooth accommodation with its limits, if and only if it receives accurate, prompt signals telling it where it is with respect to its limits, and only if it can respond to those signals quickly and accurately—or if the population and economy limit themselves well below external limits.

3. It may overshoot its limit for a while, make a correction, and undershoot, then overshoot again, in a series of oscillations, if the warning signal from the limits to the growing entity is delayed, or if the response is delayed, and if the supporting environment is not erodable when overstressed, or if it is able to recover quickly from erosion.

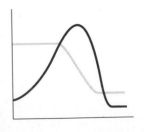

4. If the signal or response from the limit is delayed and if the environment is irreversibly eroded when overstressed, then the growing economy will overshoot its carrying capacity, degrade its resource base, and collapse. The result of this overshoot and collapse is a permanently impoverished environment and a material standard of living much lower than what could have been possible had the environment never been overstressed.

We submit that the human population and economy, drawing resources from a large but finite planet, forms a system that is structured, unless altered by human intelligence and human self-restraint, for overshoot and collapse. The prevailing industrial ethic is one of continuous growth. The resource base is both limited and erodable. The response of biological and geochemical systems to human abuse comes only after long delays. The human population acts only after further delay. And physical processes, from human population growth to forest growth to global climate change, operate with considerable momentum.

A population-economy-environment system that contains feedback delays and slow physical responses, thresholds, and erosion is literally *unmanageable*. No matter how brilliant its technologies, no matter how efficient its economy, no matter how wise its decision makers, it can't steer itself away from hazards, unless it does its best to look far forward and to test its limits very, very slowly. If it keeps its focus only on the short term and if it constantly tries to accelerate, it will overshoot.

The advent of new technologies and the flexibility of the market system are no antidotes to overshoot, and they cannot by themselves prevent collapse. In fact they themselves operate with delays that *enhance* the economy's tendency to overshoot. Technology and markets serve the values of society. If the primary goal is growth, they will produce growth, overshoot, and collapse. If the primary goals are equity and sustainability, then technology and markets can also help bring about those goals.

Overshoot is a condition in which the delayed signals from the environment aren't yet strong enough to force an end to growth. That means that a society in overshoot still has a chance, if it acts quickly, to bring itself below its limits and avoid collapse.

There even may be a recent example of the human world doing just that in its response to the destruction of the ozone layer.

THE OZONE LAYER: BACK FROM BEYOND THE LIMITS?

The ozone story illustrates all the ingredients of an overshoot and collapse system: exponential growth, an erodable environmental limit, and long response delays, both physical and political.

- *The growth:* Chlorofluorocarbons, or CFCs, are some of the most useful compounds ever invented by human beings. They were originally sold as refrigerants under the trade name Freon. Then they were found to be useful in insulation, as propellants in aerosol cans, as solvents for cleaning metals. By the early 1970s, the world was making a million tons of CFCs per year and discarding them safely into the atmosphere, or so everyone thought.

- *The limit:* High in the atmosphere, twice as high as Mount Everest, is the gossamer ozone layer that screens out a particularly harmful wavelength from the sun's incoming light—W-B, a stream of energy of just the right frequency to destroy the organic molecules that make up all life. If the ozone layer thins and more W-B light reaches the earth's surface, the results will include human skin cancers, blindness in many kinds of animals, decreased growth of green plants, and disruption of oceanic food chains. Each one percent decrease in the ozone layer is expected to produce a one percent decrease in soybean yield and a 3 percent to 6 percent increase in human skin cancer.

- *Signals:* The first scientific papers postulating that CFCs could destroy the ozone layer were published in 1974. The first measurements of a precipitous drop in ozone concentration over Antarctica were taken in the late 1970s and finally published in 1985. Since then, scientists have come up with an explanation of the "ozone hole" and have uncovered the disquieting fact that each atom of chlorine released into the stratosphere from the decay of a CFC molecule can destroy about 100,000 ozone molecules.

- *Delays:* After its manufacture, a CFC molecule may be released quickly into the atmosphere from an aerosol can, or it may remain for years in a refrigerator or air conditioner. Upon its release, it takes about fifteen years to rise up to the stratosphere. Its residence time there may vary, depending on the type of CFC, from sixty-five to five hundred years.

- *The human response:* The first official international meeting to discuss the ozone layer was convened by the United Nations Environment Program in 1985. It produced no agreement. But by 1987 the "Montreal Protocol" was signed by thirty-six nations, agreeing to cut their production of CFCs in half by

1998. Continuing ozone deterioration then spurred the world to toughen that agreement: ninety-two nations agreed in London in 1990 to phase out all CFC production by 2000. In 1991, in response to further ozone depletion, several nations unilaterally moved up their deadlines for eliminating CFCs.

It took thirteen years from the first scientific papers to the signing of the Montreal Protocol. It will take thirteen more years until the protocol, as strengthened in London, is fully implemented. The chlorine already in the stratosphere will remain there for more than a century. In fall 1991 the Antarctic ozone hole was the deepest ever measured, and in winter 1992 chlorine concentrations in the stratosphere over the Northern Hemisphere were the highest ever measured.

This is a story of overshoot and of a remarkable, worldwide human response. Whether or not it will be a story of collapse depends on how erodable or self-repairable the ozone layer is, on whether future atmospheric surprises appear, and on whether humanity has acted, and will continue to act, in time.

SIX STEPS TO AVOID COLLAPSE

Six broad measures lead to the avoidance of collapse in the World3 model and, we believe, in the world. Each of them is described here in general terms. Each can be worked out in hundreds of specific ways at all levels, from households to communities to nations to the world as a whole. Any step in any of these directions is a step toward sustainability.

- *Improve the signals.* Learn more about and monitor both the welfare of the human population and the conditions of local and planetary sources and sinks. Inform governments and the public as continuously and promptly about environmental conditions as about economic conditions. Include real environmental costs in economic prices; recast economic indicators like the GNP so that they do not confuse costs with benefits, or throughput with welfare, or the depreciation of natural capital with income.

- *Speed up response times.* Look actively for signals that indicate when the environment is stressed. Decide in advance what to do if problems appear (if possible, forecast them before they appear) and have in place the institutional and technical arrangements necessary to act effectively. Educate for flexibility and creativity, for critical thinking and for systems understanding.

- *Minimize the use of nonrenewable resources.* Fossil fuels, fossil groundwaters, and minerals should be used only with the greatest possible efficiency, recycled when possible (fuels can't be recycled, but minerals and water can), and consumed only as part of a deliberate transition to renewable resources.

- *Prevent the erosion of renewable resources.* The productivity of soils, surface waters, rechargeable groundwaters, and all living things, including forests, fish, game, should be protected and, as far as possible, restored and enhanced. These resources should only be harvested at the rate they can regenerate themselves. That requires information about their regeneration rates, and strong social sanctions or economic inducements against their overuse.

- *Use all resources with maximum efficiency.* The more human welfare can be obtained with the less throughput, the better the quality of life can be while remaining below the limits. Great efficiency gains are both technically possible and economically favorable. Higher efficiency will be essential if current and future world populations are to be supported without inducing a collapse.

- *Slow and eventually stop exponential growth of population and physical capital.* There are real limits to the extent that the first five items on this list can be pursued. Therefore this last step is the most essential. It involves institutional and philosophical change and social innovation. It requires defining desirable, sustainable levels of population and industrial output. It calls for goals defined around the idea of "enough" rather than "more." It asks, simply but profoundly, for a vision of the purpose of human existence that does not entail constant physical expansion.

This last and most daunting step toward sustainability requires solutions to the pressing problems that underlie much of the psychological and cultural commitment to growth: the problems of poverty, unemployment, and unmet nonmaterial needs. Growth as presently structured is in fact not solving those problems or is solving them only slowly and inefficiently. But until better solutions are in sight, society will never let go of its addiction to growth. Therefore there are three problems for which completely new thinking is urgently needed.

- *Poverty.* "Sharing" is a forbidden word in political discourse, probably because of the deep fear that real equity would mean not enough for anyone. "Sufficiency" and "Solidarity" are concepts that can help structure new approaches to ending poverty. Everyone needs assurance that sufficiency is possible for everyone and that there is a high social commitment to ensure it. And everyone needs to understand that the world is tied together both ecologically and economically. There is enough to go around, if we manage well. If we don't manage well, no one will escape the consequences.

- *Employment.* Human beings need to work, to have the satisfaction of personal productivity, and to be accepted as responsible members of their society. That need should not be left unfulfilled, and it should not be filled by degrading or harmful work. At the same time, employment should not be a requirement for the ability to subsist. An economic system is needed that uses and supports the contributions that all people are able and willing to make, that shares work and leisure equitably, and that does not abandon people who for reasons temporary or permanent cannot work.

- *Nonmaterial needs.* People don't need enormous cars; they need respect. They don't need closetsful of clothes; they need to feel attractive and they need excitement, variety, and beauty. People need identity, community, challenge, acknowledgment, love, joy. To try to fill these needs with material things is to set up an unquenchable appetite for false solutions to real and never-satisfied problems. The resulting psychological emptiness is one of the major forces behind the desire for material growth. A society that can admit and articulate its nonmaterial needs and find nonmaterial ways to satisfy them would require much lower material and energy throughputs and would provide much higher levels of human fulfillment.

THE SUSTAINABLE SOCIETY

A sustainable society is one that can persist over generations, one that is far-seeing enough, flexible enough, and wise enough not to undermine either its physical or its social systems of support. It is, in the words of the World Commission on Environment and Development, a society that "meets the needs of the present without compromising the ability of future generations to meet their own needs."

In a sustainable society population, capital and technology would be balanced so that the per capita material living standard is adequate and so that the society's material and energy throughputs meet three conditions:

1. Its rates of use of renewable resources do not exceed their rates of regeneration.
2. Its rates of use of nonrenewable resources do not exceed the rate at which sustainable renewable substitutes are developed.
3. Its rates of pollution emission do not exceed the assimilative capacity of the environment.

A sustainable society is not necessarily a "zero growth" society. That concept is as primitive as is the concept of "perpetual growth." Rather a sustainable society would discriminate among kinds of growth and purposes for growth. It would ask what growth is for, who would benefit, what it would cost, how long it would last, and whether it could be accommodated by the sources and sinks of the earth.

That is to say, a sustainable society would be less interested in *growth* than in *development.* As a recent World Bank report says: "Following the dictionary distinction . . . To 'grow' means to increase in size by the assimilation or accretion of materials. To 'develop' means to expand or realize the potentialities of; to bring to a fuller, greater, or better state. When something grows it gets quantitatively bigger; when it develops it gets qualitatively better."

A sustainable society would not paralyze the poor in their poverty. To do so would not be sustainable for two reasons. First, the poor would not and should not stand for it. Second, keeping any part of the population in poverty would not, except under dire coercive measures, allow the population to be stabilized. For both practical and moral reasons any sustainable society would have to be just, fair, and equitable.

A sustainable society would not experience the despondency and stagnancy, high unemployment and bankruptcy that current market systems undergo when their growth is interrupted. The difference between the transition to a sustainable society and a present-day economic recession is like the difference between stopping an automobile with the brakes and stopping it by crashing into a brick wall. A deliberate transition to sustainability would take place slowly enough so that people and businesses could find their proper places in the new society.

There is no reason why a sustainable society need be technically or culturally primitive. Freed from both material anxiety and material greed, human society could have enormous possibilities for the expansion of creativity.

A sustainable world need not be a rigid one, or a centrally controlled one. It would need rules, laws, standards, boundaries, and social agreements, of course, as does every human culture. Rules for sustainability, like every workable social rule, would be put into place not to remove freedoms but to create them or to protect them against those who would destroy them. *They could permit many more freedoms than would ever be possible in a world that continues to crowd against its limits.*

Diversity is both a cause of and a result of sustainability in nature, and therefore a sustainable human society would be diverse in both nature and culture.

A sustainable society could and should be democratic, evolving, technically advanced, and challenging. It would have plenty of problems to solve and plenty of ways for people to prove themselves, to serve each other, to realize their abilities, and to live good lives—perhaps more satisfying lives than any available today.

THE NEXT REVOLUTION

We don't underestimate the gravity of the changes that will take the present world down from overshoot and into sustainability. We think a transition to a sustainable world is technically and economically possible, but we know it is psychologically and politically daunting. The necessary changes would constitute a revolution, not in the sense of the American or French political revolutions, but in the much more profound sense of the Agricultural and Industrial Revolutions.

Like those revolutions, a sustainability revolution would change the face of the land and the foundations of human self-definitions, institutions, and cultures. It is not a revolution that can be planned or dictated. It won't follow a list of fiats from a government or from computer models. The sustainability revolution, if it happens, will be organic and evolutionary. It will arise from the visions, insights, experiments, and actions of billions of people. It will require every human quality and skill, from technical ingenuity, economic entrepreneurism, and political leadership to honesty, compassion, and love.

Are any of the necessary changes, from resource efficiency to human compassion, really possible? Can the world actually ease down below the limits and avoid collapse? Is there time? Is there enough money, technology, freedom, vision, community, responsibility, foresight, discipline, and love on a global scale?

The general cynicism of the day would say there is not a chance. That cynicism, of course, is a mental model. The truth is that no one knows. The world faces not a preordained future, but a choice. The choice is between models. One model says that this finite world for all practical purposes has no limits. Choosing that model will take us even further beyond the limits and, we believe, to collapse within the next half century.

Another model says that the limits are real and close, and that there is not enough time and that people cannot be moderate or responsible or compassionate. That model is self-fulfilling. If we choose to believe it, we will get to be right.

A third model says that the limits are real and close, and there is just exactly enough time, with no time to waste. There is just exactly enough energy, enough material, enough money, enough environmental resilience, and enough human virtue to bring about a revolution to a better world.

That model might be wrong. All the evidence we have seen, from the world data to the global computer models, suggests that it might be right. There is no way of knowing for sure, other than to try it.

For Further Discussion

1. Would you categorize these authors as "messengers of doom" and "alarmists" such as described by Easterbrook, Bast, Hill, and Rue? Why or why not?

2. How do you think that Meadows, Meadows, and Randers might answer ecorealists?

3. What do these authors mean by "overshoot"? Have they made a convincing case that such a scenario might be occurring?

4. Do you believe that the "six steps to avoid collapse" are realistic? Are likely to occur?

DISCUSSION AND STUDY QUESTIONS FOR CHAPTER 3

1. Julian Simon suggests that we should think of resources as services, meaning that we value them not in themselves but for how we use them. Compare this view to such resources as oil and coal, trees, wilderness areas, land, domestic and

wild animals. Do you agree with Simon? Why or why not?

2. In your own experience, is the environment getting cleaner? Are environmental problems more or less of an issue than when you first became aware of them? Are some problems getting worse? Which? Some getting better? Which? Do you have any thoughts about why some are improving and others not?

3. Do moderate positions such as ecorealism and ecosanity serve only to reinforce the status quo, or can they be helpful in advancing environmentally sound policies? Is realism as reasonable a position for those living outside the United States?

4. The readings in this chapter suggest very different visions for the future. What do you think the world will be like in 100 years? On what do you base your beliefs?

PART II *Basic Concepts of Environmental Ethics*

The four chapters in Part II introduce several fundamental topics that are pertinent for our understanding and analysis of environmental policy issues. Any environmental controversy involves at least one of these topics, and most involve many of them. You can only participate fully in the debates concerning these controversies if you are familiar with the topics introduced in these chapters. To return to the metaphor introduced in the introduction to Part I, if this book is a journey through environmental ethics and policy and Part I is the large-scale map, Part II provides the essential equipment and supplies you will need to pack for this journey.

Economics plays a central role in almost every environmental issue. Often, environmental controversies arise precisely because one side claims that pursuing environmental goals is too costly. In many other cases, economic solutions are proposed as the means for attaining environmental goals. Unless you are familiar with the basic concepts and categories of economics, as well as its underlying ethical and philosophical assumptions, you will not fully understand many contemporary environmental debates. The selections in Chapter 4 provide varied perspectives on the relevance of economic analysis to environmental issues.

Besides questions of economic value, environmental issues also raise many questions of aesthetic, spiritual, and religious value. For many people, the value of the natural environment rests on judgments concerning the beauty, the majesty, the awe-inspiring character of nature. Many people find in natural environments an opportunity for solitude, peace, inspiration, and meaning. For others, the natural world allows occasions for connecting to God. Chapter 5 acquaints you with several aesthetic, spiritual, and religious perspectives on the environment.

As witnessed in many of the readings in Chapter 2, the ethical status of natural objects other than human beings was either ignored or denied by most thinkers within the Western philosophical tradition. Within this tradition, natural objects were merely things to be used for human ends. A true turning point in this tradition occurred with the rise of environmental concerns. For the first time in history, widespread philosophical attention has turned to the question of the moral standing of animals, plants, and natural areas. Is it reasonable to extend direct moral concern to natural objects? Chapter 6 introduces the debates surrounding this question with some of the classic essays written on this topic.

Finally, a wide range of environmental concerns involves the science of ecology. The development of an environmental consciousness paralleled the development of ecology throughout much of the twentieth century. Ecological concepts and values are at the heart of many environmental policies and assessments. Familiarity with these concepts and values is therefore an essential element of environmental literacy. Chapter 7 introduces these topics with selections from both classic and contemporary sources.

Each chapter in the next two parts of this book will end with several short discussion cases. These cases, all taken from real-life situations, are intended to help introduce the issues and controversies covered in the chapter, to foster classroom discussions and debates, and to provide occasions for applying your own thinking to specific questions, perhaps serving as topics for written assignments or classroom presentations. Even as the readings in these two parts become more practical and more focused on policy issues, they remain somewhat abstract and general. The discussion cases should help anchor the readings in more practical affairs.

CHAPTER 4

Economics and Environmental Policy

From the local to the international level, economics and economic analysis play a major role in setting environmental policy. The common thread of economic analysis winds through local zoning ordinances aimed at protecting property values from environmental regulation to monetary policies of the World Bank that influence the development of rain forests in the Third World.

This influence is not surprising. Many environmental problems concern basic economic decisions: the allocation of scarce resources, the distribution of risks and benefits, property rights, costs of cleanup, and the production of a clean and healthy environment. The connection of ecology and economics is even reflected in the origins of each word. Both *ecology* and *economics* have their roots in the Greek word *oikos*, meaning household. *Ecology* translates as the study, or understanding of the household (the Greek word *logos* refers to a principled or rational understanding). *Economics* appends the Greek word *nomos* and translates as the rules or management of the household. Many environmental critics of economics argue that it is time to manage the household (economics) with the guidance of an understanding of the household (ecology).

Although some defend economics as a scientifically valid and value-neutral approach to policy analysis, on closer examination it is easy to see that economic analysis involves numerous ethical and philosophical assumptions. Because economics is such a powerful tool in environmental policy making, and because it does involve many philosophical aspects, it is crucial that we share a basic understanding of these issues early in this textbook. This chapter will introduce us to the central ethical and philosophical aspects of economic analysis.

Three general economic models are relevant for our survey of economics and the environment. The *neoclassical free market approach* traces its roots to Adam Smith and argues that a market free from government regulation is best suited for directing environmental policy. *Environmental economics* recognizes several serious shortcomings of the free market approach and offers a revised version of market economics. *Ecological* or *sustainable economics* takes its cue from ecology and seeks an approach that can sustain economic activity into the indefinite future. Our reading by Richard Stroup and John Baden represents traditional neoclassical economic analysis. The reading by A. Myrick Freeman is an example of environmental economics. Mark Sagoff's reading evaluates the environmental implications of two versions of free market economics. Herman Daly's essay is taken from his important book on sustainable economics.

A first step in understanding the ethical basis of economic analysis is to distinguish between ends and means. Economic analysis, and indeed all economic activity, is not an end in itself but simply a means to some other end. The end of economic analysis, as traditionally understood, is the maximum satisfaction of individual desires, or maximum happiness. According to this model, the ideal situation occurs when as many people get as much of what they want as possible. In economic terms, an economy is efficient, or Pareto optimal (named for an Italian economist Wilfredo Pareto), when it is impossible to improve the position of anyone without imposing a greater or corresponding loss on anyone else. Conversely, an economy is inefficient whenever it would be possible to improve the position of at least one person without cost.

Explained in this way, we can begin to see a utilitarian basis for economic analysis. The goal for economics, like the utilitarian goal, is maximum happiness. Happiness, in turn, is defined as the satisfaction of individual desires. People are happy when they get what they want. However, because not all people can get everything they want, the best we can hope for is getting as many people as possible as much of what they want as possible. Or, in short, maximally satisfying wants.

But how do we attain this end? Taking their lead from Adam Smith, neoclassical economists argue that the working of a free market will continuously work toward this goal. With just a few noncontroversial assumptions about human nature, we can arrange social and economic institutions in such a way that guarantees maximum satisfaction of individual interests. The noncontroversial assumptions are that humans are rationally self-interested: That is, people are motivated by their own interests, and they are generally capable of reasoning their way to that end. People can learn from their mistakes and will continuously seek the easiest, most efficient means to their ends.

Given these assumptions, we can arrange society in such a way that the pursuit of individual self-interest will be directed (as if led by an "invisible hand" in the words of Adam Smith) to the end of maximum overall happiness. These social arrangements—the well-known free market—include a system of private property rights that allows and protects the free exchange of property and protects these exchanges from fraud and coercion. Under such conditions, society will eventually reach a point of equilibrium at which resources are efficiently allocated and fairly distributed. Scarce resources go to those who most value them (those who are willing to pay the most for them), thereby bringing about greater happiness. Individuals are free to decide for themselves what they most want (their preferences), thereby prioritizing those interests that produce greatest happiness.

However, economists have long recognized that real markets seldom operate as do the ideal markets of economic theory. A variety of market failures prevent real-world markets from attaining the maximum satisfaction of individual desires: Some resources are unowned, some goods (for example, a scenic vista) do not have a market that establishes an economic value, and some costs fall on individuals who are unable to affect the market price or who do not benefit from the exchange (what economists call externalities). In light of these market failures, environmental economists recommend policies that mimic what a market would accomplish if the real world operated according to theory. Because these recommendations often include government regulations and policies, environmental economists move beyond neoclassical free market policies.

Many critics charge that even efficient markets fail to provide an ethical basis for environmental policy. Although economic evaluation might be quite good at measuring and satisfying preferences, it is not at all clear that the satisfaction of preferences is an ethical good. Fundamental questions of justice and fairness can be raised even by efficient distributions. (A distribution in which a single individual possessed all the wealth and millions of others lived in abject poverty could, in principle, be efficient on Pareto grounds.) This challenge mirrors the justice challenge to utilitarianism described in Chapter 1.

Further, economic decision making can ignore other important value questions. Many other values—aesthetic, historical, symbolic, spiritual, moral—are crucial in setting public policy and would be seriously distorted if subjected to economic evaluation.

Property Rights: The Real Issue

Richard Stroup
John Baden
with David Fractor

In the following reading, Chapter 2 from Natural Re-
*sources, economists Richard Stroup and John Baden argue
that a system of private property rights is better suited
than public ownership for protecting scarce environmental
resources. After a brief historical survey of conceptions of
property on the frontier, Stroup and Baden show how private ownership and economic markets provide an efficient
and equitable system for managing natural resources.*

*Stroup and Baden argue that a system of private own-
ership of resources provides owners with an incentive to
find the highest valued use of that resource. Individual
owners who misuse or abuse their property do so at their
own economic peril. Open markets and freely transferable
property rights ensure that resources will be allocated to
those who most value them. Further, such a system preserves individual freedom and encourages a wide diversity
of choices.*

Property rights to a resource, whether a tract of land, a
coal mine, or a spring creek, consist of having control
over that resource. Such rights are most valuable when
ownership is outright and when property can be easily
exchanged for other goods and services. Although an
important feature of a property right is the power to exclude others from using it, even limited command over
access to a resource confers status and power to the
holder. Governments typically exercise at least some
discretionary command in this regard.[1] As Douglass
North wrote, "One cannot develop a useful analysis
of the state divorced from property rights."[2] Indeed, a
theory of property rights can become a theory of the
state.

It is a common misconception that every citizen
benefits from his share of the public lands and the
resources found thereon. Public ownership of many
natural resources lies at the root of resource control
conflicts. With public ownership resources are held
in common; that is, they are owned by everyone and,
therefore, can be used by everyone. But public ownership by no means guarantees public benefits. Individuals make decisions regarding resource use, not
large groups or societies. Yet, with government control, it is not the owners who make decisions, but politicians and bureaucrats. The citizen as beneficiary is
often a fiction.

It is useful to characterize an institutional arrangement by describing how it defines and defends property rights and makes them transferable. Over the
past few years, economists and others have developed
a property rights paradigm that examines these institutional characteristics, making it possible to analyze events based on actual or proposed institutional
changes.

The property rights paradigm provides important
analytical leverage that is useful for understanding
how individuals interact in institutional contexts. The
paradigm helps us to understand history, to predict
the consequences of modern institutions, and to compare the likely outcomes of alternative arrangements.
Given the growing pressures from larger populations
and from technologies that enable us to acquire and
process more natural resources, such predictive and
analytical capabilities take on increasing importance.
We *must* manage with care. The costs of failure are
increasing.

PROPERTY RIGHTS AND
ALLOCATION OF RESOURCES

Most economists begin their analyses by assuming that
decision makers seek to maximize profits, income, or
even wealth. Property rights theorists assume that the
decision maker's goals or utility function must first

be specified in each case. It is assumed that the decision maker will maximize his own utility—not that of some institution or state—in whatever situation he finds himself.[3]

Individuals seek their own advantage within prevailing institutional arrangements. Nevertheless, they may attempt to change the "rules of the game" or the institutions themselves. When privately held property rights to urban land are attenuated by building height restrictions, for example, landowners may gain by changing the rules or by influencing their administration. Since others will fight these changes or seek similar advantages for themselves, the resulting competition may be a negative sum game. In such economic situations, the winners gain less than the losses suffered or investments made by their competitors.[4] In effect, then, negative sum games result in a net economic loss to society.

Institutional rules always allow governmental officials some discretion in determining access to resources. Claimants, therefore, have an incentive to invest in activities that might produce administrative outcomes favorable to themselves. Under these circumstances, some corruption exists in every political system. Informational lobbying, potential shifts of campaign support, actual or threatened lawsuits, and even bribery can all be brought to bear—at a cost—by those who wish to gain favorable decisions from governmental policymakers who control the rights to resources.

Economic growth and efficiency are greatly affected by the way in which existing institutions allow property rights to be traded and allocated.[5] When rights are both privately held and easily transferable, decision makers have easy access to information through bid and asked prices, as well as an incentive to move resources to higher valued uses. But if a person can gain by blocking socially useful resource moves through governmental means, then his gain is society's loss. Similarly, if potential users can gain access to the resource through government without paying the opportunity costs of the resource, then low-valued uses may dominate at the expense of more highly valued uses.

THE EVOLUTION OF PROPERTY RIGHTS TO NATURAL RESOURCES

In analyzing the effects of alternative institutional arrangements for resource allocation, it is useful to look at the evolution of those institutions that established and protected property rights to natural resources. As part of that effort, an examination of the western American frontier allows us to compare methods of defining and enforcing these property rights.[6] Two basic themes in the evolution of natural resource property rights emerge from a reading of *The Frontier in American History* by Frederick Jackson Turner.[7] First, American institutions were formed as pioneers ventured into the West and resource constraints changed. Second, opportunities provided by the frontier placed a lower limit on wages. In a general way, these observations help explain how property rights evolved as a response to changing resource prices.

We can theorize that as more information was obtained about natural resources and as their value rose, potential rents to the resources induced decision makers to develop them and thus to define and enforce the property rights governing them. Further, when voluntary associations of resource users developed and enforced these rights, a strong incentive arose to allocate the resource efficiently. Finally, as the geographic frontier closed and the number of unclaimed resources declined, individual options for increasing wealth became limited to increasing productivity, confiscating resources through legitimate or illegitimate means, or both.

History of Frontier Development

The western American frontier in the late eighteenth century contained abundant natural resources with virtually no institutions or conventions to govern their use. As long as the expected value of the marginal product of labor combined with frontier resources was less than the opportunity cost of labor in other areas, the frontier resources were not exploited. But as the demand for outputs produced from natural resources increased and the opportunity cost of employing those resources fell, settlers moved to the frontier.

When free to choose their own process for defining property rights, settlers had incentives to reduce bargaining costs. Because native American claims to resources often were not enforced by the government, their conflicts with white settlers over the control of land were settled by force. Though this was a rather expensive activity, there generally remained a strong incentive to economize in property rights definition.

Actual or potential owners have incentives to use their resources efficiently. In contrast, agents with no stake in the residuals of the bargaining process (e.g., a bureaucrat at the Environmental Protection Agency

[EPA] or the attorney in a divorce suit) have no direct and personal interest in reducing the cost of that process. Of course, when such third parties are easily monitored *and* there is competition for their jobs, incentives do appear, but this seldom occurs in the public sector. On the American frontier, the first efforts to settle resource allocation claims emerged through the establishment of voluntary associations and the development of informal property rights to resources. Land clubs, claims associations, cattlemen's associations, wagon trains, and mining camps within which individuals grappled with the allocation of water, land, livestock, minerals, timber, and even personal property all represent attempts to mitigate the problems associated with common ownership.

The forms of voluntary association varied, but each sought to bring order to competing multiple claims before the formal claims process applied to the land. These groups often had bylaws, a constitution, a management pact, a leadership selection process, and a procedure for handling disputes. In addition, outsiders who attempted to interfere with a claim held by a member of the group were confronted by the association's considerable enforcement power.

> From successive frontiers of our American history have developed needed customs, laws and organizations. The era of fur-trading produced its hunters, its barter, and the great fur companies; on the mining frontier came the staked claims and the vigilante committees; the camp meeting and the circuit rider were heard on the religious outposts; on the margins of settlement the claims clubs protected the rights of the squatter farmers; on the ranchmen's frontier the millions of cattle, the vast ranches, and the cattle companies produced pools and local, district, territorial, and national cattle associations.[8]

This process of defining individual property rights resulted in relatively low-cost methods. Claims associations in Iowa, for example, required that their members contribute from zero to fifty dollars in value each six months the claim was held. Likewise, cattlemen and livestock associations throughout the West sought to define and enforce property rights while conserving the resources necessary to the definition and enforcement process.

Two basic institutional methods were used to allocate property rights to the vast frontier resources: (1) "squatter sovereignty," or preemption, and (2) the acquisition of water rights. Typically, simple settlement was sufficient to enforce squatters' claims since the abundance of natural resources reduced the seeds of conflict. Later, however, range rights were established through crude advertising and were enforced by livestock associations. Water rights as a claims method typically entailed homesteading adjacent to water or filing claims where state or territorial laws prevailed.

Throughout this period, these extralegal, voluntary associations economized on definition and enforcement techniques. They recognized that high bargaining costs consumed resources that they intuitively knew could be put to better use.

The Frontiersman Versus the Law

> The Easterner, with his background of forest and farm, could not always understand the man of the cattle kingdom. One went on foot, the other went on horseback; one carried his law in books, the other carried his strapped around his waist. One represented tradition, the other represented innovation; one responded to convention, the other responded to necessity and evolved his own conventions. Yet the man of the timber and the town made the law for the man of the plain; the plainsman, finding this law unsuited to his needs, broke it and was called lawless.[9]

In the late nineteenth century, as statutory mandates dating from 1785 focused on the rapid disposition and transfer of government lands to private parties and the promotion of the family farm, the growing conservation movement began charging private enterprise and private ownership with exploitation. The movement to preserve the public lands intensified.

To the extent that initial disposal schemes recognized scarcities, efficiencies were realized. But the change in policy evidenced by the Homestead Act of 1862 reversed eight decades of relatively unfettered land disposal and minimal transaction costs by requiring often economically unrealistic labor and capital expenditures to retain land ownership. This reversal, of course, produced economic inefficiency, considerable human suffering, and even death.

As third party agents, the government increasingly interfered in resource allocations, setting the stage for the blossoming and growth of transfer activity. It is this institutional setting, now matured, that dominates property rights considerations in natural resource development, allocation, sale, and preservation.

The Growth of Transfer Activity

When the rule of willing consent applies, the transfer (exchange) of property is expected to benefit both parties. In contrast, when the coercive power of government is employed to transfer property from one party to another, neither equity nor efficiency can be assumed. It was these politically determined transfers that disturbed Sitting Bull—with serious consequences for Custer and his party. Politically enforced transfers require the use of the coercive power of government to transfer rights from one individual or group to another.[10] Since individual wealth is a direct function of the property rights held, an institutional environment that allows coercive transfer activity will greatly increase the marginal benefit of using resources to generate transfers.

The closing of the American frontier compounded the effect of increased returns from transfer activity. Whereas the frontier had permitted individuals to increase their wealth by establishing property rights to previously unclaimed resources, the closing of the frontier enhanced the benefits to be gained from expending resources for transfer activity. This is not the zero sum game claimed by many economists—that one's gain must come at a loss to another—but rather a negative sum game in which the resources spent by some in generating and by others in opposing transfer activity lead to a *net* loss. The private gain of one is more than offset by the other's loss coupled with the unproductive waste of resources used in the process.

Beginning with the Slaughterhouse cases (1873) and continuing through *Munn* v. *Illinois* (1877) and *Muller* v. *Oregon* (1908), the institutional setting of property rights definition began to change. Third parties were increasingly granted a "property right" in what had previously been another's exclusive resource. This historical foundation provides the context within which conflicts over the definition of current property rights regarding natural resources can be judged.

RIGHTS, MARKETS, AND RESOURCE MANAGEMENT

Another common but mistaken belief is that without social regulation, resources will be managed for profit, not people. When a natural resource is privately owned, it is often thought that the owner has only his conscience to tell him to pay attention to the desires of others. Normally, however, it is the *absence* of private, transferable ownership that leads to the resource user's lack of concern for others' desires. Private ownership holds the individual owner responsible for allocating a resource to its highest valued use, whether or not the resource is used by others. If the buffalo is not mine until I kill it and I cannot sell my interest in the living animal to another, I have no incentive—beyond altruism—to investigate others' interest in it. I will do with it as I wish. But if the buffalo is mine and I may sell it, I am motivated to consider others' value estimates of the animal. I will misuse the buffalo only at my economic peril.[11] How does this work?

Privately owned resources that are freely transferable generate decentralized decisions regarding resource uses. The market rations scarce resources and coordinates individual plans. For example, the owner of a copper mine receives information on the value of alternative uses, as well as the incentive to supply the highest valued use, through bids for copper ore or offers to buy the mine.[12] The market enables the owner to minimize the social opportunity cost of exploiting his resource simply by minimizing his total costs. Bid and asked prices of resources provide owners with information as well as the incentives to use that information for allocating resources efficiently, thereby serving others. Similarly, consumers are informed by prices of the value *others* place on a given resource. In an open market, no one consumes or controls a good desired more by others, as measured by the size of the others' bids.

The benefits of diversity, individual freedom, adaptiveness to changing conditions, the production of information, and even a certain equity derive from this market system. Diversity flourishes because there is no single, centralized decision maker. Instead, many asset owners and entrepreneurs, making their own individual decisions, compete over resource allocations. Those who correctly anticipate people's desires are rewarded the most. Those who envisioned a retail market for television in the 1930s and 1940s and acted on their vision, for example, may have been amply rewarded in the 1950s.

This system preserves individual freedom since those who support and wish to participate in each activity may do so on the basis of willing consent. If I want more logs for my log house, my neighbor need not be concerned. I must pay at least as much for the logs as anyone else would, and in so doing I give up purchasing power that could be used to buy other items. No shortage will result. Adaptiveness is encouraged in both management and consumer activities, since prices provide immediate information and incentives for action as soon as changes are seen. In the political arena any change in resource policy requires convincing a majority of the voters, or the bureaucracy, of the benefits such change would generate. The market system, however, permits individuals who envision scarcities or opportunities in the future to buy or sell resources and develop expertise that may redirect resource use. If their expectations about the future prove correct, they will profit. Losses from foolish diversions of resources, on the other hand, ultimately channel those resources away from inefficient or little valued ventures. From this marvelous adaptive quality, we get both television and Edsels. Successes quickly draw imitators. Losers are quickly dropped, and unsuccessful planners are disciplined by losses.

Information, produced as a byproduct of bids offered and prices asked, is vital to the coordination of individual plans.[13] The market pricing system provides a tangible measure of how individuals evaluate a particular product or service relative to others that use the same resources. Without a market system of exchange such assessments are virtually impossible to make, thereby rendering rational management of nonmarketed activities difficult if not impossible. No manager can make productive resource allocation decisions without knowing input and output values, without knowing, for example, how much people are willing to sacrifice for a thousand board feet of lumber. When rights are privately held and transferable, prices yield the necessary information about the relative value of alternative resource uses—information that is concise, measurable, comparable, and largely devoid of distortion.

A management system based on private property rights also provides a certain equity by having those people who use a resource or who wish to reserve it pay for it by sacrificing some of their wealth. Where natural resources are publicly owned but used by only a few, the sale of those resources could provide a new measure of equity. The proceeds from the sale of assets now in the public domain could be distributed widely in cash or in lower taxes. Alternatively, they could be invested to reduce the national debt or used to cope with the actuarial deficit of the social security system. . . . In a market, those who use the resources would receive from sale proceeds their share of the wealth and would then be required to pay for resources they use, whether for recreation, timber harvest, or research.

PRIVATE AND TRANSFERABLE PROPERTY RIGHTS

Markets can generate both equity and efficiency. Their very essence requires decentralized decision making that can promote flexibility and individual freedom, as well as the information from which rational management of resources is made possible. Yet these advantages will materialize only when property rights to each resource are privately held and easily transferable, ensuring that decision makers will have an incentive to identify the highest value of their resources, including their value to others. In the absence of such clearly defined and enforceable rights, resources may be utilized by individuals who need not compensate or outbid anyone for their use, resulting in substantial waste.

Private ownership of property rights alone is insufficient to secure efficient resource use. If rights are not easily transferable, owners may, for example, have little incentive to conserve resources for which others might be willing to pay if transfer were possible. Transferability ensures that a resource owner must reject all bids for the resource in order to continue ownership and use. Thus, if ownership is retained, the cost to others is made real and explicit.

Private and transferable property rights mandate consideration of alternative users' interests. Any failure to do so imposes economic penalties on the owner. On the other hand, nonprivate or nontransferable property rights often result in inefficiency and waste, as well as a potential indifference to others' interests. When rights are private and transferable, a decentralized market provides diversity, individual freedom, flexibility, information, and equity, since the interests of nonowners are unavoidably observed and respected.[14]

MARKET FAILURE AND POTENTIAL REMEDIES

It is important to acknowledge and describe market failures, their root causes, and their potential remedies in order to proceed to compare market with nonmarket outcomes. Market failure occurs when property rights are inadequately specified or are not controlled by those who can benefit personally by putting the resources to their most highly valued uses. Though both champions and critics of markets have long recognized the potential pitfalls of monopoly, externalities, public goods and common pool problems, transaction costs, and inequities, seldom are they traced to their origins.[15] Yet remedies cannot be developed, nor can the role of government management be properly assessed, without first understanding the causes of market failure.

Many have argued that unfettered market operations would produce monopolies in which single individuals or firms controlling the entire supply of a resource would limit output in order to increase price. In the absence of any satisfactory substitutes, the resource owner would benefit from these restrictions on production. Under these circumstances, production of additional units could be sustained at prices lower than those set by the monopoly, yet sufficiently high that other investors, were they able to gain access to the resource in question, would be willing to develop it. In this case, monopoly would result in pronounced inefficiency.[16]

A second market failure, that of externalities, is especially prominent in discussions lamenting the market misallocation of natural resources. Externalities refer to the separation of responsibility from authority in resource decision making. In other words, an externality exists when some results of a decision do not affect the decision maker.

Both negative and positive externalities can develop. Perhaps the most often condemned negative externality is air pollution. Why should some individuals, for example, suffer harm from smoke produced by others? Such pollution is often both inequitable as well as inefficient in the event that the costs of reducing the pollution are actually less than the damages such a reduction would eliminate. But as long as the decision maker is shielded from the costs or damages of his actions, negative externalities will abound.

Positive externalities present a symmetrical problem, appearing when a decision maker's actions yield benefits that cannot be captured by the decision maker.[17] If my neighbor continues to grow wheat or raise livestock on his land rather than strip-mine the coal below, I enjoy the view without having to pay him. Therefore, of course, he need not consider my values when negotiating with coal buyers and deciding how to use his land. In general, activities leading to positive externalities tend to be underproduced.

Both negative and positive externalities result from imperfectly defined property rights. If runoff from a farmer's land pollutes a stream, a negative externality occurs simply because the stream is not owned by anyone. If rights to the stream were privately held, a polluter would be liable for pollution damages in the same way he would be liable for damaging his neighbor's house.[18] The courts would enforce existing rights if damage could be proven. A resource whose rights are unassigned is likely to be abused.[19]

Positive externalities, such as the inability to charge a neighbor for the aesthetic pleasure provided by one's apple orchard, for example, also represent an absence of the right to control and exclude others from the enjoyment of all output from the land resource. Since no compensation for providing the view is received, the aesthetic values of others play no part in the owner's decision to retain or bulldoze the orchard and put it to other uses.[20]

The Logic of the Commons

The presence of so-called public goods and common pool resources, in which every management decision has external effects on others, presents another set of market problems. Public goods are those goods that, once produced, are available for anyone to use, whether or not they have contributed to their production.[21] Individuals thus may become "free riders," benefiting from goods that others have provided. Because the benefits associated with public goods are not necessarily paid for by all who enjoy them, market behavior generally underproduces such goods.

Particularly problematic regarding natural resources is the common pool problem. "A common pool is like a soda being drawn down by several small boys, each with a straw. The 'rule of capture' is in effect. The contents of the container belong to no one boy until he 'captures' it through his straw."[22] The relevance of this problem to current institutions that define ownership of oil (or in some cases underground water) is evident. Ownership of these resources is recognized only when someone actually extracts the resource from the ground, thus providing an incentive for different individuals each tapping the same reservoir to withdraw the resource as quickly as possible. This behavior may misallocate the oil or water over time. Furthermore, because of geological factors, such practices may even reduce the total volume that can be extracted from the well.[23]

Markets need not inevitably present public goods and common pool problems. Rather, it is the incomplete nature of existing property rights arrangements that has given rise to these problems. If those who failed to pay could be excluded from the benefits of current so-called public goods, no problem would exist. Demands for the goods would be reflected in offers made to purchase them. The same is true for resources subject to common pool problems. Establishing property rights to resources currently beset by overuse or underproduction because of public goods and common pool considerations would virtually eliminate these problems.

Transaction costs pose another problem for efficient market operations. Under ideal market conditions, no transaction costs would arise.[24] All individuals affected by a particular transaction would be included in the decision-making process. If there were no costs associated with defining and enforcing property rights, nor any costs of identifying and undertaking mutually beneficial exchange, all exchanges in which benefits outweighed costs would be undertaken. However, ideally efficient markets do not prevail, and some transaction costs do, in fact, almost always enter into market calculations. Thus, alternatives that might overcome the transaction costs associated with voluntary market exchange merit theoretical consideration.

Thus far, each of the problems discussed—monopoly, externalities, public goods, common pools, and transaction costs—has referred to failures in market efficiency. Though efficiency is certainly a significant indication of market success, the more subjective goal of equity must also be addressed. If efficiency means making the largest pie from our given resources, equity is the determination of how that pie is divided among the population. But which division of the economic pie is most equitable? Some may equate equity with equality, arguing that a more equal distribution of income is more equitable, or that equity requires a more equal distribution of the means of production. Even for those who do not equate equality with equity, how the pie is sliced will still be of primary importance.[25] The private property paradigm emphasizes that individuals tend to seek control over the largest possible piece of pie. Thus, how that pie is initially distributed and what rules govern the distribution have significant implications for the overall equity of the system.

If the market system fails to distribute costs and benefits in ways that are perceived as equitable, individuals may seek alternative methods of influencing that distribution. Efforts emerge to redistribute bene-

fits to low income citizens, or pleas are voiced to curb what are considered windfall profits from crude oil. Both are symptomatic of general concerns about the equity of markets.

Whether in pursuit of greater equity or efficiency, many persons have increasingly turned to governmental institutions to achieve preferred allocation of natural resources. This search for government solutions persists despite the government's tarnished performance as a natural resource manager.

PUBLIC, NONTRANSFERABLE RIGHTS IN A GOVERNMENTAL SETTING

An activist government has been lauded as the last line of defense between the bulldozer and the bald eagle. A corollary to this belief contends that those involved with governmental regulation are motivated by incentives that are incorrectly analyzed by economists, whose theories apply only to a perfectly competitive market. It is thought that even though economic analysis and economic principles can explain behavior in markets, they have little bearing on governmental actions, since motives change when people enter government service. Yet these contentions are incorrect. Economic principles apply in all settings, including bureaucracies and among primitive tribes. . . .

Though current market systems allocate resources imperfectly, even many critics of markets agree that governmental efforts to correct market imperfections have ensured neither efficiency nor equity.[26] Furthermore, there is a growing awareness that self-interest is not absent in the public sector and that economic analysis is applicable and even necessary if the actions and goals of the public sector are to be understood.

The pioneering contributions of Anthony Downs, James Buchanan and Gordon Tullock, Mancur Olson, and William Niskanen have led to a developing awareness of the problems of representative government.[27] Their analyses show some promise of approaching the rigor and predictive capacity of the economic theory of the firm.[28] While precise predictions regarding public sector decision making may not be possible, some general statements may be made with confidence.

Using the property rights approach in which decision makers act to advance their own perceived interests, we observe that, as in market systems with imperfect property rights and transaction costs, public-sector decision makers are not held fully accountable for their actions.

The public sector provides no incentives for politicians and bureaucrats to resist pressures from special interests or to manage natural resources efficiently. On the contrary, such resistance may even hinder the public decision maker's career. Presumably, those turning to government to resolve natural resource problems earnestly seek more efficient and equitable management. Why, then, are public officials not held more accountable for managing resources accordingly?

Five factors tend to undermine that accountability. First, no citizen has either the time or the resources to analyze every policy issue. Nor is every citizen able personally to influence decisions regarding most complex public policy issues. Given these constraints, the intelligent citizen's ignorance about most public policy matters is understandable.

Second, although no individuals attempt to analyze or influence *all* government policies, some do attempt to influence specific policies in which they have a pronounced interest. The result is a medley of narrowly focused, highly self-interested groups that wield tremendous influence, each over its particular policy domain. These special interest groups are able to dominate a particular policy domain precisely because others with more diffuse interests regarding such policies have little incentive to articulate their views.[29]

The system of political representation further limits the accountability of policymakers. Voters themselves have no direct input on individual issues. They merely elect representatives who, in turn, make the decisions regarding all policy issues. Such a system obviously records individual preferences imperfectly, if at all.[30]

A congressman or senator votes on hundreds of issues each year. The message sent (the vote cast) by even a thoroughly informed, decisive voter is garbled. While the voter agrees with his favorite candidate on some important issues, there may be serious disagreement on many other issues that are judged less important by that voter. By contrast, in the private sector the citizen can "vote for" tires made by one company and toasters made by another. His choices in a market are precisely recorded.

A fourth factor further weakens the incentive for efficient natural resource management. Many resource issues have significant long-term implications for future generations. How and which resources are developed today will affect future generations. Yet precisely what those future costs and benefits might be are gen-

erally poorly understood, especially by the average citizen, who remains ill-informed about such issues. Thus, most individuals evaluate the performance of politicians and bureaucrats not according to how well they shepherd resources for the future, but according to whether or not their decisions have produced current *net* benefits.

A government decision maker can seldom gain political support by locking resources away from voters to benefit the unborn. Charitable instincts toward future resource users are unaided by (and are, in fact, countered by) self-interest if the resources are publicly owned. Charity toward others at the expense of voting constituents does not usually contribute to political survival. So we can expect governmental policy to be shortsighted, especially in comparison to the long time-frames necessary for carrying out many natural resource policies. Future benefits are more difficult to measure and future costs are easy to ignore. In addition, there is no "voice of the future" in government equivalent to the rising market price of an increasingly valuable resource. The wise public resource manager who forgoes current benefits cannot personally profit from doing so.

Yet another widespread—and incorrect—assumption is that governmental intervention in private markets can be expected to produce farsighted decisions, whereas actions taken to enlarge profits or individual wealth are normally shortsighted. This belief springs from the conviction that future generations have a property right in current resources; that is, there are transgenerational property rights for the unborn. The major implication of this and similar thinking is that a market mechanism, unlike collective control, deprives future generations of resources. This belief arises in turn from the conviction that resources are being rapidly depleted, although prices for many natural minerals are in fact declining. This obvious inconsistency aside, let us assume that future generations are in some fashion granted a property right in this generation's available resources. Can the government protect the interests of these future generations better than the market?

Enter the speculator. Even though "speculator" is a term often used derisively, it merely describes a market participant who performs a service consistent with the desires of his harshest critics: He defers consumption and thereby saves for the future by paying a market price higher than any other bidder who seeks resources for present use. Indeed, exploitation will occur only when all speculative bids have been overcome.

In effect, farsighted outcomes are made possible through the activities of speculators who have an incentive to protect resources for future sale and use because such preservation may benefit them. For example, the owner of an Indiana woodlot may think that his old-growth white oak trees should be saved rather than harvested. Any concern he has for the future is powerfully influenced by how much he can gain by hoarding his resource, which is becoming more scarce and valuable, and selling it later to other hoarders or speculators. The speculator thus acts as a middleman between the present and the future.[31] His position is similar to that of another middleman, a broker of Florida oranges. By purchasing the oranges and shipping them to Montana, the fruit broker acts *as if* he cares about the desires of Montanans. He bids oranges away from Florida buyers in order to send them to Montana. The woodlot owner may act in the same way, taking wood off the market and "transporting" it into the future by failing to cut it. Since the property rights are transferable, the speculator can do well for himself while he does good for future resource users. He can profit from the transaction if he has guessed correctly. This incentive to look to the future contrasts sharply with the incentives faced in the public sector.

Private ownership allows the owner to capture the full capital value of his resource, and thus economic incentive directs him to maintain its long-term capital value. The owner of the resource, be it a fishery, a mine, or a forest, wants to produce today, tomorrow, and ten years from now; and with a renewable resource he will attempt to maintain a sustained yield. Do farmers consume their seed corn or slaughter the last of their prime breeding cattle even when prices are especially high? In contrast, when a resource is owned by everyone, the only way in which individuals can capture its economic value is to exploit the resource before someone else does.

A final factor dictates against efficient resource management within the public sector: There is no tangible internal measure of efficiency. Private sector firms whose use of resources (as measured by cost) exceeds the value of what they produce (as measured by revenue) lose money and go out of business. Their failure benefits society by removing control of scarce resources from those who use them inefficiently. Eliminating this mechanism from the market system is probably the most significant cost of government bailouts of private firms.

Government decision makers operate without this internal check on efficiency. The funding for government bureaus derives not from profits resulting from efficient use of resources, but from federal treasury "bailouts" and budget allocations. Public sector decision makers operate with no concrete measure of efficiency because their survival does not depend on the difference between costs and benefits. Indeed, the incentive is to expand rather than to economize.[32] An expanding bureau allows its administrators more opportunities for advancement and achievement, greater scope for power, and less need for unpleasant budget trimming and layoff decisions.

A REALISTIC ANALYSIS?

Our analysis has resolutely criticized decision making in the public sector. Such criticism is not overly cynical. Rather, our assessment simply recognizes that individuals, not organizations or societies, make decisions and that individuals tend to act according to their own perceived interests.

Charitable instincts are important, but the forces of simple self-interest are relentless. In order to be useful and beneficial to society as a whole, an institution must relate authority—that is, command over resources—to personal responsibility for the costs and benefits that flow from decisions.

The market relies on private property rights to hold individuals responsible for their actions. Imperfectly defined, enforced, or transferable rights generate market failures. Government is different. Representative democracy depends on informed voters and their elected representatives to ensure the accountability of governmental decision makers. Yet existing institutions and incentives provide neither informed voters nor accountability. A general understanding of the model presented above—an understanding shared by the authors of the *Federalist Papers*—can foster reforms that are likely to improve social welfare.

Notes

1. See Robert Dorfman, "The Technical Basis for Decision Making," in *The Governance of Common Property Resources,* ed. Edwin T. Haefele (Baltimore: Johns

Hopkins University Press, 1974); and Richard Stroup and John Baden, "Property Rights and Natural Resource Management," *Literature of Liberty* (October–December 1979), pp. 5–17.

2. Douglass North, "A Framework for Analyzing the State in Economic History," *Explorations in Economic History* 16 (1979): 250.

3. For an excellent review, see Eirik Furubotn and Svetozar Pejovich, "Property Rights and Economic Theory: A Survey of Recent Literature," *Journal of Economic Literature* 10 (1972): 1137–62.

4. See Anne O. Krueger, "The Political Economy of the Rent Seeking Society," *American Economic Review* 64 (June 1974): 291–303; and Gordon Tullock, "The Welfare Costs of Tariffs, Monopolies, and Theft," *Western Economic Journal* 5 (June 1967): 224–32.

5. See, for example, Steven N. S. Cheung, "The Structure of a Contract and the Theory of a Non-Exclusive Resource," *Journal of Law and Economics* 13 (April 1970): 49–70; Harold Demsetz, "Toward a Theory of Property Rights," *American Economic Review* 57 (May 1967): 347–59; and Furubotn and Pejovich, "Property Rights and Economic Theory."

6. For a more detailed account, see Terry L. Anderson and Peter J. Hill, "Property Rights as a Common Pool Resource," in *Bureaucracy vs. Environment,* ed. John Baden and Richard Stroup (Ann Arbor: University of Michigan Press, 1981), pp. 22–43.

7. Frederick Jackson Turner, *The Frontier in American History* (New York: Krieger, 1920), p. 343.

8. Louis Pelzer, *The Cattlemen's Frontier* (Glendale, Calif.: Russell Sage, 1936), p. 87.

9. Walter Prescott Webb, *The Great Plains* (Boston: Grosset and Dunlap, 1931), p. 206.

10. For more information, see Terry Anderson and Peter J. Hill, *The Birth of a Transfer Society* (Stanford, Calif.: Hoover Institution Press, 1980).

11. See John Hanner, "Government Response to the Buffalo Hide Trade—1871–1883," *Journal of Law and Economics* 24 (October 1981): 239–71.

12. For a more thorough treatment of markets in a resource setting, see Richard Stroup and John Baden, "Externality, Property Rights, and the Management of Our National Forests," *The Journal of Law and Economics* 16 (1973): 303–12.

13. Economists of the Austrian school emphasize the role of the entrepreneur who, in his search for profit, finds higher valued uses for resources. See, for example, Ludwig von Mises, *Human Action* (New Haven, Conn.: Yale University Press, 1949); and Israel Kirzner, *Competition and Entrepreneurship* (Chicago: University of Chicago Press, 1973).

14. The workings of the market are explained from a property rights approach, with a minimum of jargon, in Armen Alchian and William Allen, *University Eco-*

nomics, 3d ed. (Belmont, Calif.: Wadsworth, 1972); James Gwartney and Richard Stroup, *Economics: Private and Public Choice,* 2d ed. (New York: Academic Press, 1980); Paul Heyne, *The Economic Way of Thinking,* 3d ed. (Chicago: SRA, 1980); and Svetozar Pejovich, *Fundamentals of Economics: A Property Rights Approach* (Dallas: The Fisher Institute, 1979).

15. For a systematic treatment of the "accepted wisdom" on market failure, see Francis M. Bator, "The Anatomy of Market Failure," *The Quarterly Journal of Economics* (August 1958), pp. 351–79.

16. Nearly all introductory economics texts cover the general problem of monopoly, including the four cited in note 14.

17. The following discussion of positive externalities is drawn from Richard Stroup and John Baden, "Property Rights and Natural Resource Management," *Literature of Liberty* 2 (September–December 1979): 11.

18. See *Corpus Juris Secundum,* vol. 66, p. 9461. This common law approach is being supplemented by statutory laws that proclaim the mere existence of a pollution source a nuisance, apart from demonstrated damage. Such laws are currently being challenged in the courts.

19. Note that if property rights to clean air were easily enforced, pollution would still be produced, but only in efficient amounts. Polluters would compensate those damaged and would reduce pollution until further reductions were more costly than if they fully compensated all those harmed. A different approach considers negative externalities as a failure of law regarding liability. The implications of alternative liability laws are examined in Roland McKean, "Products Liability: Implications of Some Changing Property Rights," *Quarterly Journal of Economics* 84 (November 1970): 611–26.

20. See Ronald Coase, "The Problem of Social Cost," *The Journal of Law and Economics* 4 (October 1960): 1–44. Coase shows that in the absence of transaction costs it does not matter *who* owns the resource, only that wealth will change.

21. In his original definition of a *public good,* Paul Samuelson stated that one individual's consumption of a public good led to no reduction in others' consumption. See Paul Samuelson, "The Pure Theory of Public Expenditures," *Review of Economics and Statistics* 36 (1954): 347–53.

22. This discussion of the common pool problem derives largely from Richard Stroup and John Baden, "Property Rights and Natural Resource Management." The quote is from p. 12.

23. If many well-owners pump more rapidly from many pools, ignoring the "user cost" or reduced availability from each pool later, then oil market prices can be depressed. When this happened in the United States

in the 1930s, the government gained control of oil well production. See Edward Mitchell, *U.S. Energy Policy: A Primer* (Washington, D.C.: American Enterprise Institute, 1974). For a general treatment of problems associated with common pool resources, see Garrett Hardin and John Baden, eds., *Managing the Commons* (San Francisco: W. H. Freeman and Company, 1977), especially Hardin's "The Tragedy of the Commons," pp. 16–30.

24. For further discussions on transaction costs (the cost of reaching a final bargain among parties), see Furubotn and Pejovich, "Property Rights and Economic Theory"; and Cheung, "The Structure of a Contract."

25. The growing importance of equity is discussed in Fred Hirsch, *Social Limits to Growth* (Cambridge: Harvard University Press, 1976); Robert A. Nisbet, *Twilight of Authority* (New York: Oxford University Press, 1975); and Daniel Bell, *Cultural Contradictions of Capitalism* (New York: Basic Books, 1976).

26. Economists are still struggling with the theory of regulations, but not fruitlessly. For a technical approach see, for example, George Stigler, "The Theory of Economic Regulation," *Bell Journal of Economics and Management Science* 2 (1971): 3–21; and Sam Peltzman, "Toward a More General Theory of Regulation," *The Journal of Law and Economics* 11 (1976): 211–40. The problems of governmental (bureaucratic) control of resources are analyzed in William A. Niskanen, Jr., *Bureaucracy and Representative Government* (Chicago: Aldine-Atherton, 1971); and Thomas Borcherding, ed., *Budgets and Bureaucrats* (Durham, N.C.: Duke University Press, 1977). These problems are discussed in a natural resource context in John Baden and Richard Stroup, "The Environmental Costs of Government Action," *Policy Review* 4 (1978): 23–26.

27. Anthony Downs, *An Economic Theory of Democracy* (New York: Harper, 1957); James Buchanan and Gordon Tullock, *The Calculus of Consent* (Ann Arbor: University of Michigan Press, 1962); Mancur Olson, Jr., *The Logic of Collective Action* (New York: Schocken Books, 1965); and Niskanen, *Bureaucracy and Representative Government.*

28. For a relatively nontechnical presentation of the economics of government failure, see Gwartney and Stroup, *Economics: Private and Public Choice,* chap. 32. See also Richard B. McKenzie and Gordon Tullock, *Modern Political Economy: An Introduction to Political Economy* (New York: McGraw Hill, 1978), chaps. 5 and 6; and William Mitchell, *The Anatomy of Government Failure* (Los Angeles: International Institute for Economic Research, 1979).

29. For more rigor and detail on this and related aspects of the political process, see Gordon Tullock, *Toward a Mathematics of Politics* (Ann Arbor: University of Michigan Press, 1967).

30. On this point, see Gordon Tullock, *Private Wants and Public Means* (New York: Basic Books, 1970), pp. 107–14.

31. This discussion of speculation generally follows a similar discussion in Stroup and Baden, "Property Rights and Natural Resource Management," p. 17.

32. The public choice literature, taking a property rights approach, is developing an increasingly sophisticated set of models to explain bureaucratic behavior. See, for example, Jean-Luc Mique and Gerard Belanger, "Toward a General Theory of Managerial Discretion," *Public Choice* 17 (1974); Niskanen, *Bureaucracy and Representative Government;* Gordon Tullock, *The Politics of Bureaucracy* (Washington, D.C.: Public Affairs Press, 1965); and Oliver Williamson, *The Economics of Discretionary Behavior: Managerial Objectives in a Theory of the Firm* (Englewood Cliffs, N.J.: Prentice-Hall, 1964).

For Further Discussion

1. How important is the value of private property rights? Would you describe property rights as natural human rights or as legal rights that exist only within a particular legal and political system?

2. Property rights are often spoken of as a "bundle of rights," suggesting that there are a variety of specific rights associated with ownership of property. How many different rights do you see associated with the ownership of land? How many different things do you have a right to do with your land?

3. Do you believe that private owners would have a greater incentive to protect natural resources than is provided with social ownership?

4. What do the authors mean by equity? Efficiency? Market failure? Externalities?

The Ethical Basis of the Economic View of the Environment

A. Myrick Freeman III

In this reading, A. Myrick Freeman surveys many of the ethical issues involved in environmental economics. Freeman recognizes the limitations of the neoclassical model of economics, but he believes that this model, and the Pareto criterion that it endorses, can be a helpful tool for evaluating environmental policy. Without abandoning the general approach of economic analysis, these limitations can be overcome—sometimes with the help of government regulation and subsidies—and an economically rational policy can be developed. Freeman examines several such policy recommendations in the final section of his essay.

I. INTRODUCTION

At least in some circles, economists' recommendations for a policy concerning pollution and other environmental problems are regarded with a good deal of skepticism and perhaps even distrust. For example, when we suggest that economic factors such as cost should be taken into account in setting ambient air quality standards, we are told that it is wrong to put a price on human life or beauty. And when we argue that placing a tax or charge on the emissions of pollutants would be more effective than the present regulatory approach, we are told that this would simply create "licenses to pollute" and pollution is wrong.

I am not sure how much of this type of reaction stems from a misunderstanding or lack of familiarity with the arguments for the economists' policy recommendations, and how much is due to a rejection of the premises, analysis, and value judgments on which these recommendations are based. And I will not attempt to answer this question here. Rather, I will limit myself to making clear the rationale for some of our recommendations concerning policy and the value judgments on which they are based.

To the economist, the environment is a scarce resource which contributes to human welfare. The economic problem of the environment is a small part of the overall economic problem: how to manage our ac-

tivities so as to meet our material needs and wants in the face of scarcity. The economists' recommendations concerning the environment flow out of out analysis of the overall economic problem. It will be useful to begin with a brief review of the principal conclusions of economic reasoning concerning the allocation of scarce resources to essentially unlimited needs and wants. After reviewing some basic economic principles and the criteria that economists have used in the evaluation of alternative economic outcomes, I will explain the economic view of the environment and some of the major policy recommendations which follow from that view. I will conclude by identifying some of the major questions and possible sources of disagreement about the validity and usefulness of economic reasoning as a way of looking at environmental problems.

II. SOME BASIC ECONOMICS

We begin with the basic premises that the purpose of economic activity is to increase the well-being of the individuals who make up the society, and that each individual is the best judge of how well off he or she is in a given situation. To give this premise some operational content, we assume that each individual has preferences over alternative bundles of economic goods and services. In other words, the individual can rank all of the alternative combinations of goods and services he can consume from most preferred to least preferred. Of course there may be ties in this ranking. We assume that individuals act so as to obtain the most preferred (to them) bundles given the constraints imposed by technology and the availability of the means of production.

These preferences of individuals are assumed to have two properties which are important for our purposes: substitutability among the components of bundles, and the absence of limits on wants. Substitutability simply means that preferences are not lexicographic. Consider a consumption bundle labeled *A* with specified quantities of food, clothing, shelter, and so forth. Now consider alternative bundle *B* which contains 10 percent less clothing and the same quantities of all other goods. Since *B* contains less clothing, it is less desirable to the individual. In other words, bundle *A* is preferred to bundle *B*. But substitutability means that it is possible to alter the composition of bundle *B* by increasing the quantities of one or more of the other goods in the bundle to the point where the individual will consider *A* and *B* as equally preferred. That is to say, the individual can be compensated for the loss of some quantity of one good by increases in the quantities of one or more of the other goods. The value of the lost clothing to this individual can be expressed in terms of the quantities of the other goods which must be added to the bundle to substitute for it. This principle is the basis of the economic theory of value. In a market economy where all goods and services can be bought and sold at given prices in markets, the necessary amount of substitution can be expressed in money terms.

The significance of the substitution principle for the economic view of the environment should be apparent. If the substitution principle applies to good things that are derived from a clean environment, then it is possible to put a price on those things. The price is the money value of the quantities of other goods that must be substituted to compensate for the loss of the environmental good. Whether the substitution principle applies to those things derived from the environment is essentially an empirical question about human behavior. It is possible to think of examples that violate the substitution principle. The slogan printed on all license plates issued in New Hampshire ("Live Free or Die") shows a lexicographic preference for freedom. If the statement is believed, there is no quantity of material goods that can compensate for the loss of freedom. It is not clear that all individuals have lexicographic preferences for freedom. And the question for our purpose is whether there are similar examples in the realm of environmental goods.

By unlimited wants, I mean that for any conceivable bundle *A*, it is possible to describe another bundle *B* with larger quantities of one or more goods such that an individual would prefer *B* to *A*. Is this property plausible? It is possible to imagine some upper limit on the gross consumption of food as measured by calories or weight. But quality and variety are also goods over which individuals have preferences. And it may always be possible to conceive of a bundle containing a more exotic dish or one with more careful preparation with higher quality ingredients. Again, whether this property is plausible is an empirical question about human behavior. But its significance for anti-growth arguments is apparent.

Much of economic theory is concerned with understanding how individuals with given preferences interact as they seek to attain the highest level of satisfaction. Many societies have developed systems of markets for guiding this interaction; and historically the bulk of economists' effort has gone to the study of market systems. In part this can be explained by the historic fact that economics as a separate discipline emerged during a period of rapid industrialization, economic change, and growth in the extent of the market system. But it is also true that as early as Adam Smith's time, it was recognized that a freely functioning market system had significant advantages over alternative means of organizing and coordinating economic activity. Even in more primitive societies, markets facilitate exchange whereby an individual can attain a more preferred bundle by giving up less preferred goods in exchange for more preferred goods. And in more developed economies, markets also facilitate the specialization of productive activities and the realization of economies of scale in production.

A market system can be said to have advantages only in terms of some criterion and in comparison with some alternative set of economic institutions. It is time now to make the criterion explicit. The criterion is economic efficiency, or after the man who first developed the concept in formal terms, Pareto Optimality. An economy has reached a state of economic efficiency if it is not possible to rearrange production and consumption activity so as to make at least one person better off except by making one or more other individuals worse off. To put it differently, an economy is in an inefficient position if it is possible to raise at least one individual to a more preferred consumption bundle while hurting no one. If an economy is in an inefficient position, it is possible to achieve a sort of "free lunch" in the form of an improvement for at least one individual *at no cost* to anyone.

One of the fundamental conclusions of economic reasoning is that given certain conditions a market system will always reach a position of economic efficiency. The conditions are that: (*a*) all goods that matter to individuals (that is, all goods over which individuals have preference orderings) must be capable of being bought and sold in markets; and (*b*) all such markets must be perfectly competitive in the sense that there are large numbers of both buyers and sellers no one of which has any influence over market price. The extensiveness and competitiveness of markets are sufficient to assure that economic efficiency in the allocation of resources will be achieved. This conclusion provides much of the intellectual rationale for *laissez faire* capitalism as well as the justification for many forms of government intervention in the market, for example, anti-monopoly policies, the regulation of the prices charged by monopolies such as electric utilities, and, as we shall see, the control of pollution.

The ideal of efficiency and the perfectly competitive market economy which guarantees its attainment acts as a yardstick by which the performance of real world economies can be measured. If there is monopoly power in a market, the yardstick shows that there is a shortfall in the performance of the economy. It would be possible by eliminating monopoly and restoring perfect competition to the market to increase output in such a way that no one would be made worse off and at least one person would be made better off. How monopoly power is to be eliminated without making at least the monopolist worse off is a difficult question in practice. But I will return to this point below.

The ideal of perfect competition and economic efficiency is a powerful one. But it is not without its limitations. Perhaps the most important of these is that there is no single, unique Pareto Optimum position. Rather there is an infinite number of alternative Pareto Optimums, each different from the others in the way in which it distributes economic well-being among the members of the society.

A society in which one individual owned all of the capital, land, and resources could achieve a Pareto Optimum position. It would likely be one in which all but one of the individuals lived in relative poverty. But it would not be possible to make any of the workers better off without making the rich person worse off. This Pareto Optimum position would be quite different from the Pareto Optimum which would be achieved by an economy in which each individual owned equal shares of the land, capital, and so forth. Which Pareto Optimum position is attained by an economy depends upon the initial distribution of the entitlements to receive income from the ownership of factor inputs such as land and capital. Each conceivable distribution of rights of ownership has associated with it a different Pareto Optimum. And each Pareto Optimum position represents the best that can be done for the members of society *conditioned* upon acceptance of the initial distribution of entitlements. Since the ranking of different Pareto Optimums requires the comparison of alternative distributions of well-being, it is inherently an ethical question. There is nothing more that economic reasoning can contribute to this issue.

III. POLICY EVALUATION

Given the fact that the real world economy is characterized by many market imperfections and failures and that for a variety of reasons it is not possible to create the perfect, all encompassing market system of the Pareto ideal, we must consider piecemeal efforts to make things better at the margin. The question is: what criterion should be used to evaluate policy proposals which would alter the outcomes of existing market processes?

The Pareto Criterion says to accept only those policies that benefit some people while harming no one. In other words, this criterion rules out any policy which imposes costs on any individual, no matter how small the cost and no matter how large the benefits to any other members of the society. This is a very stringent criterion in practice. There are very few policy proposals which do not impose some costs on some members of the society. For example, a policy to curb pollution reduces the incomes and welfares of those who find it more profitable to pollute than to control their waste. The Pareto Criterion is not widely accepted by economists as a guide to policy. And it plays no role in what might be called "mainstream" environmental economics.

The most widely accepted criterion asks whether the aggregate of the gains to those made better off measured in money terms is greater than the money value of the losses of those made worse off. If the gains exceed the losses, the policy is accepted by this criterion. The gains and losses are to be measured in terms of each individual's willingness-to-pay to receive the gains or to prevent the policy-imposed losses. Thus this criterion draws on the substitutability principle discussed earlier. If the gains or losses came in the form of goods over which individuals have lexicographic preferences, this criterion could not be utilized.

This criterion is justified on ethical grounds by observing that if the gains outweigh the losses, it would be possible for the gainers to compensate fully the losers with money payments and still themselves be better off with the policy. Thus if the compensation were actually paid, there would be no losers, only gainers.

This criterion is sometimes referred to as the potential compensation criterion. This criterion is the basis of the benefit-cost analysis of public policy. Benefits are the money values of the gains to individuals and costs are the money values of the losses to individuals. If benefits exceed costs, the gainers could potentially compensate the losers.

There are two observations concerning the potential compensation criterion. First, the criterion is silent on the question of whether compensation should be paid or not. If society decides that compensation shall always be paid, compensation becomes a mechanism for assuring that there are never any losers and that all adopted policies pass the Pareto Criterion. On the other hand, if society decides that compensation should never be paid, the potential compensation criterion becomes a modern form of utilitarianism in which the aggregate of utilities is measured by the sum of the money values of all goods consumed by all individuals. Finally, society may decide that whether compensation should be paid or not depends upon the identity and relative deservingness of the gainers and losers. If this is the case, then society must adopt some basis for determining relative deservingness, that is, some ethical rule concerning the justness of creating gains and imposing losses on individuals.

The second observation concerns the measurement of gains and losses in money terms. Willingness to pay for a good is constrained by ability to pay. Economic theory shows that an individual's willingness to pay for a good depends on his income and that for most goods, higher income means higher willingness to pay, other things equal. As a consequence, the potential compensation criterion has a tendency to give greater weight to the preferences of those individuals with higher incomes. As a practical matter there are reasons to doubt that this bias is quantitatively significant in most cases. But the question is often raised when benefit-cost analysis is applied to environmental goods. And it is well to keep this point in mind.

IV. ENVIRONMENTAL ECONOMICS

The environment is a resource which yields a variety of valuable services to individuals in their roles as consumers and producers. The environment is the source of the basic means of life support—clean air and clean water. It provides the means for growing food. It is a source of minerals and other raw materials. It can be used for recreation. It is the source of visual amenities. And it can be used as a place to deposit the wastes from production and consumption activities. The economic problem of the environment is that it is a scarce resource. It cannot be called upon to provide all of the desired quantities of all of the services at the same time. Greater use of one type of environmental service usually means that less of some other type of service is available. Thus the use of the environment involves trade-offs. And the environment must be managed as an economic resource. But unlike other resources such as land, labor, or capital, the market does not perform well in allocating the environment to its highest valued uses. This is primarily because individuals do not have effective property rights in units of the environment.

For example, if a firm wishes to use one hour of labor time in production, it must find an individual who is willing to provide one hour of labor and it must pay that individual an amount at least equal to the value to the individual of that time in an alternative use. If a voluntary exchange of labor for money takes place, it is presumed that neither party is made worse off, and it is likely that both parties benefit from the exchange. Otherwise they would not have agreed to it. But if a firm wishes to dump a ton of sulfur dioxide into the atmosphere, it is under no obligation to determine whose health or whose view might be impaired by this use of the environment and to obtain their voluntary agreement through the payment of money. Thus firms need not take into account the costs imposed on others by their uses of the environment. Because there is no market for environmental services, the decentralized decision making of individuals and firms will result in a misallocation of environmental resources. The market fails. And the economy does not achieve a Pareto Optimum allocation.

Where markets have failed, economists have made two kinds of suggestions for dealing with market failure. The first is to see if markets can be established through the creation of legally transferable property rights in certain environmental services. If such property rights can be created, then markets can assume their proper role in achieving an efficient allocation of environmental services. Because of the indivisible nature of many aspects of the environment, for example, the urban air shed, there is limited scope for this solution. The second approach is to use various forms of government regulations, taxes, and subsidies to create incentives which replicate the incentives and outcomes that a perfectly functioning market would produce. Activities under this approach could include the setting of ambient air quality standards, placing limits on discharges from individual polluters, imposing taxes on pollution, and so forth. In the next section, I take up several specific applications of this approach to dealing with the environment in an economically rational manner.

V. APPLICATIONS

Environment Quality Standards

An environmental quality standard is a legally established minimum level of cleanliness or maximum level of pollution in some part of the environment, for example, an urban air shed or a specific portion of a river. A standard, once established, can be the basis for enforcement actions against a polluter whose discharges cause the standard to be violated. The principle of Pareto Optimality provides a basis for determining at what level an environmental quality standard should be set. In general, Pareto Optimality requires that each good be provided at the level for which the marginal willingness to pay for the good (the maximum amount that an individual would be willing to give up to get one more unit of the good) is just equal to the cost of providing one more unit of the good (its marginal cost).

Consider for example an environment which is badly polluted because of existing industrial activity. Consider making successive one-unit improvements in some measure of environment quality. For the first unit, individuals' marginal willingnesses to pay for a small improvement are likely to be high. The cost of the first unit of clean-up is likely to be low. The difference between them is a net benefit. Further increases in cleanliness bring further net benefits as long as the marginal willingness to pay is greater than the marginal cost. But as the environment gets cleaner, the willingness to pay for additional units of cleanliness decreases, while the additional cost of further cleanliness rises. At that point where the marginal willingness to pay just equals the marginal cost, the net benefit of further cleanliness is zero, and the total benefits of environmental improvement are at a maximum. This is the point at which the environmental quality standard should be set, if economic reasoning is followed.

There are two points to make about this approach to standard setting. First, an environmental quality standard set by this rule will almost never call for complete elimination of pollution. As the worst of the pollution is cleaned up, the willingness to pay for additional cleanliness will be decreasing, while the extra cost of further clean-up will be increasing. The extra cost of going from 95 percent clean-up to 100 percent clean-up may often be several times larger than the total cost of obtaining the first 95 percent clean-up. And it will seldom be worth it in terms of willingness to pay. Several economists have argued that the air quality standards for ozone that were first established in 1971 were too stringent in terms of the relationship between benefits and costs. If this is true, then the resources devoted to controlling ozone could be put to better use in some other economic activity. Many economists have urged Congress to require that costs be compared with benefits in the setting of ambient air quality standards.

The second point is that the logic of benefit-cost analysis does not require that those who benefit pay for those benefits or that those who ultimately bear the cost of meeting a standard be compensated for those costs. It is true that if standards are set so as to maximize the net benefits, then the gainers could fully compensate the losers and still come out ahead. But when beneficiaries do not compensate losers, there is a political asymmetry. Those who benefit call for ever more strict standards and clean-up, because they obtain the gross benefits and bear none of the costs, while those who must control pollution call for less strict standards.

Charging for Pollution

One way to explain the existence of pollution is in terms of the incentives faced by firms and others whose activities generate waste products. Each unit of pollution discharged imposes costs or damages on other individuals. But typically the dischargers are not required to compensate the losers for these costs. Thus there is no economic incentive for the discharger to take those costs into account. This is the essence of the market failure argument.

If it is impractical to establish a private market in rights to clean air, it may be possible to create a pseudo-market by government regulation. Suppose that the government imposed a charge or tax on each unit of pollution discharged and set the tax equal to the money value of the damage that pollution caused to others. Then each discharger would compare the tax cost of discharging a unit of pollution with the cost of controlling or preventing that discharge. As long as the cost of control were less than the tax or charge, the firm would prevent the discharge. In fact it would control pollution back to the point where its marginal cost of control was just equal to the marginal tax and by indirection equal to the marginal damage the pollution would cause. The properly set tax or charge would cause the firm to undertake on its own accord the optimum amount of pollution control. By replicating a market incentive, the government regulation would bring about an efficient allocation of resources.

Since the firm would likely find that some level of discharges would be more preferred to a zero discharge level, it would be paying taxes to the government equal to the damages caused by the remaining discharges. In principle, the government could use the tax revenues to compensate those who are damaged by the remaining discharges.

Risk and the Value of Life

Because some forms of pollution are harmful to human health and may increase mortality, economists have had to confront the question of the economic value of life. It turns out that the "value of life" is an unfortunate phrase which does not really reflect the true nature of the question at hand. This is because pollutants do not single out and kill readily identifiable people. Rather, they result in usually small increases in the *probability* of death to exposed *groups* of individuals. So what is really at issue is the economic value of reductions in the risk of death. This is a manageable question and one on which we have some evidence.

People in their daily lives make a variety of choices that involve trading off changes in the risk of death with other economic goods whose values we can measure in money terms. For example, some people travel to work in cars rather than by bus or by walking because of the increased convenience and lower travel time, even though they increase the risk of dying prematurely. Also, some people accept jobs with known higher risks of accidental death because those jobs pay higher wages. The "value" of saving a life can be calculated from information on individuals' trade-offs between risk and money.

Suppose there were a thousand people each of whom has a probability of .004 of dying during this next year. Suppose an environmental change would reduce that probability to .003, a change of .001. Let us ask each individual to state his or her maximum willingness to pay for that reduction in risk. Suppose for simplicity that each person states the same willingness to pay, $100. The total willingness to pay of the group is $100,000. If the policy is adopted, there will on average be one less death during this next year, (.001 × 1000). The total willingness to pay for a change that results in one fewer deaths is $100,000. This is the "value of life" that is revealed from individual preferences. Efforts to estimate the value of life from data on wage premiums for risky jobs have led to values in the range of $500,000 to $5 million.

If an economic approach is to be used in setting standards for toxic chemicals, hazardous air pollutants, and so forth, then some measure of the value of reductions in risk must be the basis for computing the benefits of pollution control. There are immense practical difficulties in providing accurate, refined estimates of this value. But these are not my concern here. Rather I am concerned with the ethical issues of even attempting to employ this approach to environmental decision making.

I think that the principal ethical issue here is compensation. Suppose that a standard has been set for an air pollutant such that even with the standard being met the population has a higher probability of death than if the pollutant were fully controlled. The standard was presumably set at this level because the cost of eliminating the remaining risk exceeded the individuals' willingness to pay to eliminate the risk. Many people would argue that the risk should be reduced to zero regardless of cost. After all, some people are being placed at risk while others are benefiting by avoiding the cost of controlling pollution. But suppose the population is compensated for bearing this risk with money from, for example, a charge on the polluting substance. Is there then any reason to argue for reducing pollution to zero? If the pollution were reduced to zero and the compensation withdrawn, the people at risk would be no better off in their own eyes than they are with the pollution and compensation. But some people would be made worse off because of the additional costs of eliminating the pollution.

Future Generations

Some environmental decisions impose risks on future generations in order to achieve present benefits. In standard benefit-cost analysis based on the economic efficiency criterion, a social rate of discount is used to weight benefits and costs occurring at different points in time. There have been long debates about the appropriateness of applying a discount rate to effects on future generations. It is argued that ethically unacceptable damage imposed on future generations may be made to appear acceptably small, from today's perspective, by discounting.

Consider the case where this generation wishes to do something which will yield benefits today worth $B. This act will also set in motion some physical process which will cause $D of damages 100,000 years from now. Assume that the events are certain and that the values of benefits and damages based on individual preferences can be accurately measured.

In brief, the argument against discounting is: at any reasonable (nonzero) discount rate, r, the present value of damages

$$\$P = \frac{\$D}{(1+r)100,000}$$

will be trivial and almost certainly will be outweighed by present benefits. The implication of discounting is that we care virtually nothing about the damages that we inflict on future generations provided that they are postponed sufficiently far into the future. Therefore, the argument goes, we should discard the discounting procedure. Instead, since the real issue is intergenerational equity, a zero discount rate should be used. This would represent the most appropriate value judgment about the relative weights to be attached to the consumption of present and future generations.

I believe this argument is confused. Certainly, the problem is equity; but that has nothing to do with discounting. Rather, the equity question revolves around the distinction between actual and potential compensation.

In order to separate the compensation and discounting issues, consider a project for which both benefits and costs are realized today. Whenever benefits are greater than costs, the efficiency criterion says that the project should be undertaken, even if the benefits and costs accrue to different groups. This is because there is at least the *possibility* of compensation. Whether compensation should be paid or not is a value judgment hinging on equity considerations.

Now consider the intergenerational case. If $B is greater than $P (the discounted present value of future damages), the project is worthwhile and should be undertaken if the objective is economic efficiency. If the trivial sum of $P is set aside now at interest, it will grow to

$$(1+r)100,000_{\$P}$$

which of course is the same as $D and therefore by definition will just compensate the future generation for the damages our actions will have imposed on them. If actual compensation is provided for, no one, present or future, will be made worse off, and some will benefit.

Some may wish to adhere to the principle that compensation should *always* be paid. The principle would apply to losers in the present as well as future generations. The discount rate would help them to calculate the amount to be set aside for future payment. Others may wish to say that whether compensation should be paid or not depends on the relative positions of potential gainers and losers. Finally some will choose to ignore the compensation question entirely. But no matter how they resolve the compensation question, they should discount future damages.

Ecological Effects

Suppose that an accidental spill of a toxic chemical or crude oil wipes out the population of some marine organism in a certain area. What is the economic value of this damage? If the organism is a fish that is sought by sports or commercial fishermen, then there are standard economic techniques for determining the willingness to pay for or value of fish in the water. If the organism is part of the food chain which supports a commercially valuable fishery, then it is also possible, at least conceptually, to establish the biological link between the organism and the economic system. The value of the organism is based on its contribution to maintaining the stock of the commercially valued fish. But if there is no link between the organism and human production or consumption activity, there is no basis for establishing an economic value. Those species that lie completely outside of the economic system also are beyond the reach of the economic rubric for establishing value.

Some people have suggested alternative bases for establishing values, for example, cost of replacing the organisms, or cost of replacing biological functions such as photosynthesis and nitrogen fixation. But if those functions have no economic value to man, for example, because there are substitute organisms to perform them, then we would not be willing to pay the full cost of replacement. And this signifies that the economic value is less, perhaps much less, than replacement cost.

Rather than introduce some arbitrary or biased method for imputing a value to such organisms, I prefer to be honest about the limitations of the economic approach to determining values. This means that we should acknowledge that certain ecological effects are not commensurable with economic effects measured in dollars. Where trade-offs between noncommensurable magnitudes are involved, choices must be made through the political system.

VI. CONCLUSIONS

The argument for the adoption of the economists' point of view concerning environmental policy can be summarized as follows. Given the premises about individual preferences and the value judgment that satisfying these preferences should be the objective of policy, the adoption of the economists' recommendations concerning environmental policy will always lead to a potential Pareto improvement, that is, it will always be possible through taxes and compensating payments to make sure that at least some people are better off and that no one loses. Society could choose not to make these compensating payments; but this choice should be on the basis of some ethical judgment concerning the deservingness of the gainers and losers from the policy.

It might be helpful at this point to review and summarize these premises and value judgments so that they might be in the focus of discussion:

1. Should individual preferences matter? If not individual preferences, then whose preferences should matter? What about ecological effects that have no perceptible effect on human welfare, that is, that lie outside of the set of things over which individuals have preferences?

2. Does the substitution principle hold for environmental services? Or are individuals' preferences for environmental goods lexicographic? This is an empirical question. Economists have developed a substantial body of evidence that people are willing to make trade-offs between environmental goods such as recreation, visual amenities, and healthful air and other economic goods.

3. Are preferences characterized by unlimited wants? This is also an empirical question. But I think that most economists would agree that if there are such limits, we have not begun to approach them for the vast bulk of the citizens of this world. A related question is whether it should be the objective of economic activity to satisfy wants without limits? But this question is more closely related to question (1) concerning the role of individual preferences.

4. Is achieving an efficient allocation of resources that important? Or, as Kelman (1981) has argued, should we be willing to accept less economic efficiency in order to preserve the idea that environmental values are in some sense superior to economic values? An affirmative answer to the latter question implies a lexicographic preference system and a rejection of the substitution principle for environmental goods.

5. Should compensation always be paid? Paid sometimes? Never? This is an ethical question. But as I have indicated, I think it plays a central role in judging the ethical implications of economists' environmental policy recommendations. Not only is there the question of whether compensation should be paid, but also the question of who should be compensated. For example, should compensation be paid to those who are damaged by the optimal level of pollution? Or should compensation be paid to those who lose because of the imposition of pollution control requirements?

References

Freeman, A. Myrick, III. "Equity, Efficiency, and Discounting: The Reasons for Discounting Intergenerational Effects," *Futures* (October, 1977), 375–376.

Kelman, Steven. "Economists and the Environmental Muddle," *The Public Interest* 641 (Summer, 1981), 106–123.

Peacock, Alan T., and Charles K. Rowley. *Welfare Economics: A Liberal Restatement*, London, M. Robertson, 1975.

For Further Discussion

1. Some people claim that using economic analysis to help set air and water quality standards is ethically wrong because it puts a price on human life and health. Why does Freeman believe that this view rests on a misunderstanding of economics? What does he mean by the "economic value of reductions in the risk of death"?

2. How does Freeman define the economic problem of the environment? Is this how you understand environmental problems?

3. According to Freeman, a "basic premise" of economics is that "each individual is the best judge of how well off he or she is in a given situation." What, precisely, does he mean by this? Do you agree?

4. What does Freeman mean by "Pareto Optimality"?

Free-Market Versus Libertarian Environmentalism

Mark Sagoff

Philosopher Mark Sagoff offers a detailed assessment of environmental economics in this selection. He believes that the utilitarian logic that underlies economic thinking is unable to provide a strong foundation for environmentalism. This selection, originally published as a review of the book Free Market Environmentalism *by Terry Anderson and Donald Leal, analyzes the ability of free market economics to attain environmental goals. Sagoff distinguishes libertarian markets—a deontological defense of markets that emphasizes free choice, property rights, and treating individuals as ends—from "free market" or utilitarian approaches. Anderson and Leal call for a utilitarian approach, relying on free markets to most efficiently attain the optimal satisfaction of individual preferences. Sagoff believes that there can be some convergence between the interests of libertarians and environmentalists, but he doubts that free market economics can provide a sound basis for environmental policy.*

Within utilitarian approaches, some (Anderson and Leal) argue that the utilitarian goal is best attained through the working of a free market. Others, those whom Sagoff identifies as "resource economists," argue that this goal is best attained by reliance on experts from economics, science, and government applying the tools of cost–benefit analysis. Sagoff rejects both types of utilitarian economics, largely because of their systematic confusion about the ethical nature of subjective preferences. Either approach is likely to result in widespread environmental destruction for the sake of economic efficiency.

Terry Anderson and Donald Leal of the Political Economy Research Center begin their book *Free Market Environmentalism* (San Francisco: Pacific Research Institute, 1991) by admitting that for many people, "the very notion of free market environmentalism is an oxymoron" (1). And one can see why. Knowing that markets transfer resources to those who pay the most for them, environmentalists worry that the highest bidder too often coincides with the lowest denominator. What free markets do best—create and satisfy consumer demand—dismays environmentalists. They

do not regard efficiency as an unmixed blessing. The more efficient markets become, the more quickly they can turn America into a shopping mall and nature into a theme park.

Abandoned strip mines, derelict factories, eroded farms, and clear-cut forests attest as surely as suburban malls and tract developments to the power of the almighty dollar to replicate itself at the expense of the natural environment. Nobody, it seems, ever went broke turning dells into delis, arcadias into arcades, or, in the words of the popular song, paradises into parking lots. Environmentalists know that effluents, emissions, and eyesores, even if efficient from an economic point of view, can still outrage us from a moral, aesthetic, cultural, and political point of view. That is why environmentalists look to politics to keep markets, however efficient, from replacing our natural birthright with bowls of consumer porridge.

THE CULT OF MICROECONOMIC EFFICIENCY

"The view that markets and the environment do not mix," Anderson and Leal observe, "is buttressed by the conception that resource exploitation and environmental degradation are inextricably linked to economic growth" (1). Environmentalists do not oppose economic development, however, if one understands it in macroeconomic terms, that is, employment, higher wages, lower inflation, in short, prosperity. Environmentalists recognize that poverty often forces people to pillage resources that prosperity permits them to protect. To the extent that free markets promote prosperity, environmentalists are likely to agree with Anderson and Leal that "market processes can be compatible with good resource stewardship and environmental quality" (6).

The proposals found in *Free Market Environmentalism,* however, have nothing to do with prosperity, employment, inflation, or any other concern associated with macroeconomics. They are restricted to the entirely separate set of issues one encounters in texts on microeconomics, particularly resource economics. These texts define the goal of resource policy in relation to microeconomic efficiency, not macroeconomic performance. Efficiency is the reason Anderson and Leal favor free markets over political choice. "In the private sector, efficiency matters"; in the political sector, "the bottom line depends on the electoral process where votes matter, not efficiency" (16).

This is true. In the political process, at least in theory, representatives debate policies on their merits and, after deliberation, determine where the public interest lies. Democracy at its best invites people to take the point of view of the community, to ponder what best expresses its values, not just to pursue their prior interests as individuals. For example, people might vote to preserve a wilderness because they believe that course is correct for the nation, although they would pay to visit the place as individuals only if it were made into a ski resort. The political result, then, depends on what people agree after deliberation is best for the community, not what is efficient in satisfying their preexisting consumer demands.

Anderson and Leal observe, again correctly, that market exchange is much more likely than political deliberation to satisfy "the subjective values of citizens" (16). This is unsurprising. Theorists since Aristotle tell us that political life aims to define the goals of the community—to elevate or educate preferences, not simply to satisfy them. Political deliberation and economic aggregation, philosophers such as Rousseau remind us, seek different ends and rest on different conceptions of public choice.

The founders of American democracy warned that republican government should not mire itself in making tradeoffs between special interests or "factions," as Madison called them. Instead, the democratic process must be seized with moral, aesthetic, and cultural questions, whether the issue is peace and war, race, gender, religion, family values, or social and environmental policy. The view opposed to Madison's—that the principal function of politics is to satisfy the preexisting preferences of individuals, however arbitrary or ill-considered—is a position for which there is no plausible argument. It is also the political theory Anderson and Leal assume is correct.

Anderson and Leal take it for granted that the main goal of environmental policy must be economic efficiency, which is to say, the satisfaction of subjective preferences, weighted by willingness to pay, to the extent that resources will allow. This is where environmentalists are likely to part company with them. Environmentalists do not assume in advance, as Anderson and Leal do, that economic efficiency is the purpose of environmental policy. Rather, they would give other concepts and values a hearing within constitutionally protected political processes bounded by personal rights.

Environmentalists, as we shall see, understand the value of liberty. They may well defend the sanctity of property rights. And they recognize the importance of economic prosperity. They may nevertheless see little virtue in microeconomic efficiency. Environmentalists may deny, indeed, that efficiency connects—as Anderson and Leal assume it does—with the goals everyone agrees are important, such as liberty, prosperity, and the protection of personal and property rights.

Environmentalists, indeed, are likely to ask why allocative efficiency is a good thing—why it is a goal worth pursuing. It is no answer that efficiency promotes economic growth or prosperity, since economists have never shown that microeconomic efficiency bears any clear relationship, conceptual or empirical, to macroeconomic performance.[1] Anderson and Leal apparently believe that efficiency promotes prosperity. As we shall see presently, however, no basis, empirical or theoretical, exists for this belief.

One cannot answer that allocative efficiency is a good thing because it promotes social welfare, since economists use "social welfare" as just a stand-in or proxy for "allocative efficiency," both of which are defined in terms of the satisfaction of preferences weighted by willingness to pay. Common wisdom ("money does not buy happiness") supports the conclusion of social science research that once basic needs are met, preference satisfaction, economic efficiency, and "social welfare" bear no significant relation to happiness or well-being.[2] Anderson and Leal themselves ridicule the collectivist notion of social welfare—as indeed they should—when they castigate natural resource economics for positing the importance of "a 'socially efficient' allocation of resources" (9).

Environmentalists, then, may see little reason to favor the single goal Anderson and Leal propose, namely, efficiency in the allocation of resources. They may, indeed, reject this goal. To understand why, consider an example. Lange's Metalmark, a beautiful and endangered butterfly, inhabits sand dunes near Los Angeles for the use of which developers are willing to pay more than $100,000 per acre. Keeping the land from development would not be efficient from a microeconomic point of view, since developers would easily outbid environmentalists. Environmentalists are likely to argue, however, that preserving the butterfly is the right thing morally, legally, and politically— even if it is not economically efficient.

Environmentalists, in short, may cast environmental problems in moral, aesthetic, cultural, political, and even religious terms. The Anderson–Leal approach, which envisions the environment strictly in microeconomic terms, may appear to environmentalists, then, as beside the point. Environmentalists, in other words, may see environmental degradation as an aesthetic, cultural, political, and moral failure but not necessarily as a market failure. Accordingly, environmentalists will not agree with Anderson and Leal that the way to solve environmental problems is to allow markets to allocate resources to the highest bidders.

FREE-MARKET VS. COST-BENEFIT ECONOMICS

Anderson and Leal begin their book by soundly attacking the cost-benefit approach of mainstream resource economics, which they rightly portray as a kind of collectivism. They summarize their criticism as follows:

> To counter market failures, centralized planning is seen as a way of aggregating information about social costs and social benefits in order to maximize the value of natural resources. Decisions based on this aggregated information are to be made by disinterested resource managers whose goal is to maximize social welfare. (9)

The cost-benefit approach, Anderson and Leal point out, "has been premised on the assumption that markets are responsible for resource misallocation and environmental degradation" (9). Actually, the approach Anderson and Leal reject is premised on two assumptions. First, mainstream resource economists define "social welfare" in terms of "preference satisfaction," and measure "preference" in terms of willingness to pay. Then they argue tautologously that those who are willing to pay the most for resources should own them, since that allocation maximizes social welfare. Anderson and Leal accept the idea that resources should go to the highest bidders apparently because they, too, define "social welfare" to coincide with allocative efficiency. They show no more interest than do the welfare economists they criticize in the

question of whether "social welfare" as they define it has anything to do with human well-being as ordinarily understood.

The second assumption—the one that Anderson and Leal challenge—is this. According to the cost-benefit theory, free markets, because of various familiar kinds of problems, fail systematically and pervasively to allocate resources to the highest would-be bidders. Mainstream resource economists then call upon the government to "correct" these market failures by setting prices in ways that will reflect the "true" worth of those resources. Mainstream resource economists conclude, then, that expert managers (presumably themselves) rather than free markets ought ultimately to be in charge of allocating environmental goods and services.

Anderson and Leal do not reject—as environmentalists might—the first premise of resource economics, which asserts the "ideal" of allocative efficiency. Anderson and Leal accept the postulate that there are myriad exogenous or pre-existing preferences "out there" and that the goal is maximally to satisfy them. Having swallowed this camel, Anderson and Leal strain at a gnat—they deny that "scientific management" and "centralized planning" will succeed better than markets in allocating resources in ways that satisfy those subjective pre-existing preferences.

The issue between Anderson–Leal and the cost-benefit position, then, comes down to this. Both hypostatize in the minds of individuals a set of preferences that exist exogenously from—and therefore are not simply artifacts of—the methodology or mechanism used to identify and to satisfy them. Anderson and Leal assert and cost-benefit theorists deny that even in instances of putative market failure—for example, when transaction costs are high—free markets will succeed better than will centralized scientific management in satisfying those exogenous preferences on a willingness-to-pay basis.

Discovering which position is correct should be no problem. One need only determine independently what these exogenous preferences are and then see whether the market or the government does the better job of meeting them. The difficulty, though, is that no method exists to get behind or away from market and cost-benefit methodologies to identify or describe the preferences against which both the accuracy and the efficacy of these mechanisms or methodologies must be tested.

Preferences, unlike cabbages, are theoretical constructs, not observable things. That is why they are so hard to measure. Subjective values or preferences exist, if anywhere, inside the minds of the people who have them; they are not part of the public world. They are hypothetical mental entities—dispositions to act in certain ways—that economists posit to explain and, presumably, predict behavior. In the old days, social scientists used to posit demons that made you act as you do; they recommended ritual exorcisms to get rid of them. Nowadays, they posit preferences, and like true scientists, obtain grants to develop methodologies to measure and to satisfy them.

Attempts to measure preferences and to predict behavior on the basis of them, however, have succeeded only in developing empirical and conceptual anomalies and difficulties, such as preference reversal and inconsistency, and still more methodologies to try to resolve *them*. Preferences do not seem to be consistent, or enduring, or to be independent of unrelated alternatives, or to have any of the qualities we would think they must have if they are to lead to behavior that is "rational." That is why dozens of major academic disciplines and industries have arisen to try to figure out how to figure out preferences. We can only hope that some day preferences, like demons, will be tossed on the dung heap of discarded-because-useless hypostatized theoretical entities. That might be real scientific progress.

Until that time comes, we can say that no independent "objective" method exists to identify, measure, or assess preferences or subjective values. The technique you use to search will largely determine which preferences you find. If you go with markets, you will get one set of preferences at any time; if you go with contingent valuation methodology, you will get another; voting behavior will get you a third; and other instruments will get other rankings or orderings. It is silly to think that preferences really exist independently of the theories that posit and techniques that identify them. It is silly to suppose that we can get "behind" these theories and techniques, moreover, to determine which of them gets preferences "right." Any theory that hypostatizes the existence of preferences—or demons or whatever—measures them correctly for the purposes of that theory. When theoretical approaches differ, they may be equally correct or incorrect, for no logical way exists to adjudicate between them.

Anderson and Leal analogize the cost-benefit or "marginalist" approach in resource economics to communism. Marginalism and Marxism, indeed, coincide in a fundamental respect. Each appeals to *hypothetical* rather than to *actual* choice. Both would set aside the results of legitimate political and economic institutions, then, in order to achieve outcomes that more

nearly represent what people would choose under ideal albeit hypothetical conditions.

The cost-benefit theorist argues that because of transaction costs and other difficulties, markets systematically fail to allocate resources to those who are willing to pay the most for them. Market failure, then, prevents the outcomes people choose from coinciding with those they prefer—in other words, those they would choose in the absence of problems like transaction costs. In order to determine what people would choose—what they prefer—economists are developing methodologies based on surveys and other scientific instruments. Empowered by these methodologies, resource economists advise public officials to "correct" the results of legitimate but imperfect economic and political processes by replacing them with outcomes that more nearly reflect what people prefer and therefore would choose in ideal conditions.[3] In this way, cost-benefit theorists insist on *preference* as distinct from and in opposition to *choice*.

Marxism likewise bases allocation on hypothetical, not actual, choice. For the Marxist, what is important is not the choices people *do* make but the choices people *would* make if they were free of their corrupt bourgeois ideology. The marginalist and the Marxist, then, both second-guess the outcomes of legitimate market and political institutions; both attempt to see through actual behavior (which is only "appearance") to measure the "true" subjective values that lurk beneath the phenomena. And both therefore argue that society should depend on a vanguard of experts, namely, themselves, to allocate resources according to their scientific methodology, which alone represents the correct view of social well-being and progress.

Environmentalists will enjoy the way Anderson and Leal present cost-benefit resource economics as a kind of centralized planning or collectivism. Environmentalists will find this aspect of their book appealing, at least on the principle that the enemy of thine enemy is thy friend.

LIBERTARIANS VS. FREE MARKETEERS

Environmentalists will reject the mainstream approach in resource economics, however, for reasons quite different from those that persuade Anderson and Leal. Environmentalists are concerned about saving magnificent landscapes and species, keeping the air and water clean, and in general getting humanity to tread more lightly on the earth. They are not concerned, as are Anderson and Leal, about satisfying preferences on a willing-to-pay basis or about maximizing a conception of social welfare defined as a function of the satisfaction of those preferences.

Accordingly, environmentalists will reject the cost-benefit approach because it takes allocative efficiency as its goal. Anderson and Leal, in contrast, accept the goal, but argue that cost-benefit management is less likely than free markets to achieve it. Why do Anderson and Leal believe that allocative efficiency is a goal worth achieving? Why do they believe it is compatible with the goals espoused by environmentalists? Indeed, why, do they associate efficient resource allocation with the functioning of free markets in which property rights are strictly enforced and well defined?

Consider, first, the concept of property rights. "At the heart of free market environmentalism," Anderson and Leal assert, "is a system of well-specified property rights" (3). They explain: "Free market environmentalism depends on a voluntary exchange of property rights between consenting owners and promotes cooperation and compromise" (8).

To see what Anderson and Leal mean by "cooperation" and "compromise," imagine a situation in which an oysterman owns or has leased and thus has the exclusive right to occupy the oyster beds he plants with oysters. Now suppose his neighbor's sewage pollutes and renders the oyster beds useless. The oysterman may have two different legal remedies. First, he might have the legal power to enjoin the nuisance, that is, to stop his neighbor from dumping effluent on the oysterbeds. The neighbor would then either meet the oysterman's price or cease polluting his property. This would seem to ensure compromise and would also protect the environment.

A second kind of "remedy" appeals to resource economists and others who would not let property rights stand in the way of an efficient allocation of resources. On this second approach, the oysterman could not sue for injunctive relief but only for compensation or damages in the amount of his lost income. Thus the polluter would not have to pay the price the oysterman demands. Rather the polluter could take whatever property rights he likes as long as he is willing to pay the price an agency of the state— usually a court—says those rights are "objectively" worth.

Libertarians know how perniciously people behave when they can take what they want, bound only by the need to pay compensation. During the early nine-teenth century, for example, owners of whaling and merchant vessels used to shanghai passersby to crew their ships. The owners did not obtain the consent of those they impressed—usually for about three years— but they did pay these people the going rate for their time as determined by labor markets. Perhaps the ship owners might have used fewer workers or raised wages to get more volunteers—but why should they have bothered? It is easier to shanghai workers and then compensate them at rates their labor is "objectively" worth.

Shall we similarly allow polluters to shanghai persons and their property as catchment areas for their emissions, provided they can compensate their victims at rates set for risks in labor or other markets? What difference in principle could possibly exist between the practice of captains of whaleboats and that of captains of industry if both are free to use the bodies of people without their consent, provided they pay compensation at rates some expert says the property, labor, or risk is worth?

The first question one must ask of free-market environmentalism, then, is who is free? Are polluters free to pollute or are individuals free to enjoin pollution? Anyone who takes liberty and property seriously must defend the right of injunctive relief in nuisance cases. Anything less simply gives polluters the power of eminent domain over any person or property they wish to violate or invade.

Anderson and Leal assert that "when waste intrudes into another's physical environment without his consent . . . that garbage becomes pollution" (135–6). One might infer that these authors require, then, that polluters gain the consent of those upon whose persons and property they would trespass. The result of such bargaining would be very favorable to environmentalists—many of whom, for ideological and other reasons, might refuse to tolerate any more than *de minimis* levels of pollution.

Anderson and Leal, however, abandon libertarian principles at this early and crucial point. They deny victims of pollution the power to obtain injunctive relief; they allow them only the right to receive compensation, in an amount determined on some "objective" basis, for the costs they incur as a result of the pollution or garbage foisted upon them. "Compensation gives the person who generates the garbage an incentive to compare the benefits of garbage production with the costs of disposal to arrive at an 'optimal' level of both" (135). In other words, Anderson and Leal put a cost-benefit test in place of injunctive relief and thus destroy the meaning and substance of property rights.

Environmentalists may view pollution more as an evil than as a cost; they may criticize effluents in ethical more than in economic terms. Yet unlike other moral wrongs—rape, murder, and theft, for example—pollution, to some extent, can be a necessary evil, since eliminating all of it would close the economy down. At some point, then, the protection of property rights becomes too onerous, where "too onerous" does not mean inefficient in economic terms but supererogatory from a moral point of view.

Environmentalists, therefore, argue for laws that limit pollution to the lowest feasible levels. These laws have an ethical, not an economic, rationale. They do not seek optimal levels of pollution; *they seek the protection of property rights.* Anderson and Leal, in spite of their putative libertarianism, apparently have no respect for the right of property owners to be free of trespass. They would give polluters power routinely to make offers their victims cannot refuse.

Environmentalists who wish to protect property rights may call for technology-forcing programs to minimize the pollutants that the economy cannot feasibly eliminate. The only alternative to minimizing this kind of pollution would be to enjoin it—if we care about protecting property rights. Thus, libertarian environmentalists must tell us which they prefer: public laws that ratchet pollution down to technologically achievable minimums, or injunctions issued against polluters so routinely that the economy may not function. The Anderson–Leal recommendation—forced sale of property rights to polluters at prices determined by the government—makes a mockery of libertarian principles—and may not protect the environment.

RESOURCE STEWARDSHIP AND ENVIRONMENTAL QUALITY

Free-market processes, Anderson and Leal write, "can be compatible with good resource stewardship and environmental quality" (6). To find out what they may mean by "good resource stewardship and environmental quality," consider an example they give. In the area of the Great Lakes, as Anderson and Leal observe,

the Kingston Plains has never recovered from logging done a hundred years ago. Efforts have been made to replant the area, but the soil is too infer-

tile and sandy. It took hundreds of years for the forest to grow, and it will take hundreds more for the area to recover. (47)

Yet Anderson and Leal note that even if clear-cutting the forest produced a nasty moonscape for centuries, this does not mean it was a bad decision. On the contrary, trees cut from this area fetched about $20 per acre which, "had it been invested in bonds or some other form of savings at the time . . . would now be worth approximately . . . $2.8 billion." Anderson and Leal plausibly argue that because "the land in this area is not worth anything close to this, we must infer that harvesting trees was the correct choice" (47). It was correct in the sense of "efficient"—which appears to be the only objective value Anderson and Leal recognize. That a policy could be good or bad in any other sense, for example, morally, aesthetically, culturally, legally, or politically, would be a surprise to them.

Environmentalists generally define "stewardship" and "environmental quality" in moral and aesthetic rather than economic terms. They dream of a land where wildlife abounds, waters flow free of pollution, wildlands and habitat are preserved, extinctions are rare, and people live in landscapes that retain the particular characteristics of their history and location. Environmentalists in the United States are proud of the political decisions their nation has made to protect its natural heritage—decisions that go to the morality of the community, not the "preferences" of individuals. When citizens and their representatives deliberate in moral and aesthetic terms about the fate of the environment, they often reach conclusions utterly at odds with those they would reach if they considered only the economic aspects—the investment potential—of things. That is why we have the Metalmark in place of another shopping mall.

We are now in a position to understand what Anderson and Leal mean by "environmental quality" and "good resource stewardship." They mean exactly what their cost-benefit cousins intend by these terms. They mean the kind of stewardship that permits any amount of strip mining, clear cutting, erosion, or degradation that produces a profit that, if invested, would generate more income than would practices of conservation. Thus, even the devastation of the forests near the Great Lakes, however it may affront environmentalist values, is compatible with good resource stewardship. "When judged against prudent investment criteria, nineteenth-century lumber markets were efficient" (47).

Environmentalists might agree that these markets were efficient. They might concede that the extinction of species, the mining of sanctuaries, the pollution of bays and estuaries and, indeed, the desecration of virtually everything of moral, aesthetic, or cultural significance all make perfect sense from the point of view of economic efficiency—the point of view Anderson and Leal absolutely take for granted and as given. Environmentalists contend, however, that the protection of what is beautiful, authentic, charming, historic, and inspiring is right and good *whether or not* it is economically efficient.

OBJECTIVE VALUES VS. SUBJECTIVE PREFERENCES

To this assertion Anderson and Leal reply: "None of these values is right or wrong; each simply represents a *special interest*" (82). How do Anderson and Leal know that none of the values environmentalists espouse is right or wrong? On what moral theory do they base this assertion? And if this theory is sound, then what shall we say about the value they endorse, that is, economic efficiency? This seems to be a policy preference—a special interest—of resource economists and two free-market economists. How much are Anderson and Leal willing to pay for resources to be allocated efficiently? Why do they believe that their preference for allocative efficiency represents an objective or intrinsic moral good while the values environmentalists hold—love of beauty, nature, purity, whatever—are just special interests? Anderson and Leal never explain why resources should go to the highest bidders, nor why this policy principle, rather than those environmentalists propose, should govern the allocation of resources. Why does the principle of allocative efficiency—or the concept of "social welfare" which is its logical equivalent—have the character of an objective normative truth, whereas every other policy prescription merely states a subjective interest or preference?

It is important not to confuse this question with another, namely, whether people should be free to make their own choices under rules fair and congenial to all. Environmentalists are likely to endorse *that* principle. They will point out that through the democratic political process of the United States, the nation has freely chosen to protect endangered species even if it is inefficient to do so. People consistently favor—and their political representatives consistently choose—environmental policies that make no sense whatever, as Anderson and Leal know, in terms of economic efficiency. Indeed, environmental statutes often explicitly rule out a cost-benefit test. Perhaps Anderson and Leal reject the results of democratic political processes, condemn the constitutional form of government, and deny that many basic political rights and freedoms are anything more than subjective preferences. But they can hardly do so in the name of freedom of choice.

It would be one thing if Anderson and Leal attempted to say why efficiency is a good thing—why it is better, all else being equal, that resources go to the highest bidder. But they do not even bother; they assume it is so; and they tell the rest of us, environmentalists especially, that other views about resource policy merely express special interests. That is the reason there is so little argument in *Free Market Environmentalism*. Its authors assume that everyone agrees allocative efficiency is the objective of resource policy, so the only question is whether markets or bureaucracies are better at achieving it. This speaks volumes about the present state of resource economics.

THE FALLACY OF DISPARATE COMPARISON

Many of the most convincing passages of *Free Market Environmentalism* show how governmental ineptitude, corruption, and well-meaning stupidity have destroyed environmental values, particularly regarding public land. The libertarian critique of below-market timber sales, subsidized water to farmers, and subsidies for nuclear power plants is a brilliant and telling one, and environmentalists cannot be too grateful for it. Likewise, free-market economists point to hopelessly complex and arcane regulations, like those associated with Superfund, that can be charitably described as full-employment acts for lawyers. These economists make their case convincingly when they tell stories of regulatory rigor mortis, bureaucratic bottleneck, and governmental gridlock. Environmental regulations, as they show in example after example, often fail even to make environmental sense.

No one can question the accuracy of these war stories; they are too true; but one may be skeptical of the conclusion free-market economists draw from them. They conclude that a free market, in which property rights are well defined and enforced, will be kinder and gentler to the environment than even a well-intentioned government actually is. In political decision making, as these war stories suggest, interest groups tend in fact to oppose rather than to accommodate each other. With private control of resources, in contrast, "each side tends to gain by satisfying the other's desires. Cooperation replaces political conflict as both sides prospect for harmony" (84).

A free market with inviolable property rights, low transaction costs, and so on, may, indeed, treat nature better than does an often bumbling and occasionally corrupt bureaucracy beset by special interests. However, this kind of argument, which is standard in the literature, commits the fallacy of disparate comparison. It compares what the perfect market would do in theory with what imperfect governmental agencies, at their worst, have done in fact.

It is easy to turn the tables on this kind of argument—to tell war stories about the malfunction of free markets and then compare these stories to the theory of Madisonian republicanism. Recently, it has been reported that a natural resource company published false statistics about the value of its holdings to run up the price of its stock. Many other examples of fraud, duplicity, rapacity, greed, and indifference to the well-being of others grace the annals of business. Compare them to the outcome of perfect Madisonian deliberation in which public officials put aside their individual interests—as legislators did when they debated the Gulf War—to ponder and then converge upon the public interest. You can defend virtually any method of decision making against its opposite by contrasting the theory of one against the practice of the other.

Thus, Anderson and Leal, after explaining how in theory a perfectly competitive market will function efficiently, observe that "the political sector operates by externalizing costs" (14). In a perfectly functioning political sector, however, officials, guided by civic virtue and brilliant economic and social analyses, would get everything right. Meanwhile, examples abound in which it is industry that operates by externalizing costs—say, by polluting the atmosphere. Externalized costs, fraud, stupidity, chicanery, special pleading, and so on, pose problems for both private- and public-sector approaches to public policy. An ideal market, moreover, presupposes an ideal government to perfectly define and defend property rights.

Free-market economists often respond, in effect, by arguing that perverse incentives operate in the public sector to polarize positions and widen conflicts. Yet perverse incentives also operate in the private sector.

Some of these are well known. For example, managers have a perverse incentive to squander resources for quick profits in order to get reputations as money-makers and thus attract a higher salary at a competing firm. And as long as quick-profit schemes exist, they present an incentive to liquidate natural resources.

The most general perverse incentive that vitiates the free market may be illustrated by an example presented by Arthur Pigou. Suppose a firm wishes to build a factory in a quiet residential area. The residents would have to grin and bear the ugly face of the factory or pay the owners not to build in their neighborhood, since nuisance law does not generally recognize amenity rights. Presumably, a free-market advocate would argue that the residents could express their wishes by compensating the factory owner for the marginal advantage between building in their neighborhood and going to the next best location. To be sure, the residents would have to overcome free-rider costs, but perhaps that is a problem they can somehow solve. Environmentalists, however, see in this situation a much darker difficulty. The developer, having received a bribe not to build in one area, has an incentive to make the same threat in another. He may profit most by never building the factory. The developer can extort payments from neighborhood after neighborhood in which he threatens construction.

One might argue that by zoning property for particular uses the government cuts off perverse market incentives of this kind. Yet libertarians are not fond of zoning in general and, in particular, they follow Richard Epstein in arguing that the government should not "take" property rights through regulation without paying compensation to the owners of those rights.[4] If a property owner has a right to build an unsightly factory on his land—even if the purpose is to extort payments or "rents" from appalled neighbors—then on the libertarian account the government may not intervene unless it is itself prepared to make those payments. The problem this example suggests is that the invisible hand of the market, in theory, becomes an invisible foot in the absence of the power of governments to enact far-reaching environmental regulations.[5]

LIBERTARIAN ENVIRONMENTALISM

In spite of these and other reservations environmentalists will entertain, there is a lot to be said for libertarian environmentalism. The problem is that Anderson and

Leal do not say it. They are libertarian north by northwest; when the wind blows southerly, they revert to cost-benefit economics. To understand the important advantages private property rights offer environmentalism, we have to go to the work of real libertarians, who offer environmentalists a much better deal than Anderson and Leal do. These writers show why libertarian environmentalism is not an oxymoron.

Libertarians other than Anderson and Leal have little patience with the argument that pollution is acceptable as long as it "pays its way"—in other words, as long as its benefits to society as a whole outweigh its costs to individuals. As libertarian philosopher Tibor Machan has pointed out, a free-market approach "requires that pollution be punished as a legal offense that violates individual rights."[6] The libertarian economist Murray Rothbard agrees: "The remedy is simply to enjoin anyone from injecting pollutants into the air, and thereby invading the rights of persons and property."[7] Jane Shaw and Richard Stroup, colleagues of Anderson and Leal at the Political Economy Research Center, add that pollution "is an invasion of person and property just as personal assault is."[8]

Libertarians insist most fundamentally that people are to be treated as ends in themselves and not merely as means to collective ends, especially collective ends as dubious as economic efficiency. Accordingly, libertarians do not tolerate forced sales—for example, they do not want the government to determine the price at which polluters may buy the right to enter your organs and contaminate your precious bodily fluids. If you cannot enjoin polluters from this kind of assault or trespass, property rights mean nothing. Risk-benefit analysis is the *alternative* to the protection of property rights, substituting centralized planning for the rule of law.

The problem with allowing routine injunctive relief in nuisance cases, however, is that individuals who refuse to be bought off could close the economy down. Recent studies using "contingent valuation methodology" suggest that a majority of Americans, indeed, would prefer injunctive relief against polluters to receiving cash compensation in *any* amount.[9] It would be convenient to stigmatize these individuals as "rent-seekers," that is, as engaging in strategic bargaining behavior to maximize the prices they get relative to what others demand. In reality, though, their behavior has nothing to do with rent-seeking. It is ethical and principled choice—free choice of exactly the kind libertarians are in business to honor and protect.

To protect property rights while avoiding the extreme consequence of closing down the economy, the preponderance of our environmental statutes require industry to *minimize* emissions it cannot feasibly *eliminate*. Critics constantly point out that these statutes often fail a cost-benefit test. Such criticisms miss the point of the laws, however, which is not to allocate resources efficiently, to "optimize" levels of pollution, or to maximize social wealth. Rather, these laws, by conceiving pollution as a nuisance or trespass rather than as a cost, strive to protect property rights.

Environmentalists could follow the lead of libertarians, therefore, by refusing to regard pollution as an economic externality and by insisting instead that it be treated as a tort. This way of characterizing pollution—in deontological rather than utilitarian terms—should sit well with environmentalists. And it has a clear moral appeal and rationale—rather than depending on the empty dogmas of welfare economics.

In a series of well-argued articles, moreover, libertarian writers have argued—in the spirit of work by Ronald Coase—"that it does not matter very much who owns a resource when it comes to determining how that resource will be used."[10] They urge, then, that "large chunks of federal land be given in fee simple to established conservation groups."[11] Anderson and Leal's colleagues John Baden and Richard Stroup write: "It would be especially beneficial if areas with both ecological and economic importance were managed by groups with the expertise to weigh the potential damage to the environment against the potential profits. Making environmental groups the owners . . . of the holdings would accomplish this end."[12]

The advantages of this suggestion are obvious. If environmental groups owned the lands they now (often in vain) implore the government to protect, they could make their own bargains with industry. They could bond petroleum companies, for example, to make a "surgical strike" on the oil in the "1001" tract, a grim area of tundra which constitutes a small fraction of the Alaska National Wildlife Refuge. The environmental groups could then invest the considerable proceeds in purchasing rain forest and other ecological wonderlands that may be far more desirable than this tract of tundra for purposes of conservation.

By contrast, environmentalists will find the Anderson–Leal approach problematical insofar as it recommends that the government privatize, that is, sell off, national forests, parks, and other public lands. Environmentalists know they will have far more difficulty than, say, oil companies in concentrating the buying power of all those who are interested in a particular disposition of a wilderness area. Therefore environmentalists are likely to view with alarm the recommendation by Anderson and Leal that "environmental groups could compete to purchase or lease public resources" (93).

Libertarian arguments offer environmentalists many advantages insofar as these arguments are not wedded—as Anderson and Leal's are—to the central dogmas of welfare economics. Not only do libertarian economists recommend that the state donate to environmentalists ecologically sensitive lands, but in putting the emphasis on choice rather than utility, libertarians are willing to back up property rights with injunctive relief, thereby prohibiting the forced sales Anderson and Leal would tolerate. Libertarians also reject the cost-benefit approach because it opposes freedom of choice—which is the right reason to reject it—rather than because public officials, as Anderson and Leal charge, can't shoot straight.

This is by no means to say that there is nothing good in free-market environmentalism apart from libertarianism. Everyone knows that the destruction of nature often results from perverse incentives—for example, the absence of property rights in a commons. Society may more effectively reach environmental goals, therefore, by restructuring market incentives than by directly allocating resources. Incentives for reducing pollution—market-based schemes for trading pollution allowances, for example—may at least in principle succeed far better than "command-and-control" approaches to reducing pollution. As long as they have a say in the preliminary question of the level of pollution to be permitted, most serious environmentalists now endorse free-market schemes like the one included in the acid rain provisions of the Clean Air Act, which allow polluters to trade pollution allowances under a "cap." Free-market advocates who argue on theoretical grounds for market-based incentives for environmental protection have by now convinced virtually everyone: at this point they are pushing against an open door. The problems worth discussing are no longer theoretical but have to do with implementation.[13]

Free-market environmentalists also do a great service by observing that when industry and other special-interest groups "capture" governmental agencies, regulation can be environmentally devastating. For example, by selling timber below cost the Forest Service causes enormous environmental damage a free market might not allow.[14] Environmentalists complained that the Office of Management and Budget during Republican administrations, while pressing free-market ideology against environmental regulation, had no problem with Department of Interior giveaways to industry. A free-market approach does not necessarily coincide with a pro-industry position. On the contrary, it provides a powerful argument against the use of government agencies to promote industry interests.

Environmentalists may object to libertarian ideology for atmospheric rather than for logical reasons: libertarians appear to be radical right-wingers who care more about property rights than about the fate of the earth. Yet what libertarians care about most—the separateness and inviolability of the individual—is not all that different from what environmentalists care about. Libertarians argue that the best way to preserve the environment is to protect property rights. Environmentalists argue that the best way to protect property rights is to preserve the environment. It may be time to examine the convergence—rather than continue to assume the opposition—of these two views.

Notes

1. The scholarly literature concludes that there is little or no relation between macroeconomic prosperity and microeconomic efficiency. For discussion, see David Shepherd, Jeremy Turk, and Aubrey Silberston, eds., *Microeconomic Efficiency and Macroeconomic Performance* (Oxford: Philip Allen, 1983).

2. For discussion, see A. Campbell, P. E. Converse, and W. Rodgers, *The Quality of American Life: Perceptions, Evaluations, Satisfactions* (New York: Russell Sage Foundation, 1976); and Michael Argyle, *The Psychology of Happiness* (London: Methuen & Co., 1987).

3. For an amusing example of this kind of economic deconstruction of the Prevention of Significant Deterioration provisions of the 1977 Clean Air Act Amendments, see Robert D. Rowe, Ralph C. D'Arge, and David Brookshire, "An Experiment on the Economic Value of Visibility," *Journal of Environmental Economics and Management* 7 (1980).

4. Richard Epstein, *Takings: Private Property and the Power of Eminent Domain* (Cambridge, Mass.: Harvard University Press, 1985).

5. The classic statement of this argument is found in E. K. Hunt, "A Radical Critique of Welfare Economics," in Edward J. Nell, ed., *Growth, Profits, and*

Property (Cambridge: Cambridge University Press, 1980), 239–49.

6. Tibor Machan, "Pollution and Political Theory," in Tom Regan, ed., *Earthbound: New Introductory Essays in Environmental Ethics* (New York: Random House, 1984): 74–106; quotation at 97.

7. See John Hospers, "What Libertarianism Is," in Tibor Machan, ed., *The Libertarian Alternative: Essays in Political and Social Philosophy* (Chicago: Nelson Hall, 1974), quoting Rothbard, 15.

8. Jane S. Shaw and Richard L. Stroup, "A Skeptical Twist," *EPA Journal* 18, no. 4 (September/October 1992): 54.

9. See Robert D. Rowe and Lauraine G. Chestnut, *The Value of Visibility: Economic Theory and Applications for Air Pollution Control* (Cambridge, Mass.: Abt Books, 1982), esp. 81.

10. Richard L. Stroup and John A. Baden, *Natural Resources: Bureaucratic Myths and Environmental Management* (San Francisco: Pacific Institute, 1983), 123.

11. Ibid. For a sample of these essays, see John Baden and Richard Stroup, eds., *Bureaucracy vs. Environment: The Environmental Costs of Bureaucratic Governance* (Ann Arbor: University of Michigan Press, 1981).

12. John A. Baden and Richard L. Stroup, "Saving the Wilderness: A Radical Proposal," *Reason* 13 (July 1981): 28–36.

13. For a good brief review of the literature on this question, see Joel A. Mintz, "Economic Reform of Environmental Protection: A Brief Comment on a Recent Debate," *Harvard Environmental Law Review* 15, no. 1

(1991): 149–64 (arguing that economic incentive and market schemes make beautiful sense in theory but often confront intractable technical, scientific, and legal difficulties).

14. For a classic study of the "capture" of air pollution control requirements by an alliance of dirty coal and clean air interests, see Bruce Ackerman and W. T. Hassler, *Clean Coal/Dirty Air, or How the Clean Air Act Became a Multibillion-Dollar Bail-Out for High Sulfur Coal Producers and What Should Be Done about It* (New Haven: Yale University Press, 1981).

For Further Discussion

1. In your own words, what is the ethical difference between libertarian, free market, and resource economics?

2. Sagoff tells us that political theorists since Aristotle have distinguished the social goal "to elevate or educate preferences" from the goal of simply satisfying preferences. What's the difference? What is most appropriate in a democracy?

3. Sagoff distinguishes between economic value and "moral, aesthetic, cultural, political, and even religious" values. Identify examples of each. Are there similarities among these categories? What are the differences?

4. What is the difference between objective values and subjective preferences?

Moving to a Steady-State Economy

Herman E. Daly

This final selection of Chapter 4 introduces the economics of sustainable development. Economist Herman Daly, perhaps the best-known champion of sustainable economics, argues that we need to replace the traditional economic goal of growth with the qualitative term development. The selection that follows, Chapter 1 from Daly's book Beyond Growth, argues that the goal of economics should be to improve the quality of life rather than simply to increase the amount of goods and services produced. In light of increasing population and finite resources, economic growth cannot increase indefinitely. Economic activity must occur at a rate and scale that can be sustained into the indefinite future.

Sustainable development, I argue, necessarily means a radical shift from a growth economy and all it entails to a steady-state economy, certainly in the North, and eventually in the South as well. My first task has to be to elaborate the case for that theoretical and practical shift in worldview. What are the main theoretical and moral anomalies of the growth economy, and how are they resolved by the steady state? And what are the practical failures of the growth economy, viewed as forced first steps toward a steady state?

It is necessary to define what is meant by the terms "steady-state economy" (SSE) and "growth economy." Growth, as here used, refers to an increase in the physical scale of the matter/energy throughput that sustains the economic activities of production and consumption of commodities. In an SSE the aggregate throughput is constant, though its allocation among competing uses is free to vary in response to the market. Since there is of course no production and consumption of matter/energy itself in a physical sense, the throughput is really a process in which low-entropy raw materials are transformed into commodities and then, eventually, into high-entropy wastes. Throughput begins with depletion and ends with pollution. Growth is quantitative increase in the physical scale of throughput. Qualitative improvement in the use made of a given scale of throughput, resulting either from improved technical knowledge or from a deeper understanding of purpose, is called "development." An SSE therefore can develop, but cannot grow, just as the planet earth, of which it is a subsystem, can develop without growing.

The steady state is by no means static. There is continuous renewal by death and birth, depreciation and production, as well as qualitative improvement in the stocks of both people and artifacts. By this definition, strictly speaking, even the stocks of artifacts or people may occasionally grow temporarily as a result of technical progress that increases the durability and repairability (longevity) of artifacts. The same maintenance flow can support a larger stock if the stock becomes longer-lived. The stock may also decrease, however, if resource quality declines at a faster rate than increases in durability-enhancing technology.

The other crucial feature in the definition of an SSE is that the constant level of throughput must be ecologically sustainable for a long future for a population living at a standard or per capita resource use that is sufficient for a good life. Note that an SSE is not defined in terms of gross national product. It is not to be thought of as "zero growth in GNP."

Ecological sustainability of the throughput is not guaranteed by market forces. The market cannot by itself register the cost of its own increasing scale relative to the ecosystem. Market prices measure the scar-

city of individual resources relative to each other. Prices do not measure the absolute scarcity of resources in general, of environmental low entropy. The best we can hope for from a perfect market is a Pareto-optimal allocation of resources (i.e., a situation in which no one can be made better off without making someone else worse off). Such an allocation can be achieved at any scale of resource throughput, including unsustainable scales, just as it can be achieved with any distribution of income, including unjust ones. The latter proposition is well known, the former less so, but equally true. Ecological criteria of sustainability, like ethical criteria of justice, are not served by markets. Markets singlemindedly aim to serve allocative efficiency. Optimal *allocation* is one thing; optimal *scale* is something else.

Economists are always preoccupied with maximizing something: profits, rent, present value, consumers' surplus, and so on. What is maximized in the SSE? Basically the maximand is life, measured in cumulative person-years ever to be lived at a standard of resource use sufficient for a good life. This certainly does not imply maximizing population growth, as advocated by Julian Simon, because too many people simultaneously alive, especially high-consuming people, will be forced to consume ecological "capital" and thereby lower the carrying capacity of the environment and the cumulative total of future lives. Although the maximand is human lives, the SSE would go a long way toward maximizing cumulative life for all species by imposing the constraint of a constant throughput at a sustainable level, thereby halting the growing takeover of habitats of other species, as well as slowing the rate of drawdown of geological capital otherwise available to future generations.

I do not wish to put too fine a point on the notion that the steady state maximizes cumulative life over time for all species, but it certainly would do better in this regard than the present value-maximizing growth economy, which drives to extinction any valuable species whose biological growth rate is less than the expected rate of interest, as long as capture costs are not too high.

Of course many deep issues are raised in this definition of the SSE that, in the interests of brevity, are only touched on here. The meanings of "sufficient for a good life" and "sustainable for a long future" have to be left vague. But any economic system must give implicit answers to these dialectical questions, even when it refuses to face them explicitly. For example, the growth economy implicitly says that there is no such thing as sufficiency because more is always better, and that a twenty-year future is quite long enough if the discount rate is 10%. Many would prefer explicit vagueness to such implicit precision.

MOVING FROM GROWTHMANIA TO THE STEADY STATE IN THOUGHT:
Theoretical and Moral Anomalies of the Growth Paradigm that Are Resolved by the Steady State

The growth economy runs into two kinds of fundamental limits: the biophysical and the ethicosocial. Although they are by no means totally independent, it is worthwhile to distinguish between them.

BIOPHYSICAL LIMITS TO GROWTH

The biophysical limits to growth arise from three interrelated conditions: finitude, entropy, and ecological interdependence. The economy, in its physical dimensions, is an open subsystem of our finite and closed ecosystem, which is both the supplier of its low-entropy raw materials and the recipient of its high-entropy wastes. The growth of the economic subsystem is limited by the fixed size of the host ecosystem, by its dependence on the ecosystem as a source of low-entropy inputs and as a sink for high-entropy wastes, and by the complex ecological connections that are more easily disrupted as the scale of the economic subsystem (the throughput) grows relative to the total ecosystem. Moreover, these three basic limits interact. Finitude would not be so limiting if everything could be recycled, but entropy prevents complete recycling. Entropy would not be so limiting if environmental sources and sinks were infinite, but both are finite. That both are finite, plus the entropy law, means that the ordered structures of the economic subsystem are maintained at the expense of creating a more-than-offsetting amount of disorder in the rest of the system. If it is largely the sun that pays the disorder costs, the entropic costs of throughput, as it is with traditional peasant economies, then we need not worry. But if these entropic costs (depletion and pollution) are mainly inflicted on the terrestrial environment, as in a modern industrial economy, then they interfere with complex ecological life-support services rendered to the economy by nature. The loss of these services should surely be counted as a cost of growth, to be weighed against benefits at the margin. But our national accounts emphatically do not do this.

Standard growth economics ignores finitude, entropy, and ecological interdependence because the concept of throughput is absent from its preanalytic vision, which is that of an isolated circular flow of exchange value . . . , as can be verified by examining the first few chapters of any basic textbook. The physical dimension of commodities and factors is at best totally abstracted from (left out altogether) and at worst assumed to flow in a circle, just like exchange value. It is as if one were to study physiology solely in terms of the circulatory system without ever mentioning the digestive tract. The dependence of the organism on its environment would not be evident. The absence of the concept of throughput in the economists' vision means that the economy carries on no exchange with its environment. It is, by implication, a self-sustaining isolated system, a giant perpetual motion machine. The focus on exchange value in the macroeconomic circular flow also abstracts from use value and any idea of purpose other than maximization of the circular flow of exchange value.

But everyone, including economists, knows perfectly well that the economy takes in raw material from the environment and gives back waste. So why is this undisputed fact ignored in the circular flow paradigm? Economists are interested in scarcity. What is not scarce is abstracted from. Environmental sources and sinks were considered infinite relative to the demands of the economy, which was more or less the case during the formative years of economic theory. Therefore it was not an unreasonable abstraction. But it is highly unreasonable to continue omitting the concept of throughput after the scale of the economy has grown to the point where sources and sinks for the throughput are obviously scarce, even if this new absolute scarcity does not register in relative prices. The current practice of ad hoc introduction of "externalities" to take account of the effects of the growing scale of throughput that do not fit the circular flow model is akin to the use of "epicycles" to explain the departures of astronomical observations from the theoretical circular motion of heavenly bodies.

Nevertheless, many economists hang on to the infinite-resources assumption in one way or another, because otherwise they would have to admit that economic growth faces limits, and that is "unthinkable."

The usual ploy is to appeal to the infinite possibilities of technology and resource substitution (ingenuity) as a dynamic force that can continuously outrun depletion and pollution. This counterargument is flawed in many respects. First, technology and infinite substitution mean only that one form of low-entropy matter/energy is substituted for another, within a finite and diminishing set of low-entropy sources. Such substitution is often very advantageous, but we never substitute high-entropy wastes for low-entropy resources in net terms. Second, the claim is frequently made that reproducible capital is a near-perfect substitute for resources. But this assumes that capital can be produced independently of resources, which is absurd. Furthermore, it flies in the teeth of the obvious complementarity of capital and resources in production. The capital stock is an agent for transforming the resource flow from raw material into a product. More capital does not substitute for less resources, except on a very restricted margin. You cannot make the same house by substituting more saws for less wood.

The growth advocates are left with one basic argument: resource and environmental limits have not halted growth in the past and therefore will not do so in the future. But such logic proves too much, namely, that nothing new can ever happen. A famous general survived a hundred battles without a scratch, and that was still true when he was blown up.

Earl Cook offered some insightful criticism of this faith in limitless ingenuity in one of his last articles. The appeal of the limitless-ingenuity argument, he contended, lies not in the scientific grounding of its premises nor in the cogency of its logic but rather in the fact that

> the concept of limits to growth threatens vested interests and power structures; even worse, it threatens value structures in which lives have been invested. . . . Abandonment of belief in perpetual motion was a major step toward recognition of the true human condition. It is significant that "mainstream" economists never abandoned that belief and do not accept the relevance to the economic process of the Second Law of Thermodynamics; their position as high priests of the market economy would become untenable did they do so. (Cook 1982, p. 198)

Indeed it would. Therefore, much ingenuity is devoted to "proving" that ingenuity is unlimited. Julian Simon, George Gilder, Herman Kahn, and Ronald Reagan trumpeted this theme above all others. Every technical accomplishment, no matter how ultimately insignificant, is celebrated as one more victory in an infinite series of future victories of technology over nature. The Greeks called this hubris. The Hebrews were warned to "beware of saying in your heart, 'My own strength and the might of my own hand won this power for me'" (Deut. 8:17). But such wisdom is drowned out in the drumbeat of the see-no-evil "optimism" of growthmania. All the more necessary is it then to repeat Earl Cook's trenchant remark that "without the enormous amount of work done by nature in concentrating flows of energy and stocks of resources, human ingenuity would be onanistic. What does it matter that human ingenuity may be limitless, when matter and energy are governed by other rules than is information?"

ETHICOSOCIAL LIMITS

Even when growth is, with enough ingenuity, still possible, ethicosocial limits may render it undesirable. Four ethicosocial propositions limiting the desirability of growth are briefly considered below.

1. *The desirability of growth financed by the drawdown of geological capital is limited by the cost imposed on future generations.* In standard economics the balancing of future against present costs and benefits is done by discounting. A time discount rate is a numerical way of expressing the value judgment that beyond a certain point the future is not worth anything to presently living people. The higher the discount rate, the sooner that point is reached. The value of the future to future people does not count in the standard approach.

Perhaps a more discriminating, though less numerical, principle for balancing the present and the future would be that the basic needs of the present should always take precedence over the basic needs of the future but that the basic needs of the future should take precedence over the extravagant luxury of the present.

2. *The desirability of growth financed by takeover of habitat is limited by the extinction or reduction in number of sentient subhuman species whose habitat disappears.* Economic growth requires space for growing stocks of

artifacts and people and for expanding sources of raw material and sinks for waste material. Other species also require space, their "place in the sun." The instrumental value of other species to us, the life-support services they provide, was touched on in the discussion of biophysical limits above. Another limit derives from the intrinsic value of other species, that is, counting them as sentient, though probably not self-conscious, beings which experience pleasure and pain and whose experienced "utility" should be counted positively in welfare economics, even though it does not give rise to maximizing market behavior.

The intrinsic value of subhuman species should exert some limit on habitat takeover in addition to the limit arising from instrumental value. But it is extremely difficult to say how much. Clarification of this limit is a major philosophical task, but if we wait for a definitive answer before imposing any limits on takeover, then the question will be rendered moot by extinctions which are now occurring at an extremely rapid rate relative to past ages.

3. *The desirability of aggregate growth is limited by its self-canceling effects on welfare.* Keynes argued that absolute wants (those we feel independently of the condition of others) are not insatiable. Relative wants (those we feel only because their satisfaction makes us feel superior to others) are indeed insatiable, for, as Keynes put it, "The higher the general level, the higher still are they." Or, as J. S. Mill expressed it, "Men do not desire to be rich, but to be richer than other men." At the current margin of production in rich countries it is very likely that welfare increments (increments in well-being) are largely a function of changes in relative income (insofar as they depend on income at all). Since the struggle for relative shares is a zero-sum game, it is clear that aggregate growth cannot increase aggregate welfare. To the extent that welfare depends on relative position, growth is unable to increase welfare in the aggregate. It is subject to the same kind of self-canceling trap that we find in the arms race.

Because of this self-canceling effect of relative position, aggregate growth is less productive of human welfare than we heretofore thought. Consequently, other competing goals should rise relative to growth in the scale of social priorities. Future generations, subhuman species, community, and whatever else has been sacrificed in the name of growth should henceforth be sacrificed less simply because growth is less productive of general happiness than used to be the case when marginal income was dedicated mainly to the satisfaction of absolute rather than relative wants.

4. *The desirability of aggregate growth is limited by the corrosive effects on moral standards resulting from the very attitudes that foster growth, such as glorification of self-interest and a scientistic-technocratic worldview.* On the demand side of commodity markets, growth is stimulated by greed and acquisitiveness, intensified beyond the "natural" endowment from original sin by the multibillion-dollar advertising industry. On the supply side, technocratic scientism proclaims the possibility of limitless expansion and preaches a reductionistic, mechanistic philosophy which, in spite of its success as a research program, has serious shortcomings as a worldview. As a research program it very effectively furthers power and control, but as a worldview it leaves no room for purpose, much less for any distinction between good and bad purposes. "Anything goes" is a convenient moral slogan for the growth economy because it implies that anything also sells. To the extent that growth has a well-defined purpose, then it is limited by the satisfaction of that purpose. Expanding power and shrinking purpose lead to uncontrolled growth for its own sake, which is wrecking the moral and social order just as surely as it is wrecking the ecological order.

The situation of economic thought today can be summarized by a somewhat farfetched but apt analogy. Neoclassical economics, like classical physics, is relevant to a special case that assumes that we are far from limits—far from the limiting speed of light or the limiting smallness of an elementary particle in physics—and far from the biophysical limits of the earth's carrying capacity and the ethicosocial limits of satiety in economics. Just as in physics, so in economics: the classical theories do not work well in regions close to limits. A more general theory is needed to embrace both normal and limiting cases. In economics this need becomes greater with time because the ethic of growth itself guarantees that the close-to-the-limits case becomes more and more the norm. The nearer the economy is to limits, the less can we accept the practical judgment most economists make, namely, that "a change in economic welfare implies a change in total welfare in the same direction if not in the same degree." Rather, we must learn to define and explicitly count the other component of total welfare that growth inhibits and erodes when it presses against limits.

MOVING FROM GROWTHMANIA TO THE STEADY STATE IN PRACTICE:
Failures of Growth as Forced First Steps Toward a Steady-State Economy

No doubt the biggest growth failure is the continuing arms race, where growth has led to less security rather than more and has raised the stakes from loss of individual lives to loss of life itself in wholesale ecocide. Excessive population growth, toxic wastes, acid rain, climate modification, devastation of rain forests, and the loss of ecosystem services resulting from these aggressions against the environment represent case studies in growth failure. Seeing them as first steps toward a steady-state economy requires the conscious willing of a hopeful attitude.

All the growth failures mentioned above are failures of the growth economy to respect the biophysical limits of its host. I would like also to consider some symptoms of growthmania within the economy itself. Three examples will be considered: money fetishism and the paper economy, faulty national accounts and the treachery of quantified success indexes, and the ambivalent "information economy."

MONEY FETISHISM AND THE PAPER ECONOMY

Money fetishism is a particular case of what Alfred North Whitehead called "the fallacy of misplaced concreteness," which consists in reasoning at one level of abstraction but applying the conclusions of that reasoning to a different level of abstraction. It argues that, since abstract exchange value flows in a circle, so do the physical commodities constituting real GNP. Or, since money in the bank can grow forever at compound interest, so can real wealth, and so can welfare. Whatever is true for the abstract symbol of wealth is assumed to hold for concrete wealth itself.

Money fetishism is alive and well in a world in which banks in wealthy countries make loans to poor countries and then, when the debtor countries cannot make the repayment, simply make new loans to enable the payment of interest on old loans, thereby avoiding taking a loss on a bad debt. Using new loans to pay interest on old loans is worse than a Ponzi scheme, but the exponential snowballing of debt is expected to be

offset by a snowballing of real growth in debtor countries. The international debt impasse is a clear symptom of the basic disease of growthmania. Too many accumulations of money are seeking ways to grow exponentially in a world in which the physical scale of the economy is already so large relative to the ecosystem that there is not much room left for growth of anything that has a physical dimension.

Marx, and Aristotle before him, pointed out that the danger of money fetishism arises as a society progressively shifts its focus from use value to exchange value, under the pressure of increasingly complex division of labor and exchange. The sequence is sketched below in four steps, using Marx's shorthand notation for labels.

1. *C-C'*. One commodity (*C*) is directly traded for a different commodity (*C'*). The exchange values of the two commodities are by definition equal, but each trader gains an increased use value. This is simple *barter*. No money exists, so there can be no money fetishism.

2. *C-M-C'*. *Simple commodity circulation* begins and ends with a use value embodied in a commodity. Money (*M*) is merely a convenient medium of exchange. The object of exchange remains the acquisition of an increased use value. *C'* represents a greater use value to the trader, but *C'* is still a use value, limited by its specific use or purpose. One has, say, a greater need for a hammer than a knife but has no need for two hammers, much less for fifty. The incentive to accumulate use values is very limited.

3. *M-C-M'*. As simple commodity circulation gave way to *capitalist circulation*, the sequence shifted. It now begins with money capital and ends with money capital. The commodity or use value is now an intermediary step in bringing about the expansion of exchange value by some amount of profit, $\Delta M = M' - M$. Exchange value has no specific use or physical dimension to impose concrete limits. One dollar of exchange value is not as good as two, and fifty dollars is better yet, and a million is much better, etc. Unlike concrete use values, which spoil or deteriorate when hoarded, abstract exchange value can accumulate indefinitely without spoilage or storage costs. In fact, exchange value can grow by itself at compound interest. But as Frederick Soddy pointed out, "You cannot permanently pit an absurd human convention [compound interest] against a law of nature [entropic decay]." "Permanently," however, is not the same as "in the

meantime," during which we have, at the micro level, bypassed the absurdity of accumulating use values by accumulating exchange value and holding it as a lien against future use values. But unless future use value, or real wealth, has grown as fast as accumulations of exchange value have grown, then at the end of some time period there will be a devaluation of exchange value by inflation or some other form of debt repudiation. At the macro level limits will reassert themselves, even when ignored at the micro level, where the quest for exchange value accumulation has become the driving force.

4. *M-M'*. We can extend Marx's stages one more step to the *paper economy,* in which, for many transactions, concrete commodities "disappear" even as an intermediary step in the expansion of exchange value. Manipulations of symbols according to arbitrary and changing tax rules, accounting conventions, depreciation, mergers, public relations imagery, advertising, litigation, and so on, all result in a positive ΔM for some, but no increase in social wealth, and hence an equal negative ΔM for others. Such "paper entrepreneurialism" and "rent-seeking" activities seem to be absorbing more and more business talent. Echoes of Frederick Soddy are audible in the statement of Robert Reich that "the set of symbols developed to represent real assets has lost the link with any actual productive activity. Finance has progressively evolved into a sector all its own, only loosely connected to industry." Unlike Soddy, however, Reich does not appreciate the role played by biophysical limits in redirecting efforts from manipulating resistant matter and energy toward manipulating pliant symbols. He thinks that, as more flexible and information-intensive production processes replace traditional mass production, somehow financial symbols and physical realities will again become congruent. But it may be that as physical resources become harder to acquire, as evidenced by falling energy rates of return on investment, the incentive to bypass the physical world by moving from $M-C-M'$ to $M-M'$ becomes ever greater. We may then keep growing on paper, but not in reality. This illusion is fostered by our national accounting conventions. It could be that we are moving toward a nongrowing economy a bit faster than we think. If the cost of toxic waste dumps were subtracted from the value product of the chemical industry, we might discover that we have already attained zero growth in value from that sector of the economy.

FAULTY NATIONAL ACCOUNTING AND THE TREACHERY OF QUANTIFIED SUCCESS INDICATORS

Our national accounts are designed in such a way that they cannot reflect the costs of growth, except by perversely counting the resulting defensive expenditures as further growth. It is by now a commonplace to point out that GNP does not reveal whether we are living off income or capital, off interest or principal. Depletion of fossil fuels, minerals, forests, and soils is capital consumption, yet such unsustainable consumption is treated no differently from sustainable yield production (true income) in GNP. But not only do we decumulate positive capital (wealth), we also accumulate negative capital (illth) in the form of toxic-waste deposits and nuclear dumps. To speak so insouciantly of "economic growth" whenever produced goods accumulate, when at the same time natural wealth is being diminished and man-made illth is increasing represents, to say the least, an enormous prejudgment about the relative size of these changes. Only on the assumption that environmental sources and sinks are infinite does such a procedure make sense.

Another problem with national accounts is that they do not reflect the "informal" or "underground" economy. Estimates of the size of the underground economy in the United States range from around 4% to around 30% of GNP, depending on the technique of estimation. The underground economy has apparently grown in recent times, probably as a result of higher taxes, growing unemployment, and frustration with the increasing complexity and arbitrariness of the paper economy. Like household production, of which they are extensions, none of these informal productive activities are registered in GNP. Their growth represents an adaptation to the failure of traditional economic growth to provide employment and security. As an adaptation to growth failure in the GNP sector, the underground economy may represent a forced first step toward an SSE. But not everything about the underground economy is good. Many of its activities (drugs, prostitution) are illegal, and much of its basic motivation is tax evasion, although in today's world there may well be some noble reasons for not paying taxes.

The act of measurement always involves some interaction and interference with the reality being measured. This generalized Heisenberg principle is especially relevant in economics, where the measurement of a success index on which rewards are based, or taxes calculated, nearly always has perverse repercussions on the reality being measured. Consider, for example, the case of management by quantified objectives applied to a tuberculosis hospital, as related to me by a physician. It is well known that TB patients cough less as they get better. So the number of coughs per day was taken as a quantitative measure of the patient's improvement. Small microphones were attached to the patients' beds, and their coughs were duly recorded and tabulated. The staff quickly perceived that they were being evaluated in inverse proportion to the number of times their patients coughed. Coughing steadily declined as doses of codeine were more frequently prescribed. Relaxed patients cough less. Unfortunately the patients got worse, precisely because they were not coughing up and spitting out the congestion. The cough index was abandoned.

The cough index totally subverted the activity it was designed to measure because people served the abstract quantitative index instead of the concrete qualitative goal of health. Perversities induced by quantitative goal setting are pervasive in the literature on Soviet planning: set the production quota for cloth in linear feet, and the bolt gets narrower; set it in square feet, and the cloth gets thinner; set it by weight, and it gets too thick. But one need not go as far away as the Soviet Union to find examples. The phenomenon is ubiquitous. In universities a professor is rewarded according to number of publications. Consequently the length of articles is becoming shorter as we approach the minimum publishable unit of research. At the same time the frequency of coauthors has increased. More and more people are collaborating on shorter and shorter papers. What is being maximized is not discovery and dissemination of coherent knowledge but the number of publications on which one's name appears.

The purpose of these examples of the treachery of quantified success indexes is to suggest that, like them, GNP is not only a passive mismeasure but also an actively distorting influence on the very reality that it aims only to reflect. GNP is an index of throughput, not welfare. Throughput is positively correlated with welfare in a world of infinite sources and sinks, but in a finite world with fully employed carrying capacity, throughput is a *cost*. To design national policies to maximize GNP is just not smart. It is practically equivalent to maximizing depletion and pollution.

The usual reply to these well-known criticisms of GNP is, "So it's not perfect, but it's all we have. What would you put in its place?" It is assumed that we *must* have some numerical index. But why? Might we not be better off without the GNP statistic, even with nothing to "put in its place"? Were not the TB patients better off without the cough index, when physicians and administrators had to rely on "soft" qualitative judgment? The world before 1940 got along well enough without calculating GNP. Perhaps we could come up with a better system of national accounts, but abandoning GNP need not be postponed until then. Politically we are not likely to abandon the GNP statistic any time soon. But in the meantime we can start thinking of it as "gross national cost."

THE AMBIVALENT "INFORMATION ECONOMY"

The much-touted "information economy" is often presented as a strategy for escaping biophysical limits. Its modern devotees proclaim that "whereas matter and energy decay according to the laws of entropy . . . information is . . . immortal." And, further, "The universe itself is made of information—matter and energy are only simple forms of it." Such half-truths forget that information does not exist apart from physical brains, books, and computers, and, further, that brains require the support of bodies, books require library buildings, computers run on electricity, etc. At worst the information economy is seen as a computer-based explosion of the symbol manipulations of the paper economy. More occult powers are attributed to information and its handler, the computer, by the silicon gnostics of today than any primitive shaman ever dared claim for his favorite talisman. And this in spite of the enormous legitimate importance of the computer, which needs no exaggeration.

Other notions of the information economy are by no means nonsensical. When the term refers to qualitative improvements in products to make them more serviceable, longer-lasting, more repairable, and better-looking, then we have what was earlier referred to as "development." To think of qualitative improvement as the embodiment of more information in a product is not unreasonable.

But the best question to ask about the information economy is that posed by T. S. Eliot in "Choruses from 'The Rock'":

Where is the wisdom we have lost in knowledge?
Where is the knowledge we have lost in information?

Why stop with an information economy? Why not a knowledge economy? Why not a wisdom economy?

Knowledge is structured, organized information rendered intelligible and understandable. It is hard to imagine embodying a bit of isolated information (in the sense of communications theory) in a product. What is required for qualitative improvement of products is knowledge—an understanding of the purpose of the item, the nature of the materials, and the alternative designs that are permitted within the restrictions of purpose and nature of the materials. Probably many writers on the subject use the term "information" synonymously with "knowledge," and what they have in mind is really already a "knowledge economy." The important step is to go to a "wisdom economy."

Wisdom involves a knowledge of techniques plus an understanding of purposes and their relative importance, along with an appreciation of the limits to which technique and purpose are subject. To distinguish a real limit from a temporary bottleneck, and a fundamental purpose from a velleity, requires wise judgment. Growthmania cannot be checked without wise judgment. Since events are forcing us to think in terms of an information economy, it is perhaps not too much to hope that we will follow that thrust all the way to a wisdom economy, one design feature of which, I submit, will be that of a dynamic steady state.

The main characteristics of such a wisdom economy were adumbrated by Earl Cook in his list of nine "Beliefs of a Neomalthusian," and I will conclude by listing them:

1. "Materials and energy balances constrain production."

2. "Affluence has been a much more fecund mother of invention than has necessity." That is, science and technology require an economic surplus to support them, and a few extra but poor geniuses provided by rapid population growth will not help.

3. "Real wealth is by technology out of nature," or, as William Petty would have said, technology may be the father of wealth, but nature is the mother.

4. "The appropriate human objective is the maximization of psychic income by conversion of natural resources to useful commodities and by the use of those commodities as efficiently as possible," and "the appropriate measure of efficiency in the conversion of resources to psychic income is the human life-hour, with the calculus extended to the yet unborn."

5. "Physical laws are not subject to repeal by men," and of all the laws of economics the law of diminishing returns is closest to a physical law.

6. "The industrial revolution can be defined as that period of human history when basic resources, especially nonhuman energy, grew cheaper and more abundant."

7. "The industrial revolution so defined is ending."

8. "There are compelling reasons to expect natural resources to become more expensive."

9. "Resource problems vary so much from country to country that careless geographic and commodity aggregation may confuse rather than clarify." That is, "it serves no useful purpose to combine the biomass of Amazonia with that of the Sahel to calculate a per capita availability of firewood."

Earl Cook would have been the last person to offer these nine points as a complete blueprint for a wisdom economy. But I think that he got us off to a good start.

For Further Discussion

1. What exactly is the distinction between economic growth and economic development? What difference would this distinction make in your life?

2. What does Daly mean by the "biophysical limits to growth"? Contrast this to Simon's claim in Chapter 3 that natural resources are infinite. Why does Daly reject the "infinite resources assumption"? Do you agree with Simon or Daly?

3. Daly tells us that "the desirability of aggregate growth is limited by its self-canceling effects on welfare." What does he mean by this? Do you agree?

4. Daly offers a description of a "wisdom economy" in contrast to the "information economy." What's the difference between wisdom and information?

5. Compare Daly's criticism of aggregate growth to Sagoff's rejection of free market economics. How are they alike? Are there any differences?

DISCUSSION AND STUDY QUESTIONS FOR CHAPTER 4

1. How should government agencies decide what is in the public interest? Is "willingness to pay" a reasonable method? Can there be experts who know what is best for the public?
2. Would you protest if a Disney-like ski resort were planned for a scenic wilderness area near your home? If the resort were built, would you visit it? Are your interests as a consumer identical to your interests as a citizen?

3. International disputes have arisen as fish populations have decreased along the coasts of many countries. Years ago countries extended their borders from 3 miles to as much as 200 miles from shore. Canadian naval ships have fired shots at foreign fishing vessels, and a fierce battle rages between U.S. and Canadian salmon fishermen. Should individuals or countries be allowed to own the oceans and the fish therein?
4. What do you think is the difference between economic growth and economic development? How could a country grow without developing? How could it develop without growing?

DISCUSSION CASES

Mineral King Valley

Mineral King Valley lies in the Sierra Nevada Mountains just outside Sequoia National Park. It has long been managed as a wilderness area by the U.S. Forest Service. In the 1960s the Forest Service began to consider plans to develop Mineral King as a recreational area. Among the interested developers was Walt Disney Enterprises, whose bid included plans for a complex of ski resorts, motels, restaurants, and other tourist facilities. These development plans incurred widespread criticism, including a lawsuit filed by the Sierra Club that was aimed at preventing the development.

Mineral King Valley is publicly owned land. Unfortunately, the public express interest in widely incompatible uses for the land: Some people want ski slopes and a recreation area, others want the land left in its undeveloped wild state. How should government agents decide such a case?

Defenders of the ski resort plan argue that citizens have expressed their values most clearly in their willingness to pay for a ski resort. Disney could finance the planned development only because there is clear evidence that this development would earn a profit and thereby Disney could repay their loans. The Sierra Club could never finance

such a purchase because, simply, the public is not willing to pay for wilderness areas. Because more people are willing to pay for a ski resort than are willing to pay for a wilderness area, democratic principles seem to suggest that the area be developed. Do you agree?

What recommendations concerning Mineral King Valley might be offered by Stroup and Baden? Mark Sagoff likely would disagree with the plan to sell development rights to the Disney Company. Why? Use Sagoff's distinction between objective values and subjective preferences to evaluate this case. Does this case involve something other than the efficient allocation of public resources?

Whales for Sale

Whales and most other endangered species are wild animals, unowned by anyone. Might private ownership be a means for protecting species from extinction?

Consider animal species that are far from extinction. Cats, dogs, chickens, cows, and horses come quickly to mind. When animals have value to people, as both whales and chickens do, private ownership seems to secure the survival of the

continued

species better than common ownership. When the animals are owned, the owner has a strong incentive to ensure that they stay around. When the animals are unowned, as is the case with whales, people have an incentive to kill as many as possible as quickly as possible before others get to them.

A system of private ownership could be administered by the United Nations and might include licenses granting exclusive rights to harvest whales. Whalers from countries like Japan and Russia would purchase these exclusive rights. These rights would prevent others from harvesting whales and thereby provide the owners with an incentive to preserve whales. Much the way farmers slaughter their chickens and steers only at a sustainable rate, whalers would have an incentive to ensure that whales survive into the indefinite future. Do you agree that private owners would have a strong incentive to protect and preserve endangered species? Rely on the discussion by Stroup and Baden to develop your answer. Does the right of private property include the right to destroy that property? If ownership of whales was granted to private individuals or groups, would the owners be within their rights to destroy all the whales? How might Stroup and Baden answer this question? Is the supply of whales infinite? How might Julian Simon (Chapter 3) and Herman Daly (Chapter 4) answer that question?

"Takings" and Property Rights

The Fifth Amendment of the U.S. Constitution grants the government the authority, known as eminent domain, to take possession of private land for the public good. When this is done, however, the Constitution requires that just compensation be paid to the landowner. The police power of government also allows seizure of property to punish wrongdoing and allows restriction of property use to prevent harms.

In recent decades, the use of wetlands has been significantly regulated. Landowners are prevented from destroying wetlands, filling in wetlands, or otherwise significantly affecting their ecological functioning. Because wetlands provide such critical functions as ensuring water quality and providing flood control, these restrictions are based, in part, on the police power of government to prevent harms. But what of cases where land use is restricted for the public good?

Environmental regulations have recently been challenged on the grounds that they are unconstitutional "takings" of private property. Consider a city that passes a zoning ordinance to protect the few remaining oak savannas or prairies within its boundaries. Development is now prohibited in areas where owners previously could have sold the land for shopping malls and housing developments. Does society owe compensation to the landowners when environmental regulations deny them profits that they could have made without the regulation? Is the right to develop property an essential part of private ownership? Compare the meaning of development used in this case to the meaning of development used by Herman Daly. What exactly is the difference? If government seeks to protect land for the "moral, aesthetic, cultural, political, and religious values" mentioned by Sagoff, should it be required to pay landowners for the lost economic value that results?

Valuing Eagles and Solitude

What is the value of wilderness? Public officials are often charged with balancing competing public interests, as when forest officials must balance logging and recreational uses for a national forest. How do you compare the value of timber with the value of quiet and solitude?

Canoeists in Canada's Quetico Park were asked to answer such questions as part of a survey conducted by economists. At present, canoeists are required to purchase permits to visit the park. Because there is no market for permits—they are allocated by the government, their availability is artificially restricted, and they cannot be resold—the price of the permits provides little information about how valuable they are. To determine this, and to provide an objective measure when comparing competing uses, economists asked canoeists to place a monetary value on their wilderness experience. How much would you be willing to pay for a permit if you could be assured that no other canoeists would be in the area? How much would you be willing to pay to see bald eagles? To see moose in the wild? What might the positions of Stroup and Baden, Freeman, and Sagoff be on this practice? Do you think that a system of private ownership would better manage parks like Quetico? Why or why not? Is willingness to pay a good measure of the public value of eagles and solitude? Why or why not?

CHAPTER 5

Aesthetic and Spiritual Values

People value the environment for many reasons, not all of which can be expressed in economic terms. As we have seen in the previous chapter, the goal of economic analysis is to translate various values, or preferences, into the common language of money. But this clearly misrepresents the nature of many of our values, or so critics charge. A complete ethics must account for all of the values that guide how we live, especially if these values are more than mere personal preferences.

Why do people value natural objects and natural areas? Even a brief survey indicates a variety of reasons: People value the environment for aesthetic, spiritual, symbolic, historical, and cultural reasons. Consider an example that highlights this point.

Why would people object to the destruction of natural areas for the construction of, say, a strip mall? What, if anything, would be wrong with building a mall with fast-food restaurants, arcades, souvenir shops, and motels along the south rim of the Grand Canyon? Presumably, a strong argument against such construction is that it would destroy the natural beauty of the Grand Canyon. The Canyon is beautiful; strip malls and arcades are ugly. The Grand Canyon, like many scenic and wild areas, also plays an important spiritual, symbolic, cultural, and historical role in American life. These are places where many people go to experience a sense of renewal, a sense of awe, a sense of connection with other people from other times. These are places where many people find meaning and value. Commercial development would desecrate such areas and destroy these values.

But value questions, at least in recent times, are thought to be hopelessly controversial. Beauty is in the eye of the beholder, according to the old saying. If beauty is in the eye of the beholder, then whether or not the Grand Canyon is beautiful depends on who is looking at it. Some people may find the Grand Can-

yon beautiful and awe inspiring. Others may feel the same way about a strip mall. Thus, according to this perspective, arguments that attribute value to natural areas really are about mere personal preferences. You prefer mountains, I prefer malls. If values are merely personal preferences, then policy decisions ought to be based on more objective standards such as voting, public opinion polls, or price.

So it seems that environmental ethics must address a wide range of value questions. The challenge is to articulate these values and defend them in ways that can legitimately influence public policy.

The first selection in this chapter is by John Muir, one of the giants of American environmentalism. Muir, a founder of the Sierra Club, was a well-known defender of preservationist policies. Muir's view was that natural areas ought to be preserved exactly because they hold value that cannot be measured by economic markets or public opinion polls. Natural areas are not commodities that should be bought and sold. In his reading, Muir addresses some scientific, aesthetic, and spiritual values of nature as he looked out across Yosemite Valley.

To the degree that natural areas give rise to experiences of beauty, awe, or reverence, it might seem our true interest lies in the experiences themselves, rather than in the natural areas that give rise to them. That is, if we could provide people with experiences of beauty or reverence, without actually visiting a wilderness, would we lose anything? If we can give people a wilderness experience by driving them through a wild safari park or through an African jungle in Disneyland, should we be concerned with preserving the original wilderness? What would be wrong with replacing trees lost to a clear-cut with plastic trees? As the title of our second reading asks, "What's wrong with plastic trees?"

In a classic article, Martin Krieger examines a range of rationales offered to protect natural environments. Although he recognizes the shortcomings of economic and technological thinking, he believes that attitudes and preferences are the keys to understanding environmental preservation. Perhaps we can change attitudes (as advertising aims to do) to make artificial environments more attractive. What's wrong with plastic trees? Krieger's guess is that "there is very little wrong with them." (While reading Krieger's essay, written in 1973, it might be fun to keep in mind the possibility of virtual reality tours of wilderness areas.)

Krieger's thesis suggests that there is an important distinction between appearance and reality. What a tree is *really,* what it is made out of, is less important than how it *appears* to us. If we believe it is real, if it gives rise to the experience of nature, then it is for all practical purposes a tree, even if it is made of plastic. In a similar vein, some argue that much environmental damage can be overcome by the practice of restoring land to its original condition. On this view nothing is lost if land that is destroyed—for example, through an oil spill or clear-cutting—can later be restored to its original condition.

In our third reading, Robert Elliot responds to this "restoration thesis." He argues that even if such restoration were possible, and he doubts that it truly is, there are strong ethical objections to it. Something of great value is lost, irretrievably, when pristine natural areas are disrupted. The causal origin of natural areas, where and how they have originated, is an important part of their value.

Elliot's view is that *knowing* something about an object can affect our experience of it and how we value

it. This suggests that there is an intellectual component to value that is often overlooked in contemporary discussions. When we say that "beauty is in the eye of the beholder," we are saying that beauty is a matter of sensory experience, and this makes the person rather passive in the experience. One views a work of art or tastes a bite of food or listens to a symphony and immediately one either likes it or not. On this subjective view, values are merely a matter of personal opinion or personal tastes: "I don't know much about art, but I know what I like."

But can knowing something about an object change our perception of it? Can knowing something about a food change how it tastes? Can knowing something about music change how it sounds? Can knowing something about the origin of a forest change how it is perceived? Can knowing something about ecology change how a natural area is perceived? Philosopher Holmes Rolston addresses this question in our fourth reading.

Aesthetic values are not the only values involved in protecting the environment; spiritual and religious values also play a major role in the lives of many people. Although some critics charge that the biblical tradition has supported an ethic of environmental destruction (see the essay by Lynn White in Chapter 2), others see in that tradition a basis for environmental stewardship.

Rosemary Radford Ruether finds in the biblical tradition several arguments for viewing nature as part of the covenant between God and humanity. Although she favors an approach to nature that is neither romantic nor merely instrumental, her essay introduces us to a variety of religious and spiritual perspectives on the natural world.

A View of the High Sierra

John Muir

John Muir (1838–1914) is one of the most influential figures in modern environmentalism. He traveled, often walked, throughout North America. His essays about the beauty of natural areas inspired generations of readers. He was involved in many of the key controversies of early environmentalism, including the fight to preserve the Hetch

Hetchy Valley near Yosemite, the creation of the national park system, and the preservation of Yosemite National Park. He was a founder and president of the Sierra Club.

In this essay, Muir describes a trip into Yosemite Valley and up Mount Ritter. Muir's description of this journey attributes a variety of values to this scenic natural area:

aesthetic, spiritual, scientific, symbolic. He speaks of God as freely as he does of geology. This trip is more than just a trip of physical discovery; one senses that Muir treats it as a trip of self-discovery as well.

Early one bright morning in the middle of Indian summer, while the glacier meadows were still crisp with frost crystals, I set out from the foot of Mount Lyell, on my way down to Yosemite Valley, to replenish my exhausted store of bread and tea. I had spent the past summer, as many preceding ones, exploring the glaciers that lie on the head waters of the San Joaquin, Tuolumne, Merced, and Owen's rivers; measuring and studying their movements, trends, crevasses, moraines, etc., and the part they had played during the period of their greater extension in the creation and development of the landscapes of this alpine wonderland. . . .

To artists, few portions of the High Sierra are, strictly speaking, picturesque. The whole massive uplift of the range is one great picture, not clearly divisible into smaller ones; differing much in this respect from the older, and what may be called, riper mountains of the Coast Range. All the landscapes of the Sierra, as we have seen, were born again, remodeled from base to summit by the developing ice-floods of the last glacial winter. But all these new landscapes were not brought forth simultaneously: some of the highest, where the ice lingered longest, are tens of centuries younger than those of the warmer regions below them. In general, the younger the mountain-landscapes—younger, I mean, with reference to the time of their emergence from the ice of the glacial period—the less separable are they into artistic bits capable of being made into warm, sympathetic, lovable pictures with appreciable humanity in them.

Here, however, on the head waters of the Tuolumne, is a group of wild peaks on which the geologist may say that the sun has but just begun to shine, which is yet in a high degree picturesque, and in its main features so regular and evenly balanced as almost to appear conventional—one somber cluster of snow-laden peaks with gray pine-fringed granite bosses braided around its base, the whole surging free into the sky from the head of a magnificent valley, whose lofty walls are beveled away on both sides so as to embrace it all without admitting anything not strictly belonging to it. The foreground was now aflame with autumn colors, brown and purple and gold, ripe in the mellow sunshine; contrasting brightly with the deep, cobalt blue of the sky, and the black and gray, and

pure, spiritual white of the rocks and glaciers. Down through the midst, the young Tuolumne was seen pouring from its crystal fountains, now resting in glassy pools as if changing back again into ice, now leaping in white cascades as if turning to snow; gliding right and left between granite bosses, then sweeping on through the smooth, meadowy levels of the valley, swaying pensively from side to side with calm, stately gestures past dipping willows and sedges, and around groves of arrowy pine; and throughout its whole eventful course, whether flowing fast or slow, singing loud or low, ever filling the landscape with spiritual animation, and manifesting the grandeur of its sources in every movement and tone.

Pursuing my lonely way down the valley, I turned again and again to gaze on the glorious picture, throwing up my arms to inclose it as in a frame. After long ages of growth in the darkness beneath the glaciers, through sunshine and storms, it seemed now to be ready and waiting for the elected artist, like yellow wheat for the reaper; and I could not help wishing that I might carry colors and brushes with me on my travels, and learn to paint. In the mean time I had to be content with photographs on my mind and sketches in my note-books. At length, after I had rounded a precipitous headland that puts out from the west wall of the valley, every peak vanished from sight, and I pushed rapidly along the frozen meadows, over the divide between the waters of the Merced and Tuolumne, and down through the forests that clothe the slopes of Cloud's Rest, arriving in Yosemite in due time—which, with me, is *any* time. And, strange to say, among the first people I met here were two artists who, with letters of introduction, were awaiting my return. They inquired whether in the course of my explorations in the adjacent mountains I had ever come upon a landscape suitable for a large painting; whereupon I began a description of the one that had so lately excited my admiration. Then, as I went on further and further into details, their faces began to glow, and I offered to guide them to it, while they declared that they would gladly follow, far or near, whithersoever I could spare the time to lead them.

Since storms might come breaking down through the fine weather at any time, burying the colors in snow, and cutting off the artists' retreat, I advised getting ready at once.

I led them out of the valley by the Vernal and Nevada Falls, thence over the main dividing ridge to the Big Tuolumne Meadows, by the old Mono trail, and thence along the upper Tuolumne River to its head.

This was my companions' first excursion into the High Sierra, and as I was almost always alone in my mountaineering, the way that the fresh beauty was reflected in their faces made for me a novel and interesting study. They naturally were affected most of all by the colors—the intense azure of the sky, the purplish grays of the granite, the red and browns of dry meadows, and the translucent purple and crimson of huckleberry bogs; the flaming yellow of aspen groves, the silvery flashing of the streams, and the bright green and blue of the glacier lakes. But the general expression of the scenery—rocky and savage—seemed sadly disappointing; and as they threaded the forest from ridge to ridge, eagerly scanning the landscapes as they were unfolded, they said: "All this is huge and sublime, but we see nothing as yet at all available for effective pictures. Art is long, and art is limited, you know; and there are foregrounds, middle-grounds, backgrounds, all alike; bare rock-waves, woods, groves, diminutive flecks of meadow, and strips of glittering water." "Never mind," I replied, "only bide a wee, and I will show you something you will like."

At length, toward the end of the second day, the Sierra Crown began to come into view, and when we had fairly rounded the projecting headland before mentioned, the whole picture stood revealed in the flush of the alpenglow. Their enthusiasm was excited beyond bounds, and the more impulsive of the two, a young Scotchman, dashed ahead, shouting and gesticulating and tossing his arms in the air like a madman. Here, at last, was a typical alpine landscape.

After feasting awhile on the view, I proceeded to make camp in a sheltered grove a little way back from the meadow, where pine-boughs could be obtained for beds, and where there was plenty of dry wood for fires, while the artists ran here and there, along the river-bends and up the sides of the cañon, choosing foregrounds for sketches. After dark, when our tea was made and a rousing fire had been built, we began to make our plans. They decided to remain several days, at the least, while I concluded to make an excursion in the meantime to the untouched summit of Ritter.

It was now about the middle of October, the springtime of snow-flowers. The first winter-clouds had already bloomed, and the peaks were strewn with fresh crystals, without, however, affecting the climbing to any dangerous extent. And as the weather was still profoundly calm, and the distance to the foot of the mountain only a little more than a day, I felt that I was running no great risk of being storm-bound.

Mount Ritter is king of the mountains of the middle portion of the High Sierra, as Shasta of the north and Whitney of the south sections. Moreover, as far as I know, it had never been climbed. I had explored the adjacent wilderness summer after summer; but my studies thus far had never drawn me to the top of it. Its height above sea-level is about 13,300 feet, and it is fenced round by steeply inclined glaciers, and cañons of tremendous depth and ruggedness, which render it almost inaccessible. But difficulties of this kind only exhilarate the mountaineer.

Next morning, the artists went heartily to their work and I to mine. Former experiences had given good reason to know that passionate storms, invisible as yet, might be brooding in the calm sun-gold; therefore, before bidding farewell, I warned the artists not to be alarmed should I fail to appear before a week or ten days, and advised them, in case a snow-storm should set in, to keep up big fires and shelter themselves as best they could, and on no account to become frightened and attempt to seek their way back to Yosemite alone through the drifts.

My general plan was simply this: to scale the cañon wall, cross over to the eastern flank of the range, and then make my way southward to the northern spurs of Mount Ritter in compliance with the intervening topography; for to push on directly southward from camp through the innumerable peaks and pinnacles that adorn this portion of the axis of the range, however interesting, would take too much time, besides being extremely difficult and dangerous at this time of year.

All my first day was pure pleasure; simply mountaineering indulgence, crossing the dry pathways of the ancient glaciers, tracing happy streams, and learning the habits of the birds and marmots in the groves and rocks. Before I had gone a mile from camp, I came to the foot of a white cascade that beats its way down a rugged gorge in the cañon wall, from a height of about nine hundred feet, and pours its throbbing waters into the Tuolumne. I was acquainted with its fountains, which, fortunately, lay in my course. What a fine traveling companion it proved to be, what songs it sang, and how passionately it told the mountain's own joy! Gladly I climbed along its dashing border, absorbing its divine music, and bathing from time to time in waftings of irised spray. Climbing higher, higher, new beauty came streaming on the sight: painted meadows, late-blooming gardens, peaks of rare architecture, lakes here and there, shining like silver, and glimpses of the forested middle region and the yellow lowlands far in the west. Beyond the range I saw the so-called Mono Desert, lying dreamily silent in thick purple light—a desert of heavy sun-glare beheld from a desert of ice-burnished granite. Here the waters divide, shouting in glorious enthusiasm, and falling eastward to vanish in the volcanic sands and dry sky of the Great Basin, or westward to the Great Valley of California, and thence through the Bay of San Francisco and the Golden Gate to the sea.

Passing a little way down over the summit until I had reached an elevation of about 10,000 feet, I pushed on southward toward a group of savage peaks that stand guard about Ritter on the north and west, groping my way; and dealing instinctively with every obstacle as it presented itself. Here a huge gorge would be found cutting across my path, along the dizzy edge of which I scrambled until some less precipitous point was discovered where I might safely venture to the bottom and then, selecting some feasible portion of the opposite wall, reascend with the same slow caution. Massive, flat-topped spurs alternate with the gorges, plunging abruptly from the shoulders of the snowy peaks, and planting their feet in the warm desert. These were everywhere marked and adorned with characteristic sculptures of the ancient glaciers that swept over this entire region like one vast ice-wind, and the polished surfaces produced by the ponderous flood are still so perfectly preserved that in many places the sunlight reflected from them is about as trying to the eyes as sheets of snow.

God's glacial-mills grind slowly, but they have been kept in motion long enough in California to grind sufficient soil for a glorious abundance of life, though most of the grist has been carried to the lowlands, leaving these high regions comparatively lean and bare; while the post-glacial agents of erosion have not yet furnished sufficient available food over the general surface for more than a few tufts of the hardiest plants, chiefly carices and eriogonæ. . . . In so wild and so beautiful a region was spent my first day, every sight and sound inspiring, leading one far out of himself, yet feeding and building up his individuality.

Now came the solemn, silent evening. Long, blue, spiky shadows crept out across the snow-fields, while a rosy glow, at first scarce discernible, gradually deepened and suffused every mountain-top, flushing the glaciers and the harsh crags above them. This was the alpenglow, to me one of the most impressive of all the terrestrial manifestations of God. At the touch of this divine light, the mountains seemed to kindle to a rapt, religious consciousness, and stood hushed and waiting like devout worshipers. Just before the alpenglow began to fade, two crimson clouds came streaming across the summit like wings of flame, rendering the sublime scene yet more impressive; then came darkness and the stars. . . .

I made my bed in a nook of the pine-thicket, where the branches were pressed and crinkled overhead like a roof, and bent down around the sides. These are the best bedchambers the high mountains afford—snug as squirrel-nests, well ventilated, full of spicy odors, and with plenty of wind-played needles to sing one asleep. I little expected company, but, creeping in through a low side-door, I found five or six birds nestling among the tassels. The night-wind began to blow soon after dark; at first only a gentle breathing, but increasing toward midnight to a rough gale that fell upon my leafy roof in ragged surges like a cascade, bearing wild sounds from the crags overhead. The waterfall sang in chorus, filling the old ice-fountain with its solemn roar, and seeming to increase in power as the night advanced—fit voice for such a landscape. I had to creep out many times to the fire during the night, for it was biting cold and I had no blankets. Gladly I welcomed the morning star.

The dawn in the dry, wavering air of the desert was glorious. Everything encouraged my undertaking and betoken success. There was no cloud in the sky, no storm-tone in the wind. Breakfast of bread and tea was soon made. I fastened a hard, durable crust to my belt by way of provision, in case I should be compelled to pass a night on the mountain-top; then, securing the remainder of my little stock against wolves and woodrats, I set forth free and hopeful.

How glorious a greeting the sun gives the mountains! To behold this alone is worth the pains of any excursion a thousand times over. The highest peaks burned like islands in a sea of liquid shade. Then the lower peaks and spires caught the glow, and long lances of light, streaming through many a notch and pass, fell thick on the frozen meadows. The majestic form of Ritter was full in sight, and I pushed rapidly on over rounded rock-bosses and pavements, my iron-shod shoes making a clanking sound, suddenly hushed now and then in rugs of bryanthus, and sedgy lake-margins soft as moss. . . .

On the southern shore of a frozen lake, I encountered an extensive field of hard, granular snow, up which I scampered in fine tone, intending to follow it to its head, and cross the rocky spur against which it leans, hoping thus to come direct upon the base of the main Ritter peak. The surface was pitted with oval hollows, made by stones and drifted pine-needles that had melted themselves into the mass by the radiation of absorbed sun-heat. These afforded good footholds, but the surface curved more and more steeply at the head, and the pits became shallower and less abundant, until I found myself in danger of being shed off like avalanching snow. I persisted, however, creeping on all fours, and shuffling up the smoothest places on my back, as I had often done on burnished granite, until, after slipping several times, I was compelled to retrace my course to the bottom, and make my way around the west end of the lake, and thence up to the summit of the divide between the head waters of Rush Creek and the northernmost tributaries of the San Joaquin.

Arriving on the summit of this dividing crest, one of the most exciting pieces of pure wilderness was disclosed that I ever discovered in all my mountaineering. There, immediately in front, loomed the majestic mass of Mount Ritter, with a glacier swooping down its face nearly to my feet, then curving westward and pouring its frozen flood into a dark blue lake, whose shores were bound with precipices of crystalline snow; while a deep chasm drawn between the divide and the glacier separated the massive picture from everything else. I could see only the one sublime mountain, the one glacier, the one lake; the whole veiled with one blue shadow—rock, ice, and water close together without a single leaf or sign of life. After gazing spellbound, I began instinctively to scrutinize every notch and gorge and weathered buttress of the mountain, with reference to making the ascent. The entire front above the glacier appeared as one tremendous precipice, slightly receding at the top, and bristling with spires and pinnacles set above one another in formidable array. Massive lichen-stained battlements stood forward here and there, hacked at the top with angular notches, and separated by frosty gullies and recesses that have been veiled in shadow ever since their creation; while to right and left, as far as I could see, were huge, crumbling buttresses, offering no hope to the climber. . . .

I could not distinctly hope to reach the summit from this side, yet I moved on across the glacier as if driven by fate. Contending with myself, the season is too far spent, I said, and even should I be successful, I might be storm-bound on the mountain; and in the cloud-darkness, with the cliffs and crevasses covered with snow, how could I escape! No; I must wait till next summer. I would only approach the mountain now, and inspect it, creep about its flanks, learn what I could of its history, holding myself ready to flee on the approach of the first storm-cloud. But we little know until tried how much of the uncontrollable there is in us, urging across glaciers and torrents, and up dangerous heights, let the judgment forbid as it may.

I succeeded in gaining the foot of the cliff on the eastern extremity of the glacier, and there discovered the mouth of a narrow avalanche gully, through which I began to climb, intending to follow it as far as possible, and at least obtain some fine wild views for my pains. . . .

I thus made my way into a wilderness of crumbling spires and battlements, built together in bewildering combinations, and glazed in many places with a thin coating of ice, which I had to hammer off with stones. The situation was becoming gradually more perilous; but, having passed several dangerous spots, I dared not think of descending; for, so steep was the entire ascent, one would inevitably fall to the glacier in case a single misstep were made. . . .

At length, after attaining an elevation of about 12,800 feet, I found myself at the foot of a sheer drop in the bed of the avalanche channel I was tracing, which seemed absolutely to bar further progress. It was only about forty-five or fifty feet high, and somewhat roughened by fissures and projections; but these seemed so slight and insecure, as footholds, that I tried hard to avoid the precipice altogether, by scaling the wall of the channel on either side. But, though less steep, the walls were smoother than the obstructing rock, and repeated efforts only showed that I must either go right ahead or turn back. The tried dangers beneath seemed even greater than that of the cliff in front; therefore, after scanning its face again and again, I began to scale it, picking my holds with intense caution. After gaining a point about half-way to the top, I was suddenly brought to a dead stop, with arms outspread, clinging close to the face of the rock, unable to move hand or foot either up or down. My doom appeared fixed. I *must* fall. There would be a moment of bewilderment, and then a lifeless rumble down the one general precipice to the glacier below.

When this final danger flashed upon me, I became nerve-shaken for the first time since setting foot on the mountains, and my mind seemed to fill with a stifling smoke. But this terrible eclipse lasted only a moment, when life blazed forth again with preternatural clearness. I seemed suddenly to become possessed of a new sense. The other self, bygone experiences, Instinct, or

Guardian Angel—call it what you will—came forward and assumed control. Then my trembling muscles became firm again, every rift and flaw in the rock was seen as through a microscope, and my limbs moved with a positiveness and precision with which I seemed to have nothing at all to do. Had I been borne aloft upon wings, my deliverance could not have been more complete.

Above this memorable spot, the face of the mountain is still more savagely hacked and torn. It is a maze of yawning chasms and gullies, in the angles of which rise beetling crags and piles of detached boulders that seem to have been gotten ready to be launched below. But the strange influx of strength I had received seemed inexhaustible. I found a way without effort, and soon stood upon the topmost crag in the blessed light.

How truly glorious the landscape circled around this noble summit!—giant mountains, valleys innumerable, glaciers and meadows, rivers and lakes, with the wide blue sky bent tenderly over them all. But in my first hour of freedom from that terrible shadow, the sunlight in which I was laving seemed all in all.

Looking southward along the axis of the range, the eye is first caught by a row of exceedingly sharp and slender spires, which rise openly to a height of about a thousand feet, above a series of short, residual glaciers that lean back against their bases; their fantastic sculpture and the unrelieved sharpness with which they spring out of the ice rendering them peculiarly wild and striking. These are "The Minarets." Beyond them you behold a sublime wilderness of mountains, their snowy summits towering together in crowded abundance, peak beyond peak, swelling higher, higher as they sweep on southward, until the culminating point of the range is reached on Mount Whitney, near the head of the Kern River, at an elevation of nearly 14,700 feet above the level of the sea. . . .

Lakes are seen gleaming in all sorts of places—round, or oval, or square, like very mirrors; others narrow and sinuous, drawn close around the peaks like silver zones, the highest reflecting only rocks, snow, and the sky. But neither these nor the glaciers, nor the bits of brown meadow and moorland that occur here and there, are large enough to make any marked impression upon the mighty wilderness of mountains. The eye, rejoicing in its freedom, roves about the vast expanse, yet returns again and again to the fountain peaks. Perhaps some one of the multitude excites special attention, some gigantic castle with turret and battlement, or some Gothic cathedral more abundantly spired than Milan's. But, generally,

when looking for the first time from an all-embracing standpoint like this, the inexperienced observer is oppressed by the incomprehensible grandeur, variety, and abundance of the mountains rising shoulder to shoulder beyond the reach of vision; and it is only after they have been studied one by one, long and lovingly, that their far-reaching harmonies become manifest. Then, penetrate the wilderness where you may, the main telling features, to which all the surrounding topography is subordinate, are quickly perceived, and the most complicated clusters of peaks stand revealed harmoniously correlated and fashioned like works of art—eloquent monuments of the ancient ice-rivers that brought them into relief from the general mass of the range. The cañons, too, some of them a mile deep, mazing wildly through the mighty host of mountains, however lawless and ungovernable at first sight they appear, are at length recognized as the necessary effects of causes which followed each other in harmonious sequence—Nature's poems carved on tables of stone—the simplest and most emphatic of her glacial compositions.

Could we have been here to observe during the glacial period, we should have overlooked a wrinkled ocean of ice as continuous as that now covering the landscapes of Greenland; filling every valley and cañon with only the tops of the fountain peaks rising darkly above the rock-encumbered ice-waves like islets in a stormy sea—those islets the only hints of the glorious landscapes now smiling in the sun. Standing here in the deep, brooding silence all the wilderness seems motionless, as if the work of creation were done. But in the midst of this outer steadfastness we know there is incessant motion and change. Ever and anon, avalanches are falling from yonder peaks. These cliff-bound glaciers, seemingly wedged and immovable, are flowing like water and grinding the rocks beneath them. The lakes are lapping their granite shores and wearing them away, and every one of these rills and young rivers is fretting the air into music, and carrying the mountains to the plains. Here are the roots of all the life of the valleys, and here more simply than elsewhere is the external flux of nature manifested. Ice changing to water, lakes to meadows, and mountains to plains. And while we thus contemplate Nature's methods of landscape creation, and, reading the records she has carved on the rocks, reconstruct, however imperfectly, the landscapes of the past, we also learn that as these we now behold have succeeded those of the preglacial age, so they in turn are withering and vanishing to be succeeded by others yet unborn.

But in the midst of these fine lessons and land-scapes, I had to remember that the sun was wheeling far to the west, while a new way down the mountain had to be discovered to some point on the timber line where I could have a fire; for I had not even burdened myself with a coat. I first scanned the western spurs, hoping some way might appear through which I might reach the northern glacier, and cross its snout; or pass around the lake into which it flows, and thus strike my morning track. This route was soon sufficiently un-folded to show that, if practicable at all, it would require so much time that reaching camp that night would be out of the question. I therefore scrambled back east-ward, descending the southern slopes obliquely at the same time. Here the crags seemed less formidable, and the head of a glacier that flows northeast came in sight, which I determined to follow as far as possible, hoping thus to make my way to the foot of the peak on the east side, and thence across the intervening cañons and ridges to camp.

The inclination of the glacier is quite moderate at the head, and, as the sun had softened the *névé,* I made safe and rapid progress, running and sliding, and keep-ing up a sharp outlook for crevasses. . . .

Night drew near before I reached the eastern base of the mountain, and my camp lay many a rugged mile to the north; but ultimate success was assured. It was now only a matter of endurance and ordinary mountain-craft. The sunset was, if possible, yet more beautiful than that of the day before. The Mono land-scape seemed to be fairly saturated with warm, purple light. The peaks marshaled along the summit were in shadow, but through every notch and pass streamed vivid sun-fire, soothing and irradiating their rough, black angles, while companies of small luminous clouds hovered above them like very angels of light. . . . I discovered the little pine thicket in which my nest was, and then I had a rest such as only a tired moun-taineer may enjoy. After lying loose and lost for a while, I made a sunrise fire, went down to the lake, dashed water on my head, and dipped a cupful for tea. The revival brought about by bread and tea was as complete as the exhaustion from excessive enjoyment and toil. Then I crept beneath the pine-tassels to bed. The wind was frosty and the fire burned low, but my sleep was none the less sound, and the evening con-stellations had swept far to the west before I awoke.

After thawing and resting in the morning sunshine, I sauntered home—that is, back to the Tuolumne camp—bearing away toward a cluster of peaks that hold the fountain snows of one of the north tributaries of Rush Creek. Here I discovered a group of beautiful glacier lakes, nestled together in a grand amphitheater. Toward evening, I crossed the divide separating the Mono waters from those of the Tuolumne, and entered the glacier basin that now holds the fountain snows of the stream that forms the upper Tuolumne cascades. This stream I traced down through its many dells and gorges, meadows and bogs, reaching the brink of the main Tuolumne at dusk.

A loud whoop for the artists was answered again and again. Their camp-fire came in sight, and half an hour afterward I was with them. They seemed unrea-sonably glad to see me. I had been absent only three days; nevertheless, though the weather was fine, they had already been weighing chances as to whether I would ever return, and trying to decide whether they should wait longer or begin to seek their way back to the lowlands. Now their curious troubles were over. They packed their precious sketches, and next morn-ing we set out homeward bound. . . .

For Further Discussion

1. Muir uses a wide variety of adjectives to describe the Sierra landscape. Some—"intense azure" and "translucent purple"—describe colors; others—"savage," "passionate," and "majestic"—describe natural objects such as rocks, storms, and moun-tains. Is there any way to distinguish descriptions that are accurate, true, or objective from descrip-tions that are inaccurate or subjective embel-lishments?

2. How many different types of value can you iden-tify in Muir's descriptions of the Sierra country?

3. Muir describes part of his trip in terms that sug-gest something of a religious or spiritual experi-ence. Are such experiences more likely to occur in wilderness areas than in cathedrals?

4. Muir suggests that the inexperienced observer might be overwhelmed by the vistas from high in the mountains, but that the "harmonies" of such scenes can be perceived only by someone who has studied them "long and lovingly." Does the knowl-edge and attitude with which you view a land-scape determine what you see?

What's Wrong with Plastic Trees?

Martin H. Krieger

In this essay, Martin Krieger examines a variety of reasons that are offered for preserving natural environments. He believes that we should find a middle ground between preservationist policies and conservationist-utilitarian policies. Such a middle ground would permit flexible preservationism yet also allow many opportunities for experiencing natural environments. Krieger concludes that the wilderness, as we now understand it, is the product of social and political decisions; the meaning we attribute to the wilderness is social, not biological. He believes that social justice—justice for human beings, not for trees—should be the ultimate goal of environmental policy. Thus, economic calculations must be a part of this decision, although symbolic and social values must also be included. To this end, we should recognize that, "Much more can be done with plastic trees . . . to give most people the feeling that they are experiencing nature."

A tree's a tree. How many more [redwoods] do you need to look at? If you've seen one, you've seen them all.—Attributed to Ronald Reagan, then candidate for governor of California.

A tree is a tree, and when you've seen one redwood, given your general knowledge about trees, you have a pretty good idea of the characteristics of a redwood. Yet most people believe that when you've seen one, you haven't seen them all. Why is this so? What implications does this have for public policy in a world where resources are not scarce, but do have to be manufactured; where choice is always present; and where the competition for resources is becoming clearer and keener? In this article, I attempt to explore some of these issues, while trying to understand the reasons that are given, or might be given, for preserving certain natural environments.

THE ECOLOGY MOVEMENT

In the past few years, a movement concerned with the preservation and careful use of the natural environment in this country has grown substantially. This ecology movement, as I shall call it, is beginning to have genuine power in governmental decision-making and is becoming a link between certain government agencies and the publics to which they are responsible. The ecology movement should be distinguished from related movements concerned with the conservation and wise use of natural resources. The latter, ascendant in the United States during the first half of this century, were mostly concerned with making sure that natural resources and environments were used in a fashion that reflected their true worth to man. This resulted in a utilitarian conception of environments and in the adoption of means to partially preserve them—for example, cost-benefit analysis and policies of multiple use on federal lands.

The ecology movement is not necessarily committed to such policies. Noting the spoliation of the environment under the policies of the conservation movement, the ecology movement demands much greater concern about what is done to the environment, independently of how much it may cost. The ecology movement seeks to have man's environment valued in and of itself and thereby prevent its being traded off for the other benefits it offers to man.

It seems likely that the ecology movement will have to become more programmatic and responsive to compromise as it moves into more responsible and bureaucratic positions vis-à-vis governments and administrative agencies. As they now stand, the policies of the ecology movement may work against resource-conserving strategies designed to lead to the movement's desired ends in 20 or 30 years. Meier has said:

> The best hope, it seems now, is that the newly evolved ideologies will progress as social movements. A number of the major tenets of the belief system may then be expected to lose their centrality and move to the periphery of collective attention. Believers may thereupon only "satisfice" with respect to these principles; they are ready to consider compromises.

What is needed is an approach midway between the preservationist and conservationist-utilitarian policies.

It is necessary to find ways of preserving the opportunity for experiences in natural environments, while having, at the same time, some flexibility in the alternatives that the ecology movement could advocate.

A new approach is needed because of the success of economic arguments in the past. We are now more concerned about social equity and about finding arguments from economics for preserving "untouched" environments. Such environments have not been manipulated very much by mankind in the recent past (hundreds or thousands of years). Traditional resource economics has been concerned not as much with preservation as with deciding which intertemporal (the choice of alternative times at which one intervenes) use of natural resources over a period of years yields a maximum return to man, essentially independent of considerations of equity. If one believes that untouched environments are unlikely to have substitutes, then this economics is not very useful. In fact, a different orientation toward preservation has developed and is beginning to be applied in ways that will provide powerful arguments for preservation. At the same time, some ideas about how man experiences the environment are becoming better understood, and they suggest that the new economic approach will be in need of some modification, even if most of its assumptions are sound.

I first examine what is usually meant by natural environments and rarity; I will then examine some of the rationales for preservation. It is important to understand the character and the weak points of the usual arguments. I also suggest how our knowledge and sophistication about environments and our differential access to them are likely to lead to levers for policy changes that will effectively preserve the possibility of experiencing nature, yet offer alternatives in the management of natural resources.

One limitation of my analysis should be made clear. I have restricted my discussion to the nation-state, particularly to the United States. If it were possible to take a global view, then environmental questions would be best phrased in terms of the world's resources. If we want undisturbed natural areas, it might be best to develop some of them in other countries. But we do not live in a politically united world, and such a proposal is imperialistic at worst and unrealistic at best. Global questions about the environment need to be considered, but they must be considered in terms of controls that can exist. If we are concerned about preserving natural environments, it seems clear that, for the moment, we will most likely have to preserve them in our own country.

THE AMERICAN FALLS: KEEPING IT NATURAL

For the last few thousand years, Niagara Falls has been receding. Water going over the Falls insinuates itself into crevices of the rock, freezes and expands in winter, and thereby causes cracks in the formation. The formation itself is a problem in that the hard rock on the surface covers a softer substratum. This weakness results not only in small amounts of erosion or small rockfalls, but also in very substantial ones when the substratum gives way. About 350,000 cubic yards (1 cubic yard equals 0.77 cubic meter) of talus lie at the base of the American Falls.

The various hydroelectric projects that have been constructed during the years have also affected the amount of water that flows over the Falls. It is now possible to alter the flow of water over the American Falls by a factor of 2 and, consequently, to diminish that of the Horseshoe (Canadian) Falls by about 10 percent.

As a result of these forces, the quality of the Falls—its grandeur, its height, its smoothness of flow—changes over the millenia [sic] and the months.

There is nothing pernicious about the changes wrought by nature; the problem is that Americans' image of the Falls does not change. Our ideal of a waterfall, an ideal formed by experiences with small, local waterfalls that seem perfect and by images created by artists and photographers, is not about to change without some effort.

When one visits the Falls today, he sees rocks and debris at the base, too much or too little water going over the edge, and imperfections in the flow of water. These sights are not likely to make anyone feel that he is seeing or experiencing the genuine Niagara Falls. The consequent effects on tourism, a multimillion-dollar-per-year industry, could be substantial.

At the instigation of local forces, the American Falls International Board has been formed under the auspices of the International Joint Commission of the United States and Canada. Some $5 to $6 million are being spent to investigate, by means of "dewatering" the Falls and building scale models, policies for intervention. That such efforts are commissioned suggests that we, as a nation, believe that it is proper and possible to do something about the future evolution of the Falls. A "Fallscape" committee, which is especially concerned with the visual quality of the Falls, has been formed. It suggests that three strategies, varying in degree of intervention, be considered.

1. The Falls can be converted into a monument. By means of strengthening the structure of the Falls, it is possible to prevent rockfalls. Also, excess rock from the base can be removed. Such a strategy might cost tens of millions of dollars, a large part of this cost being for the removal of talus.

2. The Falls could become an event. Some of the rocks at the base could be removed for convenience and esthetics, but the rockfalls themselves would not be hindered. Instead, instruments for predicting rockfalls could be installed. People might then come to the Falls at certain times, knowing that they would see an interesting and grand event, part of the cycle of nature, such as Old Faithful.

3. The Falls might be treated as a show. The "director" could control the amount of water flowing over the Falls, the size of the pool below, and the amount of debris, thereby producing a variety of spectacles. Not only could there be *son et lumière,* but it could take place on an orchestrated physical mass.

Which of these is the most nearly natural environment? Current practice, exemplified by the National Park Service's administration of natural areas, might suggest that the second procedure be followed and that the Falls not be "perfected." But would that be the famous Niagara Falls, the place where Marilyn Monroe met her fate in the movie *Niagara*? The answer to this question lies in the ways in which efforts at preservation are presented to the public. If the public is seeking a symbolic Falls, then the Falls has to be returned to its former state. If the public wants to see a natural phenomenon at work, then the Falls should be allowed to fall.

Paradoxically, the phenomena that the public thinks of as "natural" often require great artifice in their creation. The natural phenomenon of the Falls today has been created to a great extent by hydroelectric projects over the years. Esthetic appreciation of the Falls has been conditioned by the rather mundane considerations of routes of tourist excursions and views from hotel windows, as well as the efforts of artists.

I think that we can provide a smooth flow of water over the Falls and at the same time not be completely insensitive to natural processes if we adopt a procedure like that described in the third proposal. Niagara Falls is not virgin territory, the skyscrapers and motels will not disappear. Therefore, an aggressive attitude toward the Falls seems appropriate. This does not imply heavy-handedness in intervention (the first proposal), but a willingness to touch the "sacred" for esthetic as well as utilitarian purposes.

The effort to analyze this fairly straightforward policy question is not trivial. Other questions concerning preservation have fuzzier boundaries, less clear costs (direct and indirect), and much more complicated political considerations. For these reasons, it seems worthwhile to examine some of the concepts I use in this discussion.

NATURAL ENVIRONMENTS

What is considered a natural environment depends on the particular culture and society defining it. It might be possible to create for our culture and society a single definition that is usable (that is, the definition would mean the same thing to many people), but this, of course, says nothing about the applicability of such a definition to other cultures. However, I restrict my discussion to the development of the American idea of a natural environment.

The history of the idea of the wilderness is a good example of the development of one concept of natural environment. I follow Nash's discussion in the following.

A wilderness may be viewed as a state of mind, as an attitude toward a collection of trees, other plants, animals, and the land on which they all exist. The idea that a wilderness exists as a product of an intellectual movement is important. A wilderness is not discovered in the sense that some man from a civilization looked upon a piece of territory for the first time. It is the meanings that we attach to such a piece of territory that convert it to a wilderness.

The Romantic appreciation of nature, with its associated enthusiasm for the "strange, remote, solitary and mysterious," converted territory that was a threatening wildland into a desirable area capable of producing an invigorating spirit of wilderness. The "appreciation of the wilderness in this form began in cities," for whose residents the wildland was a novelty. Because of the massive destruction of this territory for resources (primarily timber), city dwellers, whose livelihood did not depend on these resources and who were not familiar with the territory, called for the preservation of wildlands. At first, they did not try to keep the most easily accessible, and therefore most economically useful, lands from being exploited, but noted that Yellowstone and the Adirondacks were rare wonders and had no other utility. They did not think of these areas as wilderness, but as untouched lands. Eventually, a battle developed between conservationists and preservationists. The conservationists (Pinchot, for example), were concerned with the wise use of lands, with science and civilization and forestry; the preservationists (Muir, for example) based their argument on art and wilderness. This latter concept of wilderness is the significant one. The preservationists converted wildland into wilderness—a good that is indivisible and valuable in itself.

This capsule history suggests that the wilderness, as we think of it now, is the product of a political effort to give a special meaning to a biological system organized in a specific way. I suspect that this history is the appropriate model for the manner in which biological systems come to be designated as special.

But it might be said that natural environments can be defined in the way ecosystems are—in terms of complexity, energy and entropy flows, and so on. This is true, but only because of all the spadework that has gone into developing in the public a consensual picture of natural environments. What a society takes to be a natural environment is one.

Natural environments are likely to be named when there are unnatural environments and are likely to be noted only when they are outnumbered by these unnatural environments. The wildlands of the past, which were frightening, were plentiful and were not valued. The new wilderness, which is a source of revitalization, is rare and so valued that it needs to be preserved.

WHEN IS SOMETHING RARE?

Something is considered to be rare when there do not exist very many objects or events that are similar to it. It is clear that one object must be distinguishable from another in order to be declared rare, but the basis for this distinction is not clear.

One may take a realist's or an idealist's view of rarity. For the realist, an object is unique within a purview: given a certain boundary, there exists no other object like it. Certainly the Grand Canyon is unique within the United States. Perhaps Niagara Falls is also unique. But there are many other waterfalls throughout the world that are equally impressive, if not of identical dimensions.

For the idealist, a rare object is one that is archetypal: it is the most nearly typical of all the objects it represents, having the most nearly perfect form. We frequently preserve archetypal specimens in museums and botanical gardens. Natural areas often have these qualities.

A given object is not always rare. Rather, it is designated as rare at one time and may, at some other time, be considered common. How does this happen? Objects become rare when a large number of people change their attitudes toward them. This may come about in a number of ways, but it is necessary that the object in question be noticed and singled out. Perhaps one individual discovers it, or perhaps it is common to everyone's experience. Someone must convince the public that the object is something special. The publicist must develop in others the ability to differentiate one object from among a large number of others, as well as to value the characteristic that makes the particular object different. If he convinces a group of people influential in the society, people who are able to affect a much larger group's beliefs, then he will have succeeded in his task. Thus it may be important that some form of snob appeal be created for the special object.

In order to create the differentiations and the differential valuations of characteristics, information and knowledge are crucial. A physical object can be transformed into an instrument of beauty, pleasure, or pride, thereby developing sufficient characteristics to be called rare, only by means of changing the knowledge we have of it and of its relation to the rest of the world. In this sense, knowledge serves an important function in the creation of rare environments, very much as knowledge in society serves an important function in designating what should be considered natural resources.

Advertising is one means of changing states of knowledge—nor does such advertising have to be wholly sponsored by commercial interests. Picture post cards, for example, are quite effective:

. . . a large number of quiet beauty spots which in consequence of the excellence of their photographs had become tourist centres. . . .

The essential was to "establish" a picture, e.g., the Tower [of London] with barges in the foreground. People came to look for the barges and in the end wouldn't have the Tower without barges. Much of the public was very conservative and, though such things as high-rise building and general facade-washing had made them [the post card producers] rephotograph the whole of London recently, some people still insisted on the old skyline, and grubby facades, and liked to believe certain new roads had never happened.

Similarly, the publicity given to prices paid at art auctions spurs the rise of these prices.

As a *result* of the social process of creating a rare object, the usual indicators of rarity become important. Economically, prices rise; physically, the locations of the rare objects become central, or at least highly significant spatially; and socially, rare objects and their possessors are associated with statuses that are valued and activities that are considered to be good.

ENVIRONMENTS CAN BE AND ARE CREATED

To recapitulate, objects are rare because men decide that they are and, through social action, convince others that they are. The rarity of an object is created through four mechanisms: designating the object as rare; differentiating it from other objects of the same species; establishing its significance; and determining its position in the context of society. The last two mechanisms are especially important, for the meaning that an environment has and its relation to other things in the society are crucial to its being considered rare. That a rare environment be irreproducible or of unchanging character is usually a necessary preliminary to our desire to preserve it. Technologies, which may involve physical processes or social organization and processes, determine how reproducible an object is, for we may make a copy of the original or we may transfer to another object the significance attached to the original. (Copying natural environments may be easier than copying artistic objects because the qualities of replicas and forgeries are not as well characterized in the case of the natural environment.) Insofar as we are incapable of doing either of these, we may desire to preserve the original environment.

In considering the clientele of rare environments, one finds that accessibility by means of transportation and communication is important. If there is no means of transportation to a rare environment, then it is not likely that the public will care about that environment. An alternative to transportation is some form of com-

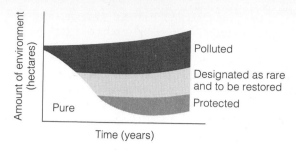

Figure 1. The development of rare environments.

munication, either verbal or pictorial, that simulates a feeling of being in the environment.

I am concerned here with the history of environments that, at first, are not considered unique. However, a similar argument could be applied to environments regarded as unique (for example, the Grand Canyon), provided they were classed with those environments most like them. Figure 1 should aid in the explanation that follows:

For example, suppose that a particular kind of environment is plentiful and that, over a period of time, frequent use causes it to become polluted. (Note that pollution need not refer just to our conventional concepts of dirtying the environment, but to a wide variety of uncleanliness and stigma as well.) Because there is a substantial amount of that environment available, man's use of it will, at first, have little effect on his perception of its rarity. As time goes on, however, someone will notice that there used to be a great deal more of that particular environment available. Suddenly, the once vast quantities of that environment begin to look less plentiful. The environment seems more special as it becomes distinguishable from the polluted environments around it. At that point, it is likely that there will be a movement to designate some fraction of the remaining environment as rare and in need of protection. There will also be a movement to restore those parts of the environment that have already been polluted. People will intervene to convert the polluted environment to a simulation of the original one.

REASONS FOR PRESERVATION

That something is rare does not imply that it must be preserved. The characteristics that distinguish it as rare must also be valued. Arguments in favor of preserving an object can be based on the fact that the object is a luxury, a necessity, or a merit.

We build temples or other monuments to our society (often by means of preservation) and believe that they represent important investments in social unity and coherence. If a forest symbolizes the frontier for a society and if that frontier is meaningful in the society's history, then there may be good reasons for preserving it. An object may also be preserved in order that it may be used in the future. Another reason, not often given but still true, for preserving things is that there is nothing else worth doing with them. For example, it may cost very little to preserve something that no one seems to have any particular reason for despoiling; therefore, we expend some small effort in trying to keep it untouched.

Natural environments are preserved for reasons of necessity also. Environments may provide ecological samples that will be useful to future generations. Recently, the long-lived bristlecone pine has helped to check radiocarbon dating and has thereby revised our knowledge of early Europe. It may be that the preservation of an environment is necessary for the preservation of an ecosystem and that our destruction of it will also destroy, as a product of a series of interactions, some highly valued aspects of our lives. Finally, it may be necessary to preserve environments in order that the economic development of the adjacent areas can proceed in a desired fashion.

Other reasons for preservation are based on merit: it may be felt by the society that it is good to preserve natural environments. It is good for people to be exposed to nature. Natural beauty is worth having, and the amenity resulting from preservation is important.

RARITY, UNIQUENESS, AND FORGERY: AN ARTISTIC INTERLUDE

The problems encountered in describing the qualities that make for "real" artistic experiences and genuine works of art are similar to those encountered in describing rare natural objects. The ideas of replica and forgery will serve to make the point.

Kubler observes that, if one examines objects in a time sequence, he may decide that some are prime objects and the others are replicas. Why should this be so? One may look at the properties of earlier objects and note that some of them serve as a source of later objects; however, since the future always has its sources in the present, any given object is a source. Therefore, one must distinguish important characteristics, perhaps arbitrarily, and say that they are seminal. Prime objects are the first to clearly and decisively exhibit important characteristics.

Why are there so few prime objects? By definition, prime objects exhibit characteristics in a clear and decisive way, and this must eliminate many other objects from the category; but why do artists not constantly create new objects, each so original that it would be prime? Not all artists are geniuses, it might be said. But this is just a restatement of the argument that most objects do not exhibit important characteristics in a clear, decisive manner. It might also be said that, if there are no followers, there will be no leaders, but this does not explain why some eras are filled with prime works and others are not.

Kubler suggests that invention, especially if too frequent, leads to chaos, which is frightening. Replication is calmer and leads only to dullness. Therefore, man would rather repair, replicate what he has done, than innovate and discard the past. We are, perhaps justifiably, afraid of what the prime objects of the future will be. We prefer natural environments to synthesized ones because we are familiar with techniques of managing the natural ones and know what the effects of such management are. Plastic trees are frightening.

What about those replicas of prime objects that are called forgeries? Something is a forgery if its provenance has been faked. Why should this bother us? If the forgery provides us with the same kind of experience we might have had with the original, except that we know it is a forgery, then we are snobbish to demand the original. But we do not like to be called snobs. Rather, we say that our opinion of the work, or the quality of our experience of it, depends on its context. History, social position, and ideology affect the way in which we experience the object. It may be concluded that our appreciation of something is only partly a product of the thing itself.

Art replicas and forgeries exist in an historical framework. So do the prime and genuine objects. And so do natural environments.

CRITERIA FOR PRESERVATION

Whatever argument one uses for preservation, there must be some criteria for deciding what to preserve. Given that something is rare and is believed to be worth preserving, rarity itself, as well as economic, ecological, or socio-historical reasons, can be used to justify preservation. I consider each of these here.

There are many economic reasons for planned intervention to achieve preservation, and I discuss two of them: one concerns the application of cost-benefit analysis to preservation; the other concerns the argument that present value should be determined by future benefits.

The work of Krutilla is an ingenious application of economics; it rescues environments from current use by arguing for their future utility. The crux of the argument follows.

Nature is irreproducible compared to the materials it provides. As Barnett and Morse have shown, there have been enough substitutions of natural materials to obviate the idea of a shortage of natural resources. It also seems likely that the value of nature and of experiences in nature will increase in the future, while the supply of natural environments will remain constant. Because it is comparatively easy to produce substitutes for the materials we get from natural environments, the cost of not exploiting an environment is small, compared to the cost of producing that environment. Finally, there is an option demand for environments: that is, there will be a demand, at a certain price, for that environment in the future. If a substantial fraction of the supply of the environment is destroyed now, it will be impossible to fill the demand in the future at a reasonable price. Therefore, we are willing to pay to preserve that option. The problem is not the intertemporal use of natural environments (as it is for natural resources), but the preservation of our options to use environments in the future, or at least the reduction of uncertainty about the availability of environments in the future.

Fisher has applied optimal investment theory, including a possibility of restoring environments to a quasi-natural state, to the problem of preservation as formulated by Krutilla. Krutilla *et al.* have applied an analysis similar to Fisher's to the preservation of Hell's Canyon.

Robinson has criticized Krutilla's argument from the following perspectives: he suggests that the amenity valued so highly by Krutilla is not necessarily that valuable; that the experiences of nature are reproducible; that refraining from current use may be costly; and that the arguments for public intervention into such environments depend on the collective consumption aspects of these environments. That is, these environments benefit everyone, and, since people cannot

be differentially charged for using them, the public must pay for these environments collectively, through government. It is well known that the users of rare environments tend to be that small fraction of the population who are better off socially and economically than the majority. However, a greater difficulty than any of these may be discerned.

It seems to me that the limitations of Krutilla's argument lie in his assumptions about how quickly spoiled environments can be restored (rate of reversion) and how great the supply of environments is. Krutilla *et al.* are sensitive to the possibility that the rate of reversion may well be amenable to technological intervention:

> Perhaps more significant, however, is the need to investigate more fully the presumption of asymmetric implications of technological progress for the value of attributes of the natural environment when used as intermediate goods, compared with their retention as assets supplying final consumption services. Irreproducibility, it might be argued, is not synonymous with irreplaceability. If reasonably good substitutes can be found, by reliance on product development, the argument for the presumption of differential effects of technological progress is weakened; or if not weakened, the value which is selected [for the reversion rate] . . . would not remain unaffected.

The supply of natural environments is affected by technology in that it can manipulate both biological processes and information and significance. The advertising that created rare environments can also create plentiful substitutes. The supply of special environments can be increased dramatically by highlighting (in ways not uncommon to those of differentiating among groups of equivalent toothpastes) significant and rare parts of what are commonly thought to be uninteresting environments.

The accessibility of certain environments to population centers can be altered to create new rare environments. Also, environments that are especially rare, or are created to be especially rare, could be very far away, since people would be willing to pay more to see them. Thus it may be possible to satisfy a large variety of customers for rare environments. The following kind of situation might result.

1. Those individuals who demand "truly" natural environments could be encouraged to fly to some isolated location where a national park with such an

environment is maintained; a substantial sum of money would be required of those who use such parks.

2. For those who find a rare environment in state parks or perhaps in small national parks, such parks could be made more accessible and could be developed more. In this way, a greater number of people could use them and the fee for using them would be less than the fee for using isolated areas.

3. Finally, for those who wish to have an environment that is just some trees, some woods, and some grass, there might be a very small park. Access would be very easy, and the rareness of such environments might well be enhanced beyond what is commonly thought possible by means of sophisticated methods of landscape gardening.

It seems to me that, as Krutilla suggests, the demand for rare environments is a learned one. It also seems likely that conscious public choice can manipulate this learning so that the environments which people learn to use and want reflect environments that are likely to be available at low cost. There is no lack of merit in natural environments, but this merit is not canonical.

THE VALUATION OF THE FUTURE

In any cost-benefit analysis that attempts to include future values, the rate which the future is discounted is crucial to the analysis. (That is, a sum of money received today is worth more to us now than the same sum received in the future. To allow for this, one discounts, by a certain percent each year, these future payments.) Changes in discount rates can alter the feasibility of a given project. If different clientele's preferences for projects correspond to different discount rates at which these projects are feasible, then the choice of a particular discount rate would place the preferences of one group over another. Preservation yields benefits that come in the future. The rich have a low rate of discount compared to the poor (say, 5 percent as opposed to 10 or 20 percent) and would impute much higher present value to these future benefits than the poor would. Baumol suggests (though it is only a hunch) that:

> . . . by and large, the future can be left to take care of itself. There is no need to lower artificially the social rate of discount in order to increase further the prospective wealth of future generations. . . . However, this does not mean that the future should in every respect be left at the mercy of the free market. . . . Investment in the preservation of such items then seems perfectly proper, but for this purpose the appropriate instrument would appear to be a set of selective subsidies rather than a low general discount rate that encourages indiscriminately all sorts of investment programs whether or not they are relevant.

Baumol is saying that the process of preserving environments may not always be fruitfully analyzed in terms of cost-benefit analyses; we are preserving things in very special cases, and each choice is not a utilitarian choice in any simple sense, but represents a balancing of all other costs to the society of having *no* preserved environments. Preservation often entails a gross change in policy, and utilitarian analyses cannot easily compare choices in which values may be drastically altered.

OTHER CRITERIA

We may decide to preserve things just because they are rare. In that case, we need to know which things are rarer than others. Leopold has tried to do this for a set of natural environments. He listed a large number of attributes for each environment and then weighted each attribute as follows. For any single attribute, determine how many environments share that attribute and assign each of them a value of $1/N$ units, where N is the number of environments that share an attribute. Then add all the weights for the environments; the environment with the largest weight is the rarest. It is clear that, if an environment has attributes which are unique, it will get one unit of weight for each attribute and thus its total weight will just equal the number of attributes. If all of the environments are about the same, then each of them will have roughly the same weight, which will equal the number of attributes divided by the number of environments. The procedure is sensitive to how differentiated we wish to make our attributes and to the attributes we choose. It is straightforward and usable, as Leopold has shown.

It seems to me that there are two major difficulties in this approach. The first, and more important, is that the accessibility of environments to their clientele, which Leopold treats as one of his 34 attributes, needs to be further emphasized in deciding what to preserve.

An environment that is quite rare but essentially inaccessible may not be as worthy of preservation as one that is fairly common but quite accessible. The other difficulty is that probably the quantity that should be used is the amount of information possessed by each environment—rather than taking $1/N$, one should take a function of its logarithm to the base 2.

An ecological argument is that environments which contribute to our stability and survival as an ecosystem should be preserved. It is quite difficult to define what survival means, however. If it means the continued existence of man in an environment quite similar to the one he lives in now, then survival is likely to become very difficult as we use part of our environment for the maintenance of life and as new technologies come to the fore. If survival means the maintenance of a healthy and rich culture, then ecology can only partially guide us in the choices, since technology has substantially changed the risk from catastrophe in the natural world. Our complex political and social organizations may serve to develop means for survival and stability sufficient to save man from the catastrophic tricks of his own technology.

If a taxonomy of environments were established, a few environments might stand out from all the rest. But what would be the criteria involved in such a taxonomy?

Another possibility is to search for relics of cultural, historical, and social significance to the nation. Such physical artifacts are preserved because the experiences they represent affect the nature of the present society. In this sense, forests are preserved to recall a frontier, and historic homes are preserved to recall the individuals who inhabited them. Of course the problem here is that there is no simple way of ordering the importance of relics and their referents. Perhaps a survey of a large number of people might enable one to assign priorities to these relics.

Finally, it might be suggested that preservation should only be used, or could sometimes be used, to serve the interests of social justice. Rather than preserving things for what they are or for the experiences they provide, we preserve them as monuments to people who deserve commemoration or as a means of redistributing wealth (when an environment is designated as rare, local values are affected). Rather than buy forests and preserve them, perhaps we should preserve slums and suitably reward their inhabitants.

All of these criteria are problematic. Whichever ones are chosen, priorities for intervention must still be developed.

PRIORITIES FOR PRESERVING THE ENVIRONMENT

Not every problem in environmental quality is urgent, nor does every undesirable condition that exists need to be improved. We need to classify environmental problems in order that we can choose from among the possible improvements.

1. There are conditions about which we must do something soon or we will lose a special thing. These conditions pertain especially to rare environments, environments we wish to preserve for their special beauty or their uniqueness. We might allocate a fixed amount of money every year to such urgent problems. Niagara Falls might be one of these, and it might cost a fraction of a dollar per family to keep it in good repair. Wilderness and monument maintenance have direct costs of a few dollars per family per year.

2. There are situations in which conditions are poor, but fairly stable. In such situations, it might be possible to handle the problem in 10 years without too much loss. However, the losses to society resulting from the delayed improvement of these facilities need to be carefully computed. For example, the eutrophied Lake Erie might be such a project. There, society loses fishing and recreational facilities. It might cost $100 per family, locally, to clean up the lake. Perhaps our environmental dollars should be spent elsewhere.

3. There are also situations in which conditions are rapidly deteriorating and in which a small injection of environmental improvement and amelioration would cause dramatic changes in a trend. Smog control devices have probably raised the cost of driving by 2 or 3 percent, yet their contribution to the relative improvement of the environment in certain areas (for example, Los Angeles) has been substantial. Fifty dollars per car per year is the estimated current cost to the car owner.

4. There may be situations in which large infusions of money are needed to stop a change. These problems are especially irksome. Perhaps the best response to them would be to change the system of production sufficiently that we can avoid such costs in the future. The costs of such change, one-time costs we hope, may be much smaller than the long-term costs of the problems themselves, although this need not be the case. The development of cleaner industrial processes is a case in point.

This is not an all-inclusive or especially inventive classification of problems, but I have devised it to suggest that many of the "urgent" problems are not so urgent.

Rare environments pose special problems and may require an approach different from that required by other environments. A poor nation is unlikely to destroy very much of its special environments. It lacks the technical and economic power to do so. It may certainly perform minor miracles of destruction through a series of small decisions or in single, major projects. These latter are often done with the aid of rich countries.

The industrialized, but not wealthy, nations have wreaked havoc with their environments in their efforts to gain some degree of wealth. It is interesting that they are willing to caution the poor nations against such a course, even though it may be a very rapid way of developing. At the U.N. Conference on the Environment this year, the poor nations indicated their awareness of these problems and their desire to develop without such havoc.

The rich nations can afford to have environments that are rare and consciously preserved. These environments are comparable to the temples of old, in that these environments will be relics of *our* time, yet this is no criterion for deciding how much should be spent on "temple building." The amount of money needed is only a small proportion of a rich country's wealth (as opposed to the cost of churches in medieval times).

Politically, the situation is complicated. There are many small groups in this country for whom certain environments are highly significant. The problem for each group is to somehow get its piece of turf, preferably uncut, unrenewed, or untouched. It seems likely that the ultimate determinant of which environments are preserved will be a process of political trade-off, in which some environments are preserved for some groups and other environments for others. Natural environments are likely to be viewed in a continuum with a large number of other environments that are especially valued by some subgroup of the society. In this sense, environmental issues will become continuous with a number of other special interests and will no longer be seen as a part of a "whole earth" movement. The power of the intellectuals, in the media, and even in union bureaucracies, with their upper middle class preferences for nature, suggests that special interest groups who are advocates for the poor and working classes will have to be wary of their own staffs.

Projects might be ranked in importance on the basis of the net benefits they provide a particular group. Marglin has suggested a means by which income redistribution could be explicitly included in cost-benefit calculations for environmental programs. If one wishes to take efficiency into account, costs minus benefits could be minimized with a constraint relating to income redistribution. This is not a simple task, however, because pricing some commodities at zero dollars, seemingly the best way of attempting a redistribution of income, may not be politically desirable or feasible. As Clawson and Knetch have pointed out, we have to be sure that in making some prices low we do not make others prohibitively high and thereby deny the persons who are to benefit access to the low-priced goods. In any case, Marglin shows that the degree to which income is redistributed will depend on how the same amount of money might have been spent in alternative activities (marginal opportunity cost). This parallels Kneese and Bower's view that the level of pollution we tolerate, or is "optimal," is that at which the marginal benefits of increasing pollution are balanced by the marginal costs of abatement measures.

In doing these cost-benefit calculations, one must consider the value of 10 years of clean lake (if we can clean up the lake now) versus 10 years of uneducated man (if we wait 10 years for a manpower training program). According to Freeman:

> . . . [the] equity characteristics of projects *within* broad classifications . . . will be roughly similar. If this surmise is correct, then the ranking of projects within these classes is not likely to be significantly affected by equity considerations. On the other hand, we would expect more marked differences in distribution patterns among classes of projects, e.g., rural recreation vs. urban air quality.

He goes on to point out that it is unlikely that such seemingly incommensurable kinds of projects will be compared with respect to equity. I suspect that it is still possible to affect specific groups in the design of a given project; furthermore, equity can be taken into consideration more concretely at this level. Careful disaggregation, in measuring effects and benefits, will be needed to ensure that minorities are properly represented.

AN ETHICAL QUESTION

I still feel quite uncomfortable with what I have said here. I have tried to show that the utilitarian and manipulative rationality inherited from the conservationist movement and currently embodied in economic analyses and modes of argument can be helpful in deciding questions of preservation and rarity. By manipulating attitudes, we have levers for intervening into what is ordinarily considered fixed and uncontrollable. But to what end?

Our ability to manipulate preferences and values tends to lead to systems that make no sense. For example, an electrical utility encourages its customers to use more electricity, and the customers proceed to do so. As a result, there are power shortages. Similarly, if we allocate resources now in order to preserve environments for future generations, their preferences for environments may be altered by this action, and there may be larger shortages.

I also fear that my own proposals might get out of hand. My purpose in proposing interventions is not to preserve man's opportunity to experience nature, although this is important, but to promote social justice. I believe that this concern should guide our attempts to manipulate, trade off, and control environments. A *summum bonum* of preserving trees has no place in an ethic of social justice. If I took this ethic seriously, I could not argue the relative merits of schemes to manipulate environments. I would argue that the ecology movement is wrong and would not answer its question about what we are going to do about the earth—I would be worried about what we are going to do about men.

CONCLUSION

With some ingenuity, a transformation of our attitudes toward preservation of the environment will take place fairly soon. We will recognize the symbolic and social meanings of environments, not just their economic utility; we will emphasize their historical significance as well as the future generations that will use them.

At the same time, we must realize that there are things we may not want to trade at all, except in the sense of letting someone else have his share of the environment also. As environments become more differentiated, smaller areas will probably be given greater significance, and it may be possible for more groups to have a share.

It is likely that we shall want to apply our technology to the creation of artificial environments. It may be possible to create environments that are evocative of other environments in other times and places. It is possible that, by manipulating memory through the rewriting of history, environments will come to

have new meaning. Finally, we may want to create proxy environments by means of substitution and simulation. In order to create substitutes, we must endow new objects with significance by means of advertising and by social practice. Sophistication about differentiation will become very important for appreciating the substitute environments. We may simulate the environment by means of photographs, recordings, models, and perhaps even manipulations in the brain. What we experience in natural environments may actually be more controllable than we imagine. Artificial prairies and wildernesses have been created, and there is no reason to believe that these artificial environments need be unsatisfactory for those who experience them.

Rare environments are relative, can be created, are dependent on our knowledge, and are a function of policy, not only tradition. It seems likely that economic arguments will not be sufficient to preserve environments or to suggest how we can create new ones. Rather, conscious choice about what matters, and then a financial and social investment in an effort to create significant experiences and environments, will become a policy alternative available to us.

What's wrong with plastic trees? My guess is that there is very little wrong with them. Much more can be done with plastic trees and the like to give most people the feeling that they are experiencing nature. We will have to realize that the way in which we experience nature is conditioned by our society—which more and more is seen to be receptive to responsible interventions.

Bentham, the father of utilitarianism, was very concerned about the uses of the dead to the living and suggested:

> If a country gentleman have rows of trees leading to his dwelling, the autoicons [embalmed bodies in an upright position] of his family might alternate with the trees; copal varnish would protect the face from the effects of rain—caoutchouc [rubber] the habiliments.

For Further Discussion

1. In speaking of Niagara Falls, Krieger suggests that although the quality of the falls changes over millennia and the months due to natural causes, "Americans' image of the Falls does not change." He seems to think that this is a problem. What, exactly, is the problem? In what ways are changes that occur over the millennia different from changes that occur over the months?

2. Krieger tells us that "what is considered a natural environment depends on the particular culture and society defining it." Do you agree? He also says that "objects are rare because men decide that they are." Is this true of rare species?

3. Consider the following: "The demand for rare environments is a learned one. It also seems likely that conscious public choice can manipulate this learning so that the environments which people learn to use and want reflect environments that are likely to be available at low cost." Do you agree or disagree?

4. Krieger claims that his ethic is to "promote social justice," not merely to promote the preservation of nature, and that a "summum bonum of preserving trees has no place in an ethic of social justice." Can you think of situations in which preserving natural areas is part of the quest for social justice?

Faking Nature

Robert Elliot

Many environmentalists object to development projects or resource uses that destroy natural areas. But could this objection be overcome if it were possible to restore natural areas? In fact, many restoration projects seek to restore woodlands, prairies, wildlife habitats, wetlands, seashores, and even rivers. If it were possible to accomplish such restoration (and this is a big if), would anything of value be lost in the process? Robert Elliot thinks that there would be something of significance lost.

In the previous reading, Martin Krieger spoke of plastic trees as providing the same experience and feeling as real trees. Elliot asks us to consider if the experience of nature truly is a matter of feelings. Even if the tree were a real one, would our experience have the same value if the tree had been replanted after an original had been harvested? Elliot believes that the history and causal origin of natural objects can play a major role in their value. By analogy with works of art, Elliot believes that there is a major value difference between originals and forgeries, even if both give rise to similar feelings and perceptions.

I

Consider the following case. There is a proposal to mine beach sands for rutile. Large areas of dune are to be cleared of vegetation and the dunes themselves destroyed. It is agreed, by all parties concerned, that the dune area has value quite apart from a utilitarian one. It is agreed, in other words, that it would be a bad thing, considered in itself, for the dune area to be dramatically altered. Acknowledging this the mining company expresses its willingness, indeed its desire, to restore the dune area to its original condition after the minerals have been extracted.[1] The company goes on to argue that any loss of value is merely temporary and that full value will in fact be restored. In other words they are claiming that the destruction of what has value is compensated for by the later creation (recreation) of something of equal value. I shall call this "the restoration thesis."

In the actual world many such proposals are made, not because of shared conservationist principles, but as a way of undermining the arguments of conservationists. Such proposals are in fact effective in defeating environmentalist protest. They are also notoriously ineffective in putting right, or indeed even seeming to put right, the particular wrong that has been done to the environment. The sand-mining case is just one of a number of similar cases involving such things as open-cut mining, clear-felling of forests, river diversion, and highway construction. Across a range of such cases some concession is made by way of acknowledging the value of pieces of landscape, rivers, forests, and so forth, and a suggestion is made that this value can be restored once the environmentally disruptive process has been completed.

Imagine, contrary to fact, that restoration projects are largely successful; that the environment is brought back to its original condition and that even a close inspection will fail to reveal that the area has been mined, clear-felled, or whatever. If this is so, then there is temptation to think that one particular environmentalist objection is defeated. The issue is by no means merely academic. I have already claimed that restoration promises do in fact carry weight against environmental arguments. Thus Mr. Doug Anthony, the Australian Deputy Prime Minister, saw fit to suggest that sand-mining on Fraser Island could be resumed once "the community becomes more informed and more enlightened as to what reclamation work is being carried out by mining companies. . . ."[2] Or consider how the protests of environmentalists might be deflected in the light of the following report of environmental engineering in the United States.

> . . . about 2 km of creek 25 feet wide has been moved to accommodate a highway and in doing so engineers with the aid of landscape architects and biologists have rebuilt the creek to the same standard as before. Boulders, bends, irregularities and natural vegetation have all been designed into the new section. In addition, special log structures have been built to improve the habitat as part of a fish development program.[3]

Not surprisingly the claim that revegetation, rehabilitation, and the like restore value has been strongly contested. J. G. Mosley reports that:

> The Fraser Island Environmental Inquiry Commissioners did in fact face up to the question of the relevance of successful rehabilitation to the decision on whether to ban exports (of beach sand minerals) and were quite unequivocal in saying that if the aim was to protect a natural area such success was irrelevant. . . . The Inquiry said: ". . . even if, contrary to the overwhelming weight of evidence before the Commission, successful rehabilitation of the flora after mining is found to be ecologically possible on all mined sites on the Island . . . the overall impression of a wild, uncultivated island refuge will be destroyed forever by mining."[4]

I want to show both that there is a rational, coherent ethical system which supports decisive objections to the restoration thesis, and that that system is not lacking in normative appeal. The system I have in mind will make valuation depend, in part, on the presence of properties which cannot survive the disruption-restoration process. There is, however, one point that needs clarifying before discussion proceeds.

Establishing that restoration projects, even if empirically successful, do not fully restore value does not by any means constitute a knock-down argument against some environmentally disruptive policy. The value that would be lost if such a policy were implemented may be just one value among many which conflict in this situation. Countervailing considerations may be decisive and the policy thereby shown to be the right one. If my argument turns out to be correct it will provide an extra, though by no means decisive, reason for adopting certain environmentalist policies. It will show that the resistance which environmentalists display in the face of restoration promises is not merely silly, or emotional, or irrational. This is important because so much of the debate assumes that settling the dispute about what is ecologically possible automatically settles the value question. The thrust of much of the discussion is that if restoration is shown to be possible, and economically feasible, then recalcitrant environmentalists are behaving irrationally, being merely obstinate or being selfish.

There are indeed familiar ethical systems which will serve to explain what is wrong with the restoration thesis in a certain range of cases. Thus preference utilitarianism will support objections to some restoration proposal if that proposal fails to maximally satisfy preferences. Likewise, classical utilitarianism will lend support to a conservationist stance provided that the restoration proposal fails to maximize happiness and pleasure. However, in both cases the support offered is contingent upon the way in which the preferences and utilities line up. And it is simply not clear that they line up in such a way that the conservationist position is even usually vindicated. While appeal to utilitarian considerations might be strategically useful in certain cases they do not reflect the underlying motivation of the conservationists. The conservationists seem committed to an account of what has value which allows that restoration proposals fail to compensate for environmental destruction despite the fact that such proposals would maximize utility. What then is this distinct source of value which motivates and underpins the stance taken by, among others, the Commissioners of the Fraser Island Environmental Inquiry?

II

It is instructive to list some reasons that might be given in support of the claim that something of value would be lost if a certain bit of the environment were destroyed. It may be that the area supports a diversity of plant and animal life, it may be that it is the habitat of some endangered species, it may be that it contains striking rock formations or particularly fine specimens of mountain ash. If it is only considerations such as these that contribute to the area's value then perhaps opposition to the environmentally disruptive project would be irrational provided certain firm guarantees were available; for instance that the mining company or timber company would carry out the restoration and that it would be successful. Presumably there are steps that could be taken to ensure the continuance of species diversity and the continued existence of the endangered species. Some of the other requirements might prove harder to meet, but in some sense or other it is possible to recreate the rock formations and to plant mountain ash that will turn out to be particularly fine specimens. If value consists of the presence of objects of these various kinds, independently of what explains their presence, then the restoration thesis would seem to hold. The environmentalist needs to appeal to some feature which cannot be replicated as a source of some part of a natural area's value.

Putting the point thus indicates the direction the environmentalist could take. He might suggest that an area is valuable, partly, because it is a natural area, one that has not been modified by human hand, one that is undeveloped, unspoilt, or even unsullied. This suggestion is in accordance with much environmentalist rhetoric, and something like it at least must be at the basis of resistance to restoration proposals. One way of teasing out the suggestion and giving it a normative basis is to take over a notion from aesthetics. Thus we might claim that what the environmental engineers are proposing is that we accept a fake or a forgery instead of the real thing. If the claim can be made good then perhaps an adequate response to restoration proposals is to point out that they merely fake nature; that they offer us something less than was taken away.[5] Certainly there is a weight of opinion to the effect that, in art at least, fakes lack a value possessed by the real thing.[6]

One way in which this argument might be nipped in the bud is by claiming that it is bound to exploit an ultimately unworkable distinction between what is natural and what is not. Admittedly the distinction between the natural and the non-natural requires detailed working out. This is something I do not propose doing. However, I do think the distinction can be made good in a way sufficient to the present need. For present purposes I shall take it that "natural" means something like "unmodified by human activity." Obviously some areas will be more natural than others according to the degree to which they have been shaped by human hand. Indeed most rural landscapes will, on this view, count as non-natural to a very high degree. Nor do I intend the natural/non-natural distinction to exactly parallel some dependent moral evaluations; that is, I do not want to be taken as claiming that what is natural is good and what is non-natural is not. The distinction between natural and non-natural connects with valuation in a much more subtle way than that. This is something to which I shall presently return. My claim then is that restoration policies do not always fully restore value because part of the reason that we value bits of the environment is because they are natural to a high degree. It is time to consider some counter-arguments.

An environmental engineer might urge that the exact similarity which holds between the original and the perfectly restored environment leaves no room for a value discrimination between them. He may urge that if they are *exactly* alike, down to the minutest detail (and let us imagine for the sake of argument that this is a technological possibility), then they must be *equally* valuable. The suggestion is that value-discriminations depend on there being intrinsic differences between the states of affairs evaluated. This begs the question against the environmentalist, since it simply discounts the possibility that events temporally and spatially outside the immediate landscape in question can serve as the basis of some valuation of it. It discounts the possibility that the manner of the landscape's genesis, for example, has a legitimate role in determining its value. Here are some examples which suggest that an object's origins do affect its value and our valuations of it.

Imagine that I have a piece of sculpture in my garden which is too fragile to be moved at all. For some reason it would suit the local council to lay sewerage pipes just where the sculpture happens to be. The council engineer informs me of this and explains that my sculpture will have to go. However, I need not despair because he promises to replace it with an exactly similar artefact, one which, he assures me, not even the very best experts could tell was not the original. The example may be unlikely, but it does have some point. While I may concede that the replica would be better than nothing at all (and I may not even concede that), it is utterly improbable that I would accept it as full compensation for the original. Nor is my reluctance entirely explained by the monetary value of the original work. My reluctance springs from the fact that I value the original as an aesthetic object, as an object with a specific genesis and history.

Alternatively, imagine I have been promised a Vermeer for my birthday. The day arrives and I am given a painting which looks just like a Vermeer. I am understandably pleased. However, my pleasure does not last for long. I am told that the painting I am holding is not a Vermeer but instead an exact replica of one previously destroyed. Any attempt to allay my disappointment by insisting that there just is no difference between the replica and the original misses the mark completely. There is a difference and it is one which affects my perception, and consequent valuation, of the painting. The difference of course lies in the painting's genesis.

I shall offer one last example which perhaps bears even more closely on the environmental issue. I am given a rather beautiful, delicately constructed, object. It is something I treasure and admire, something in which I find considerable aesthetic value. Everything is fine until I discover certain facts about its origin. I discover that it is carved out of the bone of someone killed especially for that purpose. This discovery affects me deeply and I cease to value the object in the way that I once did. I regard it as in some sense sullied, spoilt by the facts of its origin. The object itself has not changed but my perceptions of it have. I now know that it is not quite the kind of thing I thought it was, and that my prior valuation of it was mistaken. The discovery is like the discovery that a painting one believed to be an original is in fact a forgery. The discovery about the object's origin changes the valuation made of it, since it reveals that the object is not of the kind that I value.

What these examples suggest is that there is at least a prima facie case for partially explaining the value of objects in terms of their origins, in terms of the kinds of processes that brought them into being. It is easy to find evidence in the writings of people who have valued nature that things extrinsic to the present, immediate environment determine valuations of it. John Muir's remarks about Hetch Hetchy Valley are a case in point.[7] Muir regarded the valley as a place where he could have direct contact with primeval nature; he valued it, not just because it was a place of great beauty, but because it was also a part of the world that had not been shaped by human hand. Muir's valuation was conditional upon certain facts about the valley's genesis; his valuation was of a, literally, natural object, of an object with a special kind of continuity with the past. The news that it was a carefully contrived elaborate *ecological* artefact would have transformed that valuation immediately and radically.

The appeal that many find in areas of wilderness, in natural forests and wild rivers depends very much on the naturalness of such places. There may be similarities between the experience one has when confronted with the multi-faceted complexity, the magnitude, the awesomeness of a very large city, and the experience one has walking through a rain forest. There may be similarities between the feeling one has listening to the roar of water over the spillway of a dam, and the feeling one has listening to a similar roar as a wild river tumbles down rapids. Despite the similarities there are also differences. We value the forest and river in part because they are representative of the world outside our dominion, because their existence is independent of us. We may value the city and the dam because of what they represent of human achievement. Pointing out the differences is not necessarily to denigrate either. However, there will be cases where we rightly judge that it is better to have the natural object than it is to have the artefact.

It is appropriate to return to a point mentioned earlier concerning the relationship between the natural and the valuable. It will not do to argue that what is natural is necessarily of value. The environmentalist can comfortably concede this point. He is not claiming that all natural phenomena have value in virtue of being natural. Sickness and disease are natural in a straightforward sense and are certainly not good. Natural phenomena such as fires, hurricanes, volcanic eruptions can totally alter landscapes and alter them for the worse. All of this can be conceded. What the environmentalist wants to claim is that, within certain constraints, the naturalness of a landscape is a reason for preserving it, a determinant of its value. Artificially transforming an utterly barren, ecologically bankrupt landscape into something richer and more subtle may be a good thing. That is a view quite compatible with the belief that replacing a rich natural environment with a rich artificial one is a bad thing. What the environmentalist insists on is that naturalness is one factor in determining the value of pieces of the environment. But that, as I have tried to suggest, is no news. The castle by the Scottish loch is a very different kind of object, value-wise, from the exact replica in the appropriately shaped environment of some Disneyland of the future. The barrenness of some Cycladic island would stand in a different, better perspective if it were not brought about by human intervention.

As I have glossed it, the environmentalist's complaint concerning restoration proposals is that nature is not replaceable without depreciation in one aspect of its value which has to do with its genesis, its history. Given this, an opponent might be tempted to argue that there is no longer any such thing as "natural" wilderness, since the preservation of those bits of it which remain is achievable only by deliberate policy. The idea is that by placing boundaries around national parks, by actively discouraging grazing, trail-biking and the like, by prohibiting sand-mining, we are turning the wilderness into an artefact, that in some negative or indirect way we are creating an environment. There is some truth in this suggestion. In fact we need to take notice of it if we do value wilderness, since positive policies *are* required to preserve it. But as an argument against my overall claim it fails. What is significant about wilderness is its causal continuity with the past. This is something that is not destroyed by demarcating an area and declaring it a national park. There is a distinction between the "naturalness" of the wilderness itself and the means used to maintain and protect it. What remains within the park boundaries is, as it were, the real thing. The environmentalist may regret that such positive policy is required to preserve the wilderness against human, or even natural, assault.[8] However, the regret does not follow from the belief that what remains is of depreciated value. There is a significant difference between preventing damage and repairing damage once it is done. This is the difference that leaves room for an argument in favour of a preservation policy over and above a restoration policy.

There is another important issue which needs highlighting. It might be thought that naturalness only matters in so far as it is perceived. In other words it might be thought that if the environmental engineer could perform the restoration quickly and secretly, then there would be no room for complaint. Of course, in one sense there would not be, since the knowledge which would motivate complaint would be missing. What this shows is that there can be loss of value without the loss being perceived. It allows room for valuations to be mistaken because of ignorance concerning relevant facts. Thus my Vermeer can be removed and secretly replaced with the perfect replica. I have lost

something of value without knowing that I have. This is possible because it is not simply the states of mind engendered by looking at the painting, by gloatingly contemplating my possession of it, by giving myself over to aesthetic pleasure, and so on which explain why it has value. It has value because of the kind of thing that it is, and one thing that it is is a painting executed by a man with certain intentions, at a certain stage of his artistic development, living in a certain aesthetic *milieu*. Similarly, it is not just those things which make me feel the joy that wilderness makes me feel, that I value. That would be a reason for desiring such things, but that is a distinct consideration. I value the forest because it is of a specific kind, because there is a certain kind of causal history which explains its existence. Of course I can be deceived into thinking that a piece of landscape has that kind of history, has developed in the appropriate way. The success of the deception does not elevate the restored landscape to the level of the original, anymore than the success of the deception in the previous example confers on the fake the value of a real Vermeer. What has value in both cases are objects which are of the kind that I value, not merely objects which I think are of that kind. This point, it should be noted, is appropriate independently of views concerning the subjectivity or objectivity of value.

An example might bring the point home. Imagine that John is someone who values wilderness. John may find himself in one of the following situations:

1. He falls into the clutches of a utilitarian-minded super-technologist. John's captor has erected a rather incredible device which he calls an experience machine. Once the electrodes are attached and the right buttons pressed one can be brought to experience anything whatsoever. John is plugged into the machine, and, since his captor knows full well John's love of wilderness, given an extended experience as of hiking through a spectacular wilderness. This is environmental engineering at its most extreme. Quite assuredly John is being short-changed. John wants there to be wilderness and he wants to experience it. He wants the world to be a certain way and he wants to have experiences of a certain kind: veridical.

2. John is abducted, blindfolded, and taken to a simulated, plastic wilderness area. When the blindfold is removed John is thrilled by what he sees around him: the tall gums, the wattles, the lichen on the rocks. At least that is what he thinks is there. We know better: we know that John is deceived, that he is once

again being short-changed. He has been presented with an environment which he thinks is of value but isn't. If he knew that the leaves through which the artificially generated breeze now stirred were synthetic he would be profoundly disappointed, perhaps even disgusted at what at best is a cruel joke.

3. John is taken to a place which was once devastated by strip-mining. The forest which had stood there for some thousands of years had been felled and the earth torn up, and the animals either killed or driven from their habitat. Times have changed, however, and the area has been restored. Trees of the species which grew there before the devastation grow there again, and the animal species have returned. John knows nothing of this and thinks he is in pristine forest. Once again, he has been short-changed, presented with less than what he values most.

In the same way that the plastic trees may be thought a (minimal) improvement on the experience machine, so too the real trees are an improvement on the plastic ones. In fact in the third situation there is incomparably more of value than in the second, but there could be more. The forest, though real, is not genuinely that John wants it to be. If it were not the product of contrivance he would value it more. It is a product of contrivance. Even in the situation where the devastated area regenerates rather than is restored, it is possible to understand and sympathize with John's claim that the environment does not have the fullest possible value. Admittedly in this case there is not so much room for that claim, since the environment has regenerated of its own accord. Still, the regenerated environment does not have the right kind of continuity with the forest that stood there initially: that continuity has been interfered with by the earlier devastation. (In actual fact the regenerated forest is likely to be perceivably quite different to the kind of thing originally there.)

III

I have argued that the causal genesis of forests, rivers, lakes, and so on is important in establishing their value. I have also tried to give an indication of why this is. In the course of my argument I drew various analogies, implicit rather than explicit, between faking art and faking nature. This should not be taken to suggest, however, that the concepts of aesthetic evalua-

tion and judgement are to be carried straight over to evaluations of, and judgements about, the natural environment. Indeed there is good reason to believe that this cannot be done. For one thing an apparently integral part of aesthetic evaluation depends on viewing the aesthetic object as an intentional object, as an artefact, as something that is shaped by the purposes and designs of its author. Evaluating works of art involves explaining them, and judging them, in terms of their author's intentions; it involves placing them within the author's corpus of work; it involves locating them in some tradition and in some special *milieu*. Nature is not a work of art though works of art (in some suitably broad sense) may look very much like natural objects.

None of this is to deny that certain concepts which are frequently deployed in aesthetic evaluation cannot usefully and legitimately be deployed in evaluations of the environment. We admire the intricacy and delicacy of colouring in paintings as we might admire the intricate and delicate shadings in a eucalypt forest. We admire the solid grandeur of a building as we might admire the solidity and grandeur of a massive rock outcrop. And of course the ubiquitous notion of *the beautiful* has a purchase in environmental evaluations as it does in aesthetic evaluations. Even granted all this there are various arguments which might be developed to drive a wedge between the two kinds of evaluation which would weaken the analogies between faking art and faking nature. One such argument turns on the claim that aesthetic evaluation has, as a central component, a judgemental factor, concerning the author's intentions and the like in the way that was sketched above.[9] The idea is that nature, like works of art, may elicit any of a range of emotional responses in viewers. We may be awed by a mountain, soothed by the sound of water over rocks, excited by the power of a waterfall, and so on. However, the judgemental element in aesthetic evaluation serves to differentiate it from environmental evaluation and serves to explain, or so the argument would go, exactly what it is about fakes and forgeries in art which discounts their value with respect to the original. The claim is that if there is no judgemental element in environmental evaluation, then there is no rational basis to preferring real to faked nature when the latter is a good replica. The argument can, I think, be met.

Meeting the argument does not require arguing that responses to nature count as aesthetic responses. I agree that they are not. Nevertheless there are analogies which go beyond emotional content, and which may persuade us to take more seriously the claim that faked nature is inferior. It is important to make the point that only in fanciful situations dreamt up by philosophers are there no detectable differences between fakes and originals, both in the case of artefacts and in the case of natural objects. By taking a realistic example where there are discernible, and possibly discernible, differences between the fake and the real thing, it is possible to bring out the judgemental element in responses to, and evaluations of, the environment. Right now I may not be able to tell a real Vermeer from a Van Meegaran, though I might learn to do so. By the same token I might not be able to tell apart a naturally evolved stand of mountain ash from one which has been planted, but might later acquire the ability to make the requisite judgement. Perhaps an anecdote is appropriate here. There is a particular stand of mountain ash that I had long admired. The trees were straight and tall, of uniform stature, neither densely packed nor too open-spaced. I then discovered what would have been obvious to a more expert eye, namely that the stand of mountain ash had been planted to replace original forest which had been burnt out. This explained the uniformity in size, the density and so on: it also changed my attitude to that piece of landscape. The evaluation that I make now of that landscape is to a certain extent informed, the response is not merely emotive but cognitive as well. The evaluation is informed and directed by my beliefs about the forest, the type of forest it is, its condition as a member of that kind, its causal genesis and so on. What is more, the judgemental element affects the emotive one. Knowing that the forest is not a naturally evolved forest causes me to feel differently about it: it causes me to perceive the forest differently and to assign it less value than naturally evolved forests.

Val Routley has eloquently reminded us that people who value wilderness do not do so merely because they like to soak up pretty scenery.[10] They see much more and value much more than this. What they do see, and what they value, is very much a function of the degree to which they understand the ecological mechanisms which maintain the landscape and which determine that it appears the way it does. Similarly, knowledge of art history, of painting techniques, and the like will inform aesthetic evaluations and alter aesthetic perceptions. Knowledge of this kind is capable of transforming a hitherto uninteresting landscape into one that is compelling. Holmes Rolston has discussed at length the way in which an understanding and appreciation of ecology generates new values.[11] He does not claim that ecology reveals values previously unnoticed, but rather that the understanding of the complexity, diversity, and integration of the natural world which ecology affords us, opens up a new area of valuation. As the facts are uncovered, the values are generated. What the remarks of Routley and Rolston highlight is the judgemental factor which is present in environmental appraisal. Understanding and evaluation do go hand in hand; and the responses individuals have to forests, wild rivers, and the like are not merely raw, emotional responses.

IV

Not all forests are alike, not all rain forests are alike. There are countless possible discriminations that the informed observer may make. Comparative judgements between areas of the natural environment are possible with regard to ecological richness, stage of development, stability, peculiar local circumstance, and the like. Judgements of this kind will very often underlie hierarchical orderings of environments in terms of their intrinsic worth. Appeal to judgements of this kind will frequently strengthen the case for preserving some bit of the environment. Thus one strong argument against the Tasmanian Hydroelectricity Commission's proposal to dam the Lower Gordon River turns on the fact that it threatens the inundation of an exceedingly fine stand of Huon pine. If the stand of Huon pines could not justifiably be ranked so high on the appropriate ecological scale then the argument against the dam would be to that extent weakened.

One reason that a faked forest is not just as good as a naturally evolved forest is that there is always the possibility that the trained eye will tell the difference.[12] It takes some time to discriminate areas of Alpine plain which are naturally clear of snow gums from those that have been cleared. It takes some time to discriminate regrowth forest which has been logged from forest which has not been touched. These are discriminations which it is possible to make and which are made. Moreover, they are discriminations which affect valuations. The reasons why the "faked" forest counts for less, more often than not, than the real thing are similar to the reasons why faked works of art count for less than the real thing.

Origin is important as an integral part of the evaluation process. It is important because our beliefs about it determine the valuations we make. It is also important in that the discovery that something has an origin quite different to the origin we initially believe that it has, can literally alter the way we perceive that thing.[13] The point concerning the possibility of detecting fakes is important in that it stresses just how much detail must be written into the claim that environmental engineers can replicate nature. Even if environmental engineering could achieve such exactitude, there is, I suggest, no compelling reasons for accepting the restoration thesis. It is worth stressing though that, as a matter of strategy, environmentalists must argue the empirical inadequacy of restoration proposals. This is the strongest argument against restoration ploys, because it appeals to diverse value-frameworks, and because such proposals are promises to deliver a specific good. Showing that the good won't be delivered is thus a useful move to make.

Notes

1. In this case *full* restoration will be literally impossible because the minerals are not going to be replaced.
2. J. G. Mosley, "The Revegetation 'Debate': A Trap for Conservationists," *Australian Conservation Foundation Newsletter,* 12/8 (1980), 1.
3. Peter Dunk, "How New Engineering Can Work with the Environment," *Habitat Australia,* 7/5 (1979), 12.
4. See Mosley, "The Revegetation 'Debate,'" 1.
5. Offering something less is not, of course, always the same as offering nothing. If diversity of animal and plant life, stability of complex ecosystems, tall trees, and so on are things that we value in themselves, then certainly we are offered something. I am not denying this, and I doubt that many would qualify their valuations of the above-mentioned items in a way that leaves the restored environment devoid of value. Environmentalists would count as of worth programmes designed to render polluted rivers reinhabitable by fish species. The point is rather that they may, as I hope to show, rationally deem it less valuable than what was originally there.
6. See e.g. Colin Radford, "Fakes," *Mind,* 87/345 (1978) 66–76, and Nelson Goodman, *Languages of Art* (New York: Bobbs-Merrill, 1968), 99–122, though Radford and Goodman have different accounts of why genesis matters.
7. See Ch. 10 of Roderick Nash, *Wilderness and the American Mind* (New Haven: Yale University Press, 1973).
8. For example protecting the Great Barrier Reef from damage by the crown-of-thorns starfish.
9. See e.g. Don Mannison, "A Prolegomenon to a Human Chauvinist Aesthetic," in D. S. Mannison, M. A. McRobbie, R. Routley, eds., *Environmental Philosophy* (Canberra: Research School of Social Sciences, Australian National University, 1980), 212–16.
10. Val Routley, "Critical Notice of Passmore's *Man's Responsibility for Nature,*" *Australasian Journal of Philosophy,* 53/2 (1975), 171–85.
11. Holmes Rolston III, "Is There an Ecological Ethic," *Ethics,* 85/2 (1975), 93–109.
12. For a discussion of this point with respect to art forgeries, see Goodman, *Languages of Art,* esp. 103–12.
13. For an excellent discussion of this same point with respect to artefacts, see Radford, "Fakes," esp. 73–6.

For Further Discussion

1. Both Elliot and Krieger use the example of an art forgery in their essays. In what ways is a restored natural area like an art forgery? In what ways is it different?
2. Elliot summarizes his conclusions as follows: "Restoration policies do not always fully restore value because part of the reason that we value bits of the environment is because they are natural to a high degree." Could we, as Krieger suggests, manipulate this belief and teach people to value restored environments?
3. Elliot claims that "the castle by the Scottish loch is a very different kind of object, value-wise, from the exact replica" in Disneyland. What, exactly, is the value difference? The Disney parks are some of the most popular vacation destinations in the world. How do you compare the value of these parks to the value of the High Sierra as described by John Muir?
4. Elliot says that he values the forest "because there is a certain kind of causal history which explains its existence." What does he mean by this? Do you agree?

Does Aesthetic Appreciation of Landscapes Need to Be Science-Based?

Holmes Rolston III

If beauty is in the eye of the beholder, what difference does it make if the beholder knows something? Does scientific knowledge of such things as ecosystems and biological processes affect their perception as aesthetic objects? Are landscapes and wilderness areas more beautiful or more awe inspiring to people who know something about them? These are the sorts of questions that philosopher Holmes Rolston examines in the following essay.

Rolston contrasts the immediate and subjective perception of landscapes with a science-based appreciation of landscapes. Rolston believes that "the eye of the beholder is notoriously subjective, hopelessly narrow in its capacities for vision," whereas "science cultivates the habit of looking closely." Throughout this essay, Rolston develops the claim that we cannot appropriately appreciate what we do not understand.

I. MYTHS, FOLKLORE, AND NATURAL HISTORY

The lava landscapes in Hawaii's Volcanoes National Park are quite aesthetically stimulating. On a memorable evening, I watched, in the twilight, red lava roll down into the ocean. The seashore on which I stood had literally been made only a few months before. Here was more land flowing forth; I knew something of how the world was made. Next morning, overlooking a dormant crater, steaming with sulphurous fumes, I noticed flowers and a little food. These were offerings made to Pele, a goddess who dwells in Kilauea volcano, placating her to stop the flow.[1]

Contrast my understanding with this native "superstition." The native peoples gave an animistic account; I know better—about tectonic plates, magma, basaltic lava, shield volcanoes, calderas, lava plateaux, and *nuées ardentes.* Yet, in my scientific superiority, I too there experienced the sublime, a virtually religious experience, as lava out of the bowels of Earth created new landscape at the edge of the sea.

The American Indians repeatedly warned John Wesley Powell against his first trip through the Grand Canyon. The canyon once contained a trail made by the god Tavwoats for a mourning chief to go to see his wife in heaven to the West. Then the god filled up the trail with a river and forbade anyone to go there. Powell would draw Tavwoats' wrath.[2] But Powell saw the canyon geologically. He too experienced awe, but of the erosional forces of time and the river flowing. He went on to direct the US Geological Survey, and, interestingly, to head the US Bureau of Ethnology, concerned with Indian affairs. The Indian legends have only antiquarian interest; no one appreciates the canyon for what it really is, unless helped by geologists to know about the Supai formation, the Redwall limestone, the inner Precambrian gorge, and so on. That is the definitive interpretation.

The classical Chinese practiced *feng shui.*[3] The *shen* spirits were yang in character, animating heaven, the arable earth, sun, moon, stars, winds, clouds, rain, thunder, fire, mountains, rivers, seas, trees, springs, stones, and plants. The *gui* spirits were yin, especially unpredictable, and likely to be out in the evenings, in the dark, and in lonely places. Such spirits had to be considered. One avoids, for example, straight lines in buildings or roads lest they be offended, and puts an earthenware cock on one's rooftop, because the cock crowing at sunrise wards off the spirits. A life energy, *chi,* flows through the landscape and affects where one locates one's home, and what one can do in the fields. But that must make appropriate aesthetic appreciation of the real Chinese landscapes impossible.

Or consider what our great grandfathers thought about the mountains, which we now consider so scenic.[4] They were "monstrous excrescences of nature."[5] God originally made the world a smooth sphere happily habitable for the original humans; but, alas, humans sinned, and the earth was warped in punishment. Thomas Burnet is repelled by these "ruines of a broken World," "wild, vast and indigested heaps of Stones and Earth" that resulted when "confusion came into Nature."[6] John Donne called them "warts, and pock-holes in the face of th'earth."[7]

Now we know better. After geology, we are more likely to approach mountains, as did William Wordsworth a century later, as the supreme example of the permanencies amidst changes in nature, and the manifestation of "types and symbols of Eternity."[8] That may go beyond science, but it must go through science to go beyond. So do not prescientific peoples characteristically misunderstand the landscapes they inhabit?

II. IN THE EYE OF THE BEHOLDER

But then another side of the issue comes to the fore. The landscapes that we ordinarily know are not pristine nature, but cultivated landscapes, rural or pastoral, with their towns and cities. Over the centuries, people have worked out their geography with multiple kinds of industry and perception, mixing nature and culture in diverse ways, no doubt some better, some worse. But who is to say that a science-based appreciation is the only right one?[9] Nature as seen by science is just the way we Westerners currently "constitute" our world—so the phenomenologists may say. There is no reason to think this the privileged view.[10]

Aesthetics—this argument continues—is nothing that science can discover on landscapes objectively, independently of persons. Aesthetic experience of landscapes is not some pre-existing characteristic of the landscape that is found, but one that emerges when persons react to landscapes. Landscape is land-scope, land taken into human scope. "Landscape *per se* does not exist; it is amorphous—an indeterminate area of the earth's surface and a chaos of details incomprehensible to the perceptual system. A landscape requires selective viewing and a frame. The 'line' of a mountain crest, woods, or prairie silhouetted against the sky is imaginary; it lies in the eye of the beholder. Landscapes need . . . the subjectivation of nature, or interpretation in terms of human experience."[11]

The Japanese love their landscapes tamed and manicured, more parks than wilderness.[12] They like artfully to prune their pines, cultivate simple flower and rock gardens, arrange a waterfall, attract some geese, walk a path with a geometrically rising curve, look back, and enjoy the moon rising over the temple, silhouetting it all. They are hardly interested in admiring a pristine ecosystem or geological formations. Should we say that the Japanese are engaging in some aesthetic deception? Yet who are we to argue they should give up their art and learn our science? The argument is rather that humans are always the landscape architects, and even science is another cultural way of framing landscapes.

Consider my parents. My mother did not know any geomorphology or landscape ecology. Yet she enjoyed her familiar, Southern US rural landscapes. My father enjoyed the fertility of the soils in the Shenandoah Valley in Virginia; he admired a good field. On visits around, he would take a spade and turn the soil to see whether it might make a good garden. He always knew what watershed he was in, what crops were growing where. He loved a good rain. Both enjoyed the changing seasons, the dogwood and redbud in the hills in the spring, the brilliant and subtle colours of autumn.

If one is an expressionist, then whatever moods landscapes can trigger, they trigger, and that human relationship exists as surely as do the rocks or the forests on the landscape. Nature is a smorgasbord of opportunities that humans can do with as they please. No one aesthetic response is more or less correct than any other; what counts is the imaginative play, and what is remarkable is nature's richness in launching this play.

III. BEYOND THE EYE OF THE BEHOLDER

Yes, but the eye of the beholder is notoriously subjective, hopelessly narrow in its capacities for vision. One has only to consult smell or taste, for example, to realize that much more is going on than the eye can see. Science, by extending so greatly human capacities for perception, and by integrating these into theory, teaches us what is objectively there. We realize what is going on in the dark, underground, or over time. Without science, there is no sense of deep time, nor of geological or evolutionary history, and little appreciation of ecology. Science cultivates the habit of looking closely, as well as of looking for long periods of time. One is more likely to experience the landscape at multiple scales of both time and space.

Humans are the only species that can reflect outside their niche. No other animal can do this, and science greatly helps us to extend our vision. Science helps us to see the landscape as free as possible from our subjective human preferences. Science corrects for truth. There are, for example, no "badlands," as my parents might have reacted to the western Dakotas. There are no "lonely places," although there are arid landscapes with little life, where the struggle for life needs to be especially respected. Things need to be appreciated in the right categories.[13]

Daniel Boone, exploring the wild Kentucky landscape, was too uneducated to see much of what was there, supposes Aldo Leopold. "Daniel Boone's reaction depended not only on the quality of what he saw, but on the quality of the mental eye with which he saw it. Ecological science has wrought a change in our mental eye. . . . We may safely say that, as compared with the competent ecologist of the present day, Boone saw only the surface of things. The incredible intricacies of the plant and animal community . . . were as invisible to Daniel Boone as they are today to Mr. Babbitt."[14]

But then again—Leopold checks himself—science is no guarantee that one will see what is there either. "Let no man jump to the conclusion that Babbitt must take his Ph.D. in ecology before he can 'see' his own country. On the contrary, the Ph.D. may become as callous as an undertaker at the mysteries at which he officiates. . . . Perception, in short, cannot be purchased with either learned degrees or dollars; it grows at home as well as abroad." The essential perception is of "the natural processes by which the land and the living things upon it have achieved their characteristic forms . . . and by which they maintain their existence."[15] Science or no science, everyone can gain some of that sensitivity. Although one can only know the evolutionary processes in deep time with the benefit of evolutionary theory, those who reside on landscapes know, or can know, the ecological processes well enough to appreciate life coping day by day, season by season, struggling and supported on the landscape. Indeed, Boone knew existentially what it is like to live on a landscape, something the Ph.D. may have never known. Beholders need to go beyond, but this is deeper into processes in which they are already participants in the landscape. My mother and father in Virginia and Alabama, and Mr. Babbitt in rural Wisconsin, lived with a keen sense of place.

Now the argument is that we cannot appropriately appreciate what we do not *understand.* Science understands how landscapes came to be and how they now function as communities of life. But people, too, form their communities of life; humans cannot appropriately appreciate what they do not *stand-under,* that is, undergo; and the scientist *qua* scientist does not objectively undergo any such experience. That requires persons sensitively encountering landscapes, evaluating them, making a living on them, rebuilding them, responding to them.

The argument, it seems, must spiral around two foci—the one that aesthetic experience must be participatory, relating an actual beholder to a landscape; the other that nature is objective to such beholders, actually known in the physical and biological sciences. The pivotal words we use: scenery; environment; ecology; nature; and landscape form an ellipse about these foci. A richer aesthetic experience is constituted with both natural science and participatory experience in natural history.

IV. SCENERY AND ARTFORM, ECOLOGY AND EVOLUTIONARY HISTORY

Some persons enjoy landscapes rather like big art. Landscape paintings give us a taste for the real thing. What we want is not ecology, but natural art. Consider the autumn leaves in their colour, so much admired by my mother and father, indeed by us all. If one is a formalist, then it does not matter how the landscape originated. Find a vantage point where trees near and far, foreground and background, are pleasantly framed, and admire the vista. The historical genesis is irrelevant. A drive through the countryside is something like a walk through a museum of landscape paintings. In the United States, the Park Service builds pull-overs at the best selected spots, where tourists take pictures. Others buy postcards. This is appreciating the form, line, colour, texture of what we behold.

But now we can argue that to make a found art object out of a landscape is to abstract from what it ecologically is. The ecological processes are not just at the pull-over sights; they are pervasively present on the landscape. They are back home on the landscapes left behind. This organic unity in a landscape is not gained by treating it as beautiful scenery, though it might be found if one discovered its ecology.

A British visitor to the Rocky Mountains, despite the fact that his Denver hosts had urged him, "You'll love the Rockies," complained that there were too many trees of too few kinds, mostly the same monotonous evergreens, too many rocks, too much sun too high in the sky, not enough water, the scale was too big and there were not enough signs of humans, no balanced elements of form and colour, nothing like the Lake District or the Scottish lochs.[16]

Can one argue that he was wrong? One argument is that he did not have the right scientific categories. He should not have expected a homey landscape, certainly not one like his homelands. If one visits semi-arid mountains, one should expect more rocks. If one goes into the tundra, the plants will be small, and the boulders will dominate, residual from glaciation. When you understand the harshness of an arid or an alpine climate, you will find the plants' clinging to life aesthetically stimulating. One will appreciate life hunkered down low to the ground, or bent and twisted trees persisting in cold and windblown environments.

The dominant spruce in the montane zone are evergreen and shaped as they are because they can photosynthesize year round and shed the snow; needles work better than leaves in the incessant wind. Lodgepole pine replaces itself after a stand replacement fire, hence the many trees all of about the same age. A Rocky Mountain forest does not lack essences in balance, as was complained by the unappreciative visitor; to the contrary, there life persists by perpetual dialectic of the environmental resistance and conductance, wind and water, hot and cold, life and death.

An emphasis on scenic beauty may lead one to devaluate that which is not beautiful—the rotted log, or the humus, or trees that have burned, blighted, or contorted, or Burnet's "wild, vast, undigested heaps of Stones and Earth." One wants to be able to appreciate prairies, swamps, tundras, and deserts. We start looking out for a prospect that pleases us, a pastoral scene, something that photographs well, a recreational scenic view, but we end with insight into wild processes that ignore us completely. Just that insight outside our aesthetic response becomes aesthetically stimulating.

V. ENVIRONMENT, ECOLOGY, NATURE, AND LANDSCAPE

The four words "environment," "ecology," "nature," and "landscape," have different, though sometimes overlapping, logics.

1. An *environment* does not exist without some organism *environed* by the world in which it copes; the root idea is surroundings. An environment is the current field of significance for a living being, usually its home, though not always, should an animal find itself, for instance, in a strange environment. Environments are settings under which life takes place, for people, animals, plants.

2. *Ecology* is the *logic of a home;* the root idea is the interactive relationships through which an organism is constituted in its environment. Here an environment is a niche that is inhabited. There must be somebody at home, making a living there; ecology takes dwelling. One cannot visit one's ecology, though one can visit someone else's ecology. There is no ecology on the moon. But virtually over all the Earth myriads of species of fauna and flora are at home in their niches.

3. *Nature* goes back to a Greek and Latin root, *gene (g)nasi, natus,* to give birth, to generate. A "native" is born on a landscape; "pregnant" contains the same root, as does "genesis." Nature is the entire system of things, with the aggregation of all their powers, properties, processes, and products—whatever follows natural law and whatever happens spontaneously. There are two contrast classes; the supernatural, which exceeds the natural, and the cultural, where artifacts replace spontaneous nature. Ecosystems are part, though not the whole, of nature. Humans have both natural and cultural environments; landscapes are typically hybrids.

4. *Landscape* is a section of the countryside that can be seen from some place. All of nature, from quarks to cosmos, is too much for us; we can only experience nature from perspectives, sometimes with telescopes or microscopes, but usually with the unaided eye. Landscape is the scope of nature, modified by culture, from some locus, and in that sense landscape is local, located. The question arises whether landscapes exist without humans. The moon is not itself a landscape, not a moonscape, not at least without astronauts to take the surface of the moon into their scope. Landscape comes into being in the human interaction with nature. The animals, much less the plants, do not appreciate aesthetically where they are. So landscape aesthetics is something that happens when humans locate themselves. My mother constituted her landscapes. My father constituted his. I constitute mine, whether as a scientist or as an inhabitant.

The natural world is there without us. When we constitute it, we want to appreciate something of the objective geomorphology and ecosystem that exists whether or not humans are interacting with it. Realizing this, we can follow the argument that landscape perception needs to be science-based, as well as participatory. Science becomes the primary avenue for perceiving landscapes, better than any other—necessary though not sufficient for their most adequate understanding. My mother's appreciation of her landscapes would have been enriched with science, and I am the proof of that. Her son inherits her appreciation and greatly enlarges it.

VI. *MY* ENVIRONMENT AND *THE* ENVIRONMENT

A horizon is perspectival. There are no horizons without perceivers. One sense of the word "environment" has that logic, noticing the modifiers. *My* environment is rather like my horizon. I take it with me as I move through the world. Horizons require an attention span. Analogously, my environment has an owner. We can spell this "environment" with a lower case *e*.

Arnold Berleant concludes: "This is what environment *means:* a fusion of organic awareness, of meanings both conscious and unaware, of geographical location, of physical presence, personal time, pervasive movement. . . . There are no surroundings separate from my presence in that place."[17] "For nothing can be said about environment that cannot be said about its people, since environment, in the sense I am writing about it here, includes a human factor."[18] "Environment is no region separate from us. It is not only the very condition of our being but a continuous part of that being."[19] "For environments are not physical places but perceptual ones that we collaborate in making, and it is perceptually that we determine their identity and extent."[20] *My* environment is my inhabited landscape, where I work and reside; our human landscape is where we have placed our culture. There are hardly any unpossessed landscapes. Landscape is personal and cultural history made visible.

But landscapes are more public and stable than horizons; we co-inhabit them with neighbours, others in our community. So *my* environment, true in shortest scope, is rather too private a term. My environment when encountered as a landscape is a commons shared, your environment too, *our* environment. That demands another, fuller sense in which *the* environment is out there, the natural world that we move through, there before we arrive, and there after we are gone. We can spell this "Environment" with an upper case *E.* Environment is not my creation; it is the creation. I do not constitute it; it has constituted me; and now it seems arrogant and myopic to speak of foreground and background, of what I frame on *my* horizons. Environment is *the* ground of my being, and we can remove the "my" because environment is the common ground of all being.

Landscape appreciation requires stretching environment into Environment. My mother could appreciate her Alabama lived environment; she did not need science to do that. But she could appreciate only her native range sector, a residential landscape, her field of significance, though she also knew it as the setting for the fauna and flora she saw there. She treasured her mother's quiltwork depicting the plants and animals on her farmland. Science alone does not give any such regional identity with a landscape, and such identity, too, qualifies one for aesthetic experience. A scientist without love for the earth is here disqualified. One's self is extended into one's environment, into *the* Environment. The subjective self knows its objective world, the creature rejoices in the creation.

VII. SCIENTIFIC, PARTICIPATORY ENVIRONMENTAL AESTHETICS

We do not always need science to teach us what happens on landscapes, though science enriches that story. All who have had to cope in the world knew this, natives of landscapes wherever. Science brings insight into continuing organic, ecological, and evolutionary unity, dynamic genesis; but such unity may also have already been realized by pre-scientific peoples in their inhabiting of a landscape. Science can engage us with landscapes too objectively, academically, disinterestedly; landscapes are also known in participant encounter, by being embodied in them.

The Japanese, looking as they do for essences in landscapes, enjoy the transience of nature; how the cherry blossoms are here today, gone tomorrow, and will return again next year, and the next after that. Everyone who constitutes a landscape must also cope on that landscape, and in that struggle everyone who beholds landscapes can become sensitive to what is going on as the world continues on, even though they may not know its deep history in geological and evolutionary time. They know context, if not origins. They know their environment, in the lower case; but they also know dimensions of the big Environment in which we live and move and have our being, because their local experience is a puzzle piece in that bigger picture.

Living on the landscape keeps persons "tuned in," and this dimension is needed, past mere science, to appreciate what is going on on landscapes. Certainly the human coping has produced mythologies that we now find incredible—Pele extruding herself as lava, Tavwoats replacing the trail to Paradise with a forbidding canyon river, the Chinese cocks on rooftops to guard off mischievous night spirits, an angry God warping the Earth to punish iniquitous humans. Science is necessary to banish ("deconstruct") these myths, before we can understand in a corrected aesthetic.

Yet there is a check on the extent of this error, because the coping myth must minimally reflect something of the struggle to live on from generation to generation; it must give its holders some sense of adapted fit on landscapes. They cannot be altogether blind to what is going on, indeed, the more they know about this, the better they survive. Such coping in humans has aesthetic as well as cognitive components. Metaphysical fancy has to be checked by a pragmatic functioning, and this includes an operational aesthetic with some successful reference to what is there at one's location. On the ground, we have to be realists, at least enough to survive.

Animals largely lack capacities for the aesthetic appreciation of their environments, though they prefer the kinds of environments for which they are adapted. Might we expect that for humans an aesthetic appreciation of environments has any survival value? Aesthetics, some will argue, has little relationship to biological necessity. People create metaphysics, others will argue, not to map reality but to insulate themselves from a world too tough otherwise to bear. Those archaic gods and superstitions, or the Form of the Good infusing itself on recalcitrant particulars, or the yang in counterpoint to yin—these are mostly untrue, as everyone in the scientific age now knows, even though they once helped people to cope. Are we also to suppose that those worldviews that could "frame" the landscape with beauty, even though the landscape is not so, have helped people to survive? Those who see the world pleasantly (and inaccurately) leave more offspring than those who see the world grimly (and accurately). Perhaps.

But it is a simpler hypothesis to hold that persons are in fact sensitive to beauty (variously constituted

through the lenses of this or that worldview, to be sure) to the extent that beauties (or the properties that excite beauty) are those confronting humans, and come naturally. There is no need to insulate oneself by pretending that it is beautiful. Coping might sometimes require self-deception; it could more often require a self that has become sensitive to its surroundings. What if we still find the landscape has its beauty, after we see through the lenses of science? More illusion? Or better insight into truth that had already been breaking through over the millenia?

Evolutionary theory requires also that humans be an adapted fit on their landscapes. If so, humans who find their environments congenial, or even beautiful, flourish, while those who find their environments stressful, or ugly, might do less well. Such human responses can be culturally introduced, or they can have a genetic disposition, or both. Humans rebuild their environments to suit their preferences. But elements of the natural environment remain in any cultured environment; the very idea of landscapes illustrates this. If there is some harmony between nature and culture, so much the more to the human liking. Some argue that humans prefer savannah landscapes, as these are the landscapes in which humans once evolved.[21] Trees, openings, grassy fields, green space, water, a homesite with foreground and background—these elements recur in landscape paintings rather transculturally.[22] Hospital patients with such views recover from surgery more rapidly.[23]

A persistent notion in many cultures is that exposure to nature enhances psychological well-being. Scientific studies are accumulating "steadily mounting evidence that there may be considerable correspondence across Western and some non-Western cultures in terms of positive aesthetic responsiveness to natural landscapes."[24] Indeed, humans in every culture enjoy aesthetic features in their landscapes, and it is difficult for them to come under the sway of mythologies or metaphysical cosmologies (or scientific theories!) that completely erase these features. It would also be difficult for mythologies and cosmologies everywhere to create these responses as mere appearances.

Cosmological ideas must "save the appearances," and many of these are "appearances" of beauty. Some beauty breaks through these worldviews, worse and better, because there is a certain existential immediacy to inhabiting landscapes, a *Sitz-im-Leben* grounded in participatory residence where the sensory perceptions confronting us are too strong to be argued away, or cooked up, by the inferences from metaphysics. Actual landscapes keep impacting us, and our worldviews keep having to answer to this impact, willy-nilly, when we constitute our landscapes. Landscape is not passive; it acts on us. The constituting is a two-way affair.

Still, mistaken interpretative frameworks do blind us so that we cannot see what is there, they create illusions of what is not there, they leave us ignorant about what is really going on; and here science greatly educates us to what is really taking place. The native-range experience, though it has on-the-ground immediacy, lacks depth, and this deeper beauty is what science can unfold. Native-range experience to which we are genetically predisposed, or something reinforced because it produces cultural prosperity, might apply only to relatively homey-like environments, savannas, or places that we can rebuild as savannas. Science can enlarge us for the appreciation of wilder, fiercer landscapes.

So, to return to the native Hawaiians, the Southwest Indians, the Chinese, and the European theologians, we ourselves are misguided to suppose that they found nothing aesthetically positive in their landscapes, despite these aspects in their worldviews that introduced apprehension and prevented an adequate appreciation. Burnet, for instance, confesses that he was initially drawn to the mountains aesthetically: "There is nothing that I look upon with more pleasure," he first said, "than the wide Sea and the Mountains of the earth. There is something august and stately in the Air of these things, that inspires the mind with great thoughts and passions; We do naturally upon such occasions, think of God and his greatness."[25] If Burnet had pursued his studies further, in the Psalms or Job, he would have found that the Hebrews took the same delight in their promised land, mountains, valleys, and all, and interpreted it as the gift of God.

The Chinese, with their yang and yin, likewise celebrated following the natural; native Americans felt a keen sense of belonging on their landscapes; the indigenous Hawaiians lived in a community of beings where land, sea, sky, rocks, rivers, animals, plants were all alive and in the family—an enchanted world, we might say—and this view urged them to *aloha 'aina*, love for the land.[26]

Science should demythologize these views but must itself find a new myth that encourages appropriate aesthetic responses to nature, responses that will sometimes be of the sublime and the numinous. Landscape is what it is, and science can be objective about that; but landscape as phenomena is difficult to dismiss as mere phenomena, because the full story of natural history is too phenomenal, too spectacular, to be mere landscape; it is a sacrament of something noumenal. Sensitive encounter with landscape discloses dimensions of depth. And that might well have been happening before the scientists came along.

We are all aesthetic beings, first in the original, kinesthetic sense of that term; we are incarnate in flesh and blood and feel our way through the world. If science were to anaesthetize us, numb us to what is of value for our bodily well-being, we could not survive. That much we share with animals. We humans are aesthetic beings further in the philosophical sense. If science anaesthetizes us to the beauty in our landscapes, we cannot flourish. From here forward, a science-based landscape aesthetics is urgent, but it must also be a science-transcending aesthetic of participatory experience. A central feature of such an aesthetic will be the beauty of life in dialectic with its environments, the landscape as a place of satisfactory, satisfying adapted fit, on which we live, and move, and have our being. That is, ultimately, what environmental aesthetics is all about.

Once, tracking wolves in Alberta, I came upon a wolf kill. Wolves had driven a bull elk to the edge of a cliff, cornered it there, before a great pine, itself clinging to the edge. It made a good picture; the mountains on the skyline, the trees nearer in, the fallen elk at the cliff's edge. The colours were green and brown, white and grey, sombre and deep. The process, beyond the form, was still more stimulating. I was witness to an ecology of predator and prey, to population dynamics, to heterotrophs feeding on autotrophs. The carcass, beginning to decay, was already being recycled by microorganisms. All this science is about something vital, essential, and also existential, about living on the landscape. In the scene I beheld, there was time, life, death, life persisting in the midst of its perpetual perishing. My human life, too, lies in such trophic pyramids. Incarnate in this world, I saw through my environment of the moment into the Environment quintessential, and found it aesthetically exciting. As with the lava outpouring into the sea, sensitive to my location, I knew something of how the world was made.[27]

Notes

1. Michael Kiono Dudley, "Traditional Native Hawaiian Environmental Philosophy," in Lawrence S. Hamilton (ed.) *Ethics, Religion, and Biodiversity* (Cambridge: The White Horse Press, 1993), pp. 176–82.

2. John Wesley Powell, *The Exploration of the Colorado River and its Canyons* (New York: Dover Publications, 1985/1961), pp. 36–7; George Wharton James, "Indian Legends about the Grand Canyon," in *The Grand Canyon of Arizona: How to See It* (Boston: Little, Brown, and Company, 1910), pp. 225–231.

3. David S. Noss and John B. Noss, *A History of the World's Religions,* 8th edn (New York: Macmillan, 1990), p. 256; Sarah Rossbach, *Feng Shui: The Chinese Art of Placement* (New York: E. P. Dutton, 1983).

4. David Lowenthal, "Finding Valued Landscapes," *Progress in Human Geography* (London), Vol. 2, (1978), pp. 373–418; Marjorie Hope Nicolson, *Mountain Gloom and Mountain Glory: The Development of the Aesthetics of the Infinite* (Ithaca, NY: Cornell U.P., 1959).

5. Ronald Rees, "The Taste for Mountain Scenery," *History Today,* Vol. 25 (1975), 305–12, on p. 306.

6. Thomas Burnet, *The Sacred Theory of the Earth* (Carbondale, IL: Southern Illinois U.P., [1691] 1965), p. 115, p. 110.

7. John Donne, *The First Anniversary: An Anatomy of the World,* in Frank Manley (ed.), *John Donne: The Anniversaries* (Baltimore: Johns Hopkins U.P., [1611] 1963), line 300.

8. William Wordsworth, *The Prelude,* in Thomas Hutchinson (ed.), *The Poetical Works of Wordsworth,* Book VI (Oxford: Oxford U.P., [1805], 1965), line 639.

9. Yuriko Saito, "Is There a Correct Aesthetic Appreciation of Nature?," *Journal of Aesthetic Education,* Vol. 18 (1984), 35–46.

10. For related discussions, see R. W. Hepburn, "Aesthetic Appreciation of Nature," in H. Osborne (ed.), *Aesthetics and the Modern World* (New York: Weybright and Talley, 1968); and Salim Kemal and Ivan Gaskell (eds), *Landscape, Natural Beauty and the Arts* (New York: Cambridge U.P., 1993).

11. Hildegard Binder Johnson, "The Framed Landscape," *Landscape,* Vol. 23, No. 2 (1979), 26–32, on p. 27.

12. Yuriko Saiti, "The Japanese Appreciation of Nature," *British Journal of Aesthetics,* Vol. 25 (1985), pp. 239–51.

13. Allen Carlson, "Nature, Aesthetic Judgment, and Objectivity," *Journal of Aesthetics and Art Criticism,* Vol. 40 (1981), pp. 15–27.

14. Aldo Leopold, *A Sand County Almanac* (New York: Oxford U.P., 1968), pp. 173–74.

15. Ibid.

16. J. A. Walter, "'You'll Love the Rockies,'" *Landscape,* Vol. 27, No. 2 (1983), pp. 43–7.

17. Arnold Berleant, *The Aesthetics of Environment* (Philadelphia: Temple U.P., 1992), p. 34.

18. Ibid., p. 128.

19. Ibid., p. 131.

20. Ibid., p. 135.

21. Gordon H. Orians, "Habitat Selection: General Theory and Applications to Human Behaviour," in Joan S. Lockard (ed), *The Evolution of Human Social Behavior* (New York: Elsevier, North-Holland, 1980), pp. 46–66.

22. Judith H. Heerwagen and Gordon H. Orians, "Humans, Habitats, and Aesthetics," in Stephen R. Kellert and Edward O. Wilson (eds), *The Biophilia Hypothesis* (Washington: Island Press, 1993), pp. 138–172.

23. Roger S. Ulrich, "View Through a Window May Influence Recovery from Surgery," *Science,* Vol. 224 (1984), pp. 420–21.

24. Roger S. Ulrich, "Biophilia, Biophobia, and Natural Landscapes," in Stephen R. Kellert and Edward O. Wilson (eds), *The Biophilia Hypothesis* (Washington: Island Press, 1993), p. 97; Roger S. Ulrich, Robert F. Simons, Barbara D. Losito, Evelyn Fiorito, Mark A. Miles and Michael Zelson, "Stress Recovery During Exposure to Natural and Urban Environments," *Journal of Environmental Psychology,* Vol. 11 (1991), pp. 201–30.

25. Burnet, op. cit., p. 109.

26. Dudley, op. cit., p. 178.

27. This paper was presented at "Meeting in the Landscape," the First International Conference on Environmental Aesthetics, Koli, Finland, June 1994.

For Further Discussion

1. "We cannot appropriately appreciate what we do not understand." Do you agree with this statement? What do you think Rolston means by "appropriately" in this sense? How might Rolston respond to someone who claimed that "I don't know much about art, but I know what I like"?

2. According to Rolston, science can teach us things that will improve our capacity to see natural beauty. What, exactly, can science teach us about such things?

3. Rolston tells the story of a British visitor to the Rocky Mountains who complains about the monotonous scenery. Rolston asks, "Can one argue that he was wrong?" Can you?

4. Rolston contrasts the natural with the supernatural and the cultural. How would you distinguish among these three categories?

The Biblical Vision of the Ecological Crisis

Rosemary Radford Ruether

In this selection, theologian Rosemary Radford Ruether reviews several theological approaches to the ecological crisis. The first approach—what she calls the neoanimist or nature mystical approach—argues for a romantic and mystical relationship between humans and the natural world. Nature is seen as having an intrinsically divine presence within it. The second approach—the stewardship or puritan conservationist model—continues to treat nature as an object but argues that humans should be faithful and frugal stewards of God's gift to us. Ruether believes that both approaches overlook a third option: a biblically based theology that views humanity as part of creation and nature as part of the covenant between God and creation. On this view, the way humans treat nature is both symbolic of and a part of human ethics. Destruction of nature and exploitation of other humans "are profoundly understood as part of one reality, creating disaster in both." Ruether, writing in the late 1970s, was an early voice in connecting environmental concerns with social justice, a topic that we will examine at several points later in this textbook.

We need to recover an understanding of ecojustice in which the enmity or harmony of nature with humanity is part of the human historical drama of good and evil.

Two decades ago it was common to speak of the need for economic "development" among "backward" nations. The assumption behind his language was that Western-style industrialization was the model of progress, and that all nations could be judged by how far they had come along on that road. Poor nations were poor because they were at some retarded stage of this evolutionary road of development. They needed economic assistance from more "developed" nations to help them "take off" faster.

MOVEMENTS OF DISSENT

In the mid-'60s there were two major movements of dissent from this model of "developmentalism." One of them occurred primarily among social thinkers in the Third World, especially Latin America, who began to reject the idea of development for that of liberation. They contended that poor countries were poor not because they were "undeveloped," but because they were misdeveloped. They were the underside of a process in which, for five centuries, Western colonizing countries had stripped the colonized countries of their wealth, using cheap or slave labor, in order to build up the wealth which now underlies Western capitalism. One could not overcome this pattern of misdevelopment by a method of "assistance" that merely continues and deepens the pattern of pillage and dependency which created the poverty in the first place.

A few years after this critique of development from a Third World standpoint, a second dissenting movement appeared, primarily among social thinkers in advanced industrial countries. This movement focused on the issue of modern industrialized societies' ecological disharmony with the carrying capacities of the natural environment. It dealt with such issues as air, water and soil pollution; the increasing depletion of finite resources, including minerals and fossil fuels; and the population explosion.

This dissent found dramatic expression in the Club of Rome's report on *Limits to Growth,* which demonstrated that indefinite expansion of Western-style industrialization was, in fact, impossible. This system, dependent on a small affluent minority using a disproportionate share of the world's natural resources, was fast depleting the base upon which it rested: nonrenewable resources. To expand this type of industrialization would simply accelerate the impending debacle; instead, we must stop developing and try to stabilize the economic system and population where they are.

These two critiques of development—the Third World liberation perspective and the First World ecological perspective—soon appeared to be in considerable conflict with each other. The liberation viewpoint stressed pulling control over the natural resources of poor countries out from under Western power so that the developmental process could continue under autonomous, socialist political systems. The First World ecological viewpoint often sounded, whether consciously or not, as though it were delivering bad news to the hopes of poor countries. Stabilizing the world as it is seemed to suggest stabilizing its unjust relationships. The First World, having developed advanced industry at the expense of the labor and resources of the Third World, was now saying: "Sorry, the goodies have just run out. There's not enough left for you to embark on the same path." Population alarmists sounded as though Third World populations were to be the primary "targets" for reduction. Social justice and the ecological balance of humanity with the environment were in conflict. If one chose ecology, it was necessary to give up the dream of more equal distribution of goods.

RELIGIOUS RESPONSES

In the late 60s there rose a spate of what might be called theological or religious responses to the ecological crisis, again primarily in advanced industrial countries. Two major tendencies predominated among such writers. One trend, represented by books such as Theodore Roszak's *Where the Wasteland Ends,* saw the ecological crisis in terms of the entire Western Judeo-Christian reality principle. Tracing the roots of this false reality principle to the Hebrew Bible itself, Roszak, among others, considered the heart of the ecological crisis to be the biblical injunction to conquer and subdue the earth and have dominion over it. The earth and its nonhuman inhabitants are regarded as possessions or property given to "man" for "his" possession. "Man" exempts "himself" (and I use the male generic advisedly) from the community of nature, setting himself above and outside it somewhat as God "himself" is seen as sovereign over it. Humanity is God's agent in this process of reducing the autonomy of nature and subjugating it to the dominion of God and God's representative, man.

For Roszak and others, this conquest-and-dominion approach turned nature into a subjugated object and denied divine presence in it. Humanity could no longer stand in rapt contemplation before nature or enter into worshipful relations with it. A sense of ecstatic kinship between humanity and nature was destroyed. The divinities were driven out, and the rape of the earth began. In order to reverse the ecological crisis, therefore, we must go back to the root error of consciousness from which it derives. We must recover the religions of ecstatic kinship in nature that preceded and were destroyed by biblical religion. We must reimmerse God and humanity in nature, so that we can once again interact with nature as our spiritual kin, rather than as an enemy to be conquered or an object to be dominated. Only when we recover ancient animism's I-Thou relationship with nature, rather than the I-It relation of Western religion, can we recover the root principle of harmony with nature that was destroyed by biblical religion and its secular stepchildren.

This neoanimist approach to the ecological crisis was persuasive, evoking themes of Western reaction to industrialism and technological rationality that began at least as far back as the romanticism of the early 19th century. But many voices quickly spoke up in defense of biblical faith. A variety of writers took exception to romantic neoanimism as the answer, contending that biblical faith in relation to nature had been misunderstood. Most of the writers in this camp tended to come up with the "stewardship" model. Biblical faith does not mandate the exploitation of the earth, but rather commands us to be good stewards, conserving earth's goods for generations yet to come. In general, these writers did implicitly concede Roszak's point that biblical faith rejects any mystical or animist interaction with nature. Nature must be regarded as an object, not as a subject. It is our possession, but we must possess it in a thrifty rather than a profligate way.

ECONOMIC CONSIDERATIONS

One problem with both of these Western religious responses to the ecological crisis: there was very little recognition that this crisis took place within a particular economic system. The critique of the Third World liberationists was not accorded much attention or built into these responses. The ecological crisis was regarded primarily as a crisis between "man" and "nature," rather than as a crisis resulting from the way in which a particular exploitative relationship between classes, races and nations used natural resources.

The Protestant "stewardship" approach suggested a conservationist model of ecology. We should conserve resources, but without much acknowledgment that they had been unjustly used within the system that was being conserved. The countercultural approach, on the other hand, did tend to be critical of Western industrialism, but in a romantic, primitivist way. It idealized agricultural and handicraft economies but had little message for the victims of poverty who had already been displaced from that world of the preindustrial village. Thus it has little to say to the concerns of Third World economic justice, except to suggest that the inroads of Western industrialism should be resisted by turning back the clock.

Is there a third approach that has been overlooked by both the nature mystics and the puritan conservationists? Both of these views seem to me inadequate to provide a vision of the true character of the crisis and its solution. We cannot return to the Eden of the preindustrial village. However much those societies may possess elements of wisdom, these elements must be recovered by building a new society that also incorporates modern technological development. The countercultural approach never suggests ways of grappling with and changing the existing system. Its message remains at the level of dropping out into the preindustrial farm—an option which, ironically, usually depends on having an independent income!

The stewardship approach, with its mandate of thrift within the present system, rather than a recognition of that system's injustice, lacks a vision of a new and different economic order. Both the romantic and the conservationist approaches never deal with the question of ecojustice; namely, the reordering of access to and use of natural resources within a just economy. How can ecological harmony become part of a system of economic justice?

MISINTERPRETATIONS OF SCRIPTURE

To find a theology and/or spirituality of ecojustice, I would suggest that, in fact, our best foundation lies precisely in the Hebrew Bible—that same biblical vision which, anachronistically, the romantics have scapegoated as the problem and which the conservationists have interpreted too narrowly and unperceptively. Isaiah 24 offers one of the most eloquent statements of this biblical vision that is found particularly in the prophets of the Hebrew Scriptures. The puritan conservationists have too readily accepted a 19th century theology that sets history against nature—a theology which is basically western European rather than biblical. The biblical vision is far more "animistic" than they have been willing to concede. In Scripture, nature itself operates as a powerful medium of God's presence or absence. Hills leap for joy and rivers clap their hands in God's presence. Or, conversely, nature grows hostile and barren as a medium of divine wrath.

The romantics, on the other hand, have blamed Scripture for styles of thought about nature that developed in quite different circles. The concept of nature as evil and alien to humanity began basically in late apocalyptic and gnostic thought in the Christian era. The divine was driven out of nature not to turn nature into a technological instrument, but rather to make it the habitation of the devil; the religious "man" should shun it and flee from it in order to save "his" soul for a higher spiritual realm outside of and against the body and the visible, created world. Christianity and certainly Judaism objected to this concept as a denial of the goodness of God's creation, though Christianity became highly infected by this negative view of nature throughout its first few centuries, and that influence continued to be felt until well into the 17th century.

The new naturalism and science of the 17th century initially had the effect of restoring the vision of nature as good, orderly and benign—the arena of the manifestation of God's divine reason, rather than of the devil's malice. But this Deist view of nature (as the manifestation of *divine* reason) was soon replaced by a Cartesian world view that set *human* reason outside and above nature. It is this technological approach—treating nature as an object to be reduced to human control—that is the heart of modern exploitation, but it does not properly correspond to any of the earlier religious visions of nature. Any recovery of an appropriate religious vision, moreover, must be one that does not merely ignore these subsequent developments, but that allows us to review and critique where we have gone wrong in our relationship to God's good gift of the earth. In my opinion, it is precisely the vision of the Hebrew prophets that provides at least the germ of that critical and prophetic vision.

A COVENANTAL VISION

The prophetic vision neither treats nature in a romantic way nor reduces it to a mere object of human use. Rather, it recognizes that human interaction with nature has made nature itself historical. In relation to humanity, nature no longer exists "naturally," for it has become part of the human social drama, interacting with humankind as a vehicle of historical judgment and a sign of historical hope. Humanity as a part of creation is not outside of nature but within it. But this is the case because nature itself is part of the covenant between God and creation. By this covenantal view, nature's responses to human use or abuse become an ethical sign. The erosion of the soil in areas that have been abused for their mineral wealth, the pollution of air where poor people live, are not just facts of nature; what we have is an ethical judgment on the exploitation of natural resources by the rich at the expense of the poor. It is no accident that nature is most devastated where poor people live.

When human beings break their covenant with society by exploiting the labor of the worker and refusing to do anything about the social costs of production—i.e., poisoned air and water—the covenant of creation is violated. Poverty, social oppression, war and violence in society, and the polluted, barren, hostile face of nature—both express this violation of the covenant. The two are profoundly linked together in the biblical vision as parts of one covenant, so that, more and more, the disasters of nature become less a purely natural fact and increasingly become a social fact. The prophetic text of Isaiah 24 vividly portrays this link between social and natural hostility in the broken order of creation:

Behold, the Lord will lay waste the earth and make it desolate,
and he will twist its surface and scatter its inhabitants. . . .
The earth shall be utterly laid waste and utterly despoiled; . . .
The earth mourns and withers, the world languishes and withers; . . .
The earth lies polluted under its inhabitants;
for they have transgressed the laws, violated the statutes, broken the everlasting covenant.
Therefore a curse devours the earth,
and its inhabitants suffer for their guilt; . . .
The city of chaos is broken down,
every house is shut up so that none can enter. . . .
Desolation is left in the city, the gates are battered into ruins.

—Isa. 24:1, 3, 4–5, 10, 12

But this tale of desolation in society and nature is not the end of the prophetic vision. When humanity mends its relation to God, the result must be expressed not in contemplative flight from the earth but rather in the rectifying of the covenant of creation. The restoration of just relations between peoples restores peace to society and, at the same time, heals nature's enmity. Just, peaceful societies in which people are not exploited also create peaceful, harmonious and beau-

tiful natural environments. This outcome is the striking dimension of the biblical vision. The Peaceable Kingdom is one where nature experiences the loss of hostility between animal and animal, and between human and animal. The wolf dwells with the lamb, the leopard lies down with the kid, and the little child shall lead them.

They shall not hurt or destroy in all my holy mountain, for the earth
shall be full of the knowledge of the Lord.

—Is. 11:9

The biblical dream grows as lush as a fertility religion in its description of the flowering of nature in the reconciled kingdom of God's Shalom.

The wilderness and the dry land shall be glad, the desert shall rejoice and blossom;
Like the crocus it shall blossom abundantly, and rejoice with joy and singing.

—Is. 35:1–2

"The tree bears its fruit, the fig trees and vine give their full yield. . . . Rejoice in the Lord, for he has given early rain . . . The threshing floors shall be full of grain, the vats shall overflow with wine and oil."

—Joel 2:22–24

"Behold the days are coming," says the Lord, "when the plowman will overtake the reaper and the treader of grapes him who sows the seed: the mountains shall drip sweet wine, and all the hills shall flow with it."

—Amos 9:13

In the biblical view, the raping of nature and the exploitation of people in society are profoundly understood as part of one reality, creating disaster in both. We look not to the past but to a new future, brought about by social repentance and conversion to divine commandments, so that the covenant of creation can be rectified and God's Shalom brought to nature and society. Just as the fact of nature and society grows hostile through injustice, so it will be restored to harmony through righteousness. The biblical understanding of nature, therefore, inheres in a human ethical vision, a vision of ecojustice, in which the enmity or harmony of nature with humanity is part of the human historical drama of good and evil. This is indeed the sort of ecological theology we need today, not one of either romance or conservationism, but rather an ecological theology of ethical, social seriousness, through which we understand our human responsibility for ecological destruction and its deep links with the struggle to create a just and peaceful social order.

For Further Discussion

1. Ruether suggests that issues of social justice and environmental destruction are part of one reality. How does this view compare to Krieger's understanding of the relation between social justice and environmental preservation?

2. What does Ruether mean by a neoanimist approach to nature? How does it differ from the stewardship model?

3. How does Ruether's interpretation of biblical views on nature compare to the interpretation offered by Lynn White?

4. Ruether tells us that "human interaction with nature has made nature itself historical." What do you think she means by this? How is this similar to Krieger's claim that what is natural depends on the culture and society defining it? How is it different?

DISCUSSION AND STUDY QUESTIONS FOR CHAPTER 5

1. Is beauty in the eye of the beholder? How much of an aesthetic judgment is based on knowledge and how much is based on feelings?

2. Think of a beautiful nature scene. What did you think of? A mountain landscape? The ocean shore? Can a swamp be beautiful? Are farms beautiful? Why or why not?

3. Could a natural area be created in virtual reality? If aesthetic appreciation is based on feelings and perception, could those feelings and perception be created artificially?

4. What is wrong with plastic trees? Could a wilderness area ever be duplicated or restored by human means?

5. Why are natural areas valued? List as many different types of value as you can. To what degree is each the result of personal choice? To what degree is each objective?

DISCUSSION CASES

Disney and the Civil War

Manassas battlefield is a national park site in Virginia, 35 miles west of Washington, DC. The first major battle of the Civil War was fought at Manassas during the summer of 1861. The one-day battle left 900 soldiers dead. Two years later, the second battle of Manassas lasted three days and resulted in 3,400 deaths.

In 1994 a battle of a different sort was waged over plans to build a Disney theme park a few miles away from the battlefield. The park would be built on farmland and would have a historical theme, complete with villages and military re-enactments, as well as golf courses, motels, restaurants, and so forth.

Opponents argued that building such a park at such a historic site would degrade and dishonor the memory of the sacrifices made there. This land held important symbolic and historical value to the people of Virginia. Defenders argued that the project would bring economic growth—jobs, taxes, homes, business—into the area and would help all citizens. Are symbolic and historical values merely some people's personal preferences, or are they more objective than that? How would you describe the difference in experiences at Manassas and at a Disney Civil War re-creation?

Within the context of free market economics, Chapter 4 considered another proposed Disney project at Mineral King Valley in California. Explain how a defender of free market economics might decide this policy question.

In a reading in this chapter, Martin Krieger suggests that people can be taught to value environments. In another reading, Robert Elliot claims that there is a major value difference between an original landscape and one created by Disney. What Civil War experiences could people be taught to feel at a Disney theme park? How would that experience be different from a visit to the Manassas battlefield site? Could people be taught *not* to value the historic and symbolic meaning of Manassas? Should they be?

Wild Turkeys

After moving to central Minnesota from Philadelphia some years ago, I was very excited to see wild turkeys wandering across my yard one afternoon. Within the first few months of moving, I had seen bald eagles, owls, deer, and geese, as well as the turkeys, on my property. I judged these to be a very valuable part of living in the country.

It wasn't too long before I learned that the turkeys had recently been reintroduced in the area by the State Department of Natural Resources. The native turkey population had been hunted to near extinction, and these birds were part of a restocking effort to support local hunters. My previous excitement over seeing the turkeys disappeared. My subjective experience had changed with the knowledge of their origins. Was this a reasonable change?

Robert Elliot speaks of a "certain kind of causal history" when explaining the value of natural areas. How might this notion explain my differing experiences of the turkeys? Was the Department of Natural Resources wrong in their attempt to restore nature? What values, if any, could be restored by their efforts? What could not? What scientific knowledge might explain the value of native turkeys that is lacking by reintroducing the birds? Might the causal history of the growing bald eagle population add to the value of these birds?

Prairie Restoration

When European settlers first arrived in Minnesota slightly more than a century ago, much of the southern and western portion of the state was part of the great grassland prairies of the American West. Today, less than one-tenth of one percent of the original prairieland remains. The rest has been plowed, paved, and built upon.

At St. John's University, a major prairie restoration project has been underway for some years. The goal has been to restore the prairie with native species and original wetlands, as well as some

continued

surrounding oak woodlands. This prairie restoration project provides significant educational opportunities in science, history, and environmental studies. It also provides areas of beauty and solitude throughout the changing seasons. Assuming it looks indistinguishable from some of the remaining original sites (and, truth be told, it is quite beautiful), does the restored prairie have as much aesthetic value as the original value?

St. John's University is home to a religious community of Catholic monks, and part of this restoration project stems from their understanding of the responsibility to be good stewards of God's creation. What is your understanding of "stewardship"?

Martin Krieger claims that people can learn to value low-cost substitutes for natural environments. Might such restoration projects be one way to teach people about the value of original environments? What other values are served by restored environments? What values are lost forever?

English Gardens and Prairiegrass Lawns

The standard of beauty used in many U.S. neighborhoods includes carefully and neatly manicured lawns. In many cases, the grass is a nonnative species that requires significant amounts of water, fertilizer, and pesticides to maintain the weed-free and bright-green appearance.

Some homeowners have pursued a different track, planting native species of tall grass and shrubbery. The native species require less maintenance and, in the opinion of many, they are more attractive than green lawns. However, neighbors don't always agree. Some have argued that such natural yards lower property values of the surrounding homes, bring weeds into grass lawns, and generally detract from the neighborhood appearance.

Perfectly manicured lawns are modeled on classic English gardens, which have a high degree of symmetry, sharp borders, and typically a geometrical design. Those who value English gardens, as well as those who value prairiegrass lawns, would claim that they value nature and natural environments. What values do you think underlie each perspective? Is it simply a matter, as Krieger might claim, of what a particular culture and society define as beautiful? Are there different worldviews and different ethics represented by these two approaches, or is it simply an issue of personal taste? What different attitudes toward nature are represented by these two perspectives?

CHAPTER 6

Extensionism and Anthropocentrism

The debates in the previous chapters involve fairly common questions of ethics: What should we do? How should we act? What do we value? What policies will advance the well-being of our fellow humans? But there is an even more fundamental question that, perhaps because it is fundamental, may be overlooked. This question concerns the issue that philosophers call *moral standing*. What things count, ethically? What objects have moral standing?

Consider the following example. Most of us would think that there is nothing ethically wrong with splitting a piece of firewood with an axe. On the other hand, we all would think it a horrific act of immorality if I did the same thing to a human being. Why do we have such reactions? Presumably, we think this way because we believe that humans do, and pieces of firewood do not, count ethically. Humans have moral standing and firewood does not. But would your reaction change if I took my axe to a living 200-year-old oak tree? What about a gopher that was eating my garden vegetables? What about my neighbor's pet dog?

We can use a simple thought experiment to consider this issue. Imagine a continuum: On one extreme we have objects, for example, living human beings, that have full moral standing—they "count"; on the other extreme we have objects, like the piece of firewood in my backyard, that do not have moral standing. Along the continuum, depending on your own intuition, we would place other objects like plants and animals.

The philosophical issue is to translate these intuitions into a reasonable and coherent point of view. What criteria do we use to place these objects along this continuum? Why do humans count and firewood not? What characteristics do humans have that qualify them for ethical consideration? What does firewood lack that exempts it from ethical consideration? The

philosophical goal would be to identify and explain these criteria and then use them to decide more difficult cases. Are living plants more like humans or more like firewood? Are animals similar enough to humans that they should be recognized as ethical objects?

Chapter 2 introduced some traditional philosophical reflections on these questions. Until very recently, most mainstream philosophers in the Western tradition, if they considered these questions at all, thought that only humans had moral standing. As we saw in these readings, philosophers often identified something like the mind or soul as the criterion that determined standing. Humans have a mind or soul, plants and animals don't; therefore, plants and animals do not deserve status as ethical beings. At best, we might have responsibilities *regarding* plants and animals, but we have no direct responsibilities *to* them. The view that only humans have moral standing, and that therefore we have direct moral responsibilities only to humans, can be identified as anthropocentrism.

But there has always been a secondary perspective that recognized a wider membership in the class of ethical beings. Saint Francis in the Christian tradition, the utilitarian Jeremy Bentham, as well as numerous non-Western and Native American traditions, recognized to various degrees the possibility that beings other than humans should count. During the last few decades mainstream Western philosophers have begun to address this question explicitly. Spurred on by such ethical controversies as abortion, euthanasia, and responsibilities to future generations, as well as environmental issues, some philosophers have challenged the view that only free and rational adult humans have moral standing.

These issues raise a major question concerning our philosophical perspective. *Anthropocentric* (human-centered) *ethics* holds that ethical consideration should be focused solely on human well-being. Humans, and only humans, count. Although we may have reason to protect plants and animals, our reasons all ultimately have to do with human interests. Humans care for and value other creatures, and because of this we should protect and respect them. *Nonanthropocentric ethics* recognizes that objects other than humans have moral standing in their own right. From this perspective, ethical concepts and categories such as rights and welfare should be extended to include the rights and welfare of such natural objects as plants and animals.

But what criteria do we use in determining moral standing? On what terms do objects count? Several candidates suggest themselves. We might say that all living things deserve consideration, that life itself creates moral standing. Such a view would be identified as *biocentric ethics,* a life-centered ethics. This position would extend ethical consideration far beyond common boundaries.

However, there are problems with this position. If everything that is alive has standing, we would seem to face insurmountable obstacles. Have we done something ethically wrong when we swat a mosquito? When we cut down a tree? When we harvest crops for food? In response to such challenges, biocentric views offer various accounts of the ethical implications of a life-centered ethics. Typically, such views suggest that moral standing implies only that living things deserve respect and consideration, not that they have a moral standing equal to human beings.

For example, the famous humanitarian Albert Schweitzer based his ethical views on the recognition that all living beings exercise a will to live. Life is a good-in-itself, and all living things should be treated with reverence. Thus, Schweitzer's ethics would not impose strict moral obligation on us as much as it would require that in our interaction with living beings we maintain the proper attitude. We should treat life with reverence.

Other views attempt to specify a particular characteristic of life that qualifies a being for moral consideration. In answering this question, philosophers often turn to their understanding of the nature of ethics. If the purpose of ethics is to reduce suffering and produce happiness, as some utilitarians might argue, then any being who is capable of suffering pain and enjoying happiness should qualify. If the purpose is to protect interests, then beings with interests qualify. On many of these standards, some individual animals surely qualify.

Some original and interesting philosophical work in ethics developed around the suggestion that animals have moral standing. Much of this work focused on the writing of two philosophers, Australian Peter Singer and American Tom Regan. Singer has argued that sentience, the capacity for feeling pain, qualifies a being for moral standing. Echoing the utilitarian Jeremy Bentham, Singer has suggested that it is time to recognize the unethical suffering that humans inflict on animals. Regan, arguing on grounds more in line with Kantian ethics, claims that at least some animals have the cognitive capacity for conscious activities. These animals are "subjects-of-a-life" and deserve to have their interests protected in much the same way that rights protect human interests.

These arguments are often identified as *ethical extensionism,* a process by which the concepts and categories of ethics are extended beyond human beings to include a wider range of moral beings. How far should the boundaries of ethical standing be extended? Is this a process of recognizing, in a rational and logical sense, what is? Or is this more a matter of deciding to what we want to grant standing? That is, is extensionism a matter of what we choose to include in our ethical considerations? The readings in this chapter offer considerations, arguments, guidelines for extending moral standing from presently living individual humans to future generations of humans, to animals, to species, and even to plants, trees, and mountains.

The Rights of Animals and Unborn Generations

Joel Feinberg

This selection is a very influential essay written by the contemporary American philosopher, Joel Feinberg. In this 1974 essay, Feinberg argues that the criteria for having rights involve having a sake of one's own, having a good of one's own, or more generally in having interests. The interests that some things have in attaining their own good qualifies those objects as having a moral claim or "right" against us. On Feinberg's analysis, only beings with some conscious awareness of their own life can be said to have interests. Thus, it does make sense to say that many individual animals have rights. Applying this analysis, Feinberg concludes that plants and species do not, but future generations do, have moral claims upon us. These claims, generally, have to do with protecting their interests.

Every philosophical paper must begin with an unproved assumption. Mine is the assumption that there will still be a world five hundred years from now, and that it will contain human beings who are very much like us. We have it within our power now, clearly, to affect the lives of these creatures for better or worse by contributing to the conservation or corruption of the environment in which they must live. I shall assume furthermore that it is psychologically possible for us to care about our remote descendants, that many of us in fact do care, and indeed that we ought to care. My main concern then will be to show that it makes sense to speak of the rights of unborn generations against us, and that given the moral judgment that we ought to conserve our environmental inheritance for them, and its grounds, we might well say that future generations *do* have rights correlative to our present duties toward them. Protecting our environment now is also a matter of elementary prudence, and insofar as we do it for the next generation already here in the persons of our children, it is a matter of love. But from the perspective of our remote descendants it is basically a matter of justice, of respect for their rights. My main concern here will be to examine the concept of a right to better understand how that can be.

THE PROBLEM

To have a right to have a claim[1] *to* something and *against* someone, the recognition of which is called for by legal rules or, in the case of moral rights, by the principles of an enlightened conscience. In the familiar cases of rights, the claimant is a competent adult human being, and the claimee is an officeholder in an institution or else a private individual, in either case, another competent adult human being. Normal adult human beings, then, are obviously the sorts of beings of whom rights can meaningfully be predicated. Everyone would agree to that, even extreme misanthropes who deny that anyone in fact has rights. On the other hand, it is absurd to say that rocks can have rights, not because rocks are morally inferior things unworthy of rights (that statement makes no sense either), but because rocks belong to a category of entities of whom rights cannot be meaningfully predicated. That is not to say that there are no circumstances in which we ought to treat rocks carefully, but only that the rocks themselves cannot validly claim good treatment from us. In between the clear cases of rocks and normal human beings, however, is a spectrum of less obvious cases, including some bewildering borderline ones. Is it meaningful or conceptually possible to ascribe rights to our dead ancestors? to individual animals? to whole species of animals? to plants? to idiots and madmen? to fetuses? to generations yet unborn? Until we know how to settle these puzzling cases, we cannot claim fully to grasp the concept of a right, or to know the shape of its logical boundaries.

One way to approach these riddles is to turn one's attention first to the most familiar and unproblematic instances of rights, note their most salient characteristics, and then compare the borderline cases with them, measuring as closely as possible the points of similarity and difference. In the end, the way we classify the borderline cases may depend on whether we are more impressed with the similarities or the differences between them and the cases in which we have the most confidence.

It will be useful to consider the problem of individual animals first because their case is the one that has already been debated with the most thoroughness by philosophers so that the dialectic of claim and rejoinder has now unfolded to the point where disputants can get to the end game quickly and isolate the crucial point at issue. When we understand precisely what *is* at issue in the debate over animal rights, I think we will have the key to the solution of all the other riddles about rights.

INDIVIDUAL ANIMALS

Almost all modern writers agree that we ought to be kind to animals, but that is quite another thing from holding that animals can claim kind treatment from us as their due. Statutes making cruelty to animals a crime are now very common, and these, of course, impose legal duties on people not to mistreat animals; but that still leaves open the question whether the animals, as beneficiaries of those duties, possess rights correlative to them. We may very well have duties *regarding* animals that are not at the same time duties *to* animals, just as we may have duties regarding rocks, or buildings, or lawns, that are not duties *to* the rocks, buildings, or lawns. Some legal writers have taken the still more extreme position that animals themselves are not even the directly intended beneficiaries of statutes prohibiting cruelty to animals. During the nineteenth century, for example, it was commonly said that such statutes were designed to protect human beings by preventing the growth of cruel habits that could later threaten human beings with harm too. Prof. Louis B. Schwartz finds the rationale of the cruelty-to-animals prohibition in its protection of animal lovers from affronts to their sensibilities. "It is not the mistreated dog who is the ultimate object of concern," he writes. "Our concern is for the feelings of other human beings, a large proportion of whom, although accustomed to the slaughter of animals for food, readily identify themselves with a tortured dog or horse and respond with great sensitivity to its sufferings."[2] This seems to me to be factitious. How much more natural it is to say with John Chipman Gray that the true purpose of cruelty-to-animals statutes is "to preserve the dumb brutes from suffering."[3] The very people whose sensibilities are invoked in the alternative explanation, a group that no doubt now includes most of us, are precisely those who would insist that the protection belongs primarily to the animals themselves, not merely to their own tender feelings. Indeed, it would

be difficult even to account for the existence of such feelings in the absence of a belief that the animals deserve the protection in their own right and for their own sakes.

Even if we allow, as I think we must, that animals are the intended direct beneficiaries of legislation forbidding cruelty to animals, it does not follow directly that animals have legal rights, and Gray himself, for one,[4] refused to draw this further inference. Animals cannot have rights, he thought, for the same reason they cannot have duties, namely, that they are not genuine "moral agents." Now, it is relatively easy to see why animals cannot have duties, and this matter is largely beyond controversy. Animals cannot be "reasoned with" or instructed in their responsibilities; they are inflexible and unadaptable to future contingencies; they are subject to fits of instinctive passion which they are incapable of repressing or controlling, postponing or sublimating. Hence, they cannot enter into contractual agreements, or make promises; they cannot be trusted; and they cannot (except within very narrow limits and for purposes of conditioning) be blamed for what would be called "moral failures" in a human being. They are therefore incapable of being moral subjects, of acting rightly or wrongly in the moral sense, of having, discharging, or breaching duties and obligations.

But what is there about the intellectual incompetence of animals (which admittedly disqualifies them for duties) that makes them logically unsuitable for rights? The most common reply to this question is that animals are incapable of *claiming* rights on their own. They cannot make motion, on their own, to courts to have their claims recognized or enforced; they cannot initiate, on their own, any kind of legal proceedings; nor are they capable of even understanding when their rights are being violated, of distinguishing harm from wrongful injury, and responding with indignation and an outraged sense of justice instead of mere anger or fear.

No one can deny any of these allegations, but to the claim that they are the grounds for disqualification of rights of animals, philosophers on the other side of this controversy have made convincing rejoinders. It is simply not true, says W. D. Lamont,[5] that the ability to understand what a right is and the ability to set legal machinery in motion by one's own initiative are necessary for the possession of rights. If that were the case, then neither human idiots nor wee babies would have any legal rights at all. Yet it is manifest that both of these classes of intellectual incompetents have legal rights recognized and easily enforced by the courts.

Children and idiots start legal proceedings, not on their own direct initiative, but rather through the actions of proxies or attorneys who are empowered to speak in their names. If there is no conceptual absurdity in this situation, why should there be in the case where a proxy makes a claim on behalf of an animal? People commonly enough make wills leaving money to trustees for the care of animals. It is not natural to speak of the animal's right to his inheritance in cases of this kind? If a trustee embezzles money from the animal's account,[6] a proxy speaking in the dumb brute's behalf presses the animal's claim, can he not be described as asserting the animal's *rights*? More exactly, the animal itself claims its rights through the vicarious actions of a human proxy speaking in its name and in its behalf. There appears to be no reason why we should require the animal to understand what is going on (so the argument concludes) as a condition for regarding it as a possessor of rights.

Some writers protest at this point that the legal relation between a principal and an agent cannot hold between animals and human beings. Between humans, the relation of agency can take two different forms, depending upon the degree of discretion granted to the agent, and there is a continuum of combinations between the extremes. On the one hand, there is the agent who is the mere "mouthpiece" of his principal. He is a "tool" in much the same sense as is a typewriter or telephone; he simply transmits the instructions of his principal. Human beings could hardly be the agents or representatives of animals in this sense, since the dumb brutes could no more use human "tools" than mechanical ones. On the other hand, an agent may be some sort of expert hired to exercise his professional judgment on behalf of, and in the name of, the principal. He may be given, within some limited area of expertise, complete independence to act as he deems best, binding his principal to all the beneficial or detrimental consequences. This is the role played by trustees, lawyers, and ghostwriters. This type of representation requires that the agent have great skill, but makes little or no demand upon the principal, who may leave everything to the judgment of his agent. Hence, there appears, at first, to be no reason why an animal cannot be a totally passive principal in this second kind of agency relationship.

There are still some important dissimilarities, however. In the typical instance of representation by an agent, even of the second, highly discretionary kind, the agent is hired by a principal who enters into an agreement or contract with him; the principal tells his agent that within certain carefully specified boundaries "You may speak for me," subject always to the principal's approval, his right to give new directions, or to cancel the whole arrangement. No dog or cat could possibly do any of those things. Moreover, if it is the assigned task of the agent to defend the principal's rights, the principal may often decide to release his claimee, or to waive his own rights, and instruct his agent accordingly. Again, no mute cow or horse can do that. But although the possibility of hiring, agreeing, contracting, approving, directing, canceling, releasing, waiving, and instructing is present in the typical (all-human) case of agency representation, there appears to be no reason of a logical or conceptual kind why that *must* be so, and indeed that there are some special examples involving human principals where it is not in fact so. I have in mind legal rules, for example, that require that a defendant be represented at his trial by an attorney, and impose a state-appointed attorney upon reluctant defendants, or upon those tried *in absentia,* whether they like it or not. Moreover, small children and mentally deficient and deranged adults are commonly represented by trustees and attorneys, even though they are incapable of granting their own consent to the representation, or of entering into contracts, of giving directions, or waiving their rights. It may be that it is unwise to permit agents to represent principals without the latters' knowledge or consent. If so, then no one should ever be permitted to speak for an animal, at least in a legally binding way. But that is quite another thing than saying that such representation is logically incoherent or conceptually incongruous—the contention that is at issue.

H. J. McCloskey,[7] I believe, accepts the argument up to this point, but he presents a new and different reason for denying that animals can have legal rights. The ability to make claims, whether directly or through a representative, he implies, is essential to the possession of rights. Animals obviously cannot press their claims on their own, and so if they have rights, these rights must be assertable by agents. Animals, however, cannot be represented, McCloskey contends, not for any of the reasons already discussed, but rather because representation, in the requisite sense, is always of interest, and animals (he says) are incapable of having interests.

Now, there is a very important insight expressed in the requirement that a being have interests if he is to be a logically proper subject of rights. This can be appreciated if we consider just why it is that mere things cannot have rights. Consider a very precious "mere thing"—a beautiful natural wilderness, or a complex and ornamental artifact, like the Taj Mahal.

Such things ought to be cared for, because they would sink into decay if neglected, depriving some human beings, or perhaps even all human beings, of something of great value. Certain persons may even have as their own special job the care and protection of these valuable objects. But we are not tempted in these cases to speak of "thing-rights" correlative to custodial duties, because, try as we might, we cannot think of mere things as possessing interests of their own. Some people may have a duty to preserve, maintain, or improve the Taj Mahal, but they can hardly have a duty to help or hurt it, benefit or aid it, succor or relieve it. Custodians may protect it for the sake of a nation's pride and art lovers' fancy; but they don't keep it in good repair for "its own sake," or for "its own true welfare," or "well-being." A mere thing, however valuable to others, has no good of its own. The explanation of that fact, I suspect, consists in the fact that mere things have no conative life: no conscious wishes, desires, and hopes; or urges and impulses; or unconscious drives, aims, and goals; or latent tendencies, direction of growth, and natural fulfillments. Interests must be compounded somehow out of conations; hence mere things have no interests. *A fortiori*, they have no interests to be protected by legal or moral rules. Without interests a creature can have no "good" of its own, the achievement of which can be its due. Mere things are not loci of value in their own right, but rather their value consists entirely in their being objects of other beings' interests.

So far McCloskey is on solid ground, but one can quarrel with his denial that any animals but humans have interests. I should think that the trustee of funds willed to a dog or cat is more than a mere custodian of the animal he protects. Rather his job is to look out for the interests of the animal and make sure no one denies it its due. The animal itself is the beneficiary of his dutiful services. Many of the higher animals at least have appetites, conative urges, and rudimentary purposes, the integrated satisfaction of which constitutes their welfare or good. We can, of course, with consistency treat animals as mere pests and deny that they have any rights; for most animals, especially those of the lower orders, we have no choice but to do so. But it seems to me, nevertheless, that in general, animals *are* among the sorts of beings of whom rights can meaningfully be predicted and denied.

Now, if a person agrees with the conclusion of the argument thus far, that animals are the sorts of beings that *can* have rights, and further, if he accepts the moral judgment that we ought to be kind to animals, only one further premise is needed to yield the conclu-

sion that some animals do in fact have rights. We must now ask ourselves for whose sake ought we to treat (some) animals with consideration and humaneness? If we conceive our duty to be one of obedience to authority, or to one's own conscience merely, or one of consideration for tender human sensibilities only, then we might still deny that animals have rights, even though we admit that they are the kinds of beings that *can* have rights. But if we hold not only that we ought to treat animals humanely, but also that we should do so for the animals' own sake, that such treatment is something we owe animals as their due, something that can be claimed for them, something the withholding of which would be an injustice and a wrong, and not merely a harm, then it follows that we do ascribe rights to animals. I suspect that the moral judgments most of us make about animals do pass these phenomenological tests, so that most of us do believe that animals have rights, but are reluctant to say so because of the conceptual confusions about the notion of a right that I have attempted to dispel above.

Now we can extract from our discussion of animal rights a crucial principle for tentative use in the resolution of the other riddles about the applicability of the concept of a right, namely, that the sorts of beings who *can* have rights are precisely those who have (or can have) interests. I have come to this tentative conclusion for two reasons: (1) because a right holder must be capable of being represented and it is impossible to represent a being that has no interest, and (2) because a right holder must be capable of being a beneficiary in his own person, and a being without interests is a being that is incapable of being harmed or benefitted, having no good or "sake" of its own. Thus, a being without interests has no "behalf" to act in, and no "sake" to act for. My strategy now will be to apply the "interest principle," as we can call it, to the other puzzles about rights, while being prepared to modify it where necessary (but as little as possible), in the hope of separating in a consistent and intuitively satisfactory fashion the beings who can have rights from those which cannot.

VEGETABLES

It is clear that we ought not to mistreat certain plants, and indeed there are rules and regulations imposing duties on persons not to misbehave in respect to certain members of the vegetable kingdom. It is forbidden, for example, to pick wildflowers in the mountainous tundra areas of national parks, or to endanger trees by starting fires in the dry forest areas. Members of Congress introduce bills designed, as they say, to "protect" rare redwood trees from commercial pillage. Given this background, it is surprising that no one[8] speaks of plants as having rights. Plants, after all, are not "mere things"; they are vital objects with inherited biological propensities determining their natural growth. Moreover, we do say that certain conditions are "good" or "bad" for plants, thereby suggesting that plants, unlike rocks, are capable of having a "good." (This is a case, however, where "what we say" should not be taken seriously: we also say that certain kinds of paint are good and bad for the internal walls of a house, and this does not commit us to a conception of walls as beings possessed of a good or welfare of their own.) Finally, we are capable of feeling a kind of affection for particular plants, though we rarely personalize them, as we do in the case of animals, by giving them proper names.

Still, all are agreed that plants are not the kinds of beings that can have rights. Plants are never plausibly understood to be the direct intended beneficiaries of rules designed to "protect" them. We wish to keep redwood groves in existence for the sake of human beings who can enjoy their serene beauty, and for the sake of generations of human beings yet unborn. Trees are not the sorts of beings who have their "own sakes," despite the fact that they have biological propensities. Having no conscious wants or goals of their own, trees cannot know satisfaction or frustration, pleasure or pain. Hence, there is no possibility of kind or cruel treatment of trees. In these morally crucial respects, trees differ from the higher species of animals.

Yet trees are not mere things like rocks. They grow and develop according to the laws of their own nature. Aristotle and Aquinas both took trees to have their own "natural ends." Why then do I deny them the status of beings with interest of their own? The reason is that an interest, however the concept is finally to be analyzed, presupposes at least rudimentary cognitive equipment. Interests are compounded out of *desires* and *aims,* both of which presuppose something like *belief,* or cognitive awareness. . . .

WHOLE SPECIES

The topic of whole species, whether of plants or animals, can be treated in much the same way as that of individual plants. A whole collection, as such, cannot have beliefs, expectations, wants, or desires, and can flourish or languish only in the human interest-related sense in which individual plants thrive and decay. Individual elephants can have interests, but the species elephant cannot. Even where individual elephants are not granted rights, human beings may have an interest—economic, scientific or sentimental—in keeping the species from dying out, and *that* interest may be protected in various ways by law. But that is quite another matter from recognizing a right to survival belonging to the species itself. Still, the preservation of a whole species may quite properly seem to be a morally more important matter than the preservation of an individual animal. Individual animals can have rights but it is implausible to ascribe to them a right to life on the human model. Nor do we normally have duties to keep individual animals alive or even to abstain from killing them provided we do it humanely and nonwantonly in the promotion of legitimate human interests. On the other hand, we do have duties to protect threatened species; not duties to the species themselves as such, but rather duties to future human beings, duties derived from our housekeeping role as temporary inhabitants of this planet. . . .

FUTURE GENERATIONS

We have it in our power now to make the world a much less pleasant place for our descendants than the world we inherited from our ancestors. We can continue to proliferate in ever greater numbers, using up fertile soil at an even greater rate, dumping our wastes into rivers, lakes, and oceans, cutting down our forests, and polluting the atmosphere with noxious gases. All thoughtful people agree that we ought not to do these things. Most would say we have a duty not to do these things, meaning not merely that conservation is morally required (as opposed to merely desirable) but also that it is something due our descendants, something to be done for their sakes. Surely we owe it to future generations to pass on a world that is not a used up garbage heap. Our remote descendants are not yet present to claim a livable world as their right, but there are plenty of proxies to speak now in their behalf. These spokesmen, far from being mere custodians, are genuine representatives of future interests.

Why then deny that the human beings of the future have rights which can be claimed against us now in their behalf? Some are included to deny them present rights out of a fear of falling into obscure metaphysics, by granting rights to remote and unidentifiable beings who are not yet even in existence. Our unborn great-great-grandchildren are in some sense "potential" persons, but they are far more remotely potential, it may seem, than fetuses. This, however, is not the real difficulty. Unborn generations are more remotely potential than fetuses in one sense, but not in another. A much greater period of time with a far greater number of causally necessary and important events must pass before their potentiality can be actualized, it is true; but our collective posterity is just as certain to come into existence "in the normal course of events" as is any given fetus now in its mother's womb. In that sense the existence of the distant human future is no more remotely potential than that of a particular child already on its way.

The real difficulty is not that we doubt whether our descendants will ever be actual, but rather that we don't know who they will be. It is not their temporal remoteness that troubles us so much as their indeterminacy—their present facelessness and namelessness. Five centuries from now men and women will be living where we live now. Any given one of them will have an interest in living space, fertile soil, fresh air, and the like, but that arbitrarily selected one has no other qualities we can presently envision very clearly. We don't even know who his parents, grandparents, or great-grandparents are, or even whether he is related to us. Still, whoever these human beings may turn out to be, and whatever they might reasonably be expected to be like, they will have interests that we can affect, for better or worse, right now. That much we can and do know about them. The identity of the owners of these interests is now necessarily obscure, but the fact of their interest-ownership is crystal clear, and that is all that is necessary to certify the coherence of present talk about their rights. We can tell, sometimes, that shadowy forms in the spatial distance belong to human beings, though we know not who or how many they are; and this imposes a duty on us not to throw bombs, for example, in their direction. In like manner, the vagueness of the human future does not weaken its claim on us in light of the nearly certain knowledge that it will, after all, be human.

Doubts about the existence of a right to be born transfer neatly to the question of a similar right to come into existence ascribed to future generations. The rights that future generations certainly have against us are contingent rights: the interests they are sure to have when they come into being (assuming of course that they will come into being) cry out for protection from invasions that can take place now. Yet there are no actual interests, presently existent, that future generations, presently nonexistent, have now. Hence, there is no actual interest that they have in simply coming into being, and I am at a loss to think of any other reason for claiming that they have a right to come into existence (though there may well be such a reason). Suppose then that all human beings at a given time voluntarily form a compact never again to produce children, thus leading within a few decades to the end of our species. This of course is a wildly improbable hypothetical example but a rather crucial one for the position I have been tentatively considering. And we can imagine, say, that the whole world is converted to a strange ascetic religion which absolutely requires sexual abstinence for everyone. Would this arrangement violate the rights of anyone? No one can complain on behalf of presently nonexistent future generations that their future interests which give them a contingent right of protection have been violated since they will never come into existence to be wronged. My inclination then is to conclude that the suicide of our species would be deplorable, lamentable, and a deeply moving tragedy, but that it would violate no one's rights. Indeed if, contrary to fact, all human beings could ever agree to such a thing, that very agreement would be a symptom of our species' biological unsuitability for survival anyway.

CONCLUSION

For several centuries now human beings have run roughshod over the lands of our planet, just as if the animals who do live there and the generations of humans who will live there had no claims on them whatever. Philosophers have not helped matters by arguing that animals and future generations are not the kinds of beings who can have rights now, that they don't presently qualify for membership, even "auxiliary membership," in our moral community. I have tried in this essay to dispel the conceptual confusions that make such conclusions possible. To acknowledge their rights is the very least we can do for members of endangered species (including our own). But that is something.

Notes

1. I shall leave the concept of a claim unanalyzed here, but for a detailed discussion, see my "The Nature and Value of Rights," *Journal of Value Inquiry* 4 (Winter 1971): 263–277.

Reprinted with permission of the publisher from *Philosophy & Environmental Crisis,* edited by William T. Blackstone, pp. 43–68. Copyright © 1974 by the University of Georgia Press.

2. Louis B. Schwartz, "Morals, Offenses and the Model Penal Code," *Columbia Law Review* 63 (1963): 673.
3. John Chipman Gray, *The Nature and Sources of the Law,* 2d ed. (Boston: Beacon Press, 1963), p. 43.
4. And W. D. Ross for another. See *The Right and The Good* (Oxford: Clarendon Press, 1930), app. 1, pp. 48–56.
5. W. D. Lamont, *Principles of Moral Judgment* (Oxford: Clarendon Press, 1946, pp. 83–85.
6. Cf. H. J. McCloskey, "Rights," *Philosophical Quarterly* 15 (1965): 121, 124.
7. Ibid.
8. Outside of Samuel Butler's *Erewhon.*

For Further Discussion

1. Feinberg claims that unborn generations have rights against us and that we have corresponding duties to them. But because we know so little about people who will live in the distant future, does it make sense to attribute any particular rights to them? Do they, for example, have a right to live?

2. Do you think that previous generations of humans violated any of your rights? Which ones?

3. How do you understand the concept of having a "sake"? How do you understand having an "interest"?

4. Do you agree with Feinberg that our duties regarding animals are direct duties to them? If we have duties to them, does this mean that they have rights?

5. Using Feinberg's framework, do any of the following have rights: Dead humans? Fetuses? Brain-dead humans? Aliens? Corporations? Ecosystems? The earth?

All Animals Are Equal

Peter Singer

In his influential book Animal Liberation *(1975), from which the following selection is taken, philosopher Peter Singer developed a forceful argument in defense of the view that animals, at least those capable of sensation, deserve full moral consideration. Denying animals equal standing, what might be called "speciesism," is akin to denying equal standing on the basis of race or sex. The implications of this are far-reaching. Significant changes in our eating habits, in agriculture, in recreational activities, and in scientific research would be required once we recognize our direct responsibilities to animals.*

In recent years a number of oppressed groups have campaigned vigorously for equality. The classic instance is the Black Liberation movement, which demands an end to the prejudice and discrimination that has made blacks second-class citizens. The immediate appeal of the black liberation movement and its initial, if limited, success made it a model for other oppressed groups to follow. We became familiar with liberation movements for Spanish-Americans, gay people, and a variety of other minorities. When a majority group—women—began their campaign, some thought we had come to the end of the road. Discrimination on the basis of sex, it has been said, is the last universally accepted form of discrimination, practiced without secrecy or pretense even in those liberal circles that have long prided themselves on their freedom from prejudice against racial minorities.

One should always be wary of talking of "the last remaining form of discrimination." If we have learnt anything from the liberation movements, we should have learnt how difficult it is to be aware of latent prejudice in our attitudes to particular groups until this prejudice is forcefully pointed out.

A liberation movement demands an expansion of our moral horizons and an extension or reinterpretation of the basic moral principle of equality. Practices that were previously regarded as natural and inevitable come to be seen as the result of an unjustifiable prejudice. Who can say with confidence that all his or her attitudes and practices are beyond criticism? If we wish to avoid being numbered amongst the oppressors, we must be prepared to re-think even our most fundamental attitudes. We need to consider them from the point of view of those most disadvantaged by our attitudes, and the practices that follow from these attitudes. If we can make this unaccustomed mental switch we may discover a pattern in our attitudes and practices that consistently operates so as to benefit one group—usually the one to which we ourselves belong—at the expense of another. In this way we may come to see that there is a case for a new liberation movement. My aim is to advocate that we make this mental switch in respect of our attitudes and practices towards a very large group of beings: members of species other than our own—or, as we popularly though misleadingly call them, animals. In other words, I am urging that we extend to other species the basic principle of equality that most of us recognize should be extended to all members of our own species.

All this may sound a little far-fetched, more like a parody of other liberation movements than a serious objective. In fact, in the past the idea of "The Rights of Animals" really has been used to parody the case for women's rights. When Mary Wollstonecroft, a forerunner of later feminists, published her *Vindication of the Rights of Women* in 1792, her ideas were widely regarded as absurd, and they were satirized in an anonymous publication entitled *A Vindication of the Rights of Brutes.* The author of this satire (actually Thomas Taylor, a distinguished Cambridge philosopher) tried to refute Wollstonecroft's reasonings by showing that they could be carried one stage further. If sound when applied to women, why should the arguments not be applied to dogs, cats, and horses? They seemed to hold equally well for these "brutes"; yet to hold that brutes had rights was manifestly absurd; therefore the reasoning by which this conclusion had been reached must be unsound, and if unsound when applied to brutes, it must also be unsound when applied to women, since the very same arguments had been used in each case.

One way in which we might reply to this argument is by saying that the case for equality between men and women cannot validly be extended to nonhuman animals. Women have a right to vote, for instance, because they are just as capable of making rational decisions as men are; dogs, on the other hand, are incapable of understanding the significance of voting, so they cannot have the right to vote. There are many other obvious ways in which men and women resemble each other closely, while humans and other animals differ greatly. So, it might be said, men and women are similar beings, and should have equal rights, while humans and nonhumans are different and should not have equal rights.

The thought behind this reply to Taylor's analogy is correct up to a point, but it does not go far enough. There *are* important differences between humans and other animals, and these differences must give rise to *some* differences in the rights that each have. Recognizing this obvious fact, however, is no barrier to the case for extending the basic principle of equality to nonhuman animals. The differences that exist between men and women are equally undeniable, and the supporters of Women's Liberation are aware that these differences may give rise to different rights. Many feminists hold that women have the right to an abortion on request. It does not follow that since these same people are campaigning for equality between men and women they must support the right of men to have abortions too. Since a man cannot have an abortion, it is meaningless to talk of his right to have one. Since a pig can't vote, it is meaningless to talk of its right to vote. There is no reason why either Women's Liberation or Animal Liberation should get involved in such nonsense. The extension of the basic principle of equality from one group to another does not imply that we must treat both groups in exactly the same way, or grant exactly the same rights to both groups. Whether we should do so will depend on the nature of the members of the two groups. The basic principle of equality, I shall argue, is equality of consideration; and equal consideration for different beings may lead to different treatment and different rights.

So there is a different way of replying to Taylor's attempt to parody Wollstonecroft's arguments, a way which does not deny the differences between humans and nonhumans, but goes more deeply into the question of equality, and concludes by finding nothing absurd in the idea that the basic principle of equality applies to so-called "brutes." I believe that we reach this conclusion if we examine the basis on which our opposition to discrimination on grounds of race or sex ultimately rests. We will then see that we would be on shaky ground if we were to demand equality for blacks, women, and other groups of oppressed humans while denying equal consideration to nonhumans.

When we say that all human beings, whatever their race, creed or sex, are equal, what is it that we are asserting? Those who wish to defend a hierarchical, inegalitarian society have often pointed out that by whatever test we choose, it simply is not true that all humans are equal. Like it or not, we must face the fact that humans come in different shapes and sizes; they come with differing moral capacities, differing intellectual abilities, differing amounts of benevolent feeling and sensitivity to the needs of others, differing abilities to communicate effectively, and differing capacities to experience pleasure and pain. In short, if the demand for equality were based on the actual equality of all human beings, we would have to stop demanding equality. It would be an unjustifiable demand.

Still, one might cling to the view that the demand for equality among human beings is based on the actual equality of the different races and sexes. Although humans differ as individuals in various ways, there are no differences between the races and sexes *as such*. From the mere fact that a person is black, or a woman, we cannot infer anything else about that person. This, it may be said, is what is wrong with racism and sexism. The white racist claims that whites are superior to blacks, but this is false—although there are differences between individuals, some blacks are superior to some whites in all of the capacities and abilities that could conceivably be relevant. The opponent of sexism would say the same: a person's sex is no guide to his or her abilities, and this is why it is unjustifiable to discriminate on the basis of sex.

This is a possible line of objection to racial and sexual discrimination. It is not, however, the way that someone really concerned about equality would choose, because taking this line could, in some circumstances, force one to accept a most inegalitarian society. The fact that humans differ as individuals, rather than as races or sexes, is a valid reply to someone who defends a hierarchical society like, say, South Africa, in which all whites are superior in status to all blacks. The existence of individual variations that cut across the lines of race or sex, however, provides us with no defence at all against a more sophisticated opponent of equality, one who proposes that, say, the interests of those with I.Q. ratings above 100 be preferred to the interests of those with I.Q.s below 100. Would a hierarchical society of this sort really be so much better than one based on race or sex? I think not. But if we tie the moral principle of equality to the factual equality of the different races or sexes, taken as a whole, our opposition to racism and sexism does not provide us with any basis for objecting to this kind of inegalitarianism.

There is a second important reason why we ought not to base our opposition to racism and sexism on

any kind of factual equality, even the limited kind which asserts that variations in capacities and abilities are spread evenly between the different races and sexes: we can have no absolute guarantee that these abilities and capacities really are distributed evenly, without regard to race or sex, among human beings. So far as actual abilities are concerned, there do seem to be certain measurable differences between both races and sexes. These differences do not, of course, appear in each case, but only when averages are taken. More important still, we do not yet know how much of these differences is really due to the different genetic endowments of the various races and sexes, and how much is due to environmental differences that are the result of past and continuing discrimination. Perhaps all of the important differences will eventually prove to be environmental rather than genetic. Anyone opposed to racism and sexism will certainly hope that this will be so, for it will make the task of ending discrimination a lot easier; nevertheless it would be dangerous to rest the case against racism and sexism on the belief that all significant differences are environmental in origin. The opponent of, say, racism who takes this line will be unable to avoid conceding that if differences in ability did after all prove to have some genetic connection with race, racism would in some way be defensible.

It would be folly for the opponent of racism to stake his whole case on a dogmatic commitment to one particular outcome of a difficult scientific issue which is still a long way from being settled. While attempts to prove that differences in certain selected abilities between races and sexes are primarily genetic in origin have certainly not been conclusive, the same must be said of attempts to prove that these differences are largely the result of environment. At this stage of the investigation we cannot be certain which view is correct, however much we may hope it is the latter.

Fortunately, there is no need to pin the case for equality to one particular outcome of this scientific investigation. The appropriate response to those who claim to have found evidence of genetically based differences in ability between the races or sexes is not to stick to the belief that the genetic explanation must be wrong, whatever evidence to the contrary may turn up: instead we should make it quite clear that the claim to equality does not depend on intelligence, moral capacity, physical strength, or similar matters of fact. Equality is a moral ideal, not a simple assertion of fact. There is no logically compelling reason for assuming that a factual difference in ability between two people justifies any *difference in the amount of considera- tion we give to satisfying their needs and interests.* The principle of the equality of human beings is not a de- scription of an alleged actual equality among humans: it is a prescription of how we should treat humans.

Jeremy Bentham incorporated the essential basis of moral equality into his utilitarian system of ethics in the formula: "Each to count for one and none for more than one." In other words, the interests of every be- ing affected by an action are to be taken into account and given the same weight as the like interests of any other being. A later utilitarian, Henry Sidgwick, put the point in this way: "The good of any one individual is of no more importance, from the point of view (if I may say so) of the Universe, than the good of any other."[1] More recently, the leading figures in contem- porary moral philosophy have shown a great deal of agreement in specifying as a fundamental presupposi- tion of their moral theories some similar requirement which operates so as to give everyone's interests equal consideration—although they cannot agree on how this requirement is best formulated.[2]

It is an implication of this principle of equality that our concern for others ought not to depend on what they are like, or what abilities they possess—although precisely what this concern requires us to do may vary according to the characteristics of those affected by what we do. It is on this basis that the case against racism and the case against sexism must both ulti- mately rest; and it is in accordance with this principle that speciesism is also to be condemned. If possessing a higher degree of intelligence does not entitle one hu- man to use another for his own ends, how can it entitle humans to exploit nonhumans?

Many philosophers have proposed the principle of equal consideration of interests, in some form or other, as a basic moral principle; but, as we shall see in more detail shortly, not many of them have recog- nised that this principle applies to members of other species as well as to our own. Bentham was one of the few who did realize this. In a forward-looking passage, written at a time when black slaves in British domin- ions were still being treated much as we now treat nonhuman animals, Bentham wrote:

The day *may* come when the rest of the animal creation may acquire those rights which never could have been witholden from them but by the hand of tyranny. The French have already discov- ered that the blackness of the skin is no reason

why a human being should be abandoned with- out redress to the caprice of a tormentor. It may one day come to be recognized that the number of the legs, the villosity of the skin, or the termina- tion of the *os sacrum,* are reasons equally insuffi- cient for abandoning a sensitive being to the same fate. What else is it that should trace the insuper- able line? Is it the faculty of reason, or perhaps the faculty of discourse? But a full-grown horse or dog is beyond comparison a more rational, as well as a more conversable animal, than an infant of a day, or a week, or even a month, old. But suppose they were otherwise, what would it avail? The question is not, Can they reason? nor Can they *talk?* but, *Can they suffer?*[3]

In this passage Bentham points to the capacity for suffering as the vital characteristic that gives a being the *right* to equal consideration. The capacity for suf- fering—or more strictly, for suffering and/or enjoy- ment or happiness—is not just another characteristic like the capacity for language, or for higher mathe- matics. Bentham is not saying that those who try to make "the insuperable line" that determines whether the interests of a being should be considered happen to have selected the wrong characteristic. The capacity for suffering and enjoying things is a pre-requisite for having interests at all, a condition that must be satis- fied before we can speak of interests in any meaningful way. It would be nonsense to say that it was not in the interests of a stone to be kicked along the road by a schoolboy. A stone does not have interests because it cannot suffer. Nothing that we can do to it could pos- sibly make any difference to its welfare. A mouse, on the other hand, does have an interest in not being tor- mented, because it will suffer if it is.

If a being suffers, there can be no moral justifica- tion for refusing to take that suffering into considera- tion. No matter what the nature of the being, the prin- ciple of equality requires that its suffering be counted equally with the like suffering—in so far as rough comparisons can be made—of any other being. If a being is not capable of suffering, or of experiencing enjoyment or happiness, there is nothing to be taken into account. This is why the limit of sentience (using the term as a convenient, if not strictly accurate, short- hand for the capacity to suffer or experience enjoy- ment or happiness) is the only defensible boundary of concern for the interests of others. To mark this boundary by some characteristic like intelligence or rationality would be to mark it in an arbitrary way. Why not choose some other characteristic, like skin color?

The racist violates the principle of equality by giving greater weight to the interests of members of his own race, when there is a clash between their interests and the interests of those of another race. Similarly the speciesist allows the interests of his own species to override the greater interests of members of other species.[4] The pattern is the same in each case. Most human beings are speciesists. I shall now very briefly describe some of the practices that show this.

For the great majority of human beings, especially in urban, industrialized societies, the most direct form of contact with members of other species is at mealtimes: we eat them. In doing so we treat them purely as means to our ends. We regard their life and wellbeing as subordinate to our taste for a particular kind of dish. I say "taste" deliberately—this is purely a matter of pleasing our palate. There can be no defence of eating flesh in terms of satisfying nutritional needs, since it has been established beyond doubt that we could satisfy our need for protein and other essential nutrients far more efficiently with a diet that replaced animal flesh by soy beans, or products derived from soy beans, and other high-protein vegetable products.[5]

It is not merely the act of killing that indicates what we are ready to do to other species in order to gratify our tastes. The suffering we inflict on the animals while they are alive is perhaps an even clearer indication of our speciesism that the fact that we are prepared to kill them. In order to have meat on the table at a price that people can afford, our society tolerates methods of meat production that confine sentient animals in cramped, unsuitable conditions for the entire durations of their lives. Animals are treated like machines that convert fodder into flesh, and any innovation that results in a higher "conversion ratio" is liable to be adopted. As one authority on the subject has said, "cruelty is acknowledged only when profitability ceases."[6] . . .

Since, as I have said, none of these practices cater for anything more than our pleasures of taste, our practice of rearing and killing other animals in order to eat them is a clear instance of the sacrifice of the most important interests of other beings in order to satisfy trivial interests of our own. To avoid speciesism we must stop this practice, and each of us has a moral obligation to cease supporting the practice. Our custom is all the support that the meat-industry needs.

The decision to cease giving it that support may be difficult, but it is no more difficult than it would have been for a white Southerner to go against the traditions of his society and free his slaves: if we do not change our dietary habits, how can we censure those slave-holders who would not change their own way of living?

The same form of discrimination may be observed in the widespread practice of experimenting on other species in order to see if certain substances are safe for human beings, or to test some psychological theory about the effect of severe punishment on learning, or to try out various new compounds just in case something turns up. . . .

In the past, argument about vivisection has often missed this point, because it has been put in absolutist terms: Would the abolitionist be prepared to let thousands die if they could be saved by experimenting on a single animal? The way to reply to this purely hypothetical question is to pose another: *Would the experimenter be prepared to perform his experiment on an orphaned human infant, if that were the only way to save many lives?* (I say "orphan" to avoid the complication of parental feelings, although in doing so I am being overfair to the experimenter, since the nonhuman subjects of experiments are not orphans.) If the experimenter is not prepared to use an orphaned human infant, then his readiness to use nonhumans is simple discrimination, since adult apes, cats, mice and other mammals are more aware of what is happening to them, more self-directing and, so far as we can tell, at least as sensitive to pain, as any human infant. There seems to be no relevant characteristic that human infants possess that adult mammals do not have to the same or a higher degree. (Someone might try to argue that what makes it wrong to experiment on a human infant is that the infant will, in time and if left alone, develop into more than the nonhuman, but one would then, to be consistent, have to oppose abortion, since the fetus has the same potential as the infant—indeed, even contraception and abstinence might be wrong on this ground, since the egg and sperm, considered jointly, also have the same potential. In any case, this argument still gives us no reason for selecting a nonhuman, rather than a human with severe and irreversible brain damage, as the subject for our experiments.)

The experimenter, then, shows a bias in favor of his own species whenever he carries out an experiment on a nonhuman for a purpose that he would not think justified him in using a human being at an equal or lower level of sentience, awareness, ability to be self-directing, etc. No one familiar with the kind of results yielded by most experiments on animals can have the slightest doubt that if this bias were eliminated the number of experiments performed would be a minute fraction of the number performed today.

Experimenting on animals, and eating their flesh, are perhaps the two major forms of speciesism in our society. By comparison, the third and last form of speciesism is so minor as to be insignificant, but it is perhaps of some special interest to those for whom this article was written. I am referring to speciesism in contemporary philosophy.

Philosophy ought to question the basic assumptions of the age. Thinking through, critically and carefully, what most people take for granted is, I believe, the chief task of philosophy, and it is this task that makes philosophy a worthwhile activity. Regrettably, philosophy does not always live up to its historic role. Philosophers are human beings and they are subject to all the preconceptions of the society to which they belong. Sometimes they succeed in breaking free of the prevailing ideology: more often they become its most sophisticated defenders. So, in this case, philosophy as practiced in the universities today does not challenge anyone's preconceptions about our relations with other species. By their writings, those philosophers who tackle problems that touch upon the issue reveal that they make the same unquestioned assumptions as most other humans, and what they say tends to confirm the reader in his or her comfortable speciesist habits.

I could illustrate this claim by referring to the writings of philosophers in various fields—for instance, the attempts that have been made by those interested in rights to draw the boundary of the sphere of rights so that it runs parallel to the biological boundaries of the species *homo sapiens,* including infants and even mental defectives, but excluding those other beings of equal or greater capacity who are so useful to us at mealtimes and in our laboratories. I think it would be a more appropriate conclusion to this article, however, if I concentrated on the problem with which we have been centrally concerned, the problem of equality.

It is significant that the problem of *equality,* in moral and political philosophy, is invariably formulated in terms of human equality. The effect of this is that the question of the equality of other animals does not confront the philosopher, or student, as an issue itself—and this is already an indication of the failure of philosophy to challenge accepted beliefs. Still, philosophers have found it difficult to discuss the issue of human equality without raising, in a paragraph or two, the question of the status of other animals. The reason for this, which should be apparent from what I have said already, is that if humans are to be regarded as equal to one another, we need some sense of "equal" that does not require any actual, descriptive equality of capacities, talents or other qualities. If equality is to be related to any actual characteristics of humans, these characteristics must be some lowest common denominator, pitched so low that no human lacks them—but then the philosopher comes up against the catch that any such set of characteristics which covers *all* humans will not be possessed *only by humans.* In other words, it turns out that in the only sense in which we can truly say, as an assertion of fact, that all humans are equal, at least some members of other species are also equal—equal, that is, to each other and to humans. If, on the other hand, we regard the statement "All humans are equal" in some non-factual way, perhaps as a prescription, then, as I have already argued, it is even more difficult to exclude nonhumans from the sphere of equality.

This result is not what the egalitarian philosopher originally intended to assert. Instead of accepting the radical outcome to which their own reasonings naturally point, however, most philosophers try to reconcile their beliefs in human equality and animal inequality by arguments that can only be described as devious.

As a first example, I take William Frankena's well-known article "The Concept of Social Justice." Frankena opposes the idea of basing justice on merit, because he sees that this could lead to highly inegalitarian results. Instead he proposes the principle that

> . . . all men are to be treated as equals, not because they are equal, in any respect, but *simply because they are human.* They are human because they have *emotions* and *desires,* and are able to *think,* and hence are capable of enjoying a good life in a sense in which other animals are not.[7]

But what is this capacity to enjoy the good life which all humans have, but no other animals? Other animals have emotions and desires, and appear to be capable of enjoying a good life. We may doubt that they can think—although the behavior of some apes, dolphins and even dogs suggests that some of them can—but *what is the relevance of thinking?* Frankena goes on to admit that by "the good life" he means "not so much the morally good life as the happy or satisfactory life," so thought would appear to be unnecessary for enjoying the good life; in fact to emphasize the need for thought would make difficulties for the egalitarian since only some people are capable of leading intellectually satisfying lives, or morally good lives. This makes it difficult to see what Frankena's principle of equality has to do with simply being *human*. Surely every sentient being is capable of leading a life that is happier or less miserable than some alternative life, and hence has a claim to be taken into account. In this respect the distinction between humans and nonhumans is not a sharp division, but rather a continuum along which we move gradually, and with overlaps between the species, from simple capacities for enjoyment and satisfaction, or pain and suffering, to more complex ones.

Faced with a situation in which they see a need for some basis for the moral gulf that is commonly thought to separate humans and animals, but finding no concrete difference that will do the job without undermining the equality of humans, philosophers tend to waffle. They resort to high-sounding phrases like "the intrinsic dignity of the human individual";[8] they talk of the "intrinsic worth of all men" as if men (humans?) had some worth that other beings did not,[9] or they say that humans, and only humans, are "ends in themselves," while "everything other than a person can only have value for a person."[10]

This idea of a distinctive human dignity and worth has a long history; it can be traced back directly to the Renaissance humanists, for instance to Pico della Mirandola's *Oration on the Dignity of Man*. Pico and other humanists based their estimate of human dignity on the idea that man possessed the central, pivotal position in the "Great Chain of Being" that led from the lowliest forms of matter to God himself; this view of the universe, in turn, goes back to both classical and Judeo-Christian doctrines. Contemporary philosophers have cast off these metaphysical and religious shackles and freely invoke the dignity of mankind without needing to justify the idea at all. Why should we not attribute "intrinsic dignity" or "intrinsic worth" to ourselves? Fellow-humans are unlikely to reject the accolades we so generously bestow on them, and those to whom we deny the honor are unable to object. Indeed, when one thinks only of humans, it can be very liberal, very progressive, to talk of the dignity of all human beings. In so doing, we implicitly condemn slavery, racism, and other violations of human rights. We admit that we ourselves are in some fundamental sense on a par with the poorest, most ignorant members of our own species. It is only when we think of humans as no more than a small sub-group of all the beings that inhabit our planet that we may realize that in elevating our own species we are at the same time lowering the relative status of all other species.

The truth is that the appeal to the intrinsic dignity of human beings appears to solve the egalitarian's problems only as long as it goes unchallenged. Once we ask *why* it should be that all humans—including infants, mental defectives, psychopaths, Hitler, Stalin, and the rest—have some kind of dignity or worth that no elephant, pig, or chimpanzee can ever achieve, we see that this question is as difficult to answer as our original request for some relevant fact that justifies the inequality of humans and other animals. In fact, these two questions are really one: talk of intrinsic dignity or moral worth only takes the problem back one step, because any satisfactory defence of the claim that all and only humans have intrinsic dignity would need to refer to some relevant capacities or characteristics that all and only humans possess. Philosophers frequently introduce ideas of dignity, respect and worth at the point at which other reasons appear to be lacking, but this is hardly good enough. Fine phrases are the last resource of those who have run out of arguments.

In case there are those who still think it may be possible to find some relevant characteristic that distinguishes all humans from all members of other species, I shall refer again, before I conclude, to the existence of some humans who quite clearly are below the level of awareness, self-consciousness, intelligence, and sentience, of many nonhumans. I am thinking of humans with severe and irreparable brain damage, and also of infant humans. To avoid the complication of the relevance of a being's potential, however, I shall henceforth concentrate on permanently retarded humans.

Philosophers who set out to find a characteristic that will distinguish humans from other animals rarely take the course of abandoning these groups of humans by lumping them in with the other animals. It is easy to see why they do not. To take this line without rethinking our attitudes to other animals would entail that we have the right to perform painful experiments on retarded humans for trivial reasons; similarly it would follow that we had the right to rear and kill these humans for food. To most philosophers these consequences are as unacceptable as the view that we should stop treating nonhumans in this way.

Of course, when discussing the problem of equality it is possible to ignore the problem of mental defectives, or brush it aside as if somehow insignificant.[11] This is the easiest way out. What else remains? My final example of speciesism in contemporary philosophy has been selected to show what happens when a writer is prepared to face the question of human equality and animal equality without ignoring the existence of mental defectives, and without resorting to obscurantist mumbo-jumbo. Stanley Benn's clear and honest article "Egalitarianism and Equal Consideration of Interests"[12] fits this description.

Benn, after noting the usual "evident human inequalities" argues, correctly I think, for equality of consideration as the only possible basis for egalitarianism. Yet Benn, like other writers, is thinking only of "equal consideration of human interests." Benn is quite open in his defence of this restriction of equal consideration:

> . . . not to possess human shape is a disqualifying condition. However faithful or intelligent a dog may be, it would be a monstrous sentimentality to attribute to him interests that could be weighed in an equal balance with those of human beings . . . if, for instance, one had to decide between feeding a hungry baby or a hungry dog, anyone who chose the dog would generally be reckoned morally defective, unable to recognize a fundamental inequality of claims.
>
> This is what distinguishes our attitude to animals from our attitude to imbeciles. It would be odd to say that we ought to respect equally the dignity or personality of the imbecile and of the rational man . . . but there is nothing odd about saying that we should respect their interests equally, that is, that we should give to the interests of each the same serious consideration as claims to considerations necessary for some standard of well-being that we can recognize and endorse.

Benn's statement of the basis of the consideration we should have for imbeciles seems to me correct, but why should there be any fundamental inequality of claims between a dog and a human imbecile? Benn sees that if equal consideration depended on rationality, no reason could be given against using imbeciles for research purposes, as we now use dogs and guinea pigs. This will not do: "But of course we do distinguish imbeciles from animals in this regard," he says. That the common distinction is justifiable is something Benn does not question; his problem is how it is to be justified. The answer he gives is this:

> . . . we respect the interests of men and give them priority over dogs not *insofar* as they are rational, but because rationality is the human norm. We say it is *unfair* to exploit the deficiencies of the imbecile who falls short of the norm, just as it would be unfair, and not just ordinarily dishonest, to steal from a blind man. If we do not think in this way about dogs, it is because we do not see the irrationality of the dog as a deficiency or a handicap, but as normal for the species. The characteristics, therefore, that distinguish the normal man from the normal dog make it intelligible for us to talk of other men having interests and capacities and therefore claims, of precisely the same kind as we make on our own behalf. But although these characteristics may provide the point of the distinction between men and other species, they are *not* in fact the qualifying conditions for membership, or the distinguishing criteria of the class of morally considerable persons; *and this is precisely because a man does not become a member of a different species, with its own standards of normality, by reason or not possessing these characteristics.*

The final sentence of this passage gives the argument away. An imbecile, Benn concedes, may have no characteristics superior to those of a dog; nevertheless this does not make the imbecile a member of "a different species" as the dog is. *Therefore* it would be "unfair" to use the imbecile for medical research as we use the dog. But why? That the imbecile is not rational is just the way things have worked out, and the same is true of the dog—neither is any more responsible for their mental level. If it is unfair to take advantage of an isolated defect, why is it fair to take advantage of a more general limitation? I find it hard to see anything in this argument except a defence of preferring the interests of members of our own species because they are members of our own species. To those who think there might be more to it, I suggest the following mental exercise. Assume that it has been proven that there is a difference in the average, or normal, intelligence quotient for two different races, say whites and blacks. Then substitute the term "white" for every occurrence of "men" and "black" for every occurrence of "dog" in the passage quoted; and substitute "high I.Q." for "rationality" and when Benn talks of "imbeciles" replace this term by "dumb whites"—that is, whites who fall well below the normal white I.Q. score. Finally, change "species" to "race." Now re-read the passage. It has become a defence of a rigid, no-exceptions division between whites and blacks, based on I.Q. scores, *not withstanding an admitted overlap* between whites and blacks in this respect. The revised passage is, of course, outrageous, and this is not only because we have made fictitious assumptions in our substitutions. The point is that in the original passage Benn was defending a rigid division in the amount of consideration due to members of different species, despite admitted cases of overlap. If the original did not, at first reading strike us as being as outrageous as the revised version does, this is largely because although we are not racists ourselves, most of us are speciesists. Like the other articles, Benn's stands as a warning of the ease with which the best minds can fall victim to a prevailing ideology.

Notes

1. *The Methods of Ethics* (7th Ed.), p. 382.
2. For example, R. M. Hare, *Freedom and Reason* (Oxford, 1963) and J. Rawls, *A Theory of Justice* (Harvard, 1972) a brief account of the essential agreement on this issue between these and other positions, see R. M. Hare, "Rules of War and Moral Reasoning," *Philosophy and Public Affairs,* vol. 1, no. 2 (1972).
3. *Introduction to the Principles of Morals and Legislation,* ch. XVII.
4. I owe the term "speciesism" to Richard Ryder.
5. In order to produce 1 lb. of protein in the form of beef or veal, we must feed 21 lbs. of protein to the animal. Other forms of livestock are slightly less inefficient, but the average ratio in the U.S. is still $1:8$. It has been estimated that the amount of protein lost to humans in this way is equivalent to 90% of the annual world protein deficit. For a brief account, see Frances Moore Lappé, *Diet for a Small Planet* (Friends of The Earth/Ballantine, New York 1971), pp. 4–11.
6. Ruth Harrison, *Animal Machines* (Stuart, London, 1964). For an account of farming conditions, see my *Animal Liberation* (New York Review Company, 1975).
7. R. Brandt (ed.) *Social Justice* (Prentice-Hall, Englewood Cliffs, 1962), p. 19.
8. Frankena, *Op. cit,* p. 23.
9. H. A. Bedau, "Egalitarianism and the Idea of Equality" in *Nomos IX: Equality,* ed. J. R. Pennock and J. W. Chapman, New York, 1967.
10. G. Vlastos, "Justice and Equality" in Brandt, *Social Justice,* p. 48.
11. For example, Bernard Williams, "The Idea of Equality," in *Philosophy, Politics and Society* (second series), ed. P. Laslett and W. Runciman (Blackwell, Oxford, 1962), p. 118; J. Rawls, *A Theory of Justice,* pp. 509–10.
12. *Nomos IX: Equality;* the passages quoted are on p. 62ff.

For Further Discussion

1. In what ways is the animal liberation movement similar to other liberation movements? In what ways is it different?

2. Singer recognizes that there are differences between humans and other animals and that these differences entail ethical differences as well. Why does he conclude that this is no barrier to the extension of equal ethical standing to animals?

3. Do animals suffer? Are they aware of their suffering in all of the ways that humans are aware of their suffering?

4. How do you answer Singer's question: "Why should it be that all humans—including infants, mental defectives, psychopaths, Hitler, Stalin, and the rest—have some kind of dignity or worth that no elephant, pig, or chimpanzee can ever achieve?"

5. If suffering could be eliminated, do you think that Singer would accept the ethical legitimacy of eating meat?

The Case for Animal Rights

Tom Regan

Philosopher Tom Regan develops a deontological account of animal rights. In the following reading, he rejects the utilitarian approach to animal liberation because, on his view, it fails to recognize the intrinsic worth of animals. The reason it is wrong to mistreat animals is not that we inflict pain on them (what if we killed them painlessly?) but that we treat them as resources, as means to our own ends. Many animals are capable of consciously choosing their own ends. In this respect, they deserve the same moral standing, and the same moral rights, that we recognize for humans. Although Regan would disagree with Singer concerning the philosophical basis for the moral standing of animals, Regan would agree with Singer in condemning many human activities having to do with animals.

I regard myself as an advocate of animal rights—as a part of the animal rights movement. That movement, as I conceive it, is committed to a number of goals, including:

1. the total abolition of the use of animals in science

2. the total dissolution of commercial animal agriculture

3. and the total elimination of commercial and sport hunting and trapping.

There are, I know, people who profess to believe in animal rights who do not avow these goals. Factory farming they say, is wrong—violates animals' rights—but traditional animal agriculture is all right. Toxicity tests of cosmetics on animals violate their rights; but not important medical research—cancer research, for example. The clubbing of baby seals is abhorrent; but not the harvesting of adult seals. I used to think I understood this reasoning. Not any more. You don't change unjust institutions by tidying them up.

What's wrong—what's fundamentally wrong—with the way animals are treated isn't the details that vary from case to case. It's the whole system. The forlornness of the veal calf is pathetic—heart wrenching; the pulsing pain of the chimp with electrodes planted deep in her brain is repulsive; the slow, torturous death of the raccoon caught in the leg hold trap, agonizing. But what is fundamentally wrong isn't the pain, isn't the suffering, isn't the deprivation. These compound what's wrong. Sometimes—often—they make it much worse. But they are not the fundamental wrong.

The *fundamental wrong is the system that allows us to view animals as our resources,* here for us—to be eaten, or surgically manipulated, or put in our cross hairs for sport or money. Once we accept this view of animals—as our resources—the rest is as predictable as it is regrettable. Why worry about their loneliness, their pain, their death? Since animals exist for us, here to benefit us in one way or another, what harms them really doesn't matter—or matters only if it starts to bother us, makes us feel a trifle uneasy when we eat our veal scampi, for example. So, yes, let us get veal calves out of solitary confinement, give them more space, a little straw, a few companions. But let us keep our veal scampi.

But a little straw, more space, and a few companions don't eliminate—don't even touch—the fundamental wrong, the wrong that attaches to our viewing and treating these animals as our resources. A veal calf killed to be eaten after living in close confinement is viewed and treated in this way: but so, too, is another who is raised (as they say) "more humanely." To right the fundamental wrong of our treatment of farm animals requires more than making rearing methods "more human"—requires something quite different—requires the *total dissolution of commercial animal agriculture.*

How we do this—whether we do this, or as in the case of animals in science, whether and how we abolish their use—these are to a large extent political questions. People must change their beliefs before they change their habits. Enough people, especially those elected to public office, must believe in change—must want it—before we will have laws that protect the rights of animals. This process of change is very complicated, very demanding, very exhausting, calling for the efforts of many hands—in education, publicity, political organization and activity, down to the licking of envelopes and stamps. As a trained and practicing philosopher the sort of contribution I can make is limited, but I like to think, important. The currency of philosophy is ideas—their meaning and rational foundation—not the nuts and bolts of the legislative process say, or the mechanics of community organization. That's what I have been exploring over the past ten years or so in my essays and talks and, more recently, in my book, *The Case for Animal Rights*. I believe the major conclusions I reach in that book are true because they are supported by the weight of the *best arguments*. I believe the idea of animal rights has reason, not just emotion, on its side.

In the space I have at my disposal here I can only sketch, in the barest outlines, some of the main features of the book. Its main themes—and we should not be surprised by this—involve asking and answering deep foundational moral questions, questions about what morality is, how it should be understood, what is the best moral theory all considered. I hope I can convey something of the shape I think this theory is. The attempt to do this will be—to use a word a friendly critic once used to describe my work—cerebral. In fact I was told by this person that my work is "too cerebral." But this is misleading. My feelings about how animals sometimes are treated are just as deep and just as strong as those of my more volatile compatriots. Philosophers do—to use the jargon of the day—have a right side to their brains. If it's the left side we contribute or mainly should—that's because what talents we have reside there.

How to proceed? We begin by asking how the moral status of animals has been understood by thinkers who deny that animals have rights. Then we test the mettle of their ideas by seeing how well they stand up under the heat of fair criticism. If we start our thinking in this way we soon find that some people believe that we have no duties directly to animals—that we owe nothing *to them*—that we can do nothing that *wrongs them*. Rather, we can do wrong acts that involve animals, and so we have duties regarding them, though none to them. Such views may be called indirect duty views. By way of illustration:

Suppose your neighbor kicks your dog. Then your neighbor has done something wrong. But not to your dog. The wrong that has been done is a wrong to you. After all, it is wrong to upset people, and your neigh-

bor's kicking your dog upsets you. So you are the one who is wronged, not your dog. Or again: by kicking your dog your neighbor damages your property. And since it is wrong to damage another person's property, your neighbor has done something wrong—to you, of course, not to your dog. Your neighbor no more wrongs your dog than your car would be wronged if the windshield were smashed. Your neighbor's duties involving your dog are indirect duties to you. More generally, all of our duties regarding animals are indirect duties to one another—to humanity.

How could someone try to justify such a view? One could say that your dog doesn't feel anything and so isn't hurt by your neighbor's kick, doesn't care about the pain since none is felt, is as unaware of anything as your windshield. Someone could say this but no rational person will since, among other considerations, such a view will commit one who holds it to the position that no human being feels pain either—that human beings also don't care about what happens to them. A second possibility is that though both humans and your dog are hurt when kicked, it is only human pain that matters. But, again, no rational person can believe this. Pain is pain wheresoever it occurs. If your neighbor's causing you pain is wrong because of the pain that is caused, we cannot rationally ignore or dismiss the moral relevance of the pain your dog feels.

Philosophers who hold indirect duty views—and many still do—have come to understand that they must avoid the two defects just noted—avoid, that is, both the view that animals don't feel anything as well as the idea that only human pain can be morally relevant. Among such thinkers the sort of view now favored is one or another form of what is called *contractarianism*.

Here, very crudely, is the root idea: morality consists of a set of rules that individuals voluntarily agree to abide by—as we do when we sign a contract (hence the name: contractarianism). Those who understand and accept the terms of the contract are covered directly—have rights created by, and recognized and protected in, the contract. And these contractors can also have protection spelled out for others who, though they lack the ability to understand morality and so cannot sign the contract themselves, are loved or cherished by those who can. Thus young children, for example, are unable to sign and lack rights. But they are protected by the contract nonetheless because of the sentimental interests of others, most notably their parents. So we have, then, duties involving these children, duties regarding them, but no duties to them. Our duties in their case are indirect duties to other human beings, usually their parents.

As for animals, since they cannot understand the contract, they obviously cannot sign; and since they cannot sign, they have no rights. Like children, however, some animals are the objects of the sentimental interest of others. You, for example, love your dog . . . or cat. So these animals—those enough people care about: companion animals, whales, baby seals, the American bald eagle—these animals, though they lack rights themselves, will be protected because of the sentimental interests of people. I have, then, according to contractarianism, no duty directly to your dog or any other animal, not even the duty not to cause them pain or suffering; my duty not to hurt them is a duty I have to those people who care about what happens to them. As for other animals, where no or little sentimental interest is present—farm animals, for example, or laboratory rats—what duties we have grow weaker and weaker, perhaps to the vanishing point. The pain and death they endure, though real, are not wrong if no one cares about them.

Contractarianism could be a hard view to refute when it comes to the moral status of animals if it was an adequate theoretical approach to the moral status of human beings. It is not adequate in this latter respect, however, which makes the question of its adequacy in the former—regarding animals—utterly moot. For consider: morality, according to the (crude) contractarian position before us, consists of rules people agree to abide by. What people? Well, enough to make a difference—enough, that is, so that collectively they have the power to enforce the rules that are drawn up in the contract. That is very well and good for the signatories—but not so good for anyone who is not asked to sign. And there is nothing in contractarianism of the sort we are discussing that guarantees or requires that everyone will have a chance to participate equitably in framing the rules of morality. The result is that this approach to ethics could sanction the most blatant forms of social, economic, moral, and political injustice, ranging from a repressive caste system to systematic racial or sexual discrimination. Might, on this theory, does make right. Let those who are the victims of injustice suffer as they will. It matters not so long as no one else—no contractor, or too few of them—cares about it. Such a theory takes one's moral breath away . . . as if, for example, there is nothing wrong with apartheid in South Africa if too few white South Africans are upset by it. A theory with so little to recommend it at the level of the ethics of our treatment of our fellow humans cannot have anything more to recommend it when it comes to the ethics of how we treat our fellow animals.

The version of contractarianism just examined is, as I have noted, a crude variety, and in fairness to those of a contractarian persuasion it must be noted that much more refined, subtle, and ingenious varieties are possible. For example, John Rawls, in his *A Theory of Justice,* sets forth a version of contractarianism that forces the contractors to ignore the accidental features of being a human being—for example, whether one is white or black, male or female, a genius or of modest intellect. Only by ignoring such features, Rawls believes, can we insure that the principles of justice contractors would agree upon are not based on bias or prejudice. Despite the improvement a view such as Rawls's shows over the cruder forms of contractarianism, it remains deficient: it systematically denies that we have direct duties to those human beings who do not have a sense of justice—young children, for instance, and many mentally retarded humans. And yet it seems reasonably certain that, were we to torture a young child or a retarded elder, we would be doing something that wrongs them, not something that is wrong if (and only if) other humans with a sense of justice are upset. And since this is true in the case of these humans, we cannot rationally deny the same in the case of animals.

Indirect duty views, then, including the best among them, fail to command our rational assent. Whatever ethical theory we rationally should accept, therefore, it must at least recognize that we have some duties directly to animals, just as we have some duties directly to each other. The next two theories I'll sketch attempt to meet this requirement.

The first I call the *cruelty-kindness* view. Simply stated, this view says that we have a direct duty to be kind to animals and a direct duty not to be cruel to them. Despite the familiar, reassuring ring of these ideas, I do not believe this view offers an adequate theory. To make this clearer, consider kindness. A kind person acts from a certain kind of motive—compassion or concern, for example. And that is a virtue. But there is no guarantee that a kind act is a right act. If I am a generous racist, for example, I will be inclined to act kindly toward members of my own race, favoring their interests above others. My kindness would be real and, so far as it goes, good. But I trust it is too obvious to require comment that my kind acts may not be above moral approach—may, in fact, be positively wrong because rooted in injustice.

So kindness, not withstanding its status as a virtue to be encouraged, simply will not cancel the weight of a theory of right action.

Cruelty fares no better. People or their acts are cruel if they display either a lack of sympathy for or, worse, the presence of enjoyment in, seeing another suffer. Cruelty in all its guises *is* a bad thing—*is* a tragic human failing. But just as a person's being motivated by kindness does not guarantee that they do what is right, so the absence of cruelty does not assure that they avoid doing what is wrong. Many people who perform abortions, for example, are not cruel, sadistic people. But that fact about their character and motivation does not settle the terribly difficult question about the morality of abortion. The case is no different when we examine the ethics of our treatment of animals. So, yes, let us be for kindness and against cruelty. But let us not suppose that being for the one and against the other answers questions about moral right and wrong.

Some people think the theory we are looking for is *utilitarianism.* A utilitarian accepts two moral principles. The first is a principle of *equality: everyone's interests count, and similar interests must be counted as having similar weight or importance.* White or black, male or female, American or Iranian, human or animal: everyone's pain or frustration matter and matter equally with the like pain or frustration of anyone else. The second principle a utilitarian accepts is the principle of *utility: do that act that will bring about the best balance of satisfaction over frustration for everyone affected by the outcome.*

As a utilitarian, then, here is how I am to approach the task of deciding what I morally ought to do: I must ask who will be affected if I choose to do one thing rather than another, how much each individual will be affected, and where the best results are most likely to lie—which option, in other words, is most likely to bring about the best results, the best balance of satisfaction over frustration. That option, whatever it may be, is the one I ought to choose. That is where my moral duty lies.

The great appeal of utilitarianism rests with its uncompromising *egalitarianism:* everyone's interests count and count equally with the like interests of everyone else. The kind of odious discrimination some forms of contractarianism can justify—discrimination based on race or sex, for example—seems disallowed in principle by utilitarianism, as is speciesism—systematic discrimination based on species membership.

The sort of equality we find in utilitarianism, however, is not the sort an advocate of animal or human rights should have in mind. Utilitarianism has no room for the *equal moral rights of different individuals because it has no room for their equal inherent value or worth.* What has value for the utilitarian is the satisfaction of an individual's interests, not the individual whose interests they are. A universe in which you satisfy your desire for water, food, and warmth, is, other things being equal, better than a universe in which these desires are frustrated. And the same is true in the case of an animal with similar desires. But neither you nor the animal have any value in your own right. *Only your feelings do.*

Here is an analogy to help make the philosophical point clearer: a cup contains different liquids—sometimes sweet, sometimes bitter, sometimes a mix of the two. What has value are the liquids: the sweeter the better, the bitter the worse. The cup—the container—has no value. It's what goes into it, not what they go into, that has value. For the utilitarian, you and I are like the cup; we have no value as individuals and thus no equal value. What has value is what goes into us, what we serve as receptacles for; our feelings of satisfaction have positive value, our feelings of frustration have negative value.

Serious problems arise for utilitarianism when we remind ourselves that it enjoins us to bring about the best consequences. What does this mean? It doesn't mean the best consequences for me alone, or for my family or friends, or any other person taken individually. No, what we must do is, roughly, as follows: we must add up—somehow!—the separate satisfactions and frustrations of everyone likely to be affected by our choice, the satisfactions in one column, the frustrations in the other. We must total each column for each of the options before us. That is what it means to say the theory is aggregative. And then we must choose that option which is most likely to bring about the best balance of totaled satisfactions over totaled frustrations. Whatever act would lead to this outcome is the one we morally ought to perform—is where our moral duty lies. And that act quite clearly might not be the same one that would bring about the best results for me personally, or my family or friends, or a lab animal. The best aggregated consequences for everyone concerned are not necessarily the best for each individual.

That utilitarianism is an aggregative theory—that different individual's satisfactions or frustrations are added, or summed, or totaled—is the key objection to this theory. My Aunt Bea is old, inactive, a cranky, sour person, though not physically ill. She prefers to go on living. She is also rather rich. I could make a fortune if I could get my hands on her money, money she intends to give me in any event, after she dies, but which she refuses to give me now. In order to avoid a huge tax bite, I plan to donate a handsome sum of my profits to a local children's hospital. Many, many children will benefit from my generosity, and much joy will be brought to their parents, relatives, and friends. If I don't get the money rather soon, all these ambitions will come to naught. The once-in-a-lifetime-opportunity to make a real killing will be gone. Why, then, not really kill my Aunt Bea? Oh, of course I *might* get caught. But I'm no fool and, besides, her doctor can be counted on to cooperate (he has an eye for the same investment and I happen to know a good deal about his shady past). The deed can be done . . . professionally, shall we say. There is *very* little chance of getting caught. And as for my conscience being guilt ridden, I am a resourceful sort of fellow and will take more than sufficient comfort—as I lie on the beach at Acapulco—in contemplating the joy and health I have brought to so many others.

Suppose Aunt Bea is killed and the rest of the story comes out as told. Would I have done anything wrong? Anything immoral? One would have thought that I had. But not according to utilitarianism. Since what I did brought about the best balance of totaled satisfaction over frustration for all those affected by the outcome, what I did was not wrong. Indeed, in killing Aunt Bea the physician and I did what duty required.

This same kind of argument can be repeated in all sorts of cases, illustrating time after time, how the utilitarian's position leads to results that impartial people find morally callous. It *is* wrong to kill my Aunt Bea in the name of bringing about the best results for others. A good end does not justify an evil means. Any adequate moral theory will have to explain why this is so. Utilitarianism fails in this respect and so cannot be the theory we seek.

What to do? Where to begin anew? The place to begin, I think, is with the utilitarian's view of the value of the individual—or, rather, lack of value. In its place suppose we consider that you and I, for example, do have value as individuals—what we'll call *inherent value.* To say we have such value is to say that we are something more than, something different from, mere receptacles. Moreover, to insure that we do not pave the way for such injustices as slavery or sexual discrimination, we must believe that all who have inherent value have it equally, regardless of

their sex, race, religion, birthplace, and so on. Similarly to be discarded as irrelevant are one's talents or skills, intelligence and wealth, personality or pathology, whether one is loved and admired—or despised and loathed. The genius and the retarded child, the prince and the pauper, the brain surgeon and the fruit vendor, Mother Teresa and the most unscrupulous used car salesman—all have inherent value, all possess it *equally,* and *all have an equal right to be treated with respect,* to be treated in ways that do not reduce them to the status of things, as if they exist as resources for others. My value as an individual is independent of my usefulness to you. Yours is not dependent on your usefulness to me. For either of us to treat the other in ways that fail to show respect for the other's independent value is to act immorally—is to violate the individual's rights.

Some of the rational virtues of this view—what I call the rights view—should be evident. Unlike (crude) contractarianism, for example, the rights view *in principle* denies the moral tolerability of any and all forms of racial, sexual, or social discrimination; and unlike utilitarianism, this view *in principle* denies that we can justify good results by using evil means that violate an individual's rights—denies, for example, that it could be moral to kill my Aunt Bea to harvest beneficial consequences for others. That would be to sanction the disrespectful treatment of the individual in the name of the social good, something the rights view will not—categorically will not—ever allow.

The rights view—or so I believe—is rationally the most satisfactory moral theory. It surpasses all other theories in the degree to which it illuminates and explains the foundation of our duties to one another—the domain of human morality. On this score, it has the best reasons, the best arguments, on its side. Of course, if it were possible to show that only human beings are included within its scope, then a person like myself, who believes in animal rights, would be obliged to look elsewhere than to the rights view.

But attempts to limit its scope to humans only can be shown to be rationally defective. Animals, it is true, lack many of the abilities humans possess. They can't read, do higher mathematics, build a bookcase, or make *baba ghanoush.* Neither can many human beings, however, and yet we don't say—and shouldn't say—that they (these humans) therefore have less inherent value, less of a right to be treated with respect, than do others. It is the *similarities* between those human beings who most clearly, most noncontroversially have such value—the people reading this, for example—it is our similarities, not our differences, that matter most. And the really crucial, the basic similarity is simply this: *we are each of us the experiencing subject of a life, each of us a conscious creature having an individual welfare that has importance to us whatever our usefulness to others.* We want and prefer things; believe and feel things; recall and expect things. And all these dimensions of our life, including our pleasure and pain, our enjoyment and suffering, our satisfaction and frustration, our continued existence or our untimely death—all make a difference to the quality of our life as lived, as experienced by us as individuals. As the same is true of those animals who concern us (those who are eaten and trapped, for example), they, too, must be viewed as the experiencing subjects of a life with inherent value of their own.

There are some who resist the idea that animals have inherent value. "Only humans have such value," they profess. How might this narrow view be defended? Shall we say that only humans have the requisite intelligence, or autonomy, or reason? But there are many, many humans who will fail to meet these standards and yet who are reasonably viewed as having value above and beyond their usefulness to others. Shall we claim that only humans belong to the right species—the species *Homo sapiens?* But this is blatant speciesism. Will it be said, then, that all—and only—humans have immortal souls? Then our opponents more than have their work cut out for them. I am myself not ill-disposed to there being immortal souls. Personally, I profoundly hope I have one. But I would not want to rest my position on a controversial ethical issue on the even more controversial question about who or what has an immortal soul. That is to dig one's hole deeper, not climb out. Rationally, it is better to resolve moral issues without making more controversial assumptions than are needed. The question of who has inherent value is such a question, one that is more rationally resolved without the introduction of the idea of immortal souls than by its use.

Well, perhaps some will say that animals have some inherent value, only *less* than we do. Once again, however, attempts to defend this view can be shown to lack rational justification. What could be the basis of our having more inherent value than animals? Will it be their lack of reason, or autonomy, or intellect? Only if we are willing to make the same judgment in the case of humans who are similarly deficient. But it is not true that such humans—the retarded child, for example, or the mentally deranged—have less inherent value than you or I. Neither, then, can we rationally

sustain the view that animals like them in being the experiencing subjects of a life have less inherent value. *All who have inherent value have it equally, whether they be human animals or not.*

Inherent value, then, belongs equally to those who are the experiencing subjects of a life. Whether it belongs to others—to rocks and rivers, trees and glaciers, for example—we do not know. And may never know. But neither do we need to know, if we are to make the case for animal rights. We do not need to know how many people, for example, are eligible to vote in the next presidential election before we can know whether I am. Similarly, we do not need to know *how many* individuals have inherent value before we can know that some do. When it comes to the case for animal rights, then what we need to know is whether the animals who, in our culture are routinely eaten, hunted, and used in our laboratories, for example, are like us in being subjects of a life. And we *do* know this. We do *know* that many—literally, billions and billions—of these animals are subjects of a life in the sense explained and so have inherent value if we do. And since, in order to have the best theory of our duties to one another, we must recognize our equal inherent value, as individuals, *reason*—not sentiment, not emotion—*reason compels us to recognize the equal inherent value of these animals.* And, with this, their equal right to be treated with respect.

That, *very* roughly, is the shape and feel of the case for animal rights. Most of the details of the supporting argument are missing. They are to be found in the book I alluded to earlier. Here, the details go begging and I must in closing, limit myself to four final points.

The first is how the theory that underlies the case for animal rights shows that the animal rights movement is a part of, not antagonistic to, the human rights movement. The theory that rationally grounds the rights of animals also grounds the rights of humans. Thus are those involved in the animal rights movement partners in the struggle to secure respect for human rights—the rights of women, for example, or minorities and workers. The animal rights movement is cut from the same moral cloth as these.

Second, having set out the broad outlines of the rights view, I can now say why its *implications for farming and science,* for example, are both clear and uncompromising. In the case of using animals in science, the rights view is categorically abolitionist. *Lab animals are not our tasters; we are not their kings.* Because these animals are treated—routinely, systematically—as if their value is reducible to their usefulness to others, they are routinely, systematically treated with a lack of respect, and thus their rights routinely, systematically violated. This is just as true when they are used in trivial, duplicative, unnecessary or unwise research as it is when they are used in studies that hold out real promise of human benefits. We can't justify harming or killing a human being (my Aunt Bea, for example) just for these sorts of reasons. Neither can we do so even in the case of so lowly a creature as a laboratory rat. It is not just refinement or reduction that are called for, not just larger, cleaner cages, not just more generous use of anesthetic or the elimination of multiple surgery, not just tidying up the system. It is replacement—completely. The best we can do when it comes to using animals in science is—not to use them. That is where our duty lies, according to the rights view.

As for commercial animal agriculture, the rights view takes a similar abolitionist position. The fundamental moral wrong here is not that animals are kept in stressful close confinement, or in isolation, or that they have their pain and suffering, their needs and preferences ignored or discounted. *All* these *are* wrong, of course, but they are not the fundamental wrong. They are symptoms and effects of the deeper, systematic wrong that allows these animals to be viewed and treated as lacking independent value, as resources for us—as, indeed, a renewable resource. Giving farm animals more space, more natural environments, more companions does not right the fundamental wrong, any more than giving lab animals more anaesthesia or bigger, cleaner cages would right the fundamental wrong in their case. Nothing less than the total dissolution of commercial animal agriculture will do this, just as, for similar reasons I won't develop at length here, morality requires nothing less than the total elimination of commercial and sport hunting and trapping. The rights view's implications, then, as I have said, are clear—and are uncompromising.

My last two points are about philosophy—my profession. It is most obviously, no substitute for political action. The words I have written here and in other places by themselves don't change a thing. It is what we do with the thoughts the words express—our acts, our deeds—that change things. All that philosophy can do, and all I have attempted, is to offer a vision of what our deeds could aim at. And the why. But not the how.

Finally, I am reminded of my thoughtful critic, the one I mentioned earlier, who chastised me for being "too cerebral." Well, cerebral I have been: indirect duty views, utilitarianism, contractarianism—hardly the stuff deep passions are made of. I am also reminded, however, of the image another friend once set before me—the image of the ballerina as expressive of disciplined passion. Long hours of sweat and toil, of loneliness and practice, of doubt and fatigue; that is the discipline of her craft. But the passion is there, too: the fierce drive to excel, to speak through her body, to do it right, to pierce our minds. That is the image of philosophy I would leave with you; not "too cerebral," but *disciplined passion*. Of the discipline, enough has been seen. As for the passion:

There are times, and these are not infrequent, when tears come to my eyes when I see, or read, or hear of the wretched plight of animals in the hands of humans. Their pain, their suffering, their loneliness, their innocence, their death. Anger. Rage. Pity. Sorrow. Disgust. The whole creation groans under the weight of the evil we humans visit upon these mute, powerless creatures. It is our heart, not just our head, that calls for an end, that demands of us that we overcome, for them, the habits and forces behind their systematic oppression. All great movements, it is written, go through three stages: ridicule, discussion, adoption. It is the realization of this third stage—adoption—that demands both our passion and our discipline, our heart and our head. *The fate of animals is in our hands. God grant we are equal to the task.*

For Further Discussion

1. On what issues would Regan agree with Singer? On what issues would they disagree?

2. Both Singer and Regan suggest that a good test for determining the ethical status of our treatment of animals is to ask if we would accept similar treatment of humans. When, if at all, is it justified to inflict pain on or to kill humans? Why are we appalled by cannibalism?

3. On what grounds does Regan reject the utilitarian theory of social justice? Are you convinced by his defense of the rights view?

4. What does Regan mean by "inherent value"? Is this type of value discovered or created by humans?

Should Trees Have Standing? Toward Legal Rights for Natural Objects

Christopher D. Stone

In our final selection for Chapter 6, Christopher Stone extends ethical and legal consideration to trees and other natural objects. Although Stone is specifically concerned with extending legal rather than ethical rights and standing to natural objects, his approach raises important philosophical issues as well. Stone begins with an analysis of legal rights and, drawing on a parallel with nonliving objects such as corporations, argues that natural objects like trees and ecosystems can be said to have interests. Stone points out that even though corporations are not human, they are treated as legal persons and are given significant standing in law. They can be represented in court and sue to protect their own interests. So, too, with natural objects. Because they have interests, they can be harmed by our actions, and therefore they deserve legal representation and legal standing.

INTRODUCTION: THE UNTHINKABLE

In *Descent of Man*, Darwin observes that the history of man's moral development has been a continual extension in the objects of his "social instincts and sympathies." Originally each man had regard only for himself and those of a very narrow circle about him; later, he came to regard more and more "not only the welfare, but the happiness of all his fellow-men"; then "his sympathies became more tender and widely diffused, extending to men of all races, to the imbecile, maimed, and other useless members of society, and finally to the lower animals. . . ."[1]

The history of the law suggests a parallel development. Perhaps there never was a pure Hobbesian state of nature, in which no "rights" existed except in the vacant sense of each man's "right to self-defense." But it is not unlikely that so far as the earliest "families" (including extended kinship groups and clans) were concerned, everyone outside the family was suspect, alien, rightless.[2] And even within the family, persons we presently regard as the natural holders of at least some rights had none. Take, for example, children. We know something of the early rights-status of children from the widespread practice of infanticide—especially of the deformed and female.[3] (Senicide,[4] as among the North American Indians, was the corresponding rightlessness of the aged.)[5] Maine tells us that as late as the Patria Potestas of the Romans, the father had *jus vitae necisque*—the power of life and death—over his children. A fortiori, Maine writes, he had power of "uncontrolled corporal chastisement; he can modify their personal condition at pleasure; he can give a wife to his son; he can give his daughter in marriage; he can divorce his children of either sex; he can transfer them to another family by adoption; and he can sell them." The child was less than a person: an object, a thing.[6]

The legal rights of children have long since been recognized in principle, and are still expanding in practice. Witness, just within recent time, *In re Gault*,[7] guaranteeing basic constitutional protections to juvenile defendants, and the Voting Rights Act of 1970.[8] We have been making persons of children although they were not, in law, always so. And we have done the same, albeit imperfectly some would say, with prisoners,[9] aliens, women (especially of the married variety), the insane,[10] Blacks, foetuses,[11] and Indians.

Nor is it only matter in human form that has come to be recognized as the possessor of rights. The world of the lawyer is peopled with inanimate right-holders: trusts, corporations, joint ventures, municipalities, Subchapter R partnerships,[12] and nation-states, to mention just a few. Ships, still referred to by courts in the feminine gender, have long had an independent jural life, often with striking consequences.[13] We have become so accustomed to the idea of a corporation having "its" own rights, and being a "person" and "citizen" for so many statutory and constitutional purposes, that we forget how jarring the notion was to early jurists. "That invisible, intangible and artificial being, that mere legal entity" Chief Justice Marshall wrote of the corporation in *Bank of the United States v. Deveaux*[14]—could a suit be brought in *its* name? Ten years later, in the *Dartmouth College* case,[15] he was still refusing to let pass unnoticed the wonder of an entity "existing only in contemplation of law."[16] Yet, long before Marshall worried over the personifying of the modern corporation, the best medieval legal scholars had spent hundreds of years struggling with the notion of the legal nature of those great public "corporate bodies," the Church and the State. How could they exist in law, as entities transcending the living Pope and King? It was clear how a king could bind *himself*— on his honor—by a treaty. But when the king died, what was it that was burdened with the obligations of, and claimed the rights under, the treaty *his* tangible hand had signed? The medieval mind saw (what we have lost our capacity to see)[17] how *unthinkable* it was, and worked out the most elaborate conceits and fallacies to serve as anthropomorphic flesh for the Universal Church and the Universal Empire.[18]

It is this note of the *unthinkable* that I want to dwell upon for a moment. Throughout legal history, each successive extension of rights to some new entity has been, theretofore, a bit unthinkable. We are inclined to suppose the rightlessness of rightless "things" to be a decree of Nature, not a legal convention acting in support of some status quo. It is thus that we defer considering the choices involved in all their moral, social, and economic dimensions. And so the United States Supreme Court could straight-facedly tell us in *Dred Scott* that Blacks had been denied the rights of citizenship "as a subordinate and inferior class of beings, who had been subjugated by the dominant race. . . ."[19] In the nineteenth century, the highest court in California explained that Chinese had not the right to testify against white men in criminal matters because they were "a race of people whom nature has marked as inferior, and who are incapable of progress or intellectual development beyond a certain point . . . between whom and ourselves nature has placed an impassable difference."[20] The popular conception of the Jew in the 13th Century contributed to a law which treated them as "men *ferae naturae,* protected by a quasi-forest law. Like the roe and the deer, they form an order apart."[21] Recall, too, that it was not so long ago that the foetus was "like the roe and the deer." In an early suit attempting to establish a wrongful death action on behalf of a negligently killed foetus (now widely accepted practice), Holmes, then on the Massachusetts Supreme Court, seems to have thought

it simply inconceivable "that a man might owe a civil duty and incur a conditional prospective liability in tort to one not yet in being."[22] The first woman in Wisconsin who thought she might have a right to practice law was told that she did not, in the following terms:

> The law of nature destines and qualifies the female sex for the bearing and nurture of the children of our race and for the custody of the homes of the world. . . . [A]ll life-long callings of women, inconsistent with these radical and sacred duties of their sex, as is the profession of the law, are departures from the order of nature; and when voluntary, treason against it. . . . The peculiar qualities of womanhood, its gentle graces, its quick sensibility, its tender susceptibility, its purity, its delicacy, its emotional impulses, its subordination of hard reason to sympathetic feeling, are surely not qualifications for forensic strife. Nature has tempered woman as little for the juridical conflicts of the court room, as for the physical conflicts of the battle field. . . .[23]

The fact is, that each time there is a movement to confer rights onto some new "entity," the proposal is bound to sound odd or frightening or laughable. This is partly because until the rightless thing receives its rights, we cannot see it as anything but a *thing* for the use of "us"—those who are holding rights at the time.[24] In this vein, what is striking about the Wisconsin case above is that the court, for all its talk about women, so clearly was never able to see women as they are (and might become). All it could see was the popular "idealized" version of *an object it needed.* Such is the way the slave South looked upon the Black.[25] There is something of a seamless web involved: there will be resistance to giving the thing "rights" until it can be seen and valued for itself; yet, it is hard to see it and value it for itself until we can bring ourselves to give it "rights"—which is almost inevitably going to sound inconceivable to a large group of people.

The reason for this little discourse on the unthinkable, the reader must know by now, if only from the title of the paper. I am quite seriously proposing that we give legal rights to forests, oceans, rivers and other so-called "natural objects" in the environment—indeed, to the natural environment as a whole.

As strange as such a notion may sound, it is neither fanciful nor devoid of operational content. In fact, I do not think it would be a misdescription of recent developments in the law to say that we are already on the verge of assigning some such rights, although we have not faced up to what we are doing in those particular terms.[26] We should do so now, and begin to explore the implications such a notion would hold.

TOWARD RIGHTS FOR THE ENVIRONMENT

Now, to say that the natural environment should have rights is not to say anything as silly as that no one should be allowed to cut down a tree. We say human beings have rights, but—at least as of the time of this writing—they can be executed. Corporations have rights, but they cannot plead the fifth amendment; *In re Gault* gave 15-year-olds certain rights in juvenile proceedings, but it did not give them the right to vote. Thus, to say that the environment should have rights is not to say that it should have every right we can imagine, or even the same body of rights as human beings have. Nor is it to say that everything in the environment should have the same rights as every other thing in the environment.

What the granting of rights does involve has two sides to it. The first involves what might be called the legal-operational aspects; the second, the psychic and socio-psychic aspects. I shall deal with these aspects in turn.

THE LEGAL-OPERATIONAL ASPECTS

What It Means to Be a Holder of Legal Rights

There is, so far as I know, no generally accepted standard for how one ought to use the term "legal rights." Let me indicate how I shall be using it in this piece.

First and most obviously, if the term is to have any content at all, an entity cannot be said to hold a legal right unless and until *some public authoritative body* is prepared to give *some amount of review* to actions that are colorably inconsistent with that "right." For example, if a student can be expelled from a university and cannot get any public official, even a judge or administrative agent at the lowest level, either (i) to require the university to justify its actions (if only to the extent of filling out an affidavit alleging that the expulsion "was not wholly arbitrary and capricious") or (ii) to compel the university to accord the student some procedural safeguards (a hearing, right to coun-

sel, right to have notice of charges), then the minimum requirements for saying that the student has a legal right to his education do not exist.[27]

But for a thing to be *a holder of legal rights,* something more is needed than that some authoritative body will review the actions and processes of those who threaten it. As I shall use the term, "holder of legal rights," each of three additional criteria must be satisfied. All three, one will observe, go towards making a thing *count* jurally—to have a legally recognized worth and dignity in its own right, and not merely to serve as a means to benefit "us" (whoever the contemporary group of rights-holders may be). They are, first, that the thing can institute legal actions *at its behest;* second, that in determining the granting of legal relief, the court must take *injury to it* into account; and, third, that relief must run to the *benefit of it.* . . .

The Rightlessness of Natural Objects at Common Law

Consider, for example, the common law's posture toward the pollution of a stream. True, courts have always been able, in some circumstances, to issue orders that will stop the pollution. . . . But the stream itself is fundamentally rightless, with implications that deserve careful reconsideration.

The first sense in which the stream is not a rights-holder has to do with standing. The stream itself has none. So far as the common law is concerned, there is in general no way to challenge the polluter's actions save at the behest of a lower riparian—another human being—able to show an invasion of *his* rights. This conception of the riparian as the holder of the right to bring suit has more than theoretical interest. The lower riparians may simply not care about the pollution. They themselves may be polluting, and not wish to stir up legal waters. They may be economically dependent on their polluting neighbor. And, of course, when they discount the value of winning by the costs of bringing suit and the chances of success, the action may not seem worth undertaking. Consider, for example, that while the polluter might be injuring 100 downstream riparians $10,000 a year *in the aggregate,* each riparian separately might be suffering injury only to the extent of $100—possibly not enough for any one of them to want to press suit by himself, or even to go to the trouble and cost of securing co-plaintiffs to make it worth everyone's while. This hesitance will be especially likely when the potential plaintiffs consider the burdens the law puts in their way:[28] proving, *e.g.,* specific damages, the "unreasonableness" of defendant's use of the water, the fact that practicable means of abatement exist, and overcoming difficulties raised by issues such as joint casuality, right to pollute by prescription, and so forth. Even in states which, like California, sought to overcome these difficulties by empowering the attorney-general to sue for abatement of pollution in limited instances, the power has been sparingly invoked and, when invoked, narrowly construed by the courts.[29]

The second sense in which the common law denies "rights" to natural objects has to do with the way in which the merits are decided in those cases in which someone is competent and willing to establish standing. At its more primitive levels, the system protected the "rights" of the property owning human with minimal weighing of any values: *Cujus est solum, ejus est usque ad coelum et ad infernos.*[30] Today we have come more and more to make balances—but only such as will adjust the economic best interests of identifiable humans. For example, continuing with the case of streams, there are commentators who speak of a "general rule" that "a riparian owner is legally entitled to have the stream flow by his land with its quality unimpaired" and observe that "an upper owner has prima facie, no right to pollute the water."[31] Such a doctrine, if strictly invoked, would protect the stream absolutely whenever a suit was brought; but obviously, to look around us, the law does not work that way. Almost everywhere there are doctrinal qualifications on riparian "rights" to an unpolluted stream.[32] Although these rules vary from jurisdiction to jurisdiction, and upon whether one is suing for an equitable injunction or for damages, what they all have in common is some sort of balancing. Whether under language of "reasonable use," "reasonable methods of use," "balance of convenience" or "the public interest doctrine," what the courts are balancing, with varying degrees of directness, are the economic hardships on the upper riparian (or dependent community) of abating the pollution vis-à-vis the economic hardships of continued pollution on the lower riparians. What does not weigh in the balance is the damage to the stream, its fish and turtles and "lower" life. So long as the natural environment itself is rightless, these are not matters for judicial cognizance. Thus, we find the highest court of Pennsylvania refusing to stop a coal company from discharging polluted mine water into a tributary of the Lackawana River because a plaintiff's "grievance is for a mere personal inconvenience; and . . . mere private personal inconveniences . . . must yield to the necessities of a great public industry,

which although in the hands of a private corporation, subserves a great public interest."[33] The stream itself is lost sight of in "a quantitative compromise between *two* conflicting interests."[34]

The third way in which the common law makes natural objects rightless has to do with who is regarded as the beneficiary of a favorable judgment. Here, too, it makes a considerable difference that it is not the natural object that counts in its own right. To illustrate this point, let me begin by observing that it makes perfectly good sense to speak of, and ascertain, the legal damage to a natural object, if only in the sense of "making it whole" with respect to the most obvious factors. The costs of making a forest whole, for example, would include the costs of reseeding, repairing watersheds, restocking wildlife—the sorts of costs the Forest Service undergoes after a fire. Making a polluted stream whole would include the costs of restocking with fish, water-fowl, and other animal and vegetable life, dredging, washing out impurities, establishing natural and/or artificial aerating agents, and so forth. Now, what is important to note is that, under our present system, even if a plaintiff riparian wins a water pollution suit for damages, no money goes to the benefit of the stream itself to repair *its* damages. This omission has the further effect that, at most, the law confronts a polluter with what it takes to make the plaintiff riparians whole; this may be far less than the damages to the stream, but not so much as to force the polluter to desist. For example, it is easy to imagine a polluter whose activities damage a stream to the extent of $10,000 annually, although the aggregate damage to all the riparian plaintiffs who come into the suit is only $3000. If $3000 is less than the cost to the polluter of shutting down, or making the requisite technological changes, he might prefer to pay off the damages (*i.e.,* the legally cognizable damages) and continue to pollute the stream. Similarly, even if the jurisdiction issues an injunction at the plaintiffs' behest (rather than to order payment of damages), there is nothing to stop the plaintiffs from "selling out" the stream, *i.e.,* agreeing to dissolve or not enforce the injunction at some price (in the example above, somewhere between plaintiffs' damages—$3000—and defendant's next best economic alternative). Indeed, I take it this is exactly what Learned Hand had in mind in an opinion in which, after issuing an anti-pollution injunction, he suggests that the defendant "make its peace with the plaintiff as best it can."[35] What is meant is a peace between *them,* and not amongst them and the river.

I ought to make clear at this point that the common law as it affects streams and rivers, which I have been using as an example so far, is not exactly the same as the law affecting other environmental objects. Indeed, one would be hard pressed to say that there was a "typical" environmental object, so far as its treatment at the hands of the law is concerned. There are some differences in the law applicable to all the various resources that are held in common: rivers, lakes, oceans, dunes, air, streams (surface and subterranean), beaches, and so forth. And there is an even greater difference as between these traditional communal resources on the one hand, and natural objects on traditionally private land, *e.g.,* the pond on the farmer's field, or the stand of trees on the suburbanite's lawn.

On the other hand, although there be these differences which would make it fatuous to generalize about a law of the natural environment, most of these differences simply underscore the points made in the instance of rivers and streams. None of the natural objects, whether held in common or situated on private land, has any of the three criteria of a rights-holder. They have no standing in their own right; their unique damages do not count in determining outcome; and they are not the beneficiaries of awards. In such fashion, these objects have traditionally been regarded by the common law, and even by all but the most recent legislation, as objects for man to conquer and master and use—in such a way as the law once looked upon "man's" relationships to African Negroes. Even where special measures have been taken to conserve them, as by seasons on game and limits on timber cutting, the dominant motive has been to conserve them *for us*— for the greatest good of the greatest number of human beings. Conservationists, so far as I am aware, are generally reluctant to maintain otherwise.[36] As the name implies, they want to conserve and guarantee *our* consumption and *our* enjoyment of these other living things. In their own right, natural objects have counted for little, in law as in popular movements.

As I mentioned at the outset, however, the rightlessness of the natural environment can and should change; it already shows some signs of doing so.

TOWARD HAVING STANDING IN ITS OWN RIGHT

It is not inevitable, nor is it wise, that natural objects should have no rights to seek redress in their own behalf. It is no answer to say that streams and forests cannot have standing because streams and forest cannot speak. Corporations cannot speak either; nor can states, estates; infants, incompetents, municipalities or universities. Lawyers speak for them, as they customarily do for the ordinary citizen with legal problems. One ought, I think, to handle the legal problems of natural objects as one does the problems of legal incompetents—human beings who have become vegetables. If a human being shows signs of becoming senile and has affairs that he is de jure incompetent to manage, those concerned with his well being make such a showing to the court, and someone is designated by the court with the authority to manage the incompetent's affairs. The guardian (or "conservator" or "committee"—the terminology varies) then represents the incompetent in his legal affairs. Courts make similar appointments when a corporation has become "incompetent"—they appoint a trustee in bankruptcy or reorganization to oversee its affairs and speak for it in court when that becomes necessary.

On a parity of reasoning, we should have a system in which, when a friend of a natural object perceives it to be endangered, he can apply to a court for the creation of a guardianship. Perhaps we already have the machinery to do so. California law, for example, defines an incompetent as "any person, whether insane or not, who by reason of old age, disease, weakness of mind, or other cause, is unable, unassisted, properly to manage and take care of himself or his property, and by reason thereof is likely to be deceived or imposed upon by artful or designing persons."[37] Of course, to urge a court that an endangered river is "a person" under this provision will call for lawyers as bold and imaginative as those who convinced the Supreme Court that a railroad corporation was a "person" under the fourteenth amendment, a constitutional provision theretofore generally thought of as designed to secure the rights of freedmen.[38] . . .

The guardianship approach, however, is apt to raise . . . [the following objection]: a committee or guardian could not judge the needs of the river or forest in its charge; indeed, the very concept of "needs," it might be said, could be used here only in the most metaphorical way. . . .

. . . Natural objects *can* communicate their wants (needs) to us, and in ways that are not terribly ambiguous. I am sure I can judge with more certainty and meaningfulness whether and when my lawn wants (needs) water, than the Attorney General can judge whether and when the United States wants (needs) to take an appeal from an adverse judgment by a lower court. The lawn tells me that it wants water by a certain dryness of the blades and soil—immediately obvious to the touch—the appearance of bald spots, yellowing, and a lack of springiness after being walked on; how does "the United States" communicate to the Attorney General? For similar reasons, the guardian-attorney for a smog-endangered stand of pines could venture with more confidence that his client wants the smog stopped, than the directors of a corporation can assert that "the corporation" wants dividends declared. We make decisions on behalf of, and in the purported interests of, others every day; these "others" are often creatures whose wants are far less verifiable, and even far more metaphysical in conception, than the wants of rivers, trees, and land. . . .

The argument for "personifying" the environment, from the point of damage calculations, can best be demonstrated from the welfare economics position. Every well-working legal-economic system should be so structured as to confront each of us with the full costs that our activities are imposing on society. Ideally, a paper-mill, in deciding what to produce—and where, and by what methods—ought to be forced to take into account not only the lumber, acid and labor that its production "takes" from other uses in the society, but also what costs alternative production plans will impose on society through pollution. The legal system, through the law of contracts and the criminal law, for example, makes the mill confront the costs of the first group of demands. When, for example, the company's purchasing agent orders 1000 drums of acid from the Z Company, the Z Company can bind the mill to pay for them, and thereby reimburse the society for what the mill is removing from alternative uses.

Unfortunately, so far as the pollution costs are concerned, the allocative ideal begins to break down, because the traditional legal institutions have a more difficult time "catching" and confronting us with the full social costs of our activities. In the lakeside mill example, major riparian interests might bring an action, forcing a court to weigh *their* aggregate losses against the costs to the mill of installing the anti-pollution device. But many other interests—and I am speaking for the moment of recognized homocentric interests—are too fragmented and perhaps "too remote" causally to warrant securing representation and pressing for recovery: the people who own summer homes and motels, the man who sells fishing tackle and bait, the man who rents rowboats. There is no reason not to allow the lake to prove damages to them as the prima facie measure of damages to it. *By doing so, we in effect make the natural object, through its guardian, a jural entity competent to gather up these fragmented and otherwise unrep-*

resented damage claims, and press them before the court even where, for legal or practical reasons, they are not going to be pressed by traditional class action plaintiffs. Indeed, one way—the homocentric way—to view what I am proposing so far, is to view the guardian of the natural object as the guardian of unborn generations, as well as of the otherwise unrepresented, but distantly injured, contemporary humans.[39] By making the lake itself the focus of these damages, and "incorporating" it so to speak, the legal system can effectively take proof upon, and confront the mill with, a larger and more representative measure of the damages its pollution causes.

So far, I do not suppose that my economist friends (unremittent human chauvinists, every one of them!) will have any large quarrel in principle with the concept. Many will view it as a *trompe l'oeil* that comes down, at best, to effectuate the goals of the paragon class action, or the paragon water pollution control district. Where we are apt to part company is here—I propose going beyond gathering up the loose ends of what most people would presently recognize as economically valid damages. The guardian would urge before the court injuries not presently cognizable—the death of eagles and inedible crabs, the suffering of sea lions, the loss from the face of the earth of species of commercially valueless birds, the disappearance of a wilderness area. One might, of course, speak of the damages involved as "damages" to us humans, and indeed, the widespread growth of environmental groups shows that human beings do feel these losses. But they are not, at present, economically measurable losses: how can they have a monetary value for the guardian to prove in court?

The answer for me is simple. Wherever it carves out "property" rights, the legal system is engaged in the process of *creating* monetary worth. One's literary works would have minimal monetary value if anyone could copy them at will. Their economic value to the author is a product of the law of copyright; the person who copies a copyrighted book has to bear a cost to the copyright-holder because the law says he must. Similarly, it is through the law of torts that we have made a "right" of—and guaranteed an economically meaningful value to—privacy. (The value we place on gold—a yellow inanimate dirt—is not simply a function of supply and demand—wilderness areas are scarce and pretty too—, but results from the actions of the legal systems of the world, which have institutionalized that value; they have even done a remarkable job of stabilizing the price.) I am proposing we do the same with eagles and wilderness areas as we do with copyrighted works, patented inventions, and privacy: *make* the violation of rights in them to be a cost by declaring the "pirating" of them to be the invasion of a property interest.[40] If we do so, the net social costs the polluter would be confronted with would include not only the extended homocentric costs of his pollution (explained above) but also costs to the environment *per se.*

How, though, would these costs be calculated? When we protect an invention, we can at least speak of a fair market value for it, by reference to which damages can be computed. But the lost environmental "values" of which we are now speaking are by definition over and above those that the market is prepared to bid for: they are priceless.

One possible measure of damages, suggested earlier, would be the cost of making the environment whole, just as, when a man is injured in an automobile accident, we impose upon the responsible party the injured man's medical expenses. Comparable expenses to a polluted river would be the costs of dredging, restocking with fish, and so forth. It is on the basis of such costs as these, I assume, that we get the figure of $1 billion as the cost of saving Lake Erie.[41] As an ideal, I think this is a good guide applicable in many environmental situations. It is by no means free from difficulties, however.

One problem with computing damages on the basis of making the environment whole is that, if understood most literally, it is tantamount to asking for a "freeze" on environmental quality, even at the costs (and there will be costs) of preserving "useless" objects. Such a "freeze" is not inconceivable to me as a general goal, especially considering that, even by the most immediately discernible homocentric interests, in so many areas we ought to be cleaning up and not merely preserving the environmental status quo. In fact, there is presently strong sentiment in the Congress for a total elimination of all river pollutants by 1985,[42] notwithstanding that such a decision would impose quite large direct and indirect costs on us all. Here one is inclined to recall the instructions of Judge Hays, in remanding Consolidated Edison's Storm King application to the Federal Power Commission in *Scenic Hudson:*

> The Commission's renewed proceedings must include as a basic concern the preservation of natural beauty and of natural historic shrines, keeping in mind that, in our affluent society, the

cost of a project is only one of several factors to be considered.[43]

Nevertheless, whatever the merits of such a goal in principle, there are many cases in which the social price tag of putting it into effect are going to seem too high to accept. Consider, for example, an oceanside nuclear generator that could produce low cost electricity for a million homes at a savings of $1 a year per home, spare us the air pollution that comes of burning fossil fuels, but which through a slight heating effect threatened to kill off a rare species of temperature-sensitive sea urchins; suppose further that technological improvements adequate to reduce the temperature to present environmental quality would expend the entire one million dollars in anticipated fuel savings. Are we prepared to tax ourselves $1,000,000 a year on behalf of the sea urchins? In comparable problems under the present law of damages, we work out practicable compromises by abandoning restoration costs and calling upon fair market value. For example, if an automobile is so severely damaged that the cost of bringing the car to its original state by repair is greater than the fair market value, we would allow the responsible tortfeasor to pay the fair market value only. Or if a human being suffers the loss of an arm (as we might conceive of the ocean having irreparably lost the sea urchins), we can fall back on the capitalization of reduced earning power (and pain and suffering) to measure the damages. But what is the fair market value of sea urchins? How can we capitalize their loss to the ocean, independent of any commercial value they may have to someone else?

One answer is that the problem can sometimes be sidestepped quite satisfactorily. In the sea urchin example, one compromise solution would be to impose on the nuclear generator the costs of making the ocean whole somewhere else, in some other way, *e.g.,* re-establishing a sea urchin colony elsewhere, or making a somehow comparable contribution.[44] In the debate over the laying of the trans-Alaskan pipeline, the builders are apparently prepared to meet conservationists' objections halfway by re-establishing wildlife away from the pipeline, so far as is feasible.[45]

But even if damage calculations have to be made, one ought to recognize that the measurement of damages is rarely a simple report of economic facts about "the market," whether we are valuing the loss of a foot, a foetus, or a work of fine art. Decisions of this sort are always hard, but not impossible. We have increasingly taken (human) pain and suffering into account in reckoning damages, not because we think we can ascertain them as objective "facts" about the universe, but because, even in view of all the room for disagreement, we come up with a better society by making rude estimates of them than by ignoring them.[46] We can make such estimates in regard to environmental losses fully aware that what we are really doing is making implicit normative judgments (as with pain and suffering)—laying down rules as to what the society is going to "value" rather than reporting market evaluations. In making such normative estimates decision-makers would not go wrong if they estimated on the "high side," putting the burden of trimming the figure down on the immediate human interests present. All burdens of proof should reflect common experience; our experience in environmental matters has been a continual discovery that our acts have caused more long-range damage than we were able to appreciate at the outset.

To what extent the decision-maker should factor in costs such as the pain and suffering of animals and other sentient natural objects, I cannot say; although I am prepared to do so in principle.[47] Given the conjectural nature of the "estimates" in all events, and the roughness of the "balance of conveniences" procedure where that is involved, the practice would be of more interest from the socio-psychic point of view, discussed below, than from the legal-operational. . . .

THE PSYCHIC AND SOCIO-PSYCHIC ASPECTS

The strongest case can be made from the perspective of human advantage for conferring rights on the environment. Scientists have been warning of the crises the earth and all humans on it face if we do not change our ways—radically—and these crises make the lost "recreational use" of rivers seem absolutely trivial. The earth's very atmosphere is threatened with frightening possibilities: absorption of sunlight, upon which the entire life cycle depends, may be diminished; the oceans may warm (increasing the "greenhouse effect" of the atmosphere), melting the polar ice caps, and destroying our great coastal cities; the portion of the atmosphere that shields us from dangerous radiation may be destroyed. Testifying before Congress, sea explorer Jacques Cousteau predicted that the oceans (to which we dreamily look to feed our booming populations) are headed toward their own death: "The cycle

of life is intricately tied up with the cycle of water . . . the water system has to remain alive if we are to remain alive on earth."[48] We are depleting our energy and our food sources at a rate that takes little account of the needs even of humans now living.

These problems will not be solved easily; they very likely can be solved, if at all, only through a willingness to suspend the rate of increase in the standard of living (by present values) of the earth's "advanced" nations, and by stabilizing the total human population. For some of us this will involve forfeiting material comforts; for others it will involve abandoning the hope someday to obtain comforts long envied. For all of us it will involve giving up the right to have as many offspring as we might wish. Such a program is not impossible of realization, however. Many of our so-called "material comforts" are not only in excess of, but are probably in opposition to, basic biological needs. Further, the "costs" to the advanced nations is not as large as would appear from Gross National Product figures. G.N.P. reflects a social gain (of a sort) without discounting for the social *cost* of that gain, *e.g.*, the losses through depletion of resources, pollution, and so forth. As has well been shown, as societies become more and more "advanced," their real marginal gains become less and less for each additional dollar of G.N.P.[49] Thus, to give up "human progress" would not be as costly as might appear on first blush.

Nonetheless, such far-reaching social changes are going to involve us in a serious reconsideration of our consciousness towards the environment. . . .

A radical new conception of man's relationship to the rest of nature would not only be a step towards solving the material planetary problems; there are strong reasons for such a changed consciousness from the point of making us far better humans. If we only stop for a moment and look at the underlying human qualities that our present attitudes toward property and nature draw upon and reinforce, we have to be struck by how stultifying of our own personal growth and satisfaction they can become when they take rein of us. Hegel, in "justifying" private property, unwittingly reflects the tone and quality of some of the needs that are played upon:

> A person has as his substantive end the right of putting his will into any and every thing and thereby making it his, because it has no such end in itself and derives its destiny and soul from his will. This is the absolute right of appropriation which man has over all "things."[50]

What is it within us that gives us this need not just to satisfy basic biological wants, but to extend our wills over things, to object-ify them, to make them ours, to manipulate them, to keep them at a psychic distance? Can it all be explained on "rational" bases? Should we not be suspect of such needs within us, cautious as to why we wish to gratify them? When I first read that passage of Hegel, I immediately thought not only of the emotional contrast with Spinoza, but of the passage in Carson McCullers' *A Tree, A Rock, A Cloud,* in which an old derelict has collared a twelve year old boy in a streetcar cafe. The old man asks whether the boy knows "how love should be begun?"

The old man leaned closer and whispered:

"A tree. A rock. A cloud."

. . .

"The weather was like this in Portland," he said. "At the time my science was begun. I meditated and I started very cautious. I would pick up something from the street and take it home with me. I bought a goldfish and I concentrated on the goldfish and I loved it. I graduated from one thing to another. Day by day I was getting this technique. . . .

. . .

. . . "For six years now I have gone around by myself and built up my science. And now I am a master, Son. I can love anything. No longer do I have to think about it even. I see a street full of people and a beautiful light comes in me. I watch a bird in the sky. Or I meet a traveler on the road. Everything, Son. And anybody. All stranger and all loved! Do you realize what a science like mine can mean?"[51]

To be able to get away from the view that Nature is a collection of useful senseless objects is, as McCullers' "madman" suggests, deeply involved in the development of our abilities to love—or, if that is putting it too strongly, to be able to reach a heightened awareness of our own, and others' capacities in their mutual interplay. To do so, we have to give up some psychic investment in our sense of separateness and specialness in the universe. And this, in turn, is hard giving indeed, because it involves us in a flight backwards, into earlier stages of civilization and childhood in which we had to trust (and perhaps fear) our environment, for we had not then the power to master it. Yet, in doing so, we—as persons—gradually free ourselves of needs for supportive illusions. Is not this one

of the triumphs for "us" of our giving legal rights to (or acknowledging the legal rights of) the Blacks and women? . . .

. . . A few years ago the pollution of streams was thought of only as a problem of smelly, unsightly, unpotable water, *i.e.,* to us. Now we are beginning to discover that pollution is a process that destroys wondrously subtle balances of life within the water, and as between the water and its banks. This heightened awareness enlarges our sense of the dangers to us. But it also enlarges our empathy. We are not only developing the scientific capacity, but we are cultivating the personal capacities *within us* to recognize more and more the ways in which nature—like the woman, the Black, the Indian and the Alien—is like us (and we will also become more able realistically to define, confront, live with and admire the ways in which we are all different).

The time may be on hand when these sentiments, and the early stirrings of the law, can be coalesced into a radical new theory or myth—felt as well as intellectualized—of man's relationships to the rest of nature. I do not mean "myth" in a demeaning sense of the term, but in the sense in which, at different times in history, our social "facts" and relationships have been comprehended and integrated by reference to the "myths" that we are co-signers of a social contract, that the Pope is God's agent, and that all men are created equal. Pantheism, Shinto and Tao all have myths to offer. But they are all, each in its own fashion, quaint, primitive and archaic. What is needed is a myth that can fit our growing body of knowledge of geophysics, biology and the cosmos. In this vein, I do not think it too remote that we may come to regard the Earth, as some have suggested, as one organism, of which Mankind is a functional part—the mind, perhaps: different from the rest of nature, but different as a man's brain is from his lungs. . . .

. . . As I see it, the Earth is only one organized "field" of activities—and so is the *human person*—but these activities take place at various levels, in different "spheres" of being and realms of consciousness. The lithosphere is not the biosphere, and the latter not the . . . ionosphere. The Earth is not *only* a material mass. Consciousness is not only "human"; it exists at animal and vegetable levels, and most likely must be latent, or operating in some form, in the molecule and the atom; and all these diverse and in a sense hierarchical modes of activity and consciousness should be seen integrated in and perhaps transcended by an all-encompassing and "eonic" planetary Consciousness.

. . .

Mankind's function within the Earth-organism is to extract from the activities of all other operative systems within this organism the type of consciousness which we call "reflective" or "self"-consciousness—or, we may also say to *mentalize* and give meaning, value, and "name" to all that takes place anywhere within the Earth-field. . . .[52]

As radical as such a consciousness may sound today, all the dominant changes we see about us point in its direction. Consider just the impact of space travel, of world-wide mass media, of increasing scientific discoveries about the interrelatedness of all life processes. Is it any wonder that the term "spaceship earth" has so captured the popular imagination? The problems we have to confront are increasingly the world-wide crises of a global organism: not pollution of a stream, but pollution of the atmosphere and of the ocean. Increasingly, the death that occupies each human's imagination is not his own, but that of the entire life cycle of the planet earth, to which each of us is as but a cell to a body.

To shift from such a lofty fancy as the planetarization of consciousness to the operation of our municipal legal system is to come down to earth hard. Before the forces that are at work, our highest court is but a frail and feeble—a distinctly human—institution. Yet, the Court may be at its best not in its work of handing down decrees, but at the very task that is called for: of summoning up from the human spirit the kindest and most generous and worthy ideas that abound there, giving them shape and reality and legitimacy. Witness the School Desegregation Cases which, more importantly than to integrate the schools (assuming they did), awakened us to moral needs which, when made visible, could not be denied. And so here, too, in the case of the environment, the Supreme Court may find itself in a position to award "rights" in a way that will contribute to a change in popular consciousness. It would be a modest move, to be sure, but one in furtherance of a large goal: the future of the planet as we know it.

How far we are from such a state of affairs, where the law treats "environmental objects" as holders of legal rights, I cannot say. But there is certainly intriguing language in one of Justice Black's last dissents, regarding the Texas Highway Department's plan to run a six-lane expressway through a San Antonio Park.[53] Complaining of the Court's refusal to stay the plan, Black observed that "after today's decision, the people of San Antonio and the birds and animals that make their home in the park will share their quiet retreat with an ugly, smelly stream of traffic. . . . Trees, shrubs, and flowers will be mowed down."[54] Elsewhere he speaks of the "burial of public parks," of segments of a highway which "devour parkland," and of the park's heartland.[55] Was he, at the end of his great career, on the verge of saying—just saying—that "nature has 'rights' on its own account"? Would it be so hard to do?

Notes

1. C. Darwin, Descent of Man 119, 120–211 (2d ed. 1874). *See also* R. Waelder, Progress and Revolution 39 *et seq.* (1967).
2. *See* Darwin, *supra* note 1, at 113–14. . . .
3. *See* Darwin, *supra* note 1, at 113. *See also* E. Westermarck, 1 The Origin and Development of the Moral Ideas 406–12 (1912). . . .
4. There does not appear to be a word "gericide" or "geronticide" to designate the killing of the aged. "Senicide" is as close as the Oxford English Dictionary comes, although, as it indicates, the word is rare. 9 Oxford English Dictionary 454 (1933).
5. *See* Darwin, *supra* note 1, at 386–93. Westermarck, *supra* note 3, at 387–89, observes that where the killing of the aged and infirm is practiced, it is often supported by humanitarian justification; this, however, is a far cry from saying that the killing is *requested* by the victim as his right.
6. H. Maine, Ancient Law 153 (Pollock ed. 1930).
7. 387 U.S. 1 (1967).
8. 42 U.S.C. §§ 1973 *et seq.* (1970).
9. *See* Landman v. Royster, 40 U.S.L.W 2256 (E.D. Va., Oct. 30, 1971). . . .
10. *But see* T. Szasz, Law, Liberty and Psychiatry (1963).
11. *See* note 22. The trend toward liberalized abortion can be seen either as a legislative tendency back in the direction of rightlessness for the foetus—or toward increasing rights of women. This inconsistency is not unique in the law of course; it is simply support for Hohfeld's scheme that the "jural opposite" of someone's right is someone else's "no-right." W. Hohfeld, Fundamental Legal Conceptions (1923). . . .
12. Int. Rev. Code of 1954, § 1361 (repealed by Pub. L. No. 89-389, effective Jan. 1, 1969).
13. For example, *see* United States v. Cargo of the Brig Melek Adhel, 43 U.S. (2 How.) 210 (1844). There, a ship had been seized and used by pirates. All this was

done without the knowledge or consent of the owners of the ship. After the ship had been captured, the United States condemned and sold the "offending vessel." The owners objected. In denying release to the owners, Justice Story cited Chief Justice Marshall from an earlier case: "This is not a proceeding against the owner; it is a proceeding against the vessel for an offense committed by the vessel; which is not the less an offense . . . because it was committed without the authority and against the will of the owner." 43 U.S. at 234, quoting from United States v. Schooner Little Charles, 26 F Cas. 979 (No. 15,612) (C.C.D. Va. 1818).

14. 9 U.S. (5 Cranch) 61, 86 (1809).

15. Trustees of Dartmouth College v. Woodward, 17 U.S. (4 Wheat.) 518 (1819).

16. *Id.* at 636.

17. Consider, for example, that the claim of the United States to the naval station at Guantanamo Bay, at $2000-a-year rental, is based upon a treaty signed in 1903 by José Montes for the President of Cuba and a minister representing Theodore Roosevelt; it was subsequently ratified by two-thirds of a Senate no member of which is living today. Lease [from Cuba] of Certain Areas for Naval or Coaling Stations, July 2, 1903, T.S. No. 426; C. Bevans, 6 Treaties and Other International Agreements of the United States 1776–1949, at 1120 (U.S. Dep't of State Pub. 8549, 1971).

18. O. Gierke, Political Theories of the Middle Age (Maitland transl. 1927), especially at 22–30. . . .

19. Dred Scott v. Sandford, 60 U.S. (19 How.) 396, 404–05 (1856). . . .

20. People v. Hall, 4 Cal. 399, 405 (1854). . . .

21. Schechter, *The Rightlessness of Mediaeval English Jewry,* 45 Jewish Q. Rev. 121. 135 (1954) quoting from M. Bateson, *Medieval England* 139 (1904). . . .

22. Dietrich v. Inhabitants of Northampton, 138 Mass. 14, 16 (1884).

23. *In re* Goddell, 39 Wisc. 232, 245 (1875). The court continued with the following "clincher":

> And when counsel was arguing for this lady that the word, person, in sec. 32, ch. 119 [respecting those qualified to practice law], necessarily includes females, her presence made it impossible to suggest to him as *reductio ad absurdum* of his position, that the same construction of the same word . . . would subject woman to prosecution for the paternity of a bastard, and . . . prosecution for rape. *Id.* at 246.
>
> The relationship between our attitudes toward woman, on the one hand, and, on the other, the more central concern of this article—land—is captured in an unguarded aside of our colleague, Curt Berger: ". . . after all, land, like woman, was meant to be possessed. . . ." Land Ownership and Use 139 (1968).

24. Thus it was that the Founding Fathers could speak of the inalienable rights of all men, and yet maintain a society that was, by modern standards, without the most basic rights for Blacks, Indians, children and women. There was no hypocrisy; emotionally, no one *felt* that these other things were men.

25. The second thought streaming from . . . the older South [is] the sincere and passionate belief that somewhere between men and cattle, God created a *tertium quid,* and called it a Negro—a clownish, simple creature, at times even lovable within its limitations, but straitly foreordained to walk within the Veil. W. E. B. DuBois, The Souls of Black Folk 89 (1924).

26. The statement in text is not quite true; *cf.* Murphy, *Has Nature Any Right to Life?,* 22 Hast. L. J. 467 (1971). An Irish court, passing upon the validity of a testamentary trust to the benefit of someone's dogs, observed in dictum that "'lives' means lives of human beings, not of animals or trees in California." Kelly v. Dillon, 1932 Ir. R. 255, 261. (The intended gift over on the death of the last surviving dog was held void for remoteness, the court refusing "to enter into the question of a dog's expectation of life," although prepared to observe that "in point of fact neighbor's [sic] dogs and cats are unpleasantly long-lived. . . ." *Id.* at 260–61).

27. *See* Dixon v. Alabama State Bd. of Educ., 294 F.2d 150 (5th Cir.), *cert. denied,* 368 U.S. 930 (1961).

28. The law in a suit for injunctive relief is commonly easier on the plaintiff than in a suit for damages. *See* J. Gould, Law of Waters § 206 (1883).

29. However, in 1970 California amended its Water Quality Act to make it easier for the Attorney General to obtain relief, *e.g.,* one must no longer allege irreparable injury in a suit for an injunction. Cal. Water Code § 13350(b) (West 1971).

30. To whomsoever the soil belongs, he owns also to the sky and to the depths. *See* W. Blackstone, 2 Commentaries *18.

31. *See* Note, *Statutory Treatment of Industrial Stream Pollution,* 24 Geo. Wash. L. Rev. 302, 306 (1955); H. Farnham, 2 Law of Waters and Water Rights § 461 (1904); Gould, *supra* note 32, at § 204.

32. For example, courts have upheld a right to pollute by prescription, Mississippi Mills Co. v. Smith, 69 Miss. 299, 11 So. 26 (1882), and by easement, Luama v. Bunker Hill & Sullivan Mining & Concentrating Co., 41 F.2d 358 (9th Cir. 1930).

33. Pennsylvania Coal Co. v. Sanderson, 113 Pa. 126, 149, 6 A. 453, 459 (1886).

34. Hand, J. in Smith v. Staso Milling Co., 18 F.2d 736, 738 (2d Cir. 1927) (emphasis added). *See also* Harrisonville v. Dickey Clay Co., 289 U.S. 334 (1933) (Brandeis, J.).

35. Smith v. Staso, 18 F.2d 736, 738 (2d Cir. 1927).

36. By contrast, for example, with humane societies.

37. Cal. Prob. Code § 1460 (West Supp. 1971). . . .

38. Santa Clara County v. Southern Pac. R.R., 118 U.S. 394 (1886). . . .

39. *Cf.* Golding, *Ethical Issues in Biological Engineering,* 15 U.C.L.A. L. Rev. 143, 451–63 (1968).

40. Of course, in the instance of copyright and patent protection, the creation of the "property right" can be more directly justified on homocentric grounds.

41. See Schrag, *Life on a Dying Lake,* in The Politics of Neglect 167, at 173 (R. Meek & J. Straayer eds. 1971).

42. On November 2, 1971, the Senate, by a vote of 86–0, passed and sent to the House the proposed Federal Water Pollution Control Act Amendments of 1971, 117 Cong. Rec. S17464 (daily ed. Nov. 2, 1971). Sections 101(a) and (a)(1) of the bill declare it to be "national policy that, consistent with the provisions of this Act—(1) the discharge of pollutants into the navigable waters be eliminated by 1985." S.2770, 92d Cong., 1st Sess., 117 Cong. Rec. S17464 (daily ed. Nov. 2, 1971).

43. 334 F.2d 608, 624 (2d Cir. 1965).

44. Again, there is a problem involving what we conceive to be the injured entity.

45. N.Y. Times, Jan. 14, 1971, § 1, col. 2, and at 74, col. 7.

46. Courts have not been reluctant to award damages for the destruction of heirlooms, literary manuscripts or other property having no ascertainable market value. In Willard v. Valley Gas Fuel Co., 171 Ca. 9, 151 Pac. 286 (1915), it was held that the measure of damages for the negligent destruction of a rare old book written by one of plaintiff's ancestors was the amount which would compensate the owner for all detriment including sentimental loss proximately caused by such destruction. . . .

47. It is not easy to dismiss the idea of "lower" life having consciousness and feeling pain, especially since it is so difficult to know what these terms mean even as applied to humans. *See* Austin, *Other Minds,* in *Logic and Language* 342 (S. Flew ed. 1965); Schopenhauer, *On the Will in Nature,* in Two Essays by Arthur Schopenhauer 193, 281–304 (1889). Some experiments on plant sensitivity—of varying degrees of extravagance in their claims—include Lawrence, *Plants Have Feelings, Too . . . ,* Organic Gardening & Farming 64 (April 1971); Woodlief, Royster & Huang, *Effect of Random Noise on Plant Growth,* 46 J. Acoustical Soc. Am. 481 (1969); Backster, *Evidence of a Primary Perception in Plant Life,* 10 Int'l J. Parapsychology 250 (1968).

48. Cousteau, *The Oceans: No Time to Lose,* L.A. Times, Oct. 24, 1971, § (opinion), at 1, col. 4.

49. *See* J. Harte & R. Socolow, Patient Earth (1971).

50. G. Hegel, Hegel's Philosophy of Right 41 (T. Knox transl. 1945).

51. C. McCullers, The Ballad of the Sad Cafe and Other Stories 150–51 (1958).

52. D. Rudhyar, Directives for New Life 21–23 (1971).

53. 136. San Antonio Conservation Soc'y v. Texas Highway Dep't, *cert. denied,* 400 U.S. 968 (1970) (Black, J. dissenting to denial of certiorari).

54. *Id.* at 969.

55. *Id.* at 971.

For Further Discussion

1. In the previous selection, Tom Regan focuses on moral rights. In this selection, Christopher Stone speaks of legal rights. What's the difference? Does one offer a preferable strategy in pursuing public policy questions?

2. Nonliving institutions such as corporations and trusts are granted legal rights. Should they? How are they different from nonliving natural objects?

3. In the Dr. Suess story, it is the Lorax who claims to speak for the trees. If Stone's proposal is accepted, who should speak for the trees legally?

4. In law, an individual's family becomes the beneficiary deserving of compensation for the negligent killing of an individual. If trees were given legal standing, who or what should be compensated when trees or other natural objects are negligently destroyed?

DISCUSSION AND STUDY QUESTIONS FOR CHAPTER 6

1. What criteria do you use to establish moral standing? Where do you draw the line between objects that count and those that do not?

2. Is moral standing and moral value something that we humans confer on things, or is it something that we recognize as being there already? Can we choose to grant moral standing the way we can choose to grant legal standing?

3. Is there a moral difference between domestic animals such as cats and dogs and wild animals such as sharks and grizzly bears? Are grizzly bears more valuable than deer?

4. If suffering could be eliminated, are there still ethical concerns with eating animals? With hunting?

DISCUSSION CASES

Nuclear Waste and Future People

Nuclear waste from both electric power plants and military applications is among the most toxic substances on earth. Some of the waste from power plants is designated as "low-level" waste because the radiation emitted occurs at levels not particularly harmful to humans. High-level waste emits radiation that can have immediate and deadly consequences. Some of the most toxic radioactive waste, such as plutonium and uranium-235, will remain highly radioactive for hundreds of thousands of years. Plutonium remains highly toxic for 250,000 years; uranium-235, a by-product of plutonium decay, remains toxic for 710,000 years.

Debates concerning the storage of nuclear waste focus on our ability to guarantee safety. Although some storage systems might be secure into the indefinite future, this is seldom thought to be more than a few hundred years. Does it make sense to talk about responsibilities to people living 100,000 years in the future? What might those responsibilities be? What can we know about the people who might live 100,000 years in the future? On Feinberg's analysis, what rights might they have against us and what duties do we have to them?

Animal Research—LD50

Various government agencies, including the Environmental Protection Agency and the Food and Drug Administration, have used animal tests to determine the toxicity of various substances. One of the most infamous such tests was the LD50— "lethal dose 50 percent."

The LD50 test administers the substance being tested in increasing dosages to determine the level of exposure at which 50 percent of the animals die. This test, therefore, provides a common standard to determine the relative risks of various substances. With highly toxic substances, 50 percent die rather quickly and the rest are significantly poi-

soned. With less toxic substances, enormous quantities must be force fed and injected before 50 percent die. In all cases, animals that don't die still suffer, and because the scientific validity of these tests require controlled experiments, suffering animals cannot be euthanized.

What do you think of such tests? Do your views on animal testing vary depending on the animal being tested? Would your views change depending on the substance being tested or the purpose of the substance? What ethical restrictions are placed on research that uses human subjects? Would the same restrictions work for animal subjects? Why or why not? Would it make a difference in your ethical evaluation if research animals could be anesthetized?

Factory Farms

Modern factory farming is far from the image many of us have of bucolic farm life. Food animals such as chickens, calves, and pigs can be raised in conditions that appear brutal and cruel. Hens are often confined into tightly packed coops with wings clipped to prevent flapping and beaks cut to prevent pecking. Veal calves are confined in tight pens to prevent muscle growth and systematically malnourished to keep their flesh pink (due to iron deficiency). Many animals in such conditions are fed mixtures of growth hormones, vitamins, and antibiotics to accelerate growth and promote, relatively speaking, health. What ethical issues are involved in such practices?

Would Singer and Regan agree in their ethical evaluation of factory farming? Could animal farming ever be justified according to either Singer or Regan? What policy should be followed if we are convinced by Singer and Regan? Should all the animals presently living on farms be freed? Do we have an obligation to keep and feed such animals? Could we eat them, ethically, if they die a natural death?

continued

Mineral King Revisited

The Mineral King Valley dispute (see Chapter 4) raised a number of ethical issues. Christopher Stone (in the previous reading) worked as an attorney representing the Sierra Club. He argued that the law should give due consideration to the interests of the natural objects that would be disrupted or destroyed in the development plans. Is there a difference between the interests of living beings—animals and plants—and the interests of mountains or rivers? Can such nonliving natural objects be said to have interests?

Suppose the Supreme Court was convinced by Stone's argument and granted trees and natural objects legal standing. What would follow from this decision? Do we have reason to believe that the Sierra Club would be the best guardian of the trees' interests? Would you accept that a lumber company could speak for the trees? Would restoration be an adequate legal remedy to compensate for environmental destruction? Why or why not? Could humans be allowed to sue natural objects? Imagine a farmer who suffers economic loss from wolves killing his livestock. Could the farmer sue the wolves?

Species and Individuals

Not all actions that protect animals protect species, and not all actions aimed at protecting species serve the interests of individual members of that species. For example, to preserve a species it might be necessary to capture, confine (for breeding purposes), and in some cases even kill (for research purposes) individual members of the species. Do we have responsibilities to species that are different from responsibilities to individual animals? Does a species have interests? Explain how Feinberg, Singer, and Regan might evaluate captive breeding programs aimed at preserving endangered species.

There also can be situations in which individual animals are killed to benefit an entire population, as when herds are thinned to combat overpopulation. How might Feinberg, Singer, and Regan evaluate this practice?

In your opinion, do we have different responsibilities to individual animals than we have to their species? If these responsibilities conflict, which should take priority? Are species more important than individual animals?

CHAPTER 7

Holism: Ecology and Ethics

Two major philosophical questions raised by environmental issues are *extensionism,* the movement from anthropocentric to nonanthropocentric ethics, and *holism.* Chapter 6 introduced ethical extensionism, and this chapter examines holism. Under the influence of ecology, holistic ethics argues that ecological wholes, such as ecosystems and species, deserve equal, if not primary, ethical consideration. Holism holds that an ethical concern for ecosystems, relationships (for example, predator–prey), and species involves a different ethical perspective than a concern for the individual animals, trees, and plants that comprise the ecosystem or species. The individualistic approach assumed by most of the perspectives examined in the previous chapter, with its emphasis on individual living things, literally fails to see the forest for the trees.

This shift from an ethical emphasis on individuals to an emphasis on systems, relationships, and wholes, is sometimes referred to as a shift from a biocentric (life-centered) to an ecocentric ethics. The science of ecology has played a major role in the development of ecocentric ethics.

The shift from individualistic to holistic ethics has major implications for public policy. Whereas an emphasis on the welfare of individual animals might treat all mammals equally, an ecocentric ethics would more likely give preference to members of endangered species, to native species, to predators, to wild animals. The value of any particular plant or animal would depend on how that individual functions within its ecological surroundings.

Ecological holism raises intriguing philosophical questions on several levels. First, holism invites us to rethink our understanding of the nature of things (metaphysics). What is more real or fundamental—individual animals or the ecological niche that they fill? Are physiochemical processes involved in the carbon, water, nitrogen, and photosynthesis cycles more basic than the individual beings created through these cycles? What, after all, is an individual? A plant? A species? An ecosystem?

Ecological holism also raises interesting questions of epistemology. Given the discoveries of ecology, one might conclude that adequate understanding can occur only when we've understood relations and functions. This suggests that reductionistic approaches, common to physics and chemistry, miss something important when they try to reduce individuals to collections of molecules. A functional approach, reminiscent more of Aristotle than Galileo, may be the appropriate scientific method for biology.

But it is the normative and ethical questions of holism that are key for us. From the earliest days of ecology, people have drawn policy conclusions from the scientific descriptions offered by ecologists. Surely the most famous such normative conclusion was articulated by Aldo Leopold in his classic *A Sand County Almanac;* the first selection of this chapter, "The Land Ethic," is taken from this book. Leopold tells us that "a thing is right when it tends to preserve the integrity, stability, and beauty of the biotic community. It is wrong when it tends otherwise."

Ecocentric approaches face two major challenges. First, it is unclear what, if any, ethical implications can be drawn from ecology. On one hand, the science of ecology itself has not offered an unambiguous and unified theory of ecosystems. Ecologists themselves disagree, often dramatically, about the facts of ecology. On the other hand, even if there were widespread agreement on the facts, it is not clear that any ethical conclusions would obviously follow.

Consider two scenarios. Suppose ecology establishes that ecosystems naturally tend toward a state of balance and stability. What public policy would follow from this? Some might argue that given this natural stability, ecosystems can withstand significant human intervention. Others might argue that this stability gives us reason to protect the natural balance of nature and not interfere. Or, consider what might follow from ecology if it concluded that ecosystems naturally tend toward chaos rather than harmony. Some might argue that because ecosystems are chaotic anyway, we need not be concerned with how humans interact with an ecosystem. Others might argue that this chaos gives us even greater reason to tread carefully on the earth for fear that any human intervention could lead to disastrous results. So, whether we see ecosystems as balanced or chaotic, we can draw diverse policy conclusions from ecological observations.

The second major challenge focuses on the philosophical and ethical status of the relation between ecological wholes and their individual members. Such concepts as systems, relationships, and species are abstractions, seemingly removed from the real world of living, breathing individual plants and animals. According to some critics, although it is at least plausible to claim that individual animals and plants have interests and should count morally, it is unreasonable to make the same claim for abstract entities such as a species.

One particularly important criticism that follows from this points out that holistic approaches seem to subordinate the welfare of individuals—living, breathing animals, for example—to the good of the whole. Politically, such views are often totalitarian and fascist because they appear willing to sacrifice the individual for the collective. The Western tradition in ethics argues against this when it emphasizes the dignity of individuals, be they human or not.

The science of ecology, particularly its discoveries concerning the interrelatedness of the natural world, has had major impact on environmental policy. It is difficult to imagine any policy debate concerning the environment that is not deeply influenced by ecology. This chapter introduces some of the basic ethical issues involved in this approach to environmental policy.

The Land Ethic

Aldo Leopold

One of the true classics of environmentalism is Aldo Leopold's A Sand County Almanac. As J. Baird Callicott mentions in a later reading, this book is often regarded as a holy book within environmental circles, and Leopold is seen as an environmental prophet. Leopold's own career paralleled the development of ecology.

Leopold (1887–1948) spent much of his life working for government agencies as a wildlife and conservation agent. His early views were clearly within the conservationist camp. Game and forests are resources to be used, albeit in a reasonable and prudent manner. In an early essay he spoke of wolves, lions, coyotes, and bobcats as "varmints" that need to be eliminated to make room for a more efficient production of game animals, the "crop" of game management. But as he learned more about natural systems, as he learned more about ecology, he began to see the folly of such an approach.

"The Land Ethic," taken from A Sand County Almanac, provides Leopold's most mature thoughts about natu- *ral systems and the ethical conclusions that should be drawn from ecological knowledge.*

When god-like Odysseus returned from the wars in Troy, he hanged all on one rope a dozen slave-girls of his household whom he suspected of misbehavior during his absence.

This hanging involved no question of propriety. The girls were property. The disposal of property was then, as now, a matter of expediency, not of right and wrong.

Concepts of right and wrong were not lacking from Odysseus' Greece: witness the fidelity of his wife through the long years before at last his black-prowed galleys clove the wine-dark seas for home. The ethical structure of that day covered wives, but had not yet been extended to human chattels. During the three thousand years which have since elapsed, ethical criteria have been extended to many fields of conduct, with corresponding shrinkages in those judged by expediency only.

no lognness
upon eachother

THE ETHICAL SEQUENCE

This extension of ethics, so far studied only by philosophers, is actually a process in ecological evolution. Its sequences may be described in ecological as well as in philosophical terms. An ethic, ecologically, is a limitation on freedom of action in the struggle for existence. An ethic, philosophically, is a differentiation of social from anti-social conduct. These are two definitions of one thing. The thing has its origin in the tendency of interdependent individuals or groups to evolve modes of co-operation. The ecologist calls these symbioses. Politics and economics are advanced symbioses in which the original free-for-all competition has been replaced, in part, by co-operative mechanisms with an ethical content.

The complexity of co-operative mechanisms has increased with population density, and with the efficiency of tools. It was simpler, for example, to define the anti-social uses of sticks and stones in the days of the mastodons than of bullets and billboards in the age of motors.

The first ethics dealt with the relation between individuals; the Mosaic Decalogue is an example. Later accretions dealt with the relation between the individual and society. The Golden Rule tries to integrate the individual to society; democracy to integrate social organization to the individual.

There is as yet no ethic dealing with man's relation to land and to the animals and plants which grow upon it. Land, like Odysseus' slave-girls, is still property. The land-relation is still strictly economic, entailing privileges but not obligations.

The extension of ethics to this third element in human environment is, if I read the evidence correctly, an evolutionary possibility and an ecological necessity. It is the third step in a sequence. The first two have already been taken. Individual thinkers since the days of Ezekiel and Isaiah have asserted that the despoliation of land is not only inexpedient but wrong. Society, however, has not yet affirmed their belief. I regard the present conservation movement as the embryo of such an affirmation.

An ethic may be regarded as a mode of guidance for meeting ecological situations so new or intricate, or

involving such deferred reactions, that the path of social expediency is not discernible to the average individual. Animal instincts are modes of guidance for the individual in meeting such situations. Ethics are possibly a kind of community instinct in-the-making.

THE COMMUNITY CONCEPT

All ethics so far evolved rest upon a single premise: that the individual is a member of a community of interdependent parts. His instincts prompt him to compete for his place in the community, but his ethics prompt him also to co-operate (perhaps in order that there may be a place to compete for).

The land ethic simply enlarges the boundaries of the community to include soils, waters, plants, and animals, or collectively: the land.

This sounds simple: do we not already sing our love for and obligation to the land of the free and the home of the brave? Yes, but just what and whom do we love? Certainly not the soil, which we are sending helter-skelter downriver. Certainly not the waters, which we assume have no function except to turn turbines, float barges, and carry off sewage. Certainly not the plants, of which we exterminate whole communities without batting an eye. Certainly not the animals, of which we have already extirpated many of the largest and most beautiful species. A land ethic of course cannot prevent the alteration, management, and use of these "resources," but it does affirm their right to continued existence, and, at least in spots, their continued existence in a natural state.

In short, a land ethic changes the role of *Homo sapiens* from conqueror of the land-community to plain member and citizen of it. It implies respect for his fellow-members, and also respect for the community as such.

In human history, we have learned (I hope) that the conqueror role is eventually self-defeating. Why? Because it is implicit in such a role that the conqueror knows, *ex cathedra*, just what makes the community clock tick, and just what and who is valuable, and what and who is worthless, in community life. It always turns out that he knows neither, and this is why his conquests eventually defeat themselves.

In the biotic community, a parallel situation exists. Abraham knew exactly what the land was for: it was to drip milk and honey into Abraham's mouth. At the present moment, the assurance with which we regard *no doing know it all*

this assumption is inverse to the degree of our education.

The ordinary citizen today assumes that science knows what makes the community clock tick; the scientist is equally sure that he does not. He knows that the biotic mechanism is so complex that its workings may never be fully understood.

That man is, in fact, only a member of a biotic team is shown by an ecological interpretation of history. Many historical events, hitherto explained solely in terms of human enterprise, were actually biotic interactions between people and land. The characteristics of the land determined the facts quite as potently as the characteristics of the men who lived on it.

Consider, for example, the settlement of the Mississippi valley. In the years following the Revolution, three groups were contending for its control: the native Indian, the French and English traders, and the American settlers. Historians wonder what would have happened if the English at Detroit had thrown a little more weight into the Indian side of those tipsy scales which decided the outcome of the colonial migration into the cane-lands of Kentucky. It is time now to ponder the fact that the cane-lands, when subjected to the particular mixture of forces represented by the cow, plow, fire, and axe of the pioneer, became bluegrass. What if the plant succession inherent in this dark and bloody ground had, under the impact of these forces, given us some worthless sedge, shrub, or weed? Would Boone and Kenton have held out? Would there have been any overflow into Ohio, Indiana, Illinois, and Missouri? Any Louisiana Purchase? Any transcontinental union of new states? Any Civil War?

Kentucky was one sentence in the drama of history. We are commonly told what the human actors in this drama tried to do, but we are seldom told that their success, or the lack of it, hung in large degree on the reaction of particular soils to the impact of the particular forces exerted by their occupancy. In the case of Kentucky, we do not even know where the bluegrass came from—whether it is a native species, or a stowaway from Europe.

Contrast the cane-lands with what hindsight tells us about the Southwest, where the pioneers were equally brave, resourceful, and persevering. The impact of occupancy here brought no bluegrass, or other plant fitted to withstand the bumps and buffetings of hard use. This region, when grazed by livestock, reverted through a series of more and more worthless grasses, shrubs, and weeds to a condition of unstable equilibrium. Each recession of plant types bred erosion; each increment to erosion bred a further recession of plants. The result today is a progressive and mutual deterioration, not only of plants and soils, but of the animal community subsisting thereon. The early settlers did not expect this: on the ciénegas of New Mexico some even cut ditches to hasten it. So subtle has been its progress that few residents of the region are aware of it. It is quite invisible to the tourist who finds this wrecked landscape colorful and charming (as indeed it is, but it bears scant resemblance to what it was in 1848).

This same landscape was "developed" once before, but with quite different results. The Pueblo Indians settled the Southwest in pre-Columbian times, but they happened *not* to be equipped with range livestock. Their civilization expired, but not because their land expired.

In India, regions devoid of any sod-forming grass have been settled, apparently without wrecking the land, by the simple expedient of carrying the grass to the cow, rather than vice versa. (Was this the result of some deep wisdom, or was it just good luck? I do not know.)

In short, the plant succession steered the course of history; the pioneer simply demonstrated, for good or ill, what successions inhered in the land. Is history taught in this spirit? It will be, once the concept of land as a community really penetrates our intellectual life.

THE ECOLOGICAL CONSCIENCE

Conservation is a state of harmony between men and land. Despite nearly a century of propaganda, conservation still proceeds at a snail's pace; progress still consists largely of letterhead pieties and convention oratory. On the back forty we still slip two steps backward for each forward stride.

The usual answer to this dilemma is "more conservation education." No one will debate this, but is it certain that only the *volume* of education needs stepping up? Is something lacking in the *content* as well?

It is difficult to give a fair summary of its content in brief form, but, as I understand it, the content is substantially this: obey the law, vote right, join some organizations, and practice what conservation is profitable on your own land; the government will do the rest.

Is not this formula too easy to accomplish anything worth-while? It defines no right or wrong, assigns no

woods & farm interme

obligation, calls for no sacrifice, implies no change in the current philosophy of values. In respect of land-use, it urges only enlightened self-interest. Just how far will such education take us? An example will perhaps yield a partial answer.

By 1930 it had become clear to all except the ecologically blind that southwestern Wisconsin's topsoil was slipping seaward. In 1933 the farmers were told that if they would adopt certain remedial practices for five years, the public would donate CCC labor to install them, plus the necessary machinery and materials. The offer was widely accepted, but the practices were widely forgotten when the five-year contract period was up. The farmers continued only those practices that yielded an immediate and visible economic gain for themselves.

This led to the idea that maybe farmers would learn more quickly if they themselves wrote the rules. Accordingly the Wisconsin Legislature in 1937 passed the Soil Conservation District Law. This said to farmers, in effect: *We, the public, will furnish you free technical service and loan you specialized machinery, if you will write your own rules for land-use. Each county may write its own rules, and these will have the force of law.* Nearly all the counties promptly organized to accept the proffered help, but after a decade of operation, *no county has yet written a single rule.* There has been visible progress in such practices as strip-cropping, pasture renovation, and soil liming, but none in fencing woodlots against grazing, and none in excluding plow and cow from steep slopes. The farmers, in short, have selected those remedial practices which were profitable anyhow, and ignored those which were profitable to the community, but not clearly profitable to themselves.

When one asks why no rules have been written, one is told that the community is not yet ready to support them; education must precede rules. But the education actually in progress makes no mention of obligations to land over and above those dictated by self-interest. The net result is that we have more education but less soil, fewer healthy woods, and as many floods as in 1937.

The puzzling aspect of such situations is that the existence of obligations over and above self-interest is taken for granted in such rural community enterprises as the betterment of roads, schools, churches, and baseball teams. Their existence is not taken for granted, nor as yet seriously discussed, in bettering the behavior of the water that falls on the land, or in the preserving of the beauty or diversity of the farm landscape. Land-use ethics are still governed wholly by economic self-interest, just as social ethics were a century ago.

To sum up: we asked the farmer to do what he conveniently could to save his soil, and he has done just that, and only that. The farmer who clears the woods off a 75 per cent slope, turns his cows into the clearing, and dumps its rainfall, rocks, and soil into the community creek, is still (if otherwise decent) a respected member of society. If he puts lime on his fields and plants his crops on contour, he is still entitled to all the privileges and emoluments of his Soil Conservation District. The District is a beautiful piece of social machinery, but it is coughing along on two cylinders because we have been too timid, and too anxious for quick success, to tell the farmer the true magnitude of his obligations. Obligations have no meaning without conscience, and the problem we face is the extension of the social conscience from people to land.

No important change in ethics was ever accomplished without an internal change in our intellectual emphasis, loyalties, affections, and convictions. The proof that conservation has not yet touched these foundations of conduct lies in the fact that philosophy and religion have not yet heard of it. In our attempt to make conservation easy, we have made it trivial.

SUBSTITUTES FOR A LAND ETHIC

When the logic of history hungers for bread and we hand out a stone, we are at pains to explain how much the stone resembles bread. I now describe some of the stones which serve in lieu of a land ethic.

One basic weakness in a conservation system based wholly on economic motives is that most members of the land community have no economic value. Wildflowers and songbirds are examples. Of the 22,000 higher plants and animals native to Wisconsin, it is doubtful whether more than 5 per cent can be sold, fed, eaten, or otherwise put to economic use. Yet these creatures are members of the biotic community, and if (as I believe) its stability depends on its integrity, they are entitled to continuance.

When one of these non-economic categories is threatened, and if we happen to love it, we invent subterfuges to give it economic importance. At the beginning of the century songbirds were supposed to be disappearing. Ornithologists jumped to the rescue with some distinctly shaky evidence to the effect that insects would eat us up if birds failed to control them. The evidence had to be economic in order to be valid.

how truly

Currently regulation Based only on economics

It is painful to read these circumlocutions today. We have no land ethic yet, but we have at least drawn nearer the point of admitting that birds should continue as a matter of biotic right, regardless of the presence or absence of economic advantage to us.

A parallel situation exists in respect of predatory mammals, raptorial birds, and fish-eating birds. Time was when biologists somewhat overworked the evidence that these creatures preserve the health of game by killing weaklings, or that they control rodents for the farmer, or that they prey only on "worthless" species. Here again, the evidence had to be economic in order to be valid. It is only in recent years that we hear the more honest argument that predators are members of the community, and that no special interest has the right to exterminate them for the sake of a benefit, real or fancied, to itself. Unfortunately this enlightened view is still in the talk stage. In the field the extermination of predators goes merrily on: witness the impending erasure of the timber wolf by fiat of Congress, the Conservation Bureaus, and many state legislatures.

Some species of trees have been "read out of the party" by economics-minded foresters because they grow too slowly, or have too low a sale value to pay as timber crops: white cedar, tamarack, cypress, beech, and hemlock are examples. In Europe, where forestry is ecologically more advanced, the non-commercial tree species are recognized as members of the native forest community, to be preserved as such, within reason. Moreover some (like beech) have been found to have a valuable function in building up soil fertility. The interdependence of the forest and its constituent tree species, ground flora, and fauna is taken for granted.

Lack of economic value is sometimes a character not only of species or groups, but of entire biotic communities: marshes, bogs, dunes, and "deserts" are examples. Our formula in such cases is to relegate their conservation to government as refuges, monuments, or parks. The difficulty is that these communities are usually interspersed with more valuable private lands; the government cannot possibly own or control such scattered parcels. The net effect is that we have relegated some of them to ultimate extinction over large areas. If the private owner were ecologically minded, he would be proud to be the custodian of a reasonable proportion of such areas, which add diversity and beauty to his farm and to his community.

In some instances, the assumed lack of profit in these "waste" areas has proved to be wrong, but only after most of them had been done away with. The present scramble to reflood muskrat marshes is a case in point.

There is a clear tendency in American conservation to relegate to government all necessary jobs that private landowners fail to perform. Government ownership, operation, subsidy, or regulation is now widely prevalent in forestry, range management, soil and watershed management, park and wilderness conservation, fisheries management, and migratory bird management, with more to come. Most of this growth in governmental conservation is proper and logical, some of it is inevitable. That I imply no disapproval of it is implicit in the fact that I have spent most of my life working for it. Nevertheless the question arises: What is the ultimate magnitude of the enterprise? Will the tax base carry its eventual ramifications? At what point will governmental conservation, like the mastodon, become handicapped by its own dimensions? The answer, if there is any, seems to be in a land ethic, or some other force which assigns more obligation to the private landowner.

Industrial landowners and users, especially lumbermen and stockmen, are inclined to wail long and loudly about the extension of government ownership and regulation to land, but (with notable exceptions) they show little disposition to develop the only visible alternative: the voluntary practice of conservation on their own lands.

When the private landowner is asked to perform some unprofitable act for the good of the community, he today assents only with outstretched palm. If the act costs him cash this is fair and proper, but when it costs only fore-thought, open-mindedness, or time, the issue is at least debatable. The overwhelming growth of land-use subsidies in recent years must be ascribed, in large part, to the government's own agencies for conservation education: the land bureaus, the agricultural colleges, and the extension services. As far as I can detect, no ethical obligation toward land is taught in these institutions.

To sum up: a system of conservation based solely on economic self-interest is hopelessly lopsided. It tends to ignore, and thus eventually to eliminate, many elements in the land community that lack commercial value, but that are (as far as we know) essential to its healthy functioning. It assumes, falsely, I think, that the economic parts of the biotic clock will function without the uneconomic parts. It tends to relegate to government many functions eventually too large, too complex, or too widely dispersed to be performed by government.

An ethical obligation on the part of the private owner is the only visible remedy for these situations.

THE LAND PYRAMID

An ethic to supplement and guide the economic relation of land presupposes the existence of some mental image of land as a biotic mechanism. We can be ethical only in relation to something we can see, feel, understand, love, or otherwise have faith in.

The image commonly employed in conservation education is "the balance of nature." For reasons too lengthy to detail here, this figure of speech fails to describe accurately what little we know about the land mechanism. A much truer image is the one employed in ecology: the biotic pyramid. I shall first sketch the pyramid as a symbol of land, and later develop some of its implications in terms of land-use.

Plants absorb energy from the sun. This energy flows through a circuit called the biota, which may be represented by a pyramid consisting of layers. The bottom layer is the soil. A plant layer rests on the soil, an insect layer on the plants, a bird and rodent layer on the insects, and so on up through various animal groups to the apex layer, which consists of the larger carnivores.

The species of a layer are alike not in where they came from, or in what they look like, but rather in what they eat. Each successive layer depends on those below it for food and often for other services, and each in turn furnishes food and services to those above. Proceeding upward, each successive layer decreases in numerical abundance. Thus, for every carnivore there are hundreds of his prey, thousands of their prey, millions of insects, uncountable plants. The pyramidal form of the system reflects this numerical progression from apex to base. Man shares an intermediate layer with the bears, raccoons, and squirrels which eat both meat and vegetables.

The lines of dependency for food and other services are called food chains. Thus soil-oak-deer-Indian is a chain that has now been largely converted to soil-corn-cow-farmer. Each species, including ourselves, is a link in many chains. The deer eats a hundred plants other than oak, and the cow a hundred plants other than corn. Both, then, are links in a hundred chains. The pyramid is a tangle of chains so complex as to seem disorderly, yet the stability of the system proves it to be a highly organized structure. Its functioning depends on the co-operation and competition of its diverse parts.

In the beginning, the pyramid of life was low and squat; the food chains short and simple. Evolution has added layer after layer, link after link. Man is one of thousands of accretions to the height and complexity of the pyramid. Science has given us many doubts, but it has given us at least one certainty: the trend of evolution is to elaborate and diversify the biota.

Land, then, is not merely soil; it is a fountain of energy flowing through a circuit of soils, plants, and animals. Food chains are the living channels which conduct energy upward; death and decay return it to the soil. The circuit is not closed; some energy is dissipated in decay, some is added by absorption from the air, some is stored in soils, peats, and long-lived forests; but it is a sustained circuit, like a slowly augmented revolving fund of life. There is always a net loss by downhill wash, but this is normally small and offset by the decay of rocks. It is deposited in the ocean and, in the course of geological time, raised to form new lands and new pyramids.

The velocity and character of the upward flow of energy depend on the complex structure of the plant and animal community, much as the upward flow of sap in a tree depends on its complex cellular organization. Without this complexity, normal circulation would presumably not occur. Structure means the characteristic numbers, as well as the characteristic kinds and functions, of the component species. This interdependence between the complex structure of the land and its smooth functioning as an energy unit is one of its basic attributes.

When a change occurs in one part of the circuit, many other parts must adjust themselves to it. Change does not necessarily obstruct or divert the flow of energy; evolution is a long series of self-induced changes, the net result of which has been to elaborate the flow mechanism and to lengthen the circuit. Evolutionary changes, however, are usually slow and local. Man's invention of tools has enabled him to make changes of unprecedented violence, rapidity, and scope.

One change is in the composition of floras and faunas. The larger predators are lopped off the apex of the pyramid; food chains, for the first time in history, become shorter rather than longer. Domesticated species from other lands are substituted for wild ones, and wild ones are moved to new habitats. In this worldwide pooling of faunas and floras, some species get out of bounds as pests and diseases, others are extinguished. Such effects are seldom intended or foreseen;

they represent unpredicted and often untraceable readjustments in the structure. Agricultural science is largely a race between the emergence of new pests and the emergence of new techniques for their control.

Another change touches the flow of energy through plants and animals and its return to the soil. Fertility is the ability of soil to receive, store, and release energy. Agriculture, by overdrafts on the soil, or by too radical a substitution of domestic for native species in the superstructure, may derange the channels of flow or deplete storage. Soils depleted of their storage, or of the organic matter which anchors it, wash away faster than they form. This is erosion.

Waters, like soil, are part of the energy circuit. Industry, by polluting waters or obstructing them with dams, may exclude the plants and animals necessary to keep energy in circulation.

Transportation brings about another basic change: the plants or animals grown in one region are now consumed and returned to the soil in another. Transportation taps the energy stored in rocks, and in the air, and uses it elsewhere; thus we fertilize the garden with nitrogen gleaned by the guano birds from the fishes of seas on the other side of the Equator. Thus the formerly localized and self-contained circuits are pooled on a world-wide scale.

The process of altering the pyramid for human occupation releases stored energy, and this often gives rise, during the pioneering period, to a deceptive exuberance of plant and animal life, both wild and tame. These releases of biotic capital tend to becloud or postpone the penalties of violence.

This thumbnail sketch of land as an energy circuit conveys three basic ideas:

1. That land is not merely soil.

2. That the native plants and animals kept the energy circuit open; others may or may not.

3. That man-made changes are of a different order than evolutionary changes, and have effects more comprehensive than is intended or foreseen.

These ideas, collectively, raise two basic issues: Can the land adjust itself to the new order? Can the desired alterations be accomplished with less violence?

Biotas seem to differ in their capacity to sustain violent conversion. Western Europe, for example, carries a far different pyramid than Caesar found there. Some large animals are lost; swampy forests have become

meadows or plowland; many new plants and animals are introduced, some of which escape as pests; the remaining natives are greatly changed in distribution and abundance. Yet the soil is still there and, with the help of imported nutrients, still fertile; the waters flow normally; the new structure seems to function and to persist. There is no visible stoppage or derangement of the circuit.

Western Europe, then, has a resistant biota. Its inner processes are tough, elastic, resistant to strain. No matter how violent the alterations, the pyramid, so far, has provided some new *modus vivendi* which preserves its habitability for man, and for most of the other natives.

Japan seems to present another instance of radical conversion without disorganization.

Most other civilized regions, and some as yet barely touched by civilization, display various stages of disorganization, varying from initial symptoms to advanced wastage. In Asia Minor and North Africa diagnosis is confused by climatic changes, which may have been either the cause or the effect of advanced wastage. In the United States the degree of disorganization varies locally; it is worst in the Southwest, the Ozarks, and parts of the South, and least in New England and the Northwest. Better land-uses may still arrest it in the less advanced regions. In parts of Mexico, South America, South Africa, and Australia a violent and accelerating wastage is in progress, but I cannot assess the prospects.

This almost world-wide display of disorganization in the land seems to be similar to disease in an animal, except that it never culminates in complete disorganization or death. The land recovers, but at some reduced level of complexity, and with a reduced carrying capacity for people, plants, and animals. Many biotas currently regarded as "lands of opportunity" are in fact already subsisting on exploitative agriculture, i.e. they have already exceeded their sustained carrying capacity. Most of South America is overpopulated in this sense.

In arid regions we attempt to offset the process of wastage by reclamation, but it is only too evident that the prospective longevity of reclamation projects is often short. In our own West, the best of them may not last a century.

The combined evidence of history and ecology seems to support one general deduction: the less violent the man-made changes, the greater the probability of successful readjustment in the pyramid. Violence, in turn, varies with human population density; a dense population requires a more violent conversion. In this respect, North America has a better chance for permanence than Europe, if she can contrive to limit her density.

This deduction runs counter to our current philosophy, which assumes that because a small increase in density enriched human life, that an indefinite increase will enrich it indefinitely. Ecology knows of no density relationship that holds for indefinitely wide limits. All gains from density are subject to a law of diminishing returns.

Whatever may be the equation for men and land, it is improbable that we as yet know all its terms. Recent discoveries in mineral and vitamin nutrition reveal unsuspected dependencies in the up-circuit: incredibly minute quantities of certain substances determine the value of soils to plants, of plants to animals. What of the down-circuit? What of the vanishing species, the preservation of which we now regard as an esthetic luxury? They helped build the soil; in what unsuspected ways may they be essential to its maintenance? Professor Weaver proposes that we use prairie flowers to reflocculate the wasting soils of the dust bowl; who knows for what purpose cranes and condors, otters and grizzlies may some day be used?

LAND HEALTH AND
THE A-B CLEAVAGE

A land, ethic, then, reflects the existence of an ecological conscience, and this in turn reflects a conviction of individual responsibility for the health of the land. Health is the capacity of the land for self-renewal. Conservation is our effort to understand and preserve this capacity.

Conservationists are notorious for their dissensions. Superficially these seem to add up to mere confusion, but a more careful scrutiny reveals a single plane of cleavage common to many specialized fields. In each field one group (A) regards the land as soil, and its function as commodity-production; another group (B) regards the land as a biota, and its function as something broader. How much broader is admittedly in a state of doubt and confusion.

In my own field, forestry, group A is quite content to grow trees like cabbages, with cellulose as the basic forest commodity. It feels no inhibition against violence; its ideology is agronomic. Group B, on the other hand, sees forestry as fundamentally different from

agronomy because it employs natural species, and manages a natural environment rather than creating an artificial one. Group B prefers natural reproduction on principle. It worries on biotic as well as economic grounds about the loss of species like chestnut, and the threatened loss of the white pines. It worries about a whole series of secondary forest functions: wildlife, recreation, watersheds, wilderness areas. To my mind, Group B feels the stirrings of an ecological conscience.

In the wildlife field, a parallel cleavage exists. For Group A the basic commodities are sport and meat; the yardsticks of production are ciphers of take in pheasants and trout. Artificial propagation is acceptable as a permanent as well as a temporary recourse— if its unit costs permit. Group B, on the other hand, worries about a whole series of biotic side-issues. What is the cost of predators of producing a game crop? Should we have further recourse to exotics? How can management restore the shrinking species, like prairie grouse, already hopeless as shootable game? How can management restore the threatened rarities, like trumpeter swan and whooping crane? Can management principles be extended to wildflowers? Here again it is clear to me that we have the same A-B cleavage as in forestry.

In the larger field of agriculture I am less competent to speak, but there seem to be somewhat parallel cleavages. Scientific agriculture was actively developing before ecology was born, hence a slower penetration of ecological concepts might be expected. Moreover the farmer, by the very nature of his techniques, must modify the biota more radically than the forester or the wildlife manager. Nevertheless, there are many discontents in agriculture which seem to add up to a new vision of "biotic farming."

Perhaps the most important of these is the new evidence that poundage or tonnage is no measure of the food-value of farm crops; the products of fertile soil may be qualitatively as well as quantitatively superior. We can bolster poundage from depleted soils by pouring on imported fertility, but we are not necessarily bolstering food-value. The possible ultimate ramifications of this idea are so immense that I must leave their exposition to abler pens.

The discontent that labels itself "organic farming," while bearing some of the earmarks of a cult, is nevertheless biotic in its direction, particularly in its insistence on the importance of soil flora and fauna.

The ecological fundamentals of agriculture are just as poorly known to the public as in other fields of land-use. For example, few educated people realize that the marvelous advances in technique made during recent decades are improvements in the pump, rather than the well. Acre for acre, they have barely sufficed to offset the sinking level of fertility.

In all of these cleavages, we see repeated the same basic paradoxes: man the conqueror *versus* man the biotic citizen; science the sharpener of his sword *versus* science the searchlight on his universe; land the slave and servant *versus* land the collective organism. Robinson's injunction to Tristram may well be applied, at this juncture, to *Homo sapiens* as a species in geological time:

> Whether you will or not
> You are a King, Tristram, for you are one
> Of the time-tested few that leave the world,
> When they are gone, not the same place it was.
> Mark what you leave.

THE OUTLOOK

It is inconceivable to me that an ethical relation to land can exist without love, respect, and admiration for land, and a high regard for its value. By value, I of course mean something far broader than mere economic value; I mean value in the philosophical sense.

Perhaps the most serious obstacle impeding the evolution of a land ethic is the fact that our educational and economic system is headed away from, rather than toward, an intense consciousness of land. Your true modern is separated from the land by many middlemen, and by innumerable physical gadgets. He has no vital relation to it; to him it is the space between cities on which crops grow. Turn him loose for a day on the land, and if the spot does not happen to be a golf links or a "scenic" area, he is bored stiff. If crops could be raised by hydroponics instead of farming, it would suit him very well. Synthetic substitutes for wood, leather, wool, and other natural land products suit him better than the originals. In short, land is something he has "outgrown."

Almost equally serious as an obstacle to a land ethic is the attitude of the farmer for whom the land is still an adversary, or a taskmaster that keeps him in slavery. Theoretically, the mechanization of farming ought to cut the farmer's chains, but whether it really does is debatable.

One of the requisites for an ecological comprehension of land is an understanding of ecology, and this is by no means co-extensive with "education"; in fact, much higher education seems deliberately to avoid ecological concepts. An understanding of ecology does not necessarily originate in courses bearing ecological labels; it is quite as likely to be labeled geography, botany, agronomy, history, or economics. This is as it should be, but whatever the label, ecological training is scarce.

The case for a land ethic would appear hopeless but for the minority which is in obvious revolt against these "modern" trends.

The "key-log" which must be moved to release the evolutionary process for an ethic is simply this: quit thinking about decent land-use as solely an economic problem. Examine each question in terms of what is ethically and esthetically right, as well as what is economically expedient. A thing is right when it tends to preserve the integrity, stability, and beauty of the biotic community. It is wrong when it tends otherwise.

It of course goes without saying that economic feasibility limits the tether of what can or cannot be done for land. It always has and it always will. The fallacy the economic determinists have tied around our collective neck, and which we now need to cast off, is the belief that economics determines *all* land-use. This is simply not true. An innumerable host of actions and attitudes, comprising perhaps the bulk of all land relations, is determined by the land-users' tastes and predilections, rather than by his purse. The bulk of all land relations hinges on investments of time, forethought, skill, and faith rather than on investments of cash. As a land-user thinketh, so is he.

I have purposely presented the land ethic as a product of social evolution because nothing so important as an ethic is ever "written." Only the most superficial student of history supposes that Moses "wrote" the Decalogue; it evolved in the minds of a thinking community, and Moses wrote a tentative summary of it for a "seminar." I say tentative because evolution never stops.

The evolution of a land ethic is an intellectual as well as emotional process. Conservation is paved with good intentions which prove to be futile, or even dangerous, because they are devoid of critical understanding either of the land, or of economic land-use. I think it is a truism that as the ethical frontier advances from the individual to the community, its intellectual content increases.

The mechanism of operation is the same for any ethic: social approbation for right actions; social disapproval for wrong actions.

By and large, our present problem is one of attitudes and implements. We are remodeling the Alhambra with a steam-shovel, and we are proud of our yardage. We shall hardly relinquish the shovel, which after all has many good points, but we are in need of gentler and more objective criteria for its successful use.

For Further Discussion

1. Leopold speaks of an extension of ethics to land. How does this view compare to the ethical extensionism found in earlier essays by Singer, Reagan, and Stone?

2. Leopold suggests that, like Odysseus's slaves, land is still considered to be mere property. How does Leopold's view on land compare to the understanding of property rights defended in the earlier essay by Stroup and Baden?

3. The concept of a community plays an important role in Leopold's land ethic. What do you understand about this idea? How is a community different from a society? Are humans naturally communal beings? What does Leopold mean when he tells us to change our role from conquerer of the land community to "plain member and citizen of it"?

4. The most influential line in this essay is the conclusion that "a thing is right when it tends to preserve the integrity, stability, and beauty of the biotic community." How do you interpret this line? What does "integrity" mean when applied to biotic communities? What does "stability" mean in this context? When is a biotic community beautiful?

The Conceptual Foundations of the Land Ethic

J. Baird Callicott

Although Leopold was a giant in ecological science and policy, he was not a philosopher. If his writings are fertile sources for philosophical and ethical insights, they are soil that needs to be worked to produce its abundant harvest. Over the past two decades, philosopher J. Baird Callicott has developed a sustained and philosophically more sophisticated interpretation of Leopold. In this essay, Callicott develops and defends a sympathetic interpretation of Leopold's land ethic. He attempts to situate Leopold within a philosophical context and defend the normative implications that can be drawn from Leopold's work.

As Wallace Stegner observes, *A Sand County Almanac* is considered "almost a holy book in conservation circles," and Aldo Leopold a prophet, "an American Isaiah." And as Curt Meine points out, "The Land Ethic" is the climatic essay of *Sand County,* "the upshot of 'The Upshot.'"[1] One might, therefore, fairly say that the recommendation and justification of moral obligations on the part of people to nature is what the prophetic *A Sand County Almanac* is all about. . . .

Here I first examine and elaborate the compactly expressed abstract elements of the land ethic and expose the "logic" which binds them into a proper, but revolutionary, moral theory. I then discuss the controversial features of the land ethic and defend them against actual and potential criticism. I hope to show that the land ethic cannot be ignored as merely the groundless emotive exhortations of a moonstruck conservationist or dismissed as entailing wildly untoward practical consequences. It poses, rather, a serious intellectual challenge to business-as-usual moral philosophy.

"The Land Ethic" opens with a charming and poetic evocation of Homer's Greece, the point of which is to suggest that today land is just as routinely and remorsely enslaved as human beings then were. A panoramic glance backward to our most distant cultural origins, Leopold suggests, reveals a slow but steady moral development over three millennia. More of our relationships and activities ("fields of conduct") have fallen under the aegis of moral principles ("ethical criteria") as civilization has grown and matured. If moral growth and development continue, as not only a synoptic review of history, but recent past experience suggest that it will, future generations will censure today's casual and universal environmental bondage as today we censure the casual and universal human bondage of three thousand years ago.

A cynically inclined critic might scoff at Leopold's sanguine portrayal of human history. Slavery survived as an institution in the "civilized" West, more particularly in the morally self-congratulatory United States, until a mere generation before Leopold's own birth. And Western history from imperial Athens and Rome to the Spanish Inquisition and the Third Reich has been a disgraceful series of wars, persecutions, tyrannies, pogroms, and other atrocities.

The history of moral practice, however, is not identical with the history of moral consciousness. Morality is not descriptive; it is prescriptive or normative. In light of this distinction, it is clear that today, despite rising rates of violent crime in the United States and institutional abuses of human rights in Iran, Chile, Ethiopia, Guatemala, South Africa, and many other places, and despite persistent organized social injustice and oppression in still others, moral consciousness is expanding more rapidly now than ever before. Civil rights, human rights, women's liberation, children's liberation, animal liberation, etc., all indicate, as expressions of newly emergent moral ideals, that ethical consciousness (as distinct from practice) has if anything recently accelerated—thus confirming Leopold's historical observation.

Leopold next points out that "this extension of ethics, so far studied only by philosophers"—and therefore, the implication is clear, not very satisfactorily studied—"is actually a process in ecological evolution" (202).* What Leopold is saying here, simply, is that we may understand the history of ethics, fancifully alluded to by means of the Odysseus vignette, in biological as well as philosophical terms. From a biological point of view, an ethic is "a limitation on freedom of action in the struggle for existence" (202).

I had this passage in mind when I remarked that Leopold manages to convey a whole network of ideas in a couple of phrases. The phrase "struggle for existence" unmistakably calls to mind Darwinian evolution as the conceptual context in which a biological account of the origin and development of ethics must

ultimately be located. And at once it points up a paradox: Given the unremitting competitive "struggle for existence" how could "limitations on freedom of action" ever have been conserved and spread through a population of *Homo sapiens* or their evolutionary progenitors?

For a biological account of ethics, as Harvard social entomologist Edward O. Wilson has recently written, "the central theoretical problem . . . [is] how can altruism [elaborately articulated as morality or ethics in the human species], which by definition reduces personal fitness, possibly evolve by natural selection?" [2] According to modern sociobiology, the answer lies in kinship. But according to Darwin—who had tackled this problem himself "exclusively from the side of natural history" in *The Descent of Man*—the answer lies in society. [3] And it was Darwin's classical account (and its divers variations), from the side of natural history, which informed Leopold's thinking in the late 1940s.

Let me put the problem in perspective. How, we are asking, did ethics originate and, once in existence, grow in scope and complexity?

The oldest answer in living human memory is theological. God (or the gods) imposes morality on people. And God (or the gods) sanctions it. . . .

Western philosophy, on the other hand, is almost unanimous in the opinion that the origin of ethics in human experience has somehow to do with human reason. . . .

An evolutionary natural historian, however, cannot be satisfied with either of these general accounts of the origin and development of ethics. The idea that God gave morals to man is ruled out in principle—as any supernatural explanation of a natural phenomenon is ruled out in principle in natural science. And while morality might *in principle* be a function of human reason (as, say, mathematical calculation clearly is), to suppose that it is so *in fact* would be to put the cart before the horse. Reason appears to be a delicate, variable, and recently emerged faculty. It cannot, under any circumstances, be supposed to have evolved in the absence of complex linguistic capabilities which depend, in turn, for their evolution upon a highly developed social matrix. But we cannot have become social beings unless we assumed limitations on freedom of action in the struggle for existence. Hence we must have become ethical before we became rational.

Darwin, probably in consequence of reflections somewhat like these, turned to a minority tradition of modern philosophy for a moral psychology consistent with and useful to a general evolutionary account of ethical phenomena. A century earlier, Scottish phi-

---

*Page references are to Aldo Leopold's *A Sand County Almanac with Sketches Here and There* (New York: Oxford University Press, 1949).

losophers David Hume and Adam Smith had argued that ethics rest upon feelings or "sentiments"—which, to be sure, may be both amplified and informed by reason.[4] And since in the animal kingdom feelings or sentiments are arguably far more common or widespread than reason, they would be a far more likely starting point for an evolutionary account of the origin and growth of ethics.

Darwin's account, to which Leopold unmistakably (if elliptically) alludes in "The Land Ethic," begins with the parental and filial affections common, perhaps, to all mammals.[5] Bonds of affection and sympathy between parents and offspring permitted the formation of small, closely knit social groups, Darwin argued. Should the parental and filial affections bonding family members chance to extend to less closely related individuals, that would permit an enlargement of the family group. And should the newly extended community more successfully defend itself and/or more efficiently provision itself, the inclusive fitness of its members severally would be increased, Darwin reasoned. Thus, the more diffuse familial affections, which Darwin (echoing Hume and Smith) calls the "social sentiments," would be spread throughout a population.[6]

Morality, properly speaking—i.e., morality as opposed to mere altruistic instinct—requires, in Darwin's terms, "intellectual powers" sufficient to recall the past and imagine the future, "the power of language" sufficient to express "common opinion," and "habituation" to patterns of behavior deemed, by common opinion, to be socially acceptable and beneficial.[7] Even so, ethics proper, in Darwin's account, remains firmly rooted in moral feelings or social sentiments which were—no less than physical faculties, he expressly avers—naturally selected, by the advantages for survival and especially for successful reproduction, afforded by society.[8]

The protosociobiological perspective on ethical phenomena, to which Leopold as a natural historian was heir, leads him to a generalization which is remarkably explicit in his condensed and often merely resonant rendering of Darwin's more deliberate and extended paradigm: Since "the thing [ethics] has its origin in the tendency of interdependent individuals or groups to evolve modes of cooperation, . . . all ethics so far evolved rest upon a single premise: that the individual is a member of a community of interdependent parts" (202–3).

Hence, we may expect to find that the scope and specific content of ethics will reflect both the perceived boundaries and actual structure or organization of a cooperative community or society. *Ethics and society or community are correlative.* This single, simple principle constitutes a powerful tool for the analysis of moral natural history, for the anticipation of future moral development (including, ultimately, the land ethic), and for systematically deriving the specific precepts, the prescriptions and proscriptions, of an emergent and culturally unprecedented ethic like a land or environmental ethic.

Anthropological studies of ethics reveal that in fact the boundaries of the moral community are generally coextensive with the perceived boundaries of society.[9] And the peculiar (and, from the urbane point of view, sometimes inverted) representation of virtue and vice in tribal society—the virtue, for example, of sharing to the point of personal destitution and the vice of privacy and private property—reflects and fosters the life way of tribal peoples.[10] Darwin, in his leisurely, anecdotal discussion, paints a vivid picture of the intensity, peculiarity, and sharp circumscription of "savage" mores: "A savage will risk his life to save that of a member of the same community, but will be wholly indifferent about a stranger."[11] As Darwin portrays them, tribespeople are at once paragons of virtue "within the limits of the same tribe" and enthusiastic thieves, manslaughterers, and torturers without.[12]

For purposes of more effective defense against common enemies, or because of increased population density, or in response to innovations in subsistence methods and technologies, or for some mix of these or other forces, human societies have grown in extent or scope and changed in form or structure. Nations—like the Iroquois nation or the Sioux nation—came into being upon the merger of previously separate and mutually hostile tribes. Animals and plants were domesticated and erstwhile hunter-gatherers became herders and farmers. Permanent habitations were established. Trade, craft, and (later) industry flourished. With each change in society came corresponding and correlative changes in ethics. The moral community expanded to become coextensive with the newly drawn boundaries of societies and the representation of virtue and vice, right and wrong, good and evil, changed to accommodate, foster, and preserve the economic and institutional organization of emergent social orders. . . .

Most educated people today pay lip service at least to the ethical precept that all members of the human species, regardless of race, creed, or national origin, are endowed with certain fundamental rights which it is wrong not to respect. According to the evolutionary scenario set out by Darwin, the contemporary moral ideal of human rights is a response to a perception—however vague and indefinite—that mankind worldwide is united into one society, one community—however indeterminate or yet institutionally unorganized. As Darwin presciently wrote:

As man advances in civilization, and small tribes are united into larger communities, the simplest reason would tell each individual that he ought to extend his social instincts and sympathies to all the members of the same nation, though personally unknown to him. This point being once reached, there is only an artificial barrier to prevent his sympathies extending to the men of all nations and races. If, indeed, such men are separated from him by great differences of appearance or habits, experience unfortunately shows us how long it is, before we look at them as our fellow-creatures.[13]

According to Leopold, the next step in this sequence beyond the still incomplete ethic of universal humanity, a step that is clearly discernible on the horizon, is the land ethic. The "community concept" has, so far, propelled the development of ethics from the savage clan to the family of man. "The land ethic simply enlarges the boundary of the community to include soils, waters, plants, and animals, or collectively: the land" (204).

As the foreword to *Sand County* makes plain, the overarching thematic principle of the book is the inculcation of the idea—through narrative description, discursive exposition, abstractive generalization, and occasional preachment—"that land is a community" (viii). The community concept is "the basic concept of ecology" (viii). Once land is popularly perceived as a biotic community—as it is professionally perceived in ecology—a correlative land ethic will emerge in the collective cultural consciousness.

Although anticipated as far back as the mid-eighteenth century—in the notion of an "economy of nature"—the concept of the biotic community was more fully and deliberately developed as a working model or paradigm for ecology by Charles Elton in the 1920s.[14] The natural world is organized as an intricate corporate society in which plants and animals occupy "niches," or as Elton alternatively called them "roles" or "professions," in the economy of nature.[15] As in a feudal community, little or no socioeconomic mobility

(upward or otherwise) exists in the biotic community. One is born to one's trade.

Human society, Leopold argues, is founded, in large part, upon mutual security and economic interdependency and preserved only by limitations on freedom of action in the struggle for existence—that is, by ethical constraints. Since the biotic community exhibits, as modern ecology reveals, an analogous structure, it too can be preserved, given the newly amplified impact of "mechanized man," only by analogous limitations on freedom of action—that is, by a land ethic (viii). A land ethic, furthermore, is not only "an ecological necessity," but an "evolutionary possibility" because a moral response to the natural environment—Darwin's social sympathies, sentiments, and instincts translated and codified into a body of principles and precepts—would be automatically triggered in human beings by ecology's social representation of nature (203).

Therefore, the key to the emergence of a land ethic is, simply, universal ecological literacy.

The land ethic rests upon three scientific cornerstones: (1) evolutionary and (2) ecological biology set in a background of (3) Copernican astronomy. Evolutionary theory provides the conceptual link between ethics and social organization and development. It provides a sense of "kinship with fellow-creatures" as well, "fellow-voyagers" with us in the "odyssey of evolution" (109). It establishes a diachronic link between people and nonhuman nature.

Ecological theory provides a synchronic link—the community concept—a sense of social integration of human and nonhuman nature. Human beings, plants, animals, soils, and waters are "all interlocked in one humming community of cooperations and competitions, one biota."[16] The simplest reason, to paraphrase Darwin, should, therefore, tell each individual that he or she ought to extend his or her social instincts and sympathies to all the members of the biotic community though different from him or her in appearance or habits.

And although Leopold never directly mentions it in *A Sand County Almanac*, the Copernican perspective, the perception of the Earth as "a small planet" in an immense and utterly hostile universe beyond, contributes, perhaps subconsciously, but nevertheless very powerfully, to our sense of kinship, community, and interdependence with fellow denizens of the Earth household. It scales the Earth down to something like a cozy island paradise in a desert ocean.

Here in outline, then, are the conceptual and logical foundations of the land ethic: Its conceptual elements are a Copernican cosmology, a Darwinian protosociobiological natural history of ethics, Darwinian ties of kinship among all forms of life on Earth, and an Eltonian model of the structure of biocenoses all overlaid on a Humean-Smithian moral psychology. Its logic is that natural selection has endowed human beings with an affective moral response to perceived bonds of kinship and community membership and identity; that today the natural environment, the land, is represented as a community, the biotic community; and that, therefore, an environmental or land ethic is both possible—the biopsychological and cognitive conditions are in place—and necessary, since human beings collectively have acquired the power to destroy the integrity, diversity, and stability of the environing and supporting economy of nature. In the remainder of this essay I discuss special features and problems of the land ethic germane to moral philosophy.

The most salient feature of Leopold's land ethic is its provision of what Kenneth Goodpaster has carefully called "moral considerability" for the biotic community per se, not just for fellow members of the biotic community: [17]

> In short, a land ethic changes the role of *Homo sapiens* from conqueror of the land-community to plain member and citizen of it. It implies respect for his fellow-members, *and also respect for the community as such.* (204, emphasis added)

The land ethic, thus, has a holistic as well as an individualistic cast.

Indeed, as "The Land Ethic" develops, the focus of moral concern shifts gradually away from plants, animals, soils, and waters severally to the biotic community collectively. Toward the middle, in the subsection called Substitutes for a Land Ethic, Leopold invokes the "biotic rights" of *species*—as the context indicates—of wildflowers, songbirds, and predators. In The Outlook, the climactic section of "The Land Ethic," nonhuman natural entities, first appearing as fellow members, then considered in profile as species, are not so much as mentioned in what might be called the "summary moral maxim" of the land ethic: "A thing is right when it tends to preserve the integrity, stability, and beauty of the biotic community. It is wrong when it tends otherwise" (224–25).

By this measure of right and wrong, not only would it be wrong for a farmer, in the interest of higher prof-

its, to clear the woods off a 75 percent slope, turn his cows into the clearing, and dump its rainfall, rocks, and soil into the community creek, it would also be wrong for the federal fish and wildlife agency, in the interest of individual animal welfare, to permit populations of deer, rabbits, feral burros, or whatever to increase unchecked and thus to threaten the integrity, stability, and beauty of the biotic communities of which they are members. The land ethic not only provides moral considerability for the biotic community per se, but ethical consideration of its individual members is preempted by concern for the preservation of the integrity, stability, and beauty of the biotic community. The land ethic, thus, not only has a holistic aspect; it is holistic with a vengeance.

The holism of the land ethic, more than any other feature, sets it apart from the predominant paradigm of modern moral philosophy. It is, therefore, the feature of the land ethic which requires the most patient theoretical analysis and the most sensitive practical interpretation.

As Kenneth Goodpaster pointed out, mainstream modern ethical philosophy has taken egoism as its point of departure and reached a wider circle of moral entitlement by a process of generalization: [18] I am sure that *I,* the enveloped ego, am intrinsically or inherently valuable and thus that *my* interests ought to be considered, taken into account, by "others" when their actions may substantively affect *me.* My own claim to moral consideration, according to the conventional wisdom, ultimately rests upon a psychological capacity—rationality or sentiency were the classical candidates of Kant and Bentham, respectively—which is arguably valuable in itself and which thus qualifies *me* for moral standing.[19] However, then I am forced grudgingly to grant the same moral consideration I demand from others, on this basis, to those others who can also claim to possess the same general psychological characteristic.

A *criterion* of moral value and consideration is thus identified. Goodpaster convincingly argues that mainstream modern moral theory is based, when all the learned dust has settled, on this simple paradigm of ethical justification and logic exemplified by the Benthamic and Kantian prototypes.[20] If the criterion of moral values and consideration is pitched low enough—as it is in Bentham's criterion of sentiency—a wide variety of animals are admitted to moral entitlement.[21] If the criterion of moral value and consideration is pushed lower still—as it is in Albert Schweitzer's reverence-for-life ethic—all minimally conative things (plants as well as animals) would be extended moral considerability.[22] The contemporary animal liberation/rights, and reverence-for-life/life-principle ethics are, at bottom, simply direct applications of the modern classical paradigm of moral argument. But this standard modern model of ethical theory provides no possibility whatever for the moral consideration of wholes—of threatened *populations* of animals and plants, or of endemic, rare, or endangered *species,* or of biotic *communities,* or most expansively, of the *biosphere* in its totality—since wholes per se have no psychological experience of any kind.[23] Because mainstream modern moral theory has been "psychocentric," it has been radically and intractably individualistic or "atomistic" in its fundamental theoretical orientation.

Hume, Smith, and Darwin diverged from the prevailing theoretical model by recognizing that altruism is as fundamental and autochthonous in human nature as is egoism. According to their analysis, moral value is not identified with a natural quality objectively present in morally considerable beings—as reason and/or sentiency is objectively present in people and/or animals—it is, as it were, projected by valuing subjects.[24]

Hume and Darwin, furthermore, recognize inborn moral sentiments which have society as such as their natural object. Hume insists that "we must renounce the theory which accounts for every moral sentiment by the principle of self-love. We must adopt a more *public affection* and allow that the *interests of society* are not, *even on their own account,* entirely indifferent to us."[25] And Darwin, somewhat ironically (since "Darwinian evolution" very often means natural selection operating exclusively with respect to individuals), sometimes writes as if morality had no other object than the commonweal, the welfare of the community as a corporate entity:

> We have now seen that actions are regarded by savages, and were probably so regarded by primeval man, as good or bad, solely as they obviously affect the welfare of the tribe,—not that of the species, nor that of the individual member of the tribe. This conclusion agrees well with the belief that the so-called moral sense is aboriginally derived from social instincts, for both relate at first exclusively to the community.[26]

Theoretically then, the biotic community owns what Leopold, in the lead paragraph of The Outlook,

calls "value in the philosophical sense"—i.e., direct moral considerability—because it is a newly discovered proper object of a specially evolved "public affection" or "moral sense" which all psychologically normal human beings have inherited from a long line of ancestral social primates (223). . . .[27]

[T]he conceptual foundations of the land ethic provide a well-formed, self-consistent theoretical basis for including both fellow members of the biotic community and the biotic community itself (considered as a corporate entity) within the purview of morals. The preemptive emphasis, however, on the welfare of the community as a whole, in Leopold's articulation of the land ethic, while certainly *consistent* with its Humean-Darwinian theoretical foundations, is not *determined* by them alone. The overriding holism of the land ethic results, rather, more from the way our moral sensibilities are informed by ecology.

Ecological thought, historically, has tended to be holistic in outlook.[28] Ecology is the study of the *relationships* of organisms to one another and to the elemental environment. These relationships bind the *relata*—plants, animals, soils, and waters—into a seamless fabric. The ontological primacy of objects and the ontological subordination of relationships, characteristic of classical Western science, is, in fact, reversed in ecology.[29] Ecological relationships determine the nature of organisms rather than the other way around. A species is what it is because it has adapted to a niche in the ecosystem. The whole, the system itself, thus, literally and quite straightforwardly shapes and forms its component parts.

Antedating Charles Elton's community model of ecology was F. E. Clements' and S. A. Forbes' organism model.[30] Plants and animals, soils and waters, according to this paradigm, are integrated into one superorganism. Species are, as it were, its organs; specimens its cells. Although Elton's community paradigm (later modified, as we shall see, by Arthur Tansley's ecosystem idea) is the principal and morally fertile ecological concept of "The Land Ethic," the more radically holistic superorganism paradigm of Clements and Forbes resonates in "The Land Ethic" as an audible overtone. In the peroration of Land Health and the A-B Cleavage, for example, which immediately precedes The Outlook, Leopold insists that

> in all of these cleavages, we see repeated the same basic paradoxes: man the conqueror *versus* man

the biotic citizen; science the sharpener of his sword *versus* science the searchlight on his universe; land the slave and servant *versus* land the collective organism. (223)

And on more than one occasion Leopold, in the latter quarter of "The Land Ethic," talks about the "health" and "disease" of the land—terms which are at once descriptive and normative and which, taken literally, characterize only organisms proper.

In an early essay, "Some Fundamentals of Conservation in the Southwest," Leopold speculatively flirted with the intensely holistic superorganism model of the environment as a paradigm pregnant with moral implications:

> It is at least not impossible to regard the earth's parts—soil, mountains, rivers, atmosphere, etc.—as organs or parts of organs, of *a coordinated whole,* each part with a definite function. And if we could see *this whole, as a whole,* through a great period of time, we might perceive not only organs with co-ordinated functions, but possibly also that process of consumption and replacement which in biology we call metabolism, or growth. In such a case we would have all the visible attributes of a living thing, which we do not realize to be such because it is too big, and its life processes too slow. And there would also follow that invisible attribute—a soul or consciousness—which . . . many philosophers of all ages ascribe to all living things and aggregates thereof, including the "dead" earth.
>
> Possibly in our intuitive perceptions, which may be truer than our science and less impeded by words than our philosophies, we realize the indivisibility of the earth—its soil, mountains, rivers, forests, climate, plants, and animals—and *respect it collectively* not only as a useful servant but as a living being, vastly less alive than ourselves, but vastly greater than ourselves in time and space. . . . Philosophy, then, suggests one reason why we cannot destroy the earth with moral impunity; namely, that the "dead" earth is an organism possessing a certain kind and degree of life, which we intuitively respect as such.[31]

Had Leopold retained this overall theoretical approach in "The Land Ethic," the land ethic would doubtless have enjoyed more critical attention from philosophers. The moral foundations of a land or, as he might then have called it, "earth" ethic, would rest upon the hypothesis that the Earth is alive and en-

souled—possessing inherent psychological characteristics, logically parallel to reason and sentiency. This notion of a conative whole Earth could plausibly have served as a general criterion of intrinsic worth and moral considerability, in the familiar format of mainstream moral thought.

Part of the reason, therefore, that "The Land Ethic" emphasizes more and more the integrity, stability, and beauty of the environment as a whole, and less and less the "biotic right" of individual plants and animals to life, liberty, and the pursuit of happiness, is that the superorganism ecological paradigm invites one, much more than does the community paradigm, to hypostatize, to reify the whole, and to subordinate its individual members.

In any case, as we see, rereading "The Land Ethic" in light of "Some Fundamentals," the whole Earth organism image of nature is vestigially present in Leopold's later thinking. Leopold may have abandoned the "earth ethic" because ecology had abandoned the organism analogy, in favor of the community analogy, as a working theoretical paradigm. And the community model was more suitably given moral implications by the social/sentimental ethical natural history of Hume and Darwin. . . .

The Land Pyramid is the pivotal section of "The Land Ethic"—the section which effects a complete transition from concern for "fellow-members" to the "community as such." It is also its longest and most technical section. A description of the "ecosystem" ([1935 British ecologist Arthur] Tansley's deliberately nonmetaphorical term) begins with the sun. Solar energy "flows through a circuit called the biota" (215). It enters the biota through the leaves of green plants and courses through plant-eating animals, and then on to omnivores and carnivores. At last the tiny fraction of solar energy converted to biomass by green plants remaining in the corpse of a predator, animal feces, plant detritus, or other dead organic material is garnered by decomposers—worms, fungi, and bacteria. They recycle the participating elements and degrade into entropic equilibrium any remaining energy. According to this paradigm

> land, then, is not merely soil; it is a fountain of energy flowing through a circuit of soils, plants, and animals. Food chains are the living channels which conduct energy upward; death and decay return it to the soil. The circuit is not closed; . . . but it is a sustained circuit, like a slowly augmented revolving fund of life. (216)

In this exceedingly abstract (albeit poetically expressed) model of nature, process precedes substance and energy is more fundamental than matter. Individual plants and animals become less autonomous beings than ephemeral structures in a patterned flux of energy. . . . The maintenance of "the complex structure of the land and its smooth functioning as an energy unit" emerges in The Land Pyramid as the *summum bonum* of the land ethic (216).

From this good Leopold derives several practical principles slightly less general, and therefore more substantive, than the summary moral maxim of the land ethic distilled in The Outlook. "The trend of evolution [not its "goal," since evolution is ateleological] is to elaborate and diversify the biota" (216). Hence, among our cardinal duties is the duty to preserve what species we can, especially those at the apex of the pyramid—the top carnivores. "In the beginning, the pyramid of life was low and squat; the food chains short and simple. Evolution has added layer after layer, link after link" (215–16). Human activities today, especially those, like systematic deforestation in the tropics, resulting in abrupt massive extinctions of species, are in effect "devolutionary"; they flatten the biotic pyramid; they choke off some of the channels and gorge others (those which terminate in our own species).[32]

The land ethic does not enshrine the ecological status quo and devalue the dynamic dimension of nature. Leopold explains that "evolution is a long series of self-induced changes, the net result of which has been to elaborate the flow mechanism and to lengthen the circuit. Evolutionary changes, however, are usually slow and local. Man's invention of tools has enabled him to make changes of unprecedented violence, rapidity, and scope" (216–17). "Natural" species extinction, i.e., species extinction in the normal course of evolution, occurs when a species is replaced by competitive exclusion or evolves into another form.[33] Normally speciation outpaces extinction. Mankind inherited a richer, more diverse world than had ever existed before in the 3.5 billion-year odyssey of life on Earth.[34] What is wrong with anthropogenic species extirpation and extinction is the *rate* at which it is occurring and the *result:* biological impoverishment instead of enrichment.

Leopold goes on here to condemn, in terms of its impact on the ecosystem, "the world-wide pooling of faunas and floras," i.e., the indiscriminate introduction of exotic and domestic species and the dislocation of native and endemic species; mining the soil for its stored biotic energy, leading ultimately to diminished fertility and to erosion; and polluting and damming water courses (217).

According to the land ethic, therefore: Thou shalt not extirpate or render species extinct; thou shalt exercise great caution in introducing exotic and domestic species into local ecosystems, in extracting energy from the soil and releasing it into the biota, and in damming or polluting water courses; and thou shalt be especially solicitous of predatory birds and mammals. Here in brief are the express moral precepts of the land ethic. They are all explicitly informed—not to say derived—from the energy circuit model of the environment.

The living channels—"food chains"—through which energy courses are composed of individual plants and animals. A central, stark fact lies at the heart of ecological processes: Energy, the currency of the economy nature, passes from one organism to another, not from hand to hand, like coined money, but, so to speak, from stomach to stomach. Eating *and being eaten,* living *and dying* are what make the biotic community hum.

The precepts of the land ethic, like those of all previous accretions, reflect and reinforce the structure of the community to which it is correlative. Trophic asymmetries constitute the kernel of the biotic community. It seems unjust, unfair. But that is how the economy of nature is organized (and has been for thousands of millions of years). The land ethic, thus, affirms as good, and strives to preserve, the very inequities in nature whose social counterparts in human communities are condemned as bad and would be eradicated by familiar social ethics, especially by the more recent Christian and secular egalitarian exemplars. A "right to life" for individual members is not consistent with the structure of the biotic community and hence is not mandated by the land ethic. This disparity between the land ethic and its more familiar social precedents contributes to the apparent devaluation of individual *members* of the biotic community and augments and reinforces the tendency of the land ethic, driven by the systemic vision of ecology, toward a more holistic or community-per-se orientation. . . .

Today, two processes internal to civilization are bringing us to a recognition that our renunciation of our biotic citizenship was a mistaken self-deception. Evolutionary science and ecological science, which certainly are products of modern civilization now supplanting the anthropomorphic and anthropocentric myths of earlier civilized generations, have rediscovered our integration with the biotic community. And the negative feedback received from modern civilization's technological impact upon nature—pollution, biological impoverishment, etc.—forcefully reminds us that mankind never really has, despite past assumptions to the contrary, existed apart from the environing biotic community.

This reminder of our recent rediscovery of our biotic citizenship brings us face to face with the paradox posed by Peter Fritzell:[35] Either we are plain members and citizens of the biotic community, on a par with other creatures, or we are not. If we are, then we have no moral obligations to our fellow members or to the community per se because, as understood from a modern scientific perspective, nature and natural phenomena are amoral. Wolves and alligators do no wrong in killing and eating deer and dogs (respectively). Elephants cannot be blamed for bulldozing acacia trees and generally wreaking havoc in their natural habitats. If human beings are natural beings, then human behavior, however destructive, is natural behavior and is as blameless, from a natural point of view, as any other behavioral phenomenon exhibited by other natural beings. On the other hand, we are moral beings, the implication seems clear, precisely to the extent that we are civilized, that we have removed ourselves from nature. We are more than natural beings; we are metanatural—not to say, "supernatural"—beings. But then our moral community is limited to only those beings who share our transcendence of nature, i.e., to human beings (and perhaps to pets who have joined our civilized community as surrogate persons) and to the human community. Hence, have it either way—we are members of the biotic community or we are not—a land or environmental ethic is aborted by either choice.

But nature is *not* amoral. The tacit assumption that we are deliberating, choice-making ethical beings only to the extent that we are metanatural, civilized beings, generates this dilemma. The biosocial analysis of human moral behavior, in which the land ethic is grounded, is designed precisely to show that in fact intelligent moral behavior *is* natural behavior. Hence, we are moral beings not in spite of, but in accordance with, nature. To the extent that nature has produced at least one ethical species, *Homo sapiens,* nature is not amoral.

Alligators, wolves, and elephants are not subject to reciprocal interspecies duties or land ethical obligations themselves because they are incapable of conceiving and/or assuming them. Alligators, as mostly solitary, entrepreneurial reptiles, have no apparent moral sentiments or social instincts whatever. And while wolves and elephants certainly do have social instincts and at least protomoral sentiments, as their social behavior amply indicates, their conception or imagination of community appears to be less culturally plastic than ours and less amenable to cognitive information. Thus, while we might regard them as ethical beings, they are not able, as we are, to form the concept of a universal biotic community, and hence conceive an all-inclusive, holistic land ethic.

The paradox of the land ethic, elaborately noticed by Fritzell, may be cast more generally still in more conventional philosophical terms: Is the land ethic prudential or deontological? Is the land ethic, in other words, a matter of enlightened (collective, human) self-interest, or does it genuinely admit nonhuman natural entities and nature as a whole to true moral standing?

The conceptual foundations of the land ethic, as I have here set them out, and much of Leopold's hortatory rhetoric, would certainly indicate that the land ethic is deontological (or duty oriented) rather than prudential. In the section significantly titled The Ecological Conscience, Leopold complains that the then-current conservation philosophy is inadequate because "it defines no right or wrong, assigns no obligation, calls for no sacrifice, implies no change in the current philosophy of values. In respect of land-use, it urges *only* enlightened self-interest" (207–8, emphasis added). Clearly, Leopold himself thinks that the land ethic goes beyond prudence. In this section he disparages mere "self-interest" two more times, and concludes that "obligations have no meaning without conscience, and the problem we face is the extension of the social conscience from people to land" (209).

In the next section, Substitutes for a Land Ethic, he mentions rights twice—the "biotic right" of birds to continuance and the absence of a right on the part of human special interest to exterminate predators.

Finally, the first sentences of The Outlook read: "It is inconceivable to me that an ethical relation to land can exist without love, respect, and admiration for land, and a high regard for its value. By value, I of course mean something far broader than mere economic value; I mean value in the philosophical sense" (223). By "value in the philosophical sense," Leopold can only mean what philosophers more technically call "intrinsic value" or "inherent worth."[36] Something that has intrinsic value or inherent worth is valuable in and of itself, not because of what it can do for us. "Obligation," "sacrifice," "conscience," "respect," the ascription of rights, and intrinsic value—all of these are consistently opposed to self-interest and seem to indicate decisively that the land ethic is of the deontological type.

Some philosophers, however, have seen it differently. Scott Lehmann, for example, writes,

> Although Leopold claims for communities of plants and animals a "right to continued existence," his argument is homocentric, appealing to the human stake in preservation. Basically it is an argument from enlightened self-interest, where the self in question is not an individual human being but humanity—present and future—as a whole. . . .[37]

Lehmann's claim has some merits, even though it flies in the face of Leopold's express commitments. Leopold does frequently lapse into the language of (collective, long-range, human) self-interest. Early on, for example, he remarks, "in human history, we have learned (I hope) that the conqueror role is eventually *self*-defeating" (204, emphasis added). And later, of the 95 percent of Wisconsin's species which cannot be "sold, fed, eaten, or otherwise put to economic use," Leopold reminds us that "these creatures are members of the biotic community, and if (as I believe) its stability depends on its integrity, they are entitled to continuance" (210). The implication is clear: the economic 5 percent cannot survive if a significant portion of the uneconomic 95 percent are extirpated; nor may *we,* it goes without saying, survive without these "resources."

Leopold, in fact, seems to be consciously aware of this moral paradox. Consistent with the biosocial foundations of his theory, he expresses it in sociobiological terms:

> An ethic may be regarded as a mode of guidance for meeting ecological situations so new or intricate, or involving such deferred reactions, that the path of social expediency is not discernible to the average individual. Animal instincts are modes of guidance for the individual in meeting such situations. Ethics are possibly a kind of community instinct-in-the-making. (203)

From an objective, descriptive sociobiological point of view, ethics evolve because they contribute to the inclusive fitness of their carriers (or, more reductively still, to the multiplication of their carriers' genes); they are expedient. However, the path of self-interest (or to the self-interest of the selfish gene) is not discernible to the participating individuals (nor, certainly, to their genes). Hence, ethics are grounded in instinctive feeling—love, sympathy, respect—not in self-conscious calculating intelligence. Somewhat like the paradox of hedonism—the notion that one cannot achieve happiness if one directly pursues happiness per se and not other things—one can only secure self-interest by putting the interests of others on a par with one's own (in this case long-range collective human self-interest and the interest of other forms of life and of the biotic community per se).

So, is the land ethic deontological or prudential, after all? It is both—self-consistently both—depending upon point of view. From the inside, from the lived, felt point of view of the community member with evolved moral sensibilities, it is deontological. It involves an affective-cognitive posture of genuine love, respect, admiration, obligation, self-sacrifice, conscience, duty, and the ascription of intrinsic value and biotic rights. From the outside, from the objective and analytic scientific point of view, it is prudential. "There is no other way for land to survive the impact of mechanized man," nor, therefore, for mechanized man to survive his own impact upon the land (viii).

Notes

1. Wallace Stegner, "The Legacy of Aldo Leopold"; Curt Meine, "Building 'The Land Ethic.'" The oft-repeated characterization of Leopold as a prophet appears traceable to Roberts Mann, "Aldo Leopold: Priest and Prophet," *American Forests* 60, no. 8 (August 1954): 23, 42–43; it was picked up, apparently, by Ernest Swift, "Aldo Leopold: Wisconsin's Conservationist Prophet," *Wisconsin Tales and Trails* 2, no. 2 (September 1961): 2–5; Roderick Nash institutionalized it in his chapter, "Aldo Leopold: Prophet," in *Wilderness and the American Mind* (New Haven: Yale University Press, 1967; revised edition, 1982).

2. Edward O. Wilson, *Sociobiology: The New Synthesis* (Cambridge: Harvard University Press, 1975), 3. See also W. D. Hamilton, "The Genetical Theory of Social Behavior," *Journal of Theoretical Biology* 7 (1964): 1–52.

3. Charles R. Darwin, *The Descent of Man and Selection in Relation to Sex* (New York: J. A. Hill and Company, 1904). The quoted phrase occurs on p. 97.

4. See Adam Smith, *Theory of the Moral Sentiments* (London and Edinburgh: A. Millar, A. Kinkaid, and J. Bell, 1759) and David Hume, *An Enquiry Concerning the Principles of Morals* (Oxford: The Clarendon Press, 1777; first published in 1751). Darwin cites

both works in the key fourth chapter of *Descent* (pp. 106 and 109, respectively).

5. Darwin, *Descent,* 98ff.

6. Ibid., 105f.

7. Ibid., 113ff.

8. Ibid., 105.

9. See, for example, Elman R. Service, *Primitive Social Organization: An Evolutionary Perspective* (New York: Random House, 1962).

10. See Marshall Sahlins, *Stone Age Economics* (Chicago: Aldine Atherton, 1972).

11. Darwin, *Descent,* 111.

12. Ibid., 117ff. The quoted phrase occurs on p. 118.

13. Ibid., 124.

14. See Donald Worster, *Nature's Economy: The Roots of Ecology* (San Francisco: Sierra Club Books, 1977).

15. Charles Elton, *Animal Ecology* (New York: Macmillan, 1927).

16. Aldo Leopold, *Round River* (New York: Oxford University Press, 1953), 148.

17. Kenneth Goodpaster, "On Being Morally Considerable," *Journal of Philosophy* 22 (1978): 308–25. Goodpaster wisely avoids the term *rights,* defined so strictly albeit so variously by philosophers, and used so loosely by nonphilosophers.

18. Kenneth Goodpaster, "From Egoism to Environmentalism" in *Ethics and Problems of the 21st Century,* ed. K. E. Goodpaster and K. M. Sayre (Notre Dame, Ind.: University of Notre Dame Press, 1979), 21–35.

19. See Immanuel Kant, *Foundations of the Metaphysics of Morals* (New York: Bobbs-Merrill, 1959; first published in 1785); and Jeremy Bentham, *An Introduction to the Principles of Morals and Legislation,* new edition (Oxford: The Clarendon Press, 1823).

20. Goodpaster, "Egoism to Environmentalism." Actually Goodpaster regards Hume and Kant as the cofountainheads of this sort of moral philosophy. But Hume does not reason in this way. For Hume, the other-oriented sentiments are as primitive as self-love.

21. See Peter Singer, *Animal Liberation: A New Ethics for Our Treatment of Animals* (New York: Avon Books, 1975) for animal liberation; and see Tom Regan, *All That Dwell Therein: Animal Rights and Environmental Ethics* (Berkeley: University of California Press, 1982) for animal rights.

22. See Albert Schweitzer, *Philosophy of Civilization: Civilization and Ethics,* trans. John Naish (London: A. & C. Black, 1923). For a fuller discussion see J. Baird Callicott, "On the Intrinsic Value of Nonhuman Species," in *The Preservation of Species,* ed. Bryan Norton (Princeton: Princeton University Press, 1986), 138–72.

23. Peter Singer and Tom Regan are both proud of this circumstance and consider it a virtue. See Peter Singer, "Not for Humans Only: The Place of Nonhumans in Environmental Issues" in *Ethics and Problems of the 21st Century,* 191–206; and Tom Regan, "Ethical Vegetarianism and Commercial Animal Farming" in *Contemporary Moral Problems,* ed. James E. White (St. Paul, Minn.: West Publishing Co., 1985), 279–94.

24. See J. Baird Callicott, "Hume's Is/Ought Dichotomy and the Relation of Ecology of Leopold's Land Ethic," *Environmental Ethics* 4 (1982): 163–74, and "Non-anthropocentric Value Theory and Environmental Ethics," *American Philosophical Quarterly* 21 (1984): 299–309, for an elaboration.

25. Hume, *Enquiry,* 219.

26. Darwin, *Descent,* 120.

27. I have elsewhere argued that "value in the philosophical sense" means "intrinsic" or "inherent" value. See J. Baird Callicott, "The Philosophical Value of Wildlife," in *Valuing of Wildlife: Economic and Social Values of Wildlife,* ed. Daniel J. Decker and Gary Goff (Boulder, Col.: Westview Press, 1986), 214–221.

28. See Worster, *Nature's Economy.*

29. See J. Baird Callicott, "The Metaphysical Implications of Ecology," *Environmental Ethics* 8 (1986): 300–315, for an elaboration of this point.

30. Robert P. McIntosh, *The Background of Ecology: Concept and Theory* (Cambridge: Cambridge University Press, 1985).

31. Aldo Leopold, "Some Fundamentals of Conservation in the Southwest," *Environmental Ethics* 1 (1979): 139–40, emphasis added.

32. I borrow the term "devolution" from Austin Meredith, "Devolution," *Journal of Theoretical Biology* 96 (1982): 49–65.

33. Holmes Rolston, III, "Duties to Endangered Species," *Bioscience* 35 (1985): 718–26. See also Geerat Vermeij, "The Biology of Human-Caused Extinction," in Norton, *Preservation of Species,* 28–49.

34. See D. M. Raup and J. J. Sepkoski, Jr., "Mass Extinctions in the Marine Fossil Record," *Science* 215 (1982): 1501–3.

35. Peter Fritzell, "The Conflicts of Ecological Conscience," in *Companion to A Sand County Almanac,* edited by J. Baird Callicott (Madison: University of Wisconsin Press, 1987).

36. See Worster, *Nature's Economy.*

37. Scott Lehmann, "Do Wildernesses Have Rights?" *Environmental Ethics* 3 (1981): 131.

For Further Discussion

1. Callicott speaks of a biological account of ethics, suggesting that ethics may have contributed to the ability of humans to adapt and survive. Does this sound plausible to you? What implications are drawn from this concerning our relationship to land?

2. Callicott tells us that the land ethic has both an individualistic and a holistic cast. What do these concepts mean in this context? What is a holistic ethic?

3. What, according to Callicott, is the criterion of moral value and consideration? How does this differ from the views of Feinberg, Singer, Regan, and Stone?

4. Callicott claims that "the maintenance of the complex structure of the land and its smooth function-

ing as an energy unit emerges as the *summum bonum* of the land ethic." How do you interpret what Callicott is saying?

5. Callicott concludes that the land ethic is both "deontological" and "prudential." Explain what he means by this. Do you agree with this conclusion?

Environmental Holism and the Individual

Don E. Marietta, Jr.

Philosopher Don Marietta examines a variety of charges against holistic environmentalism. Marietta pays close attention to the charge that holistic approaches are totalitarian or fascist in subordinating the good of the individual to the good of the whole. He concludes that although some versions of holism might be susceptible to such charges, more moderate versions can be made consistent with the demands of social justice.

CONCERNS ABOUT HOLISTIC ENVIRONMENTALISM

. . . A number of other philosophers have espoused a philosophy that plays down the significance of humans in favor of nonhuman animals, or of all living things, or of the natural environment as a whole. Various names, such as ecological holism, deep ecology, biocentrism, and ecocentrism, indicate different but somewhat similar approaches to environmental philosophy, at least similar in that they all seem to deny a special place for humans in the scheme of things. Holistic philosophies, which incorporate in various ways the concept of the moral equality of humans and other species, are growing in acceptance by environmental ethicists and scientists and others with an interest in the environment. Biologists, especially, are acutely aware of the dependence of human life on the natural environment, and many of them find in nature precious but threatened values. Because humans cannot

live well or very long without a healthy ecosystem, treatment of the environment is clearly seen as a matter of moral significance; for some this is a moral priority. Many people feel a moral commitment to the natural environment that transcends responsibility to provide for human biological need.

This acceptance of holism is not universal, however; some moral philosophers have serious reservations about the consequences of practicing this new approach to ethics. . . .

Some ethicists consider holism a threat to the rights of individual human beings or to individual animals. William Aiken found the practical implications of holism "astounding, staggering," because of the loss of individual rights. Tom Regan claims that "what holism gives us is a fascist understanding of the environment." Marti Kheel called ecological holism "totalitarian." [1]

Eric Katz writes about "the substitution problem" which is a result of failure on the part of environmental holists to understand the difference between thinking of the natural environment as an organism and as a community. If nature is understood in terms of an organismic model, the individual loses significance; one individual can be substituted for another if it makes no difference to the whole environment. This might happen when one species of animal can fill a biological niche and replace another species. Katz is disturbed by the possibility of this kind of substitution occurring when a matter of moral significance is at issue. To Katz, it appears that individuals are being valued for their instrumental roles in nature rather than for their intrinsic worth.[2] If the possibility of the substitution of one nonhuman animal for another or one nonhuman species for another is disturbing, certainly the thought of substituting one human for another as pawns in a natural system is horrifying. It will not be morally acceptable to most people to take into account only the role of humans as members of an ecosystem.

A number of feminists think that holistic environmentalism has failed to realize the historical connection between abuse of the environment and the denial of women's rights. Some fear that holism will actually encourage the continuation of the subjugation of women. Sara Ann Ketchum brought to my attention the fear that holism might tend to deny freedom in reproduction, which would be a special concern to women, who might be forced to have children, because both the bearing of children and the caring for children fall most heavily on them. . . .

The concern of supporters of animal rights, such as Tom Regan, is that individual animals will be sacrificed to the good of the whole natural system. Holistic environmentalists have shown more interest in preserving animals in their natural role in natural habitats than they have shown in individual animals, which are the primary concern of animal advocates. Animal rights advocates talk about the pain animals are made to suffer and their loss of freedom. Holists talk about niches in natural environments; they do not express much concern for pet animals and farm animals. The holists do not show the same interest in conscious animals that can feel pain as do the animal activists. For the holists, not only are the animals with consciousness important, but also the lowliest decomposers in a food chain because the natural system needs the activity of worms and insects, even of microorganisms. The species that are most threatened are of most concern. The reason for this concern is not pain, which is a necessary part of natural systems in which every species eats and is eaten; the concern is the preservation of a dynamic balance within the natural system. The attractive, conscious animals that tend to look a bit like people are not favored more than the unattractive and even the inconspicuous forms of life. . . . Environmental holists and defenders of animal rights have different approaches to such things as hunting and hospitals for wild animals. . . .

These reservations about holistic environmentalism are serious. Are they well founded? Are such dangers real? Is holism a genuine threat to individual rights and recognition of the worth of persons? . . .

DIFFERENT KINDS OF HOLISM

My defense of holism will require a clarification of environmental holism, which will show that there are several types of holism, and I will further try to show that the types that could lead to antihumanistic practices are not acceptable on logical grounds, not simply because they are morally offensive to humanistic thinkers. There are several ways in which the ethical aspects of holism can be presented, and holistic duties can be presented in moderate or in extreme versions. Most of the literature about holism has dealt with it as though it were a rather simple doctrine. . . . The distinctions I am making are relevant to that debate, but they approach holism from a different angle and are more inclusive.

First let us consider holism as a scientific doctrine, which I refer to as scientific holism. Scientific holism is based on those findings of ecology that show the interrelatedness of organisms in natural systems and the connections between those systems. . . . The concept of nature as a vast economy, with different organisms making their own contributions to the whole system of nature, doing their own work, has been a basic theme of biology since the time of Linnaeus. Living things are related by trophic webs or food chains. . . . Within these webs, organisms of various species play roles of production, consumption, and decomposition. There have been several interpretations of this theme and disagreement about a number of points.[3] The basic concept, however, has been an important aspect of ecological science, which has explored in detail the interdependence of living things on each other and on the nonliving parts of the biosphere. The concept of everything being connected to everything else is a popular idea with environmentalists. Much environmental education makes this a basic theme. The interrelatedness of organisms is often expressed in figures of speech using social analogies such as the family relationship or kinship or citizenship. The concept has been expressed in the poetry of Gary Snyder, Robinson Jeffers, and Francis Thompson. . . .

It is important to see that not all versions of environmental holism that claim to be based on ecological science have the same relationship to scientific studies, even when we consider only those ecological researches that study ecosystems. Scientific holism can be a carefully stated doctrine that claims only what biological research has actually discovered about the relationships between specified plant and animal species and between them and their environments. It can give an understanding of specific ecosystems by showing how various parts of the system are related to other parts of the system. . . .

Out of this study of positively identified relationships, holism as a general notion is developed; it then becomes a conceptual model, a very general guiding principle that structures inquiry, determines what is significant, and gives a framework around which the science is systematically organized. As a conceptual model or scientific paradigm, holism can provide an overall way of seeing the world for environmental education. By pedagogical use of this model, various bits of information are made to fit together and biological knowledge becomes more striking and understandable.

The relatedness of organisms can be stated in an exaggerated way, as in the poem by Francis Thompson that says "thou canst not stir a flower/Without troubling of a star." When an environmental educator says "every living thing is dependent upon every other living thing," it should be understood that this ambiguous statement cannot be a claim that each thing is known to be dependent upon each and every other thing, as Thompson's star upon the plucked flower. The statement has been generalized beyond what research has actually demonstrated. Philosophers can also make exaggerated holistic statements; in fact Paul W. Taylor admits to having overstated the idea of dependency in his paper, "Ethics of Respect for Nature."[4]

Exaggeration is not the only fault we must be careful to avoid. The form a scientific holism takes must be attuned to the best understanding of ecology available at the time. We must not go ahead as though the theories of Frederic L. Clements and Eugene P. Odum were universally accepted ecological science. We must recognize the effect of chaos theory, patch dynamics, hierarchy theory, and other new approaches to ecology. We cannot escape the complexity and the developing nature of ecology. Environmental philosophers are concerned about the effect of such new concepts as hierarchy theories on ethical and metaphysical doctrines, as well they should be.[5] A philosophical definition of holism must be open to development as ecological science develops. It must avoid becoming rigid and dogmatic. I believe that hierarchy theory shows that there are several heuristic ways to look at natural systems. There may well be several ways conceptually to model the roles of humans in natural systems. I do not see this as indicating that holistic per-

spectives on humans and nature are to be avoided, but it does warn us against exaggerating our statements of the oneness of humans and nature, and it warns us against proclaiming a particular version of holism as *the* scientific model of humans in nature. . . .

Holism can also be stated in terms of what is valuable. I will call this axiological holism. This type of holism ascribes value on the basis of usefulness to the biosphere. An organism plays a role in the ecosystem, and that is what makes the organism valuable. Anything that in turn is useful to the useful organism is also valuable to the natural order. Axiological holism appears in a number of papers, such as J. Baird Callicott's 1979 paper on animal liberation.[6]

Axiological holism can be expressed in various degrees, from a modest claim that one source of value, among others, is the contribution made to the biosphere to an extreme claim that contribution to the biosphere is the only significant value. Between these two expressions of axiological holism is the strong, but more moderate, claim that contribution to the biosphere is the main basis for weighing moral obligations. This claim would imply that an important contribution to the ecosystem could be an overriding consideration in ethics. Written expressions of axiological holism have not been sufficiently explicit to determine exactly which version of holism was intended. There may be no defender of the most extreme version at this time. Statements that might be read as extreme axiological holism, when read out of context, do not seem to intend such an extreme claim when seen in relation to other statements of the writer. Some statements in literary works, especially poems and essays quoted with approval by some holistic philosophers, might imply an extreme concept of value, but they are usually not expressed in specific terms of value.

Environmental holism can be presented in terms of moral duty. I will call this deontic holism. This approach to holism claims that there is a moral duty to be a responsible member of the biospherical community. The moral standing of nonhuman entities also comes from membership in the biotic community. A modest approach to deontic holism would see duty to the ecosystem as one source of moral obligation, among others. Treating duty to the biotic community as primary and overriding would be a stronger kind of holism. To treat duty to the ecosphere as the only significant moral obligation would be the most extreme version. I am not sure any ethicist at this time accepts an extreme approach that would reject the concerns of humanistic ethics. Such strong defenders of holistic environmentalism as Devall, Taylor, and Sessions do not seem to adopt the severe positions that are theoretically possible; even though some of their statements could be read as expressions of an extreme approach, other statements clearly indicate a more moderate view. Callicott, with his recognition of different levels of moral obligation, clearly does not think that the land ethic calls for an extreme approach.

Holistic writers seem to accept axiological and deontic claims as mutually supportive insights. At this early stage in the development of environmental ethics, the use of both kinds of holism does not seem very systematic, but some work has been done in explicating the connection between axiological and deontic aspects of ethics.

Peter Miller has a theory of value as richness that he relates to ethical obligation, but he does not give much explanation of the way richness relates to obligation.[7]

J. Baird Callicott has written several papers on value theory. He is not very explicit about the way values are related to ethics, but it seems clear that he sees the connection in terms of his ethical theory, which builds on David Hume, Charles Darwin, and sociobiological theory. It is a moral-sense approach that stresses the value to humans of social sentiments. The benevolent and social sentiments can be directly related to the valuing of the good of other persons and the natural world.[8]

Holmes Rolston, III has written extensively on values in nature. His main concern is the nature of values, and he does an extensive taxonomy of value, including both levels of value and types of value. He deals with the effect of values on ethics. Rolston suggests a view on the way value judgments affect one's sense of moral obligation. He speaks of value in nature generating duty toward nature. He says that "we follow what we love, and the love of an intrinsic value is always a moral relationship."[9]

I think that my treatment of value and the basis of duty, based on the effect of a person's individual world view on ethical beliefs, along with the volitional element in the person's constitution of the world, helps explain the mechanism by which axiological and deontic claims are connected. . . .

ANTIHUMANISTIC HOLISM

The different kinds of holism I have mentioned are some possible versions of the doctrine. I am not claiming that each of the possible positions I have indicated has had an advocate. I am interested in them as views that have been supported or that might be defended in the future. They provide a sort of spectrum that will help make the whole picture clearer. My main concern is assessing the possible positions to determine which of the deontic and axiological approaches are sound and which are not.

An examination of the spectrum of holistic positions indicates that what the humanistic critic fears is the extreme versions of deontic and axiological holism. One approach to holism that threatens humanistic ethics is the extreme deontic claim that duty to the biosphere overrides all other duties. The claim that all duty can be reduced to responsibility to the planet is even more threatening because it denies much that we value in human life. The other threatening approach to holism is the extreme axiological approach, which denies any source of value except contribution to the stability of the ecosystem. This denies humanistic and personalistic values by reducing human beings to their ecological role. Is the critic correct in fearing that these threatening forms of holism are the logical outcome of holistic ethics?

I do not believe that the extreme versions are the logical outcome of environmental holism. My reason is a logical one. In the first place, the scientific evidence does not fully justify these extreme versions. They may seem at first to be based on ecological science, but they are not. I am not basing this on the impossibility of deducing moral obligations from factual knowledge; I support an approach to ethical theory that holds that a deductive justification of moral claims is not necessary. What is lacking in the extreme versions of holism is adequate justification for claiming a connection between the factual knowledge provided by ecology and the reductive and exclusivist claims made by the extreme versions of holism. The extreme claims lack the fittingness that is the necessary link between ecology and ethics. . . . The lack of fittingness is related to two logical flaws of the extreme versions: they are extremely reductionistic, in that they reduce the significance of human life to involvement in the biosphere, and abstract, in that they do not rest on specific concrete and observable relationships and interdependencies. As I will explain, this is not just a matter of distaste for abstraction and this type of reduction. These are logical flaws in the extreme forms of holism. There are flaws of several types. Abstraction is flawed because its claims are not supported by evidence; it is theoretical formulation that has run too far ahead of its factual base. Reduction is flawed because it simplifies things by ignoring matters that are evident and that have been held important by generations of responsible thinkers.

The extreme forms of holism are logically flawed by a lack of internal consistency; in their attempt to support holism, they are contrary to holism. It is not holism to ignore all values except the value of an organism to the ecosystem, as the extreme axiological version does. It is not holism to reduce all obligation to the one obligation to promote the stability of the natural environment, as the extreme deontic version does. Holism must be inclusive to be whole. Holism simply cannot be so reductionistic as to leave important matters out of consideration. Holism cannot be satisfied with a partial picture of humanity.

THE LOGICAL CONCLUSION

It is easy to see how a person who has read the writings of leading holistic ethicists might assume that an extreme approach to holism would be the correct position. There is implicit in environmental holism a turn away from a position that rests on a narrow concept of human interests. Callicott wrote of a connection between holism and misanthropy. Several representatives of the deep ecology perspective have cited literary works that express strong antipathy toward human beings. Callicott cites as an example of misanthropy a statement in *Desert Solitaire* by Edward Abbey, who said he would sooner shoot a man than a snake.[10] The poet Robinson Jeffers explains that he wrote the first part of *The Double Axe* "to present a certain philosophical attitude, which might well be called Inhumanism, a shifting of emphasis and significance from man to not-man."[11]

. . . Strong statements of devotion to the environment and nonhuman animals is understandable when we see the harm people have done and continue to do to the system of nature. We have good reason to fear for the ability of the planet to support the forms of life we have come to love in the face of human destruction. Anger and dismay are understandable emotional responses to the situation. This is not the same, however, as an intellectually sound response. I do not believe that logical consistency will lead holists in the future to abandon the humanistic and personalistic aspects of ethics. Logical consistency will not force a rejection of concern for civil rights, respect for individual persons, and a recognition of the duties we have

to our families, friends, and the several communities of which we are a part. Environmental ethicists in future generations might adopt extreme protective measures, but it will not be logical consistency with holism that brings this about. The severity of environmental problems might force the acceptance of draconian measures, but a properly understood holism will not be the basis of this. Even nonholistic moralists might turn to extreme measures in the face of great threats to human survival. Not holism, but fear and desperation, would account for these actions.

We need to see clearly when extreme types of holism are logically unacceptable. A careful look at the extreme forms of holism reveals that they are extremely reductionistic, reducing human life to the barest aspects of relationship to the natural environment. These extreme forms also tend to be so abstract that they do not speak of the actual entities that we experience in our lived worlds and that our sciences study. Such an extreme holism would substitute for the richness of our lives in the world a world of abstract relationships and incomplete images of persons, animals, and plants.

When environmental holism makes the system of nature and the individuals within it into abstractions, it has gone away from the ecological science it claims as its basis. It has lost sight of the actual individual plants and animals that compose the system. In place of the relationships between actual organisms and the land community in which they live, we now have an abstract notion of interdependence. Symbols have replaced substances, and the study of nature is in danger of losing the control of its experiential grounding. An adequate ethical system cannot be informed by such an abstract conception of nature, one in which the parts of the natural system become hazily perceived symbolic ciphers. The individual person whose interests we would consider sacrificing for the common good is without individuality or personal character. Under these conditions of thought, the substitution problem that worries Eric Katz would be almost inevitable.[12]

Extreme abstraction goes hand in hand with extreme reduction. When the only value accorded to the individual person is that which comes from serving a useful function in the ecosystem, the person has significance only as a part of a greater whole, and only a part of the individual person is treated as the whole person. The rest of the person has been disallowed; the person has been reduced. When the only significant behavior of the individual person is that which makes a contribution to the whole, the ecosystem, the same reduction takes place. This kind of abstraction and reduction reduces each individual person to those aspects of being human that we all have in common. When all persons are alike, what reason is there for not substituting one for another?

What is wrong with a reductionist approach? Why is it not just one alternative among others, the right approach for those who prefer it? I hold that extreme abstraction and reductionism are not only unacceptable to our present state of moral sensibility but also unjustifiable assumptions. They become meaningless by losing touch with our knowledge and experience. The reduction and abstractness are serious faults that prevent the extreme versions from being acceptable on logical grounds. It is not a matter of one's preference. These extreme versions are too seriously flawed to be the basis of an adequate ethic.

This sort of reduction has the same faults we have seen in other reductionistic views, such as defining people in terms of sex, religion, or position in society. It is like the mistake of thinking of humans only in terms of rationality or in terms of the person's relation to a deity and ignoring all of the other aspects of what it is to be human. The abstract reductionism of extreme holism ignores many things that are morally significant, but that is only part of its failure. It ignores vast areas of human life and experience of the world. Its understanding of human beings is so incomplete that it cannot be correct. We are biological organisms who are part of an ecosystem, but to define us only in biological terms is to give an inadequate understanding of what we are. We are not fully defined unless social, cognitive, psychological, moral, and aesthetic aspects of our existence are taken fully into account. It is possible that all these aspects of human life have a biological basis, but this does not justify defining human beings in the most narrow biological concepts, those that can be applied to any other member of the biotic community. That people are biological beings must broaden biology, not narrow our concept of people.

Reductionistic views are not acceptable as alternative versions of environmental holism. They must be rejected. This is not a matter of preference. An adequate world view must be scientific and inclusive of our experience of our common world. The extreme versions of holism are so reductionistic that they are the antithesis of holism. They do not really consider the whole because they leave out elements of vital importance. They do not recognize actualities of human life. Any adequate holism must recognize all the important biological, individual, and social aspects of being human. Reductionistic holism is not scientific, complete, or factually adequate because it attempts to treat some things as being what they are not and fails to recognize them for what they are.

The logical inadequacy of extreme versions of holism is paralleled by another kind of inadequacy, moral inadequacy. An ethical approach that does not take into account all factors that might be morally relevant is inadequate. It is marred by the incompleteness that comes with reductionistic and overly abstract ethical systems. In order to deal with all that belongs in the determination of an ethical issue, moral thinking must be grounded in an adequate world view. . . . An adequate world view must incorporate an understanding of humanness in its fullness. Every aspect of what is involved in being human must be considered. All of the qualities of human life must be recognized and their moral significance weighed.

Human vulnerabilities as well as human capacities must be recognized. The human susceptibility to pain is one of those features of human life that has great moral relevance. It is ethically significant that human beings can feel many different types of pain, psychological as well as physical. Humans can suffer disappointment, grief, loneliness, a sense of rejection or of being scorned, a feeling of having been treated unjustly, and many kinds of frustration. We do not know with certainty what capacity other animals have for feeling these many types of pain, but to whatever degree these organisms have a capacity for pain, it is morally relevant in determining ethical treatment of animals. To dismiss this aspect of ethics as placing too much importance on self-consciousness, claiming that it is a kind of chauvinism based on a human characteristic, is a mistake. One need not, and I think should not, make self-consciousness the only matter of ethical significance; it is, however, something that must not be left out of ethical consideration. Utilitarians may be in error in making the experience of pleasure and pain the only guide to ethical decisions, but no sound ethical system can deny the moral relevance of pain.

It is ethically significant that humans are rational and self-conscious beings. Some ethicists seem to treat this as a matter of little or no importance, rejecting attention to human self-consciousness as a kind of "speciesism." This happens largely in reaction to an equally inadequate approach to ethics that gives this aspect of humanness too great a role, even seeing it as an absolute distinction between humans and all other forms of life and making it a requirement for being morally considerable. To hold that humans do not have duties to beings that do not share this human quality is a reductionism that ignores the moral significance of a number of morally relevant matters, such as the capacity to feel pain, being alive, and being a part of an ecosystem. It is not sound moral thinking, however, to substitute one reductionist approach for another. Self-awareness is morally significant, but not the only matter of significance. Exactly how important it is in relation to other considerations needs to be studied carefully in each case.

As we criticize traditional ethical systems, we should remember that they were developed in response to certain common needs of human individuals and of human societies. This is one reason for our being critical of them, because our ancestors may have had a faulty understanding of some aspects of human life, but it is also a reason why we must be very careful not to discard them carelessly; they were responses to real concerns that might be important still. To the traditional concerns about human personal and social life, we must add some broader concerns as we seek to discover the morally right responses to our impact on the natural environment.

As we take account of our role in the system of nature, we do not need to ignore the concerns of humanistic ethics. It would be an intellectual and moral mistake to reduce our understanding of ourselves to that of organisms in an ecosystem. We do have a large impact on our environment, and this is a matter of moral concern, but we should not reduce our range of ethical concerns to environmental concerns. It would be a mistake to ignore concerns about the natural environment and a mistake to reduce all of morality to extreme holistic environmental ethics. We must acknowledge the many levels of moral obligation, from family responsibilities and obligations to close associates to duties to distant peoples, future generations, and the biosphere. . . .

Trying to take into account everything that is morally relevant forces us to face complexity, conflicts, and confusions, but there is no justifiable alternative to hard study and the making of hard choices. We can expect our traditional ethical concerns to conflict with our responsibility toward the natural environment, but there is no morally acceptable way to escape difficult decisions. We cannot justify making our decisions easier by being reductionistic. We must bring together concerns about the environment, concerns about liberty and justice, and concerns about the quality of human life. In doing this we must maintain a rational consistency, recognizing that what applies to humans will apply to many other animals. . . .

A nonreductionistic and concrete holistic approach seeks to see every part of the biosphere clearly and completely. Ecological science and other natural and social sciences must be employed to provide as adequate a picture as possible of organisms, species, and the nonliving parts of the environment. A concept of the whole biosphere must incorporate knowledge of the individual members. This use of science should not be a return to outmoded concepts of science or to questionable metaphysics. If we talk about knowing things as they really are, we must keep in mind that we are talking loosely, in respect to both science and metaphysics. Knowing that all knowledge must be grasped within a conceptual model must keep us from thinking that we can claim to have knowledge that is not colored by our world views. The concept of "objective" knowledge of "brute facts" lost its credibility and usefulness long ago. We can, however, avoid excessive abstraction and reduction in the way we think about the members of the biosphere.

A nonreductive holism requires a responsibly scientific basis in order to have an ethical wholeness, one that does not set aside any part of a person's moral responsibility. . . .

Will future generations of moral philosophers abandon the humanistic ethics that have guided humanity for many centuries? They might do so. Predicting the thoughts and actions of future generations of moral philosophers is not something I am competent to do, nor is it the concern of this book. We do not know what sorts of situations they will face. We might have some idea of what the environmental problems will be, but we do not know with any certainty how severe some of the familiar problems will be or what new threats may arise. There is much we do not know about future social and political situations, and we do not know what ideologies will be shaping human thought when generations to come must make important decisions. It may be necessary for them to take draconian measures to secure the survival of life on Earth. The point I wish to make is that if moral thinkers adopt ethical principles we would now consider unnecessary and even misanthropic, it will not be the result of sound, nonreductionistic holism.

This does not mean that we can wash our hands of all concern about the future. Draconian measures might result from our failure to take moderate measures today, such as reasonable action to control human population. When facing a crisis that threatens the existence of humanity, moral thinkers might sacrifice many of the humanistic values that are important to us now. This could happen even if we do not adopt a holistic moral perspective; even with a philosophy of nature that does not acknowledge that humans are citizens and members of the biosphere, future generations could take strong measures to preserve the human race. My point is that these harsh actions would be their response to an environmental crisis; the abandonment of humanistic ethics would not follow an inexorable logic of holism.

I advocate a humanistic holism, on the grounds that it is logically sounder than abstract and reductionistic versions of holism and that it is morally superior in that it takes into account important matters that should be considered in making moral judgments. It will not do to reject this notion of humanistic holism as an oxymoron. As amusing as they might be as word play, oxymorons do not signal logical impossibility. On a serious level of thought, we can see that the concept of holism and the concept of humanism modify each other when used in one phrase, indicating what concept of humanism and what concept of holism are being used.

A humanistic holism, which avoids the reductionism that ignores important aspects of our lives and the abstractness that has lost touch with the actualities of people and the natural system in which they live, is the intellectually and morally responsible approach to our lives and to the world. *Person/Planet,* the title of a book by Theodore Roszak, gives us a clear word picture of the goal of our moral endeavor. A person-planetary approach will preserve the evolving ethical standards by which we have guided our lives over centuries. At the same time, we will be able to bring into our ethical perspective the findings of ecological science and recognize ourselves as inhabitors of the planet. We need a moral perspective that will enable us to preserve life on the planet and to produce an enduring human society. If we avoid abstractionism, which loses sight of the concrete relationships that make up our biological lives on Earth, and reductionism, which loses sight of what we are as persons, we can act responsibly toward the planet and toward persons.

Notes

1. William Aiken, "Ethical Issues in Agriculture," in *Earthbound,* ed. Tom Regan (New York: Random House, 1984), 269; Tom Regan, *The Case for Animal Rights* (Berkeley: University of California Press, 1983), 372; Marti Kheel, "The Liberation of Nature: A Circular Affair," *Environmental Ethics* 7.2 (Summer 1985): 135.

2. Eric Katz, "Organicism, Community, and the 'Substitution Problem,'" *Environmental Ethics* 7.3 (Fall 1985): 241.

3. For a good and readable overview of the development of ecology, see Donald Worster, *Nature's Economy* (Garden City, NY: Anchor, 1979).

4. Paul W. Taylor, "In Defense of Biocentrism," *Environmental Ethics* 5.3 (Fall 1983): 239.

5. See, for example, Donald Worster, "The Ecology of Order and Chaos," *Environmental Review* 14.1–2 (1990): 1; Karen J. Warren and Jim Cheney, "Ecosystem Ecology and Metaphysical Ecology: A Case Study," *Environmental Ethics* 15.2 (Summer 1993): 99; and J. Baird Callicott, "On Warren and Cheney's Critique of Callicott's Ecological Metaphysics," *Environmental Ethics* 15.4 (Winter 1993): 373.

6. J. Baird Callicott, "Animal Liberation: A Triangular Affair," *Environmental Ethics* 2.4 (Winter 1980): 324.

7. Peter Miller, "Value as Richness: Toward a Value Theory for the Expanded Naturalism in Environmental Ethics," *Environmental Ethics* 4.2 (Summer 1982): 101.

8. J. Baird Callicott, "On the Intrinsic Value of Nonhuman Species," in *The Preservation of Species: The Value of Biological Diversity,* ed. Bryan G. Norton (Princeton, NJ: Princeton University Press, 1986), 138–72; "Hume's *Is/Ought* Dichotomy and the Relation of Ecology to Leopold's Land Ethic," *Environmental Ethics* 4.2 (Summer 1982): 163; and "Intrinsic Value, Quantum Theory, and Environmental Ethics," *Environmental Ethics* 7.3 (Fall 1985): 257.

9. Holmes Rolston III, "Values in Nature," *Environmental Ethics* 3.2 (Summer 1981): 113; "Are Values in Nature Subjective or Objective?" *Environmental Ethics* 4.2 (Summer 1982): 125; "Values Gone Wild," *Inquiry* 23 (1983): 181; "Valuing Wildlands," *Environmental Ethics* 7 (Spring 1985): 23; and *Environmental Ethics: Duties to and Values in the Natural World* (Philadelphia: Temple University Press, 1988), 32, 41.

10. Callicott, "Animal Liberation," 326.

11. Robinson Jeffers, *The Double Axe and Other Poems* (New York: Liveright, 1977), xxi.

12. Katz, "Organicism," 241.

For Further Discussion

1. Outline the philosophical challenges to holism discussed at the beginning of this essay. Which do you consider most troubling?

2. Some critics charge that holistic environmentalism is fascist because it sacrifices the individual rights

for the good of the whole. Are we ever ethically justified in sacrificing the rights of individuals for the good of the whole community?

3. What does Marietta mean by "scientific holism"? What does he mean by "axiological holism"? What relationships exist between these two ideas?

4. Why does Marietta believe that reductionistic views are not acceptable alternatives to holism? Do you agree with his reasoning?

5. Explain the concept of humanistic holism. How does this version answer the charge of environmental fascism?

The Ecology of Order and Chaos

Donald Worster

In this essay, historian Donald Worster examines the development of the science of ecology with an eye toward the ethical implications drawn from this science. He points out that early models from ecology supported a view of ecosystems as stable and harmonious. For many people, this leads to specific policy conclusions concerning human responsibility to the natural world. However, Worster points out that in recent decades a more chaotic picture has emerged from ecology. Worster seems to doubt that any particular ethical conclusions can be drawn from the science of ecology.

The science of ecology has had a popular impact unlike that of any other academic field of research. Consider the extraordinary ubiquity of the word itself: it has appeared in the most everyday places and the most astonishing, on day-glo T-shirts, in corporate advertising, and on bridge abutments. It has changed the language of politics and philosophy—springing up in a number of countries are political groups that are self-identified as "Ecology Parties." Yet who ever proposed forming a political party named after comparative linguistics or advanced paleontology? On several continents we have a philosophical movement termed "Deep Ecology," but nowhere has anyone announced a movement for "Deep Entomology" or "Deep Polish Literature." Why has this funny little word, ecology, coined by an obscure 19th-century German scientist, acquired so powerful a cultural resonance, so widespread a following?

Behind the persistent enthusiasm for ecology, I believe, lies the hope that this science can offer a great deal more than a pile of data. It is supposed to offer a pathway to a kind of moral enlightenment that we can call, for the purposes of simplicity, "conservation." The expectation did not originate with the public but first appeared among eminent scientists within the field. For instance, in his 1935 book *Deserts on the March*, the noted University of Oklahoma, and later Yale, botanist Paul Sears urged Americans to take ecology seriously, promoting it in their universities and making it part of their governing process. "In Great Britain," he pointed out,

> the ecologists are being consulted at every step in planning the proper utilization of those parts of the Empire not yet settled, thus . . . ending the era of haphazard exploitation. There are hopeful, but all too few signs that our own national government realizes the part which ecology must play in a permanent program.[1]

Sears recommended that the United States hire a few thousand ecologists at the county level to advise citizens on questions of land use and thereby bring an end to environmental degradation; such a brigade, he thought, would put the whole nation on a biologically and economically sustainable basis.

In a 1947 addendum to his text, Sears added that ecologists, acting in the public interest, would instill in the American mind that "body of knowledge," that "point of view, which peculiarly implies all that is meant by conservation."[2] In other words, by the time of the 1930s and 40s, ecology was being hailed as a much needed guide to a future motivated by an ethic of conservation. And conservation for Sears meant restoring the biological order, maintaining the health of the land and thereby the well-being of the nation, pursuing by both moral and technical means a lasting equilibrium with nature.

While we have not taken to heart all Sears's suggestions—have not yet put any ecologists on county payrolls, with an office next door to the tax collector and sheriff—we have taken a surprisingly long step in his direction. Every day in some part of the nation, an ecologist is at work writing an environmental impact report or monitoring a human disturbance of the landscape or testifying at a hearing.

Twelve years ago I published a history, going back to the 18th century, of this scientific discipline and its ideas about nature.[3] The conclusions in that book still strike me as being, on the whole, sensible and valid: that this science has come to be a major influence on our perception of nature in modern times; that its ideas, on the other hand, have been reflections of ourselves as much as objective apprehensions of nature; that scientific analysis cannot take the place of moral reasoning; that science, including the science of ecology, promotes, at least in some of its manifestations, a few of our darker ambitions toward nature and therefore itself needs to be morally examined and critiqued from time to time. Ecology, I argued, should never be taken as an all-wise, always trustworthy guide. We must be willing to challenge this authority, and indeed challenge the authority of science in general; not be quick to scorn or vilify or behead, but simply, now and then, to question.

During the period since my book was published, there has accumulated a considerable body of new thinking and new research in ecology. In this essay I mean to survey some of that recent thinking, contrasting it with its predecessors, and to raise a few of the same questions I did before. Part of my argument will be that Paul Sears would be astonished, and perhaps dismayed, to hear the kind of advice that ecological experts have to give these days. Less and less do they offer, or even promise to offer, what he would consider to be a program of moral enlightenment—of "conservation" in the sense of a restored equilibrium between humans and nature.

There is a clear reason for that outcome, I will argue, and it has to do with drastic changes in the ideas that ecologists hold about the structure and function of the natural world. In Sears's day ecology was basically a study of equilibrium, harmony, and order, it had been so from its beginnings. Today, however, in

many circles of scientific research, it has become a study of disturbance, disharmony, and chaos, and coincidentally or not, conservation is often not even a remote concern.

At the time *Deserts on the March* appeared in print, and through the time of its second and even third edition, the dominant name in the field of American ecology was that of Frederic L. Clements, who more than any other individual introduced scientific ecology into our national academic life. He called his approach "dynamic ecology," meaning it was concerned with change and evolution in the landscape. At its heart Clements's ecology dealt with the process of vegetational succession—the sequence of plant communities that appear on a piece of soil, newly made or disturbed, beginning with the first pioneer communities that invade and get a foothold.[4] Here is how I have defined the essence of the Clementsian paradigm:

> Change upon change became the inescapable principle of Clements's science. Yet he also insisted stubbornly and vigorously on the notion that the natural landscape must eventually reach a vaguely final climax stage. Nature's course, he contended, is not an aimless wandering to and fro but a steady flow toward stability that can be exactly plotted by the scientist.[5]

Most interestingly, Clements referred to that final climax stage as a "superorganism," implying that the assemblage of plants had achieved the close integration of parts, the self-organizing capability, of a single animal or plant. In some unique sense, it had become a live, coherent thing, not a mere collection of atomistic individuals, and exercised some control over the nonliving world around it, as organisms do.

Until well after World War II Clements's climax theory dominated ecological thought in this country.[6] Pick up almost any textbook in the field written forty, or even thirty, years ago, and you will likely find mention of the climax. It was this theory that Paul Sears had studied and took to be the core lesson of ecology that his county ecologists should teach their fellow citizens: that nature tends toward a climax state and that, as far as practicable, they should learn to respect and preserve it. Sears wrote that the chief work of the scientist ought to be to show "the unbalance which man has produced on this continent" and to lead people back to some approximation of nature's original health and stability.[7]

But then, beginning in the 1940s, while Clements and his ideas were still in the ascendant, a few scientists began trying to speak a new vocabulary. Words like "energy flow," "trophic levels," and "ecosystem" appeared in the leading journals, and they indicated a view of nature shaped more by physics than botany. Within another decade or two nature came to be widely seen as a flow of energy and nutrients through a physical or thermodynamic system. The early figures prominent in shaping this new view included C. Juday, Raymond Lindeman, and G. Evelyn Hutchinson. But perhaps its most influential exponent was Eugene P. Odum, hailing from North Carolina and Georgia, discovering in his southern saltwater marshes, tidal estuaries, and abandoned cotton fields the animating, pulsating force of the sun, the global flux of energy. In 1953 Odum published the first edition of his famous textbook, *The Fundamentals of Ecology*.[8] In 1966 he became president of Ecological Society of America.

By now anyone in the United States who regularly reads a newspaper or magazine has come to know at least a few of Odum's ideas, for they furnish the main themes in our popular understanding of ecology, beginning with the sovereign idea of the ecosystem. Odum defined the ecosystem as "any unit that includes all of the organisms (i.e., the 'community') in a given area interacting with the physical environment so that a flow of energy leads to clearly defined trophic structure, biotic diversity, and material cycles (i.e., exchange of materials between living and nonliving parts) within the system."[9] The whole earth, he argued, is organized into an interlocking series of such "ecosystems," ranging in size from a small pond to so vast an expanse as the Brazilian rainforest.

What all those ecosystems have in common is a "strategy of development," a kind of game plan that gives nature an overall direction. That strategy is, in Odum's words, "directed toward achieving as large and diverse an organic structure as is possible within the limits set by the available energy input and the prevailing physical conditions of existence."[10] Every single ecosystem, he believed, is either moving toward or has already achieved that goal. It is a clear, coherent, and easily observable strategy; and it ends in the happy state of order.

Nature's strategy, Odum added, leads finally to a world of mutualism and cooperation among the organisms inhabiting an area. From an early stage of competing against one another, they evolve toward a more symbiotic relationship. They learn, as it were, to work together to control their surrounding environment, making it more and more suitable as a habitat, until at last they have the power to protect themselves from its stressful cycles of drought and flood, winter and summer, cold and heat. Odum called that point "homeostasis." To achieve it, the living components of an ecosystem must evolve a structure of interrelatedness and cooperation that can, to some extent, manage the physical world—manage it for maximum efficiency and mutual benefit.

I have described this set of ideas as a break from the past, but that is misleading. Odum may have used different terms than Clements, may even have had a radically different vision of nature at times; but he did not repudiate Clements's notion that nature moves toward order and harmony. In the place of the theory of the "climax" stage he put the theory of the "mature ecosystem." His nature may have appeared more as an automated factory than as a Clementsian superorganism, but like its predecessor it tends toward order.

The theory of the ecosystem presented a very clear set of standards as to what constituted order and disorder, which Odum set forth in the form of a "tabular model of ecological succession." When the ecosystem reaches its end point of homeostasis, his table shows, it expends less energy on increasing production and more on furnishing protection from external vicissitudes; that is, the biomass in an area reaches a steady level, neither increasing nor decreasing, and the emphasis in the system is on keeping it that way—on maintaining a kind of no-growth economy. Then the little, aggressive, weedy organisms common at an early stage in development (the r-selected species) give way to larger, steadier creatures (K-selected species), who may have less potential for fast growth and explosive reproduction but also better talents at surviving in dense settlements and keeping the place on an even keel.[11] At that point there is supposed to be more diversity in the community—i.e., a greater array of species. And there is less loss of nutrients to the outside; nitrogen, phosphorous, and calcium all stay in circulation within the ecosystem rather than leaking out. Those are some of the key indicators of ecological order, all of them susceptible to precise measurement. The suggestion was implicit but clear that if one interfered too much with nature's strategy of development, the effects might be costly: a serious loss of nutrients, a decline in species diversity, an end to biomass stability. In short, the ecosystem would be damaged.

The most likely source of that damage was no mystery to Odum: it was human beings trying to force up the production of useful commodities and stupidly risking the destruction of their life support system.

Man has generally been preoccupied with obtaining as much "production" from the landscape as possible, by developing and maintaining early successional types of ecosystems, usually monocultures. But, of course, man does not live by food and fiber alone: he also needs a balanced CO_2-O_2 atmosphere, the climatic buffer provided by oceans and masses of vegetation, and clean (that is, unproductive) water for cultural and industrial uses. Many essential life-cycle resources, not to mention recreational and esthetic needs, are best provided man by the less "productive" landscapes. In other words, the landscape is not just a supply depot but is also the *oikos*—the home—in which we must live.[12]

Odum's view of nature as a series of balanced ecosystems, achieved or in the making, led him to take a strong stand in favor of preserving the landscape in as nearly natural a condition as possible. He suggested the need for substantial restraint on human activity—for environmental planning "on a rational and scientific basis." For him as for Paul Sears, ecology must be taught to the public and made the foundation of education, economics, and politics; America and other countries must be "ecologized."

Of course not every one who adopted the ecosystem approach to ecology ended up where Odum did. Quite the contrary, many found the ecosystem idea a wonderful instrument for promoting global technocracy. Experts familiar with the ecosystem and skilled in its manipulation, it was hoped in some quarters, could manage the entire planet for improved efficiency. "Governing" all of nature with the aid of rational science was the dream of these ecosystem technocrats.[13] But technocratic management was not the chief lesson, I believe, the public learned in Professor Odum's classroom; most came away devoted, as he was, to preserving large parts of nature in an unmanaged state and sure that they had been given a strong scientific rationale, as well as knowledge base, to do it. We must defend the world's endangered ecosystems, they insisted. We must safeguard the integrity of the Greater Yellowstone ecosystem, the Chesapeake Bay ecosystem, the Serengeti ecosystem. We must protect

species diversity, biomass stability and calcium recycling. We must make the world safe for K-species.[14]

That was the rallying cry of environmentalists and ecologists alike in the 1960s and early 1970s, when it seemed that the great coming struggle would be between what was left of pristine nature, delicately balanced in Odum's beautifully rational ecosystems, and a human race bent on mindless, greedy destruction. A decade or two later the situation has changed considerably. There are still environmental threats around, to be sure, and they are more dangerous than ever. The newspapers inform us of continuing disasters like the massive 1989 oil spill in Alaska's Prince William Sound, and reporters persist in using words like "ecosystem" and "balance" and "fragility" to describe such disasters. So do many scientists, who continue to acknowledge their theoretical indebtedness to Odum. For instance, in a recent British poll, 447 ecologists out of 645 questioned ranked the "ecosystem" as one of the most important concepts their discipline has contributed to our understanding of the natural world; indeed, "ecosystem" ranked first on their list, drawing more votes than nineteen other leading concepts.[15] But all the same, and despite the persistence of environmental problems, Odum's ecosystem is no longer the main theme in research or teaching in the science. A survey of recent ecology textbooks shows that the concept is not even mentioned in one leading work and has a much diminished place in the others.[16]

Ecology is not the same as it was. A rather drastic change has been going on in this science of late—a radical shifting away from the thinking of Eugene Odum's generation, away from its assumptions of order and predictability, a shifting toward what we might call a new *ecology of chaos*.

In July 1973, the *Journal of the Arnold Arboretum* published an article by two scientists associated with the Massachusetts Audubon Society, William Drury and Ian Nisbet, and it challenged Odum's ecology fundamentally. The title of the article was simply "Succession," indicating that old subject of observed sequences in plant and animal associations. With both Frederic Clements and Eugene Odum, succession has been taken to be the straight and narrow road to equilibrium. Drury and Nisbet disagreed completely with that assumption. Their observations, drawn particularly from northeastern temperate forests, strongly suggested that the process of ecological succession does not lead anywhere. Change is without any determinable direction and goes on forever, never reaching a point of stability. They found no evidence of any progressive development in nature: no progressive increase over time in biomass stabilization, no progressive diversification of species, no progressive movement toward a greater cohesiveness in plant and animal communities, nor toward a greater success in regulating the environment. Indeed, they found none of the criteria Odum had posited for mature ecosystems. The forest, they insisted, no matter what its age, is nothing but an erratic, shifting mosaic of trees and other plants. In their words, "most of the phenomena of succession should be understood as resulting from the differential growth, differential survival, and perhaps differential dispersal of species adapted to grow at different points on stress gradients."[17] In other words, they could see lots of individual species, each doing its thing, but they could locate no emergent collectivity, nor any strategy to achieve one.

Prominent among their authorities supporting this view was the nearly forgotten name of Henry A. Gleason, a taxonomist who, in 1926, had challenged Frederic Clements and his organismic theory of the climax in an article entitled, "The Individualistic Concept of the Plant Association." Gleason had argued that we live in a world of constant flux and impermanence, not one tending toward Clements's climaxes. There is no such thing, he argued, as balance or equilibrium or steady-state. Each and every plant association is nothing but a temporary gathering of strangers a clustering of species unrelated to one another, here for a brief while today, on their way somewhere else tomorrow. "Each . . . species of plant is a law unto itself," he wrote.[18] We look for cooperation in nature and we find only competition. We look for organized wholes, and we can discover only loose atoms and fragments. We hope for order and discern only a mishmash of conjoining species, all seeking their own advantage in utter disregard of others.

Thanks in part to Drury and Nisbet, this "individualistic" view was reborn in the mid-1970s and, during the past decade, it became the core idea of what some scientists hailed as a new, revolutionary paradigm in ecology. To promote it, they attacked the traditional notion of succession; for to reject that notion was to reject the larger idea that organic nature tends toward order. In 1977 two more biologists, Joseph Connell and Ralph Slatyer, continued the attack, denying the old claim that an invading community of pioneering species, the first stage in Clements's sequence, works to prepare the ground for its successors, like a group of Daniel Boones blazing the trail for civilization. The first comers, Connell and Slatyer maintained, manage in most cases to stake out their claims and successfully defend them; they do not give way to a later, superior group of colonists. Only when the pioneers die or are damaged by natural disturbances, thus releasing the resources they have monopolized, can latecomers find a foothold and get established.[19]

As this assault on the old thinking gathered momentum, the word "disturbance" began to appear more frequently in the scientific literature and be taken far more seriously. "Disturbance" was not a common subject in Odum's heyday, and it almost never appeared in combination with the adjective "natural." Now, however, it was as though scientists were out looking strenuously for signs of disturbance in nature—especially signs of disturbance that were not caused by humans—and they were finding it everywhere. During the past decade those new ecologists succeeded in leaving little tranquility in primitive nature. Fire is one of the most common disturbances they noted. So is wind, especially in the form of violent hurricanes and tornadoes. So are invading populations of microorganisms and pests and predators. And volcanic eruptions. And invading ice sheets of the Quaternary Period. And devastating droughts like that of the 1930s in the American West. Above all, it is these last sorts of disturbances, caused by the restlessness of climate, that the new generation of ecologists have emphasized. As one of the most influential of them, Professor Margaret Davis of the University of Minnesota, has written: "For the last 50 years or 500 or 1,000—as long as anyone would claim for 'ecological time'—there has never been an interval when temperature was in a steady state with symmetrical fluctuations about a mean. . . . Only on the longest time scale, 100,000 years, is there a tendency toward cyclical variation, and the cycles are asymmetrical, with a mean much different from today."[20]

One of the most provocative and impressive expressions of the new post-Odum ecology is a book of essays edited by S. T. A. Pickett and P. S. White, *The Ecology of Natural Disturbance and Patch Dynamics* (published in 1985). I submit it as symptomatic of much of the thinking going on today in the field. Though the final section of the book does deal with ecosystems, the word has lost much of its former meaning and implications. Two of the authors in fact open their contribution with a complaint that many scientists assumed that "homogeneous ecosystems are a reality," when in truth "virtually all naturally occurring and man-disturbed ecosystems are mosaics of environmental conditions." "Historically," they write, "ecologists have been slow to recognize the importance of disturbances and the heterogeneity they generate." The reason for this slowness? "The majority of both theoretical and empirical work has been dominated by an equilibrium perspective."[21] Repudiating that perspective, these authors take us to the tropical forests of South and Central America and to the Everglades of Florida, showing us instability on every hand: a wet, green world of continual disturbance—or as they prefer to say, "of perturbations." Even the grasslands of North America, which inspired Frederic Clements's theory of the climax, appear in this collection as regularly disturbed environments. One paper describes them as a "dynamic, fine-textured mosaic" that is constantly kept in upheaval by the workings of badgers, pocket gophers, and mound-building ants, along with fire, drought, and eroding wind and water.[22] The message in all these papers is consistent: The climax notion is dead, the ecosystem has receded in usefulness, and in their place we have the idea of the lowly "patch." Nature should be regarded as a landscape of patches, big and little, patches of all textures and colors, a patchwork quilt of living things, changing continually through time and space, responding to an unceasing barrage of perturbations. The stitches in that quilt never hold for long.

Now, of course, scientists have known about gophers and winds, the Ice Age and droughts for a considerable time. Yet heretofore they have not let those disruptions spoil their theories of balanced plant and animal associations, and we must ask why that was so. Why did Clements and Odum tend to dismiss such forces as climatic change, at least of the less catastrophic sort, as threats to the order of nature? Why have their successors, on the other hand, tended to put so much emphasis on those same changes, to the point that they often see nothing but instability in the landscape?

One clue comes from the fact that many of these disturbance boosters are not and have never been ecosystem scientists; they received their training in the subfield of population biology and reflect the growing confidence, methodological maturity, and influence of the subfield.[23] When they look at a forest, the population ecologists see only the trees. See them and count them—so many white pines, so many hemlocks, so many maples and birches. They insist that if we know all there is to know about the individual species that constitute a forest, and can measure their lives in precise, quantitative terms, we will know all there is to

know about that forest. It has no "emergent" or organismic properties. It is not some whole greater than the sum of its parts, requiring "holistic" understanding. Outfitted with computers that can track the life histories of individual species, chart the rise and fall of populations, they have brought a degree of mathematical precision to ecology that is awesome to contemplate. And what they see when they look at population histories for any patch of land is wildly swinging oscillations. Populations rise and populations fall, like stock market prices, auto sales, and hemlines. We live, they insist, in a non-equilibrium world.[24]

There is another reason for the paradigmatic shift I have been describing, though I suggest it quite tentatively and can offer only sketchy evidence for it. For some scientists, a nature characterized by highly individualistic associations, constant disturbance, and incessant change may be more ideologically satisfying than Odum's ecosystem, with its stress on cooperation, social organization, and environmentalism. A case in point is the very successful popularizer of contemporary ecology, Paul Colinvaux, author of *Why Big Fierce Animals Are Rare* (1978). His chapter on succession begins with these lines: "If the planners really get hold of us so that they can stamp out all individual liberty and do what they like with our land, they might decide what whole counties full of inferior farms should be put back into forest." Clearly, he is not enthusiastic about land-use planning or forest restoration. And he ends that same chapter with these remarkably revealing and self-assured words:

> We can now . . . explain all the intriguing, predictable events of plant successions in simple, matter of fact, Darwinian ways. Everything that happens in successions comes about because of the different species go about earning their livings as best they may, each in its own individual manner. What look like community properties are in fact the summed results of all these bits of private enterprise.[25]

Apparently, if this example is any indication, the social Darwinists are back on the scene, and at least some of them are ecologists, and at least some of their opposition to Odum's science may have to do with a revulsion toward its political implications, including its attractiveness for environmentalists. Colinvaux is very clear about the need to get some distance between himself and groups like the Sierra Club.

I am not alone in wondering whether there might be a deeper, half-articulated ideological motive generating the new direction in ecology. The Swedish historian of science, Thomas Söderqvist, in his recent study of ecology's development in his country, concludes that the present generation of evolutionary ecologists

> seem to do ecology for fun only, indifferent to practical problems, including the salvation of the nation. They are mathematically and theoretically sophisticated, sitting indoors calculating on computers, rather than traveling out in the wilds. They are individualists, abhorring the idea of large-scale ecosystem projects. Indeed, the transition from ecosystem ecology to evolutionary ecology seems to reflect the generational transition from the politically consciousness generation of the 1960s to the "yuppie" generation of the 1980s.[26]

That may be an exaggerated characterization, and I would not want to apply it to every scientist who has published on patch dynamics or disturbance regimes. But it does draw our attention to an unmistakable attempt by many ecologists to disassociate themselves from reform environmentalism and its criticisms of human impact on nature.

I wish, however, that the emergence of the new post-Odum ecology could be explained so simply in those two ways: as a triumph of reductive population dynamics over holistic consciousness, or as a triumph of social Darwinist or entrepreneurial ideology over a commitment to environmental preservation. There is, it seems, more going on than that, and it is going on all through the natural sciences—biology, astronomy, physics—perhaps going on through all modern technological societies. It is nothing less than the discovery of chaos. Nature, many have begun to believe is *fundamentally* erratic, discontinuous, and unpredictable. It is full of seemingly random events that elude our models of how things are supposed to work. As a result, the unexpected keeps hitting us in the face. Clouds collect and disperse, rain falls or doesn't fall, disregarding our careful weather predictions, and we cannot explain why. Cars suddenly bunch up on the freeway, and the traffic controllers fly into a frenzy. A man's heart beats regularly year after year, then abruptly begins to skip a beat now and then. A ping pong ball bounces off the table in an unexpected direction. Each little snowflake falling out of the sky turns out to be completely unlike any other. Those are ways in which nature seems, in contrast to all our previous theories and methods, to be chaotic. If the ultimate test of any body of scientific knowledge is its ability to predict events, then all the sciences and pseudo-sciences—physics, chemistry, climatology, economics, ecology—fail the test regularly. They all have been announcing laws, designing models, predicting what an individual atom or person is supposed to do; and now, increasingly, they are beginning to confess that the world never quite behaves the way it is supposed to do.

Making sense of this situation is the task of an altogether new kind of inquiry calling itself the science of chaos. Some say it portends a revolution in thinking equivalent to quantum mechanics or relativity. Like those other 20th-century revolutions, the science of chaos rejects tenets going back as far as the days of Sir Isaac Newton. In fact, what is occurring may be not two or three separate revolutions but a single revolution against all the principles, laws, models, and applications of classical science, the science ushered in by the great Scientific Revolution of the 17th century.[27] For centuries we have assumed that nature, despite a few appearances to the contrary, is a perfectly predictable system of linear, rational order. Give us an adequate number of facts, scientists have said, and we can describe that order in complete detail—can plot the lines along which everything moves and the speed of that movement and the collisions that will occur. Even Darwin's theory of evolution, which in the last century challenged much of the Newtonian worldview, left intact many people's confidence that order would prevail at last in the evolution of life; that out of the tangled history of competitive struggle would come progress, harmony, and stability. Now that traditional assumption may have broken down irretrievably. For whatever reason, whether because empirical data suggests it or because extrascientific cultural trends do—the experience of so much rapid social change in our daily lives—scientists are beginning to focus on what they had long managed to avoid seeing. The world is more complex than we ever imagined, they say, and indeed, some would add, ever can imagine.[28]

Despite the obvious complexity of their subject matter, ecologists have been among the slowest to join the cross-disciplinary science of chaos. I suspect that the influence of Clements and Odum, lingering well into the 1970s, worked against the new perspective, encouraging faith in linear regularities and equilibrium in the interaction of species. Nonetheless, eventually there arrived a day of conversion. In 1974 the Princeton mathematical ecologist Robert May published a paper with the title, "Biological Populations with Non-overlapping Generations: Stable Points, Stable Cycles, and Chaos."[29] In it he admitted that the mathematical models he and others had constructed were inadequate approximations of the ragged life histories of organisms. They did not fully explain, for example, the aperiodic outbreaks of gypsy moths in eastern hardwood forests or the Canadian lynx cycles in the subarctic. Wildlife populations do not follow some simple Malthusian pattern of increase, saturation, and crash.

More and more ecologists have followed May and begun to try to bring their subject into line with chaotic theory. William Schaefer is one of them; though a student of Robert MacArthur, a leader of the old equilibrium school, he has been lately struck by the same anomaly of unpredictable fluctuations in populations as May and others. Though taught to believe in "the so-called 'Balance of Nature,'" he writes, ". . . the idea that populations are at or close to equilibrium," things now are beginning to look very different.[30] He describes himself as having to reach far across the disciplines, to make connections with concepts of chaos in the other natural sciences, in order to free himself from his field's restrictive past.

The entire study of chaos began in 1961, with efforts to simulate weather and climate patterns on a computer at MIT. There, meteorologist Edward Lorenz came up with his now famous "Butterfly Effect," the notion that a butterfly stirring the air today in a Beijing park can transform storm systems next month in New York City. Scientists call this phenomenon "sensitive dependence on initial conditions." What it means is that tiny differences in input can quickly become substantial differences in output. A corollary is that we cannot know, even with all our artificial intelligence apparatus, every one of the tiny differences that have occurred or are occurring at any place or point in time; nor can we know which tiny differences will produce which substantial differences in output. Beyond a short range, say, of two or three days from now, our predictions are not worth the paper they are written on.

The implications of this "Butterfly Effect" for ecology are profound. If a single flap of an insect's wings in China can lead to a torrential downpour in New York, then what might it do to the Greater Yellowstone Ecosystem? What can ecologists possibly know about all the forces impinging on, or about to impinge on, any piece of land? What can they safely ignore and what must they pay attention to? What distant, invisible, minuscule events may even now be happening that will change the organization of plant and animal life in our back yards? This is the predicament, and the

challenge, presented by the science of chaos, and it is altering the imagination of ecologists dramatically.

John Muir once declared, "When we try to pick out anything by itself, we find it hitched to everything else in the universe."[31] For him, that was a manifestation of an infinitely wise plan in which everything functioned with perfect harmony. The new ecology of chaos, though impressed like Muir with interdependency, does not share his view of "an infinitely wise plan" that controls and shapes everything into order. There is no plan, today's scientists say, no harmony apparent in the events of nature. If there is order in the universe—and there will no longer be any science if all faith in order vanishes—it is going to be much more difficult to locate and describe than we thought.

For Muir, the clear lesson of cosmic complexity was that humans ought to love and preserve nature just as it is. The lessons of the new ecology, in contrast, are not at all clear. Does it promote, in Ilya Prigogine and Isabelle Stenger's words, "a renewal of nature," a less hierarchical view of life, and a set of "new relations between man and nature and between man and man"?[32] Or does it increase our alienation from the world, our withdrawal into post-modernist doubt and self-consciousness? What is there to love or preserve in a universe of chaos? How are people supposed to behave in such a universe? If such is the kind of place we inhabit, why not go ahead with all our private ambitions, free of any fear that we may be doing special damage? What, after all, does the phrase "environmental damage" mean in a world of so much natural chaos? Does the tradition of environmentalism to which Muir belonged, along with so many other nature writers and ecologists of the past—people like Paul Sears, Eugene Odum, Aldo Leopold, and Rachel Carson—make sense any longer? I have no space here to attempt to answer those questions or to make predictions but only issue a warning that they are too important to be left for scientists alone to answer. Ecology today, no more than in the past, can be assumed to be all-knowing or all-wise or eternally true.

Whether they are true or false, permanent or passingly fashionable, it does seem entirely possible that these changes in scientific thinking toward an emphasis on chaos will not produce any easing of the environmentalist's concern. Though words like ecosystem or climax may fade away and some new vocabulary take their place, the fear of risk and danger will likely become greater than ever. Most of us are intuitively aware, whether we can put our fears into mathematical formulae or not, that the technological power we have accumulated is *destructively* chaotic; not irrationally, we fear it and fear what it can to do us as well as the rest of nature.[33] It may be that we moderns, after absorbing the lessons of today's science, find we cannot love nature quite so easily as Muir did; but it may also be that we have discovered more reason than ever to respect it—to respect its baffling complexity, its inherent unpredictability, its daily turbulence. And to flap our own wings in it a little more gently.

Notes

1. Paul Sears, *Deserts on the March,* 3rd ed. (Norman: University of Oklahoma Press, 1959), p. 162.

2. Ibid., p. 177.

3. Donald Worster, *Nature's Economy: A History of Ecological Ideas* (New York: Cambridge University Press, 1977).

4. This is the theme in particular of Clements's book *Plant Succession* (Washington: Carnegie Institution, 1916).

5. Worster, p. 210.

6. Clements's major rival for influence in the United States was Henry Chandler Cowles of the University of Chicago, whose first paper on ecological succession appeared in 1899. The best study of Cowles's ideas is J. Ronald Engel, *Sacred Sands: The Struggle for Community in the Indiana Dunes* (Middletown, CT: Wesleyan University Press, 1983), pp. 137–59. Engel describes him as having a less deterministic, more pluralistic notion of succession, one that "opened the way to a more creative role for human beings in nature's evolutionary adventure" (p. 150). See also Ronald C. Tobey, *Saving the Prairies: The Life Cycle of the Founding School of American Plant Ecology, 1895–1955* (Berkeley: University of California, 1981).

7. Sears, p. 142.

8. This book was co-authored with his brother Howard T. Odum, and it went through two more editions, the last appearing in 1971.

9. Eugene P. Odum, *Fundamentals of Ecology* (Philadelphia: W. B. Saunders, 1971), p. 8.

10. Odum, "The Strategy of Ecosystem Development," *Science,* 164 (18 April 1969): 266.

11. The terms "K-selection" and "r-selection" came from Robert MacArthur and Edward O. Wilson, *Theory of Island Biogeography* (Princeton: Princeton University Press, 1967). Along with Odum, MacArthur was the leading spokesman during the 1950s and 60s for the view of nature as a series of thermodynamically balanced ecosystems.

12. Odum, "Strategy of Ecosystem Development," p. 266. See also Odum, "Trends Expected in Stressed Ecosystems," *BioScience,* 35 (July/August 1985): 419–422.

13. A book of that title was published by Earl P. Murphy, *Governing Nature* (Chicago: Quadrangle Books, 1967). From time to time, Eugene Odum himself seems to have caught that ambition or leant his support to it, and it was certainly central to the work of his brother, Howard T. Odum. On this theme see Peter J. Taylor, "Technocratic Optimism, H. T. Odum, and the Partial Transformation of Ecological Metaphor after World War II," *Journal of the History of Biology,* 21 (Summer 1988): 213–244.

14. A very influential popularization of Odum's view of nature (though he is never actually referred to in it) is Barry Commoner's *The Closing Circle: Nature, Man, and Technology* (New York: Alfred A. Knopf, 1971). See in particular the discussion of the four "Laws" of ecology, pp. 33–46.

15. Communication from Malcolm Cherrett, *Ecology,* 70 (March 1989): 41–42.

16. See Michael Begon, John L. Harper, and Colin R. Townsend, *Ecology: Individuals, Populations, and Communities* (Sunderland, Mass.: Sinauer, 1986). In another textbook, Odum's views are presented critically as the traditional approach: R. J. Putnam and S. D. Wratien, *Principles of Ecology* (Berkeley: University of California Press, 1984). More loyal to the ecosystem model are Paul Ehrlich and Jonathan Roughgarden, *The Science of Ecology* (New York: Macmillan, 1987); and Robert Leo Smith, *Elements of Ecology,* 2nd ed. (New York: Harper & Row, 1986), though the latter admits that he has shifted from an "ecosystem approach" to more of an "evolutionary approach" (p. xiii).

17. William H. Drury and Ian C. T. Nisbet, "Succession," *Journal of the Arnold Arboretum,* 54 (July 1973): 360.

18. H. A. Gleason, "The Individualistic Concept of the Plant Association," *Bulletin of the Torrey Botanical Club,* 53 (1926): 25. A later version of the same article appeared in *American Midland Naturalist,* 21 (1939): 92–110.

19. Joseph H. Connell and Ralph O. Slatyer, "Mechanisms of Succession in Natural Communities and Their Role in Community Stability and Organization," *The American Naturalist,* 111 (November–December 1977): 1119–1144.

20. Margaret Bryan Davis, "Climatic Instability, Time Lags, and Community Disequilibrium," in *Community Ecology,* ed. Jared Diamond and Ted J. Case (New York: Harper & Row, 1986), p. 269.

21. James R. Karr and Kathryn E. Freemark, "Disturbance and Vertebrates: An Integrative Perspective," *The Ecology of Natural Disturbance and Patch Dynamics,* eds. S. T. A. Pickett and P. S. White (Orlando, Fla.: Academic Press, 1985), pp. 154–55. The Odum school of thought is, however, by no means silent. Another recent compilation has been put together in his honor, and many of its authors express a continu-

ing support for his ideas: L. R. Pomeroy and J. J. Alberts, eds., *Concepts of Ecosystem Ecology: A Comparative View* (New York: Springer-Verlag, 1988).

22. Orie L. Loucks, Mary L. Plumb-Menties, and Deborah Rogers, "Gap Processes and Large-Scale Disturbances in Sand Prairies," *ibid.,* pp. 72–85.

23. For the rise of population ecology see Sharon E. Kingsland, *Modeling Nature: Episodes in the History of Population Ecology* (Chicago: University of Chicago Press, 1985).

24. An influential exception to this tendency is F. H. Bormann and G. E. Likens, *Pattern and Process in a Forested Ecosystem* (New York: Springer-Verlag, 1979), which proposes in Chap. 6 the model of a "shifting mosaic steady-state." See also P. Yodzis, "The Stability of Real Ecosystems," *Nature,* 289 (19 February 1981): 674–76.

25. Paul Colinvaux, *Why Big Fierce Animals Are Rare: An Ecologist's Perspective* (Princeton: Princeton University Press, 1978), pp. 117, 135.

26. Thomas Söderqvist, *The Ecologists: From Merry Naturalists to Saviours of the Nation. A Sociologically Informed Narrative Survey of the Ecologization of Sweden, 1895–1975.* (Stockholm: Almqvist & Wiksell International, 1986), p. 281.

27. This argument is made with great intellectual force by Stengers, *Order Out of Chaos* (Boulder: Shambala/New Science Library, 1984). Prigogine won the Nobel Prize in 1977 for his work on the thermodynamics of nonequilibrium systems.

28. An excellent account of the change in thinking is James Gleick, *Chaos: The Making of a New Science* (New York: Viking, 1987). I have drawn on his explanation extensively here. What Gleick does not explore are the striking intellectual parallels between chaotic theory in science and post-modern discourse in literature and philosophy. Post-Modernism is a sensibility that has abandoned the historic search for unity and order in nature, taking an ironic view of existence and debunking all established faiths. According to Todd Gitkin, "Post-Modernism reflects the fact that a new moral structure has not yet been built and our culture has not yet found a language for articulating the new understandings we are trying, haltingly, to live with. It objects to all principles, all commitments, all crusades—in the name of an unconscientious evasion." On the other hand, and more positively, the new sensibility leads to emphasis on democratic coexistence: "a new 'moral ecology'—that in the preservation of the other is a condition for the preservation of the self." Gitkin, "Post-Modernism: The Stenography of Surfaces," *New Perspectives Quarterly,* 6 (Spring 1989): 57, 59.

29. The paper was published in *Science,* 186 (1974): 645–647. See also Robert M. May, "Simple Mathematical Models with Very Complicated Dynamics," *Nature,* 261 (1976): 459–67. Gleick discusses May's work in *Chaos,* pp. 69–80.

30. W. M. Schaeffer, "Chaos in Ecology and Epidemiology," in *Chaos in Biological Systems,* ed. H. Degan, A. V. Holden, and L. F. Olsen (New York: Plenum Press, 1987), p. 233. See also Schaeffer, "Order and Chaos in Ecological Systems," *Ecology,* 66 (February 1985): 93–106.

31. John Muir, *My First Summer in the Sierra* (1911; Boston: Houghton Mifflin, 1944), p. 157.

32. Prigogine and Stengers, pp. 312–13.

33. Much of the alarm that Sears and Odum, among others, expressed has shifted to a global perspective, and the older equilibrium thinking has been taken up by scientists concerned about the geo- and biochemical condition of the planet as a whole and about human threats, particularly from the burning of fossil fuels, to its stability. One of the most influential texts in this new development is James Lovelock's *GAIA: A New Look at Life on Earth* (Oxford: Oxford University Press, 1979). See also Edward Goldsmith, "Gaia: Some Implications for Theoretical Ecology," *The Ecologist,* 18, nos. 2/3 (1988): 64–74.

For Further Discussion

1. There is a long tradition in philosophy of insisting that no ethical conclusions can be drawn from scientific facts alone. This tradition argues that there is a significant logical distinction between statements of fact and statements of value. Would Worster agree with this view? What reasons might he offer in support of that conclusion?

2. How does the climax theory differ from the chaos theory in ecology? How does Aldo Leopold's discussion of the biotic community compare with the ecological views of Clements and Odum as described in this essay?

3. What ethical conclusions might you draw from the ecological claim that "change is without any determinable direction and goes on forever, never reaching a point of stability"? What ethical conclusions would you draw from the contrary: "Ecological change is always directed toward a point of stability"?

4. Examine the quotation from Paul Colinvaux on page 251. How do you distinguish words and phrases that are scientific and factual from those that are evaluative and normative?

5. What is meant by the butterfly effect? What conclusions might John Muir draw from such phenomena? What conclusions might Paul Colinvaux

draw from the same phenomena? What advice might Worster offer us in light of these views?

DISCUSSION AND STUDY QUESTIONS FOR CHAPTER 7

1. Leopold advises us to consider the "integrity, stability, and beauty" of the biotic community. How do you understand these words when they are applied to an ecosystem?
2. Some animal rights advocates criticize ecocentric ethics as fascist. This suggests that ecocentric ethics, like political fascism, is willing to sacrifice the well-being of the individual for the good of the whole. Politically, this is seen as a grave injustice.

Is it unjust to sacrifice the well-being of individual animals for the good of a species? For the good of an ecosystem?
3. What ethical conclusions can be drawn from ecological data? Suppose ecology establishes that natural ecosystems tend toward a natural balance or harmony. What ethical conclusions would follow from that? Suppose ecology establishes that chaos rather than harmony is the norm. What follows from this?
4. Are ecosystems more real than individuals? What is an individual? Darwininan theory suggests that species rather than individuals evolve. Does this mean that species are real? Should species have moral standing?

DISCUSSION CASES

Yellowstone Fires

During the summer of 1988 massive fires burned throughout Yellowstone National Park. Nearly 1 million acres within the park had burned by the time a September snowstorm helped bring the fires under control. During recent decades, the U.S. Park Service had reversed earlier policies that had encouraged fire fighting and had adopted a policy of allowing natural fires to burn themselves out. This was regarded as sound ecological policy because fire has always been a part of the natural ecological process. Fires contribute to biotic diversity and natural plant growth, they provide benefits to many plant and animal species, and they can help recycle soil nutrients. Should fires in national parks, national forests, and national wilderness areas be vigorously fought?

Aldo Leopold advises us that "a thing is right when it tends to preserve the integrity, stability and beauty of the biotic community." According to this criterion, what policy should the Park Service adopt concerning forest fires? Is a charred and burned-out forest beautiful? Based on the essay on

aesthetic value in Chapter 5, how might Holmes Rolston evaluate the beauty of a forest fire?

Many animals and plants died in these Yellowstone fires. Do humans have an obligation to rescue wild animals from forest fires? How might Peter Singer answer this question? How might Leopold? Following the recommendations of Christopher Stone's essay, should the Park Service be sued on behalf of the plants and animals destroyed in these fires?

Reintroducing Wolves

Aldo Leopold, writing in the early decades of the twentieth century, once referred to predators such as wolves and coyotes as "varmints" and encouraged their elimination. Due to hunting and to loss of habitat, wolves were no longer in most of the American West by the 1940s. In the early 1980s, the U.S. Fish and Wildlife Service, following a policy most in line with Leopold's later writings, began plans to reintroduce wolves into Yellowstone Park. The plan met with fierce criticism from farm-
continued

ers and ranchers in the area who believed that the wolves would threaten their livestock. In 1995, after surviving several court challenges, the reintroduction plan was enacted. By the late 1990s a viable wolf population had been established in Yellowstone. Do you think this is a good thing? Should ranchers be compensated when their livestock is killed by wolves? Some plant and animal species become extinct independently of human action. Does your view on reintroducing wolves into an ecosystem depend on why the wolves are no longer there?

In reviewing various ethical theories, Callicott suggests that ethical status might ultimately rest on feelings of affection or kinship. For many ranchers, such feelings are reserved for their livestock but certainly not for wolves. Can people be taught to love wolves?

Olympia Goats

The ecological perspective defended by people like Aldo Leopold places great value on native species and discounts the value of nonnative "exotics." During the 1920s park rangers at Olympia National Park in Washington State introduced mountain goats into the park primarily as a game species for hunters. The goats adapted quite well, increasing to a population of over twelve hundred by the mid-1980s.

At that time the U.S. Park Service concluded that the goats posed a threat to the native ecosystem, causing erosion along steep hillsides and eating endangered native plants. After various attempts to sterilize and capture these goats failed to rid the area of all goats, the Park Service began a plan to shoot the remaining goats. Citing ecological reasons, various environmental groups agreed with the Park Service plan. Animal welfare groups opposed these plans by arguing, in part, that the goats

had once been native to the area. In your mind, what value implications follow from the distinction between native and nonnative species?

Consider how Leopold's concept of ecosystem integrity might be used to evaluate this case. Do native species contribute more to an ecosystem's integrity than nonnative species? Why might native species be more highly valued than nonnative species? How might the various models of ecological theory described by Worster differ on their value of native species?

Whitetail Deer and St. John's University

St. John's University is located among 2,400 acres of pine and hardwood forests. The woods around St. John's provide habitat for many animals species, including whitetail deer. Like many places around the country, the deer population has outgrown its limited habitat. Protected from hunters and with natural predators like the wolf long gone from the area, deer overpopulation has become a concern for land managers at St. John's. Deer destroy many native plant species, including oak saplings and other tree species. Harsh winters result in many deer dying of starvation. Deer that wander outside of the local woods destroy crops and often are involved in traffic accidents. As a measure to control the deer population, a limited deer hunt was allowed within the woods during the fall of 1997. Are there more attractive alternatives to hunting as a means of population control?

Can the integrity of an ecosystem be managed by humans or is something of integrity lost in the process? Compare what Leopold might say about this situation to the animal welfare views of Singer and Regan from Chapter 5. Are such deer hunts equivalent to ethical fascism in that they sacrifice the lives of individual deer for the good of the woods?

PART III *Policies and Controversies in Environmental Ethics*

Over the next seven chapters we turn directly to specific environmental policy issues. The goal of the previous chapters was to provide the context and the philosophical tools necessary for a reasoned and sophisticated analysis of such issues. It is now time to address the ethical dimension of environmental policy.

At this point it may be helpful to return to the discussion of ethics and public policy introduced in Chapter 1. As traditionally understood, ethical questions are raised from two perspectives: (1) Individual moral questions deal with how we, as individuals, should live our lives; (2) public policy questions deal with how we, as a society, should live together. Although many questions of the first sort will be raised in the following chapters, our primary emphasis will be on questions of public policy.

In Chapter 1 we described public policy as addressing three fundamental questions: What should we do? Why should we do it? How should we do it? In the readings that follow, we will find many cases in which people disagree about the goals of public policy. Some think we should increase preservation of wilderness areas; some believe we have too much land already preserved. Some believe that hunting is unethical; others see it as an ecologically beneficial activity. Debates concerning the goals of policy raise essential ethical questions, and many of the controversies that follow raise such questions.

We also shall find instances when authors disagree about the reasons that justify particular policies. For example, there is disagreement over the value of preserving wilderness areas even among people who believe that such areas should be preserved. Some argue on behalf of the ecosystem (ethical holists), others on behalf of the plants and animals that dwell therein (ethical extensionists), while still others argue on behalf of the aesthetic and historical values or on behalf of conserving economic resources.

Finally, we will examine cases in which the disagreement lies at the level of procedures or policies adopted as means to common goals. One example concerns management of national forests. Even among people who agree that forests should be managed on the basis of multiple use (including timber harvesting and various forms of recreation), people disagree about how best to do that. Some argue that scientific experts trained in forestry should manage this land according to the best management practices established by science. Others want to let the market sift through the competing interests to find the best balance of interests.

Another example of procedural disagreement can be found in the chapter on pollution. Although there is a widespread and strong consensus that pollution ought, at least, to be reduced to levels that are not harmful to humans, there is disagreement concerning the goals and justification of pollution policy. Some defend general, across-the-board standards that each and every industry must meet. Some argue that cost–benefit analysis should be used in a case-by-case manner. Others defend the use of tradable "pollution permits" that provide rewards and incentives to least-polluting industries.

Remember that ethical issues are often hidden within these debates about procedures and means. Particular policy approaches are defended as scientific or rational or objective, with the assumption that whatever policy recommendations emerge from such means are obviously correct. You should always be on guard in the face of such claims, looking for underlying value and philosophical assumptions. Many of the selections chosen for the following chapters seek to make these value assumptions explicit, bringing them into the light of day for careful examination.

Once again, each chapter in Part III ends with a variety of short discussion cases, drawn from real-life situations. These cases are not intended to be exhaustive treatments of particular policy cases but simply to illustrate the type of controversies associated with the issues raised in the chapter. The cases should highlight issues by connecting the readings to practical policy events. I also hope that they will provoke further discussion and debate, both within and outside of the classroom.

CHAPTER 8

Pollution

At first glance, pollution does not seem to raise any major ethical controversies. There has been wide agreement that air, water, and soil pollution is bad, that polluting activities need to be reduced, and that polluted resources need to be cleaned up. Further, as we saw in the readings by Easterbrook and by Bast, Hill, and Rue in Chapter 3, there is evidence for concluding that environmental pollution is decreasing, that the environment is actually getting cleaner. The only questions that remain seem to be the relatively minor ones of how to do this and who should pay for it.

Despite these appearances, controversies remain. We still need to think about our vision of the society in which we live. Is any level of pollution acceptable? An argument could be made on economic grounds that all pollution involves wasted resources and therefore is inefficient. Other values also are at stake. Pollution threatens other living beings, and it may affect ecosystems in unknown ways. We should also think about different kinds of pollution and what they say about the type of people we are. Do nuclear waste, petrochemicals, smog, insecticides, and industrial pollutants raise the same ethical issues as untreated sewage, unclean water, and smoke from burning rain forests? Are the ethical issues of pollution faced by residents of Los Angeles similar to or different from the issues faced by residents of Malaysia or Calcutta?

Even when there is agreement concerning the goals of pollution policy, disagreement remains concerning the appropriate means for achieving those goals. Consider the development of U.S. law in this regard.

Many of the most important environmental laws were enacted during the early 1970s. The Clean Air Act of 1970 and the Federal Water Pollution Act of 1972 (amended as the Clean Water Act of 1977) adopted similar strategies for addressing environmental pollution. Prior to these laws, individuals who suffered a harm from pollution could address this harm through the laws of civil liability. That is, the legal burden rested on the victims to prove that a harm occurred, that they deserved compensation for that harm, and that some identifiable person was responsible for causing that harm. These environmental laws changed this approach by shifting the burden from victim to polluter. Under these laws, government regulators established standards of clean air and clean water, and both individuals and industries were held to these standards. The goal was to prevent harm from occurring rather than compensating for the harm after the fact.

The process here suggests that a political judgment was made concerning the type of world in which we wish to live. It is better to live in a relatively clean world that prevents harm than a world in which harms are allowed and compensation is given if the harm unjustly occurs. In our terms, the goal of public policy was set through the political process (in fact, passed by a Democratic Congress and signed into law by a Republican president). The means for attaining that goal involved using experts who provided scientific testimony to regulatory agencies who then established the standards.

Under the Reagan administration during the 1980s, both the ends and the means of these policies came under criticism. Opponents argued that cost–benefit analysis should be used both for establishing the proper standards for air and water quality and for determining the proper procedures for attaining those standards. This approach suggests that the question of pollution is really a question of acceptable risk. Consider the question: Is exposure to one part per billion of lead in drinking water dangerous? On one approach, this is a question to be answered by experts in science and health. On another approach, this is a question of relative safety, answerable by individuals who decide how much risk they are willing to face.

The original approach taken by many environmental laws was to address such questions scientifically and to bring in economic considerations only when deciding the appropriate policies for attaining the scientifically established standards of safety. The approach defended by the Reagan administration was to use economic considerations to determine the standards themselves. Acceptable risk, after all, involves comparisons between alternatives and trade-offs, exactly the sort of thing that cost–benefit analysis is designed to do. According to the first approach, economic calculations are relevant only for judging means; political processes determine our ends. According to the Reagan administration's approach, economic calculations are relevant for judging both means and ends.

These are some of the important ethical and policy issues that you should keep in mind as your read and evaluate the selections that follow.

Assessing Environmental Health Risks

Ann Misch

In the first reading of this chapter, Ann Misch, a researcher with the Worldwatch Institute in Washington, DC, details many of the actual and potential problems that result from our modern reliance on a wide range of chemicals. Scientific assessment of the environmental health risks associated with many everyday-contact chemicals is very limited. Although science has been quite good at identifying immediate and short-term harm caused by pollutants, it is less helpful in identifying the long-term effects of chemical exposure. This problem is compounded by the very limited scientific study of exposure to many combinations of chemicals. Science is also just beginning to understand the effects of low-level exposure to toxic chemicals. Misch reviews many environmental health risks and potential health hazards associated with exposure to chemicals. She also points out that scientific study of chemical exposure typically relies on "average" exposure levels and thus ignores the unequal distribution of risks, both nationally and internationally.

Throughout the second half of this century, chemical companies have trumpeted the miraculous powers of chemicals on billboards, in magazine ads, and on television: "Better Things for Better Living Through Chemicals," appealed a long-running Du Pont ad. "Without chemicals, life itself would be impossible," claimed Monsanto. And they did bring us miracles— antibiotics, penicillin, and a vast selection of creature comforts far beyond anything our forebears could have imagined. Synthetic fibers, dry cleaning, spoil-proof food, crop-saving pesticides, contraceptives, contact lenses . . . the list is endless. All in all, chemical manufacturers have heaped tens of thousands of compounds on the bandwagon of progress, creating every possible convenience—and chasing every imagined ache or emptiness from our lives.

But along with all their benefits, these new creations have generated a long list of problems, including serious health consequences. Our enthusiasm for new chemicals and the products and services they allow has outstripped our attention to their long-term effects. While billions of dollars have been lavished on product development, marketing, promotion, and advertising, little has been spent on observing chemicals' interactions with living things and the environment. And these effects can never be thoroughly tested; the combinations of chemicals now in our food, water, clothing, and homes defy measurement.

Historically, research on chemicals' effects has led environmental health experts to one conclusion: there is an indisputable link between exposure to certain industrial substances and specific serious diseases, particularly cancer. Leukemia has been linked to benzene, an ingredient in gasoline, for example, and mesothelioma, a form of cancer, is considered a signature of asbestos exposure. The share of total chronic disease due to environmental pollution is still vigorously debated, however.

But some scientists are also beginning to look beyond the obvious—cancer and other easily diagnosable problems—to other health consequences of the chemical age. What they are finding puts a different face on claims of harmlessness accepted by consumers in a less questioning era. In the summer of 1991, for example, 21 scientists gathered at the Wingspread Conference Center in Wisconsin presented fresh evidence that a wide assortment of environmental pollutants had the potential to undermine biological functioning and so affect the overall competence of animals studied both in the lab and in the wild. Many of the substances they investigated had caused broad yet subtle damage by disrupting vital physiological systems, including the nervous system, the endocrine system (responsible for regulating hormones), and the immune system (which defends the body against infectious disease and cancer).

What the industrial chemicals discussed at Wingspread have in common—beyond, in some cases, their cancer-causing potential—is the ability to wreak silent havoc at much lower levels of exposure than those typically associated with cancer. Because most of the damage observed by the scientists reporting at Wingspread was so insidious, the group determined that similar effects in people exposed to the same contaminants might go unnoticed unless researchers specifically hunted for them. They called for a major investigation to better assess the extent of subtle chemical damage to human health.

Although the suspicion that chemicals can cause toxic effects more subtle than cancer, acute poisoning, or birth defects is not new, toxicologists and other scientists until recently lacked the tools to investigate many less obvious health effects. U.S. regulatory agencies have emphasized avoiding cancer; in the process, they presumed that a public protected against cancer was also a public protected against other toxic outcomes of chemical exposure. But environmental health research is proving that assumption false with each revelation that environmental pollutants can impair functioning and overall biological competence—in the absence of overt signs of disease.

The findings of scientists like those gathered at Wingspread do not mean that our previous focus on avoiding cancer was wrong. Yet the recent research on noncancer health effects certainly implies that our view of chemical risks is incomplete. The discovery that a whole universe of other health effects may be associated with the products of our industrial age has profound implications for public health and regulatory policy. The continuous appearance of toxic effects at lower and lower levels of exposure is especially troubling, since low-level exposure to some chemicals is practically universal.

The growing repertoire of toxic effects that scientists are beginning to document begs for a much more conservative approach to chemical regulation. To include these fresh findings in chemical risk assessment, regulatory agencies must look beyond cancer to assess the effect of chemicals on overall biological competence. This task involves, at a minimum, careful consideration of potential toxic effects on the nervous system, the endocrine and reproductive systems, and the immune system.

THE CHEMICAL LOAD

Society has understood for centuries the dangers posed by many natural and synthetic substances, often through casual observation of diseases that have beset workers in various "dirty" industries. Both Hippocrates in late fourth century B.C. and Charles Dickens in the mid-nineteenth century noted cases of lead poisoning among workers, for example. But with the Industrial Revolution came a new era of broad population exposure to natural and synthetic substances, as large-scale chemical production began and naturally occurring chemical substances were refined and altered.

By the twenties, industry began to produce synthetic chemicals, some of which began to escape into the environment. Polychlorinated biphenyls (PCBs), for instance, were introduced in the United States in 1930 but banned there and in Canada beginning in the late seventies; they were used for decades in electrical transformers, plastics, paints, varnishes, and waxes. PCB residues have been found in the tissues of people and animals in disparate parts of the globe—even in the fat of polar bears in the Arctic.

By the fifties, industry had invented thousands of important industrial compounds, some of which were as-yet-unrecognized toxicants. Among them were a number that contained chlorine, an important element in the production of many synthetic chemicals. Chlorine proved useful to the chemical industry, since it bonds readily with carbon compounds. Between 1920 and 1990, U.S. production of chlorine rose hundredfold, and now stands at roughly 10 million tons a year—roughly a third of the world total.

Unfortunately, thanks to their chlorine-carbon bonds, many industrial compounds are quite stable, and only break down slowly. Many organochlorine substances (compounds containing carbon and chlorine) are not water-soluble, but do dissolve in fat. Like filings drawn to a magnet, these substances migrate to the reserves of fat stored in the tissues of fish, birds, mammals, and people. DDT, an organochlorine pesticide, caused eggshell thinning among American bald eagles, leading to the death of unhatched eaglets and to broad population declines among these birds. Two recent studies in the United States have linked tissue concentrations of DDT with an increased risk of breast cancer in women.

In addition to creating brand-new hazards, twentieth-century industry has coaxed many other naturally toxic substances from rock and soil to use in manufacturing, and thereby released them into the environment. Industry's reliance on these metals has pushed more than 300 times the amount of lead, 20 times as much cadmium, and four times as much arsenic into the atmosphere than is naturally present. Gold mining in the Amazon Basin pollutes the region with some 90–120 tons of mercury annually as miners collect gold from the riverbeds of the region. Elevated levels of mercury have been found in fish, river waters, and people, suggesting that the pursuit of gold has caused a vast public health problem in the basin. Global emissions of mercury to the atmosphere are estimated at 4,500 tons a year. Minute quantities of cadmium, lead, and mercury have proved poisonous to the central nervous system.

The world's understanding of how all these substances affect human beings is still elementary. A few years ago, the National Research Council (NRC) looked into just how much is actually known: it found no information at all on the possible toxic effects of more than 80 percent of the 50,000 or so industrial chemicals (a category that excludes pesticides, food additives, cosmetics, and drugs) used in the United States. And there are still many important unanswered questions about the remaining 20 percent. Occupational exposure limits have been set for fewer than 700 of these 50,000 chemicals. For those produced in amounts exceeding 1 million pounds a year, for example, the NRC found that virtually no testing had been done on the potential for neurobehavioral damage, birth defects, or toxic effects that might span several generations by passing from parents to offspring.

The NRC report is not really so surprising, since even the U.S. Environmental Protection Agency (EPA) in the vast majority of cases does not require manufacturers of industrial chemicals to run specific tests to determine whether their products have adverse effects before putting them on the market. Most chemicals, says Erik Olson, an attorney at the National Resources Defense Council, "are innocent until proven guilty."

Pesticides' toxic effects are slightly better understood than those of industrial chemicals. The U.S. government began to make some progress with a 1988 law requiring that manufacturers submit health data on 620 active ingredients in older pesticides, although it turns out that much of the information EPA wants simply does not exist. Unfortunately, there still is not enough information to determine the health effects of more than 60 percent of the pesticides currently used in the United States.

Chemical manufacturers usually do not submit full reports on their products' potential toxic effects, but independent researchers have documented some of these. Based on current scientific literature, the NRC estimates that a third of the 197 substances to which a million or more American workers are exposed have the potential to be neuro-toxic or to damage the central nervous system and the brain. A partial list of these products includes many solvents, pesticides, and several metals.

Many common industrial substances—including benzene, dioxin, certain pesticides, and some metals—also have the ability to interfere with the immune system. Many organochlorines (including dioxins, furans, and PCBs), as well as the pesticides chlordane, DDT, heptachlor, and hexachlorobenzene, disrupt the endocrine system and impair reproductive abilities. Some toxic organochlorine substances have been banned in the United States and other countries, but they can remain in the environment for decades. And many, like DDT, are still used in other parts of the world.

Although workers often have the highest exposures to these toxic compounds, other people are exposed through consumer products, through drinking water and indoor air, or by virtue of living next to a factory or hazardous waste site. Some 53 million Americans apply herbicides to their lawns, for instance, and millions use commercial bug sprays inside the home. Shoe polish, glues, household cleaners, varnishes, and other everyday consumer products stored in the home contain neurotoxic chemicals. Some hazardous waste sites harbor neurotoxic chemicals that threaten to contaminate the drinking-water supplies of nearby communities. Outside the workplace, however, little is known about the precise extent to which people are exposed to single toxic chemicals, let alone this dangerous mix of hazardous substances.

CANCER AND THE ENVIRONMENT

One of the most feared and most investigated consequences of exposure to environmental pollutants is cancer. Like industry, cancer and chronic diseases are prominent features of the twentieth century. In industrial countries, cancer causes 20 percent of all deaths. By contrast, infectious and parasitic diseases account for less than 5 percent of all deaths in these nations. The reverse holds true in developing countries, where infectious and parasitic diseases cause far more deaths.

Explanations for the higher rates of cancer in industrial countries often invoke the vast differences in diet, smoking habits, and methods of preserving food. Compared with these factors, pollution may play a smaller part. Nonetheless, the role of industrial pollutants in cancer is not negligible. Indeed, for some cancers, toxic substances may make an important contribution. But it is difficult to quantify the role of toxicants in human cancer for three reasons: little is known about the toxicity of most chemicals, most exposures occur at very low levels, and people are exposed to countless substances. As a result, estimates of the share of cancers attributable to toxic exposure range from 7 percent to more than 20.

Until recently, a heated dispute among cancer experts and epidemiologists surrounded the question of whether the already high rates of cancer in industrial countries were increasing. The debate was compounded when cancer related to the use of tobacco, a risk factor that experts hold responsible for roughly 30 percent of all tumors, was included. Smoking's overwhelming impact on the prevalence of cancer means that increases in smoking-related cancers can drive overall cancer trends upward.

But recent data show that one of the most important cancers related to smoking, lung cancer, has levelled off in some industrial countries. The rates of death from lung cancer among men from Finland and the United Kingdom, for example, peaked in the seventies and are now falling. In other industrial countries, including Canada, Germany, Sweden, Switzerland, and the United States, deaths from lung cancer are not increasing as fast as previously. (They are falling, for example, for American men under the age of 45.) This suggests that if cancer is increasing overall, then lung cancer is not the explanation.

In recent studies, a number of researchers have found increases in the incidence and mortality rate of cancers with no known links to smoking, as well as increases in overall cancer. Many of these increases have occurred in older people. A 1990 study of trends in countries reporting mortality data to the World Health Organization found that between 1968 and 1987, death rates from cancer of the central nervous system (including brain cancer), breast cancer, kidney cancer, multiple myeloma, non-Hodgkin's lymphoma,

and skin cancer had increased in people over age 54 in six different countries. (Roughly two thirds of all deaths from cancer occur after the age of 60.) Cancers of the brain and central nervous system more than doubled in people between the ages of 75 and 84 and almost doubled among those aged 65 to 74. Total cancer among people 55 or older in all six countries also increased when lung and stomach cancer (both of which account for a large share of cancer and are linked to well-known risk factors) were excluded.

But the increases are not confined to older people. For example, the incidence of brain cancer and cancer of the central nervous system in American boys under age 20 rose 16 percent between 1973–77 and 1983–87, while the incidence of lymphoma rose 15 percent. Leukemia also rose during this period.

Studies in Europe and the United States have also revealed two- to fourfold increases in the incidence of testicular cancer during the last 50 years. One investigation in the United States found an "apparent epidemic increase over time in the risk of testicular cancer for young men aged 15 to 44." The authors based their observation on rates of testicular cancer recorded in the Connecticut Tumor Registry between 1935 and 1979, during which the incidence of testicular cancer among men aged 25 to 44 rose more than two and a half times. Although they offered no specific explanations for the increase, the authors commented that "the greater rate of increases seen for the recent [generations] suggests either the introduction of new carcinogens during the 1950s or increased exposure to already present carcinogens."

Cancer statistics are far from straightforward. One of the first questions experts often ask is whether the increases measured are "real." Many factors could cause deceptive increases in the spread of cancer. One of these is the reliability of national registries in which cancer deaths are recorded. If cancer deaths are incompletely reported to the registry, for example, better reporting causes an apparent increase. Artificial peaks in cancer incidence can also follow new techniques for diagnosis, new programs for cancer screening, and gains in access to health care. A portion of the increase in the incidence of breast cancer in American women, for example, is probably due to the increased use of mammography.

A number of cancer researchers speculate that much if not all of the rise in cancer stems from improvements in diagnosis. But researchers Devra Davis, currently a senior scientific advisor in the U.S. Department of Health and Human Services, and David Hoel of the Medical University of South Carolina, in Charleston, contend that for improved accuracy to explain the increase, there should be a corresponding decline in recorded cases of poorly diagnosed cancer. Traditionally, undiagnosed cancers are recorded as deaths due to unknown cancer or other unknown causes. As diagnoses sharpened, reason Davis and Hoel, then some of the cancers in these categories should shift into ones for which the specific causes of death are identified.

But what a number of researchers found when they looked closely at the mortality statistics in different countries was not a shift in the distribution of deaths (those attributed to nonspecific causes versus a specific cancer), but a rise in both well-diagnosed and poorly diagnosed cases. This means that cancer itself must be on the increase. Further, they argue that for better diagnosis to explain the consistent increase in cancer in many different industrial countries, advances in diagnosis would need to be spread evenly across the industrial world. Such uniform improvements in detection both among different countries and among men and women are unlikely.

If cancer truly is increasing in many industrial countries, as these recent studies suggest, then improvements in treatment may not be enough to deflect the impact of the disease. Researchers thus will need to pay renewed attention to the origins of cancer in order to better prevent it.

Despite being the focus of intense medical research in industrial countries, cancer is still a complex and elusive disease. It is startling that the specific causes of most cancers still evade medical understanding. But researchers do know that many factors, including alcohol, diet, genes, and environmental pollution, can all play a role.

Certain genes, for example, confer a unique susceptibility to cancer. Recently, a section of DNA containing a gene that predisposes individuals to develop colon cancer was discovered by researchers at Johns Hopkins University. The malfunctioning of genes that normally suppress tumors may account for the appearance of some cases of cancer. Researchers have also identified around 50 so-called oncogenes. Proto-oncogenes are genes that are normally involved in cell growth and specialization. When mutated, they transform into oncogenes and orchestrate the appearance and growth of tumors.

Yet the apparent rises in cancer incidence in the United States and other industrial countries happened too suddenly to make broader genetic susceptibility a plausible explanation. This means that an answer must be sought in the environment.

Experts estimate that environmental factors—which include diet, smoking, drinking alcohol, viruses, occupation, and geographical location—account for at least 60 percent of all cancer. Preserving food through salting, smoking, or pickling raises the risk of oral and stomach cancer, for example. Eating a diverse variety of vegetables may actually block certain kinds of cancer; the disappearance of these ingredients from modern diets may, in turn, lead to higher rates of these cancers in industrial countries.

Environmental cues clearly interact with genes. Dietary fat, for instance, may act as a cancer promoter, fanning the spread of malignant tissue. The fat-laden diet of typical Americans, who derive more than 40 percent of their calories from fat, has been tied to colon cancer and implicated by some researchers in breast cancer.

Chemical substances, too, can spur the development of cancer. But very few substances have been tested for their cancer-causing potential. Most of those known to cause cancer in mice or rats are not regulated by U.S. agencies. The International Agency for Research on Cancer has identified 60 environmental agents that can cause cancer in people. These include chemicals, groups of related chemicals, mixtures of different chemicals, radiation, drugs, and industrial processes or occupational exposures linked to cancer.

Exposure to toxic substances can begin a cascade of events that ultimately leads to cancer. This can happen in a number of ways. One mechanism is by directly damaging the DNA and causing mutations. Other toxicants may cause cancer by suppressing immunity. There is no direct evidence linking chemically lowered immunity to cancer, but a number of substances that are known carcinogens also appear capable of suppressing the immune system. Benzo(a)pyrene, a chemical found in automobile exhaust and coal smoke, and asbestos, a natural mineral fiber, are two examples.

A third way exposure can lead to cancer is by having the toxic substance act like a hormone. Diethylstilbesterol (DES), a drug prescribed to American women in the fifties and sixties to prevent miscarriages, is a synthetic version of estrogen, a female sex hormone. Like estrogen, DES promotes tumors in laboratory animals. It has also been associated with vaginal cancer in the daughters of women who took the drug. A number of other environmental pollutants appear to have the potential to perturb the endocrine system, which is responsible for the regulation of hormones (as discussed later in this chapter).

Other toxic substances disable the enzymes that normally break down toxicants, or interfere with the body's ability to mend flaws in the DNA. Under normal circumstances, these flaws constantly arise but are promptly repaired by processes in the cell.

Epidemiological studies in a broad sample of industrial countries have found higher rates of certain cancers among farmers and other agricultural workers who are exposed to a wide array of natural toxins and synthetic hazards. . . . In most countries, farmers are on the whole healthier than other people. They smoke less and get more exercise than the average person, so their rates for overall cancers and heart disease are lower. But despite their better health status, farmers across the industrial world appear to be at greater risk for developing some forms of cancer. Most of these have been associated with exposure to one or more industrial toxicants.

Many of the cancers that are rising in farmers and in industrial countries have also been observed in patients that have suppressed immune systems. Evidence from laboratory animals confirms that some environmental pollutants, such as pesticides, can suppress the immune system (as discussed in a later section), which leads researchers to speculate that environmental pollutants may also promote cancer by dampening immunity. Aaron Blair and Sheila Hoar Zahm reviewed various studies linking farming and cancer risks and suggested that farmers might be the vanguard of a population at risk for cancers linked to environmental pollutants. Though it is undoubtedly not a role they wanted, farmers may be the "canaries" of environmental health—providing a signal that all is not well, much as the canaries kept in cages in mines did by dying when there was not enough oxygen to breathe.

ENDANGERED NERVOUS SYSTEMS

The human nervous system may be the most sensitive target of environmental pollutants. A number of pollutants that are ubiquitous in the environment, home, or workplace—including lead, solvents, and pesticides—have the potential to injure the nervous system.

The central nervous system's sensitivity is due in part to the fact that nerve cells do not replenish themselves when they die, unlike most other cells in the body. This fixed endowment of nerve cells can only shrink, not expand: as people age, they lose neurons. Healthy 90-year-olds, for example, have about 75 percent of their original set of neurons. Fortunately, since these nerve cells number in the billions, the brain has a large "reserve capacity" of them. Because of this, old people occasionally retain much of the intellectual crispness they had in their twenties. But exposure to certain toxicants increases the rate at which nerve cells are lost. So people who lose an additional one tenth of a percent a year might in their sixties have the same number of neurons as healthy nonagenarians.

Very few industrial chemicals have been tested for their neurotoxic potential, as already mentioned. Tests of 197 chemicals that at least a million American workers are exposed to found that 65 of them could cause neurological damage. Roughly 20 million U.S. workers are exposed to neurotoxic chemicals, many of which are solvents. . . .

The most infamous example of a substance poisonous to the nervous system—albeit not a synthetic one—is lead, recognized for its far-reaching toxicity in workers since ancient Greece. Modern standards for occupational exposure to lead still fall short of protecting workers in the United States and other countries, in the eyes of a number of experts. Lead is now ubiquitous in the environment. In the late sixties, it was found in the snow of Greenland at 200 times ancient levels. The metal is considered a major environmental health threat in Eastern Europe, the United States, and many other countries.

Lead is one of the best studied neurotoxins. The more closely researchers have looked at the metal's toxicity, the more concerned they have grown about its effects at very low levels. Between 1972 and 1991, the amount of lead concentration in blood that U.S. federal agencies deemed "safe" dropped by 75 percent. In 1991, the Centers for Disease Control announced that subtle effects on the central nervous system of children began at blood-lead concentrations above 10 micrograms per deciliter. The previous guideline had established blood-lead levels over 25 micrograms per deciliter of blood as cause for concern.

The burning of leaded automobile fuel is an important source of children's exposure to lead in many countries. A second source is leaded paint, which still lines millions of American homes as well as those elsewhere. When the United States began removing lead from gasoline (which is scheduled to be finished by 1996), concentrations of lead in blood also dropped. Many countries in Europe, Latin America, and Asia still rely heavily on leaded automobile fuel, however. . . .

U.S. researchers are beginning to appreciate the hazards of some other neurotoxins, one group of which is solvents. These can cause short-term memory problems, dizziness, fatigue, irritability, and an inability to concentrate, as well as structural changes in the brain and nervous system. Some 10 million American workers and 1–2 million German workers are exposed to solvents.

Many ordinary items that can be bought at any hardware store or drugstore contain solvents. Shoe polish, rubber cement, nail polish remover, furniture polish, bathroom tile cleaners, disinfectants, paints, and paint thinner all contain varying amounts of solvents. Clothes brought back from the neighborhood dry cleaners release percholorethylene, a chlorinated solvent, as they hang in the closet.

In the eighties a series of studies, mostly Scandinavian, tied neurological damage to chronic solvent exposure among painters, woodworkers, and other workers. In some studies, investigators found lower performance among workers who were exposed at levels within the legal limits set by the government. For example, a Swedish study of car and industrial spray painters found a statistically significant rise in psychiatric symptoms such as irritability and difficulties concentrating. The investigators also found slower reaction and perception times, reduced manual dexterity, and poorer short-term memory. Workers with solvent-related symptoms accounted for the majority of patients in Sweden's occupational medicine clinics in the mid-eighties. Psychiatric problems that can be traced to occupational solvent exposure are compensated in Sweden by the Swedish National Social Insurance Board.

A New York State Department of Health study of apartment buildings that also housed dry cleaning shops found that the indoor levels of solvents sometimes exceeded workplace standards by a wide margin. Since few, if any, studies of nonoccupational exposure to solvents other than for glue sniffers have been done, it is hard to know what these exposures mean. Judith Schreiber, a toxicologist with the New York Department of Health, points out that health effects of using solvents outside the workplace may in some cases be more serious than those associated with some exposures on the job since evaluations of the risks posed by workplace exposures are based on studies of healthy middle-aged white workers. Workers are exposed only during the working day, but people living in solvent-contaminated apartments may spend most of their time at home.

In addition, infants and children are probably more susceptible than the typical worker to the neurotoxic effects of solvents. Infants lack the fully developed "blood-brain barrier" that protects adults from some toxicants and, because of their high metabolism, both infants and young children assimilate more of an airborne toxicant than adults do.

Kaye Kilburn, a cardiologist with training in neurology, has investigated subtle changes in certain nervous system functions among communities exposed to trichloroethylene in their drinking water. "Balance is very sensitive to solvents," says Kilburn. He also found that people with solvent exposure had diminished or absent blink reflexes and were slower to react on perception and coordination tests. Kilburn thinks that such tests of nervous system performance are the "future of neurotoxicity because all normal people test the same on them. It doesn't matter whether or not they were dull or brilliant."

Another possible consequence of exposure to neurotoxic substances is neurodegenerative disease. Parkinson's disease, for example, is caused by the deterioration of nerve cells in regions of the brain governing movement. Telltale signs of the disease include a shuffling walk, tremor, and rigidity. But because the brain has so much extra capacity, the disease can progress quite far before any warning signs occur. The hallmark symptoms of Parkinson's appear only after the level of a brain chemical called dopamine drops 70–80 percent. Dopamine, one of a handful of neurotransmitters that shuttle messages between nerve cells, is produced by a specific group of motor neurons.

Some evidence suggests that dopamine-producing neurons can be assailed by various environmental agents. In the early eighties, researchers stumbled across cases of a syndrome very similar to Parkinson's among California drug addicts who had manufactured batches of a heroin-like drug in their own basements. The drug contained a chemical known as MPTP. When metabolized, MPTP converted into a toxic chemical, MPP+, that turned out to be responsible for the mysterious cases like Parkinson's found among these young adults.

MPP+ bears a close resemblance to the herbicide paraquat, which has been banned in many industrial countries but which is still allowed in many developing nations. A number of epidemiological studies have detected higher levels of Parkinson's disease in farmers and other workers exposed to herbicides. One study reported in *Neurology* in 1992 found a threefold higher risk among people with herbicide exposure compared with controls. Conditions resembling Parkinson's disease have also appeared in workers exposed to carbon disulfide, carbon monoxide, and manganese.

The case for an environmental cause of neurodegenerative illnesses is complicated by the contribution of genes to some forms of these diseases. It may turn out that genes create an underlying susceptibility to nerve loss, in effect setting the stage for a disease's later appearance when prompted by environmental cues.

GENDER, REPRODUCTION, AND DEVELOPMENT

Recent research has stirred intense interest in the potential effects of pollutants on fertility and the normal development of the fetus. In the United States, a fifth of all pregnancies end in spontaneous abortion before the fifth month. Fifteen percent of all babies are born prematurely or with low birth weight. And close to 14 percent of all married couples in their childbearing years (aged 18–54) have trouble conceiving. The origin of an estimated 60 percent of reproductive and developmental disorders is unknown, leaving ample room for environmental pollutants to play a contributing role.

Epidemiological studies have recently detected adverse reproductive effects associated with a group of chemicals known as glycol ether solvents. These are used in a wide variety of products and industrial processes, including the manufacture of paints, varnishes, primers, film, and electrical wire insulation, and as printing inks, cleaners, and deicers on aircraft. A series of studies among semiconductor workers has revealed higher rates of spontaneous abortions and delayed conception among women who worked closely with these chemicals. Two of the studies came up with remarkably similar estimates of raised risk among women in a process called photolithography and among those with the greatest exposure to short-chain glycol ethers.

The U.S. Occupational Health and Safety Administration (OSHA) recently proposed lowering the maximum airborne concentrations of four of the glycol ethers to a fraction of their current legal levels. If OSHA adopts this rule, it will be the first time a substance or group of substances has been regulated by the agency because of its reproductive toxicity.

While glycol ethers pose a hazard in the workplace, a number of researchers are pursuing the theory that a whole menu of environmental pollutants have a broad influence on reproductive behavior and function by modulating the effects of hormones. Independently, many scientists have observed these effects in laboratory animals and in wildlife populations that live in polluted environments.

By enhancing or dulling the effects of hormones, these pollutants have a far-reaching effect on fertility,

gender, and sex-linked behavior. Traits that seem fixed or inherent are surprisingly malleable when changes in the internal hormonal environment occur. A group of Canadian scientists, for example, studied the offspring of gerbils; they discovered that the litter's sex ratio, a trait thought to be under the master control of genes, was strongly influenced by the hormonal environment of the mother at an early age. Developing females that shared the womb with male siblings got exposed to male hormones produced by their brothers. Unerringly, female gerbils with such exposure gave birth to more males when they themselves bore litters. Conversely, female fetuses that had been sandwiched between two fetal sisters produced more female offspring.

Recently, attention has focused on chemicals that duplicate or interfere with the effects of estrogen, the female sex hormone. Tampering with estrogen levels can have profound effects on the normal development of sexual organs of the fetus, and on later sexual function. In male laboratory rats, estrogen reduces the number of sperm and also inhibits the descent of the testes from the abdomen. (Male rat pups, like male human infants, are born with undescended testes.)

These effects are mirrored in studies of the only estrogen inadvertently tested in people—diethylstilbesterol. A higher than average number of men exposed to DES while in their mother's womb have testes that failed to descend. They also have a higher incidence of abnormal urethral tubes and reduced sperm counts and semen. Undescended testes are a major risk factor for testicular cancer. Two small studies also found that DES-exposed women (daughters of women who took DES during their pregnancies) were three to five times as likely to have a homosexual or bisexual orientation as women who were not exposed to the hormone through their mothers.

Exposure to other environmental estrogens is constant. Natural estrogens, known as phytoestrogens, exist in plants and fungi. Oral contraceptives, taken by millions of women, contain synthetic estrogens. High-fat diets also appear to increase circulating levels of estrogen in women, while increasing dietary fiber may have the opposite impact. Finally, a whole array of chemicals are estrogenic in their effects: PCBs, DBCP (a nematicide), and kepone and methoxychlor (both organochlorine pesticides), as well as some components of plastics, all have effects that resemble estrogen.

Recent evidence suggests that this load of environmental estrogens could have a pervasive influence on human reproduction. A 1992 study in the *British Medical Journal* by a group of Danish researchers reported a 50-percent decline in the quantity of sperm in human semen in 20 different countries. In an article several months later in *The Lancet,* Niels Skakkebaek, the lead author of the study and an endocrinologist at the University of Copenhagen, hypothesized that this dramatic spread of subfertility might be due to the "sea of estrogens" bathing the twentieth-century environment. The increase in subfertility that Skakkebaek and his colleagues discovered coincides with documented increases around the world in other reproductive defects, such as testicular cancer, failed testicular descent, and abnormalities of the urethra.

One mystery is how potent the different environmental estrogens are. For the most part, we do not know. Many environmental estrogens seem only weakly estrogenic. In theory, however, continuous exposure to weak estrogens or exposure at a critical point during development could lead to reproductive abnormalities and changes in sex-linked behavior.

A number of scientists believe that the natural estrogens found in plants are not the most likely explanation for widespread reproductive problems seen by scientists in wildlife populations in North America and in marine mammals around Scandinavian countries. Theodora Colborn, a zoologist at the World Wildlife Fund, claims that chemicals that interfere with hormones are responsible for the widespread reproductive dysfunction many wildlife biologists have observed among populations of birds and mammals around the Great Lakes in the United States and Canada. A list of the hormonally active chemicals in the Great Lakes includes PCBs, dioxin, and an assortment of pesticides.

A second mystery is how such a diverse group of structurally unrelated chemicals have practically identical biological effects. "It's been an enigma for years. We still don't know what makes an estrogen an estrogen from a chemical standpoint," says John McLachlan, scientific director of the U.S. National Institute of Environmental Health Sciences and an expert on environmental estrogens. McLachlan, for one, expects to find more and more substances that interfere with natural estrogen levels.

THE IMMUNE SYSTEM

AIDS has made clear the role of a healthy immune system in fending off disease by showing what happens when immunity is weakened or abolished. AIDS patients die of bacterial and viral infections, such as pneumonia and cytomegalovirus, that surmount low-ered immune barriers. Similarly, people who received organ transplants and take immunosuppressive drugs in order to lull their immune system into accepting alien tissue run a greater risk of developing cancer.

Researchers are still feeling their way around the main contours of the immune system, however. Basic questions, such as how large the immune system's reserve forces are and whether a well-functioning component can compensate for a disabled one, remain unresolved. Immunotoxicology is such a recent offshoot of toxicology that few chemicals have been banned or regulated because of their adverse effects on the immune system. "Not until the mid-eighties were there enough people on board to fully assess immunotoxicity of a variety of chemicals," says Loren Koller, dean of the Oregon State University School of Veterinary Medicine in Corvallis.

Many industrial chemicals and pollutants have an effect on one or more parts of the immune system when they are tested in animals. Benzene, dioxin, lead, mercury, ozone, nitrogen dioxide, PCBs, pesticides, and chemical mixtures like those found in contaminated groundwater all perturb the immune system. But it is difficult to say what these perturbations mean. Some epidemiological studies, for example, have found that people exposed to polybrominated biphenyls (used as fire retardants), asbestos, and certain metals experience changes in the populations of certain cells of the immune system (such as B cells and T cells). But scientists still have not determined the significance of these changes to overall immune functioning and health.

The best evidence for a link between broad environmental pollution and immune dysfunction may be the effect of air pollution on asthma and respiratory illnesses. Deaths as a result of asthma are rising in a number of industrial countries, including Australia, Canada, Denmark, New Zealand, Sweden, the United Kingdom, and the United States. In the last two, the prevalence of asthma also seems to be increasing, especially among children. Between 1982 and 1991, for example, the prevalence of asthma rose 56 percent in Americans under the age of 18, compared with 36 percent in the general population. Between 1979 and 1987, Americans under the age of four were the fastest growing group of people entering the hospital because of asthma attacks.

Many cases of asthma are due to an abnormal immune response known as allergic hypersensitivity. This lies at the opposite end of the spectrum from immune suppression. A suppressed immune system is less capable of mounting a response to invading bacteria and viruses. A hypersensitive immune system, on the other hand, is extra-vigilant, overreacting to the slightest provocation. Asthmatic attacks may be a manifestation of this underlying vigilance.

Certain kinds of air pollution common in industrial countries, such as acidic aerosols, nitrogen dioxide (NO_2), ozone, and sulfur dioxide (SO_2), worsen asthma. Acidic aerosols are formed from SO_2 and NO_2 by a chemical reaction that recruits ground-level ozone. (A different reaction, involving sunlight and the volatile organic chemicals present in automobile exhaust, produces ground-level ozone.) Studies of the effects of acidic aerosols and sulfur dioxides on asthma patients show that both pollutants increase spasms in the airways. Acidic aerosols are currently not included as one of the six "criteria" air pollutants (carbon monoxide, lead, nitrogen dioxide, ozone, particulates, and sulfur oxides) regulated under EPA's national air quality standards.

Particles are another pollutant whose role in triggering asthma may be substantial. They are made up of a variety of substances and come in many different sizes, which complicates the task of regulating them. Some particles, for example, incorporate acidic aerosols; others do not. Studies of the health effects of fine particles in Canada and in U.S. cities have found a connection between concentrations of particles in the air and hospital visits for asthma. A recent study in Seattle found an increased risk of asthma attacks (measured by emergency admissions) even when particulate concentrations in the air met federal air quality standards.

Scientists do not fully understand the causes of asthma and propose diverse explanations for its increase. Some of the theories include lack of physician training, indoor air pollution (especially cigarette smoke and house mites), and poorly regulated acidic and nonacidic particles. Whatever role these other factors play in asthma, there is little question that air pollution increases both the incidence and severity of asthma and that air pollutants linked to asthma exist at health-threatening levels in many urban areas.

PROOF AND THE LIMITS OF SCIENCE

Establishing an industrial pollutant or pollutants as the cause of a specific health outcome in people is no easy job. The task requires a mix of toxicological and epidemiological evidence. Data showing that a substance causes cancer in laboratory animals, for example, are generally not regarded as sufficient to prove carcinogenicity in people. EPA and the International Agency for Research on Cancer require both experimental and epidemiological data before listing substances as known human carcinogens.

Epidemiological evidence is considered more reliable than toxicological data, since it provides direct information on human health effects. But the very terms used by epidemiologists betray the limitations of their science: they speak of "associations" instead of "causes."

As a general rule, epidemiology is a blunt tool. Epidemiological studies often miss the small contribution of a pollutant among all the competing explanations for higher disease rates. Investigations of cancer clusters in communities in the United States, for example, have mostly failed to pinpoint a specific cause for the unusual outcroppings of the disease.

This is partly due to the fact that cancer is a broad outcome influenced by many factors, including alcohol, diet, and genes. But it is also the result of another weakness of many epidemiological studies—the lack of "statistical power." Rarer toxic effects that would occur only in a very large study population often fail to appear in small communities. Detecting small increases in risk, such as a one-in-a-million increase in the incidence of cancer, is practically impossible in most cases, since epidemiologists would have to study a very large population for a long period of time. (Cancer latency is often 20 years or more.) The chances that all the members of such a large population would stay in place long enough to let investigators count any excess cases of cancer are very slight. Epidemiologists also find it difficult to assess the health effects of slight exposures to toxicants, since the outcome of such exposures can be subtle and since it is difficult to reconstruct exposures that happened many years ago.

In general, the minimum increased risk that epidemiological studies can detect is somewhere between 10 and 20 percent. Yet U.S. regulations generally allow no more than a one-in-a-million risk of cancer from lifetime exposure to any chemical carcinogen. As a result, even risks that are many times above those that laws supposedly allow might exist but remain undetected by epidemiological studies, and unassigned to any environmental cause.

But the biggest disadvantage of epidemiological studies is that they only measure ill effects once they have occurred, and so furnish no predictions for future effects given present exposures. This limitation can be offset by toxicological studies, which can provide some warning of risk to humans before exposure occurs. But there are many gaps in toxicological evaluations of chemicals as well. Toxicologists are often forced to test at high doses, meaning that they also collect little direct information on health effects at low levels of exposure. High doses are used to ensure that any toxic effects that might exist are evoked in small populations of laboratory animals. Health outcomes at high levels of exposure are then extrapolated to effects at lower levels, using mathematical models. Finally, scientists make educated guesses as to the human health effects that might parallel theoretical low-level effects in laboratory animals.

Low-level exposures could be extremely important in terms of subtle and persistent toxicity. Human epidemiological studies of populations exposed to low levels of PCBs and lead have suggested that levels of exposures entirely within the range experienced by the general population can be detrimental to the central nervous system—slowing growth, delaying development, and creating IQ deficits.

Toxicology falls short of reflecting reality in another way. While most substances are tested individually, in isolation, people rarely encounter chemicals one at a time. Most exposure to solvents, for example, is to mixtures rather than single chemicals. Researchers have garnered little information on the health effects of various combinations of chemicals, but what little there is suggests that chemical interactions may affect toxicity profoundly. Adding chemicals together can produce synergistic, additive, or antagonistic effects. Alcohol, for example, enhances the toxicity of carbon disulfide and other solvents.

Pesticides, too, interact synergistically. Malathion and a pesticide known as EPN enhance each other's effects. EPA permitted both pesticides to be applied to 33 different crops. (The permit to use EPN was cancelled by EPA in 1987.) According to the National Research Council's latest report on the health effects of pesticides in children, "the existence of synergism at low levels of exposure cannot be assessed directly. It is conceivable that two compounds, innocuous by themselves, might interact chemically even at low doses to form a new substance that is toxic."

There is also some evidence suggesting that part of the population is more sensitive than the average person to the effects of a whole range of chemicals. For these people, who suffer from a syndrome known as multiple chemical sensitivity (MCS), the risks posed by even very slight exposure to chemicals might be much greater than currently is appreciated. MCS is an extreme example of the general problem of how to

tie chemical causes to real or apparent health outcomes. There is little evidence to suggest so far that chemically sensitive people actually have been injured. No adverse effects are generally believed to exist at such low concentrations, making MCS extremely controversial within the medical community. Nonetheless, people who have chemical sensitivity often claim their memory and other intellectual abilities have been damaged.

However the debate on MCS is resolved, the condition adds to the evidence that people differ remarkably in their susceptibility to the toxic effects of individual chemicals, for a variety of reasons. Genetic defects, poor nutrition, and lowered immunity can exacerbate certain toxic effects. Low-income children who lack iron and calcium in their diet, for example, tend to be more severely affected by lead exposure than better-nourished children exposed to similar lead levels.

The wide range in individual sensitivity to chemicals and scientific uncertainties about the potency of many substances mean that no neat formulas for the health risks posed by environmental pollutants exist. Even the most painstakingly performed "risk assessments" are plagued by genuine scientific controversy.

Yet in most countries where chemical emissions and pollution levels are controlled, regulations are based on the assumption that people are fairly similar in their susceptibility to toxicants and that numbers can be plugged into risk formulas. Regulatory agencies charged with evaluating the hazards posed by pollutants use a factor of 10 to account for individual differences that might make some people more vulnerable to toxic effects than others. This estimate of individual differences may not be appropriate. Risk assessments of suspected toxicants also generally fail to consider the extra danger many chemicals pose to the fetus during critical stages of development and children's weaker defenses against toxic hazards.

Just as worrisome is the fact that some portions of the population are exposed to more pollutants than others. Many workers, for example, are routinely exposed to health-threatening levels of toxic substances. U.S. laws often permit workers to be exposed to health risks from chemicals that are 100 times greater than those allowed for the general population—such as a one-in-a-thousand risk of cancer instead of one-in-a-million. In many developing countries, protections for workers are nonexistent or not enforced.

In the United States, environmental health risks are widely acknowledged to be more severe in low-income and minority communities. Traditional civil rights groups, such as the National Association for the Advancement of Colored People, have begun to turn their attention to "environmental justice." Studies by public interest groups have documented that three out of five African Americans and Hispanic Americans live in communities with toxic waste sites. Some 55 percent of poor black children have blood-lead levels associated with adverse effects on the nervous system. One quarter of poor white children run this risk, compared with 7 percent of affluent white children.

Studies in the United States and in other industrial countries have shown higher rates of cancer in farmworkers and other groups who work with pesticides, as mentioned earlier. Pesticides are also well-known neurotoxins. But little is known about rates of cancer or neurological illness among certain groups of farmworkers, such as migrants in the United States, and farm laborers in developing countries, who have high exposure to pesticides. (Only 1–2 percent of pesticide-related illnesses that occur in the United States are thought to be reported.)

The poor are at higher risk from chemical exposure for a number of reasons: they are more likely to work in dirty professions, to live close to polluted sites, to suffer from inadequate nutrition (which can exacerbate toxic effects), and to have less access to health care. A recent study found measurable impacts of low levels of lead on children's cognitive performance. The authors note that had they studied poor children in place of middle-class children, such effects might have been difficult to distinguish from other variables affecting development, despite the fact that the effects in poor children may loom even larger than those among affluent children.

Unfortunately, most risk assessments ignore the fact that exposure to toxic chemicals is unequal and rely instead on estimates of "average" exposure levels. But just as hunger can exist as a problem of serious proportions even while average food consumption is adequate, levels of toxic exposure may be a problem in many local regions without being a problem on a national level.

In order to really assess the risks posed by chemicals, regulators must act not only as toxicologists but as sociologists. "Who has poor access to health care, who lives in dirty neighborhoods, and who suffers from bad nutrition? From a public health perspective, these are all relevant answers to the question, Who is at risk from toxic exposure?" comments Robert Ginsburg, an environmental health consultant in Chicago.

He notes that one-part-per-million benzene exposure in Winnetka, Illinois, represents an entirely different risk than in a southeast Chicago neighborhood where 80 percent of the residents are poor and live close to a major highway, a steel plant, a landfill, and a sewage treatment plant.

Translating such realities into policies that better safeguard public health is not exactly straightforward. The most desirable, long-term solution is preventing exposure in the first place. This is best accomplished by lowering our reliance on synthetic chemicals, perhaps through a combination of pushes and prompts, such as tax incentives and federally mandated phase-outs. Farmers can be weaned off pesticides, for example, if they move toward integrated pest management, which combines nonchemical methods of pest control, such as rotating crops and introducing pests' natural enemies, with selective use of chemicals. In fact, the Clinton administration recently proposed just such an approach.

Implementing similar measures in industry is far more complicated, but many substitute chemicals or new processes may remove the need for a number of toxic chemicals currently in use. The semiconductor industry, for example, originally dependent on such toxic solvents as methylene acid and hydrochloric acid for its manufacturing processes, during the last few years has substituted more-benign water-based processes in a number of instances. Other substances, such as lead and mercury, could be removed from many products. Until recently, for example, mercury was allowed in interior latex paints in the United States. The death of a toddler in Michigan due to evaporation of mercury fumes from fresh paint belatedly led to an EPA ban.

Unfortunately, more modest, interim measures will probably govern the exposure to toxic substances that communities and workers endure for the next few generations. But some intermediate steps nonetheless have the potential to make a significant contribution to protecting the public. The first one is education: both workers and the general public need much more information about the chemical hazards they routinely face. A second step involves more research into groups likely to live in hot spots of contamination, such as around the Great Lakes region in North America, in the Amazon Basin, and in inner cities everywhere.

Caution also clearly calls for the thorough evaluation of suspect substances for their ability to cause a wide range of health effects in different groups of people. Safeguarding public health is presumably the ultimate goal of regulatory agencies. Adopting one of the central principles of the public health field—prevention of disease—would amount to a revolution in the way most governments now regulate chemicals. Government, industry, and the public have implicitly regarded synthetic chemicals as benign until epidemiological studies provide evidence to the contrary. By definition, action at this stage comes too late.

Questions no doubt remain about the precise contribution of industrial pollutants to human disease and dysfunction. But there is ample evidence that cumulatively they are causing significant harm to humans. Given the shadow this casts over these "conveniences" of modern life, overturning the presumption of innocence about chemicals is long overdue.

For Further Discussion

1. In light of Misch's observations, what should be our public policy in the face of incomplete information about potential harms? Where does your answer fit within the optimistic, moderate, and pessimistic alternatives described in the readings of Chapter 3?

2. Who benefits and who pays the costs of our contemporary reliance on chemicals? Do the same people who benefit bear the risks?

3. Is risk assessment a technical question best left to environmental and health scientists, or is it a political question that deserves public discussion and debate? What are the costs and benefits of each alternative?

4. Should we continue to live in ways that rely so heavily on the widespread use of chemicals? What social and political changes would be needed before we could reduce chemical use significantly?

The Case for Optimal Pollution

William Baxter

In this classic essay, William Baxter lays out clearly the free-market approach to pollution. Baxter acknowledges his philosophical commitments: environmental policy should serve people, not penguins; freedom of choice is a central value; humans should be treated as ends, not means; waste is a bad thing; individuals should have incentives and opportunities to pursue their own interests and preferences. With these starting points, Baxter argues that public policy should aim for an optimal level of pollution, rather than for the total elimination of pollution, and that economic markets are the best means for attaining this goal.

I start with the modest proposition that, in dealing with pollution, or indeed with any problem, it is helpful to know what one is attempting to accomplish. Agreement on how and whether to pursue a particular objective, such as pollution control, is not possible unless some more general objective has been identified and stated with reasonable precision. We talk loosely of having clean air and clean water, of preserving our wilderness areas, and so forth. But none of these is a sufficiently general objective: each is more accurately viewed as a means rather than as an end.

With regard to clean air, for example, one may ask, "how clean?" and "what does clean mean?" It is even reasonable to ask, "why have clean air?" Each of these questions is an implicit demand that a more general community goal be stated—a goal sufficiently general in its scope and enjoying sufficiently general assent among the community of actors that such "why" questions no longer seem admissible with respect to that goal.

If, for example, one states as a goal the proposition that "every person should be free to do whatever he wishes in contexts where his actions do not interfere with the interests of other human beings," the speaker is unlikely to be met with a response of "why." The goal may be criticized as uncertain in its implications or difficult to implement, but it is so basic a tenet of our civilization—it reflects a cultural value so broadly shared, at least in the abstract—that the question "why" is seen as impertinent or imponderable or both.

I do not mean to suggest that everyone would agree with the "spheres of freedom" objective just stated. Still less do I mean to suggest that a society could subscribe to four or five such general objectives that would be adequate in their coverage to serve as testing criteria by which all other disagreements might be measured. One difficulty in the attempt to construct such a list is that each new goal added will conflict, in certain applications, with each prior goal listed; and thus each goal serves as a limited qualification on prior goals.

Without any expectation of obtaining unanimous consent to them, let me set forth four goals that I generally use as ultimate testing criteria in attempting to frame solutions to problems of human organization. My position regarding pollution stems from these four criteria. If the criteria appeal to you and any part of what appears hereafter does not, our disagreement will have a helpful focus: which of us is correct, analytically, in supposing that his position on pollution would better serve these general goals. If the criteria do not seem acceptable to you, then it is to be expected that our more particular judgments will differ, and the task will then be yours to identify the basic set of criteria upon which your particular judgments rest.

My criteria are as follows:

1. The spheres of freedom criterion stated above.

2. Waste is a bad thing. The dominant feature of human existence is scarcity—our available resources, our aggregate labors, and our skill in employing both have always been, and will continue for some time to be, inadequate to yield to every man all the tangible and intangible satisfactions he would like to have. Hence, none of those resources, or labors, or skills, should be wasted—that is, employed so as to yield less than they might yield in human satisfactions.

3. Every human being should be regarded as an end rather than as a means to be used for the betterment of another. Each should be afforded dignity and regarded as having an absolute claim to an even-handed application of such rules as the community may adopt for its governance.

4. Both the incentive and the opportunity to improve his share of satisfactions should be preserved to every individual. Preservation of incentive is dictated by the "no-waste" criterion and enjoins against the continuous, totally egalitarian redistribution of satisfactions, or wealth; but subject to that constraint, everyone should receive, by continuous redistribution if necessary, some minimal share of aggregate wealth so as to avoid a level of privation from which the opportunity to improve his situation becomes illusory.

The relationship of these highly general goals to the more specific environmental issues at hand may not be readily apparent, and I am not yet ready to demonstrate their pervasive implications. But let me give one indication of their implications. Recently scientists have informed us that use of DDT in food production is causing damage to the penguin population. For the present purposes let us accept that assertion as an indisputable scientific fact. The scientific fact is often asserted as if the correct implication—that we must stop agricultural use of DDT—followed from the mere statement of the fact of penguin damage. But plainly it does not follow if my criteria are employed.

My criteria are oriented to people, not penguins. Damage to penguins, or sugar pines, or geological marvels is, without more, simply irrelevant. One must go further, by my criteria, and say: Penguins are important because people enjoy seeing them walk about rocks; and furthermore, the well-being of people would be less impaired by halting use of DDT than by giving up penguins. In short, my observations about environmental problems will be people-oriented, as are my criteria. I have no interest in preserving penguins for their own sake.

It may be said by way of objection to this position, that it is very selfish of people to act as if each person represented one unit of importance and nothing else was of any importance. It is undeniably selfish. Nevertheless I think it is the only tenable starting place for analysis for several reasons. First, no other position corresponds to the way most people really think and act—i.e., corresponds to reality.

Second, this attitude does not portend any massive destruction of nonhuman flora and fauna, for people depend on them in many obvious ways, and they will be preserved because and to the degree that humans do depend on them.

Third, what is good for humans is, in many respects, good for penguins and pine trees—clean air for example. So that humans are, in these respects, surrogates for plant and animal life.

Fourth, I do not know how we could administer any other system. Our decisions are either private or collective. Insofar as Mr. Jones is free to act privately, he may give such preferences as he wishes to other forms of life: he may feed birds in winter and do with less himself, and he may even decline to resist an advancing polar bear on the ground that the bear's appetite is more important than those portions of himself that the bear may choose to eat. In short my basic premise does not rule out private altruism to competing life-forms. It does rule out, however, Mr. Jones' inclination to feed Mr. Smith to the bear, however hungry the bear, however despicable Mr. Smith.

Insofar as we act collectively on the other hand, only humans can be afforded an opportunity to participate in the collective decisions. Penguins cannot vote now and are unlikely subjects for the franchise—pine trees more unlikely still. Again each individual is free to cast his vote so as to benefit sugar pines if that is his inclination. But many of the more extreme assertions that one hears from some conservationists amount to tacit assertions that they are specially appointed representatives of sugar pines, and hence that their preferences should be weighted more heavily than the preferences of other humans who do not enjoy equal rapport with "nature." The simplistic assertion that agricultural use of DDT must stop at once because it is harmful to penguins is of that type.

Fifth, if polar bears or pine trees or penguins, like men, are to be regarded as ends rather than means, if they are to count in our calculus of social organization, someone must tell me how much each one counts, and someone must tell me how these life-forms are to be permitted to express their preferences, for I do not know either answer. If the answer is that certain people are to hold their proxies, then I want to know how those proxy-holders are to be selected: self-appointment does not seem workable to me.

Sixth, and by way of summary of all the foregoing, let me point out that the set of environmental issues under discussion—although they raise very complex technical questions of how to achieve any objective—ultimately raise a normative question: what *ought* we to do? Questions of *ought* are unique to the human mind and world—they are meaningless as applied to a nonhuman situation.

I reject the proposition that we *ought* to respect the "balance of nature" or to "preserve the environment" unless the reason for doing so, express or implied, is the benefit of man.

I reject the idea that there is a "right" or "morally correct" state of nature to which we should return. The word "nature" has no normative connotation. Was it "right" or "wrong" for the earth's crust to heave in contortion and create mountains and seas? Was it "right" for the first amphibian to crawl up out of the primordial ooze? Was it "wrong" for plants to reproduce themselves and alter the atmospheric composition in favor of oxygen? For animals to alter the atmosphere in favor of carbon dioxide both by breathing oxygen and eating plants? No answers can be given to these questions because they are meaningless questions.

All this may seem obvious to the point of being tedious, but much of the present controversy over environment and pollution rests on tacit normative assumptions about just such nonnormative phenomena: that it is "wrong" to impair penguins with DDT, but not to slaughter cattle for prime rib roasts. That it is wrong to kill stands of sugar pines with industrial fumes, but not to cut sugar pines and build housing for the poor. Every man is entitled to his own preferred definition of Walden Pond, but there is no definition that has any moral superiority over another, except by reference to the selfish needs of the human race.

From the fact that there is no normative definition of the natural state, it follows that there is no normative definition of clean air or pure water—hence no definition of polluted air—or of pollution—except by reference to the needs of man. The "right" composition of the atmosphere is one which has some dust in it and some lead in it and some hydrogen sulfide in it—just those amounts that attend a sensibly organized society thoughtfully and knowledgeably pursuing the greatest possible satisfaction for its human members.

The first and most fundamental step toward solution of our environmental problems is a clear recognition that our objective is not pure air or water but rather some optimal state of pollution. That step immediately suggests the question: How do we define and attain the level of pollution that will yield the maximum possible amount of human satisfaction?

Low levels of pollution contribute to human satisfaction but so do food and shelter and education and music. To attain ever lower levels of pollution, we must pay the cost of having less of these other things. I contrast that view of the cost of pollution control with the more popular statement that pollution control will "cost" very large numbers of dollars. The popular statement is true in some senses, false in others; sorting out the true and false senses is of some importance. The first step in that sorting process is to achieve a clear understanding of the difference between dollars and resources. Resources are the wealth of our nation; dollars are merely claim checks upon those resources. Resources are of vital importance; dollars are comparatively trivial.

Four categories of resources are sufficient for our purposes: At any given time a nation, or a planet if you prefer, has a stock of labor, of technological skill, of capital goods, and of natural resources (such as mineral deposits, timber, water, land, etc.). These resources can be used in various combinations to yield goods and services of all kinds—in some limited quantity. The quantity will be larger if they are combined efficiently, smaller if combined inefficiently. But in either event the resource stock is limited, the goods and services that they can be made to yield are limited; even the most efficient use of them will yield less than our population, in the aggregate, would like to have.

If one considers building a new dam, it is appropriate to say that it will be costly in the sense that it will require x hours of labor, y tons of steel and concrete, and z amount of capital goods. If these resources are devoted to the dam, then they cannot be used to build hospitals, fishing rods, schools, or electric can openers. That is the meaningful sense in which the dam is costly.

Quite apart from the very important question of how wisely we can combine our resources to produce goods and services is the very different question of how they get distributed—who gets how many goods? Dollars constitute the claim checks which are distributed among people and which control their share of national output. Dollars are nearly valueless pieces of paper except to the extent that they do represent claim checks to some fraction of the output of goods and services. Viewed as claim checks, all the dollars outstanding during any period of time are worth, in the aggregate, the goods and services that are available to be claimed with them during that period—neither more nor less.

It is far easier to increase the supply of dollars than to increase the production of goods and services—printing dollars is easy. But printing more dollars doesn't help because each dollar then simply becomes a claim to fewer goods, i.e., becomes worth less.

The point is this: many people fall into error upon hearing the statement that the decision to build a dam, or to clean up a river, will cost $X million. It is regrettably easy to say: "It's only money. This is a wealthy country, and we have lots of money." But you cannot build a dam or clean a river with $X million—unless you also have a match, you can't even make a fire. One builds a dam or cleans a river by diverting labor and steel and trucks and factories from making one kind of goods to making another. The cost in dollars is merely a shorthand way of describing the extent of the diversion necessary. If we build a dam for $X million, then we must recognize that we will have $X million less housing and food and medical care and electric can openers as a result.

Similarly, the costs of controlling pollution are best expressed in terms of the other goods we will have to give up to do the job. This is not to say the job should not be done. Badly as we need more housing, more medical care, and more can openers, and more symphony orchestras, we could do with somewhat less of them, in my judgment at least, in exchange for somewhat cleaner air and rivers. But that is the nature of the trade-off, and analysis of the problem is advanced if that unpleasant reality is kept in mind. Once the trade-off relationship is clearly perceived, it is possible to state in a very general way what the optimal level of pollution is. I would state it as follows:

People enjoy watching penguins. They enjoy relatively clean air and smog-free vistas. Their health is improved by relatively clean water and air. Each of these benefits is a type of good or service. As a society we would be well advised to give up one washing machine if the resources that would have gone into that washing machine can yield greater human satisfaction when diverted into pollution control. We should give up one hospital if the resources thereby freed would yield more human satisfaction when devoted to elimination of noise in our cities. And so on, trade-off by trade-off, we should divert our productive capacities from the production of existing goods and services to the production of a cleaner, quieter, more pastoral nation up to—and no further than—the point at which we value more highly the next washing machine or hospital that we would have to do without than we value the next unit of environmental improvement that the diverted resources would create.

Now this proposition seems to me unassailable but so general and abstract as to be unhelpful—at least unadministerable in the form stated. It assumes we can measure in some way the incremental units of human satisfaction yielded by very different types of goods. The proposition must remain a pious abstraction until I can explain how this measurement process can occur. . . .

But I insist that the proposition stated describes the result for which we should be striving—and again, that it is always useful to know what your target is even if your weapons are too crude to score a bull's-eye.

For Further Discussion

1. How do the policy recommendations offered by Baxter flow from the value assumptions of free market economics?

2. Explain the distinction between clean air and water and safe air and water. According to Baxter,

who are the best judges for determining relative levels of safety? Do you agree?

3. What does Baxter mean by "waste"? Apply this definition to such resources as water, soil, trees, wilderness areas, rare species. Do you agree with Baxter's understanding?

4. Baxter suggests that the assumption that people are selfish is the only tenable starting place for policy analysis because no other position better corresponds to reality. Do you agree? Are all people naturally selfish? Should public policy be based on this assumption?

5. How might someone such as Christopher Stone (Chapter 6) respond to Baxter's challenge that

"someone must tell me how these life forms are to be permitted to express their preferences"?

6. Do you agree with Baxter that "what is good for humans is, in many respects, good for penguins and pine trees"?

7. Do you agree with the claim that "the word 'nature' has no normative connotation"?

8. Baxter suggests that "every person should be free to do whatever he wishes in contexts where his actions do not interfere with the interests of other human beings." When someone's actions do interfere with the interests of others, should those actions be prohibited, or should an optimal level of interference be sought?

The Morality of Pollution Permits

Paul Steidlmeier

One challenge to economic solutions to environmental quality is that within competitive markets firms have little incentive to reduce pollution beyond present industry standards. Voluntarily allocating resources to reduce pollution beyond what is being done by one's competitors places an individual firm at a competitive disadvantage. Further, if the government sets mandatory standards above those set within an industry, some firms (the ones that can no longer compete when these added costs are factored in) may be forced out of business. In turn, this leads to less competition and higher costs all around.

To meet these challenges, a system of tradable permits in pollution allowances was established by the Clean Air Act of 1990. This system works by first establishing a level of acceptable pollution (in fact, but not necessarily, using economic criteria for this). After this level is established, firms are granted permits, literally a license to pollute, which allows each firm a certain level of emissions. If every firm meets its permit requirements, the overall pollution goal is met. The key part of this policy is that firms are allowed to trade and sell these permits. This means that a firm that is efficient enough to reduce its pollution beyond the permit levels is not put at a competitive disadvantage. Firms are allowed to trade or sell the "leftover" permission to pollute, presumably to a firm that cannot otherwise meet

its goals. Thus, there is an incentive to reduce pollution even beyond the required level whereas firms unable to meet these levels have some leeway to remain competitive.

Significant controversy has surrounded this policy. Is it a license to pollute? Does it abandon the goal of eliminating pollution altogether? In this selection, management professor Paul Steidlmeier reviews a wide range of ethical issues raised by the policy of pollution permits.

I. INTRODUCTION

The question of whether the end justifies the means has been one of the staples of ethical discourse over the centuries. In the Clean Air Act of 1990 marketable permits in pollution allowances have emerged as one of policy makers' favorite means to the end of environmental integrity. Many observers have disputed whether such allowances are at all effective in improving the environment. More important than their effectiveness, however, people have challenged such permits on the grounds that they are not a morally acceptable means to the end because they do not eliminate pollution, but rather set levels of "acceptable" risk. If marketable pollution permits are a means to an end, that end itself stands in need of clarification. Specifically, is all pollution to be eliminated or might there be acceptable levels of pollution that are deemed negligible in that no one is clearly harmed? In assessing the appropriateness of pollution permits, I examine both of these alternatives.

. . .

III. MARKETABLE POLLUTION PERMITS AND PUBLIC POLICY

There is a tangle of different positions regarding ecology and the environment and the use of economic incentives. A person's position combines (1) an ecological world view and methods of moral reasoning, (2) an assessment of empirical fact, and (3) an assessment of manageability in view of the public policy balance of power. For purposes of discussion I have summarized the various positions under three general *typologies*: (1) resourcism's legal/market ecology strategies, (2) the preservationists' public policy/regulatory ecology strategies, and (3) postmodern ecology strategies. Although these typologies (especially the third) represent oversimplifications of the positions of representative groups, they are, nonetheless, adequate to reveal some important differences in ethical, empirical, and social power analyses on the part of various protagonists. . . .

Marketable Pollution Allowances

The first step in implementing a system of marketable permits involves setting a target level for environmental quality. After the level is set, this environmental quality level then has to be more precisely defined in terms of total allowable emissions. Currently an allowance is a license to emit a single ton of sulfur dioxide or nitrogen oxide per year. Permits, which function like property rights to emissions, are allocated to firms. Firms are allowed to trade these permits either internally, within or between plants of the same firm, or between firms: buying them when firms exceed allowable emissions and selling them when they accumulate unused emission rights. In economic theory, *assuming* that firms minimize their overall production costs and that permit markets are competitive, marketable allowances represent a cost-effective economic way of limiting the aggregate pollution emissions level.

. . .

Legal/Market Economism

In 1990 Congress instigated a public debate by announcing its intentions to pass a new Clean Air Act. Central to this debate are the views of the legal/market ecologists. Although they tend to deplore the degradation of the environment, they frame the issues within an individualist (atomistic) view of society, as well as an atomistic view of nature, and try to limit environmental policy to legal or market mechanisms. The legal approach is centered around the establishment of environmental property rights while the market approach focuses upon reducing costs and maximizing profits. Advocates of this position hold that individuals and free associations of people who cause harm to others by polluting the environment should be held liable for damages to third parties.

The world view of legal/market ecologists embodies a quasi-libertarian and contractual approach to ecology. According to this view, people act egotistically. The relation of people to nature is governed by the free choice of individuals, conditioned by the fact that all enjoy due liberty. The key concern is whether the self-interest of one party also respects the legitimate self-interest of others. Market incentives and legal redress of abuses are favored as the best means to ensure that these interests are respected.

Legal/market ecologists consider public policy interventions not only unnecessarily coercive, but also wasteful and inefficient. For them the medicine is worse than the disease. These ecologists oppose any "big government" coercion while at the same time insisting upon the honoring of contracts between free agents.

Government, they argue, has a role to facilitate the functioning of legal and market mechanisms. It should not, however, be the principal actor. Groups of this persuasion, such as the Heritage Foundation and the CATO Institute, have been highly vocal critics of the Environmental Protection Agency, because, in principle, they view the government's regulatory role as both being ethically coercive and inefficient. Furthermore, they believe that EPA programs do more harm than good. In particular, by deterring private enterprise from voluntarily clean ups, these programs inhibit the emergence of ecological market forces.

In their moral critique, these groups concentrate on five issues: the ranking system of environmental harms, the financial mechanisms, the approach to liability, inefficiency, and the selection of improper means to the end. They hold that the Environmental

Protection Agency represents the wrong means to the end. They argue that building in economic incentives is a better approach. If the main threat to private companies were the law of torts (common law), businesses would be confronted with a much more effective threat to their livelihood. It is improper to rely on the government when private solutions are available. Relying on common law instead of statutory regulation would leave control to the market place, allowing for private parties to have more leeway than excessive regulation by the government permits.

These groups clearly link their non-interventionist market approach with the legal framework. In doing so, they make a number of assumptions. For example, they believe that the marketplace is best equipped to handle the environment on the grounds that the profit incentive combined with the severe penalties involved in a possible adverse common law suit are enough to resolve these problems.

One possible solution to our environmental problems involves an extension of property rights. For example, there is currently little or no recognition of property rights to the atmosphere. As a result, no private entity or individual has an incentive to monitor the condition of the air and protect it against pollution. Trusting the protection of the atmosphere to private sources, the legal/market ecologists argue, would be far more effective than trusting it to various groups of bureaucrats. Air rights could be established in the same way that property rights for oil are. Once air was no longer seen as a "free good," there would be an entirely different attitude toward those who contaminate it. Every property owner, in effect, would then be a police officer, overseeing his or her own rights. In this way, creating property rights in the atmosphere would vastly increase the number of people with a strong incentive to protect our environment.

Although the provision to allow market trading in pollution permits finds a hospitable reception among the legal/market ecologists, they in fact would prefer no regulatory EPA role at all. If all regulation were eliminated, they would then rely on market forces and the coherent logic of the law. Whether these operational assumptions will work, however, is an open question. Although common law provides a logical framework for handling environmental problems, up to this point, it has not proven to be adequate in practice. There are a number of reasons why it has not yet worked. (1) Rulings by various courts have proven to be inconsistent among themselves. (2) The legal system has historically emphasized compensa-

tion over prevention and future protection of the public. (3) Bringing suit involves very high expenses and this deters potential plaintiffs. (4) Lawsuits are invariably brought after the fact. They are, therefore, not preventive except insofar as a ruling may be assumed to establish legal precedent and, in this way, function as a deterrent. (5) The technological complexity of environmental problems leaves few courts equipped to handle them. This complexity leads judges to transfer problems to government agencies which supposedly have expertise. (6) The question of the moral standing of the environment itself is not explicitly considered.

The belief in a market approach is, likewise, logically based and cannot be facilely discounted. The federal government has allocated billions of dollars to clean up the environment. These expenditures have stimulated technological innovation and spurred many new companies to enter this new market. For these new companies, there is a ready economic demand to clean up pollution caused by others or to develop means to prevent it in the first place.

Companies and industries that have traditionally caused pollution also present a market argument, but with a different focus. They want anti-pollution regulations to be postponed until they become economically viable. . . .

The Traditional Public Policy Approach

Although the legal/market approach is logically coherent, its operational assumptions have not yet been borne out. Because of the above noted problems associated with common law solutions to environmental pollution and because the markets have not brought the problem under control, public policy ecologists feel that federal and state statutory law is needed to protect U.S. citizens against the dangers posed by environmental pollution.

The public policy ecologists see the environment as a common heritage and insist that economic concerns be subordinated to maintaining the holistic integrity of the environment. In many ways, this camp insists that good ecology and sound economics are not in conflict (the argument turning on the definitions of "good" and "sound").

According to the public policy ecologists, nature is a common possession which is not only useful but holistic, and which, therefore, cannot be left either to a utilitarian economic calculus or libertarian caprice. They recognize that there are market fundamentalists who would subdivide Yellowstone park because doing so would be efficient economically in terms of supply and demand. Public environmental policy in the United States has clearly been preservationist. Proponents of this preservationist position suggest a higher norm than either economic utilitarianism or individual liberty. They treat the environment as a public good that is to be managed in the human interest. This interest is broadly defined to include not only economic interests, but also cultural and aesthetic interests.

The public policy ecologists are "interventionist" because they favor strict regulation through public policy and government institutions so as to secure the public interest. They believe that public policy is the only way to get everyone to act properly: to neglect public policy means to ignore our obligations to present and future generations who will suffer from pollution.

Advocates of the public policy approach tend to see polluters as driven by greed and narrow self-interest such that they must be forcibly restrained. At the same time, they argue that market forces place coercive limits upon freedom such that one must pollute in order to survive. They view public policy as a market improvement and an enhancement of the freedom to act responsibly. Moreover, with respect to conscience, they emphasize the collective bonds uniting all of humanity and the necessity for public action to preserve the public good.

The social ethical program that this group favors is evident from recent history. Congressional action during the 1970s (an era that came to be called the "environmental" decade) included several statutes intended to narrow the range of acceptable disposal options. Although a clean air act was passed in 1963, the 1970 Clean Air Act was the first comprehensive air pollution law. It has been amended in 1977 and in 1990.

The preservationist group is generally favorable to trading in pollution permits as a practical measure to get industry on board to solve the problem. This approach is consistent with the traditional methodology of setting minimum standards of acceptable environmental risks which are then set in a "trade-off balance" with economic costs. At the same time, the preservationists want the EPA to retain administrative and regulatory control of the market by setting emission standards, performing environmental audits, and assessing penalties. It is just these points that the Hahn and Hester study found to have eviscerated the effectiveness of permit trading programs. Those in the legal/market camp want to eliminate these transaction costs by leaving emission controls to market forces, with disputes settled in the courts.

Toward a Postmodernist Environmental Paradigm

Despite the flurry of all this legislation, many who favor public regulation of the environment remain profoundly dissatisfied with what has come to pass. Barry Commoner, the director of the Center for the Biology of Natural Systems at Queens College, City University of New York, has stated that the EPA practices bad science and bad policy and is failing to protect the environment. Whether it is correct to group Commoner with the postmodernists is open to debate because his thought spans a number of decades. At times, his thought seems very close to traditional public policy approaches. However, recently, he has objected to the traditional approach in one important respect: the setting of allowable minimum levels of pollution and risk. In doing so, in my opinion, he sounds a postmodernist theme even though he may not articulate the philosophy of ecocentrism, deep ecology, or ecofeminism, of which he is critical. Commoner wants to eliminate pollutants rather than merely control them at some level of "acceptable risk." This section is based on a speech that Commoner gave to the EPA in January 1988. In it, he focuses upon (1) production technology and (2) the scientific basis of the regulatory process.

Commoner believes that even though environmental pollution is a nearly incurable disease, it can be prevented. Environmental degradation is built into the modern instruments of production. Most of our environmental problems are the inevitable result of the sweeping changes in the technology of production that transformed the U.S. economic system after World War II. Only in the few instances in which the technology of production has been changed has the environment been substantially improved. In most cases, the production technology remains unchanged.

Attempts have been made to trap pollutants in appended control devices. Such attempts, however, only improve the environment at best modestly, and in some cases, not at all. If a pollutant is attacked at the point of origin, it can be eliminated; once it is produced, it is too late. Unfortunately, the legislative base of the U.S. environmental program has been created without reference to the origin of the crisis that it is supposed to solve. Our environmental laws do not discuss the origin of environmental pollutants—why we are afflicted with pollutants that the laws are designed to control. Because environmental legislation has ignored the origin of the assault on environmental quality, it has only dealt with its subsequent effects. Moreover, because the legislation has defined the disease (environmental pollution) as a collection of symptoms, it mandates only mitigating measures. The notion of preventing pollution—the only measures that really work—appears but fitfully in the environmental laws and has never been given any administrative force.

Commoner believes that the great majority of the assaults on the environment are, in fact, preventable. His approach would mean sweeping changes in the major systems of production—agriculture, industry, power production, and transportation. This approach represents social (as contrasted with private) governance of the means of production. Commoner contends that such measures would also lead to greater economic efficiency. He argues that a good deal of the U.S. economic decline derives from the fact that the new, highly polluting post–World War II production technologies are based on large-scale, centralized, capital and energy intensive facilities. As a result, the country's overall economic efficiency is now heavily encumbered by low capital productivity and low energy productivity. Thus, he argues, the technological changes that reduce environmental impact can also improve economic productivity. For example, decentralized electric power systems, by reducing fuel consumption, improve not only air pollution, but the economic efficiency of power production as well.

On this point, Commoner finds support from an unexpected source, Michael Porter, a leading international expert in strategic management at Harvard Business School, who argues that:

> Stringent standards for product performance, product safety, and environmental impact pressure companies to improve quality, upgrade technology, and provide features which respond to consumer and social demands. Easing standards, however tempting, is counterproductive.

According to Commoner, today's largely unsuccessful regulatory effort is based on a defective but now well-established process. First, EPA estimates the degree of harm represented by different levels of numerous environmental pollutants. Next, it chooses some "acceptable" level of harm and establishes emission and/or ambient concentration standards that can presumably achieve that risk level. Polluters are then expected to respond by introducing control measures that will bring emissions or ambient concentrations to the required levels. If the regulation survives the inevitable challenges from industry, the polluters will invest in the appropriate control systems. If all goes well, and it frequently does not, at least some areas of the country and some production facilities will eventually be in compliance with the regulation. The net result, however, is that the "acceptable" pollution level is frozen in place because the industries, having heavily invested in the equipment designed to just reach the required level, are unlikely to invest in further improvements. Moreover, the public, having been told that the accompanying hazard is "acceptable," is likely to be equally satisfied.

Clearly, this process is the inverse of a preventative, public health approach. Because it strives not for the continuous improvement of environmental health, but for the social acceptance of some, hopefully low, risk to health, some level of pollution, and some risk to health is the "unavoidable" price that must be paid for the material benefits of modern technology. This approach involves an empirical scientific question which needs to be resolved on a case-by-case basis. Depending on the pollutant, it is possible that there may be no real risk to health. Thus, the methodology which aggregates all pollutants together may lead to false conclusions regarding risk. Nevertheless, in most cases, in which there are substantial risks to health, the present regulatory approach, by setting a standard of "acceptable" exposure to the pollutant, erects an administrative barrier that blocks further improvement in environmental quality. In contrast, the preventative approach at least aims at progressively reducing the risk to health.

How does the EPA decide when to stop and at what level to set standards? Since the pollutants' ultimate effect can often be assessed by the number of lives lost, risk-benefit analysis requires that a value be placed on a human life. Doing so, however, inhibits society's duty to deal with important moral questions (i.e., how many people should be allowed to become ill or even die as a result of specific policy decisions) by permitting the risk-benefit equation to masquerade as science rather than as an ethic displaying little regard for individual human life and welfare. In this way, regulatory agencies have been driven into positions that seriously diminish the force of social morality, eroding the integrity of regulation and diminishing public faith in the meaning of environmental legislation.

A related area of concern is the impact of the current process on science itself. Although the scientific participants may be convinced that their decisions are evenhanded and objective, the consequences are not.

Each such decision means that some people will save a good deal of money and others will spend more, that some people will be more concerned about their children's health and others less. For a few people, the decision creates a political problem and, for a few others, a welcome political opportunity. These are simply the facts of regulatory life. In this situation, there is an understandable tendency to find purely scientific grounds, which appear to be free of economic or other judgments, for unequivocal standards of exposure—a firm line below which there is a simple message: "healthy." The no-effect level (NOEL) is such a standard, a threshold level below which it can be said, on presumably objective and scientific grounds, that all is well. On the other hand, if there is a linear relationship between dosage and effect, determining the allowable standard moves from the seemingly solid realm of science to the more arguable domain of policy judgment. Since every level of exposure, no matter how small, results in a comparable degree of medical risk, the choice of an "acceptable" standard must somehow balance the expected harm against some other value—the supposed worth of a human life, or the cost of controlling or cleaning up the pollutant. However, if the risk assessment is changed—for example, by increasing the dosage that is expected to generate a particular risk level—the standard can be altered without changing the social judgment, thus avoiding the contentious area of discourse.

According to Commoner, there is an inherent contradiction between "science" and "policy." In the present context, the relevant attributes of science are, first, its demand for rigorous, validated methods that are independent of the expected results, and second, its objectivity, or the independence of the data and analysis used to reach the results from the interests of those affected by them. In contrast, policy is usually defined as "prudence or wisdom in the management of affairs" and "management or procedure based primarily on material interest." Currently, none of us is ready to prescribe what should be done to remedy specific environmental problems, for to do so would require the courage to challenge the taboo against even questioning the present dominance of private interests over the public interest. Such questioning calls for both good science *and* wise politics.

In advocating a "no-effect level," Commoner hints at a postmodernist vision of economic society. He is not, however, in the same camp as the ecocentrists, the deep ecologists, or the ecofeminists. These groups would find themselves in a more profound philosophical dissent over tradable permits because the

well-being interests of sentient nonhumans and of ecosystems are not absolutely respected—that is, their moral claims are subject to trade-offs based on human interests.

IV. THE BALANCE OF POWER IN THE PUBLIC POLICY PROCESS AND THE LEGITIMACY OF "SECOND-BEST" SOLUTIONS

Congress, the EPA, private enterprise, and the public are all challenged to seek reliable long-term solutions to the problem of managing the environment in a morally responsible way. Sound policy is multifaceted. It must address the economic effects of policy, improve the functioning of markets and the law, and increase the efficiency of public agencies. There are some recent indications that society's "ecological efficiency" is growing. Indeed, ends and means are in much sharper focus today than thirty years ago, and polluters can less easily escape culpability.

We live in a pluralistic society. To get anything done, the active cooperation of countless people with different ecological world views must be elicited. For something to work, it must be empirically comprehensive, value critical, technologically feasible, and systems manageable. When it comes to clean air, none of these issues is ever fully resolved, as each possesses a dynamic, developmental characteristic. Empirical knowledge of the ecology as well as ecological technologies are constantly developing. A critical value-based assessment of the ecology, likewise, shows continual refinement. The last point, systems manageability in a sociopolitical sense, raises the issue of the balance of power in public policy processes and is especially difficult to deal with.

The greatest resistance to sound ecological practices emanates from economic interests. While the cause of pollution may be fairly clear, solutions are not all that easy. For instance, simply forcing a business to pay the clean up costs may make it uneconomical for it to continue either in the industry or to operate the plant in question. The increased costs bear economic consequences. Recognition of these consequences has led to proposals for economic incentives, such as permits, to directly counteract economic resistance on the part of business enterprises or local communities.

For none of the three typologies surveyed above do permits represent the moral ideal. They are, rather, a compromise second-best approach, which has the practical effect of getting a majority on board. Permits are not without problems. They may well create environmental hot spots if one locality buys them up; they must be audited by independent observers; they impose significant transaction costs through administration and regulation. Furthermore, if there were a market in allowable emissions, many environmentalists would like to enter the market, buy up the permits and take them off the market and out of the reach of polluters! At present such action is not allowed.

I think permits have a certain legitimacy as a second-best, short-term solution. The United States is not alone in adopting social compromises of minimum standard approaches. The European Environment Agency displays a similar approach. As a means to the end, however, I think they are morally flawed, for they do not eliminate the problem in terms of the valid moral interests of humans, sentient nonhumans, or ecosystems. They do, nevertheless, have the beneficial social effect of producing a good debate that, as Hazel Henderson suggests, may lead to many other measures if people are creative. Proposals include full-cost pricing of environmental pollution by companies, mass media exposés, consumer activism, application of new technologies, and, most importantly, the retooling of cultures in terms of values.

. . .

To conclude, in the short run society often settles for a second-best approach in order to build a consensus and get something done. Such a procedure, however, carries a moral risk: the second-best approach may become enshrined as the pervading norm. Practical considerations required to get anything done at all suggest a step-by-step procedure. At the same time, care must be taken to ensure that the momentum for change is not dissipated by partial solutions. Improvement in the environment depends upon the continued raising of a critical social voice by environmental activists in order to stimulate critical social engagement by all members of society. Judging by the long-term support of the environmental movement in the United States throughout the twentieth century, there is reason to hope that such momentum will continue in the long run despite setbacks. Only as society becomes more value critical, only as empirical work becomes more solid and comprehensive, and only as the horizon of technological feasibility expands will solutions that are better than second-best be perceived as viable options and be implemented. To conclude

with such an expression of hope may seem naive. It is good, however, to recall that less than thirty years ago Rachel Carson also seemed hopelessly naive to many people. Nevertheless, the seed that she and others planted has taken root.

For Further Discussion

1. Are pollution permits simply a license to pollute? What, if anything, is wrong with that?

2. How does the policy process used by pollution permits differ from the policy recommendations offered by Baxter? How does the concept of a "target level for environmental quality" differ from the "optimal level of pollution"?

3. What is the difference between the legal/market approach to public policy and the traditional public policy approach? What are the ethical assumptions of each? Which do you favor?

4. What problems associated with the traditional public policy approach are thought to be solved by establishing pollution permits? Do you think permits will solve these problems?

DISCUSSION AND STUDY QUESTIONS FOR CHAPTER 8

1. Should polluters be required to prove that their pollutants are safe before being allowed to discharge this material, or should they be allowed to pollute until the material is shown to be harmful?

2. Are pollution credits a license to pollute? Are compensatory payments to harms caused by pollution a license to pollute? Is there anything wrong with such strategies?

3. Is there an optimal level of pollution? Should this level be established by markets or by scientific experts?

4. Is all pollution a form of waste? Shouldn't economic incentives encourage business to reduce waste?

5. Should every business that contributes hazardous waste to a Superfund site be required to pay all costs of cleanup? Should the government (using tax dollars) clean up hazardous sites first and seek payment from those liable later? Or should those responsible for dumping the toxins be identified first?

DISCUSSION CASES

Global Warming and Greenhouse Gases

Global warming is one predicted consequence of an increasing amount of greenhouse gases in the atmosphere. These gases—mainly carbon dioxide, methane, nitrous oxides, and chlorofluorocarbons—act the way that panes of glass function in greenhouses. They allow sunlight in to warm the earth but prevent the heat from radiating back into space. This greenhouse effect is the scientifically accepted explanation of atmospheric warming. There is also widespread agreement that the amount of these gases in the atmosphere has increased significantly due to human activities, most notably burning of fossil fuels. But there is disagreement over the consequences of this increase. Some models predict an increase in global temperatures that will result in climate change, significant agricultural disruption, and massive flooding. Others suggest that temperature changes will remain within normal ranges of fluctuation and that both humans and nature will be able to adapt to these changes. In the face of such uncertainty, is there a most reasonable public policy? Explain how a system of international pollution permits might be used to address global warming. If such a policy were adopted, should each country be granted the same permitted level of pollution to begin with?

During the fall of 1997, all the countries of the world met in Kyoto, Japan, to negotiate a treaty on global warming. Less industrialized countries opposed setting strict standards, arguing that such standards would place them at an unfair economic disadvantage because the industrialized world has already reaped the advantages of the economic growth that fueled global warming in the first place. They argued that industrialized countries such as the United States should be required to meet stricter target levels than the less industrialized world. Should industrialized countries be held to higher pollution control standards than less industrialized countries?

Superfund Payments

In 1980 the U.S. Congress passed the Comprehensive Environmental Response, Compensation and Liability Act, commonly known as the Superfund law. The law was aimed at cleaning up hazardous waste sites throughout the country. Part of this law made polluters "jointly and severally liable" for cleanup. This means that any business that contributes to the pollution can be held liable for all of the cleanup costs associated with the site. The rationale behind this standard was the difficulty in establishing exactly who was responsible for what part of the hazardous waste. To some observers, this represented a very unfair standard.

Superfund cleanup was administered by the Environmental Protection Agency (EPA). The most common cleanup procedure involved transfer of toxic waste to specially designed and EPA-approved landfills. Environmental critics charge that this does not solve the problem of hazardous waste. How would you design a hazardous waste cleanup program?

Consider how William Baxter might evaluate responsibility for toxic waste. Is there an optimal level of toxic wastes? How would the market set public policy in this case? Should industries that produce toxic waste be required to create the means to detoxify the waste, before being allowed to sell their products? What responsibility does the chemical industry have for the products it creates?

Boomer v. Atlantic Cement Co.

A very early legal case involving pollution clearly captures the tension between utilitarian and deontological ethics. *Boomer* v. *Atlantic Cement Co.* involved neighbors suing a cement plant outside of Albany, New York. Some neighbors charged that airborne pollutants from the plant were causing damage to both health and property. They filed suit to stop the pollution.

In deciding the case, the court reasoned that the costs involved in closing the plant outweighed the costs of the harms done to residents. The court concluded that it would be more reasonable to keep the plant open but require the company to pay, on an ongoing basis, costs associated with the harms done. To some critics, this created a license to pollute. (*continued*)

Explain how a system of pollution permits such as the one described by Paul Steidlmeier might work in a case such as *Boomer* v. *Atlantic Cement*. How would the target levels of pollution be established? How, if at all, would this differ from the setting of "optimal" pollution levels as described by Baxter? Baxter suggests that "every person should be free to do whatever he wishes in contexts where his actions do not interfere with the interests of other human beings." What decision would the court have made in this case if it had followed this principle? When someone's actions do interfere with the interests of others, should those actions be prohibited, or should an optimal level of interference be sought?

Reserve Mining Company

Reserve Mining Company was located along the north shore of Lake Superior in Silver Bay, Minnesota. Reserve Mining processed low-grade iron ore, called taconite, into a form suitable for steel production. Processing involved crushing taconite rock into fine-grain dust, mixing this dust with water, and then flushing this water through a series of powerful magnets that filtered out most of the iron particles. The remaining dust and water was discharged into Lake Superior.

During the first decades of operation, Reserve Mining operated without controversy and within the law. Because no provable harm was associated with the discharge of rock dust and water, the company seemed environmentally responsible. By the late 1960s, however, standards had changed. Lawsuits were filed on behalf of local residents and communities seeking to prevent the continued discharge into Lake Superior.

Initially, the legal burden of proof rested on those who claimed damages from pollution. Individuals could recover compensation and prevent continued pollution if they could prove that they were harmed by the negligent discharge of pollution. The Reserve Mining case reversed this presumption. The courts shifted the burden in pollution cases from those claiming to be harmed onto the polluters. The courts concluded that Reserve Mining had to prove that the discharge into Lake Superior was safe. What reasons can be given to support either burden of proof option?

Should the same standards for establishing the burden of proof be used in all pollution cases? Baxter suggests that one's freedom ends where it interferes with the interests of others. Given this standard, who should have to prove their claim—the person acting or the persons claiming that they are being interfered with? Should individuals be free to act until it is shown that their actions interfere with others? Should the same standard that applies to individual freedom be applied to the actions of institutions like corporations, cities, and countries?

CHAPTER 9

Ethics and Animals

Issues such as air and water pollution first focused attention on the ethical questions surrounding environmental interests. These issues required little significant change in common approaches to ethics. Standard ethical concepts and principles were simply applied to new social concerns. But as the field of environmental ethics developed, it became clear that standard approaches for ethics needed to be extended beyond normal boundaries. The ethical responsibilities of humans *regarding* and *to* animals instigated many of these developments.

The assumption that humans have ethical responsibilities regarding animals has been recognized for a long time in the Western philosophical tradition. Mistreating animals could give rise to habits such as cruelty and insensitivity that could lead to unethical behavior toward humans. Humans also had interests in animals, such as property rights, that could be adversely affected by the mistreatment of animals. Such concerns have had implications for public policy. Laws prohibiting cruelty to animals as well as hunting and fishing regulations result, in part, from these concerns. But a wider range of policy issues arose once people began to consider the possibility that we have direct ethical responsibilities to animals.

Three general policy areas merit close attention when we consider ethical standing for animals. The use of animals as a major food source, and the specific practices of animal agriculture, raise significant ethical questions. The use of animals in commercial and scientific research, particularly when it involves inflicting bodily harm or death, raises another set of ethical questions. Finally, hunting and fishing activities, whether they are done for food or for sport, are the focus of a third range of policy issues.

Readings by Peter Singer and Tom Regan in Chapter 6 addressed some of the ethical issues involved in using animals for food. Singer's book, *Animal Libera-* *tion,* called public attention to many of the abusive practices of factory farming. Many times the ways in which animals are raised and slaughtered entail a significant amount of pain and suffering. But, as Regan pointed out, even when the suffering is minimal, serious ethical questions can be raised about the very practice of using animals as food.

Animals are also used in a wide variety of research situations—ranging from the use of rabbits for experiments testing the safety of eye mascara to the use of primates for research in psychology and medicine. There have been challenges concerning the effectiveness of such studies, the need for such studies, and the morality of their painful and sometimes deadly consequences.

Finally, humans use animals in a wide range of sport and recreational activities. Many of these activities, most obviously hunting and fishing, result in the death of animals for human entertainment. Weighing the interests of animals—whether in terms of the suffering they endure or in terms of the rights they have—against the recreational interests of humans is a major ethical concern.

The shift from individualistic to holistic ethics, particularly the shift encouraged by the study of ecology, raises an entire range of different issues. From an ecological perspective, individual animals are not all created equal. There is a major difference between domestic and wild animals. Farm-raised chickens and cows, for example, are not as interesting or as valuable as bald eagles and bison. Members of an endangered species may be more valuable than common species such as starlings or dogs. Scarce predators such as grizzly bears and sharks may be seen as more valuable than overpopulated species such as deer.

The tension between animal welfare ethics and ecocentric ethics is perhaps most obvious in debates

287

concerning the use of hunting as a means of population control and ecosystem management. In many areas of the country, deer population has increased to the point where local habitats are threatened. Destruction of plant species (due to overgrazing), starvation, disease, and deer increasingly being involved in highway collisions have led some wildlife managers to support an increase in hunting as one means for thinning deer population.

Such activities horrify those who defend the ethical treatment of animals. These activities are seen as on a par with Jonathan Swift's *Modest Proposal.* Just as surely as it would be horrific to hunt, kill, and eat humans as a humane management solution to human overpopulation, it is equally inhumane to do so with animals. In the minds of many people, there is a wide gap between the concerns of animal welfare ethics and ecologically based environmental ethics.

Five Arguments for Vegetarianism

William O. Stephens

In this reading, philosopher William Stephens examines five arguments supporting vegetarianism as an ethical responsibility. Some of Stephens's arguments are explicitly connected to environmental concerns, others are not. Stephens also considers objections that have been raised against each of these arguments. His conclusion is that a strong ethical case can be offered in support of vegetarianism.

Stephens considers the following five arguments: meat-eating is unjust because it involves an unjust distribution of scarce food resources; the livestock industry leads to great ecological damage; vegetarianism is connected to a lifestyle that battles social and political oppression; treating animals as food violates our ethical responsibilities to them; and meat-eating causes serious harm to human health.

I. INTRODUCTION

In this paper I will examine five different arguments for adopting a vegetarian diet. These "arguments" can be viewed as various persuasive strategies directed towards different audiences. Many readers will be familiar with some of these arguments, but I think it is useful to bring them all together and to see them as presenting a cumulative case for vegetarianism. Taken as a whole they lead one in a certain direction regarding the choice of one's diet. Although these arguments may have different degrees of logical persuasiveness and different rhetorical audiences, it is worthwhile to ask whether virtuous persons find in themselves some

moral trait that inclines them to respond sympathetically to each argument, and even more sympathetically to the persuasive case taken as a whole. That is, I want to bring these arguments together in order to challenge an otherwise serious, mature human being who wants to be a morally good person. What kind of person would be unmoved by the cumulative case for vegetarianism? Wouldn't the case at least make one more sensitive to dietary choices concerning meat? And would such sensitivity lead naturally in the direction of vegetarianism?

While the conclusions of these arguments may prescribe nonidentical and overlapping scopes of dietary restriction, these restrictions all include abstaining from intensively raised, grain-fed, factory-farmed sentient animals such as cattle, pigs, and poultry, and perhaps also lambs (sheep). Thus the dietary goal towards which the arguments lead is not strict veganism which excludes consumption of all animal products, including dairy food and eggs. Nor do the arguments exclude consumption of those fish, crustaceans, mollusks, and other organisms whose sentience is doubtful and which have not been bred, raised, and slaughtered by intensive, factory-farming methods. Moreover, these arguments do not apply universally to all people in all agricultural circumstances. These arguments do not apply to those few people who, out of genuine necessity, must, in order to survive, hunt and/or trap wild animals in remote areas that are unsuitable for raising crops. Nor for that matter do these arguments rule out either passive cannibalism, that is, eating the corpses of humans who died natural deaths, or passive carnivo-rousness, that is, eating the corpses of nonhuman ani-mals who either died natural deaths or were killed accidentally like roadkill. Rather, these arguments ad-dress the usual situation of North Americans and Euro-peans who live in agriculturally-wealthy communities that enjoy ample dietary alternatives to grain-fed, fac-tory-farmed beef, pork, chicken, turkey and lamb.

The five arguments I will discuss are the Argument from Distributive Justice, the Argument from Environ-mental Harm, the Feminist Argument from Sexual Politics, the Argument from Moral Consideration for Animals, and the Prudential Argument from Health. After presenting these arguments I will offer a critical discussion of them. I will then conclude with a brief analysis of the moral character of the person who re-sists the cumulative case aimed at persuading consum-ers of affluent nations to set a meatless diet as a virtu-ous goal.

II. THE ARGUMENTS FOR VEGETARIANISM

A. The Argument from Distributive Justice

This first argument was advanced as early as 1971 by Frances Moore Lappé, and has been repeated by such philosophers as Peter Singer, James Rachels, Ste-phen R. L. Clark, Mary Midgley, and mentioned in passing by still others. The argument can be recon-structed as follows:

1. 16 to 21 lbs. of grain and soy are needed to pro-duce 1 lb. of beef. 6 to 8 lbs. of grain and soy are needed to produce 1 lb. of pork. 4 lbs. of grain and soy are needed to produce 1 lb. of turkey meat. 3 lbs. of grain and soy are needed to produce 1 lb. of chicken meat.

2. Therefore, converting grain and soy to meat is a very wasteful means of producing food. (From 1)

3. Every day millions of human beings in the world suffer and die from lack of sufficient grains and legumes for a minimally decent diet.

4. By choosing to eat meat when sufficient grains and vegetables are available for a healthy diet for oneself, one participates in and perpetuates a very wasteful means of producing food.

5. If one eats meat knowing 3 and 4, then one en-dorses a very wasteful means of producing food,

and shows an insensitivity to malnourished and starving human beings.

6. By knowingly participating in and perpetuating a very wasteful means of producing food, the meat eater shows a selfish refusal to share with starving human beings food that could have been made available to them, and thereby shows disregard for the principle of distributive justice.

7. Developing nations mimic the dietary habits of Americans, and Americans are setting a harm-ful, irresponsible example by wasting grain to produce and consume meat.

8. Therefore, members of affluent nations ought to adopt vegetarian diets and boycott meat so as not to be implicated in the wasteful and un-just system of meat production, and to show concern for the welfare of unfortunate human beings.

Basically, the idea here is that eating meat perpetu-ates a system which indirectly harms other human be-ings. Therefore, to choose to be a part of this system indicates a disregard for those people, and this in effect contaminates one's moral character.

The Worldwatch Institute reports that:

Large areas of the world's cropland now produce grains for animals. Roughly 38 percent of the world's grain—especially corn, barley, sorghum, and oats—is fed to livestock, up from 35 percent in 1960. Wealthy meat-consuming regions dedi-cate the largest shares of their grain to fattening livestock, while the poorest regions use the least grain as feed. In the United States, for example, animals account for 70 percent of domestic grain use, while India and sub-Saharan Africa offer just 2 percent of their cereal harvest to livestock.

By consuming intensively-raised, grain-fed meat, the few who are affluent indulge in a luxury produced by wasting grain that is desperately needed by the many who are poor. Since those suffering from mal-nutrition and starvation surely do not deserve to be without adequate nutrition, the principle of distribu-tive justice dictates that we who are fortunate enough to live in agriculturally-wealthy nations at least ought to boycott the luxury of meat, and instead adopt a vegetarian diet. By choosing to forego meat, lower the demand for it, and thereby exert pressure to reduce meat production, we contribute to the possibility of many more people being fed by the freed up grains and vegetables, or equivalently, by the freed up acreage of fertile land. The criticism of the moral character of meat eaters that is embedded in this argument is that they are selfishly squandering our agricultural wealth to support their luxurious food preference instead of resting content with a modest yet healthy vegetarian diet in order to share the fruits (and grains and vegetables) of our abundant agricultural wealth with those who, by accident of birth, live in agriculturally poor areas.

Jeremy Rifkin adds to this argument by observing that in order to make room for cattle grazing, the cattle industry (and the spreading desertification it has caused) has displaced millions of people in developing nations from their ancestral lands, forcing them to migrate to squalid urban areas where, suffering from chronic hunger, they succumb to diet-deficiency diseases. I shall discuss the environmentally harmful results of the global meat industry in the next argument. Here I need only observe that the Worldwatch Institute also has reported that "rising meat consumption among the fortunate in developing societies sometimes squeezes out food production for the poor and boosts imports of feed grains." The upshot of these considerations is that the factory farming of animals indirectly harms human beings.

B. The Argument from Environmental Harm

This argument is motivated by the interest many who are sensitive to environmental issues have in "treading lightly on the planet." The sources of my reconstruction of this argument are Jeremy Rifkin, Frances Moore Lappé, and the Worldwatch Institute, but it has been mentioned by many others. This argument runs as follows:

1. Livestock manure mixed with nitrogen from artificial fertilizers produces harmful nitrates which pollute groundwater and cause nervous system impairments, cancer, and methemoglobinemia ("blue baby" syndrome).

2. Cattle feedlots are a dangerous source of organic pollutants, accounting for more than half the toxic organic pollutants found in fresh water.

3. Livestock cause considerable amounts of soil compaction and erosion; each pound of feedlot steak costs about 35 lbs. of eroded topsoil.

4. The destruction of thousands of species of tropical plants, insects, birds, reptiles, and mammals through deforestation is to a great extent caused by the creation of livestock pasture land which in only a few years loses its fertility.

5. Livestock production contributes considerably to the depletion of soil fertility.

6. Livestock are a major cause of the depletion of fresh water aquifers; 3,000 liters of water are used to produce a single kilogram of American beef.

7. Cattle play a prominent role in global desertification as a primary factor in all four causes of it: (a) overgrazing, (b) overcultivation of the land, (c) deforestation, (d) improper irrigation techniques.

8. The Bureau of Land Management has exterminated to near extinction mountain lions, bears, lynx, bobcats, and eagles in order to expand pasture land for livestock.

9. Livestock are a significant cause of damage to the narrow streambank habitats vital to arid-land ecology. These "riparian zones" are in the worst condition in history. This riparian zone damage has, for example, resulted in the depopulation of fresh water fish species.

10. Livestock have degraded and drastically transformed plant ecosystems in the western U.S., causing depopulations of songbirds, elk, bighorn sheep, and pronghorn antelope.

11. Livestock production consumes considerable amounts of nonrenewable energy; producing the red meat and poultry eaten each year by a typical American uses the equivalent of 190 liters of gasoline.

12. The grain-fed cattle complex is a significant factor in the emission of three of the four global "greenhouse" warming gases—methane, carbon dioxide, and nitrous oxides. Livestock account for 15 to 20 percent of global methane emissions.

13. One-third of the value of all raw materials consumed for all purposes in the United States is consumed in feed for livestock.

14. Consequently, livestock are one of the most serious causes of environmental harm, and livestock production and meat-eating are at

odds with sustainable development. In contrast, plant agriculture and vegetarian diets are sustainable, environmentally benign practices.

15. Therefore, it is ecologically beneficial to boycott livestock by adopting a vegetarian diet.

The Worldwatch Institute reports that:

> Cattle and other ruminant livestock such as sheep and goats graze one-half of the planet's total land area. Ruminants, along with pigs and poultry, also eat feed and fodder raised on one-fourth of the cropland. Ubiquitous and familiar, livestock exert a huge, and largely unrecognized, impact on the global environment.

Basically, the argument is that with over a billion cows, bulls, and steers worldwide, the cattle industry is responsible, either directly or indirectly, for considerable ecological devastation.

This Argument from Environmental Harm appeals to the value of ecosystems as more worthy of preserving than the system of raising cattle and other sentient animals that causes manifold harms to the environment in order to generate meat. Notice that one need not establish whether ecosystems have inherent value independent of human valuing, or whether environmental damage is wrong only because it harms human beings. All one needs to accept is that the environmental damage to which the commercial meat industry greatly contributes is *bad*. The Worldwatch Institute states that "If livestock are to live in balance with the environment again, First World consumers will have to eat less meat, while Third World citizens will need to keep their meat consumption low."

C. The Feminist Argument from Sexual Politics

This third argument is that there is an intimate connection between vegetarianism and feminism, and between male dominance and meat eating. Carol J. Adams argues that "to talk about eliminating meat is to talk about displacing one aspect of male control and demonstrates the ways in which animals' oppression and women's oppression are linked together." Adams calls this connection "the sexual politics of meat." She claims it is overtly acknowledged when we hear that men, and especially soldiers, athletes, and other "working men," need meat to be strong and virile, or when wives report that they could give up meat, but prepare it for their husbands who insist on it. Adams tries to reveal the more covert associations between meat eating and male dominance that she claims are deeply embedded within our patriarchal culture. She writes:

> By speaking of the *texts of meat* we situate the production of meat's meaning within a political-cultural context. None of us chooses the meanings that constitute the texts of meat, we adhere to them. Because of the personal meaning meat eating has for those who consume it, we generally fail to see the social meanings that have actually predetermined the personal meaning. Recognizing the texts of meat is the first step in identifying the sexual politics of meat.

Adams argues that it is meaningful to speak of "texts of meat" for three reasons. First, meat carries a recognizable message since it is seen as an essential and nutritious item of food. Second, meat's meaning is unchangeable because it recurs continuously at mealtimes, in advertisements, and in conversations. Third, meat is comprised of a system of relations having to do with food production, attitudes toward animals, and, by extension, acceptable violence toward them.

The Feminist Argument from Sexual Politics can be reconstructed, at length, as follows:

1. In the Bible, the male prerogative for meat is exhibited in Leviticus 6, according to which "The meat so delicately cooked by the priests, with wood and coals in the altar, in clean linen, no woman was permitted to taste, only the males among the children of Aaron." The fused oppression of women and animals through the power of naming can be traced to the story of the Fall in Genesis in which women and an animal, the serpent, are blamed for the Fall, and Adam is entitled to name both Eve (after the Fall) and the other animals (before the Fall).

2. According to the ancient Greek myth, Zeus, the patriarch of patriarchs, desires Metis, chases her, coaxes her to a couch with "honeyed words," subdues her, rapes her, and then swallows her, but he claims that he receives her counsel from his belly, where she remains. This myth collapses together sexual violence against women and meat eating and exhibits the masculine consumption of female language.

3. Fairy tales exhibit meat eating generally as the male's role. The King in his countinghouse ate four-and-twenty blackbirds in a pie, while the Queen ate bread and honey. Folktales of all nations, including

Jack and the Beanstalk, depict giants as male and "fond of eating human flesh."

4. In most nontechnological cultures, obtaining meat was performed by men. In societies with animal-based economies, men hunt and control meat distribution, thus wielding economic and social power typically used to dominate women. In contrast, societies with plant-based economies in which women gather vegetables tend to be egalitarian since women gain an essential economic and social role without abusing it.

5. The language of the hunt implies that it is a variation of rape, since the word "venison" (which originally meant the flesh of any animal killed in the chase or by hunting) derives from the Latin word *venari*, to hunt, and is akin to the Sanskrit term meaning "he desires, attacks, gains." "According to the *American Heritage Dictionary,* the word 'venery' had two definitions (now both archaic): Indulgence in or the pursuit of sexual activity (from *Venus,* love), and also the act, art, or sport of hunting, the chase (from *vener,* to hunt)."

6. In many nontechnological societies women are forbidden to eat meat. They may not eat pork in the Solomon Islands, fish, seafood, chicken, duck, and eggs in some cultures in Asia, and chicken, goat, partridge, or other game birds in equatorial Africa. The Kufa of Ethiopia punished women who ate chicken by enslaving them, while the Walamo put to death women who violated the restriction of eating fowl.

7. In famine situations (e.g. in Ethiopia) women engage in deliberate self-deprivation, serving men meat at the expense of their own nutritional needs.

8. In 19th-century British working-class families, where poverty forced a conscious distribution of meat, men received it, not women.

9. Dr. George Beard, a 19th-century advocate of white superiority, endorsed meat as superior food that more highly evolved, more civilized white men eat. The beef-eating English, Beard said, keep in subjection "the rice-eating Hindoo and Chinese" peasant and "the potato-eating Irish peasant."

10. Hegel wrote: "The difference between men and women is like that between animals and plants. Men correspond to animals, while women correspond to plants because their development is more placid."

11. Originally "men" was a generic term for all humans, and "meat" was a generic term for all solid foods. But "meat" no longer means all food, and "men" no longer includes women. Today "meat" represents the essence or principal part of something, whereas "vegetable" represents passivity and monotonous existence. Colloquially, "vegetable" is a synonym for a person severely brain-damaged or comatose. "To vegetate is to lead a passive existence; just as to be feminine is to lead a passive existence."

12. Twentieth-century meat textbooks proclaim meat to be a virile food. In technological societies, cookbooks reflect the presumption that men eat meat (e.g. the *New McCall's Cookbook* states that London Broil is a man's favorite dinner).

13. In our society, football players drink beer because it's a man's drink, and eat steak because it's a man's meal. The emphasis is on "man-sized portions" and "hero" sandwiches. Meat-and-potatoes men are our stereotypical strong and hearty, rough and ready, able males. Hearty beef stews are named "Manhandlers." The ex-head football coach of the Chicago Bears, Mike Ditka, operates a restaurant that features "he-man food" such as steaks and chops.

14. Men who batter women have often used the absence of meat from their meal as a pretext for violence against women.

15. Animals' lives precede and enable the existence of meat. Through butchering, the live animal is replaced by a dead body, thus transforming it into food. So animals in name and body are made absent referents for meat to exist. Since women are also made absent referents through pornographic pictures of pigs, there is an intersection between sexual violence against women and meat eating.

16. The coherence meat achieves as a meaningful item of food arises from the "patriarchal attitudes" that the end justifies the means, that the objectification of other beings is a necessary part of life, and that violence can and should be masked.

17. Therefore, meat's recognizable message is closely associated with the male role in our patriarchal, meat-advocating cultural discourse, and so the oppression of women and the other animals is interdependent.

Having reached this conclusion about the sexual politics of meat, Adams extends her argument as follows:

18. A meal is an amalgam of food dishes, each introduced in precise order. Each course is seen as leading up to and then coming down from the central

entree that is meat. This pattern is evidence of stability. Thus to remove meat, as the centerpiece of a meal, is to threaten the structure of the larger patriarchal culture.

19. Since meat eating is a measure of a virile culture and individual, our society equates vegetarianism with emasculation or femininity.

20. The fact that people do not often closely scrutinize their own meat eating is an example of the prerogative of those in the dominant order determining what is worthy of conversation and critique.

21. Consequently, vegetarians become trapped by this dominant patriarchal worldview, and fail to perceive that in a meat-eating culture, the ill health, death of animals, and ecological spoilage caused by meat eating do not really matter.

22. It is a very important fact that the hidden majority of this world has been primarily vegetarian. The dietary history of most cultures indicates that complete protein dishes were made of vegetables and grains.

23. As a result, what is most threatening to our cultural discourse is self-determined vegetarianism in cultures where meat is plentiful. Vegetarianism acts as a sign of autonomous female being and signals a rejection of male control and violence.

24. Since some vegetarians, vegetarian groups, and vegetarian cultures are sexist, adopting an overt feminist perspective is necessary for rebuking a meat eating *and* patriarchal world.

25. Therefore, feminism AND vegetarianism ought to be embraced by members of our patriarchal culture in order to transform it from within.

To build her case Adams uses linguistic and etymological analyses, mythology, scripture, folklore, anthropological, sociological, and historical studies, cultural observations, and literary analysis of Mary Shelley's book *Frankenstein,* drawing throughout from a wide range of sources. For example, she notes that in the English tradition it is female hares that are hunted (as in Playboy bunnies) and it is female chickens that are eaten because the flesh of males is believed to be of poor quality, and that by using chickens and cows to produce eggs and dairy products before being slaughtered "we exploit their femaleness as well." In summary, then, the basic gist of Adams' Feminist Argument from Sexual Politics is that since meat is a symbol of patriar-

chal oppression, domination, and violence perpetrated against both nonhuman animals and women, vegetarianism represents an explicit rejection of our "Meat is king" patriarchal culture.

D. The Argument from Moral Consideration for Animals

This argument is probably the most familiar one to philosophical audiences and has several different formulations, but since the object of moral concern in each is the sentient animals themselves, whether couched in terms of the value of their lives, their moral rights, or their suffering, I group them all together as arguments appealing to moral consideration for animals. I will only briefly reconstruct the two most influential versions: Peter Singer's utilitarian argument from suffering and Tom Regan's deontological argument from inherent value.

Singer's argument can be reconstructed as follows:

1. The interests of every sentient being affected by an action ought to be taken into account and given the same weight as the like interest of any other sentient being.

2. Practices which inflict suffering on sentient beings without good reason are morally wrong.

3. Factory farming inflicts considerable suffering on cattle, pigs, sheep, turkeys, and chickens, all of which are sentient beings.

4. Humans do not need meat for a healthy diet (see section E below).

5. Sentient beings have a serious interest in not being made to suffer.

6. Humans have only a trivial interest in meat since it is a dietary luxury. (From 4)

7. Therefore, the trivial interest humans have in eating meat is outweighed by the serious interest factory-farmed animals have in not being made to suffer. (From 1, 3, 6)

8. Therefore, factory farming inflicts suffering on sentient beings without good reason. (From 3, 7)

9. Therefore, the practice of factory farming is morally wrong. (From 2, 8)

10. We ought neither to participate in, nor perpetuate, morally wrong practices.

11. Therefore, we ought to boycott factory farming by becoming vegetarians. (From 9, 10)

Singer's utilitarian contention here is that through vegetarianism, decreasing the demand for factory-farmed meat will reduce animal suffering.

Regan's argument can be reconstructed as follows:

1. Experiencing subjects of a life are living, conscious beings who have beliefs and desires, perception, memory, and a sense of the future, including their own future, have an emotional life together with feelings of pleasure or pain, have preference and welfare interests, have the ability to initiate action in pursuit of their desires and goals, have a psychophysical identity over time, and have an individual welfare in the sense that their experiential life fares well or ill for them, logically independently of their utility for others and logically independently of their being the object of anyone else's interests.

2. Cattle, pigs, sheep, chickens, and turkeys are experiencing subjects of a life.

3. All experiencing subjects of a life have inherent value.

4. All beings with inherent value have equal inherent value, and a right to be treated respectfully. All moral agents have a duty to respect the rights of all such beings.

5. We fail to treat beings with inherent value respectfully if we treat them in ways that detract from their welfare, that is, in ways that harm them.

6. Raising and slaughtering cattle, pigs, sheep, chickens, and turkeys harms them and treats them as mere resources.

7. Therefore, raising and slaughtering cattle, pigs, sheep, chickens, and turkeys violates their right to be treated respectfully, and so is fundamentally unjust. (From 4, 5)

8. Therefore, we moral agents have a duty to boycott factory-farmed products and become vegetarians so as not to be causally implicated in this unjust practice. (From 3, 6)

Regan's argument rests on no utilitarian calculation weighing the interests farm animals have in not suffering against the interests meat eaters have in eating meat. Instead, he argues that animals are experiencing subjects of a life, have inherent value, and thus have a *prima facie* right not to be harmed by being raised and slaughtered when we can become vegetarians without being made worse off by doing so.

Both Regan and Singer appeal to the moral consideration we owe the animals themselves; Singer enjoins us to reduce their pain and suffering, while Regan enjoins us to respect them as beings with inherent value equal to our own. Both contend that we are wronging these animals whom we breed into existence, make to suffer, and slaughter.

E. The Prudential Argument from Health

The last argument for vegetarianism is perhaps the least philosophically interesting argument because it turns on a simple appeal to self-interest, but it is probably the argument that has succeeded in persuading most people. In recent years, many nutritionists have judged eating meat to be unhealthy.

The argument can be reconstructed as follows:

1. The Eskimos, the Laplanders, the Greenlanders, and the Russian Kurgi tribes are populations with the highest animal flesh consumption in the world; they are also among the populations with the lowest life expectancy, often only about 30 years.

2. The Russian Caucasians, the Yucatan Indians, the East Indian Todas, and the Pakistan Hunzakuts are other peoples who live in harsh conditions, but they subsist with little or no animal flesh and have life expectancies of 90 to 100 years, some of the highest in the world.

3. The United States has the most sophisticated medical technology in the world, and one of the most temperate climates, yet it is also one of the highest consumers of meat and animal products in the world, and has one of the lowest life expectancies of industrialized nations.

4. The cultures with the longest life spans in the world are the Vilcambas, who live in the Andes of Ecuador, the Abkhasians, who live on the Black Sea, and the Hunzas, who live in the Himalayas of Northern Pakistan. These people also enjoy full, active lives, working and playing at 80 and beyond. All three groups are either totally vegetarian or close to it; meat and dairy products combined account for only 1½% of the total calories of the Hunzas, the largest of the three groups.

5. Several different studies have shown that the stamina and strength of vegetarians is superior to that of meat eaters. A number of world-class athletes are vegetarians.

6. Meat eaters risk serious and sometimes fatal illness from trichinosis, salmonella, mercury poisoning, and clostridium perfringens gastroenteritis.

7. Consuming meat (eggs, dairy food) and animal fat increases one's chances of suffering heart disease, atherosclerosis (hardened and narrowed arteries), high cholesterol, stroke, peptic ulcers, colon cancer, breast cancer, uterine cancer, cervical cancer, prostate cancer, osteoporosis, kidney disease, and even lung cancer.

8. Those who suffer from angina and other cardiac diseases, rheumatoid arthritis, kidney stones, diverticulitis, gall bladder disease, peptic ulcers, diabetes, asthma, and hypertension have been shown to benefit by switching to a vegetarian diet.

9. The human intestine is anatomically very different from that of natural carnivores, such as dogs and cats. Carnivore bowels are short and straight with smooth walls that guarantee short transit times. Human bowels are long and winding and full of pouches with deeply puckered walls. Wolves and other natural carnivores have highly acidic saliva and digestive secretions designed to dissolve the bones of their prey, whereas human saliva is highly alkaline, and human digestive secretions are far less acidic.

10. The dentition, facial structure, and digestive system of humans do not closely resemble those of natural omnivores, such as bears. Rather, our teeth appear to be designed for the grinding of grains, vegetables, and fruits, and our intestines for their digestion. So human physiology suggests the evolutionary history of an herbivorous species.

11. Therefore, a balanced vegetarian diet tends to be healthier than a diet containing meat and animal fat.

Some evidence also suggests that a strict vegan diet is the healthiest diet of all. Thus prudence would seem to dictate eliminating at least all beef, pork, lamb, and poultry from one's diet, and preferably all fish, seafood, eggs, and dairy products as well. As conservative a group as the American Dietetic Association, having reviewed the current literature on the nutritional status of vegetarians, has concluded that "vegetarian diets are healthful and nutritionally adequate when appropriately planned." Thus, since a balanced, meatless diet is

healthier than a diet containing meat, there appear to be strong prudential reasons for becoming a vegetarian.

III. CRITICAL DISCUSSION

A critic could object that if each one of these arguments is flawed by weak reasoning, then all I have presented are five poor arguments for vegetarianism. So let's now look briefly at each argument from the standpoint of its relation to a wider persuasive strategy. That is, consider each argument as a logical or rhetorical moment that attempts to build a stronger and stronger case for the *prima facie* virtue of vegetarianism.

A. Distributive Justice or Gustatory Guilt?

First let's consider the Argument from Distributive Justice. It is certainly true that if an individual American in Omaha refuses to buy and eat a particular hamburger, the hamburger will not magically transform itself into a bowl of porridge large enough to sustain ten hungry Rwandans. The lines of causation that stretch out between the boycotting of meat by Americans, the market effect this will have on international agribusinesses, rising surpluses of grain worldwide, and political decisions to export such surpluses to famine-plagued areas, are without question long, complicated, difficult to establish, and even more difficult to predict. One could argue that such tenuous, convoluted causal lines are too easily severed by unforeseen or uncontrollable circumstances. But even granting this in no way concedes that such causal lines are *unreal*. Such a boycott would not be a mere symbolic gesture. Coupled with political action, it could exert real market pressure to undercut the meat industry.

But even if the effects of one discrete human action are negligible in a utilitarian sense, the ethics of individual boycotting can never be generated merely by appeals to overall consequences. If everyone were to act in a certain manner, good consequences might occur. However, boycotts may fail as collective action dissipates or never gets energized in the first place. Action remains the expression of individual virtue. If some action or practice produces widespread suffering or injustice, a virtuous person will not be insensitive to this. By making the *personal* choice to abstain from meat, a virtuous individual would be actively expressing her compassion for famine victims. She would be making a moral exemplar of herself, whether others rally to follow her example in sufficient number to achieve the hoped for market effects or not. It is a matter of moral integrity, not empirical, utilitarian calculations of probable consequences.

Consider the fact that Americans lead the world in meat consumption, with 112 kilograms per capita in 1990; that averages out to over 2 kilograms per week, whereas in India on average 2 kilograms per capita are consumed per *year*. Our view of meat consumption might be transformed if we bear in mind that if Americans were to reduce their meat consumption by only 10 percent for one year, it would free at least 12 million tons of grain for human consumption—or enough to feed 60 million starving people. We can see factory-farmed meat as a luxury indulged in predominantly by Americans and Europeans at the expense of the poor of developing nations. As such it is a form of wastefulness and selfishness at odds with distributive justice. Thus it betrays a lack of compassion for those who deserve decent food.

B. Environmental Harm or Harmless Heifers?

Peter Singer observed some time ago that "there would be environmental benefits from ending factory farming, which is energy intensive and leads to problems in disposing of the huge quantities of animal wastes which it concentrates on one site." Here Singer is only concerned with the environmental harm resulting from factory-farming animals on land which could be put to other agricultural uses for humans. He grants that "If a calf, say, grazes on rough pasture land that grows only grass and could not be planted with corn or any other crop that provides food edible by human beings, the result will be a net gain of protein for human beings, since the grown calf provides us with protein that we cannot—yet—extract economically from grass." This suggests a counterargument to the Argument from Distributive Justice. The defender of meat eating could argue that by limiting consumption to those animals (e.g. goats) that graze on unfarmable "rough pasture land that grows only grass" as Singer describes it (e.g. mountain slopes), meat eaters would *not* be depriving hungry people of any grain protein at all.

The first reply to this argument is that it already concedes that farmable land should not be used to support meat production. But the deeper reply is that this argument too quickly assumes that all land "that grows only grass" can and rightly should be used to produce animal protein for humans. This assumption can be challenged by asserting the ecological value such grassland (say, open prairie) has independent of human agricultural use. Peter S. Wenz has argued that if healthy ecosystems are of value, and the value of an ecosystem is positively related to its degree of health, then people have *prima facie* obligations to avoid harming, to repair damage to, and to improve the health of ecosystems. He reasons that using land to grow large quantities of food impairs the health of the ecosystems involved, so people have a *prima facie* obligation to meet their nutritional needs through minimal use of land. Because vegetarianism enables people to do this, Wenz infers, we have a *prima facie* obligation to be vegetarians. Wenz contends that for healthy people in our society, the countervailing considerations are generally of little weight. He concludes that people have an obligation that is not merely *prima facie* to try vegetarianism for a length of time sufficient to become habituated to it.

A second objection to the Argument from Environmental Harm might be that abstaining from pork chops and steak fajitas may to some degree aid in preserving natural ecosystems, but can hardly reverse the wholesale, widespread ecological devastation that is occurring on a global scale. In short, this criticism is that a vegetarian diet won't be enough to heal the planet. My reply is that vegetarianism is not being advanced as a cure-all solution to the plethora of ecological ills. Rather, I claim only that it embodies one concrete example of what it means in practice to "tread lightly on the planet." By consuming less rather than more of the planet's agricultural resources three times a day, each vegetarian definitely contributes to positive environmental change.

Moreover, those who, for ecological reasons, choose to recycle glass, plastic, and paper, use public transportation or a bicycle instead of commuting alone by car, or who choose to install more energy efficient devices in their homes, should be just as sympathetic to this argument for vegetarianism. If one is motivated to take steps to decrease the amount of garbage one generates or the harmful emissions of one's car, then it is only consistent to make dietary choices that reflect a desire to decrease the amount of energy, pollution, and agricultural resources required to eat. If a person sees the value of healthy ecosystems and feels the urgency to preserve them, then integrity of character would also dictate making the appropriate choices of what to consume. Human beings are just as dependent on the ecosystems they inhabit as nonhuman animals and plants are, and as fellow creations of evolution are no less natural organisms. This perspective can instill in us an attitude of proper humility toward the rest of nature. Such considerations have led one moral philosopher to advocate what he calls a morality without hubris. Mature human adults can also reflect on the ethics of their diet, and they have the moral freedom and physiological ability to adapt their nutritional choices so as to exact a lighter toll on the planet's renewable resources. To claim that "the meat eater symbolizes his sense of solidarity with the ecological cycles within which he locates the human race" betrays an ignorance of the central location of the meat industry within the realm of ecological destruction.

C. Patriarchy of Pork or Feminist Fuss?

When I explained Adams' views to one of my colleagues, he related to me a story about the dinner ritual of his wife's family in western Nebraska. There were a total of nine sons and seven daughters. The father (i.e. patriarch) would be seated in the middle of one long side of an 8 foot by 4 foot formica-covered dining table. To his left his sons would be seated clockwise around the table. After preparing the meal, his wife would sit to his right, with her daughters seated to her right, counter-clockwise around the table. In serving the meat, the father would always serve first himself and then his sons, going clockwise around the table. This would mean there would be either poorer choices of meat, or else no meat at all, left for the daughters. Often the daughters would have to have peanut butter and jelly sandwiches. But even when some meat remained for the daughters, there was usually no meat left for the mother, since she was invariably the last to be served.

Moreover, when I have discussed arguments for vegetarianism with my undergraduate students, my experience has consistently been that a significantly greater number of young women are receptive to vegetarianism than young men. Admittedly, this is not a scientific survey with a sample guaranteed to be representative. Nevertheless, my colleague's story and my own experience, though anecdotal, do lend some support to Adams' argument.

Additional evidence can be gleaned from media advertising. Television commercials for steak sauces feature robust, hefty men announcing their zealous appetites for thick, juicy steaks. While the actress Cybill Shepherd has done a few beef ads, women models selling steak or steak sauce are quite the exception. Consider the beef industry's familiar slogan, "Beef: real food for real people." The clear message here is that vegetarians are really Unpeople.

In contrast to Adams' Feminist Argument from Sexual Politics, Jack Weir has offered what he calls a sociocultural appeal to excuse (not justify) the eating of meat by Americans. Like Adams, Weir takes the ubiquity of meat in our culture to be significant, yet he draws conclusions contrary to hers. The first two "sociocultural factors" which Weir says are "relevant," "although not idealistically unavoidable," are that "beliefs about animals are often religiously based and dogmatically implacable," and that "agribusiness is the most powerful and wealthy multinational industry and is unlikely to stop meat production." Similar sociocultural appeals could be made by white supremacists that their beliefs about African-Americans are scientifically-based (on their own genetic theories) and dogmatically implacable, and by sexists that their beliefs about women are anthropologically-based on gender differences and dogmatically implacable. Moreover, to call the observation that "agribusiness is the most powerful and wealthy multinational industry and is unlikely to stop meat production" a "factor" that makes vegetarianism nonobligatory is simply a veiled way of citing it as a reason for not giving up meat. This strikes me as analogous to citing the observation that the tobacco industry is a powerful and wealthy multinational industry and is unlikely to stop cigarette production as a reason for not giving up smoking. Weir concludes that "Because meat-eating is so deeply entrenched in our culture, moderation and reform are probably best . . . in our implacably carnivorous society." I doubt Adams could be very receptive to this attempt at rationalization. On her analysis, she would probably interpret this to be saying, in effect, that "Because the oppression of women is so deeply entrenched in our culture, moderate sexism and reform of gender exclusive language are probably best . . . in our implacably patriarchal society."

However, some of Adams' assumptions are suspicious. Take for example her labeling of the ideas that "the end justifies the means," "objectification of other beings is a necessary part of life," and "violence can and should be masked" as "patriarchal attitudes." Gandhi and Martin Luther King were pacifists and were men, but surely that would not make pacifism a "patriarchal attitude." Some of the connections Adams tries to establish between meat-advocating discourse and patriarchal culture seem rather strained and somewhat farfetched. If the Feminist Argument from Sexual Politics were the only argument for vegetarianism, it might not sway the hardened skeptic who could object that there is no logically necessary connection between meat eating and patriarchy. Yet Adams' argument does, I think, retain an interesting degree of plausibility in its own right, and it adds another rhetorical dimension to the cumulative case for vegetarianism.

D. Concern for Animals or Soppy Sentimentalism?

First, one could reply to Singer and Regan that their arguments would not prohibit eating nonhuman animals that have been accidentally killed by automobiles on the highway (i.e., roadkill) or that have died "natural" deaths from old age. Steve Sapontzis writes:

> . . . even if not morally objectionable, the prospect of our becoming scavengers in order to satisfy our lust for meat strikes me, at least, as bizarre. The prospect of our raising cattle, sheep, hogs, and other animals until they die of old age to satisfy that same lust seems almost equally bizarre. I know of no one who has become convinced of the moral obligation to liberate animals from human exploitation who has also retained such a craving for meat that he or she has resorted to or even seriously contemplated either of these two activities. . . . [E]xcept for the eating of (biological) animals that the eater feels confident are not sentient, vegetarianism is a consequence of animal liberation.

These remarks strike me as correct.

Hud Hudson has argued that we must distinguish between the actual consumption of factory-farmed meat on the one hand, and directly supporting the factory-farming industry economically through the purchase of meat for oneself, or through allowing meat to be purchased by another on one's behalf, or through one's purchase of meat wholly on another's behalf. Hudson writes: ". . . even if we accepted Regan's argument, we have no *moral* reason to regard the *eating* of some portion of a factory-farmed animal, which has fallen off a carelessly driven delivery truck and into our hands, never to be paid for and never to be missed during inventory, as morally impermissible." This raises the question of whether finding and wearing, but not purchasing, a necklace made from the finger bones of a murdered man, or a jacket made from the tanned skin of a murdered woman, would be morally objectionable. What would it indicate about the character of a person who wore items with this kind of morally problematic history? Perhaps a compassionate person would feel moral discomfort, or even revulsion, enjoying something made possible only by the suffering of another. Even though enjoying the carelessly-driven delivery truck meat would not be causally linked to *economically* supporting the factory-farming system that produced it, since no human being ever paid money for it, I suggest that a person sympathetic to the Argument from Moral Consideration for Animals would feel morally tainted deriving pleasure from eating a portion of an animal that paid for that pleasant consumption with its own pain, suffering, and sentient life.

A common criticism of the utilitarian argument for vegetarianism is that as long as farm animals experience a greater balance of pleasure over pain while they exist, then breeding them into existence, treating them on balance decently, and then killing and eating them to increase the gustatory utility of meat eaters, yields greater net utility than a vegetarian world devoid of all farm animals. One could object that this argument fails to include the loss of utility that would have accrued from the balance of the farm animals' lives had they not been slaughtered. Yet this objection can be countered by the "replaceability argument" discussed by Singer. If one is sympathetic to Regan's view that animals have *inherent* value, then one can reject the very idea that animal lives are "replaceable" at all. Here I do find Regan's position more appealing than Singer's since it strikes me as wrong to view animals as our resources to create, manipulate, slaughter, consume, and replace in the name of maximizing the utility of the class of sentient beings. Perhaps a better criticism of the replaceability argument is that it fails to factor in the number of wild animals that could come into existence on their own once we stopped breeding so many domesticated farm animals into existence. Given the fact that farm animals today are the product of dozens of years of selective breeding by humans, these animals are sentient artifacts that humans have manufactured for illegitimate purposes. That is why at this point I would part company with Regan and maintain that battery chickens and grain-fed steers have less inherent value than bald eagles and grizzly bears.

Moreover, I am suspicious of Frederick Ferré's inference that having respect for inherent value means benevolently bringing into existence as many bearers of inherent value as is compatible with their collective well being. Respect for beings with inherent value could well mean treating benevolently all existing beings with inherent value, rather than creating new ones. For example, I think we have an obligation to care for the many *existing* human babies, not to create *more* babies who *might* have pleasant lives. Given the enormous problems of human overpopulation, I contend that today we have an obligation to reduce the current misery of existing humans (and other sentient beings). Yet I think we have no obligation to create new humans who might experience pleasant lives in the future, since their existence would probably further compound the problems of human overpopulation.

Another persistent criticism of the utilitarian argument is that a compassionate person can be more effective adopting any number of other tactics designed to reduce the suffering of factory-farmed animals without becoming a vegetarian. Hudson has responded to this skepticism about the market impact an individual vegetarian has on the leviathan meat industry by developing an argument for vegetarianism which appeals to collective responsibility. Hudson asserts that:

> . . . certain individuals, by virtue of their membership in a loosely-structured group, are at least partially morally responsible for not collectively preventing certain harms by committing themselves to modified, moral (conditional) vegetarianism along with other members of that group, even though none of the individuals could have prevented the harm by acting independently.

Hudson reasons that since the collective inactivity of the group of nonvegetarians contributes to the demand for factory-farmed meat, the members of this group are collectively causally implicated in a morally-abhorrent chain of events, and in order to extricate themselves from that chain, members of this group have reason to, and are morally obligated to, abstain from purchasing factory-farmed products, and to the extent that this affects their eating habits, act as if eating meat with that sort of history is in itself a moral wrong. Hudson's argument grounds the wrong of the meat industry in the harm done to the sentient animals. But if his appeal to collective responsibility is legitimate, then as we have seen in this paper the "loosely-structured group" of nonvegetarians is also partially morally responsible for not collectively preventing harm to poor famine victims in developing nations, harm to the environment, and harm to oppressed women in our patriarchal culture. Just as it seems plausible to think that we are collectively responsible for the suffering of factory-farmed animals, it seems equally plausible to think we are collectively responsible for world hunger, ecological diversity and preservation, and hierarchical, oppressive institutions of all kinds.

Many other objections to the Argument from Moral Consideration for Animals, including the "replace-ment argument," have been made by R. G. Frey and, in my view, fairly refuted by S. F. Sapontzis, so I will not rehearse them here. Instead I wish to focus on the virtue of compassion. As Hume observed long ago, compassion and sympathy actually move people to act much more than carefully constructed pieces of philosophical reasoning which hinge upon contentious interpretations of theoretical principles. I suggest that compassionate persons who had to breed, raise, and slaughter by their own hands the animals they would eat would be greatly disinclined to do so. The anguished cries, terrified struggles, and spurting blood of the farm animals would no doubt deter many people from cutting off the animals' heads in order to make a meal of them. The gory, visceral experience of slaughtering a breathing, feeling animal may trigger the sensitive person's latent compassionate impulse enough to make the prospect of a fleshy meal quite unappetizing. The suggestion here is that if a person would be unwilling to perform the labor necessary for producing an item she wants to have (or consume), then that realization should deter the person from having (or consuming) that item even when it is produced by the labor of another. I take it that we have a strong intuition that just as it would be wrong to murder an innocent person, it would also be wrong to hire a hitman to murder an innocent person. Thus, if we would have moral qualms about slaughtering a helpless sentient animal, even one bred into existence for that purpose, then consistency would require that we extend those moral qualms to, in effect, hiring workers in abattoirs to kill sentient animals.

Here my critic could reply that this argument would force me to do without shoes, my automobile, and the housing material manufactured for the construction of my home since I would not enjoy producing such things myself (assuming I had the required skills). But this objection fails because I *would* be willing to manufacture my own shoes precisely because I would *not* be morally repelled from performing the labors needed to do so. My critic may then reply that those who feel no such moral repulsion in slaughtering nonhuman animals have no moral compulsion to be vegetarians. But if this is the case, then I suggest that we have *prima facie* grounds for doubting the depth of compassion those people have for the following reason. We can and do fault racists for having little or no compassion for members of other races. We can and do fault sexists for having little or no compassion for members of the other sex. A virtuous person will have and will show compassion to Africans, African-Americans, Hispanics, Asians, AmerInds, Caucasians, females, and males alike. Similarly, a person possessed of the virtue of integrity will have compassion for the living, breathing, feeling animals who are bred into existence, raised and made to suffer in unpleasant conditions, and slaughtered as cheaply, not as humanely, as possible, all for the sole purpose of satisfying the luxurious preference for the taste of their flesh.

E. Prudential Health or Dietary Delusion?

A criticism that has been leveled at all moral arguments for vegetarianism is that while it may be ethically "pure" or "ideal" to abstain from meat, we can have no general duty to become vegetarians because meat is necessary for a healthful diet. Jack Weir has argued that abstaining from meat is at best supererogatory and at worst dangerous to one's health, and Steve Sapontzis has offered a critical response to Weir. A more protracted exchange has ensued between Kathryn Paxton George and Evelyn Pluhar, with Gary E. Varner joining the fray. It seems to me that Sapontzis and Pluhar have had the better of these exchanges. Worries about deficiencies from strict vegan diets have ranged from protein, calcium, and iron to zinc and riboflavin. The majority of nutritionists seem to agree that vegans can amply satisfy their needs for these nutrients by eating a variety of grains, legumes, vegetables, and fruits. B^{12} (cobalamin or cyanocobalamin) is the only vitamin which strict vegetarians may not be able to obtain from a balanced plant diet, but only 1 mcg of this vitamin is needed per day, and it can be stored in the body for days at a time. Some packaged food, particularly breakfast cereals, are enriched with B^{12}, health food stores carry vegetarian B^{12} supplements, usually made from algae, and nearly all common multivitamin tablets also contain B^{12}. Finally, George's worries about detrimental effects of a vegan diet on pregnant and lactating women and on children appear to be misplaced. The Preventive Medicine Research Institute is convinced that "Plant-based diets provide a good balance of nutrients to support a healthy pregnancy and are superior to diets containing milk or other animal products." Barnard and his associates hold that "A vegan menu is preferred for nursing women, too. A plant-based diet reduces levels of environmental contaminants in breast milk, compared to that of meat eaters." Barnard and his associates also maintain that "The New Four Food Groups are great for kids. Vegetarian children grow up to be slimmer and healthier, and to live longer than their meat-eating friends."

While it is true that the specific dietary needs of individuals vary, and that some people are allergic to some plant foods, no one has yet established that even a strict vegan diet cannot be adapted to fulfill each person's dietary needs. Carol Adams' claim that anthropological evidence suggests that humans have predominantly been vegetarians has been confirmed by other authors: "Studies of tribal Australian aborigines and the Kung-San of South Africa—groups that live under conditions similar to those of our ancestors—show that only about one fourth of their caloric intake derives from animal products. Nuts, seeds, fruits, and vegetables are the staple foods of these groups. A view of early humans as *gatherers* rather than hunters is a more accurate portrayal."

IV. CONCLUSION

A final, more global counterargument could be constructed as a *reductio ad absurdum*. It would run like this. If we adopt vegetarianism to help feed the hungry, preserve the environment, resist sexist oppression, save sentient animals, and be kind to our colons, then we'll be led to give up more and more of our enjoyments out of guilt that we're doing harm. We'll quit our academic jobs to join UNICEF or the Peace Corps in order to work full time helping to end starvation. We'll get rid of our cars, eschew all use of fossil fuels and nuclear energy, build environmentally-friendly, efficient homes to live in, and tend organic gardens to live on. We'll revolt against all patriarchal institutions, razing all oppressive bodies to the ground, and building new, truly egalitarian, truly gender-neutral, non-exploitative societies in their place. We'll eliminate all animal products, including all those used in cosmetics, clothing, food, fuel, building materials, and even baseball mitts. Finally, we'll wear sealed masks and body suits that protect us from all airborne viruses, bacteria, and disease-carrying substances, breathing only filtered oxygen from respirators, drinking only purified distilled water, and eating only organically grown, irradiated plant-food, free of all toxic residue. Thus, according to this argument, if we judge that eating factory-farmed meat is so problematic that we opt for vegetarianism, then consistency will force us to give up our jobs, our cars, our homes, our relationships, and virtually our entire way of life, all in order to strive to achieve an utterly innocuous, totally benign existence. An utterly innocuous, totally benign existence is impossible for human beings in this world, and so striving for such an existence would be futile. Therefore, accepting the five grounds for vegetarianism presented here commits one to a life of complete futility.

This attempted *reductio* contains a disguised false dilemma. The assumption underlying this objection is that either one must do *everything possible* to produce beneficial results, or else one should do nothing at all to produce beneficial results (and so one may eat as one pleases). The choice is not between being a moral saint (cum-health fanatic) and being a heartless egoist. I do not think that people who have deliberately chosen to become vegetarians are moral saints. But I do think that receptiveness to the first four arguments is linked to the character trait of compassion. I suggest that the five different arguments for abstaining from intensively raised, grain-fed birds and mammals constitute reasons for vegetarianism that are at least as strong as the reasons for many daily actions that are routinely accepted by most people. My conclusion is that this cumulative case for vegetarianism succeeds in establishing that vegetarianism is, in at least five different respects, a virtuous dietary commitment. If I am correct, then this shifts the burden of proof to meat eaters who believe their dietary choice is without moral taint.

What motivates the objections to adopting a vegetarian diet? What inclines consumers of affluent, industrial nations to continue to eat meat despite familiarity with criticisms of it? The inertia of habit, the custom of food choices and learned preferences passed down by our parents from their parents, ubiquitous cultural conditioning, and nutritional ignorance are all formidable forces that resist philosophical argument. As Cato said, "It is a difficult task, O citizens, to make speeches to the belly which has no ears." But even if we do slowly modify our eating habits over time, gradually eating less and less meat, where do we stop after giving up meat? I leave that question open for future discussion. Here I conclude by suggesting that given the various virtues of vegetarianism I have discussed, working toward a meatless diet is a worthy endeavor for a person who values compassion, humility, and integrity.

For Further Discussion

1. Which of Stephens's five arguments do you find most convincing? Which do you find least convincing? Why?

2. Stephens talks of logical persuasiveness and rhetorical effectiveness. What is the difference? How do you respond to the logical and rhetorical force of his arguments?

3. Stephens suggests that certain types of meat-eating are not ruled out by his arguments. Do you disagree with any of the exceptions? Why?

4. In an earlier selection, William Baxter discusses waste as underused resources. Are meat-eating and animal agriculture wasteful in this sense? Is waste ever ethically responsible? Do you think Baxter might be convinced by Stephens's argument from distributive justice?

5. The argument from environmental harm claims that "if livestock are to live in balance with the environment again, First World consumers will have to eat less meat." What public policies might be advanced to promote this goal?

The Case for the Use of Animals in Biomedical Research

Carl Cohen

In this essay, philosopher Carl Cohen defends the practice of using animals for biomedical research. Directly addressing the arguments of Singer and Regan (see Chapter 6), Cohen argues that animals lack the capacities that would qualify them for moral standing. They have no rights against us, and their interests as sentient beings are not equal to human interests. Besides his critical response to the views of Singer and Regan, Cohen offers reasons to support the use of animals in biomedical research. Although he acknowledges that there are humane (that is, human-centered) reasons to refrain from unnecessary experimentation on animals, Cohen concludes that there are good reasons to increase the use of animals in research.

Using animals as research subjects in medical investigations is widely condemned on two grounds: first, because it wrongly violates the *rights* of animals, and second, because it wrongly imposes on sentient creatures much avoidable *suffering*. Neither of these arguments is sound. The first relies on a mistaken understanding of rights; the second relies on a mistaken calculation of consequences. Both deserve definitive dismissal.

WHY ANIMALS HAVE NO RIGHTS

A right, properly understood, is a claim, or potential claim, that one party may exercise against another. The target against whom such a claim may be registered can be a single person, a group, a community, or (perhaps) all humankind. The content of rights claims also varies greatly: repayment of loans, nondiscrimination by employers, noninterference by the state, and so on. To comprehend any genuine right fully, therefore, we must know *who* holds the right, *against whom* it is held, and *to what* it is a right.

Alternative sources of rights add complexity. Some rights are grounded in constitution and law (e.g., the right of an accused to trial by jury); some rights are moral but give no legal claims (e.g., my right to your keeping the promise you gave me); and some rights (e.g., against theft or assault) are rooted both in morals and in law.

The differing targets, contents, and sources of rights, and their inevitable conflict, together weave a tangled web. Notwithstanding all such complications, this much is clear about rights in general: they are in every case claims, or potential claims, within a community of moral agents. Rights arise, and can be intelligibly defended, only among beings who actually do, or can, make moral claims against one another. Whatever else rights may be, therefore, they are necessarily human; their possessors are persons, human beings.

The attributes of human beings from which this moral capability arises have been described variously by philosophers, both ancient and modern: the inner consciousness of a free will (Saint Augustine); the grasp, by human reason, of the binding character of moral law (Saint Thomas); the self-conscious participation of human beings in an objective ethical order (Hegel); human membership in an organic moral community (Bradley); the development of the human self through the consciousness of other moral selves (Mead); and the underivative, intuitive cognition of the rightness of an action (Prichard). Most influential has been Immanuel Kant's emphasis on the universal

human possession of a uniquely moral will and the autonomy its use entails. Humans confront choices that are purely moral; humans—but certainly not dogs or mice—lay down moral laws, for others and for themselves. Human beings are self-legislative, morally *auto-nomous*.

Animals (that is, nonhuman animals, the ordinary sense of that word) lack this capacity for free moral judgment. They are not beings of a kind capable of exercising or responding to moral claims. Animals therefore have no rights, and they can have none. This is the core of the argument about the alleged rights of animals. The holders of rights must have the capacity to comprehend rules of duty, governing all including themselves. In applying such rules, the holders of rights must recognize possible conflicts between what is in their own interest and what is just. Only in a community of beings capable of self-restricting moral judgments can the concept of a right be correctly invoked.

Humans have such moral capacities. They are in this sense self-legislative, are members of communities governed by moral rules, and do possess rights. Animals do not have such moral capacities. They are not morally self-legislative, cannot possibly be members of a truly moral community, and therefore cannot possess rights. In conducting research on animal subjects, therefore, we do not violate their rights, because they have none to violate.

To animate life, even in its simplest forms, we give a certain natural reverence. But the possession of rights presupposes a moral status not attained by the vast majority of living things. We must not infer, therefore, that a live being has, simply in being alive, a "right" to its life. The assertion that all animals, only because they are alive and have interests, also possess the "right to life" is an abuse of that phrase, and wholly without warrant.

It does not follow from this, however, that we are morally free to do anything we please to animals. Certainly not. In our dealings with animals, as in our dealings with other human beings, we have obligations that do not arise from claims against us based on rights. Rights entail obligations, but many of the things one ought to do are in no way tied to another's entitlement. Rights and obligations are not reciprocals of one another, and it is a serious mistake to suppose that they are.

Illustrations are helpful. Obligations may arise from internal commitments made: physicians have obligations to their patients not grounded merely in their patients' rights. Teachers have such obligations to their students, shepherds to their dogs, and cowboys to their horses. Obligations may arise from differences of status: adults owe special care when playing with young children, and children owe special care when playing with young pets. Obligations may arise from special relationships: the payment of my son's college tuition is something to which he may have no right, although it may be my obligation to bear the burden if I reasonably can; my dog has no right to daily exercise and veterinary care, but I do have the obligation to provide these things for her. Obligations may arise from particular acts or circumstances: one may be obliged to another for a special kindness done, or obliged to put an animal out of its misery in view of its condition—although neither the human benefactor nor the dying animal may have had a claim of right.

Plainly, the grounds of our obligations to humans and to animals are manifold and cannot be formulated simply. Some hold that there is a general obligation to do no gratuitous harm to sentient creatures (the principle of nonmaleficence); some hold that there is a general obligation to do good to sentient creatures when that is reasonably within one's power (the principle of beneficence). In our dealings with animals, few will deny that we are at least obliged to act humanely—that is, to treat them with the decency and concern that we owe, as sensitive human beings, to other sentient creatures. To treat animals humanely, however, is not to treat them as humans or as the holders of rights.

A common objection, which deserves a response, may be paraphrased as follows:

> If having rights requires being able to make moral claims, to grasp and apply moral laws, then many humans—the brain-damaged, the comatose, the senile—who plainly lack those capacities must be without rights. But that is absurd. This proves [the critic concludes] that rights do not depend on the presence of moral capacities.

This objection fails: it mistakenly treats an essential feature of humanity as though it were a screen for sorting humans. The capacity for moral judgment that distinguishes humans from animals is not a test to be administered to human beings one by one. Persons who are unable, because of some disability, to perform the full moral functions natural to human beings are certainly not for that reason ejected from the moral community. The issue is one of kind. Humans are of such

a kind that they may be the subject of experiments only with their voluntary consent. The choices they make freely must be respected. Animals are of such a kind that it is impossible for them, in principle, to give or withhold voluntary consent or to make a moral choice. What humans retain when disabled, animals have never had.

A second objection, also often made, may be paraphrased as follows:

> Capacities will not succeed in distinguishing humans from the other animals. Animals also reason; animals also communicate with one another; animals also care passionately for their young; animals also exhibit desires and preferences. Features of moral relevance—rationality, interdependence, and love—are not exhibited uniquely by human beings. Therefore [this critic concludes], there can be no solid moral distinction between humans and other animals.

This criticism misses the central point. It is not the ability to communicate or to reason, or dependence on one another, or care for the young, or the exhibition of preference, or any such behavior that marks the critical divide. Analogies between human families and those of monkeys, or between human communities and those of wolves, and the like, are entirely beside the point. Patterns of conduct are not at issue. Animals do indeed exhibit remarkable behavior at times. Conditioning, fear, instinct, and intelligence all contribute to species survival. Membership in a community of moral agents nevertheless remains impossible for them. Actors subject to moral judgment must be capable of grasping the generality of an ethical premise in a practical syllogism. Humans act immorally often enough, but only they—never wolves or monkeys—can discern, by applying some moral rule to the facts of a case, that a given act ought or ought not to be performed. The moral restraints imposed by humans on themselves are thus highly abstract and are often in conflict with the self-interest of the agent. Communal behavior among animals, even when most intelligent and most endearing, does not approach autonomous morality in this fundamental sense.

Genuinely moral acts have an internal as well as an external dimension. Thus, in law, an act can be criminal only when the guilty deed, the actus reus, is done with a guilty mind, mens rea. No animal can ever commit a crime; bringing animals to criminal trial is the mark of primitive ignorance. The claims of moral right are similarly inapplicable to them. Does a lion have a right to eat a baby zebra? Does a baby zebra have a right not to be eaten? Such questions, mistakenly invoking the concept of right where it does not belong, do not make good sense. Those who condemn biomedical research because it violates "animal rights" commit the same blunder.

IN DEFENSE OF "SPECIESISM"

Abandoning reliance on animal rights, some critics resort instead to animal sentience—their feelings of pain and distress. We ought to desist from the imposition of pain insofar as we can. Since all or nearly all experimentation on animals does impose pain and could be readily forgone, say these critics, it should be stopped. The ends sought may be worthy, but those ends do not justify imposing agonies on humans, and by animals the agonies are felt no less. The laboratory use of animals (these critics conclude) must therefore be ended—or at least very sharply curtailed.

Argument of this variety is essentially utilitarian, often expressly so; it is based on the calculation of the net product, in pains and pleasures, resulting from experiments on animals. Jeremy Bentham, comparing horses and dogs with other sentient creatures, is thus commonly quoted: "The question is not, Can they reason? nor Can they talk? but, Can they suffer?"

Animals certainly can suffer and surely ought not to be made to suffer needlessly. But in inferring, from these uncontroversial premises, that biomedical research causing animal distress is largely (or wholly) wrong, the critic commits two serious errors.

The first error is the assumption, often explicitly defended, that all sentient animals have equal moral standing. Between a dog and a human being, according to this view, there is no moral difference; hence the pains suffered by dogs must be weighed no differently from the pains suffered by humans. To deny such equality, according to this critic, is to give unjust preference to one species over another; it is "speciesism." The most influential statement of this moral equality of species was made by Peter Singer:

> The racist violates the principle of equality by giving greater weight to the interests of members of his own race when there is a clash between their interests and the interests of those of another race. The sexist violates the principle of equality by favoring the interests of his own sex. Similarly the

speciesist allows the interests of his own species to override the greater interests of members of other species. The pattern is identical in each case.

This argument is worse than unsound; it is atrocious. It draws an offensive moral conclusion from a deliberately devised verbal parallelism that is utterly specious. Racism has no rational ground whatever. Differing degrees of respect or concern for humans for no other reason than that they are members of different races is an injustice totally without foundation in the nature of the races themselves. Racists, even if acting on the basis of mistaken factual beliefs, do grave moral wrong precisely because there is no morally relevant distinction among the races. The supposition of such differences has led to outright horror. The same is true of the sexes, neither sex being entitled by right to greater respect or concern than the other. No dispute here.

Between species of animate life, however—between (for example) humans on the one hand and cats or rats on the other—the morally relevant differences are enormous, and almost universally appreciated. Humans engage in moral reflection; humans are morally autonomous; humans are members of moral communities, recognizing just claims against their own interest. Human beings do have rights; theirs is a moral status very different from that of cats or rats.

I am a speciesist. Speciesism is not merely plausible; it is essential for right conduct, because those who will not make the morally relevant distinctions among species are almost certain, in consequence, to misapprehend their true obligations. The analogy between speciesism and racism is insidious. Every sensitive moral judgment requires that the differing natures of the beings to whom obligations are owed be considered. If all forms of animate life—or vertebrate animal life?—must be treated equally, and if therefore in evaluating a research program the pains of a rodent count equally with the pains of a human, we are forced to conclude (1) that neither humans nor rodents possess rights, or (2) that rodents possess all the rights that humans possess. Both alternatives are absurd. Yet one or the other must be swallowed if the moral equality of all species is to be defended.

Humans owe to other humans a degree of moral regard that cannot be owed to animals. Some humans take on the obligation to support and heal others, both humans and animals, as a principal duty in their lives; the fulfillment of that duty may require the sacrifice of many animals. If biomedical investigators abandon the effective pursuit of their professional objectives because they are convinced that they may not do to animals what the service of humans requires, they will fail, objectively, to do their duty. Refusing to recognize the moral differences among species is a sure path to calamity. (The largest animal rights group in the country is People for the Ethical Treatment of Animals; its codirector, Ingrid Newkirk, calls research using animal subjects "fascism" and "supremacism." "Animal liberationists do not separate out the *human* animal," she says, "so there is no rational basis for saying that a human being has special rights. A rat is a pig is a dog is a boy. They're all mammals.")

Those who claim to base their objection to the use of animals in biomedical research on their reckoning of the net pleasures and pains produced make a second error, equally grave. Even if it were true—as it is surely not—that the pains of all animate beings must be counted equally, a cogent utilitarian calculation requires that we weigh all the consequences of the use, and of the nonuse, of animals in laboratory research. Critics relying (however mistakenly) on animal rights may claim to ignore the beneficial results of such research, rights being trump cards to which interest and advantage must give way. But an argument that is explicitly framed in terms of interest and benefit for all over the long run must attend also to the disadvantageous consequences of not using animals in research, and to all the achievements attained and attainable only through their use. The sum of the benefits of their use is utterly beyond quantification. The elimination of horrible disease, the increase of longevity, the avoidance of great pain, the saving of lives, and the improvement of the quality of lives (for humans and for animals) achieved through research using animals is so incalculably great that the argument of these critics, systematically pursued, establishes not their conclusion but its reverse: to refrain from using animals in biomedical research is, on utilitarian grounds, morally wrong.

When balancing the pleasures and pains resulting from the use of animals in research, we must not fail to place on the scales the terrible pains that would have resulted, would be suffered now, and would long continue had animals not been used. Every disease eliminated, every vaccine developed, every method of pain relief devised, every surgical procedure invented, every prosthetic device implanted—indeed, virtually every modern medical therapy is due, in part or in whole, to experimentation using animals. Nor may we ignore, in the balancing process, the predictable gains

in human (and animal) well-being that are probably achievable in the future but that will not be achieved if the decision is made now to desist from such research or to curtail it.

Medical investigators are seldom insensitive to the distress their work may cause animal subjects. Opponents of research using animals are frequently insensitive to the cruelty of the results of the restrictions they would impose. Untold numbers of human beings—real persons, although not now identifiable—would suffer grievously as the consequence of this well-meaning but shortsighted tenderness. If the morally relevant differences between humans and animals are borne in mind, and if all relevant considerations are weighed, the calculation of long-term consequences must give overwhelming support for biomedical research using animals.

CONCLUDING REMARKS

Substitution

The humane treatment of animals requires that we desist from experimenting on them if we can accomplish the same result using alternative methods—in vitro experimentation, computer simulation, or others. Critics of some experiments using animals rightly make this point.

It would be a serious error to suppose, however, that alternative techniques could soon be used in most research now using live animal subjects. No other methods now on the horizon—or perhaps ever to be available—can fully replace the testing of a drug, a procedure, or a vaccine, in live organisms. The flood of new medical possibilities being opened by the successes of recombinant DNA technology will turn to a trickle if testing on live animals is forbidden. When initial trials entail great risks, there may be no forward movement whatever without the use of live animal subjects. In seeking knowledge that may prove critical in later clinical applications, the unavailability of animals for inquiry may spell complete stymie. In the United States, federal regulations require the testing of new drugs and other products on animals, for efficacy and safety, before human beings are exposed to them. We would not want it otherwise.

Every advance in medicine—every new drug, new operation, new therapy of any kind—must sooner or later be tried on a living being for the first time. That trial, controlled or uncontrolled, will be an experiment. The subject of that experiment, if it is not an animal, will be a human being. Prohibiting the use of live animals in biomedical research, therefore, or sharply restricting it, must result either in the blockage of much valuable research or in the replacement of animal subjects with human subjects. These are the consequences—unacceptable to most reasonable persons—of not using animals in research.

Reduction

Should we not at least reduce the use of animals in biomedical research? No, we should increase it, to avoid when feasible the use of humans as experimental subjects. Medical investigations putting human subjects at some risk are numerous and greatly varied. The risks run in such experiments are usually unavoidable, and (thanks to earlier experiments on animals) most such risks are minimal or moderate. But some experimental risks are substantial.

When an experimental protocol that entails substantial risk to humans comes before an institutional review board, what response is appropriate? The investigation, we may suppose, is promising and deserves support, so long as its human subjects are protected against unnecessary dangers. May not the investigators be fairly asked, Have you done all that you can to eliminate risk to humans by the extensive testing of that drug, that procedure, or that device on animals? To achieve maximal safety for humans we are right to require thorough experimentation on animal subjects before humans are involved.

Opportunities to increase human safety in this way are commonly missed; trials in which risks may be shifted from humans to animals are often not devised, sometimes not even considered. Why? For the investigator, the use of animals as subjects is often more expensive, in money and time, than the use of human subjects. Access to suitable human subjects is often quick and convenient, whereas access to appropriate animal subjects may be awkward, costly, and burdened with red tape. Physician-investigators have often had more experience working with human beings and know precisely where the needed pool of subjects is to be found and how they may be enlisted. Animals, and the procedures for their use, are often less familiar to these investigators. Moreover, the use of animals in place of humans is now more likely to be the target of zealous protests from without. The upshot is that humans are sometimes subjected to risks that animals could have borne, and should have borne, in their

place. To maximize the protection of human subjects, I conclude, the wide and imaginative use of live animal subjects should be encouraged rather than discouraged. This enlargement in the use of animals is our obligation.

Consistency

Finally, inconsistency between the profession and the practice of many who oppose research using animals deserves comment. This frankly ad hominem observation aims chiefly to show that a coherent position rejecting the use of animals in medical research imposes costs so high as to be intolerable even to the critics themselves.

One cannot coherently object to the killing of animals in biomedical investigations while continuing to eat them. Anesthetics and thoughtful animal husbandry render the level of actual animal distress in the laboratory generally lower than that in the abattoir. So long as death and discomfort do not substantially differ in the two contexts, the consistent objector must not only refrain from all eating of animals but also protest as vehemently against others eating them as against others experimenting on them. No less vigorously must the critic object to the wearing of animal hides in coats and shoes, to employment in any industrial enterprise that uses animal parts, and to any commercial development that will cause death or distress to animals.

Killing animals to meet human needs for food, clothing, and shelter is judged entirely reasonable by most persons. The ubiquity of these uses and the virtual universality of moral support for them confront the opponent of research using animals with an inescapable difficulty. How can the many common uses of animals be judged morally worthy, while their use in scientific investigation is judged unworthy?

The number of animals used in research is but the tiniest fraction of the total used to satisfy assorted human appetites. That these appetites, often base and satisfiable in other ways, morally justify the far larger consumption of animals, whereas the quest for improved human health and understanding cannot justify the far smaller, is wholly implausible. Aside from the numbers of animals involved, the distinction in terms of worthiness of use, drawn with regard to any single animal, is not defensible. A given sheep is surely not more justifiably used to put lamb chops on the supermarket counter than to serve in testing a new contraceptive or a new prosthetic device. The needless killing of animals is wrong; if the common killing of them for our food or convenience is right, the less common but more humane uses of animals in the service of medical science are certainly not less right.

Scrupulous vegetarianism, in matters of food, clothing, shelter, commerce, and recreation, and in all other spheres, is the only fully coherent position the critic may adopt. At great human cost, the lives of fish and crustaceans must also be protected, with equal vigor, if speciesism has been forsworn. A very few consistent critics adopt this position. It is the reductio ad absurdum of the rejection of moral distinctions between animals and human beings.

Opposition to the use of animals in research is based on arguments of two different kinds—those relying on the alleged rights of animals and those relying on the consequences for animals. I have argued that arguments of both kinds must fail. We surely do have obligations to animals, but they have, and can have, no rights against us on which research can infringe. In calculating the consequences of animal research, we must weigh all the long-term benefits of the results achieved—to animals and to humans—and in that calculation we must not assume the moral equality of all animate species.

For Further Discussion

1. Have you ever used animals in your own science education? Do you believe that the research you conducted or the knowledge you gained justified the treatment of the animal involved?

2. What does Cohen mean by "morally autonomous"? Why does he deny that animals lack this capacity? How does he distinguish humans who lack this capacity from animals? Are you convinced by his reasoning?

3. Cohen dismisses Singer's argument against speciesism as "unsound" and "atrocious." Do you think that talk about speciesism trivializes racism and sexism?

4. Cohen distinguishes two separate arguments given to oppose the use of animals in biomedical research. How would you summarize his response to each?

A Critical Analysis of Hunters' Ethics

Brian Luke

Is hunting animals an ethical activity? Can it be justified as part of, or independent of, an environmental ethic? Hunting is often used as a tool of wildlife management, and hunting groups have been strong advocates for policies that preserve wilderness areas and other animal habitats. Aldo Leopold, one of the giants of environmental policy and ethics, was a lifelong avid hunter. Hunters are often concerned with the ethics of hunting, insisting on responsible hunting practices.

In this essay, Brian Luke examines the ethical values implicit in codes of ethics that govern sport hunting. These codes commit hunters to an ethical view, a view that suggests direct ethical responsibilities to animals. But this ethics would also suggest that hunting itself is ethically wrong. Luke develops his case against hunting by arguing that the ethical commitments of hunters themselves establish a strong case for the immorality of hunting.

My father told me more than once that there were killers and there were hunters, good and bad ways to kill animals, worst men and best men.

—Robert Franklin Gish [1]

HOW TO HUNT

A survey of hunting literature reveals a high degree of consensus regarding what constitutes ethical sport hunting.[2] The primary rules of the "Sportsman's Code" are the following:

SC1. Safety first;
SC2. Obey the law;
SC3. Give fair chase;
SC4. Harvest the game;

SC5. Aim for quick kills;

SC6. Retrieve the wounded.[3]

Attention to safety and obedience to the law (SC1 and SC2) are the bare minimum requirements expected of every shooter, while SC3 through SC6 form additional rules that one must follow to be considered a truly ethical hunter. Hunters who violate SC2 are called "poachers," while hunters who routinely disregard any or all of SC1 through SC6 are known as "slob hunters," in distinction from the conscientious followers of the code known as "true sportsmen."

What general principles are entailed by the sportsman's code? Most of the rules of the code are susceptible to a multitude of interpretations, some anthropocentric, some nonanthropocentric. Since the *prima facie* case against hunting arises from a nonanthropocentric principle of respect for animals, it is important to determine whether an anthropocentric or a nonanthropocentric reading of the sportsman's code has greater validity. Moreover, the strength of the *prima facie* case against hunting depends on the precise nature of the respect for animals (e.g., whether the respect is for nonhuman species or for nonhuman individuals). Thus, in the following discussion I attempt to be as precise as possible about what is entailed by the sportsman's code. I do so by considering the rules not as isolated formulae, but in the context of how hunters apply them and how they understand them. I conclude in this section that although some parts of the sportsman's code are straightforwardly anthropocentric, the code as a whole entails a strong principle calling for the minimization of the harm done to nonhuman individuals.

Safety is highly stressed by hunters. The purpose, of course, is to protect hunters and other humans from the dangerous weapons used by hunters. Thus, SC1 implies no ethical principles regarding the treatment of nonhuman animals. However, SC2, the injunction to obey the law, may be interpreted nonanthropocentrically if the law is understood to be protecting animals for their own sakes. Hunting regulations limit the number of animals of each kind that a licensed hunter may kill over a given time period. These regulations are promulgated mainly to protect species from overhunting. How SC2 should be interpreted thus depends on hunters' understanding of the reasons for this animal protection. Legal regulation of hunting could be taken as a furtherance of healthy ecosystemic functioning for its own sake or as a recognition of the species' own interest in survival. Nevertheless,

there are plenty of human-centered reasons to protect nonhuman species, and these seem to be foremost in hunters' understanding of SC2. In explaining the need to obey hunting regulations, hunters most often refer to obligations to future human hunters. For instance:

> As a hunter, you have a responsibility to future generations to see to the conservation of the animals you hunt. . . . Hunting seasons and bag limits are established to allow the taking of some animals while sustaining wildlife populations. In this way hunters are allowed a harvest, and breeding populations are maintained.[4]

Thus, rules SC1 and SC2 are straightforwardly anthropocentric, primarily functioning to further a human interest in pursuing hunting in a safe and sustainable fashion. The rule SC3, "give fair chase," is more ambiguous. "Fair chase" refers to the restrictions on the means of hunting that sport hunters feel they must apply to remain ethical. Fair chase, as Jim Posewitz explains it, is "a balance that allows hunters to occasionally succeed while animals generally avoid being taken."[5] Although there is disagreement at times between hunters over whether a particular restriction is necessary for a chase to be fair, the techniques most universally denounced include pursuing animals whose flight is restricted by water, deep snow, or fencing, using motorized vehicles to chase or herd animals, tracking animals electronically, employing automatic weaponry, and hunting animals placed outside of their native habitat.

Fair chase is about letting "the hunted animal have his *chance,* that he be able, in principle, to avoid capture."[6] Stated this way, with the emphasis on the animal potentially escaping, it might seem that SC3 embodies some respect for the hunted animals, that the hunters are trying to be fair *to their targets* by refraining from totally overwhelming them with technology. Although hunters do at times understand fair chase this way, as a mark of respect for their prey, it is a superficial fairness insofar as it is not in the interests of any particular game animal to be hunted "sportingly," but rather not to be hunted at all. Joy Williams sarcastically refutes the idea that sport hunting is "a balanced jolly game of mutual satisfaction between the hunter and the hunted—*Bam, bam, bam, I get to shoot you and you get to be dead.*"[7]

Indeed, some hunters insist that the rule of fair chase is not followed out of consideration for the animals themselves. Ortega chides those who assume that fair chase "arises from the pure gentlemanliness of a

Knight of the Round Table."[8] He argues that the total absence of restrictions on the means of killing wild animals would "annihilate the essential character of the hunt"[9] because

> . . . hunting is precisely the series of efforts and skills which the hunter has to exercise to dominate with sufficient frequency the countermeasures of the animal which is the object of the hunt. If these countermeasures did not exist, if the inferiority of the animal were absolute, the opportunity to put the activities involved in hunting into effect would not have occurred.[10]

The point of fair chase is to preserve the hunting experience *for the hunter,* in particular to maintain *hunting* as the development and application of certain skills in distinction from effortless *killing* via high technology. Hunting becomes dull in the absence of fair chase restrictions: "Boredom occurs when the hunt is always a failure or always a success. Neither our superiority nor our inferiority to the hunted may be absolute."[11] From this point of view, the ultimate standard used to decide how to balance human hunting technology against the elusive abilities of the prey is the maximization of the sportsman's pleasure in the hunt, particularly whether the pursuit requires the application of hunting skills or "disciplines." This standard is applied in Nugent's discussion of whether hunting hogs on private preserves violates fair chase:

> [T]he anti-hunters like to claim it is considered "slob hunting." Not true. . . . A wild hog is a wild hog, and I've found no difference in or out of an enclosure. I get a blast out of it either way (and of course they are delicious regardless). It will demand all the disciplines that all good bowhunting takes.[12]

Thus, the primary motivation for following SC3—retaining the hunt as an enjoyable exercise of certain skills—centers on the interests of human hunters. In other words, this part of the sportsman's code, like SC1 and SC2, entails no particular respect for animals and no *prima facie* case against hunting.

"Today, using what is killed is essential to ethical hunting."[13] The rule SC4, "harvest the game," is stated as something the hunter does *after* the kill, but in practice this rule affects the sportsman's behavior prior to the kill as well, since the hunter will leave off killing specific animals when he or she anticipates finding their harvest distasteful: "I will not shoot coots or mer-gansers as I can't find a way to get them to taste good enough to eat."[14] Matt Cartmill notes that only since World War II has the injunction to harvest the kill been widely applied in the West.[15] Today, however, SC4 is such an established part of the sportsman's code that even pure trophy hunters seek a way to portray themselves as satisfying it. They do so by extending the notion of "harvest" from its primary connotation of eating the flesh to include the taking of the antlers. Mounting the head is translated from a display of conquest into a "responsible use." We must "responsibly utilize the products of the game we harvest, appreciating the God-given flesh, hides, bone, *horn & antlers* with respect and dignity."[16]

In his book, *The Right to Hunt,* Whisker attempts to give a purely anthropocentric analysis of hunting practices. He thus interprets SC4 as a rule intended to safeguard other people against wasted resources. As Whisker notes, it follows from such an analysis that harvesting the kill is *not* required if there is an overabundance of game relative to people's needs: "If game is scarce, or if some are in need, and he wantonly kills and wastes he will have committed an offense against charity. If, however, game is plentiful, and if there is no known need, he may not have sinned at all."[17] It is only by thus contextualizing the application of SC4 that one can read it anthropocentrically. However, hunters today generally do not contextualize this rule; rather, they see it as applying to all hunting situations: "Under *all* circumstances, the ethical hunter cares for harvested game in a respectful manner, leaving no waste,"[18] and "my father taught me my first lesson about fishing and hunting ethics and conservation: You *always* eat what you catch and keep."[19] Hunter Robert Franklin Gish criticizes the jackrabbit hunting he tried and rejected growing up in New Mexico:

> [D]ead jackrabbits were never picked up, never treated as "game" to take home and eat. Those jackrabbits were nothing more than motion to try to stop. It was the cruelest kind of hunting. . . . Three or four guns shooting over the cab and to the side of a bouncing truck is not the kind of hunting any person who values life would confess to.[20]

There is no suggestion here that leaving the rabbit carcasses in the brush was wrong because people were in need of meat; rather, the failure to harvest is deemed cruel because it is seen as linked to a deficient perception of the value of animal life—seeing rabbits as nothing more than "motion to try to stop."

In his defense of hunting, Theodore Vitali starts with the premise that "the life of the animal is a good. For the animal in question, it is better to be alive than dead."[21] The intrinsic value of the hunted animal's life entails that there must be a good reason for the kill:

> [K]illing is an evil, . . . in such acts there is a loss of something good, in this case, the life of the animals. And for there ever to be the deliberate taking away of something good, there needs to be a proportionate good that provides an adequate reason for this deliberate loss.[22]

The use of parts of the animal's body enjoined by SC4 is an attempt to provide such an adequate reason for the deliberately inflicted loss. Sport hunters are not aiming for true mercy kills, for they generally shoot healthy animals. Rule SC4 represents the hunters' recognition that they must have a good reason for committing the *prima facie* wrong act of ending a healthy animal's life. This recognition leads naturally to the sense that the greater use they make of the animal's body, the more ethical they become as hunters. Nugent writes, "It is our ultimate legal and ethical responsibility to *maximize* the utilization of this animal at the dinner table and beyond."[23] Likewise, Cartmill writes:

> Some hunters . . . make a point of trying to eat, wear, or utilize in some other way every possible scrap of their quarry's body. "I don't waste anything," proclaims one hunter held up as a model in a 1985 National Rifle Association ad. "I process the meat, tan the buckskin, make thread and lacings from the sinew, even scrimshaw the bones."[24]

By making use of the dead body, the hunter hopes to redeem his act of prematurely ending the animal's life. The sportsman's insistence on harvesting the kill presupposes a recognition of the intrinsic value of the hunted individual's life.

The general point of the next rule, SC5, is well stated by Jim Posewitz: "When the time comes to kill an animal, your responsibility is to do it efficiently. . . . The ethical hunter will constantly work toward the ideal of making all shots on target and instantly fatal."[25] This principle is affirmed by all those who promulgate hunting ethics. The ideal of always killing instantly leads to numerous practical injunctions: ethical hunters must be familiar with their weaponry and choose weapons appropriate for the intended prey; they must practice sufficiently to become a reliable shot; they must be conscientious about shot selection, passing up those shots that may only wound the target or miss entirely.[26]

Anthropocentric interpretations of SC5 are possible, but they are not consistent with the hunters' expressed attitudes. For instance, one might read SC5 as encouraging the hunter to practice shooting and acquire familiarity with weapons so as not to unnecessarily lose game. If the concern were merely to maximize the hunter's take, then the advice would be to work diligently to improve one's shooting skill *and* to take uncertain shots when it is clear that no better opportunity is forthcoming. However, this approach is inconsistent with the sportsman's commitment always to *pass up* uncertain shots:

> [I]f the first shot is a miss, and a better shooting opportunity does not present itself, do not continue firing. . . . Continuing to shoot at fleeing targets or groups of birds, hoping to "get lucky," is blatantly unethical. It risks crippling animals and hitting the wrong animal.[27]

Posewitz's comment about crippling animals or hitting the wrong animal supports an interpretation of SC5 focused on consequences for animals rather than on hunters' interests. Maurice Wade quotes one hunter's evidently nonanthropocentric interpretation of SC5, but then suggests a deeper anthropocentric explanation for hunters' commitment to this rule:

> C. H. D. Clarke, himself a sport hunter, wrote the following: "A hunter who, deliberately, ignorantly, thoughtlessly, or through lack of the skill which an ethical sportsman would strive for, needlessly risks inflicting pain is unethical and imperils his moral right to hunt." That sport hunters accept this anticruelty ethic should not be surprising for, like other aspects of the hunting ethic, it contributes to the challenge of the hunt. Making a clean kill is often more difficult than making a dirty kill.[28]

As in the case of fair chase, while hunters do construct their code out of concern to maximize the pleasure of hunting, including particularly the challenge, it would be a mistake to understand SC5 completely in terms of hunters' self-interest. First, note that sportsmen advocate developing one's shooting skill by practicing at the shooting range, not by taking difficult and uncertain long shots at live animals, even though this latter practice could well be challenging and enjoyable for some hunters. Thus, SC5 is evidently not motivated exclusively by hunters' self-interest.[29] Further support for this interpretation comes from hunters' explanations of SC5—unlike fair chase, which hunters often explain anthropocentrically, SC5 is consistently explained in terms of the importance of minimizing animal suffering. Consider the comments above by Clarke and by any number of other hunting defenders,[30] which show the sportsman's determination to minimize the suffering that he or she is responsible for inflicting on hunted animals.

Rule SC6, "retrieve the wounded," also shows this sense of personal responsibility toward individual animals. Swan summarizes it as follows: "A hunter always hopes that his shots will kill quickly and cleanly. If not, I was taught, it is your duty to find and kill the wounded animal as quickly as possible."[31] Hunters do not like the idea of game animals suffering. This aversion can be seen in their strained arguments for the benignity of shooting: "In most cases, a well-placed, sharp arrowhead causes imperceptible pain, zero hydrostatic shock, and literally puts the animal to sleep on its feet."[32] The commitment to aim for instant kills and follow up on all woundings expresses the hunter's desire to minimize the suffering that he or she causes to wild animals by hunting. As with SC5, a prudential interpretation of SC6 in terms of maximizing the hunter's take is implausible, given the hunters' insistence on retrieving the wounded under *all* circumstances.[33] There are times at which the likelihood of retrieving a wounded animal is so small that the expected gain for the hunter is not worth continuing the search. Yet, even under such circumstances, the hunter who perseveres is valorized, as in Posewitz's tale of the bowhunter who for thirty consecutive days returned to the woods to track an elk that he had fatally wounded. His friends encouraged him to "abandon his obsession" and shoot another elk that season, but "He was no longer looking for any elk; he was looking for the elk he shot. . . . the hunter stayed with the hunt until he satisfied himself that it was over. He had mortally wounded an animal and did not rest until he sat with that animal. This is a profound expression of respect."[34] In this case, the tracking continued past the point at which any alleviation of suffering or even harvesting of flesh was likely, since well before the elk was found he was certainly dead and picked over by scavengers. Yet the perseverance was still seen as profoundly respectful by Posewitz because the hunter was acting on a sense of personal responsibility for the animal that he shot. Viewed in this way, rules SC5 and SC6 are not abstract injunctions to minimize suffering in the aggregate; rather

they represent the hunters' sense of personal responsibility for the suffering of the specific individual that they shoot.

WHETHER TO HUNT

Of the six parts of the sportsman's code, only SC1, SC2, and SC3 can be interpreted anthropocentrically. The rest of the code presumes two nonanthropocentric principles: a recognition of the intrinsic value of individual animal lives and a sense of personal responsibility for minimizing one's imposition of animal suffering. Given a prior decision to hunt, these principles give rise to the practical injunctions detailed above—avoid wanton killing, aim for instant kills, etc. Nevertheless, these principles also put into question the acceptability of hunting itself. The intrinsic value of animal life implies that we should avoid unnecessary killing altogether, not that we may kill provided that we find a use for the corpse. The commitment to avoid causing unnecessary pain implies leaving healthy animals alone, not just shooting them more carefully, the rules hunters have developed to prescribe *how* to hunt ethically presume principles that pointedly raise the question of *whether* to hunt at all. In other words, the sportsman's code entails that hunting is *prima facie* wrong.

Hunters have developed a number of arguments in various attempts to overcome this *prima facie* case against hunting. In this section, I consider the four major defenses of hunting that are usually given, not to assess these arguments on their own terms, but rather to relate them to the sportsman's code and the nonanthropocentric principles that the code entails. The key question is: have hunters successfully rebutted their own case against hunting?

Meat Procurement

The first defense of hunting is that it is an acceptable way to obtain food: "one of the more legitimate arguments for hunting is to hunt your own food and take responsibility for it."[35] Many hunters point out that death in the slaughterhouse is by no means more humane than the death imposed by hunters. This argument is directed toward Anglo-European non-hunters (most of whom eat the flesh of slaughtered animals) and is evidently telling—most Americans support hunting for meat and oppose hunting for trophies.[36] However, in the context of the sportsman's code such *ad hominem* arguments against non-hunting meat eaters are insufficient: being committed to minimizing their infliction of suffering, hunters must determine whether there is a source of nutrition entailing less harm than either game or factory-farmed flesh—i.e., they must address the possibility of vegetarianism.

Two recent defenders of hunting, Swan and Kerasote, do so. Swan writes that he tried a vegetarian diet for a year and it made him ill,[37] while Kerasote, who also abstained from meat for a while, returned to hunting after calculating that his consumption of vegetarian food imported from outside his bioregion was indirectly costing more animal lives than if he killed and ate one elk a year.[38] Although these points are useful for rendering Swan's and Kerasote's own hunting consistent with their commitment to minimize their imposition of unnecessary harm, neither point can be used as the basis of a generalized defense of North American meat hunting. American vegetarians are typically as healthy or healthier than American meat eaters, and few Americans live in that particular combination of ecological circumstances that Kerasote needs to support a utilitarian defense of his hunting. For hunters in general, killing wild animals and eating their flesh is an unnecessary imposition of harm and is thus inconsistent with the principles entailed by the sportsman's code.

Atavism

One of the most frequently presented defenses of hunting is the atavism argument. According to this argument, "Man" has been a hunter throughout most of "his" existence,[39] acquiring predatory instincts that cannot have been totally lost in the brief period of time since the development of agriculture.[40] It is suggested that modern sport hunting is an expression of these lingering instincts,[41] and a way of linking civilized man with his prehistoric origins.[42] Various conclusions are drawn from this analysis, including the claims that hunting, being instinctive, is not subject to moral evaluation,[43] that hunting today is necessary for emotional stability, fulfillment and happiness,[44] and that the abolition of hunting, by repressing an instinctive need, has led or could lead to various seriously negative consequences such as drug abuse and intrahuman violence.[45]

There are major problems with the atavism argument. Sport hunting can be explained easily without recourse to predatory instincts,[46] and the empirical evidence for the evolution of such instincts is shaky. The presumption that "Man" evolved as a hunter has been challenged by recent anthropological theory, according to which humans have been foragers, not hunters, throughout most of our existence, gathering plants, insects and perhaps a few stray small animals, so that scavenging is as likely as hunting to be the first means by which the flesh of large mammals was acquired.[47] Moreover, several writers have pointed out that the occurrence of prehistoric hunting does not necessitate the evolution of predatory instincts.[48] Finally, the presumption of an inherited hunting instinct is difficult to maintain in the face of the preponderance of nonhunters and anti-hunters today—a population that is evidently no less well-adjusted than the hunters themselves.[49]

In addition to these points, there are two other difficulties related directly to the sportsman's code that are important in assessing the atavism argument. The first is that the code indicates a significant ambivalence by hunters about their killing. Hunting is hedged by an elaborate network of restrictions, conditions, and guidelines to prevent it from lapsing into a completely unacceptable activity. Thus, when Swan suggests that "in each of us there is a leopard,"[50] or Nugent claims that he is as "much a natural predator as any Canis Lupus or Ursa Horribilus,"[51] they are evidently forgetting that wolves, bears, and leopards show none of the compunction over killing their prey that human hunters do. Even if sport hunting is an expression of some kind of predatory instinct, the sportsman's code indicates that human hunters are also disposed *against* killing and inflicting pain. Thus, humans are not *natural* predators; rather, they are *conflicted* predators. Because it is not clear that emotional adjustment, happiness, fulfillment, etc., can ever come from expressing a disposition (the predatory "instinct") in such a way that it conflicts with some other disposition (our compassion for animals) that may be just as "instinctive," this conflict undermines the atavism argument.[52]

Hunters reject the obvious resolution of this conflict—for example, such nonlethal stalking practices as wildlife photography—insisting in various ways that the intent to kill is essential to the hunt. Yet, it is precisely this part of hunting, the killing, that is most difficult to interpret as instinctive, given our frequent reluctance and resistance to harming animals. For example, Pluhar writes:

> One of my colleagues, an avid hunter, once told me in exasperation that his 13-year-old son had

just ruined his chance to "get" his first buck. Although the child was in perfect position to shoot the deer, he did not pull the trigger. "I couldn't do it, Dad," the boy explained: "he was looking right into my eyes!" [53]

The first kill is the most difficult.[54] Helping boys "work through this problem" is part of the process that Swan describes as a primary challenge facing hunters today—overcoming guilt.[55] After killing his first animal, Gish's anguish was so severe that he worried for weeks that his mother would die from plague transmitted to her from a flea jumping off the carcass. He writes:

> When I killed my first rabbit with my new Benjamin air rifle, . . . I was overcome with a . . . pervasive sadness. I shot the rabbit in the back and paralyzed it. And it took another shot to stop its wild cries and suffering. My father coaxed me into the responsibility of finishing it off myself, albeit with tears in my eyes.[56]

Pluhar's comment about another child seems apt here: "Such children do not appear to be genetically programmed to kill." [57]

The second difficulty for the atavism argument raised by the sportsman's code is that although the code greatly conditions the practice of modern sport hunting, it is not part of our evolutionary inheritance, but rather is a social construction developed within a particular time and place. Ortega writes:

> This is the reason men hunt. When you are fed up with the troublesome present, with being "very twentieth century," you take your gun, whistle for your dog, go out to the mountain, and, without further ado, give yourself the pleasure during a few hours or a few days of being "Paleolithic." [58]

While North American hunters today do often fantasize that they are acting as their Paleolithic forefathers, the reality is quite different. Note that even though the sportsman's code can be summarized with just six rules, some of the rules, particularly SC1, SC2 and SC3, are actually rubrics for whole lists of specific injunctions. These injunctions are complicated and specific to time and place: they are not instinctive; they must be carefully studied; and they greatly condition the modern experience of hunting. For example, Swan describes the process of hunting for waterfowl in California: how to sign up for the lottery that determines who may shoot, where and when you must go to be assigned a blind, which types of weapons and ammuni-

tion are allowed, the complicated determination of how many of which kinds of birds one may legally shoot ("four per day, eight in possession . . . no more than three mallards, only one of which is a female, only one pintail of either sex, and no more than two redheads and/or canvasbacks"). Swan concludes: "If I haven't got all these regulations exactly correct, I apologize. I only have a Ph.D. in natural resources, and I have a little trouble reading eight-point condensed-type manuals of regulations." [59]

One might argue that prehistoric hunting was most likely also highly ritualized and rule-bound so the regulations surrounding hunting today do not prevent it from being atavistic. However, the specific forms that the rules take differ between cultures, and those who deploy the atavistic argument never address the crucial question of which rules so significantly alter the hunting experience that their adoption decisively separates modern hunting from the primal experience. For example, it is presumed that prehistoric hunters did *not* self-impose limits on their means of hunting. The supposition is that early men hunted out of necessity, and it is generally agreed that it makes no sense for the truly needy to restrict their means of killing: "It would be as absurd for a hunter in need of food to wait until dawn to kill a deer as it would be for a cougar to wait." [60] Because today's North American hunters are generally not facing starvation regardless of the outcome of their hunt, they have the luxury to apply the rules of fair chase to make the hunting experience a more exciting challenge. How do we know that the one who staves off hunger by using any means at his disposal to kill is expressing the same "instinctive" disposition as one who self-consciously restricts his means of killing in order to construct the most thrilling hunting experience?

Conservation

Another common defense of hunting is the conservation argument, according to which "there would be few wild animals if there were no hunters" [61] because hunters as a group form "one of the most effective pressure groups in existence working to preserve" natural habitat.[62]

The premise regarding the relative contribution of hunters to conservation can be challenged.[63] Even granting this premise, there are still problems with the conservation argument given the nonanthropocentric principles behind the sportsman's code. Rules SC5 and SC6 entail a commitment to minimize the unnecessary harm that one inflicts on individual animals; however, if there are ways to generate funds for conservation other than through the sale of hunting licenses and the taxation of weapons and ammunition, then hunting is an unnecessary imposition of harm. Causey (who defends the atavism argument) agrees with anti-hunters such as Pluhar that conservation can *more* effectively be served by using general tax revenues than by charging hunters.[64]

The hunter who defends his sport with the conservation argument faces a particular difficulty at the culminating moment of the hunt just before he shoots. Because, at this point, his or her financial contribution to conservation has already been made, no further benefit will come from continuing the hunt to its fatal conclusion. The kill is in this sense an unnecessary harm for which the hunter is personally responsible. On the other hand, hunters might know that given the opportunity they will conclude the hunt with a killing shot, and simultaneously recognize that they would not contribute much to conservation except to further their hunting experiences. Thus, their contribution to conservation is impossible without the deaths and possible injuries of some wild animals.[65] Indeed, those who use the conservation argument generally recognize that hunters support conservation because they enjoy hunting and want to preserve game species so as to continue their sport.[66] However, this desire for enjoyment again puts the conservation argument into tension with the sportsman's code because rule SC4 entails the intrinsic value of the individuals, that their continued lives are a good for the animals themselves. By saying in effect that "we will conserve animals only if we can enjoy tracking and shooting them," hunters are precisely denying the value of animal lives independent of their utility as moving targets.[67]

Hunters are interested not just in the *continuation* of those species that they enjoy hunting, but also in the proliferation of those game species in sufficient numbers to maximize pleasurable hunting opportunities. The proliferation of game animals sometimes requires the complete or near extermination of natural predators—a case in point being the recent killing of wolves in Alaska to generate greater numbers of elk and caribou for human hunters to pursue.[68] This eradication underscores the inconsistency between the conservation argument and hunters' own ethics. The intrinsic value of hunted animals recognized by SC4 applies equally well to wolves and other natural predators since they also seek to continue their lives. Yet, these animals are at times subjected to extermination efforts funded by the "conservation" dollars of hunters.

Wildlife Management

Hunters often attempt to overcome the *prima facie* case against hunting through the wildlife management argument. According to this argument, the absence of natural nonhuman predators necessitates the regulated use of human hunting to maintain the populations of prey species at healthy, sustainable levels.[69] Although this argument is one of the most common defenses of hunting, it applies at most to only a few hunting situations. Deer comprise about two percent of the animals that North Americans kill each year; yet, for the remaining 98 percent (doves, rabbits, squirrels, quail, pheasant, ducks and geese),[70] no one even suggests there are overpopulation problems.[71] Deer hunters cannot realistically portray themselves as "Florence Nightingales with rifles"[72]—euthanizing deer to save them from an agonizing death by starvation—because, whether hunting for meat or for trophies, deer hunters select the healthy adults for killing, just those individuals most likely to survive a hard winter.[73] As a result, human hunters cannot be said to be replacing natural predators, since natural predators are more likely to kill the young, the old, and the unhealthy.[74] Deer hunters might grant that the individuals they kill are not really being helped, but still maintain that their killing is justified because by helping decrease the herd size they save others from starvation. This idea actually reverses reality, since it is because of hunting that U.S. deer herds are often unnaturally large—in order to boost deer herd size to please hunters, wildlife managers (both public and private) feed deer, manipulate flora, and decimate natural predators.[75] Once the herd has become unnaturally large through such measures, wildlife managers make sure it stays that way, insuring an annual "harvestable surplus" for hunters by carefully regulating how many does are killed.[76] Even in the absence of natural predators, deer herds that are not hunted by humans tend to reach and maintain a stable population level below the carrying capacity of the habitat, and below the levels fostered by states managing wildlife for hunters.[77] Therefore, since the wildlife management system functions primarily to further the interests of hunters, it is question begging to assert the importance of hunting as a wildlife management tool in response to the *prima facie* case against hunting raised by the sportsman's code.

THE PARADOX OF HUNTERS' ETHICS

The sportsman's code, particularly SC4 through SC6, enjoins hunters to hunt in a way that recognizes both the moral burden of killing and the importance of minimizing one's infliction of pain. Because hunters aim to kill and at times cannot help inflicting pain, the sportsman's code raises a strong *prima facie* case against hunting altogether. According to the analysis of the preceding section, this case against hunting has not been successfully rebutted. Thus, hunters' ethics are paradoxical: hunters become more ethical by hunting in a way that is sensitive to the animal's interests in avoiding pain and in continuing to live; nevertheless, this very sensitivity and respect for animals entails that hunting is not justifiable, that even true sportsmen are not acting ethically. There are three major responses to this paradoxical situation: embracing the paradox, renouncing the code, and renouncing hunting.

Notes

1. Robert Franklin Gish, *Songs of My Hunter Heart: A Western Kinship* (Albuquerque: University of New Mexico Press, 1992), p. 63.
2. *Sport hunting* is defined here as hunting done for its own sake, in contrast to subsistence hunting (done as a means of survival) and market hunting (done to sell parts of the animals' bodies).
3. I have attempted to express succinctly the ideas most frequently reiterated by those Westerners concerned with developing a hunting ethic. The best single source that affirms and discusses each of these rules is Jim Posewitz, *Beyond Fair Chase: The Ethic and Tradition of Hunting* (Helena, Mont.: Falcon Press, 1994). Ted Nugent includes versions of each of these rules in his "Sportsman's Creed," printed on page one of his book, *Blood Trails: The Truth about Bowhunting* (Jackson, Mich.: Ted Nugent, 1991). Other sources that articulate some or all of these rules include James Swan, *In Defense of Hunting* (New York: Harper-Collins, 1995); Ted Kerasote, *Bloodties: Nature, Culture, and the Hunt* (New York: Kodansha, 1993); James Whisker, *The Right to Hunt* (North River Press, 1981); Robert Franklin Gish, *Songs of My Hunter Heart;* and José Ortega y Gasset, *Meditations on Hunting* (New York: Charles Scribner's Sons, 1972).
4. Posewitz, *Beyond Fair Chase,* pp. 13, 27. Cf. Whisker, *The Right to Hunt,* pp. 88 and xvi.
5. Posewitz, *Beyond Fair Chase,* p. 57.
6. Ortega, *Meditations on Hunting,* p. 49.
7. Joy Williams, "The Killing Game," in *Women on Hunting,* ed. Pam Houston (Hopewell, N.J.: Ecco Press, 1995), p. 252.
8. Ortega, *Meditations on Hunting,* p. 50.
9. Ibid., p. 45.
10. Ibid., p. 49.
11. Whisker, *The Right to Hunt,* p. 6.
12. Nugent, *Blood Trails,* p. 71. Compare Leopold's statement that "the recreational value of game is inverse to the artificiality of its origin" (quoted in Posewitz, *Beyond Fair Chase,* p. 60).
13. Posewitz, *Beyond Fair Chase,* p. 90.
14. Swan, *In Defense of Hunting,* p. 155. See also Kerasote, *Bloodties,* p. 194.
15. Matt Cartmill, *A View to a Death in the Morning: Hunting and Nature through History* (Cambridge: Harvard University Press, 1993), p. 232.
16. "A Sportsman's Creed," bylaw 8. Nugent, *Blood Trails,* p. 1 (emphasis added).
17. Whisker, *The Right to Hunt,* p. 89.
18. Posewitz, *Beyond Fair Chase,* pp. 90–91 (emphasis added).
19. Swan, *In Defense of Hunting,* p. 126 (emphasis added).
20. Gish, *Songs of My Hunter Heart,* p. 63.
21. Theodore Vitali, "Sport Hunting: Moral or Immoral?" *Environmental Ethics* 12 (1990): 76.
22. Theodore Vitali, "The Ethics of Hunting: Killing as Life-Sustaining," *Reason Papers* 12 (1987): 37.
23. Nugent, *Blood Trails,* p. 101 (emphasis added). See also p. 135.
24. Cartmill, *A View to a Death in the Morning,* pp. 231–32.
25. Posewitz, *Beyond Fair Chase,* p. 35.
26. See, for example, Vitali, "The Ethics of Hunting," p. 39.
27. Posewitz, *Beyond Fair Chase,* pp. 36–37. See also Swan, *In Defense of Hunting,* p. 220.
28. Maurice Wade, "Animal Liberation, Ecocentrism and the Morality of Sport Hunting," *Journal of the Philosophy of Sport* 17 (1990): 17.
29. See Swan, *In Defense of Hunting,* p. 182.
30. For example, "Her slender, once-powerful leg was ripped nearly off. What had the pain been like? . . . The first doe's sacrifice impressed upon me that I never wanted to wound or wing an animal. To hunt must be to kill as swiftly as possible" (Gish, *Songs of My Hunter Heart,* p. 127); and Vitali, "The Ethics of Hunting," p. 39; Ann Causey, "On the Morality of Hunting," *Environmental Ethics* 11 (1989): 334–35; Swan, *In Defense of Hunting,* p. 181; Nugent, *Blood Trails,* p. iv.
31. Swan, *In Defense of Hunting,* p. 132.
32. Nugent, *Blood Trails,* p. 135. Cf. *Deer & Deer Hunting,* October 1991, p. 51.
33. Posewitz, *Beyond Fair Chase,* pp. 39–40; Swan, *In*

Defense of Hunting, p. 188; Nugent, *Blood Trails,* p. 1; etc.

34. Posewitz, *Beyond Fair Chase,* pp. 80–83.
35. Gish, *Songs of My Hunter Heart,* p. 100. Also Nugent, *Blood Trails,* p. 9.
36. Swan, *In Defense of Hunting,* p. 9.
37. Ibid., p. 15.
38. Kerasote, *Bloodties,* pp. 232–33.
39. Ortega, *Meditations on Hunting,* p. 102; Whisker, *The Right to Hunt,* pp. ix, 24.
40. Swan, *In Defense of Hunting,* p. 175; Whisker, *The Right to Hunt,* pp. 18–20, 66.
41. Ortega, *Meditations on Hunting,* p. 119; Gish, *Songs of My Hunter Heart,* p. xii; Swan, *In Defense of Hunting,* pp. 12–13; Whisker, *The Right to Hunt,* pp. 18, 30–31.
42. Posewitz, *Beyond Fair Chase,* p. 110; Nugent, *Blood Trails,* p. 116.
43. Causey, "On the Morality of Hunting," p. 338.
44. Ortega, *Meditations on Hunting,* p. 27; Swan, *In Defense of Hunting,* p. 177.
45. Paul Shepard, *The Tender Carnivore and the Sacred Game* (New York: Scribner's, 1973), p. 150; Swan, *In Defense of Hunting,* pp. 126–27; Nugent, *Blood Trails,* p. 129.
46. Evelyn Pluhar, "The Joy of Killing," *Between the Species* 7 (1991): 123.
47. See Cartmill, *A View to a Death in the Morning,* pp. 15–28; Andree Collard with Joyce Contrucci, *Rape of the Wild: Man's Violence against Animals and the Earth* (Bloomington: Indiana University Press, 1989), chap. 2; Mason, *An Unnatural Order,* chap. 2.
48. Marc Bekoff and Dale Jamieson, "Sport Hunting as an Instinct," *Environmental Ethics* 13 (1991): 375–78; Pluhar, "The Joy of Killing," p. 123.
49. See Causey, "On the Morality of Hunting," p. 338; Pluhar, "The Joy of Killing," p. 123; Cartmill, *A View to a Death in the Morning,* p. 229.
50. Swan, *In Defense of Hunting,* p. 13.
51. Nugent, *Blood Trails,* p. v.
52. Daniel Dombrowski argues that moral concern for nonhuman animals is as likely to be part of our evolutionary heritage as the urge to kill (Daniel Dombrowski, "Comment on Pluhar," *Between the Species* 7 (1991): 130–31).
53. Pluhar, "The Joy of Killing," p. 123.
54. The film [*After the First*] tells the story of a young boy who goes on his first deer hunt with his father. When the boy looked at the beauty of the deer poised in the early morning, its antlered head raised above the bushes, he had hesitated, then shot; and as father and son looked down together at the dead deer, the father had said, 'After the first it won't be so hard.'" Helen Prejean, *Dead Man Walking* (New York: Vintage Books, 1993), p. 185.
55. Swan, *In Defense of Hunting,* p. 29.

56. Gish, *Songs of My Hunter Heart,* p. 62.
57. Pluhar, "The Joy of Killing," pp. 123–24; see also Cartmill, *A View to a Death in the Morning,* pp. 229–31.
58. Ortega, *Meditations on Hunting,* p. 116.
59. Swan, *In Defense of Hunting,* pp. 259–60.
60. Vitali, "The Ethics of Hunting," pp. 38–39. Compare Swan, *In Defense of Hunting,* pp. 181–82; Gish, *Songs of My Hunter Heart,* p. 51; Kerasote, *Bloodties,* p. 77.
61. Posewitz, *Beyond Fair Chase,* p. 105.
62. Robert Loftin, "The Morality of Hunting," *Environmental Ethics* 6 (1984): 248. See also Swan, *In Defense of Hunting,* pp. 6, 147; and Whisker, *The Right to Hunt,* pp. 83–84.
63. Williams, "The Killing Game," pp. 264–65; Pluhar, "The Joy of Killing," p. 126, n 8.
64. Causey, "On the Morality of Hunting," p. 341; Pluhar, "The Joy of Killing," pp. 121–22.
65. On this basis Loftin calls animals killed by hunters "martyrs" and argues that "their deaths should be seen as a sacrificial act in the best sense" (Loftin, "The Morality of Hunting," p. 248).
66. E.g., Nugent, *Blood Trails,* pp. 14, 115; Whisker, *The Right to Hunt,* pp. 152–53.
67. Williams calls sportsman's conservation a "contradiction in terms (We protect things now so that we can kill them later)" (Williams, "The Killing Game," p. 258). Roger King points out the essentially anthropocentric attitude behind the conservation argument (Roger J. H. King, "Environmental Ethics and the Case for Hunting," *Environmental Ethics* 13 (1991): 81–82).
68. See Pluhar, "The Joy of Killing," pp. 121–22; Causey, "On the Morality of Hunting," p. 342.
69. See Vitali, "Sport Hunting," and Nugent, *Blood Trails,* p. 8, for just two examples.
70. Swan, *In Defense of Hunting,* p. 8.
71. See Pluhar, "The Joy of Killing," p. 121; Loftin, "The Mortality of Hunting," p. 244; Cartmill, *A View to a Death in the Morning,* p. 232.
72. Kerasote, *Bloodties,* p. 218.
73. See Byron Dalrymple, *Deer Hunting with Dalrymple* (New York: Arco, 1983), pp. 53–54, on the selection involved in meat hunting for deer, and Vitali, "Sport Hunting," p. 70, on the negative consequences of trophy hunting for the healthiest males.
74. Loftin, "The Morality of Hunting," p. 245; Vitali, "Sport Hunting," p. 71.
75. Swan, *In Defense of Hunting,* p. 77; Nugent, *Blood Trails,* pp. 113–14; Kerasote, *Bloodties,* p. 214; Pluhar, "The Joy of Killing," p. 121; Wenz, "Ecology, Morality, and Hunting," pp. 193–94; King, "Environmental Ethics and the Case for Hunting," p. 68.
76. Ron Baker, *The American Hunting Myth* (New York: Vantage Press, 1985), p. 81.
77. Ibid., pp. 73–77.

For Further Discussion

1. Luke begins his essay with a quote from Robert Franklin Gish. What is your reaction to this quotation?

2. Luke claims that although much of the hunter's code is anthropocentric, it is also derived from a direct respect for animals. Is it possible to respect an animal at the same time that one seeks to kill it?

3. To your mind, is there a difference between hunting deer and hunting rare big game trophies, such as elephants and grizzly bears?

4. What hunting values are attained by killing animals that could not be attained by photographing them? Could photography ever replace hunting?

5. Hunters are often aligned with environmentalists in defending policies to preserve wilderness areas, although they usually differ significantly in their reasons for supporting such policies. If they agree on the public policy question, does it really matter that they disagree on reasons?

Animal Liberation and Environmental Ethics: Bad Marriage, Quick Divorce

Mark Sagoff

In previous chapters we suggested that extension of moral standing to nonhuman animals (ethical extensionism) and an ethical concern for ecological wholes such as ecosystems and species (holism) are among the most fundamental concepts of environmental ethics. But, according to many observers there are irreconcilable disagreements between these two perspectives.

In this selection, philosopher Mark Sagoff examines the relationship between animal liberation and environmental ethics. He argues that a moral concern for the well-being of individual animals—a central issue to such thinkers as Peter Singer and Tom Regan—is incompatible with the ecological concerns of environmentalists. On Sagoff's view, "environmentalists cannot be animal liberationists. Animal liberationists cannot be environmentalists."

"The land ethic," Aldo Leopold wrote in *A Sand County Almanac,* "simply enlarges the boundaries of the community to include soils, waters, plants, and animals, or collectively, the land." What kind of community does Leopold refer to? He might mean a *moral* community, for example, a group of individuals who respect each other's right to treatment as equals or who regard one another's interests with equal respect and concern. He

may also mean an *ecological* community, that is, a community tied together by biological relationships in interdependent webs or systems of life.

Let us suppose, for a moment, that Leopold has a *moral* community in mind; he would expand our *moral* boundaries to include not only human beings, but also soils, waters, plants and animals. Leopold's view, then, might not differ in principle from that of Christopher Stone, who has suggested that animals and even trees be given legal standing, so that their interests may be represented in court. Stone sees the expansion of our moral consciousness in this way as part of a historical progress by which societies have recognized the equality of groups of oppressed people, notably blacks, women and children. Laurence Tribe eloquently makes the same point:

What is crucial to recognize is that the human capacity for empathy and identification is not static; the very process of recognizing rights in those higher vertebrates with whom we can already empathize could well pave the way for still further extensions as we move upward along the spiral of moral evolution. It is not only the human liberation movements—involving first blacks, then women, and now children—that advance in waves of increased consciousness.

Peter Singer, perhaps more than any other writer, has emphasized the analogy between human liberation movements (for example, abolitionism and sufferagism) and "animal liberation" or the "expansion of our moral horizons" to include members of other species in the "basic principle of equality." Singer differs from Stone and Tribe, however, in two respects. First, he argues that the capacity of animals to suffer pain or to enjoy pleasure or happiness places people under a moral obligation which does not need to be enhanced by a doctrine about rights. Second, while Stone is willing to speak of the interests of his lawn in being watered, Singer argues that "only a being with subjective experiences, such as the experience of pleasure or the experience of pain, can have interests in the full sense of the term." A tree, as Singer explains, may be said to have an "interest" in being watered, but all this means is that it needs water to grow properly as an automobile needs oil to function properly. Thus, Singer would not include rocks, trees, lakes, rivers or mountains in the moral community or the community of morally equal beings.

Singer's thesis, then, is not necessarily that animals have rights which we are to respect. Instead, he argues that they have utilities that ought to be treated on an equal basis with those of human beings. Whether Tribe and Stone argue a weaker or a different thesis depends upon the rights they believe animals and other natural things to have. They may believe that all animals have a right to be treated as equals, in effect, they may agree with Singer that the interests of *all* animals should receive equal respect and concern. On the other hand, Tribe, Stone or both may believe that animals have a right only to life or only to those very minimal and basic rights without which they could not conceivably enjoy any other right. I will, for the moment, assume that Tribe and Stone agree that animals have basic rights, for example, a right to live or a right not to be killed for their meat. I will consider later the possibility that environmental law might protect the rights of animals without necessarily improving their welfare or protecting their lives.

Moral obligations to animals, to their well-being or to their rights, may arise in either of two ways. First, duties to non-human animals may be based on the principle that cruelty to animals is obnoxious, a principle nobody denies. Muckraking journalists (thank God for them) who depict the horrors which all too often occur in laboratories and on farms, appeal quite properly to the conviction and intuition that people should never inflict needless pain on animals and especially not for the sake of profit. When television documentaries or newspaper articles report the horrid ways in which domestic animals are often treated, the response is, as it should be, moral revulsion. This anger is directed at human responsibility for the callous, wanton and needless cruelty human beings inflict on domestic animals. It is not simply the pain but the way it is caused which justifies moral outrage.

Moral obligations, however, might rest instead on a stronger contention, which is that human beings are obliged to prevent and to relieve animal suffering however it is caused. Now, insofar as the animal equality or animal liberation movement makes a philosophically interesting claim, it insists on the stronger thesis, that there is an obligation to serve the interests, or at least to protect the lives, of *all* animals who suffer or are killed, whether on the farm or in the wild. Singer, for example, does not stop with the stultifying platitude that human beings ought not to be cruel to animals. No; he argues the controversial thesis that society has an obligation to prevent the killing of animals and even to relieve their suffering wherever, however, and as much as it is able, at a reasonable cost to itself.

I began by supposing that Aldo Leopold viewed the community of nature as a *moral* community—one in which human beings, as members, have obligations to all other animals, presumably to minimize their pain. I suggested that Leopold, like Singer, may be committed to the idea that the natural environment should be preserved and protected only insofar as, and because, its protection satisfies the needs or promotes the welfare of individual animals and perhaps other living things. I believe, however, that this is plainly not Leopold's view. The principle of natural selection is not obviously a humanitarian principle; the predator-prey relation does not depend on moral empathy. Nature ruthlessly limits animal populations by doing violence to virtually every individual before it reaches maturity; these conditions respect animal equality only in the darkest sense. Yet these are precisely the ecological relationships which Leopold admires; they are the conditions which he would not interfere with, but protect. Apparently, Leopold does not think that an ecological system has to be an egalitarian moral system in order to deserve love and admiration. An ecological system has a beauty and an authenticity that demands respect—but plainly not on humanitarian grounds.

In a persuasive essay, J. Baird Callicott describes a number of differences between the ideas of Leopold and those of Singer—differences which suggest that Leopold's environmental ethic and Singer's humane utilitarianism lead in opposite directions. First, while Singer and other animal liberationists deplore the suffering of domestic animals, "Leopold manifests an attitude that can only be described as indifference." Second, while Leopold expresses an urgent concern about the disappearance of species, Singer, consistently with his premises, is concerned with the welfare of individual animals, without special regard to their status as endangered species. Third, the preservation of wilderness, according to Leopold, provides "a means of perpetuating, in sport form, the more virile and primitive skills. . . ." He had hunting in mind. Leopold recognized that since top predators are gone, hunters may serve an important ecological function. Leopold was himself an enthusiastic hunter and wrote unabashedly about his exploits pursuing game. The term "game" as applied to animals, Callicott wryly comments, "appears to be morally equivalent to referring to a sexually appealing young woman as a 'piece' or to a strong, young black man as a 'buck'—if animal rights, that is, are to be considered on par with women's rights and the rights of formerly enslaved races."

Singer expresses disdain and chagrin at what he calls "'environmentalists'" organizations such as the Sierra Club and the Wildlife Fund, which actively support or refuse to oppose hunting. I can appreciate Singer's aversion to hunting, but why does he place the word "environmentalist" in shudder quotes when he refers to organizations like the Sierra Club? Environmentalist and conservationist organizations traditionally have been concerned with ecological, not humanitarian issues. They make no pretense of acting for the sake of individual animals; rather, they attempt to maintain the diversity, integrity, beauty and authenticity of the natural environment. These goals are ecological, not eleemosynary. Their goals are entirely consistent, then, with licensing hunters to shoot animals whose populations exceed the carrying capacity of their habitats. Perhaps hunting is immoral; if so, environmentalism is consistent with an immoral practice, but it is environmentalism without quotes nonetheless. The policies environmentalists recommend are informed by the concepts of population biology, not the concepts of animal equality. The S.P.C.A. does not set the agenda for the Sierra Club.

I do not in any way mean to support the practice of hunting; nor am I advocating environmentalism at this time. I merely want to point out that groups like the Sierra Club, the Wilderness Society and the World Wildlife Fund do not fail in their mission insofar as they devote themselves to causes other than the happiness or welfare of individual creatures; that never was their mission. These organizations, which promote a love and respect for the functioning of natural ecosystems, differ ideologically from organizations that make the suffering of animals their primary concern—groups like the Fund for Animals, the Animal Protection Institute, Friends of Animals, the American Humane Association, and various single issue groups such as Friends of the Sea Otter, Beaver Defenders, Friends of the Earthworm, and Worldwide Fair Play for Frogs.

D. G. Ritchie, writing in 1916, posed a difficulty for those who argue that animals have rights or that we have obligations to them created simply by their capacity to suffer. If the suffering of animals creates a human obligation to mitigate it, is there not as much an obligation to prevent a cat from killing a mouse as to prevent a hunter from killing a deer? "Are we not to vindicate the rights of the persecuted prey of the stronger?" Ritchie asks. "Or is our declaration of the rights of every creeping thing to remain a mere hypocritical formula to gratify pug-loving sentimentalists?"

If the animal liberation or animal equality movement is not to deteriorate into "a hypocritical formula to gratify pug-loving sentimentalists," it must insist, as Singer does, that moral obligations to animals are justified, in the first place, by their distress, and, in the second place, by human ability to relieve that distress. The liberationist must morally require society to relieve animal suffering wherever it can and at a lesser cost to itself, whether in the chicken coop or in the wild. Otherwise, the animal liberationist thesis becomes interchangeable with the platitude one learns along with how to tie shoestrings: people ought not to be cruel to animals. I do not deny that human beings are cruel to animals, that they ought not to be, that this cruelty should be stopped and that sermons to this effect are entirely appropriate and necessary. I deny only that these sermons have anything to do with environmentalism or provide a basis for an environmental ethic.

In discussing the rights of human beings, Henry Shue describes two that are basic in the sense that "the enjoyment of them is essential to the enjoyment of all other rights." These are the right to physical security and the right to minimum subsistence. These are positive, not merely negative rights. In other words, these rights require governments to provide security and subsistence, not merely to refrain from invading security and denying subsistence. These basic rights require society, where possible, to rescue individuals from starvation; this is more than the merely negative obligation not to cause starvation. No; if people have basic rights—and I have no doubt they do—then society has a positive obligation to satisfy those rights. It is not enough for society simply to refrain from violating them.

This, surely, is true of the basic rights of animals as well, if we are to give the conception of "right" the same meaning for both people and animals. For example, to allow animals to be killed for food or to permit them to die of disease or starvation when it is within human power to prevent it, does not seem to balance fairly the interests of animals with those of human beings. To speak of the rights of animals, of treating them as equals, of liberating them, and at the same time to let nearly all of them perish unnecessarily in the most brutal and horrible ways is not to display humanity but hypocrisy in the extreme.

Where should society concentrate its efforts to provide for the basic welfare—the security and subsistence—of animals? Plainly, where animals most lack this security, when their basic rights, needs, or interests are most thwarted and where their suffering is most intense. Alas, this is in nature. Ever since Darwin, we have been aware that few organisms survive to reach sexual maturity; most are quickly annihilated in the struggle for existence. Consider as a rough but reasonable statement of the facts the following:

> All species reproduce in excess, way past the carrying capacity of their niche. In her lifetime a lioness might have 20 cubs; a pigeon, 150 chicks; a mouse, 1,000 kits; a trout, 20,000 fry, a tuna or cod, a million fry or more; an elm tree, several million seeds; and an oyster, perhaps a hundred million spat. If one assumes that the population of each of these species is, from generation to generation, roughly equal, then on the average only one offspring will survive to replace each parent. All the other thousands and millions will die, one way or another.

The ways in which creatures in nature die are typically violent: predation, starvation, disease, parasitism, cold. The dying animal in the wild does not understand the vast ocean of misery into which it and billions of other animals are born only to drown. If the wild animal understood the conditions into which it is born, what would it think? It might reasonably prefer to be raised on a farm, where the chances of survival for a year or more would be good, and to escape from the wild, where they are negligible. Either way, the animal will be eaten: few die of old age. The path from birth to slaughter, however, is often longer and less painful in the barnyard than in the woods. Comparisons, sad as they are, must be made to recognize where a great opportunity lies to prevent or mitigate suffering. The misery of animals in nature—which humans can do much to relieve—makes every other form of suffering pale in comparison. Mother Nature is so cruel to her children she makes Frank Perdue look like a saint.

What is the practical course society should take once it climbs the spiral of moral evolution high enough to recognize its obligation to value the basic rights of animals equally with that of human beings? I do not know how animal liberationists, such as Singer, propose to relieve animal suffering in nature (where most of it occurs), but there are many ways to do so at little cost. Singer has suggested, with respect to pest control, that animals might be fed contraceptive chemicals rather than poisons. It may not be beyond the reach of science to attempt a broad program of contraceptive care for animals in nature so that fewer will fall victim to an early and horrible death. The government is spending hundreds of millions of dollars to store millions of tons of grain. Why not lay out this food, laced with contraceptives, for wild creatures to feed upon? Farms which so overproduce for human needs might then satisfy the needs of animals. The day may come when entitlement programs which now extend only to human beings are offered to animals as well.

One may modestly propose the conversion of national wilderness areas, especially national parks, into farms in order to replace violent wild areas with more humane and managed environments. Starving deer in the woods might be adopted as pets. They might be fed in kennels; animals that once wandered the wilds in misery might get fat in feedlots instead. Birds that now kill earthworms may repair instead to birdhouses stocked with food, including textured soybean protein that looks and smells like worms. And to protect the brutes from cold, their dens could be heated, or shelters provided for the all too many who will otherwise freeze. The list of obligations is long, but for that reason it is more, not less, compelling. The welfare of all animals is in human hands. Society must attend not solely to the needs of domestic animals, for they are in a privileged class, but to the needs of all animals, especially those which without help, would die miserably in the wild.

Now, whether you believe that this harangue is a *reductio* of Singer's position, and thus that it agrees in principle with Ritchie, or whether you think it should be taken seriously as an ideal is of no concern to me. I merely wish to point out that an environmentalist must take what I have said as a *reductio*, whereas an animal liberationist must regard it as stating a serious position, at least if the liberationist shares Singer's commitment to utilitarianism. Environmentalists cannot be animal liberationists. Animal liberationists cannot be environmentalists. The environmentalist would sacrifice the lives of individual creatures to preserve the authenticity, integrity and complexity of ecological systems. The liberationist—if the reduction of animal misery is taken seriously as a goal—must be willing, in principle, to sacrifice the authenticity, integrity and complexity of ecosystems to protect the rights, or guard the lives, of animals.

A defender of the rights of animals may answer that my argument applies only to someone like Singer who is strongly committed to a utilitarian ethic. Those who emphasize the rights of animals, however, need not argue that society should enter the interests of animals equitably into the felicific calculus on which policy is based. For example, Laurence Tribe appeals to the rights of animals not to broaden the class of wants to be included in a Benthamite calculus but to "move beyond wants" and thus to affirm duties "ultimately independent of a desire-satisfying conception." Tribe writes:

> To speak of "rights" rather than "wants," after all, is to acknowledge the possibility that want-maximizing or utility-maximizing actions will be ruled out in particular cases as inconsistent with a structure of agreed-upon obligations. It is Kant, not Bentham, whose thought suggests the first step toward making us "different persons from the manipulators and subjugators we are in danger of becoming."

It is difficult to see how an appeal to rights helps society to "move beyond wants" or to affirm duties "ultimately independent of a desire-satisfying conception." Most writers in the Kantian tradition analyze rights as claims to something in which the claimant has an interest. Thus, rights-theorists oppose utilitarianism not to go beyond wants but because they believe that some

wants or interests are moral "trumps" over other wants and interests. To say innocent people have a right not to be hanged for crimes they have not committed, even when hanging them would serve the general welfare, is to say that the interest of innocent people not to be hanged should outweigh the general interest in deterring crime. To take rights seriously, then, is simply to take some interests, or the general interest, more seriously than other interests for moral reasons. The appeal to rights simply is a variation on utilitarianism, in that it accepts the general framework of interests, but presupposes that there are certain interests that should not be traded off against others.

A second problem with Tribe's reply is more damaging than the first. Only *individuals* may have rights, but environmentalists think in terms of protecting *collections, systems* and *communities*. Consider Aldo Leopold's oft-quoted remark: "A thing is right when it tends to preserve the integrity, stability, and beauty of the biotic community. It is wrong when it tends to do otherwise." The obligation to preserve the "integrity, stability, and beauty of the biotic community," whatever those words mean, implies no duties whatever to individual animals in the community, except in the rare instance in which an individual is important to functioning of that community. For the most part, individual animals are completely expendable. An environmentalist is concerned only with maintaining a population. Accordingly, the moral obligation Leopold describes cannot be grounded in or derived from the rights of individuals. Therefore, it has no basis in rights at all.[1]

Consider another example: the protection of endangered species. An individual whale may be said to have rights, but the species cannot; a whale does not suddenly have rights when its kind becomes endangered. No; the moral obligation to preserve species is not an obligation to individual creatures. It cannot, then, be an obligation that rests on rights. This is not to say that there is no moral obligation with regard to endangered species, animals or the environment. It is only to say that moral obligations to nature cannot be enlightened or explained—one cannot even take the first step—by appealing to the rights of animals and other natural things.

Garrett Hardin, in his "Foreword" to *Should Trees Have Standing?*, suggests that Stone's essay answers Leopold's call for a "new ethic to protect land and other natural amenities. . . ." But as one reviewer has pointed out,

Stone himself never refers to Leopold, and with good reason; he comes from a different place, and his proposal to grant rights to natural objects has emerged not from an ecological sensibility but as an extension of the philosophy of the humane movement.

A humanitarian ethic—an appreciation not of nature, but of the welfare of animals—will not help us to understand or to justify an environmental ethic. It will not provide necessary or valid foundations for environmental law.

Note

1. Tom Regan discusses this issue in *The Case for Animal Liberation* (1983):

    Because paradigmatic rights-holders are individuals, and because the dominant thrust of contemporary environmental efforts (e.g., wilderness preservation) is to focus on the whole rather than on the part (i.e., the individual), there is an understandable reluctance on the part of environmentalists to "take rights seriously," or at least a reluctance to take them as seriously as the rights view contends we should. . . . A rights-based environmental ethic . . . ought not to be dismissed out of hand by environmentalists as being in principle antagonistic to the goals for which they work. It isn't. Were we to show proper respect for the rights of individuals who make up the biotic community, would not the *community* be preserved?

    (*Id.* at 362.) I believe this is an empirical question, the answer to which is "no." The environmentalist is concerned about preserving evolutionary processes; whether these processes, e.g., natural selection, have deep enough respect for the rights of individuals to be preserved on those grounds, is a question that might best be left to be addressed by an evolutionary biologist.

For Further Discussion

1. Are all animals created equal? Do you value endangered species as much as you value such common species as chickens and cows? Identify as many different values as you can that support human concern for animals. In doing this, consider the different values attributed to: chickens, dogs, grizzly bears, bald eagles, insects, earthworms, chimpanzees, dolphins, whales, bass, salmon, and deer.

2. How might defenders of animal liberation, such as Peter Singer and Tom Regan, respond to Sagoff's conclusions? With whom do you agree?

3. Sagoff contrasts the reproduction and survival rates of numerous species. How does Sagoff use these facts in his criticism of the animal rights perspective?

4. Are there *two* positions concerning the ethical status of animals—they have standing or they don't—or are there *three*—ecologically sensitive animals do, others do not? Can you think of any way to reconcile animal rights with environmental ethics?

DISCUSSION AND STUDY QUESTIONS FOR CHAPTER 9

1. Even many nonvegetarians refuse to eat certain types of animal flesh. What, exactly, is the difference between eating beef and eating dog?

2. Is there an ethical difference between hunting for sport and hunting for food? Does it matter if an alternative food source—for example, a grocery store—is readily available?

3. Some animals are threatened because of particular traits: elephants for their ivory, rhinoceroses for their horns, various furry animals for their pelts. If it were possible to breed these traits out of these animals, in effect creating a separate subspecies, would that be right?

4. Some research involving harm to animals is done to test therapies for life-threatening diseases, whereas other research is done to test consumer products such as cosmetics and food additives. Is all scientific research ethically equal?

DISCUSSION CASES

Primate Research and Animal Liberation

Various groups such as the Animal Defense League and the Animal Liberation Front have taken strong moral stands against what they see as abuse of animals. Some groups have resorted to illegal acts in the effort to liberate animals, ranging from pouring blood on fur coats to setting fire to trucks owned by meat processing companies to breaking into university laboratories to free animals used in scientific research. Although these acts are illegal, such groups generally claim to be committed to nonviolence. In one of the most well-known cases, activists broke into laboratories that were using primates to study the medical effects of severe head trauma. The activists pointed out that primates were subjected to massive head injuries and claimed that this treatment was brutally immoral.

Various species of animals have been used in a wide variety of research projects: primates for researching therapies for illnesses such as AIDS and polio, rabbits for testing safety of eye mascara, specifically bred mice for testing genetics, and common white rats for undergraduate experimentation in psychology courses. Is every use equally justified or equally unethical?

Carl Cohen claims that animals lack the moral autonomy necessary for having rights. Is this an empirical claim that might be answered by scientific research, or is it a conceptual claim that is true by definition? How would one know if primates were autonomous?

The Ethics of Fishing

Hunting is the usual recreational activity that animal rights activists focus on. They claim that it is unethical to pursue and kill animals for sport. Fishing seldom gets the same attention even though it involves similar acts. People stalk, pursue, hook, and often kill fish merely for entertainment. Fish often seem to suffer greatly during the experience. Also, as sometimes happens with hunting, non-

native species are often introduced into an ecosystem for the benefit of sport. These exotics can sometimes have detrimental effects of the native species and overall ecosystem. Do hunting and fishing stand or fall according to the same arguments?

Are you familiar with either a formal or informal code of ethics for fishing? Are any aspects of that code aimed at the well-being of the fish, or are they all anthropocentric? Which of Stephens's arguments might apply to eating fish? Which do not?

Many government agencies that manage lakes, forests, and wilderness areas are funded in part from hunting and fishing fees. As a result, fisheries management often faces conflict between ecological and recreational goals. For example, the game fish desired for fishing can significantly alter the ecosystem of a lake or river. Stocking game fish can also be very expensive, thus diverting financial resources away from more ecological projects. What general policy would you propose for fisheries management?

Breeding Endangered Species

The Endangered Species Act not only established policies to prevent harm to endangered species, it also directed government agencies to take steps to recover viable populations. One strategy for this second goal includes captive breeding of endangered species. Animals such as the California condor have been taken from the wild, often forever, and brought into captivity for breeding purposes. In effect, the individual animal is sacrificed for the good of the species. Do we have a responsibility to ensure the survival of a species? Is this equally true for all species? Is it as true for mosquitos and bacteria as it is for condors and whales? Is biological diversity itself valuable?

Many environmental objections to human activities focus on the effects of those activities on animals. Pollution has led to the near extinction of numerous fish species, for example. If it were possible to breed animals to be less susceptible to the effects

(continued)

of pollution, would this be a reasonable alternative to reducing pollution? Would genetic engineering be a helpful tool for environmentalists?

Selective Breeding

Humans have bred plant species such as corn and wheat for thousands of years. Although many plants have become extinct as a result (think of the effort to preserve heirloom species of garden vegetables), many more species have been created by human management. Farm animals such as cows, horses, and chickens have been similarly bred for human purposes. Species of rats are bred for laboratory purposes. Is there a value or ethical difference between species created by humans and those that are not? Is there a value or ethical difference between domestic animals and wild animals?

What principles should guide the breeding of plants and animals? Many new species of food crops have been created in the laboratory. With the development of cloning, the same can be true for animals as well. If biological diversity is good, is human-created diversity good? What answer might be suggested by Leopold's prescription to preserve the integrity and stability of a biotic community? Might William Baxter's claim that no normative conclusions can be drawn from something being natural suggest that such breeding is an acceptable means for optimizing our environmental goals?

CHAPTER 10

Ethics and Land

Although humans' relationship with animals may have drawn the widest public attention of all environmental policy issues, our relationship with land may be more fundamental. It is difficult to think of an environmental controversy that does not involve how humans have understood, valued, or treated the land. The next two chapters address some of the issues of land-use policy. Chapter 10 introduces some ethical issues that arise concerning both private and publicly owned land. Chapter 11 turns to the vital case of wilderness lands.

Some pivotal environmental cases in the earliest years of this century concerned public ownership and management of land. From the earliest days of U.S. history, the federal government owned substantial amounts of land. The Louisiana Purchase was just one case in which ownership of land, at least under the present legal system, originated in the public sector. During the late nineteenth and early twentieth centuries, government decisions created both the national park system and federal forestlands.

From these earliest days, controversy surrounded the proper use of public lands. In one famous case, Sierra Club founder John Muir argued that the scenic Hetch Hetchy Valley in the Sierra Nevada mountains should be preserved as a wilderness area. Gifford Pinchot, founder of the U.S. Forest Service, argued that this land would be better used as the site for a dam and reservoir to supply water and energy to the San Francisco Bay Area. To many people, this was the major case for establishing the distinction between *preservationist* and *conservationist* approaches to land. Preservationists argue that some land should be excluded from human use; conservationists argue that the land should be used but in reasonable and conservative fashion. Deciding the best and proper use for public lands raises some of the most fundamental questions of public policy.

In general, public lands should be managed for the public interest. But what is the public interest? Some would argue (see the reading by Stroup and Baden in Chapter 4, for example) that the public interest would be best served by transferring ownership to the private sector. On this view, private citizens are the best judge of the public interest; government ought simply to get out of the business of owning land.

Assuming that the government does own land, what should it do with this land? Preservationists might argue that the government should be guided by an ecological understanding of what is best for the land itself. Thus, if the government owns seashore property, this property should be managed in a way that best preserves the ecological functioning of the shoreland. If the government owns scenic canyons, it ought to preserve the natural beauty of the area. If the government owns a forest of giant sequoias, it ought to preserve and protect these symbols of nature's majesty. Conservationists might argue that government should regulate use of the land in a way that ensures equal access to a wide variety of citizens.

These sorts of decisions assume that government has a legitimate role to play in deciding questions of value. Yet, especially in a democracy, government seems ill-positioned to do this. Liberal political theory claims that government ought to remain neutral on questions of value. Given that citizens will disagree about such decisions, government's role should be to remain the neutral arbitrator or judge of competing interests. However, evidence from the turn of the century suggests that the public interest is better served when government is involved in both the ownership and management of property, especially when that property is highly valued and productive.

During the early years of the twentieth century, President Theodore Roosevelt led a progressive political movement against the laissez-faire, monopolistic practices of the nineteenth century. These progressives argued that natural resources should exist for the benefit of all citizens and not just for the few wealthy landowners and industrialists. Roosevelt and his political allies argued for a more active federal government to fight the concentration of power in the hands of a few and to direct natural resources for the common good. One of Roosevelt's closest advisors and friends was Gifford Pinchot, head of the U.S. Forest Service.

Pinchot instituted policies that would put management of national forestlands in the hands of trained forest professionals. These scientifically trained experts were in the best position to decide how to manage these public lands for the greatest good for the greatest number. These natural lands were definitively treated as resources—Pinchot's ethics was a thoroughgoing anthropocentrism—but they were seen as resources that should benefit all citizens over the long term. Pinchot's brand of environmentalism is conservationist and utilitarian. The land is equated with resources—but resources that are to be used conservatively for the long-term public interest rather than for the short-term private interest.

This philosophy guided the U.S. Forest Service throughout most of its history. Some critics have charged, however, that government bureaucracies are terribly inefficient at distributing resources. Echoing arguments that we've examined previously (in Chapters 4 and 8), these critics argue that free markets are more appropriate means for attaining their shared utilitarian goal of optimally satisfying the public's demand for resources.

But, of course, these debates often discount the perspective that there is more than mere instrumental value to land. Land is valued in many ways and for many reasons (note the essays by Thoreau in Chapter 2, Sagoff in Chapter 4, Muir in Chapter 5, and Leopold in Chapter 7). A full understanding of the many ethical issues raised by land policy must address the range of values attributed to natural areas. In turn, appreciating these values requires some understanding of the more general philosophies or worldviews in which these values are embedded. Readings in this chapter address questions of land policy by examining some influential and fundamental values and attitudes concerning land.

Without question, the most fundamental relationship that humans have with the land involves farming the land for food. Yet, in recent decades there has been an ambiguous relationship between environmentalists and farming. Too often, when environmentalists focus on farming they do so to criticize farming practices. Farmers clearly treat the soil as a resource to be used, albeit usually according to conservationist principles. This contrasts with the preservationist perspective common among environmentalists. Farmers also are criticized for their use of chemical fertilizers and pesticides, which many see as environmentally harmful. In addition, environmentalists point to cases of groundwater contamination and depletion as evidence of the environmental hazards of modern farming. Farmers also are faulted for draining wetlands and destroying other sensitive environmental areas to increase acreage for plowing. Finally, the treatment of farm animals is a central concern for many environmentalists.

It is strange that this tension should exist between two groups who are so committed to the land. This is especially so when we consider that a major environmental figure, Aldo Leopold, wrote many of his most influential and inspiring essays about his Wisconsin farm. Perhaps the key is to neither praise nor fault all farming in general but to recognize that some farming practices are environmentally and socially good and others are not.

What is good farming? According to many, good farming must exist in an ecologically balanced relationship with the land. A variety of farming techniques—crop rotation and diversification, contour plowing, elimination of chemical fertilizers and pesticides, use of native plants, composting—often identified as sustainable agriculture are defended by both environmentalists and farmers. Such techniques are contrasted with the practices of modern agribusiness that are based on industrial models of production. This highly efficient modern farm relies heavily on complex and expensive farm machinery, chemical pesticides and fertilizers, hybrid seed, large irrigation systems, specialization, and, all too often, significant amounts of borrowed capital. The factory-farms targeted by animal rights advocates are typical of this approach.

Many farmers and environmentalists are critical of the agribusiness industry. They fault agribusiness for both the decline of small family farms and for much environmental damage. Yet defenders explain that this approach has developed for a reason. Agribusiness is the result of market pressures for greater efficiency and productivity. In short, economic markets have given rise to modern farming. The final essays in this chapter examines the question of good farming, markets, and the public good.

The Training of a Forester

Gifford Pinchot

Gifford Pinchot was head of the U.S. Forest Service under President Theodore Roosevelt. He was one of the first scientifically trained foresters in the United States and worked to bring the scientific principles of forestry into the management of national forests. His progressive politics sought to make these resources available to all citizens, and he fought to prevent resources from being exploited for the benefit of a few wealthy individuals. His ethical approach was clearly utilitarian and his environmentalism clearly conservationist. In this reading, taken from his book The Training of a Forester, *Pinchot lays out this utilitarian and conservationist agenda in characteristically direct and clear terms.*

WHAT IS A FOREST?

First, what is forestry? Forestry is the knowledge of the forest. In particularly, it is the art of handling the forest so that it will render whatever service is required of it without being impoverished or destroyed. For example, a forest may be handled so as to produce saw logs, telegraph poles, barrel hoops, firewood, tan bark, or turpentine. The main purpose of its treatment may be to prevent the washing of soil, to regulate the flow of streams, to support cattle or sheep, or it may be handled so as to supply a wide range and combination of uses. Forestry is the art of producing from the forest whatever it can yield for the service of man.

. . .

THE FOREST AND THE NATION

The position of the forest in the housekeeping of any nation is unlike that of any other great natural resource, for the forest not only furnishes wood, without which civilization as we know it would be impossible, but serves also to protect or make valuable many of the other things without which we could not get on. Thus the forest cover protects the soil from the effects

of wind, and holds it in place. For lack of it hundreds of thousands of square miles have been converted by the winds from moderately fertile, productive land to arid drifting sands. Narrow strips of forest planted as windbreaks make agriculture possible in certain regions by preventing destruction of crops by moisture-stealing dry winds which so afflict the central portions of our country.

Without the forests the great bulk of our mining for coal, metals, and the precious minerals would be either impossible or vastly more expensive than it is at present, because the galleries of mines are propped with wood, and so protected against caving in. So far, no satisfactory substitute for the wooden railroad tie has been devised; and our whole system of land transportation is directly dependent for its existence upon the forest, which supplies more than one hundred and twenty million new railroad ties every year in the United States alone.

The forest regulates and protects the flow of streams. Its effect is to reduce the height of floods and to moderate extremes of low water. The official measurements of the United States Geological Survey have finally settled this long-disputed question. By protecting mountain slopes against excessive soil wash, it protects also the lowlands upon which this wash would otherwise be deposited and the rivers whose channels it would clog. It is well within the truth to say that the utility of any system of rivers for transportation, for irrigation, for water-power, and for domestic supply depends in great part upon the protection which forests offer to the headwaters of the streams, and that without such protection none of these uses can be expected long to endure.

Of the two basic materials of our civilization, iron and wood, the forest supplies one. The dominant place of the forest in our national economy is well illustrated by the fact that no article whatsoever, whether of use or ornament, whether it be for food, shelter, clothing, convenience, protection, or decoration, can be produced and delivered to the user, as industry is not organized, without the help of the forest in supplying wood. An examination of the history of any article, including the production of the raw material, and its manufacture, transportation, and distribution, will at once make this point clear.

The forest is a national necessity. Without the material, the protection, and the assistance it supplies, no nation can long succeed. Many regions of the old world, such as Palestine, Greece, Northern Africa, and Central India, offer in themselves the most impressive object lessons of the effect upon national prosperity and national character of the neglect of the forest and its consequent destruction.

THE FORESTER'S POINT OF VIEW

The central idea of the Forester, in handling the forest, is to promote and perpetuate its greatest use to men. His purpose is to make it serve the greatest good of the greatest number for the longest time. Before the members of any other profession dealing with natural resources, the Foresters acquired the long look ahead. This was only natural, because in forestry it is seldom that a man lives to harvest the crop which he helped to sow. The Forester must look forward, because the natural resource with which he deals matures so slowly, and because, if steps are to be taken to insure for succeeding generations a supply of the things the forest yields, they must be taken long in advance. The idea of using the forest first for the greatest good of the present generation, and then for the greatest good of succeeding generations through the long future of the nation and the race—that is the Forester's point of view.

The use of foresight to insure the existence of the forest in the future, and, so far as practicable, the continued or increasing abundance of its service to men, naturally suggested the use of foresight in the same way as to other natural resources as well. Thus is was the Forester's point of view, applied not only to the forest but to the lands, the minerals, and the streams, which produced the Conservation policy. The idea of applying foresight and common-sense to the other natural resources as well as to the forest was natural and inevitable. It works out, equally as a matter of course, into the conception of a planned and orderly development of all that the earth contains for the uses of men. This leads in turn to the application of the same principle to other questions and resources. It was foreseen from the beginning by those who were responsible for inaugurating the Conservation movement that its natural development would in time work out into a planned and orderly scheme for national efficiency, based on the elimination of waste, and directed toward the best use of all we have for the greatest good of the greatest number for the longest time. It is easy to see that this principle (the Forester's principle, first brought to public attention by Foresters) is the key to national success.

For Further Discussion

1. What values should guide decisions made by professional foresters? Why? Should foresters who work for government agencies such as the U.S. Forest Service have special ethical responsibilities?

2. Pinchot tells us that forestry is the art of producing from the forest whatever it can yield for the service of man. How do forests serve human beings? How does Pinchot answer this question?

3. Have the reasons for protecting forests changed from Pinchot's time?

4. Do you agree with Pinchot that the central purpose of the forester is to make the forest serve the greatest good of the greatest number for the greatest time? What practical forest policies might follow from this principle?

Reforming the Forest Service

Randel O'Toole

*In recent years Pinchot's Forest Service has been criticized
for its management of the national forests. Many environmentalists charge that the U.S. Forest Service has not only
allowed but has actually encouraged much destruction of
forestland. The Forest Service, a section of the U.S. Department of Agriculture, has built logging roads and permitted widespread tree-cutting. Others claim that the goal
of multiple use—of balancing logging interests with recreational use—has gone too far in restricting harvesting
of timber.*

*In this selection, which is taken from Chapter 12 and
the Conclusion of his book, economist Randel O'Toole argues that market mechanisms should be used in deciding
the proper balance between competing uses of public lands.
O'Toole believes that prescriptive legislation, laws and regulations governing resource use, has failed to serve the
public interest. He believes that market mechanisms can
satisfactorily balance the interests of such diverse groups
as the Forest Service, economists, and deep ecologists. He
argues that this proposal will be "politically viable, economically efficient, and environmentally sound."*

Environmentalists who are concerned about national
forest problems have traditionally proposed to solve
them through prescriptive legislation. Such proposals
have rarely received wide support, in part because
Congress has preferred to leave technical issues to its
experts—in this case, Forest Service professionals. In
addition, laws that regulate practices rather than incentives treat the symptoms and not the causes of forest problems and thus may be doomed to fail.

The 1980s have seen two new groups whose means
and objectives appear to be quite different from each
other as well as from more traditional environmental
groups. The New Resource Economists, as represented
by the Political Economy Research Center of Bozeman,
Montana, work primarily in the academic world and
may be best known for their proposals to sell the national forests. Earth First!, a major faction of the Deep
Ecologists, primarily take direct actions against environmental destruction, such as sitting in front of

bulldozers, and strongly prefer public ownership. Although these two groups appear to represent polar
extremes, in fact there are many similarities between
them.

These three views appear to present significantly
different alternatives for reforming the Forest Service.
Individually, none of them is feasible. However, it may
be possible to blend the best elements of each view
into a proposal that is politically viable, economically
efficient, and environmentally sound.

THE FAILURE OF PRESCRIPTIVE LEGISLATION

Popular support for prescriptive legislation is probably
due to the belief that Forest Service leaders are timber
primacists who need restrictions to prevent them from
taking economically or environmentally unacceptable
actions. A number of prescriptive proposals were in
the bill sponsored by West Virginia Senator Jennings
Randolph during the debates over the National Forest
Management Act in 1976. Largely written by environmental groups, this bill was strenuously opposed by
the timber industry and the Forest Service as being too
"prescriptive." The final legislation, which focused on
a forest planning process, included few prescriptions.
Yet it is unlikely that the Randolph bill would have
corrected any of the problems with national forest
management.

Perhaps the most prescriptive section of the Randolph bill would have prohibited clearcutting in the
eastern mixed hardwood forests. Yet the bill merely required detailed environmental assessments prior to
clearcutting elsewhere. The final legislation struck the
clearcutting proscription but retained the requirement
for environmental assessments and other clearcutting
limits.

The Randolph bill would have required that trees,
or most trees in a stand, be "dead, mature, or large"
before cutting. "Mature" was defined as a tree that has

reached the peak of its vigor and is declining, and "large" was defined as trees that are the size of those in a natural mature stand. The final legislation required that trees not be cut before the stands they are in reached "culmination of mean annual increment"—their peak of growth. The difference between the two is that some individual trees may peak much later than entire stands. In this case, the final legislation was just as prescriptive as the Randolph bill, but the prescription is slightly different.

The Randolph bill would have required the Forest Service to limit timber sales to the nondeclining yield levels of each ranger district. The final legislation limited timber sales to the nondeclining yield levels of each national forest. Again, the two are equally prescriptive, but the prescriptions are slightly different.

The Randolph bill would have required that all trees be marked before cutting. The final legislation allows the Forest Service to mark only the perimeters of clearcuts before cutting. While the Randolph bill is more prescriptive, the additional prescription existed only to make clearcutting less economically efficient, and therefore less attractive, than other forms of cutting.

The Randolph bill prohibited "type conversion" of eastern hardwood forests to pine and required that other proposed type conversions receive detailed analyses and public review. The final legislation struck the proscription but retained the analysis and review requirements.

The Randolph bill required the preparation of forest plans, as did the final legislation, and prohibited higher levels of the Forest Service or USDA from establishing targets for the national forests. This was struck from the final legislation, and many people have accused the Forest Service of using RPA objectives as targets. However, many national forest plans that failed to meet these objectives have been approved.

Finally, the Randolph bill required the Forest Service to develop a cost accounting system to compare the benefits and costs of timber sales. It also required the Forest Service to report below-cost timber sales to Congress each year. The final legislation eliminated the cost accounting system but retained the reporting requirement. In 1984 Congress appropriated $400,000 for the Forest Service to develop a cost accounting system. . . . the system that was developed was strongly biased toward timber.

The Randolph bill illustrates two major problems with prescriptive legislation. First, either it is so specific that its writers must be experts on the management of every national forest or it is so vague that it is meaningless. For example, in proscribing clearcutting

from eastern hardwood forests, the Randolph bill was proclaiming Congress to be a better expert on forest management than the Forest Service. Yet by requiring just an interdisciplinary review prior to clearcutting elsewhere, the Randolph bill was creating a vague standard that could be enforced only by requiring bureaucrats to create mountains of paperwork.

Second, by trying to change forest practices without changing incentives, prescriptive legislation treats symptoms rather than causes. Recognizing that the Forest Service often sells timber at a loss, the Randolph bill would have required the Forest Service to develop a cost accounting system. But the *Timber Sale Program Information Reporting System,* . . . shows that managers with an incentive to sell below-cost timber can be trusted to produce a cost accounting system that obscures such losses.

This is demonstrated by numerous cases where prescriptive legislation was passed by Congress but ignored by the Forest Service. For example, a good case can be made that below-cost sales are already illegal. Cross-subsidized sales also appear to violate the *Forest Service Manual.* Yet such sales take place routinely in every Forest Service region. Section 14(a) of the National Forest Management Act requires that timber be sold for not less than "appraised value." . . . Legislative history and *Forest Service Manual* directives indicate that appraised value means fair market value, the price at which a buyer and seller would agree to exchange. Since no reasonable seller would agree to sell at a loss, below-cost sales appear to violate this provision.

The *Forest Service Manual* also specifically and directly proscribes cross-subsidized timber sales. Section 2403.25 of the manual states that "the appraised value of a tract of timber will not be reduced to obtain utilization of a species, size, or class." This is called the *tract value policy.* Section 2422.56(3) of the manual gives, as a specific example, a sale that includes 5 million board feet of pine, which is appraised at $22 per thousand board feet, and 5 million board feet of fir, which is appraised at minus $4 per thousand.

In a typical cross-subsidized sale, if the base rates for the two species were $9 per thousand, the value of the pine would be reduced to $9 to raise the value of the fir to $9. The total sale would then sell for $90,000. But the manual says that the sale cannot be sold for less than $110,000, which is the value of the pine alone. The manual permits adjustments if selling the pine alone would reduce the value of the pine. For example, brush disposal costs might be the same for the pine alone as for the pine together with the fir. This adjustment, however, would probably be less than

$20,000. Thus, the cross-subsidized sale violates the tract value policy.

. . . As many as 40 percent of national forest timber sales may be cross-subsidized and thus are likely to violate this policy. A CHEC review of recent timber sales on the Sierra National Forest (CA) found seventeen sales in two years violating the policy. In eight cases, high bid prices raised the total sale price above the calculated tract value. In the remaining nine cases, the high bid prices fell short of the tract value by an average of more than $100,000 per sale.

The Knutson-Vandenberg Act and other laws give managers powerful incentives to sell timber below cost and to cross-subsidize timber. Laws against below-cost sales are as unlikely to halt such sales as existing laws and provisions of the *Forest Service Manual* have halted past sales below cost.

Numerous examples exist in which the Forest Service has interpreted the law to its own benefit and has subsequently found that interpretation to be illegal. A West Virginia district court and the U.S. Court of Appeals for the Fourth Circuit found the Forest Service's practice of clearcutting to be in clear violation of a law passed seventy-five years before the court decision. In Oregon a 1903 law prohibiting entry into a municipal watershed by any member of the public other than a Forest Service official was violated by the Forest Service for seventeen years beginning in 1958. In that year the Forest Service began large-scale timber operations in the watershed, operations that ceased in 1975 when a court ruled them illegal.

Prescriptive legislation may attempt to make the Forest Service more efficient and environmentally sensitive. However, it will greatly increase centralization— nothing is more central than Congress deciding how each acre of the national forests should be managed. Prescriptive legislation will also reduce the flexibility of the Forest Service to respond to new research and changes in the supply and demand for forest resources. Because of these problems, prescriptive legislation will fail to make the Forest Service significantly more efficient or sensitive to environmental issues than it is today.

PRIVATE PROPERTY AND PUBLIC CHOICE

A major difference between traditional environmental groups and the New Resource Economists is the latter's focus on incentives rather than on outcomes. "In general, public-spirited actions regarding environmental quality, resource management, and other public issues have tended to be oriented rather simplistically toward desired outcomes rather than toward processes carefully designed to produce those outcomes," say Richard Stroup and John Baden, two of the leading New Resource Economists. Where environmental groups seek prescriptive legislation to obtain the outcome they desire, the New Resource Economists seek to use markets as the best process for achieving that outcome.

Although many people disagree with proposals to sell the public lands, the New Resource Economists have much to say that is worthwhile. Led by Baden, Stroup, and Terry Anderson, economists and political scientists who founded the Political Economy Research Center in Bozeman, Montana, the New Resource Economists believe that the marketplace is better at allocating resources than is interest-group politics.

According to Stroup and Baden, people acquire wealth in one of two ways: by purchasing less-valuable resources and converting them to more-valuable ones and by convincing the government to transfer resources from public to private use or from one person to another. "By moving toward political allocation and away from the rule of willing consent," they say, "we have moved from a society that rewards productive activity and willing exchange to one where many of a person's best investment opportunities lie in influencing transfer activities."

The marketplace is centered around the notion of private property. Property owners "have incentives to use their resources efficiently," say Stroup and Baden. People can make money in the market by taking low-value goods and converting them to high-value goods. Thus, the market distributes resources to those who place the greatest value on those resources. "Everyone can be made better off when goods and services are moved to higher valued uses," they conclude.

"The benefits of diversity, individual freedom, adaptiveness to changing conditions, the production of information, and even a certain equity derive from this market system," say Stroup and Baden. The market promotes diversity because there is no single, centralized decision maker. It preserves individual freedom since those who support and wish to participate in each activity may do so on the basis of willing consent. It is equitable because those who gain the benefits pay the costs.

The government, on the other hand, is based on coercive activity. Say Stroup and Baden, "Government, with its monopoly on sanctified coercion, has the potential for being the most efficient engine ever designed for the generation of plunder." According to the New Resource economists, timber companies, ranchers, and wilderness users find it in their interests to convince Congress to force society as a whole to pay part of the costs of timber cutting, grazing, or recreation. Thus, they become "free riders" who enjoy the benefits of the national forests with everyone else paying the costs.

New Resource Economists disagree with the traditional belief that public lands are *public goods* from which everyone benefits. "It is a common misconception that every citizen benefits from his share of the public lands and the resources found thereon," say Stroup and Baden. "But public ownership by no means guarantees public benefits. Individuals make decisions regarding resource use, not groups or societies. Yet, with government control, it is not the owners who make the decisions, but politicians and bureaucrats." . . .

Critics of the New Resource Economists say that the "perfect market" envisioned by Stroup and Baden exists only in fantasy. But the New Resource Economists respond that many of the "market failures" that make people suspicious of markets are not market failures at all but institutional failures. For the market to work, private property rights to resources must be easily transferable.

"If rights are not easily transferable, owners may, for example, have little incentive to conserve resources for which others might be willing to pay if transfer were possible," say Stroup and Baden. "Transferability ensures that a resource owner must reject all bids for the resource in order to continue ownership and use. Thus, if ownership is retained, the cost to others is made real and explicit."

The market works when rights are both privately held and easily transferable because decision makers have both easy access to information through bid and asked prices and an incentive to move resources to higher-valued uses. The New Resource Economists believe that private ownership would solve most natural resource problems.

"When a natural resource is privately owned, it is often thought that the owner has only his conscience to tell him to pay attention to the desires of others," say Stroup and Baden. "Normally, however, it is the *absence* of private, transferable ownership that leads to the resource user's lack of concern for others' desires. Private ownership holds the individual owner responsible for allocating a resource to its highest valued use, whether or not the resource is used by others.

"If the buffalo is not mine until I kill it and I cannot sell my interest in the living animal to another, I have no incentive—beyond altruism—to investigate others' interest in it. I will do with it as I wish. But if the buffalo is mine and I may sell it, I am motivated to consider others' value estimates of the animal. I will misuse the buffalo only at my economic peril." Most environmental problems, such as lack of protection for wildlife, air pollution, and poor water quality, are due to the lack of transferable property rights.

Water rights are an important example. Currently, landowners in the West can have the right to use water, but ownership is still claimed by the states. Irrigators with historic water rights defend their right to use the water even if some other use is far more valuable. If water *ownership* rights were vested with various individuals, they could sell the right to pollute the water, use it for irrigation, provide it to cities, or maintain fish habitat.

Anderson compares projected water shortages with the energy crisis of the late 1970s, which he says was partly caused by government controls of market prices. "When prices were allowed to rise to market levels, energy supplies increased and demand decreased. Before this can happen with water, ownership of the resource will have to be specified and the impediments to trade removed."

Some market failures still exist, but even with their imperfections, markets are often more efficient than governments, say the New Resource Economists. The national forests, for example, were created with the idea that "scientific foresters" employed by the public could objectively determine the method of management that best meets the public interest. But to assume that managers will be altruistic, say Stroup and Baden, it must be assumed "that culture can 'rewire' people so that the public interest becomes self-interest." Instead, "Property rights theorists assume that the decision maker will maximize his own utility—not that of some institution or state—in whatever situation he finds himself."

Stroup and Baden point out that, "for the bureaucrat, the tax base is essentially a common pool resource ripe for exploitation." Thus, agencies make their best efforts to mine the resource before another agency gets them. "A common pool is like a soda being drawn down by several small boys, each with a straw," explain Stroup and Baden. "The 'rule of capture' is in effect. The contents of the container belong to no one boy until he 'captures' it through his straw."

These problems have caused most of the environmental destruction the United States has seen in this century. "Unconstrained by the need to generate profits, bureaucrats may ignore or exaggerate the economic efficiency of the projects they administer," say Stroup and Baden. "If government agencies were required to meet standards of economic efficiency, many of their environmentally destructive practices would not occur. In the absence of such standards, the American taxpayer is, in effect, subsidizing the destruction of the environment and enhancing the welfare of bureaucrats and special interests."

Many people believe that the government can do a better job of protecting the interests of future generations than the marketplace. In fact, say the New Resource Economists, the exact reverse is true. "A government decision maker can seldom gain political support by locking resources away to benefit the unborn," say Stroup and Baden. Instead, the pressure is to use the resources as rapidly as possible to benefit constituents. The long-term costs are ignored because most terms of office are short.

In contrast, if resources will be more valuable in the future than they are today, the market will encourage conservation. Private speculators make money by deferring resource use to the future. Stroup and Baden point out that speculators "save for the future by paying a market price higher than any other bidder who seeks resources for present use." . . . A 10 percent discount rate effectively represents a ten-year planning horizon, whereas the political planning horizon is rarely much longer than two years.

DEEP ECOLOGY: A FUNDAMENTAL TRANSFORMATION

At first glance, the Deep Ecologists represent the polar extreme from the New Resource Economics. Although Deep Ecology is a broad movement which has no single representative, most people who consider themselves Deep Ecologists go beyond the prescriptive legislation proposed by traditional environmentalists to demanding "a fundamental transformation of industrial and global society." Deep Ecologists claim that their movement is a "religious and philosophical revolution of the first magnitude." A fundamental tenet of this religion is biocentrism, the view that "all things in the biosphere have an equal right to live and blossom and to reach their own individual forms."

The Deep Ecologists are, in part, responding to the endless battles being fought by more traditional environmental groups. Peter Berg likens classic environ-

mentalism to "a hospital that consists only of an emergency room. No maternity care, no podiatric clinic, no promising therapy: just mangled trauma cases. Many of them are lost or drag on in wilting protraction, and if a few are saved there are always more than can be handled jamming through the door." He adds that "no one can doubt the moral basis of environmentalism, but the essentially defensive terms of its endless struggle mitigate against ever stopping the slaughter." Deep Ecologists believe that the environmental movement is too centralized, and they prefer decentralist, local approaches to environmental problems.

Rather than getting involved in political battles, Deep Ecologists prefer to promote their general goal of biocentric equality. Of course, all life must consume other living things to survive, and Deep Ecologists do not claim that every individual plant or animal has an equal right to life. Instead, they advocate preservation of genetic diversity, species, and ecosystems.

Humans, say the Deep Ecologists, "have no right to reduce this richness and diversity [of the Earth] except to satisfy *vital* needs." They add that "the term 'vital need' is left deliberately vague to allow for considerable latitude in judgment." Although this appears vague, it is at least clear that the 1973 Endangered Species Act is a fundamental step in the direction of Deep Ecology.

At least some Deep Ecologists realize that a mass conversion to their religion is unlikely. Even if it happened, it could not last long. Writing in a series of essays on Deep Ecology, Garrett Hardin points out that "a species composed of pure altruists is impossible." If such a species existed, it would soon be replaced by a mutation that is less altruistic.

Yet many Deep Ecologists believe that a society based on their attitudes is possible and may even have once existed. The epitome of human social existence was reached, they believe, by the hunting and gathering tribes. Hunters and gatherers may have had greater social and sexual equality, less epidemic disease, and more leisure time, say Deep Ecologists.

Ironically, early hunters may have been responsible for the extinction of numerous large mammals, such as the mammoth and cave bear. Social conditions in hunting tribes were probably also less idyllic than claimed by Deep Ecologists. Yet it is significant that the hunters and gatherers represent the culture that most celebrated individuals above the social group.

Where Deep Ecology resembles a religion, it borrows its tools from the hunters and gatherers. Adherents are encouraged to make vision quests, while selected individuals are treated as shamans. Vision quests and shamans disappeared from agricultural and later societies, which often place the group higher than the individual and depend on group rites rather than individual abilities—both mystical and otherwise.

The Deep Ecologists' preference for individual freedom and action conflicts with the traditional environmental tendency toward prescriptive legislation. Prescriptive legislation is possible only through the coercive power of the government, and this, in turn, is based on the philosophical belief that society's needs are paramount to the needs of the individual. Deep Ecologists reject this belief and instead call for the promotion of an ethic or morality that will suspend the need for any such coercive activities.

Although privatization and biocentrism seem to have little in common with each other, in fact there are many similarities between the New Resource Economists and the Deep Ecologists. Each group proposes to "subvert the dominant paradigm"—a paradigm being the ideas and beliefs shared by a community of individuals. The New Resource Economists oppose the paradigm that turns to government regulation to solve environmental problems. The Deep Ecologists oppose the paradigm that uses political compromise to gain environmental protection for scarce, nonmarket resources.

In a larger sense, both groups are responding to the dominant paradigm of interest-group politics, in which environmental battles are endlessly fought but never won. The New Resource Economists and the Deep Ecologists each seek a fundamental solution that will give resource managers the incentive to protect and produce the most valuable resources. Decentralization, both groups believe, is an important part of that solution. Decentralization promotes diversity, which enables ecosystems and society to adapt to change. Decentralization also provides efficiency in resource use because it encourages innovation and the spread of innovative ideas.

The preference for decentralization is a consequence of both groups' strong belief in individual freedom. New Resource Economists see society as merely a collection of individuals and refuse to believe in the existence of any "social good" that is more than the sum of individual goods. Similarly, the Deep Ecologists favor the rights of the individual over the rights of society—with the proviso that "individual" be defined to include other species as well. Deep Ecologists, like the New Resource Economists, tend to be anarchistic.

The similarities between Deep Ecology and New Resource Economics parallel the similarities between ecology and economics. Although the language used by ecologists differs from that of economists, it frequently translates to identical concepts. Where economists discuss efficiency, decentralization, and incentives, ecologists discuss the maximum power principle, diversity, and feedback loops.

The fact that these very different terms have identical meanings underscores the obvious point that *ecology* and *economics* have the same root: *oikos*, a Greek word meaning "house." Where ecology translates to "study of the house," economics is "management of the house." Although the "houses" scrutinized by ecologists and economists differ, ecological systems are really economic systems, and economic systems are really ecological systems.

Thus, New Resource Economists and Deep Ecologists use different terminology but reach many of the same conclusions: that the current system of allocating resources is flawed and that it must be replaced by one that gives individuals the incentive to manage those resources properly. Where the New Resource Economists and Deep Ecologists disagree is in their views of public lands and markets. The New Resource Economists view public land ownership as a fundamental evil that has no value in itself and that leads to environmental destruction as well as transfers of wealth from the poor to the rich. The Deep Ecologists would apply exactly the same terms to markets.

Yet markets are compatible with public land ownership. In fact, markets are the key to reforming public land management because they most closely resemble a natural ecosystem. Unlike a centrally planned economy, a successful ecosystem doesn't require an omnipotent regulator to decide how much of each good will be produced, nor does it rely on the religion or ethics of its individual members. Instead, the ecosystem uses feedback loops to be self-regulating. Through the market system, the Forest Service can be reformed in such a way that it, too, is largely self-regulating. Such reforms can achieve the goals of both the New Resource Economists and the Deep Ecologists.

BLENDING NEW RESOURCE ECONOMICS AND DEEP ECOLOGY

Successful reforms of the Forest Service must be based on a blend of New Resource Economics and Deep Ecology. Reforms must recognize that the Forest Service is run by ordinary people who are motivated by incentives. One of the most important of these incentives is the desire of the agency to increase its budget. Reforms that try to fight the bureaucracy's natural tendency to increase its budget will be doomed to failure. Instead, they should employ this tendency so that the budget becomes part of a feedback loop: As the agency maximizes its budget, it also accomplishes social, economic, and environmental goals.

Reforms should be based on two sound ecological principles: efficiency and diversity. Just as an organism must be efficient to compete in its environment, an organization such as the Forest Service must be efficient to serve the public. The current set of laws governing the Forest Service encourages inefficiency on a massive scale. Reforms should give managers the incentives to produce the highest-valued goods or services rather than producing goods or services that are below cost or that actually have negative values. Markets can provide this incentive.

Reforms should also encourage diversity since diversity will allow local managers to respond to local situations and to develop innovative techniques to improve forest management. The Forest Service achieved its reputation as an "excellent organization" largely because it was once the most decentralized agency in the federal government. Many of the current problems with the agency are due, perhaps indirectly, to increasing centralization over the last ten to twenty years. A return to decentralization can make the agency more responsive to public demand and changing tastes.

Diversity and efficiency will make the Forest Service more environmentally sensitive. Yet reforms should also recognize that some national forest resources, such as endangered plant and wildlife species, may be valuable yet difficult to sell in a marketplace. The Deep Ecology principle that all species are valuable is important here. The protection of these resources should not be neglected in reforms designed to make the agency more efficient or more diverse. Markets can make the Forest Service more sensitive than it is now to the lack of public demand for such goods as lodgepole pine and to the increasing public demand for such goods as recreation. The Endangered Species Act can protect those species not protected by markets just as well in national forests oriented to markets as in national forests oriented to interest-group politics.

Finally, reforms should recognize the irreversibility of certain forest management actions. Changes in forest policy should be designed to maintain options for the future. The reforms themselves should be reversible so that unexpected and undesirable consequences can be corrected if necessary. This means that public lands should be retained in public ownership so that public demand can easily correct any unexpected and unintended effects of reforms.

The reforms . . . attempt to meet these standards. Markets are used to provide forest managers with feedback that will naturally lead them to manage national forests for their most valuable uses. A minimum of prescriptive legislation is required to protect endangered species and a few other truly nonmarket resources. Diversity is ensured by treating each national forest as an individual unit rather than as a part of a centrally planned organization. The resulting proposal, it is hoped, is not only economically efficient and environmentally sound but politically viable as well.

. . .

A New Environmental Agenda

Traditional environmental assessments of national forest controversies are based on a fundamental misdiagnosis of the problem. Correcting this misdiagnosis leads to a new view of the Forest Service: Not bad people intent on environmental destruction and economic waste, but ordinary people motivated by incentives, like everyone else.

The appropriation process has taught the Forest Service that Congress is most responsive to production of commodities such as timber and grazing. Laws like the Knutson-Vandenberg Act and the Public Rangelands Improvement Act also encourage commodity production. Below-cost timber sales, overgrazing, road construction in potential wilderness areas, and neglect of important recreation values, as well as fabrication of data in forest plans or reforestation reports, all result from the agency's desire to maximize its budget.

If this is true, then replacing the people who run the Forest Service, perhaps through presidential elections, will have negligible effects on the agency—as John Crowell learned to his frustration. For example, if all Forest Service foresters were replaced by wildlife biologists tomorrow, the biologists would soon propose below-cost sales to augment K-V funds for wildlife. Changing the legal prescriptions under which the agency must operate will merely lead to more double-talk and clever accounting systems. Reform of the Forest Service must come by changing the incentives that motivate national forest managers.

This necessarily means divorcing the Forest Service from congressional appropriations. Appropriations are politically determined through a process that favors those special interest groups that can gain the most income. In the long run, national forest appropriations will be oriented toward development and environmental destruction.

Changing incentives also requires charging for recreational use. Managers whose budgets increase when recreational use increases but decrease when below-cost sales occur will act very differently from the way Forest Service officials act today. Any reforms that are not accompanied by recreation fees will fail to protect national forest amenity values.

It is true that sacrifices will be needed to achieve these reforms. Recreationists must accept increased recreation fees, but in exchange they will gain a significant increase in recreation opportunities. Environmentalists must accept the fact that some national forests will be managed predominantly for timber, but in exchange they will halt development of millions of acres of below-cost timber lands. Timber purchasers must accept that many national forests will sell virtually no timber, but in exchange they will gain a stabilization of timber supplies throughout the National Forest System. The Forest Service must accept a reduction in its bureaucracy, but in exchange it will see an end to morale problems caused by polarization, red tape, and delays.

Marketization may appear revolutionary, but unlike many revolutionary ideas that are politically infeasible, marketization can be successfully promoted by a coalition of interest groups simply because so many interests will benefit. These beneficiaries include:

- Recreationists, who will gain expanded opportunities for hunting, fishing, camping, and other outdoor recreation activities;

- Counties, whose incomes from national forest management are likely to double or triple;

- Private landowners, whose timber, forage, and recreation resources will all gain in value when below-cost sales cease in the national forests;

- Wilderness advocates and wildlife lovers who are concerned about the effects of below-cost timber sales and grazing on wildlands, water quality, and other values;

- Forest managers who are tired of the pointless red tape of environmental impact statements, environmental analyses, and other paperwork; and

- Fiscal conservatives and others who are concerned about the federal deficit.

Perhaps the largest and most powerful of these interest groups is the environmental community, which has been fairly evenly matched with the timber industry in previous congressional debates. In 1976, for example, the timber industry and the Forest Service were able to convince Congress to reject most environmental reforms only because wildlife interest groups sided with the industry and agency. With the support of wildlife groups, environmentalists may have veto power over industry-supported legislation—but the industry probably would retain veto power over legislation supported only by environmentalists.

Thus, the support of counties, private landowners, and other interest groups is critical to the passage of Forest Service reforms. These groups, and particularly the counties, traditionally support the timber industry because they are not aware of what inefficient national forest management is costing them. The proposed reforms offer them many benefits and should be politically feasible if they side with the environmental community.

Environmentalists have much to gain from reforms, and reforms are not likely to take place at all without their support. Yet the dramatic changes proposed by this book will at first seem risky to most environmentalists. The dominant environmental paradigm—that recreation, wildlife, water, and other amenities are strictly nonmarket resources that will be destroyed without government regulation and prescriptive legislation—is not easy to subvert.

Yet marketization is far superior to the current system of interest group politics. It will protect millions of acres of wildlands that the Forest Service plans to road and log. It will stop below-cost pesticide spraying and bring overgrazing under control. It will provide a wider variety of recreation opportunities than is available today. It will put an end to fighting losing battles against environmentally destructive and economically absurd projects. In short, the proposals in this book should provide a blueprint for environmental action regarding the Forest Service and many other natural resource agencies.

For Further Discussion

1. In what areas would Pinchot disagree with O'Toole concerning the management of national forests? In what areas might they agree?

2. What exactly does O'Toole mean by "prescriptive legislation"? What reasons are offered to explain its failure?

3. Compare what O'Toole says about private property to the views of Stroup and Baden (Chapter 4)

and Baxter (Chapter 8). What values are promoted by the right of private property?

4. O'Toole agrees with the new resource economists in rejecting the idea that public lands are public goods. What does he mean by this, and do you agree with this conclusion?

5. Do you agree that markets are the best means for determining the proper use of public lands?

Anglo-American Land Use Attitudes

Eugene C. Hargrove

In this selection, philosopher Eugene Hargrove describes some of the philosophical ideas that underlie many contemporary attitudes toward the land. How the land is understood, a philosophical and conceptual issue, has profoundly influenced land-use values and policies. If we do not understand this fundamental question, we will fail to understand much of the contemporary land-use debate.

Hargrove traces three historical sources of contemporary American land-use attitudes: the traditions of Germanic and Saxon freemen, Thomas Jefferson's theory of allodial rights, and Locke's theory of property rights. To these beginnings, Hargrove links the strong property rights perspective characteristic of those critical of many environmental initiatives.

INTRODUCTION

Such protected areas as Yosemite, Yellowstone, and the Grand Canyon are often cited as great successes of the environmental movement in nature preservation and conservation. Yet, not all natural objects and areas worthy of special protection or management are of such national significance and these must be dealt with at state, regional, or local levels. In such cases, environmentalists almost always plead their cause before a county court, a local administrative political body, usually consisting of three judges elected by the rural community, who may or may not have legal backgrounds.

Here the environmentalists are probably in for a great shock. Inevitably, some rural landowner will defend his special property rights to the land in question. He will ask the court rhetorically, "What right do these outsiders, these so-called environmentalists, have to come in here and try to tell me what to do with my land?" and answering his own question, he will continue, "They don't have any right. I worked the land; it's my property, and no one has the right to tell me what to do with it!" The environmentalists may be surprised that the farmer does not bother to reply to any of their carefully made points, but the real shock comes at the end when the county court dismisses the environmental issues, ruling in favor of the landowner.

While the environmentalists may suspect corruption (and such dealings are not unlikely), usually both the judges and the landowner are honestly convinced that they have all acted properly. The property rights argument recited by the rural landowner is a very powerful defense, particularly when presented at this level of government. The argument is grounded in a political philosophy almost three centuries old as well as in land use practices which go back at least to Saxon and perhaps even to Celtic times in Europe and England. When the argument is presented to county court judges who share these beliefs and land use traditions, the outcome of the court decision is rarely in doubt. On the other hand, the tradition that natural objects and areas of special beauty or interest ought to be protected from landowners claiming special property rights, and from the practice of landowning in general, is of very recent origin, and without comparable historical and emotional foundations. . . .

My present purpose is to examine traditional land use attitudes. First, I examine the ancient land use practices which gave rise to these attitudes, second, the political activities and views of Thomas Jefferson which secured a place for them in American political and legal thought, and, finally, the political philosophy of John Locke which provided them with a philosophical foundation.

LANDHOLDING AMONG EARLY GERMAN AND SAXON FREEMEN

About two thousand years ago most of Europe was occupied by tribes and peoples known collectively as the Celts. At about that time, these peoples came under considerable pressure from the Romans moving up from the south and from Germanic tribes entering central Europe from the east. Five hundred years later, the Celts had either been subjugated by the German and Roman invaders or pushed back into Ireland and fringe areas of England. The Roman Empire, too, after asserting its presence as far north as England, was in decay. Roman influence would continue in the south, but in northern and central Europe as well as in most of England German influence would prevail.

The Germanic tribes which displaced the Celts and defeated the Romans were composed of four classes: a few nobles or earls, a very large class of freemen, a smaller class of slaves, and a very small class of semi-free men or serfs. Freemen were the most common people in early German society. They recognized no religious or political authority over their own activities, except to a very limited degree. As *free* men, they could, if they desired, settle their accounts with their neighbors and move to another geographical location. Each freeman occupied a large amount of land, his freehold farmstead, on which he grazed animals and, with the help of his slaves, grew crops. When necessary, he joined together with other freemen for defense or, more often, for the conquest of new territories.[1]

Freemen were the key to German expansion. When overcrowding occurred in clan villages and little unoccupied land remained, freemen moved to the border and with other freemen defeated and drove away the neighboring people. Here they established for themselves their own freehold farmsteads. Their descendants then multiplied and occupied the vacant land between the original freehold estates. When land was no longer available, clan villages began to form again and many freemen moved on once more to the new borders to start new freehold farmsteads. In this way, the Germans slowly but surely moved onward across northern and central Europe with freemen leading the way until no more land was available.

Strictly speaking, a freeman did not own his land. The idea of landownership in the modern sense was still many centuries away. In England, for example, landowning did not become a political and legal re-

ality until 1660 when feudal dues were finally abolished once and for all. Freemen, however, lived in pre-feudal times. They usually made a yearly offering to the local noble or earl, but technically this offering was a gift rather than a feudal payment and had nothing to do with their right to their land. As the term *freehold* suggests, a freeman held his land freely without any forced obligations to an overlord or to his neighbors.

In early times, when land was readily available, each freeman occupied as much land as he needed. There was no set amount that a freeman ought to have and no limit on his holdings, except that he could not hold more land than he could use. Thus, in effect, his personal dominion was restricted only by the number of animals that he had available for grazing and the number of slaves he had for agricultural labor. Sometimes, when the land began to lose its fertility, he would abandon his holdings and move to some other unoccupied location nearby. The exact location of each holding was only vaguely determined, and when disputes arose about boundaries, they were settled with the help of the testimony of neighbors or, when that failed, by armed combat between the parties involved.

Much of the unoccupied land was held in common with other freemen in accordance with various local arrangements. Sometimes the use was regulated by establishing the number of cattle that each freeman could place on the land. In other cases, plots were used by different freemen every year on a rotational basis.

When occupied border lands were no longer available for new freemen to settle, the way of life of the freemen began to change. The primary problem was one of inheritance. In the beginning, land had never been divided; rather, it has always been "multiplied" as sons moved to adjacent areas and established new freehold farmsteads. Eventually, however, it became necessary for the sons to divide the land which had been held by their father. A serious problem then developed, for, if division took place too many times, then the holdings became so small that they had little economic value, and the family as a whole slipped into poverty.

The solution was *entail*, i.e., inheritance along selected family lines. The most common form of entail was *primogeniture*, according to which the eldest son inherited everything and the others little or nothing. In this way, the family head remained powerful by keeping his landholdings intact, but most of his brothers were condemned to the semifree and poverty-stricken life of serfdom. As a result of these new inheritance practices, the number of freemen became an increasingly smaller portion of the society as a whole as most of the rest of the population, relatives included, rapidly sank to the level of serfs.

Another problem affecting freemen was taxation. The custom of giving an offering to the local noble was gradually replaced by a tax, and once established, taxes often became large burdens on many of the poorer freemen who in many instances paid taxes while other richer landholders were exempted. In such circumstances, freemen often gave up their status and their lands to persons exempted from the taxes and paid a smaller sum in rent as tenants.

Germans thus made a transition from prefeudal to feudal conditions, and freemen ceased to be an important element in the community as a whole. While freemen never disappeared altogether, most lost the economic freedom that they had formerly had. Although theoretically free to move about as they pleased, they often lacked the economic means of settling their accounts, and so in most cases were little better off than the serfs.

These feudal conditions did not appear in England until long after they were firmly established in Europe. At the time of the conquest of England by William the Conqueror most Englishmen were freemen. Thus, in England, unlike in Germanic Europe, prefeudal conditions did not slip away gradually but were abruptly replaced by a feudal system imposed on much of the native population by the victorious Normans. Under such circumstances, freemen declined in numbers, but struggled as best they could to maintain their freeman status in opposition to Norman rule and as a part of their Saxon heritage. As a result, freemen managed to maintain a presence in England no longer conceivable in Europe. Through them, memories of the heyday of the flamboyant Saxon freemen remained to shade political thought and to shape land use attitudes for centuries after the conquest. Ironically, the conquest drew attention to a class status which might otherwise have quietly passed away.

There were four major political divisions in Saxon England: the kingdom, the shire (called the *county* after the arrival of the Normans), the hundred, and the township, the last two being subdivisions of the shire or county. Throughout English history the exact nature of the government of the kingdom fluctuated, sometimes very radically. Changes occurred in the hundreds and the townships as the courts at these levels were gradually replaced by those of the local nobility, probably with the support of the government of the kingdom. The shire or county and its court or moot, however, persisted unchanged and continued to be one of the most important political units from the earliest Saxon times in England to the present day in both England and the United States.

The county court met to deal with cases not already handled by the hundred moots and with other business of common importance to the community. The meetings were conducted by three men: the alderman, representing the shire; the sheriff, representing the king; and the bishop, representing the church. All freemen in the county had the right to attend the court and participate in the decision process. Most of them, of course, were usually too busy to come except when personal interests were at stake.

There are only small differences between the county courts of Saxon and Norman times and those of modern rural America. The three judges, alderman, sheriff, and bishop, have been replaced by elected judges. Court procedure in most of these courts, however, remain as informal today as it was in pre-Norman England. In many, no record is kept by the court of its decisions and, in such cases, except for word of mouth and intermittent coverage by the news media, little is known of what goes on there. Court judges are primarily concerned with keeping the local landowners contented by resolving local differences and by providing the few community services under the administrative jurisdiction of the court, e.g., maintaining dirt or gravel roads. This casual form of government is replaced only when the county becomes urbanized, thereby, enabling residents to incorporate it and enjoy extensive new administrative and legal powers and, of course, responsibilities.

The special considerations given to the local landowner by the modern rural county court reflects the relationship of Saxon freemen to the court at the time when such courts first came into existence. The court evolved out of the freemen's custom of consulting with his neighbors during local disputes as an alternative to physical combat between the parties involved. Thus, rather than being something imposed on the freemen from above, the court was created by them for their own convenience. Since the freemen gave up little or none of their personal power, the power of the court to enforce its decisions was really nothing more than the collective power of the freemen ultimately comprising the membership of the court. From the earliest times, freemen had had absolute control over all matters pertaining to their own landholdings. When county courts were formed, freemen retained this authority over what they considered to be their own personal affairs. This limitation on the power of the court was maintained for more than a thousand years as part of the traditional conception of what a county court is, and how it is supposed to function. Today, when a landowner demands to know what right the court or anyone else has to tell him what to do with his own land he is referring to the original limitations set on the authority of the county court, and is appealing to the rights which he has informally inherited from his political ancestors, Saxon or German freemen—specifically, the right to do as he pleases without considering any interests except his own.

A modern landowner's argument that he has the right to do as he wishes is normally composed of a set series of claims given in a specific order. First, he points out that he or his father or grandfather worked the land in question. Second, he asserts that his ownership of the land is based on the work or labor put into it. Finally, he proclaims the right of uncontrolled use as a result of his ownership claim. Not all of this argument is derived directly from the freemen's world view. As mentioned above, the modern concept of ownership was unknown to freemen who were engaged in landholding rather than landowning. In other respects, however, there are strong similarities between the views of modern landowners and those of the freemen.

Landholding among German freemen was based on work. A freeman, like the nineteenth-century American homesteader, took possession of a tract of land by clearing it, building a house and barns, and dividing the land into fields for the grazing of animals and for the growing of crops. In this way, his initial work established his claim to continued use.

This emphasis on work as the basis for landholding is especially clear in connection with inheritance. When plenty of vacant land was available, landholdings were never divided among the sons, but, as described above, the sons moved to unoccupied land nearby and started their own freehold farmsteads. Thus, inheritance in those early times was not the acquisition of land itself but rather the transferral of the right to acquire land through work. This distinction is reflected in the early German word for inheritance, *Arbi* in Gothic and *Erbi* in Old High German, both of which have the same root as the modern High German word *Arbeit,* meaning work.[2]

Thus freemen were interested in land use rather than landownership. The right to land was determined by their social status as freemen and not by the fact that they or their fathers had occupied or possessed a particular piece of ground. The specific landholdings, thus, were not of major importance to the early freemen. Conceivably, they might move several times to new landholdings abandoning the old without the size of their landholdings being affected in any way. As mentioned above, it was their ability to use their holdings, the number of grazing animals, and slave workers they owned, not some form of ownership, which determined the size of their landholdings at any particular time in their lives.

Of course, once unoccupied land ceased to be readily available, freemen started paying much more attention to their land as property, encouraging the development of the idea of landownership in the modern sense. When the inheritance of sons became only the right to work a portion of their father's holdings, the transition from landholding to landowning was well on its way.

Until the time when there were no more unoccupied lands to move to, there was really no reason for freemen to be concerned with proper use or management of their land or for them to worry about possible long-term problems for themselves or their neighbors resulting from misuse and abuse of particular pieces of land. When a freeman lost his mobility, however, he did start trying to take somewhat better care of his land, occasionally practicing crop rotation and planting trees to replace those he cut down, but apparently these new necessities had little influence on his general conviction that as a freeman he had the right to use and even abuse his land as he saw fit.

Today's rural landowner finds himself in a situation not unlike that of freemen in the days when inheritance became the division of land rather than the multiplication of it. In the late eighteenth century and during most of the nineteenth, American rural landowners led a way of life much like that of prefeudal German freemen: now modern landowners face the same limitations their freemen ancestors did as feudal conditions began to develop. Although willing to take some steps toward good land management, especially those which provide obvious short-term benefit, when faced with broader issues involving the welfare of their neighbors and the local community and the protection and the preservation of the environment as a whole, they claim ancient rights which have come down to them from German freemen, and take advantage of their special influence with the local county court, a political institution as eager to please them today as it was more than a thousand years ago.

THOMAS JEFFERSON AND THE ALLODIAL RIGHTS OF AMERICAN FARMERS

When British colonists arrived in North America, they brought with them the land laws and land practices that were current in England at that time. These included entail, primogeniture, and most other aspects of the feudal tenure system which had taken hold in England after the Norman Conquest. The American Revolution called into question the right of the king of England to lands in North America which in turn led to attempts to bring about major land reform—specifically, efforts to remove all elements of the feudal system from American law and practice and replace them with the older Saxon freehold tenure system. At the forefront of this movement was a young Virginian lawyer named Thomas Jefferson. . . .

From the first moment that Jefferson began airing his land tenure opinions, however, he made it completely clear that they were based entirely on Saxon, and not on Norman, common law. Thus, he consistently spoke of allodial rights—*allodial* being the adjectival form of the Old English word *allodium* which refers to an estate held in absolute dominion without obligation to a superior—i.e., the early Germany and Saxon freehold farmstead. . . .

Noting the right of a Saxon freeman to settle his accounts and move to another realm at his own pleasure without obligation to the lord of his previous domain, Jefferson argues that this is also the case with the British citizens who moved to North America. According to this analogy, England has no more claim over residents of America than Germany has over residents of England. In accordance with Saxon tradition, the lands of North America belong to the people living there and not to the king of England.[3] . . .

It is not the king, Jefferson declares, but the individual members of a society collectively or their legislature that determine the legal status of land, and, if they fail to act, then, in accordance with the traditions of Saxon freemen, "each individual of the society may appropriate to himself such lands as he finds vacant, and occupancy will give him title."[4] . . . Jefferson, of course, did not succeed in refuting the claim of the king of England to all land in British America, but by arguing in terms of this old dispute, he gives his position a legal basis which would have strong appeal among Englishmen with Saxon backgrounds, assuring some political support of the American cause in England.

In 1776, Jefferson got the opportunity to try to turn his theory into practice. Although Jefferson is most famous for writing the *Declaration of Independence,* most of his time that year was spent working on his draft of the Virginia constitution and on the reform of various Virginia laws including the land reform laws. In his draft constitution, Jefferson included a provision which gave each person of full age the right to fifty acres of land "in full and absolute dominion." In addition, lands previously "holden of the crown in fee-simple" and all other lands appropriated in the future were to be "holden in full and absolute dominion, of no superior whatever."[5] Although these provisions were deleted, and similar bills submitted to the legislature failed to pass, Jefferson, nevertheless, did succeed in getting the legislature to abolish the feudal inheritance laws, entail and primogeniture. . . .

As for the government selling the land, Jefferson was completely opposed. "I am against selling the land at all," he writes to Pendleton, "By selling the lands to them, you will disgust them, and cause an avulsion of them from the common union. They will settle the lands in spite of every body." This prediction proved to be remarkably correct as evidenced by the fact that the next eighty years of American history was cluttered with squatters illegally occupying government land and then demanding compensation for their "improvements" through special preemption laws.[6]

In 1784, when he was appointed to head the land committee in the Congress of the Confederacy, Jefferson had a second opportunity to reestablish the Saxon landholding system. Whether Jefferson tried to take advantage of this opportunity is not known because the report of the committee, called the Ordinance of 1784, contains nothing about allodial rights to land. In addition, it even contains recommendations for the selling of western lands as a source of revenue for the government. It should be noted, however, that in one respect at least the document still has a very definite Saxon ring to it. Jefferson managed to include in his report a recommendation that settlers be permitted to organize themselves into new states on an equal footing with the original colonies. This recommendation, which was retained in the Ordinance of 1787, a revised version of the earlier ordinance, not only created the political structure necessary to turn the thirteen colonies into a much larger union of states, but also provided future generations of Americans with an independence and mobility similar to that enjoyed by the early Saxon and German freemen. In his *Summary View* of 1774, as mentioned above, Jefferson had argued that just as the Saxons invading England had had the right to set up an independent government, so British Americans had the right to an independent government in North America. The Ordinances of 1784 and 1787 extended his right to movement and self-determination of American settlers leaving the jurisdiction of established states and moving into the interior of the continent. In large measure, it is thanks to this provision that Americans today are able to move from state to state without any governmental control in the form of visas, passports, immigration quotas, or the like as unhassled by such details as were early German freemen.

The absence of any provisions specifically granting landowners full and absolute dominion over their land, however, does not mean that Jefferson abandoned this conception of landholding or ownership. Privately and in his published writings he continued to champion the right of Americans to small freehold farmsteads. The only major change seems to be that Jefferson stopped trying to justify his position in terms of historical precedents and instead began speaking in moral terms claiming that small independent landholders were the most virtuous citizens any state could ever hope to have. In a letter to John Jay in 1785, Jefferson writes:

> Cultivators of the earth are the most valuable citizens. They are the most vigorous, the most independent, the most virtuous, and they are tied to their country and wedded to it's liberty and interests by the most lasting bands.[7]

In a letter to James Madison in the same year, he adds:

> Whenever there is in any country, uncultivated lands and unemployed poor, it is clear that the laws of property have been so far extended as to violate natural right. The earth is given as a common stock for man to labour and live on. If, for the encouragement of industry we allow it to be appropriated, we must take care that other employment be furnished to those excluded from that appropriation. If we do not the fundamental right to labour the earth returns to the unemployed. It is too soon yet in our country to say that every man who cannot find employment but who

can find uncultivated land, shall be at liberty to cultivate it, paying a moderate rent. But it is not too soon to provide by every possible means that as few as possible shall be without a little portion of land. The small landholders are the most precious part of the state.[8]

. . . These remarks are probably . . . the basis for the position of rural landowners today when faced with environmental issues. They are defending the American moral virtues which they have always been told their style of life and independence represents.

Had Jefferson been alive in the late nineteenth century when his views were being cited in opposition to the preservation of Yellowstone or were he alive today to see his Saxon freemen busily sabotaging county planning and zoning, he might have become disillusioned with his faith in the virtues of independent rural landowners. Jefferson, after all, as a result of his purchase of the Natural Bridge, perhaps the first major act of nature preservation in North America, ranks as a very important figure in the history of the nature preservation movement. Unfortunately, however, Jefferson's homesteaders and their modern day descendants did not always retain his aesthetic interest in nature or his respect for sound agricultural management which he interwove with his Saxon land use attitudes to form a balanced land use philosophy.

In part, the callousness and indifference of most rural landowners to environmental matters reflects the insensitivity of ancient Saxon freemen who viewed land as something to be used for personal benefit and who, being semi-nomadic, were unconcerned about whether that use would result in irreparable damage to the particular piece of land that they held at any given point in their lives. In addition, however, it can also be traced back to the political philosophy and theory of property of John Locke, a seventeenth-century British philosopher, who had a major impact on the political views of Jefferson and most other American statesmen during the American Revolution and afterwards. This influence is the subject of the next section.

JOHN LOCKE'S THEORY OF PROPERTY

As noted above, German and Saxon freemen did not have a concept of landownership, but only of landholding. As long as there was plenty of land for everyone's use, they did not concern themselves with exact boundaries. Disputes arose only when two freemen wanted to use the same land at the same time. By the end of the Middle Ages, however, with land in short supply, landholders began enclosing their landholdings to help ensure exclusive use. Enclosure kept the grazing animals of others away and also provided a sign of the landholder's presence and authority. Although enclosure was only a small step towards the concept of landownership, it, nonetheless, proved useful as a pseudo-property concept in early seventeenth-century New England where Puritans were able to justify their occupation of Indian lands on the grounds that the lack of enclosures demonstrated that the lands were vacant. Landownership became an official legal distinction in England after 1660 with the abolishment of feudal dues. The concept of landownership was introduced into British social and political philosophy thirty years later as part of John Locke's theory of property. This theory was presented in detail in Locke's *Two Treatises of Government,* a major work in political philosophy first published in 1690.[9]

Jefferson had immense respect and admiration for Locke and his philosophical writings. On one occasion, he wrote to a friend that Locke was one of the three greatest men that had ever lived—Bacon and Newton being the other two. Jefferson's justification of the American Revolution in "The Declaration of Independence" was borrowed directly from the *Second Treatise.* Many of Jefferson's statements in the document are almost identical to remarks made by Locke. For example, when Jefferson speaks of "life, liberty, and the pursuit of happiness," he is closely paraphrasing Locke's own views. His version differs from Locke's in only one minor respect: Jefferson substitutes for Locke's "enjoyment of property" the more general phrase "the pursuit of happiness," a slight change made to recognize other enjoyments in addition to those derived from the ownership of property. . . .

In the *Second Treatise* Locke bases property rights on the labor of the individual:

> Though the Earth, and all inferior Creatures be common to all Men, yet, every Man has a *Property* in his own *Person.* This no Body has any Right to but himself. The *Labour* of his Body, and the *Work* of his Hands, we may say, are properly his. Whatsoever then he removes out of the State that Nature hath provided, and left in, he hath mixed his *Labour* with, and joyned to it something that is his own, and thereby makes it his *Property.*[10]

This theory of property served Locke's friends well since it made their property rights completely inde-

pendent of all outside interest. According to Locke, property rights are established without reference to kings, governments, or even the collective rights of other people. If a man mixes his labor with a natural object, then the product is his.

The relevance of Locke's labor theory to the American homestead land use philosophy becomes especially clear when he turns to the subject of land as property:

> But the *chief matter of Property* being now not the Fruits of the Earth, and the Beasts that subsist on it, but the *Earth it self* as that which takes in and carries with it all the rest; I think it is plain, that *Property* in that too is acquired as the former. *As much land* as a Man Tills, Plants, Improves, Cultivates, and can use the Product of, so much is his *Property.* He by his Labour does, as it were, inclose it from the Common. . . . God, when He gave the World in common to all Mankind, commanded Man also to labour, and the penury of his Condition required it of him. God and his Reason commanded him to subdue the Earth, *i.e.* improve it for the benefit of Life, and therein lay out something upon it that was his own, his labour. He that in Obedience to this Command of God, subdued, tilled, sowed any part of it, thereby annexed to it something that was his *Property,* which another had no Title to, nor could without injury take from him.[11]

In this passage, the right of use and ownership is determined by the farmer's labor. When he mixes his labor with the land, the results are *improvements,* the key term in homesteading days and even today in rural America where the presence of such improvements may qualify landowners for exemption from planning and zoning under a grandfather clause. Since property rights are established on an individual basis independent of a social context, Locke's theory of property also provides the foundation for the landowner's claim that society has little or no role in the management of his land, that nobody has the right to tell him what to do with his property.

Locke reenforces the property owner's independence from societal restraints with an account of the origins of society in which property rights are supposedly more fundamental than society itself. According to Locke, the right to the enjoyment of property is a presocietal *natural right.* It is a natural right because it is a right which a person would have in a state of nature. Locke claims that there was once, at some time in the distant past, a true state of nature in which people possessed property as a result of their labor, but, nevertheless, did not yet have societal relations with one another. This state of nature disappeared when these ancient people decided to form a society, thereby giving up some of their previous powers and rights. They did not, however, Locke emphatically insists, relinquish any of their natural rights to their own property, and the original social contract establishing the society did not give society any authority at all over personal property. In fact, the main reason that society was formed, according to Locke's account, was to make it possible for individuals to enjoy their own property rights more safely and securely. Thus, society's primary task was and allegedly still is to protect private property rights, not to infringe on them. A government which attempts to interfere with an individual's natural and uncontrolled right to the enjoyment of his property, moreover, deserves to be overthrown and the citizens of the society are free to do so at their pleasure. In effect, Locke is arguing along lines completely compatible with the early Saxon and Jeffersonian doctrine that a landowner holds his property in full and absolute dominion without any obligation to a superior.

The similarity of Locke's position to this doctrine invites the conclusion that Locke, like Jefferson, was drawing inspiration from Saxon common law and that Locke's social contract was actually the establishment of the shire or county court by Saxon freemen. Curiously, however, Locke makes no mention of the Saxons in these contexts and, even more curiously, no political philosopher ever seems to have considered the possibility that Locke might have been referring to this period of English history. In his chapter on conquest, nevertheless, Locke does demonstrate (1) that he knew what a freeman was, (2) that he was aware of the legal conflicts resulting from the Norman Conquest, and (3) that he sided with the Saxons in that controversy. In the one paragraph where he mentions the Saxons by name, he flippantly remarks that, even if they did lose their rights as freemen at the time of the conquest, as a result of the subsequent six centuries of intermarriage all Englishmen of Locke's day could claim freeman status through some Norman ancestor and it would "be very hard to prove the contrary."[12] Locke may have chosen not to mention the specifics of Saxon history fearing that if he did so, his political philosophy might have been treated as nothing more than just another call for a return to Saxon legal precedents. It is hard to imagine, nonetheless, that Locke's readers in the seventeenth century were not aware of these unstated connections considering the ease with which Jefferson saw them eighty years later in colonial North America. It is also possible, of course, that Locke may have been ignorant of the details of Saxon common law and may have simply relied on the popular land use attitudes of his day without being aware of their Saxon origin. At any rate, however, the ultimate result would be the same—a political philosophy which provides philosophical foundations for the ancient Saxon land use attitudes and traditions. . . .

Not everyone in the first half of the nineteenth century shared Jefferson's enthusiasm for land reform based on Saxon common law modified by Locke's theory of property, and for a time the idea of landholding independent of landowning continued to be influential in American political and legal thought. Early versions of the homestead bill before the beginning of the Civil War, for example, often contained inalienability and reversion clauses. According to these, a homesteader had the right to use the land, but could not subdivide it, sell it, or pass it on to his children after his death. These limitations, however, were not comparable with the wishes of potential homesteaders who wanted to be landowners, not just landholders, and, as a result, they were not included in the Homestead Act of 1862. It is unlikely that homesteading based entirely on Saxon common law ever had much chance of passing Congress because early nineteenth-century settlers squatting illegally on Western lands and demanding the enactment of special preemption laws had always had landownership as their primary objective.[13]

Because it was probably Locke's theory of property as much as Saxon common law which encouraged American citizens and immigrants to move westward, both should be given a share of the credit for the rapid settlement of the American West which ultimately established a national claim to all the lands west of the Appalachians as far as the Pacific. This past benefit to the American people, nevertheless, should not be the only standard for evaluating this doctrine's continuing value. We must still ask just how well the position is suited to conditions in twentieth-century America.

MODERN DIFFICULTIES WITH LOCKE'S POSITION

One obvious problem with Locke's theory today is his claim that there is enough land for everyone.[14] This premise is of fundamental importance to Locke's argument because, if a present or future shortage of land can be established, then any appropriation of land past or present under the procedure Locke recommends, enclosure from the common through labor, is an injustice to those who must remain unpropertied. By Locke's own estimates there was twice as much land at the end of the seventeenth century as all the inhabitants of the Earth could use. To support these calculations Locke pointed to the "in-land, vacant places of America"—places which are now occupied.[15] Since Locke's argument depends on a premise which is now false, Locke would have great difficulty advancing and justifying his position today.

Another problem is Locke's general attitude toward uncultivated land. Locke places almost no value on such land before it is improved and after improvement he says the labor is still the chief factor in any value assessment:

> . . . when any one hath computed, he will then see, how much *labour makes the far greatest part of the value* of things we enjoy in this World: And the ground which produces the materials, is scarce to be reckon'd in, as any, or at most, but a very small part of it: So little, that even amongst us, Land that is left wholly to Nature, that hath no improvement of Pasturage, Tillage, or Planting, is called, as indeed it is, *waste* and we shall find the benefit of it amount to little more than nothing.

According to Locke's calculations, 99 to 99.9 percent of the value of land even after it is improved still results from the labor and not the land. Although these absurdly high figures helped strengthen Locke's claim that labor establishes property rights over land, by making it seem that it is primarily the individual's labor mixed with the land rather than the land itself which is owned, such estimates, if presented today, would be considered scientifically false and contrary to common sense.[16]

Locke's land-value attitudes reflect a general desire prevalent in Locke's time as well as today for maximum agricultural productivity. From Locke's point of view, it was inefficient to permit plants and animals to grow naturally on uncultivated land:

> . . . I ask whether in the wild woods and uncultivated waste of America left to Nature, without any improvement, tillage, or husbandry, a thousand acres will yield the needy and wretched inhabitants as many conveniences of life as ten acres of equally fertile land doe in Devonshire where they are well cultivated?[17]

The problem, however, is not just productivity and efficiency, but also a general contempt for the quality of the natural products of the Earth. Locke writes with great conviction that "*Bread* is more worth than Acorns, *Wine* than Water, and *Cloth* or *Silk* than Leaves, Skins or Moss."[18] Even though we might be inclined to agree with Locke's pronouncements in certain contexts, the last two hundred years of the American experience have provided us with new attitudes incompatible with those of Locke and his contemporaries, and apparently completely unknown to them, which place high value on trees, water, animals, and even land itself in a wholly natural and unimproved condition. Unlike Locke, we do not always consider wilderness land or uncultivated land synonymous with waste.

At the very core of Locke's land-value attitudes is his belief that "the Earth, and all that is therein, is given to Men for the Support and Comfort of their being." In one sense, this view is very old, derived from the biblical and Aristotelian claims that the Earth exists for the benefit and use of human beings. At the same time, it is very modern because of Locke's twin emphasis on labor and consumption. Both of these activities are of central importance in communistic and capitalistic political systems, and they became so important precisely because the founders and ideologists of each system originally took their ideas about labor and consumption from Locke's philosophy. In accordance with these ideas, the Earth is nothing more than raw materials waiting to be transformed by labor into consumable products. The Greeks and Romans would have objected to this view on the grounds that labor and consumption are too low and demeaning to be regarded as primary human activities.[19] From a twentieth-century standpoint, given the current emphasis on consumption, the neglect of the aesthetic and scientific (ecological) value of nature seems to be a more fundamental and serious objection to this exploitative view.

The worst result of Locke's property theory is the amoral or asocial attitude which has evolved out of it. Locke's arguments have encouraged landowners to behave in an antisocial manner and to claim that they have no moral obligation to the land itself, or even to the other people in the community who may be affected by what they do with their land. This amoral attitude, which has been noted with dismay by Aldo Leopold, Garrett Hardin, and others, can be traced directly to Locke's political philosophy, even though Locke himself may not have intended to create this effect. The reasons why this moral apathy developed are complex.

First, the divine rights of kings had just been abolished. In accordance with this doctrine, the king had

had *ultimate* and *absolute* property rights over all the land in his dominion. He could do whatever he wanted with this land—give it away, take it back, use it himself, or even destroy it as he saw fit. Locke's new theory of property stripped the king of this power and authority and transferred these *ultimate* and *absolute* rights to each and every ordinary property owner. This transfer has been a moral disaster in large part because the king's rights involved moral elements which did not carry over to the new rights of the private landowner. As God's agent on Earth, the king was morally obligated to adhere to the highest standards of right and wrong. Furthermore, the king, as the ruler of the land, had a moral and political obligation to consider the general welfare of his entire kingdom whenever he acted. Of course, kings did not always behave as they should have, but, nevertheless, there were standards recognized by these kings and their subjects as to what constituted proper and kingly moral behavior. Private landowners, however, did not inherit these sorts of obligations. Because they were not instruments of church or state, the idea that they should have moral obligations limiting their actions with regard to their own property does not seem to have come up. The standard which landowners adopted to guide their actions was a purely selfish and egotistical one. Because it involved nothing more than the economic interest of the individual, it was devoid of moral obligation or moral responsibility. . . .

Theoretically, Locke's qualification of the right to destroy property is compatible with the American conception of checks and balances and it might have provided a *political* solution to the problem, though not a moral one. Unfortunately, however, it has not been carried over into our political and legal system as successfully as the right to destroy. A man certainly has a right in the United States to sue for damages in court after the fact, when the actions of others have clearly injured him or his property, but the right of the government to take preventive action before the damage is done has not been effectively established. It is this preventive action which private landowners are assailing when they assert their right to use and even destroy their land as they see fit without any outside interference. The success of landowners in this area is amply demonstrated by the great reluctance of most state legislatures to place waste management restrictions on small private *landowners* which have long governed the activities of rural land *developers*.

Government regulation of individual private landowners has been ineffective historically because, from the very beginnings of American government, representation at state and federal levels has nearly always been based on landownership, an approach which has usually assured rural control of the legislature even when most of the citizens in the state lived in urban population centers. Government leaders intent on acting primarily in the interests of landowners could hardly have been expected to play the preventive role which Locke recommends. The unwillingness of legislators to act in this way in the nineteenth century and most of the twentieth, moreover, further contributed to the amoral belief of rural landowners that they can do whatever they want without being concerned about the welfare or rights of others.

When Jefferson attempted to build American society on a Lockeian foundation of small landowners, he did so in large measure because he believed that small landowners would make the most virtuous citizens. He failed to foresee, however, that the independence provided by Locke's presocietal natural rights would discourage rather than encourage social responsibility, and, therefore, would contribute little to the development of moral character in American landowners. Since social responsibility is basic to our conception of morality today, the claim of landowners that their special rights relieve them of any obligation or responsibility to the community can be regarded only as both socially and morally reprehensible. The position of such rural landowners is analogous to that of a tyrannical king. Tyranny is always justified, when it is justified at all, by a claim that the tyrant has the *right* to do as he pleases regardless of the consequences. In practice, however, the impact of rural landowners more closely approaches anarchy than tyranny, but only because landowners, though sharing a common desire to preserve their special rights, do not always have common economic interests. As a result, landowners are usually more willing to promote the theoretical rights of their fellow property owners than their specific land use and development projects, which as members of society, they may find objectionable or even despicable—in spite of their Saxon and Lockeian heritage rather than because of it.

A landowner cannot justify his position morally except with the extravagant claim that his actions are completely independent and beyond any standard of right and wrong—a claim which Locke, Jefferson, and even Saxon freemen would probably have hesitated to make. Actually, there is only one precedent for such a claim. During the Middle Ages, church philosophers concluded that God was independent of all moral standards. They felt compelled to take this position because moral limitations of God's actions would have conflicted with His omnipotence. Therefore, they reasoned that God's actions created moral law—i.e., defined moral law—and that theoretically moral law could be radically changed at any moment. Descartes held this position in the seventeenth century, and in the nineteenth and twentieth centuries some atheistic existential philosophers have argued that because God is dead each man is now forced to create his own values through his individual actions. Although this position could be adopted as a defense of the landowners' extraordinary amoral rights, it would probably be distasteful to most landowners. Without it, this aspect of the rural landowners' position may be indefensible.[20]

Today, of course, whenever Locke's theory of property and the heritage of the ancient Saxon freeman surface in county courts, at planning and zoning meetings, and at state and federal hearings on conservation and land management, they still remain a formidable obstacle to constructive political action. As they are normally presented, however, they are certainly not an all-purpose answer to our environmental problems or even a marginally adequate reply to environmental criticism. When a landowner voices a Lockeian argument he is consciously or unconsciously trying to evade the land management issues at hand and to shift attention instead to the dogmatic recitation of his special rights as a property owner.

As I noted above, some of Locke's fundamental assumptions and attitudes are either demonstrably false or no longer generally held even among landowners. These difficulties need to be ironed out before the landowners can claim that they are really answering their environmental critics. Furthermore it is likely that, even if the position can be and is modernized, the moral issues will still be unresolved.

As it stands, the force of the rural landowners' arguments depends on their historical associations—their Biblical trappings, the echoes of Locke's political philosophy, the Saxon common-law tradition, the feudal doctrine of the divine rights of kings, and the spirit of the nineteenth-century American land laws. Can they be modernized? That remains to be seen. Until they are, however, landowners, environmentalists, politicians, and ordinary citizens should regard them with some suspicion.

Notes

1. The account given in this section is based most directly on Denman W. Ross, *The Early History of Land-Holding Among the Germans* (Boston: Soule and Bugbee, 1883), and Walter Phelps Hall, Robert Greenhalgh Albion, and Jennie Barnes Pope, *A History of England and the Empire-Commonwealth,* 4th ed. (Boston: Ginn and Company, 1961).

2. Ross, *Land-Holding,* p. 24.

3. Thomas Jefferson, "A Summary View of the Rights of British America," in *The Portable Thomas Jefferson,* ed. Merrill D. Peterson (New York: Viking Press, 1975), pp. 4–5.

4. Ibid., pp. 17–19.

5. Thomas Jefferson, "Draft Constitution for Virginia," in *Portable Jefferson,* p. 248.

6. Jefferson to Edmund Pendleton, 13 August 1776, in *Papers of Thomas Jefferson,* 1:492.

7. Jefferson to John Jay, 23 August 1785, in *Portable Jefferson,* p. 384.

8. Jefferson to James Madison, 28 October 1785, in *Portable Jefferson,* p. 397.

9. John Locke, *Two Treatises of Government,* ed. Thomas I. Cook (New York and London: Hafner Press, 1947).

10. Locke, *Second Treatise,* sec. 27.

11. Ibid., sec. 32.

12. Ibid., sec. 177.

13. Paul W. Gates, *History of Public Land Law Development* (Washington, D.C.: Public Land Law Commission, 1968), pp. 390–393.

14. Locke, *Second Treatise,* sec. 33.

15. Ibid., sec. 36.

16. Ibid., secs. 42–43.

17. Ibid., sec. 37.

18. Ibid., sec. 42.

19. Ibid., sec. 26: for a full discussion of labor and consumption see Hannah Arendt, *The Human Condition* (Chicago and London: University of Chicago Press, 1958), chap. 3.

20. Jean-Paul Sartre, *Existentialism and Human Emotions* (New York: Philosophical Library, 1957), pp. 13–18.

For Further Discussion

1. Prior to reading Hargrove, what was your own understanding of the origins of the private ownership of land? From where does the right to own land come?

2. Is there a difference between private ownership of personal property such as a home or a car and private ownership of land? Should there be a difference?

3. Does the value of land ownership change when a society moves from an agricultural to an industrial basis?

4. Hargrove distinguishes between land ownership and land use. What is this distinction exactly?

5. Hargove describes Locke's labor theory of property. On this theory, when labor is mixed with unowned land, the land is improved and the laborer comes to have a right to the land. Do you agree with Locke that land comes to have value only through human labor?

Good Farming and the Public Good

Donald Worster

There is no more fundamental relationship with the land than farming. But how one farms makes all the difference in the world. Some farming techniques are extremely detrimental to the land; some are respectful and deferential. In this selection, Donald Worster contrasts an approach to farming that stresses productivity and wealth with an approach that can be understood as the "art, science, and the wisdom of growing values in the soil." This latter approach, what Worster calls good farming, is the only approach that will truly serve the public good.

Rain is a blessing when it falls gently on parched fields, turning the earth green, causing the birds to sing. But when it rains and rains, for forty days and nights, as it did for Noah, then the waters rise and destroy. Life is everywhere like that. Too little is a curse, too much is a plague.

For thousands of years, the philosopher's task has been to discover an optimum point where men and women can live modestly and securely, avoiding the extremes. The philosopher may seek a point of environmental balance where there is neither too little nor too much of nature's gifts. Or he may try to define the point where private ambitions and collective needs are in harmony, where individual appetites do not overrun the commonwealth and society's demands do not cut too deeply into individual freedoms. When philosophy is applied to the definition of a society's welfare, we call that point the "public good." Farmers, more than most people, ought to be responsive to that philosophical quest for a harmonious, balanced good, for it has been their aim over a long history to seek moderation from nature and cooperation from their neighbors.

Yet it has been awhile since American agriculture, as a whole, has enjoyed a feeling of balance. The problem has not been in nature so much as in our society. We have not had a feeling of balance because we have come to hold extravagant ideas of what agriculture should contribute economically to the nation and the farmer. These days we are not a people noted for moderate thinking, so perhaps we have no reason to expect the idea of moderate farming to thrive. The most serious consequence of an immoderate culture, I will argue, is that the public good will not be well understood and therefore will not be achieved—in agriculture or in other areas. Another consequence is that farmers in the aggregate will suffer immensely and so will the practice of farming.

That has indeed happened in America, and we can blame it on our extreme dedication to the goal of maximizing agricultural productivity and wealth. In turn, that goal stems from a larger cultural conviction that wealth is our main aim in this nation and that wealth is an unlimited good. Almost everything we have celebrated as our success in farming has been defined in terms of those ends. It has now, however, become clear that our ends have been our undoing. We wanted more rain, unlimited rain, and we got a flood. It has left American farmers drowning in dreary statistics: crop reports, production charts, mortgage rates, energy bills, land and commodity prices.

For the past century or so, the nation has had to deal repeatedly with gluts of farm production, particularly in the cash grains. In 1981, a U.S. Department of Agriculture study declared an end to that nemesis; farmers, it predicted, would see no more price-depressing surpluses because we were at last able to peddle all the excess overseas. World population and demand had caught up at last with the American farmer. The study admitted that selling all our glut abroad would make the life of domestic producers more unstable than ever, that they would find themselves on a wild roller coaster ride, plunging up and down the track of international markets. But the old problem of recurrent overproduction had now been licked. There would never again be mountains of grain piled up and waiting for boxcars, never again be oranges or cotton or sugar cane going to rot.

Things have not worked out quite the way they were supposed to. In the summer of 1982, the old surplus problem appeared again: an all-time record wheat harvest of 2.1 billion bushels; not enough American buyers to absorb it; the Russians emerging once again as the main hope of salvation, but acting unpredictably as usual; crop prices declining; real farm income the lowest it had been in fifty years. Angry Colorado farmers, fighting against foreclosures on their farms, were tear-gassed by the sheriff, a replay of the turmoil in the 1920s and 1930s. Consequently, doubt begins to creep into our farm policy assumptions. Is there possibly something wrong with our approach to agriculture? How else can we explain the fact that farming, after so much attention, continues to be in so much trouble?

There can never be a perfect equilibrium in farming or any other sphere; existence might be unbearably dull if there were. But when you are flooded again and again, you are not in any danger of boredom. You naturally look up and ask, What's making so much rain? Can we turn it down a little? In America we have seldom accepted excess in nature; on the contrary, we have put our best talent and energy to work getting rid of it, making what is dry wet and wet dry. Why then do we accept excess when it is our own doing? Simply because powerful elements in our society do not allow us to recognize that there is such a thing as too much productivity, too much chasing after wealth. Despite overwhelming evidence that the idea is not working, good farming continues to mean *more*. We have not yet come to see that more of any good may, after a point, wash us away.

In the depressed 1930s, when times were even harder for farmers than they are today, American political leaders took counsel. They listened to ranchers, growers, sharecroppers, agronomists, soil experts, and marketing specialists; a few of those leaders raised basic questions of value. What, they asked, is agriculture for? What is the ultimate moral reason behind the pursuit of abundance, new farm technology, and an expanding economy—or is there one? Are our farm policies improved means to unimproved or unexamined ends? Whose ends should farming serve, those of the rich and powerful or those of the poor? How much

of nature should we spend on human desires? What are rational limits to our demands? What is the public good in agriculture and what kind of farming will most likely achieve it? Those questions are as relevant today as they were fifty years ago.

To begin a reappraisal, let us consider what the public good has come to mean in the standard discussions of agricultural policy. "Cheap and plentiful food" is the most common theme. The public deserves to eat at the lowest possible cost, so we are told, and then they will be able to spend the rest of their wages on an automobile or at the movies or on college. This, in fact, is what our policy persistently has assumed to be the chief public interest. Agriculture is supposed to contribute mainly to the wealth of Americans generally, to make possible an ever-higher level of personal consumption.

Such has long been a typical farm expert's definition of the public good. But is it what the American public really wants? According to scattered public opinion polls, people say they prefer having lots of small farms around rather than a few big ones, more little dairies and truck gardens, not so many giant agribusinesses. It might be that encouraging smaller farms would be worth more to people than saving a nickel on cheese or lettuce; a choice between the two options, however, is not clearly presented to them when they walk into a supermarket. The farm experts merely assume, on the basis of marketplace behavior, that the public wants cheapness above all else. Cheapness, of course, is supposed to require abundance, and abundance is supposed to come from greater economies of scale, more concentrated economic organization, and more industrialized methods. The entire basis for that assumption collapses if the marketplace is a poor or imperfect reflector of what people want. And it is. In matters of national defense, education, health care, and old-age assistance, we do not assume that the marketplace would be an adequate basis for public policy. Why then should we make that assumption in agriculture?

A corollary, and sometimes a rival, to the notion that good farming is farming that makes America richer through mass production, is the belief that farming is successful when it makes farmers as a special group more affluent. A common belief among policy makers is that swelling prosperity down on the farm immeasurably benefits society. No other sector of the economy has managed so fully as this one to identify its private fortunes with the public good—not dentists, not teachers, not even defense contractors. Billions of tax dollars have gone into scientific research and innovation to promote higher production, lower costs, and greater income on the farm. Heavily subsidized irrigation water flows to some very wealthy growers in the western states. Price-support programs, which have raised consumer prices by as much as 20 percent, have put money into agricultural pockets. A huge department in Washington, as well as agencies in every state, looks out for the welfare of this group. Hardly anyone begrudges the farm sector of all these gifts, though it is well established that most of them go to the wealthy few. We have been taught from Thomas Jefferson's day on down that what is good for the farmer is good for America.

That doctrine has survived some revolutionary changes in the conditions of rural folk. Currently only about six million people live on farms, and only a small fraction of them (one in ten) now falls below the poverty line. Contrast those numbers with the eleven million who were unemployed at the depth of the recent recession, roaming city streets, standing in bread lines, dreading the day their unemployment benefits ran out, or consider the several million more who have given up looking for work. Although farmers, like the unemployed and destitute, have at times experienced the indifference of other Americans to their plight, by and large they have retained an unusually sympathetic audience. When the man or woman on a tractor is in economic trouble, there is widespread worry; something, the newspapers agree, must be done. Farmers, particularly the better-off ones, maintain powerful friends and well-financed lobbyists to plead their case for governmental aid—unlike the city welfare mother who is roundly berated by the president and legislators in Congress, told to economize further on her food budget, to find a cheaper slum to live in, to forget a "handout" from Washington, to get off the public dole. In other words, the welfare and profit of this rather small group of largely middle- and upper-income farmers and ranchers have become identified with the public good in agriculture.

What is wrong with that identification? Do modern, business-oriented farmers not deserve our compassion when they go through hard times and are threatened with the loss of their assets and even their entire farms? They do, and they deserve our respect for their intelligence, hard work, fortitude, and skill. But respect and compassion should not be confused with favoritism or the private welfare of a single group with the larger welfare of the American commonwealth. When that confusion occurs, farmers are hurt;

their welfare depends on a clear, reasoned concept of public good. What is good for all Americans, we must understand, will in the long run be good for farming, too.

The source of our difficulties is not a lack of popular concern for farmers or such superficial things as stagnant overseas markets or expensive credit. Rather, it is an inadequate idea of what truly constitutes the agricultural good of this nation.

The predominant idea of the public good in agriculture takes two forms: first, forever increasing the gross farm product and, second, forever seeking to augment the wealth of the farm sector (even if it means losing most of our farmers). Both programs are tied to the ideology and pressures of the marketplace. When pushed to the extreme as we have pushed it, that market mentality becomes seriously destabilizing to rural communities. It produces a perpetually crisis-ridden farm economy. Worse, it embitters people because it cannot deliver what it says it will: a general contentment and happiness. When the marketplace is made the main idea, it diminishes other values, leads to a degrading of personal independence, social bonds, virtue, and patriotism—for those qualities cannot thrive in an unbridled culture of acquisition, which the mentality of market maximization leads to.

In an earlier America of extensive rural poverty and poor living conditions more could be said for the vigorous pursuit of wealth in the marketplace, just as more may be said for it today in Bangladesh or Haiti. But when that pursuit persists beyond the point of material sufficiency, when it becomes a dream of unlimited economic gain, troubles follow. That is what has happened to American farmers and indeed to this country in general. Farmers must run their machines nonstop to keep up with the self-aggrandizing industrialists. The faster farmers go, the more crops they harvest, the more secure their position in the marketplace may be, the more they can buy—so they hope. But what they win in that way lasts only for a brief while. A continual uncertainty is their fate in this society.

The average farmer is not altogether responsible for this predicament. He did not set up the race, and he is not leading in it but is somewhere back in the pack, straining to catch up with corporate presidents, athletes, lawyers, movie stars, and engineers. The modern farmer lives in an intensely high-pressure world of many wealth maximizers. In that milieu, growing food becomes his only defense, his sole means of competing for social position. Unfortunately for him, food has been a comparatively poor basis for income growth, for it quickly saturates its market: humans can eat only so much lettuce or beef. Unlike others in the race, the farmer must always confront the biological limits of the consumer. He cannot make more money without finding more mouths and bellies to feed. Agriculture, by its very nature, is a productive activity that deals primarily with real human needs, not the contrived wants around which the game of maximization revolves. That difference must inescapably put the farmer at a disadvantage.

In another respect agriculture is not unique. It cannot evade the bitter disappointment over shrinking promises that is endemic in marketplace societies. All individuals cannot maximize their wealth; some people have to give up something in order for others to get all they want. The social philosophy of private accumulation is a lot like Calvinism: an elect few are chosen to live in paradise, while the rest can go to hell. The number of the elect is not fixed once and for all; it decreases steadily to the vanishing point. Especially since World War II that outcome has been a familiar experience in American farming. In 1900, there were 5.7 million farms in the United States, averaging 138 acres apiece. By 1978, the number had dropped to 2.5 million, and their average size was 415 acres. Over the past thirty years the typical American farm increased its spread by 20 to 30 percent each decade, and all of the increase was taken from a neighbor's side of the fence. At that rate the promise of unlimited farm riches will someday soon be made only to a tiny privileged remnant.

Any farmer must look in the mirror each morning, like an anxious Puritan on his way out to church, and ask, "Am I to be among the saved or not?" Do what he will, the odds are clearly running against him. And if the promise of the land will tingle the ears of fewer and fewer people, if eventually it will belong to an oligopoly—and it is naive to believe agriculture alone can forever escape the corporate takeover—then the offer of unbounded wealth for farmers will turn out to have been a fraud.

The public good cannot be realized in agriculture, therefore, by the untrammeled workings of the market economy and the endless striving for private profit that it institutionalizes. The market creates wealth all right, but its wealth cannot satisfy; it holds up an ideal that is never really achieved, receding indefinitely before our eyes. A farm policy defined only in market terms inevitably must destroy the agricultural community to make it prosper. It must lead to disillusion-

ment and frustration, uprooting and alienation, wearing farmers out, then casting them off.

That is not to say there are no benefits at all in the free-market approach; rather, the arguments on its behalf can take us only a short way toward locating the optimum good of our social life, and then they become immoderate and illusionary. It is now time in the United States to try another tack, to search beyond the marketplace to serve the public good.

Suppose we begin by simply asking what it is that we as a society want out of farming in the future. Are there significant human values that agriculture can help us to realize? Once we have answered that large question, we can call in the farmers, along with the agronomists, the economists, and the fertilizer salesmen, to make the ideal real. By that strategy we might establish better control over where we are going, decide where we want to end up, and stop the drift toward rural chaos. Good farming, by that approach, would be understood as the art, the science, and the wisdom of growing values in the soil—and no calling can be more honorable than that.

Slowly, several worthy answers to that large question of the common good in agriculture have begun to emerge in public discussions. They are familiar in one form or another to us all. The task now is to make them as compelling as possible and move them out from under the deadening shadow of profit maximization.

1. *Good farming is farming that makes people healthier.* It does so by creating and delivering food of the highest attainable nutritional quality and safety. Agriculture fails in its most obvious mission when that quality of healthfulness is missing or when it becomes corrupted by such things as toxic chemical residues. One of the most serious calamities to befall modern industrial farming is that it has turned food into a suspect, potentially dangerous commodity. When people begin to bite gingerly into apples, wondering whether cancer might be lurking there, or when they hesitate to drink a cup of milk, remembering that heptachlor has been found in the dairy's cows, or when they are unsure whether chemical growth-stimulants are lingering in a chicken-salad sandwich, then agriculture has created for itself the most serious possible problem. After all, the essential point of farming is to keep people alive. No gain in export earnings or farm profit, no ease of harvesting or freedom from pests can justify risking human life, can excuse putting the public's health in danger; to act or think otherwise ought to

violate ethics as much as the willful practice of bad medicine. Yet the willingness to risk life and health has become daily news in contemporary food processing and agriculture. The problem is compounded by the fact that farmers may conscientiously harvest crops that meet the strictest standards of nutrition and safety but then must turn them over to numerous processors who, for the sake of profit, have been known to take most of the nutrition out, put additives in, turn wheat into Twinkies and corn into breakfast-table candy. The more complex and powerful the system of farm production, the more sensitive and strict must be the moral consciousness behind it and the more elaborate and expensive the system of public control overseeing it. There is no cheaper, simpler, easier way to realize this value.

2. *Good farming is farming that promotes a more just society.* For a long time in America, the land was where most people expected to go for their start in life, where they hoped to find opportunity and secure a living. The land, always the land: if not in this place, then farther west. Our society's thinking about fairness and democracy reflects even now a reliance on the land as an available, inexhaustible resource. Today, however, we are telling the majority of rural people that there is not enough farmland for them, that they will have to go someplace else for their livelihood, although it is never precisely indicated where that "someplace else" is. If agriculture passes the buck, where will it stop? Does agriculture not have an obligation to the poor and landless in its midst? An obligation to pay decent wages to its laborers and to make room for new farmers, rather than expecting the besieged, depressed cities to take the unwanted? Agriculture, through both private and public agencies, can and should give assistance to struggling racial minorities across the country: to black farmers who are living as tenants on worn-out land, to Indian farmers who need irrigation water, to small Hispanic growers who seek a fair share of attention from county extension agents, to Hawaiians who want land for taro and cultural survival. The agricultural community should work to lop the top off of the rural pyramid of wealth, which is reaching stratospheric heights; today a mere 5 percent of the nation's landowners control almost half the farm acreage, while in the Mountain West a minuscule 1 percent owns 38 percent of all agricultural land and, in the Pacific states, that same percent owns 43 percent of the land. Agriculture, however, is not doing any of these things. On the contrary, it is everywhere retreat-

ing rapidly from a commitment to justice and democracy. Meanwhile, several other nations are managing, despite the pressure of the world marketplace and industrialization, to hold onto the democratic principle; the Danes, for example, have long pursued the ideal of a rural world where few have too little and even fewer have too much. When our own farm experts and leaders rediscover that moral value, American agriculture will be stronger and more successful than it is today.

3. *Good farming is farming that preserves the earth and its network of life.* Obviously, agriculture involves the rearranging of nature to bring it more into line with human desires, but it does not require exploiting, mining, or destroying the natural world. The need for agriculture also does not absolve us from the moral duty and the common-sense advice to farm in an ecologically rational way. Good farming protects the land, even when it uses it. It does not knock down shelterbelts to squeeze a few more dollars from a field. It does not poison the animal creation wholesale to get rid of coyotes and bobcats. It does not drain entire rivers dry, causing irreversible damage to estuaries and aquatic ecosystems, in an uncontrolled urge to irrigate the desert. It does not tolerate the yearly loss of 200 tons of topsoil per acre from farms in the Palouse hills of eastern Washington. Those are the ways of violence. American agriculture of late, pushed by market forces and armed with unprecedented technology, has increasingly become a violent enterprise.

Good farming, in contrast, is a profession of peace and cooperation with the earth. It is work that calls for wise, sensitive people who are not ashamed to love their land, who will treat it with understanding and care, and who will perceive its future as their own. Many farmers and ranchers are still like that and can give us all advanced lessons in ecological ethics. But most preservation-minded farmers are not old men and women, preparing for retirement. There is great danger that they will sell out to less informed, less careful individuals or corporations, who may acquire more earth than they can know intimately and farm well. Somehow we must avoid that outcome. Agriculture's future must be oriented toward using land according to the principles of practical ecology—toward a conserving ethic and intelligence. That orientation is essential if we want to leave our children a planet as fruitful as the one we inherited.

Other public goods I have not mentioned include creating beauty in the landscape, strengthening rural families, aiding the world's hungry—especially helping them produce their own food, diverting investment capital into other sectors of the economy that now need it more than agriculture, and preserving the rural past and its traditions. All of them require us to make policy, not merely make more food. These common goods do not assume that the lot of farmers can be bettered without also considering what the entire human community requires.

Americans are often accused of being a privatizing people. We take the question of public good and break it into millions of little pieces, into every individual's private wants, and then reduce it further by trying to put a price on those pieces. This is my property, we say, and no one can tell me how to farm it. These are my cows, and I will graze them where I like and as hard as I like, until the grass is dead if I like.

There is another America, however, one that has been more open to ideas of the common welfare. That America usually can be found today in less progressive corners, often in rural neighborhoods where there is still a long memory running back to a time when farm folk got their living together and worked more as partners with the land. The future of agriculture will be determined by whether that community thinking can be nurtured and grown more abundantly. It is easily the most important crop we can raise and harvest in the United States.

Those who have forgotten what that sense of rural commonality was like in earlier periods can find it again in the pages of history and literature: in a novel, for example, like O. E. Rölvaag's *Giants in the Earth,* which depicts the settling of the Dakota prairie by a group of Norwegian immigrants. They brought little into that grassland besides themselves and their old-country habits of mutual aid. When nature gave them more than they wanted—gave them plagues of grasshoppers, droughts, and blizzards—they struggled together and came through to more moderate times. They endured and even prospered, but they did not become rich. They made homes for themselves, but they did not conquer that hard land. What they achieved was a wary peace with the prairie, an affectionate and understanding peace, a peace that reflected the fact that they were at peace among themselves. Each family had its own property and fences, its own way of doing things. Occasionally they competed against one another in friendly rivalries. But the overarching principle of their lives, as Rölvaag describes it, was the maintenance of a social bond, which finally became a bond with the strange, foreign land where they settled. Communalism of that sort, in real life as well as in fiction, is receding today, but it has not yet altogether disappeared over the horizon.

The challenge now is to retrieve that commitment to community from the past, from scattered pockets of rural life, and to find a modern expression for it in this new age of industrial agriculture.

At the heart of any nation's agricultural policy must be its ideal of a good farmer. For a number of years we have told farmers, through our colleges, agricultural magazines, government officials, and exporters, one clear thing: get as much as you possibly can out of the land. We have not told them how many farmers would have to be sacrificed to meet that instruction or how much it would deprive the few who remained of their freedom, contentment, or husbandry.

But sooner or later the prevailing ideals wear out, giving way to new ones or to new versions of even older ones. The American ideal of good farming, and the agricultural policy we have built on it, may be ready for a shift in the directions suggested here. In the not-too-distance future, farming may come to mean again a life aimed at permanence, an occupation devoted to value as well as technique, a work of moderation and balance. That is a shift in which we all have a stake.

For Further Discussion

1. Worster contrasts moderate and balanced agriculture with a style of farming that emphasizes agricultural productivity and wealth. What could the point of farming be if not maximizing agricultural productivity?

2. Worster suggests that many problems presently faced by American agriculture stem from an inadequate idea of what truly constitutes the agricultural good of the nation. What is this inadequate idea and what is the alternative?

3. "A farm policy defined only in market terms inevitably must destroy the agricultural community to make it prosper." Explain what Worster means by this sentence. Do you agree with his analysis?

4. How does Worster define good farming? What other alternative definitions can you think of? How might a defender of free markets define good farming?

5. Using Worster's account of good farming, can you develop a parallel definition of good forestry? Of good land use?

Nature as the Measure for a Sustainable Agriculture

Wes Jackson

Wes Jackson is perhaps the best-known advocate for the approach to farming that Worster would call good farming and many others would call sustainable agriculture. Jackson believes that "the best agriculture for any region is one that best mimics the region's natural ecosystems." His Land Institute, in Salina, Kansas, has led the way in promoting this ecologically sensitive, sustainable approach to agriculture. In this essay, Jackson offers a brief description and defense of this approach to farming.

At the Land Institute in Salina, Kansas, we use the prairie as our standard or measure in attempting to wed ecology and agriculture. When Wendell Berry dedicated our new greenhouse in March 1988, he traced the literary and scientific history of our work at the institute. To set the stage for understanding the institute's place in the grand scheme of things, I shall review the history he provided.

Berry first cited Job:

> . . . ask now the beasts, and they shall teach thee;
> and the fowls of the air, and they shall tell thee:
> Or speak to the earth, and it shall teach thee;
> and the fishes of the sea shall declare unto thee.

Later Berry mentioned other writings. At the beginning of *The Georgics* (36–29 B.C.), Virgil advised that

> . . . before we plow an unfamiliar patch
> It is well to be informed about the winds,
> About the variations in the sky,
> The native traits and habits of the place,
> What each locale permits, and what denies.

Toward the end of the 1500s, Edmund Spenser called nature "the equall mother" of all creatures, who "knittest each to each, as brother unto brother." Spenser also saw nature as the instructor of creatures and the ultimate earthly judge of their behavior. Shakespeare, in *As You Like It,* put the forest in the role of teacher and judge; Touchstone remarks, "You have said; but whether wisely or no, let the forest judge."

Milton had the lady in *Comus* describe nature in this way:

> She, good cateress,
> Means her provision only to the good
> That live according to her sober laws
> And holy dictate of spare Temperance.

And Alexander Pope, in his *Epistle to Burlington*, counseled gardeners to "let Nature never be forgot" and to "consult the Genius of the Place in all."

"After Pope," Berry has stated, "so far as I know, this theme departs from English poetry. The later poets were inclined to see nature and humankind as radically divided, and were no longer much interested in the issues of a *practical* harmony between the land and its human inhabitants. The romantic poets, who subscribed to the modern doctrine of the preeminence of the human mind, tended to look upon nature, not as anything they might ever have practical dealings with, but as a reservoir of symbols."

In my own region of the prairies, I think of Virgil's admonition: "Before we plow an unfamiliar patch / It is well to be informed about the winds." What if the settlers and children of settlers who gave us the dust bowl on the Great Plains in the 1930s had heeded that two-thousand-year-old advice? What if they had heeded Milton's insight that nature "means her provision only to the good / That live according to her sober laws / And holy dictate of spare Temperance"? Virgil was writing about agricultural practices, whereas Milton was writing of the spare use of nature's fruits. It is interesting that the poets have spoken of both practice in nature and harvest of nature.

Berry pointed out that this theme surfaced again among the agricultural writers, first in 1905 in a book by Liberty Hyde Bailey entitled *The Outlook to Nature.* The grand old dean at Cornell wrote, "If nature is the norm then the necessity for correcting and amending abuses of civilization become baldly apparent by very contrast. The return to nature affords the very means of acquiring the incentive and energy for ambitious and constructive work of a high order." In *The Holy*

Earth Bailey advanced the notion that "most of our difficulty with the earth lies in the efforts to do what perhaps ought not to be done." He continued, "A good part of agriculture is to learn how to adapt one's work to nature. . . . To live in right relation with his natural conditions is one of the first lessons that a wise farmer or any other wise man learns."

J. Russell Smith's *Tree Crops,* published in 1929, contributed to the tradition. Smith was disturbed with the destruction of the hills because "man has carried to the hills the agriculture of the flat plain." Smith too believed that "farming should fit the land."

In 1940 Sir Albert Howard's *An Agriculture Testament* was published. For Howard, nature was "the supreme farmer": "The main characteristic of Nature's farming can therefore be summed up in a few words. Mother earth never attempts to farm without live stock; she always raises mixed crops; great pains are taken to preserve the soil and to prevent erosion; the mixed vegetable and animal wastes are converted into humus; there is no waste; the processes of growth and the processes of decay balance one another; ample provision is made to maintain large reserves of fertility; the greatest care is taken to store the rainfall; both plants and animals are left to protect themselves against disease."

It may appear that our work at the Land Institute is part of a succession in a literary and scientific tradition, for we operate with the assumption that the best agriculture for any region is one that best mimics the region's natural ecosystems. That is why we are trying to build domestic prairies that will produce grain. We were ignorant of this literary and scientific tradition, however, when we began our work. I did have a background in botany and genetics and could see the difference between a prairie and a wheat field out my windows at the Land Institute, but as Berry said about the poets and scientists he quoted, understanding probably comes out of the familial and communal handing down of agrarian common culture rather than from any succession of teachers and students in the literary culture or in the schools. As far as the literary and scientific tradition is concerned, Berry pointed out that it is a series, not a succession. The succession is only in the agrarian common culture. I came off the farm out of a family of farmers, and apparently my "memory" of nature as measure is embedded in that agrarian common culture. George Bernard Shaw said that, "perfect memory is perfectful forgetfulness." To know something well is not to know where it came from. That is probably the nature of succession in the nonformal culture.

"UNWITTING ACCESSIBILITY TO THE WORLD"

It is always easier to think of a better way to produce either food or a consumer item than it is to propose how to avoid using food or a gadget wastefully. Of all the poets mentioned here, Milton is the only one who wrote about human consumption. Yet if nature is to be our measure, we must be attentive to the "holy dictate of spare Temperance." This is where I see humankind's split with nature widening, and therefore an examination of what is at work is in order.

I believe that we live in a fallen world. By that, I mean that to meet our food and fiber needs, we have changed the face of the earth. We employed human cleverness to make the earth yield up an unbounded technological array that has produced even more countless things. In agriculture we have hot-wired the landscape, bypassing nature's numerous control devices. What drives us to do this in the face of the evidence all around that we are destroying our habitat? Carlos Castaneda's Don Juan called it our "unwitting accessibility to the world." I should explain my interpretation of that phrase.

A few years ago on the last page of *Life* I saw a memorable photograph of a near-naked and well-muscled tribesman of Indonesian New Guinea, staring at a parked airplane in a jungle clearing. The caption noted the Indonesian government's attempt to bring such "savages" into the money economy. A stand had been set up at the edge of the jungle and was reportedly doing a brisk business in beer, soda pop, and tennis shoes.

We can imagine what must have followed for the members of the tribe, what the wages of their "sin," their "fall," must have been—decaying teeth, anxiety in a money system, destruction of their social structure. If they were like most so-called primitive peoples, then in spite of having a hierarchical structure, their society was much more egalitarian than industrialized societies today.

Unlike Adam and Eve, who partook of the tree of knowledge, the New Guinea tribesmen did not receive an explicit commandment to avoid the goodies of civilization. They were simply given unwittingly accessibility to the worldly items of beer, soda pop, and tennis shoes. In the Genesis version, the sin involves disobedience, an exercise of free will. In the latter version, the "original sin" is our unwitting accessibility to the material things of the world. I perceive that to be the largest threat to our planet and to our ability to regard nature as the standard.

In *Beyond the Hundredth Meridian,* Wallace Stegner described the breakdown of American Indian culture:

> For however sympathetically or even sentimentally a white American viewed the Indian, the industrial culture was certain to eat away at the tribal cultures like lye. One's attitude might vary, but the fact went on regardless. What destroyed the Indian was not primarily political greed, land hunger, or military power, not the white man's germs or the white man's rum. What destroyed him was the manufactured products of a culture, iron and steel, guns, needles, woolen cloth, *things that once possessed could not be done without.* [italics added]
>
> It was not the continuity of the Indian race that failed; what failed was the continuity of the diverse tribal cultures. These exist now only in scattered, degenerated reservation fragments among such notably resistant peoples as the Pueblo and Navajo of the final, persistent Indian Country. And here what has protected them is aridity, the difficulties in the way of dense white settlement, the accident of relative isolation, as much as the stability of their own institutions. Even here a Hopi dancer with tortoise shells on his calves and turquoise on his neck and wrists and a kirtle of fine traditional weave around his loins may wear down his back as an amulet a nickel-plated Ingersoll watch, or a Purple Heart medal won in a white man's war. Even here, in Monument Valley where not one Navajo in ten speaks any English, squaws may herd their sheep through the shadscale and rabbitbrush in brown and white saddle shoes and Hollywood sunglasses, or gather under a juniper for gossip and bubblegum. The lye still corrodes even the resistant cultures.

This reality—things that once possessed cannot be done without—is so powerful that it occupies our unconscious, and yet we know that nature "means her provision only to the good / That live according to her sober laws / And holy dictate of spare Temperance."

VOLUNTARY POVERTY AS A PATH TOWARD INSIGHT

Lynn White has proposed Saint Francis of Assisi as the patron saint of ecologists. Francis held the radical position that all of creation was holy. Yet nothing in the record shows that he arrived at that position because he was initially endowed with the wilderness psyche of a Henry David Thoreau or a John Muir or an Aldo Leopold. In fact, his entry point was from a nearly opposite end of the spectrum—this son of a well-to-do man chose poverty. Apparently, Francis took seriously the words of Jesus of Nazareth, "If you have done it unto the least of these my brethren, you have done it unto Me." Francis's intimate identification with the least, his joining the least, must have prepared him psychologically to be sensitively tuned to all of creation—both the living and the nonliving world. He was a Christian pantheist who believed that birds, flowers, trees, and rocks had spiritual standing.

Francis, the founder of the most heretical brand of Christianity ever, began his journey with a marriage to poverty. This poverty, voluntarily chosen, apparently was the prerequisite for making what White called "the greatest spiritual revolutionary in Western history." His marriage to poverty and his sparing use of the earth's resources led to deep ecological insight. You may remember the legend of the famous wolf of Gubbio, which had been eating livestock and people. Francis approached the wolf and asked him, in the name of Jesus Christ, to behave himself. The wolf gave signs that he understood. Francis then launched into a description of all that the wolf had done, including all the livestock and people he had killed, and told him, "You, Friar Wolf, are a thief and a murderer" and therefore "fit for the gallows." The wolf made more signs of understanding, and Francis continued, stating in effect, "I see you have a contrite heart about this matter, and if you promise to behave, I'll see to it that you are fed." The wolf showed signs that he promised, and the legend has it that the wolf came to town, went in and out of people's houses as a kind of pet, and lived that way two years before he died. The townspeople all fed him, scratched his ears, and so on. It was life at the dog food bowls of Gubbio.

In light of that story, I once thought that Saint Francis might more properly be regarded as the patron saint of domesticators. Anyone able to encourage a wolf to quit acting like a wolf should act, given everything from its enzyme system to its fangs, is not likely to be regarded as an ecologist, let alone a patron saint for such.

A few weeks ago, at about three in the morning, the barking of our two dogs woke me up. Soon there was hissing and growling and more barking right on the back porch. I went outside and saw a raccoon cowering under a step stool by the dog food bowl. I chased it away with a stick, sicced our border collie Molly on him, and went back to bed. I confidently went to sleep.

A few minutes later I heard more barking, more hissing, more growling. Once again I went outside, and the dogs and I ran the coon off. Back to bed, and sure enough the same story. This time, however, I left the porch light on for this nocturnal animal. I lay in bed and listened, and there was no more ruckus. I felt pleased with myself for having solved the problem with a light switch by means of my biological knowledge about nocturnal animals, and I went to sleep. The next morning, I headed out the back door to begin the day's work, and there in a box of tinder on a table was the coon, sleeping away. Both dogs were asleep under the table. Each time I returned to the house during the day, I expected to see that the coon had gone. But he didn't leave. And as he slept that day, the dogs would walk by, look up, and sniff, well on their way toward accepting their new fellow resident.

What was going on here? We are taught to consider recent changes when something unusual breaks a pattern. And so I might have an answer. Late in the afternoons this winter I have been going to my woods with a chain saw, some gasoline, and matches to burn brush. I am clearing out most of the box elder trees that have grown up there over the last forty years. They are early-stage succession trees that have mostly covered the area where the former tenant logged out all the walnuts and burr oaks. The box elders are the first trees to green up in the spring, and they accommodate woodpeckers. But they are not good for lumber, and though we burn them, they are very low in fuel value. I want to accelerate succession by planting some walnuts and oaks for my grandchildren. Nearly every box elder I took down was hollow. The woods are less than a quarter of a mile from my house along the Smoky Hill River, and I suspect that I destroyed the home of Friar Coon and that Friar Coon, looking for a new home, simply moved into mine.

Now back to Gubbio. Why was the thieving, murdering wolf forgiven and then given a life at the dog food bowls? I suspect that the ecological context necessary to accommodate proper wolfhood around Gubbio had been destroyed, that the usual predator-prey relationship had been somehow disrupted, and that Francis realized it. If so, Francis's deeper ecological insight was a derivative of his respect and love for nature. He could, after all, have organized a posse to eliminate the killer wolf, but instead his love and respect for nature—in turn at least partly derived from his identification with poverty—made him forgiving and compassionate.

But there is still an item on our agenda for discussion. Was Saint Francis engaged in an act of domesti-

cation? It appears so, but if harmony with nature is what we seek, should we not be willing to be in harmony any way we can? The wolf's tame behavior demonstrates that nature is not rigid. Humanity (Francis) reached out to nature (the wolf), and the wolf responded. If we insist that wild nature be so rigid, we are denying one of the most important properties of nature—resilience. The unanswered question here is, Once the wolf comes to town, can he ever return to the wild? Life at a dog bowl in Gubbio may be easier than life that relies on the fang and the occasional berry. If the wolf is unable to return, then that particular wolf is a fallen creature. Would it matter that he was made that way initially by the fallen ecological context of humanity's making? Wolves and coons, unable to return to their original context because they have gained unwittingly accessibility to the world, may be little different from the New Guinea tribesmen or the native Americans. Stegner's phrase, "things that once possessed could not be done without," can be applied beyond people. Grizzlies in the garbage at Yellowstone and elephants in African dump heaps come to mind as modern expressions of the same problem. A wolf, or any other creature, unable to return is a fallen animal dependent on fallen humanity. And so another ethical question comes on the agenda: What right do we have to create a fallen world for other species when we know that life at the dog food bowl is second best? Following that line of thinking, our crops and livestock represent fallen species that accommodate fallen humanity.

INTERPENETRATION OF THE DOMESTIC AND THE WILD

The stories about the wolf in Gubbio and the coon on the porch illustrate the problem of trying to use nature as measure, for in those examples the interpenetration of the domestic and the wild is total. During the last five hundred years or so, the ratio of the domestic to the wild has increased so much, especially in the Western Hemisphere, that wilderness is becoming an artifact of civilization. Civilization is all that can save wilderness now. The wild that produced us, that we were dependent upon, is now dependent on us. We pay homage to wildness in the United States by regarding pristine wilderness as a kind of a saint. But that presents some problems, too.

In Christianity the tradition of sainthood calls upon the faithful to stand or kneel before an image of a saint, light a candle, meditate, think, and perhaps whisper some words before departure. Often these faithful go in peace, perhaps thinking, "Well, that's covered," and they carry on more or less as before. This isolation of virtue can also be found in countless wilderness advocates who clamor to have wilderness set aside as pristine. They will stand in forest wilderness, soak up its silence, walk out, and send money to defend it. They may say little in protest, however, about the spread of lethal farm chemicals over more than half a million square miles of the best agricultural land in the world, and soil erosion may be no concern of theirs. I do not object to either saints or wilderness, but to keep the holy isolated from the rest, to treat our wilderness as a saint and to treat Kansas or East Saint Louis otherwise, is a form of schizophrenia. Either all the earth is holy, or it is not. Either every square foot deserves our respect, or none of it does.

Would Earth First! activists or Deep Ecologists be as interested in cleaning up East Saint Louis, for example, as they are in defending wilderness? Would Earth First! activists be as fervent about defending a farmer's soil conservation effort or chemical-free crop rotation as they are about spiking a tree or putting sugar in the fuel tank of a bulldozer?

It is possible to love a small acreage in Kansas as much as John Muir loved the entire Sierra Nevada. That is fortunate, for the wilderness of the Sierra will disappear unless little pieces of nonwilderness become intensely loved by lots of people. In other words, Harlem and East Saint Louis and Iowa and Kansas and the rest of the world where wilderness has been destroyed will have to be loved by enough of us, or wilderness is doomed. Suddenly we see we are dealing with a range of issues. For that reason Saint Francis's entire life becomes an important example. People who struggle for social justice by working with the poor in cities and people out to prevent soil erosion and save the family farm are suddenly on the same side as the wilderness advocate. All have joined the same fight.

NATURE AS MEASURE IN AGRICULTURAL RESEARCH

Rather than deal with problems *in* agriculture here and now, we at the Land Institute address the problem *of* agriculture, which began when agriculture began some eight to ten thousand years ago. We have seen that nature is an elusive standard. Nevertheless, it

seems to us at the Land Institute less elusive than any other standard when sustainability is our primary objective. The nature we look to at the Land Institute is the never-plowed native prairie.

We have around one hundred acres of such land at the institute, and when we compare prairie with the ordinary field of corn or wheat, important differences become apparent. From our typical agricultural fields, valuable nutrients run toward the sea, where for all practical purposes most of them are gone for good. The prairie, on the other hand, by drawing nutrients from parent rock material or subsoil, all the while returning chemicals produced by life, actually builds soil. The prairie, like nearly all of nature, runs mostly on contemporary sunlight, whereas our modern agricultural fields benefit from the stored sunlight of extinct ancient floras. Diversity does not necessarily yield stability overall; nevertheless, the chemical diversity inherent in the diverse plant species of the prairie confronts insects and pathogens, making epidemics, so common to agricultural monocultures, rare on the prairie. Because no creature has an all-consuming enzyme system, diversity yields some protection. The prairie therefore does not require the introduction of chemicals with which species have had no evolutionary experience.

So what are the basic differences between a prairie and an agricultural field? A casual examination of the ordinary differences will help us to see that the prairie features perennials in a polyculture, whereas modern agriculture features annuals in a monoculture. Our work at the Land Institute is devoted primarily to exploring the feasibility of an agriculture that features herbaceous perennials grown in a mixture for seed production—that is, domestic prairies—as substitutes for annual monocultures grown in rows on ground that can erode.

We address four basic questions in our experiments at the institute. First, can herbaceous perenniality and high seed yield go together? Because perennial plants must divert some photosynthate to belowground storage for overwintering, it may be difficult to breed perennials to produce as much seed as annual crops that die after reproducing. Perennial species differ greatly in both relative and absolute amounts of energy devoted to seed production, however, so theoretically there seems to be no reason why a fast-growing, well-adapted species could not yield adequate seed while retaining the ability to overwinter.

Before we begin to breed for stable high seed yields in an herbaceous perennial, we determine its genetic potential. Whether we start with a wild introduced species or a wild native, the development of perennial seed-producing polycultures will require that we select for varieties that perform well in polyculture. The potential improvement, therefore, depends on the range of existing genetic variability in the wild. To assess this variation requires an adequate sample drawn from the geographic range of the species and then an evaluation of the collection within a common garden.

Now our second question: Can a polyculture of perennial seed producers outyield the same species grown in monoculture? Overyielding occurs when interspecific competition in a plant community is less intense than intraspecific competition. Thus, we believe that through differences in resource use and timing of demand, multispecies fields typically yield more per unit area than do monospecific stands.

Our third question is, Can a perennial polyculture provide much of its own fertility? Specifically, can such internal factors as nitrogen fixation and weathering of primary minerals compensate for nutrients removed in harvested seed? To answer this question, we must document nutrient cycling in the soil, nutrient content of seed, and capacity of crop plants to enrich the soil.

As our fourth and final question we ask, Can a perennial mixture successfully contend with phytophagous insects, pathogens, and weeds? If we are to protect a crop, a combination of breeding for resistant lines and studies on the effects of species diversity must converge. Insect pests can be managed through a combination of attracting predators and preventing insects from locating host plants. A mixture of species, and of genotypes within species, may reduce the incidence and spread of disease. Weeds may be controlled either allelopathically or via continuous shading of the soil surface by the perennials.

Though we keep all four of these questions in mind, the most pressing biological question at the Land Institute is whether perennials and high seed yield can go together. To answer this question, we started a plant inventory that had the following steps: (1) we reviewed the literature of seed yield in winter-hardy herbaceous perennials; (2) we collected seed and plants in nature and developed an herbary of approximately three hundred species, each grown in five-meter-long rows; and (3) we planted more than forty-three hundred accessions of more than one hundred species representing seven grass genera. Our inventory continues even though we are currently focusing on five species plus a hybrid of our making.

The relationship between perennials and high yield also involves the issue of sustained production. Prairies, after all, feature perennials, but they do not feature high seed yield. Ultimately, we have to explore the optimum balance between sustainability and yield.

In addition to the inventory of potentially high-yielding species, other sorts of inventories, such as an inventory of the vegetative structure, are necessary for long-term considerations. Because perennial roots are a major feature of our work, an inventory of the soil relationships in the prairie and in our plots is also essential. The ecological inventory includes more than analysis of the phytomass ratios; it also includes an ongoing inventory of the insects and pathogens in our herbary and in our experimental plots. We always compare the results with those from our prairie, the system that represents the least departure from what was here before white settlement.

The inventory phase will probably never end. The ecological inventory is particularly long-lasting because of the countless number of interactions over time. Even research on the question of perennials and high yield will require several years, for it amounts to an investigation of long-term demographic patterns in perennial seed production, a field that is largely unexplored. Studies thus far at the Land Institute have shown increases, decreases, and oscillations in seed yield over time. But we always come back to these questions: What was here? What will nature permit us to do here? And what will nature help us to do here? Wendell Berry once wrote in a letter, "When we cut the forest and plowed our prairies we never knew what we were doing because we never knew what we were undoing." It is now a matter of practical necessity to learn what we were undoing.

In a 1986 preliminary investigation, Jon Piper, the institute's ecologist, asked, How much aboveground plant life is supported each year by the prairie, and what are the proportions of grasses, legumes, and composites? Those plant families comprise most of our temperate agricultural species. Net production of the plants at Piper's grassland sites (five hundred to seven hundred grams per square meter) was similar to that of many midwestern crops. At their peaks, grasses composed 67 to 94 percent of plant matter, and legumes and composites represented 16 and 11 percent of vegetation, respectively. Piper concluded from these encouraging results that "a sustainable agricultural system for central Kansas is feasible if perennial grasses were featured followed by nearly equal proportions of legumes and composites."

In 1985 we began to examine insects and plant pathogens qualitatively in nine experimental plots at the Land Institute. Every week from May through August, and every other week from September through October, insects were collected with a sweep net or from individual plants. All diseased plants were sent to the Disease Diagnosis Laboratory at Kansas State University for pathogen identification. All sampled plots showed a diversity of both beneficial and harmful insects. Several foliar diseases were present, but few were serious. We continued that inventory in 1986 but with important modifications. The prairie was sampled using sweep nets every third week. Over the years, we have sampled the prairie, the herbary, and the experimental plots for insects and pathogens and have made numerous comparisons.

A final example of this soft approach to sustainable-agriculture research is the design of the large polyculture experiment we intend to establish in 1991. For that experiment, we think about the species components we intend to introduce, the planting density of each species, the ratios of species to one another, and so forth.

THREE FINAL QUESTIONS

We have three final questions to consider. First, is perennial polyculture or ecosystem agriculture inherently more complicated and therefore less likely to succeed than monoculture agriculture, be it of the annual or the perennial variety?

My answer is, not necessarily. The disciplines of science are divided to explore the various levels in the hierarchy of structure from atoms to molecules, cells, tissues, organs, organ systems, and organisms. At each level of aggregation, it is the emergent qualities more than the contents that define the discipline. A physicist may have learned about the structure and workings of an atom in great detail. Though some understanding of atoms is necessary for a chemist, the chemist does not need to know the atom with the same intricacy of detail as the physicist. A chemist mostly studies reactions. On up the hierarchy, we see that chemistry is important to a cell biologist but does not define cell biology, and a good cell biologist does not need to have a chemist's detailed knowledge. Cell biology as a field is not more complicated than chemistry or physics, though cells are more complex than molecules or atoms. Likewise, ecosystem agriculture will be more complex than monoculture agriculture,

but the management of agro-ecosystems may not be more complicated. Ecosystem agriculturalists will take advantage of the natural integrities of ecosystems worked out over the millennia.

When we deal with nature's designs, a great deal of ignorance on our part is tolerable. Much error is forgiven. Ignorance is tolerable until we begin to impose our own designs on nature's landscape. Even then certain kinds of ignorance and large amounts of forgetfulness will be tolerated. (Not knowing is a kind of ignorance preferred over knowing things that just are not so. At least one does not have to unlearn what is not known.) When we impose our own designs on nature's landscape, we do so with the presumption that we know what we are doing, and we have to assume responsibility for our mistakes. By imitating nature's patterns, we should be able to reduce error by taking advantage of nature's complexity, thus minimizing complications for ourselves. Farmers and scientists alike may not know why certain associations of plants and animals grant sustainability, just that they do. And though there is little wrong with finding out why certain associations work, from the point of view of a farmer interested in running a sustainable farm, knowing why is not always necessary.

Our second question is, How critical is species diversity, and if it is necessary, how much and what kind are optimum? As mentioned earlier, diversity does not necessarily lead to stability. Numerous diverse ecosystems are less stable than simpler ones. We can raise a question about the inherent value of diversity by considering two extremes. At one extreme, we could assemble a diverse hodgepodge of species, plants that have never grown together in an ecosystem. At the other extreme, we could assemble plant species that have histories of growing together—on the prairie, let us say. In the latter case, natural integrities have evolved to the point that large numbers of genetic ensembles interact in a species mix. This area warrants much research.

Another important consideration is associated with the diversity question. As species are selected for future experimentation, we may need to determine to what extent the genetic profile is tuned to interspecific versus intraspecific complementarity. In all of our important domestic grains, the genetic assembly of an individual plant resonates against members of its own kind (intraspecific complementarity). On a prairie, that is not the case. Prairie plants are more tuned to interact with different species (interspecific complementarity).

The third question we ask is this: Is it true that, for any biotic system, internal control uses material and energy resources more efficiently than external control? In a hierarchy of structure—beginning with an individual plant, then the field (an ecosystem), then the farm (a larger ecosystem), and then the farm community (an even larger ecosystem)—it will be necessary to think about the efficient use of material and energy resources.

This philosophical consideration is of great practical importance. Consider a plant's resistance to an insect. If a plant uses its genetic code to make a chemical that is distasteful to an insect, thereby granting itself protection, we would call that internal control. If we, perhaps unknowingly, remove that ability through breeding, the plant is susceptible and we apply an insecticide on the plant's surface to grant it protection. That is external control. Yield increases, but the resource cost for protection is paid from the outside, and seemingly the total cost would be greater.

Another example is nitrogen fertility. If the feedstock for commercial nitrogen fertilizer is natural gas, the total energy cost would be higher than if the plant fixed its own nitrogen. In the first case, we are using what we might call vertical energy, or time-compressed energy; in the second case, we are using horizontal, or contemporary, energy from the sun. As our supplies of vertical energy run out and we are forced to use horizontal energy, then the answer to our major question becomes crucial, for at that point the energy source becomes a land-use problem.

A third example is weed control. If the roots of a plant produce an herbicide to keep back most weeds, then weed protection comes from within the plant (allelopathy). The plant's production of such an herbicide will come at a cost in yield. But let us say that a plant lacks the ability to produce the herbicide and that mechanical weeding is necessary. If we pay the cost on the farm the way we used to—that is, harvest biomass from a pasture or field to feed horses supplying the power for mechanical weed removal—then it seems obvious that the overall cost will be higher.

We are faced with extremely difficult choices and, I believe, extremely difficult times. Our goal must be a harmony between the human economy and nature's economy that will preserve both. In the greenhouse dedication speech mentioned earlier, Wendell Berry pointed out that such a goal is traditional: "The world is now divided between those who adhere to this ancient purpose and those who by intention do not, and this division is of far more portent for the future of the world than any of the presently recognized national or political or economic divisions."

Recalling his outline of the literary and scientific traditions, Berry concluded, "The remarkable thing about this division is its relative newness. The idea that we should obey nature's laws and live harmoniously with her as good husbanders and stewards of her gifts is old. . . . And I believe that until fairly recently our destructions of nature were more or less unwitting—the by-products, so to speak, of our ignorance or weakness or depravity. It is our present principled and elaborately rationalized rape and plunder of the natural world that is a new thing under the sun."

For Further Discussion

1. Jackson begins this essay by reviewing the works of various people who speak of our ability to learn from the earth. How would you contrast this view with Locke's notion that humans improve the land by changing it with their labor?

2. Jackson suggests that the best agriculture for any region is one that best mimics the region's natural ecosystems. Do you agree? Does this imply that humans should not attempt to use technology and science to improve agricultural production?

3. Quoting Wallace Stegner, Jackson tells us that "things once possessed cannot be done without." Explain what this means. Do you find any evidence for this in your own experiences?

4. Do you agree with Jackson that either all the earth is holy, or it is not? Does a small piece of the Kansas prairie or a small plot of land in East St. Louis have as much aesthetic, spiritual, or moral value as Yellowstone Park, the Grand Canyon, or Yosemite?

DISCUSSION AND STUDY QUESTIONS FOR CHAPTER 10

1. How should government agencies manage public lands? What is the "public interest"? Should the desires of each constituency receive equal consideration and equal weight?

2. Can private property be used by owners in any way that does not cause harm to other people? Does the right to own land imply a right to destroy it?

3. Land without any buildings is often described as undeveloped or unimproved land. One result is that such land is taxed at a lower rate than land

with buildings. How can land be improved or developed?

4. Some people mourn the decline of the family farm. Others point to the increased production that comes from centralized agribusiness. What values, both social and personal, might be involved in a family farm that would be missing in large agribusinesses?

5. What's the difference between land and property? Between dirt and soil?

DISCUSSION CASES

Who Owns the Moon?

Consider the following thought experiment to assess your understanding of property rights. In July 1969 astronaut Neil Armstrong became the first human being to walk on the moon. As Armstrong jumped to the surface, he offered the famous description: "One short step for a man, one giant leap for mankind." Does anyone own the moon? Could anyone ever come to own the moon? What would it take to own a planet? Could the United States argue that, because it was the first country to travel to the moon, and because the United States "claimed" it, the moon belongs to the United States? Suppose that, instead of unfurling an American flag, Armstrong staked his claim with the Armstrong family flag, announcing that he was claiming this unowned land for his family. Would Neil Armstrong own the moon? How could land that is unowned first come to be owned?

John Locke argued that land comes to be owned when someone mixes their labor with it as long as there is "enough and as good" remaining for others. Whose labor was mixed with the moon? Do all taxpayers have an equal claim on moon ownership? What about citizens of other countries who did not have an opportunity to participate in the lunar landing? Is there enough and as good remaining for them? Or is the moon something that shouldn't be owned at all? What would this mean?

Already there have been discussions about mining the moon for minerals. How should we decide the proper, ethical use of the moon?

Wise Use?

Several groups have formed, especially in the American West, to fight what they see as burdensome environmental regulations that restrict private property rights and prevent wide access to public lands. The so-called Sagebrush rebellion of the late 1970s and early 1980s and, more recently, the wise-use movement represent a broad coalition of people who oppose environmental regulation of private and public lands. Gifford Pinchot coined the phrase "wise use" to describe his conversationist policies aimed at managing public lands for the benefit of all citizens.

The wise-use movement includes ranchers, cattlemen, farmers, loggers, land developers, hunters, commercial and sport fishermen, off-road vehicle users, and others. Although it has broad appeal as a grassroots movement, it has also attracted financial and political support from corporations. Some wise-use groups are directly connected to industrial, logging, mining, and development interests.

The wise-use movement has two primary goals, both related to traditional American values. This movement opposes environmental regulation because it places an unfair burden on private property owners. When regulation prevents development of natural resources, private owners are denied their property rights without full compensation. Environmental regulation of public lands is also unfair in that it seems to privilege the interests of one group—the environmentalists—over the interests

(*continued*)

of others—ranchers, loggers, and so on. On this view, public land should be managed for the well-being of all, not just for the well-being of a small, elite group. Thus, both private property rights and the utilitarian goal of promoting the greatest overall good are violated by restrictive environmental laws.

Compare the interpretation of wise use offered by Gifford Pinchot and Randel O'Toole with an interpretation that might be offered by Donald Worster or Wes Jackson. Who do you think has the better understanding of wisdom in regard to land use? Why?

Lucas v. South Carolina Coastal Council

The Fifth Amendment to the U.S. Constitution prohibits private property from being "taken for public use without just compensation." The "takings" clause, as it has come to be called, traditionally has been understood to prohibit the government from physically occupying or taking ownership of private property without paying the owner just compensation. Thus, the principle of eminent domain allows government to take private property for roads and other public utilities only if just compensation is paid to the owner. Of course, the police power of government allows restrictions on property rights to prevent public nuisances and harms.

In recent years the "takings" clause has been used to protect private owners from environmental regulations that restrict the owners' use of their property. A major case in this area was *Lucas v. South Carolina Coastal Council.*

Oceanfront property in low-lying areas and especially along barrier islands is subjected to constant wind and water erosion and flooding. Storms and hurricanes have wiped out entire areas, causing billions of dollars in damage, erosion of beaches, and habitat loss, as well as jeopardizing adjacent areas by weakening jetties, seawalls, and dunes. Despite these threats, oceanfront property is among the most highly valued in the world.

Beginning in the late 1970s and continuing through the 1980s, South Carolina instituted a series of laws restricting development along its coastline. State agencies had discovered that more than a quarter of the state's coastline was eroding and recognized that development along the shoreline was hastening this erosion. In 1988 South Carolina precluded any development within a certain setback area along the ocean.

In 1986, David Lucas purchased two oceanfront lots on the barrier island Isle of Palm in South Carolina. As is typical of such land, the landscape changes over time. Between 1957 and 1962, these lots were under water, and evidence suggested that they were at least partially covered by water for twenty of the last forty years. By 1994, they were 300 feet from the ocean. When Lucas planned to build on these lots (he had previously participated in large development projects in the same area), he was informed that his lots fell within the setback area and that as a result he would be unable to build on this property.

Lucas sued South Carolina, claiming that the restrictions left his land without economic value and thus were an unconstitutional "takings" of his property. In a complex decision, the U.S. Supreme Court agreed that a regulation that leaves land without any economic value was indistinguishable from outright seizure of the land and thus did constitute a "takings." The case was sent back to South Carolina to determine if this use of the land would constitute a restrictable nuisance.

Pacific Lumber Company

Pacific Lumber was a publicly traded company operating out of Humboldt County in northern California and operated by the same family for over one hundred years. Their management philosophy was based on sound ecological principles. They harvested approximately 2 percent of their trees yearly, a rate comparable to the annual growth rate of the trees. They also preserved large stands of redwood trees.

Economically, this management philosophy had been profitable, although perhaps not as profitable as possible. The company also had good relations with its employees, stockholders, and the local community. However, this management philosophy made Pacific Lumber an inviting takeover target for outside investors. They were a profitable and debt-free business, and they possessed many unused and undervalued resources (the 98 percent of the trees left unharvested). In 1986, Charles Hurwitz's company—Maxxam Corp.—organized

(continued)

a leveraged buyout of Pacific Lumber. Almost $800 million of the $900 million purchase price was financed by high-interest loans. Pacific Lumber's resources, specifically its unused trees, were, in effect, collateral for these loans.

The results of the takeover were predictable. To repay the loans Maxxam had taken to purchase Pacific Lumber, they had to quickly sell off its assets and increase harvesting, including the remaining redwood groves. Defenders of this takeover point out that on economic grounds this was successful only because the company's resources were used more efficiently after the takeover than previously. In general terms, society was getting more of what it wanted after the takeover than it had been getting prior to the takeover.

What limits should be placed on private property rights? Should owners be free to use their property in ways that destroy the land? Is it reasonable, from an economic or ethical perspective, to use resources at a rate that cannot be sustained over the long term? Is Maxxam using its land in a wise manner?

CHAPTER 11

Wilderness Preservation

In the readings that follow, wilderness is called "a major contemporary concern," a "focal point of much contemporary debate," and "a fundamental tenet—indeed a passion—of the environmental movement." Preserving wild areas, whether they be tropical rain forests or Arctic tundra, certainly is among the primary environmental policy issues of the day.

Concern with preserving and protecting wilderness areas dates back at least to the early nineteenth century when Ralph Waldo Emerson and Henry David Thoreau wrote of the unspoiled wilderness and the natural and authentic human experiences these areas offer. In recent decades, incited by the loss of wild areas due to population growth and resource extraction, wilderness protection has come to the forefront of the environmental agenda.

But this concern has not always prevailed. The concept of the dangerous wilderness and the wild world is as old as recorded history. Throughout much of Western civilization, the wilderness is characterized as an inhospitable and threatening place. The wilderness is where Adam and Eve were sent into exile after their expulsion from Eden. It was an accursed place. Moses also wandered in the wilderness for forty years before reaching the Promised Land. The New Testament describes Jesus struggling in the wilderness for forty days and there meeting Satan himself. As described by Roderick Nash in Chapter 2, early American settlers found a "hideous and desolate wilderness," home to "hellish fiends and brutish men that devils worshipped."

This perspective surely encouraged a hostile and aggressive approach to the natural world. Wilderness was seen as something to be conquered and tamed. Natural areas were to be developed and improved. Unused wild areas were being wasted, having value only when put to human use. Inhabitants of wild areas, be they human or animal, were dangerous and brutish beasts.

This perspective is to be contrasted with what might be called the romantic model of the wilderness. Rooted in writers such as Jean-Jacques Rousseau and Thoreau, the romantic model reverses the value perspective of the earlier view. From the romantic perspective, wild areas represent what is genuine, natural, and authentic. Cities and the rest of the civilized world are artificial and corrupting. Romantics believe that humans are naturally good, becoming corrupt only under the influence of society. Romantics deny that the scientific approach to nature—an approach that reduces the natural world to physical and material objects—tells us all there is to know about the natural world. The natural world is alive with aesthetic, spiritual, and moral values. Wild areas in particular, as areas most free from human intervention, provide the best opportunities for experiencing these central human values. In the words of Thoreau, "In wildness is the preservation of the world."

It is not difficult to see how both perspectives have influenced public policy. Government policy, from North America and Australia in the eighteenth and nineteenth centuries and continuing throughout much of the world today, views wilderness as something to be conquered, controlled, settled, and developed. Native peoples who lived on these lands are to be civilized, saved, converted, or conquered. Wild animals are varmints and beasts that need to be exterminated.

The romantic model has also greatly influenced public policy. In many ways, it underlies a good deal of contemporary environmental policy. Beginning with the creation of national parks such as Yellowstone and Yosemite and continuing through the Wilderness Act of 1964, U.S. environmental policy often speaks of the wilderness as an unspoiled area that deserves protection from any human disturbance.

But these philosophical views are ideals, and public policy matters seldom fit into neat categories. Most policy debates concerning wilderness areas involve a blend of values and attitudes toward the wild. The readings in this chapter attempt to sort out the various and competing understandings of wilderness, and two of the three selections suggest that it is time for a major rethinking of the concept of wilderness.

These debates focus on a common concept of the wilderness, what one author calls the "received" view. This received view holds that wild areas are, as defined by the Wilderness Act of 1964, "where the earth and its community of life are untrammeled by man, where man himself is a visitor who does not remain." Viewed in this way, wilderness areas possess a value that can be lost when humans interfere too much. In the United

States, wilderness areas are open to such activities as hiking, camping, fishing, and some hunting, but are closed to commercial activities, building, logging, mining, and most motorized recreational activities.

But suppose that this received view of the wilderness is flawed. Suppose that there are good reasons to abandon this model of the wilderness. Might there be significant policy implications if we switch to a more ambiguous understanding of the wilderness? Are there reasons for making a sharp distinction between human culture and wild nature? These are some of the questions raised and examined in the following readings.

Rethinking Wilderness: The Need for a New Idea of Wilderness

Michael P. Nelson

According to Michael Nelson, the "received" concept of wilderness is seriously flawed. This concept views the wilderness as a place unaffected by humans, a place where humans are visitors but not residents. Nelson sees this view as the most influential and common understanding among environmentalists. This widely accepted view, however, has serious shortcomings; an uncritical acceptance of it presents troubling puzzles, paradoxes, and dilemmas.

Nelson outlines five major problems with this understanding of wilderness: it is not universalizable, it is ethnocentric, it is ecologically naive, it separates humans from nature, and its referent is nonexistent. Ignoring these problems has potentially tragic implications for those interested in protecting wilderness areas. Nelson suggests that we need to adopt a more flexible understanding of wilderness.

temporary debate. These debates presume a noncontroversial, common concept of wilderness. Yet, the concept of wilderness as we have come to view it has serious problems.

In this essay, it is my contention that a conceptual analysis and revision of our accepted notion of wilderness will better serve to protect and defend those areas that we commonly think of and legally designate as wilderness. To begin, I briefly point out what is the contemporary "received" (as I will call it) concept of wilderness. Then, I discuss the shortcomings involved with the received idea of wilderness, including the potentially tragic implications of not revising it. Along the way, I offer some suggestions for an alternative notion of wilderness.

INTRODUCTION[1]

Throughout the world, a major contemporary concern is protecting the natural environment. The environmental spectrum ranges from the natural to the nonnatural. As the extreme on one end of this spectrum, wilderness has become the focal-point of much con-

THE "RECEIVED" IDEA OF WILDERNESS

In the minds of most people in the developed West, especially in the U.S., the conception of wilderness is noncontroversial and unproblematic. After all, we received this idea of wilderness from the likes of John

Muir, Theodore Roosevelt, Henry David Thoreau, Sigurd Olson, Aldo Leopold, and Bob Marshall. Wilderness advocates of the last hundred years or so have had a common general understanding or definition of "wilderness." A few examples are the following:

> *Bob Marshall:* "a region which contains no permanent inhabitants, possesses no possibility of conveyance by any mechanical means and is sufficiently spacious that a person in crossing it must have the experience of sleeping out."[2]

> *American Heritage Dictionary:* "an unsettled, uncultivated region left in its natural condition."[3]

> *Aldo Leopold:* "the raw material out of which man has hammered the artifact called civilization."[4]

> *Aldo Leopold:* "a continuous stretch of country preserved in its natural state, open to lawful hunting and fishing, big enough to absorb a two weeks' pack trip, and kept devoid of roads, artificial trails, cottages, or other works of man."[5]

> *The Wilderness Act of 1964:* ". . . [an] area in contrast with those areas where man and his own works dominate the landscape . . . where the earth and its community of life are untrammeled by man, where man himself is a visitor who does not remain."[6]

In these definitions, the common conception of "wilderness" is that of an area unaffected by humans and human activities; an area where humans are at most only spectators or visitors; an area where environmental change is governed by natural processes and not by human-induced ones. Wilderness is pictured in opposition or antithesis to civilization. This view of wilderness I will call the "received" view or conception.

The received view of wilderness presents troubling puzzles, paradoxes, and dilemmas, to which I now turn.

SHORTCOMINGS OF THE RECEIVED CONCEPT OF WILDERNESS

What is troubling about "the forest primeval," to borrow Longfellow's phrase? This purist concept of wilderness presents five difficulties.

First, the received view of wilderness is very much an American idea and is arguably not universalizable.

Given the fact that the environmental problems we currently face are global in scope, it would seem that we would want a corresponding philosophy of conservation that would be universalizable. However, in much of the world, the application of this received American idea of wilderness makes little sense. In much of the developed world, such as western Europe, there are no remotely untouched areas: and, hence, these parts of the world would be left entirely out of this discussion, except as a negative model and antithesis to wilderness.

In other large chunks of the world, indigenous peoples are inhabiting what we think of as and hope will be designated, wilderness. For instance, humans have lived for 11,000 years in the Natak Wilderness in Alaska. According to the received concept, these areas would not qualify. We arguably need a concept we can apply globally, and the received, purist notion of wilderness cannot be applied globally.

Second, the received idea is ethnocentric—which is likely the reason why we cannot universalize it. After the northern Europeans left the "Old World" and stepped off the boat onto the western hemisphere, they were not stepping into a pristine wilderness where human influence had not already had significant effect. They mistakenly thought they were, likely because this "New World" appeared significantly different from the human-dominated landscapes in Europe.

At the time of the "discovery" by Columbus, the western hemisphere was populated by 40 to 80 million people.[7] These inhabitants, like all forms of life, had modified their environment. Native Americans had actively managed their lands—primarily with fire. The composition of the forest had been altered, grasslands had been created, erosion was severe in certain areas, wildlife had been disrupted, and such things as roads, fields, earthworks, and settlements were already widely scattered. The introduction of "Old World" diseases reduced the number of native peoples by as much as 90 percent, giving the early European immigrants the illusion that they had stumbled upon a vast and unpopulated wilderness.

The so-called "wilderness" that the Pilgrims found themselves in was one created by humans. Hence, according to the purist idea, it was not a wilderness at all.

Third, the received view of wilderness is not externally consistent. That is, it is at odds with certain aspects of theoretical ecology.

The received wilderness idea paints a picture of nature as a static landscape. This image follows from the now outdated view that ecosystems remained in, and always strove toward, a stationary state, called a climax community, until and unless disturbed by some outside force.

Recent ecological thought, however, gives a much different perspective. Ecologists now believe that ecosystems are in a constant state of flux. An ecosystem's usual condition, in other words, is to be constantly changing, regardless of whether or not the interference is anthropogenic.[8] However, the idea of preserving wilderness seems to suggest retaining wilderness as a "still-shot": preserving those conditions the landscape maintained prior to the incursion of the European settlers.

A paradox results. The only way to fulfill the purist wilderness vision of changeless preservation would be by actively and intensely managing the tracts of land. However, such actions would not only violate the innately dynamic quality of nature, but would also violate the received view of wilderness as untrammeled.

Fourth, the received view of wilderness uncritically accepts the modernist notion that there is a definite and significant distinction between humans and nature. Many naively believe that we humans exist over-and-against and apart from nature, that something qualitatively unique distinguishes our existence from that of lions, lilies, and lichens.

According to evolutionary theory and basic ecology, the boundary-line separating humans from nature is blurry and tenuous at best. As big precocious apes, humans were and are subject to the same evolutionary and ecological forces, rules, and laws as other living things. Human activities, for better or worse, are *in principle* no less natural than the activities of beavers or pitcher-plants.

Contrary to popular belief, this realization has its environmental advantages. It does not necessarily imply that clear-cuts, ozone depletion, and land development are all okay because they are natural. Quite the contrary. We humans are part of nature, and we have an appropriate role or place in nature. This does not mean that any and all of our environmental modifications are wise or permissible. In fact, many are quite harmful and bad. Just as the actions of over-browsing deer, while "natural," can be bad, many of our actions, while also "natural," can have harmful effects on the biotic community and even on the health and preservation of our own species. If the earth were populated with 5.5 billion acacia-toppling elephants instead of 5.5 billion highly-consumptive

Homo sapiens, then we would have too many elephants and an environmental crisis on our hands, albeit an environmental crisis of a different kind. Just as the environmental impact of deer and elephants can be good or bad, so too can the environmental impact of humans.

As J. Baird Callicott observes, the received view implies that all human interventions in nature are bad:

> . . . measured by the purist wilderness standard, all human impact is bad impact, not because we humans are innately bad, but because we humans are not a part of nature.[9]

But some human environmental impact is arguably good, for instance, rejuvenating a burned-out and environmentally decrepit bit of land or the biologically diversifying effects of Native Americans' pyrotechnology. It would be premature to dismiss all anthropogenic environmental alterations.

Fifth, as thought of in the received view, wilderness no longer exists, if it ever did: it's an ontological reality. There are no places on the earth untouched by human influence. Even Antarctica has not escaped human impact, and the oceans have been changed. Given the fact that our environmental effect is no longer merely on a local point-source level but rather is now global in scope, all of the earth's surface, subsurface, atmospheric, and aquatic regions have been altered by human hands. Acid rain respects no humanly created political boundaries, including wilderness areas. Rivers flow into and out of, and winds blow in and out of, wilderness areas, bringing with them the effluents of humanity. Global warming has altered the chemical composition of the oceans and eventually will cause populations of plant and animal species in wilderness areas to migrate. In other words, human impact, whether direct or indirect, can be seen and felt globally and universally. Hence, to cling to a purist notion of wilderness seems impractical and impossible.

SOME IMPLICATIONS

Unacceptable practical implications flow from these five conceptual muddles of the received wilderness idea. This consequences are problematic and troubling.

Politically speaking, adopting a conservation philosophy centered on wilderness preservation would seem to be a defensive and losing strategy. Our current wilderness areas are under increasing pressure from visitors, timber interests, exotic species invasion, land developers, oil and mining interests, and hydroelectric plans. These areas exist as small and isolated islands of highly vulnerable wild areas amidst a much larger sea of human-dominated landscapes.

As wilderness advocates will tell you, it is a constant uphill battle to have an area designated and remain as wilderness. The burden-of-proof for such designation seems always to lie with the advocate and seldom with those who would despoil the area. According to the received view, there is only an ever-shrinking number of potential wilderness areas. In the words of Aldo Leopold, "Wilderness is a resource which can shrink but not grow."[10] Accepting the received view of wilderness permits only a defensive and backward-looking strategy.

Not only is the received view bad conservation strategy politically, but it also leaves wilderness wide open to interpretation and abuse by the "enemies of wilderness." Given that the burden-of-proof seems to lie with the wilderness proponent, the enemy of wilderness need only provide convincing evidence that an area is not a pristine tract and its wilderness preservation status will be jeopardized and possibly ruined forever.

Currently, wilderness areas are designatable as such because they possess certain qualities. These qualities ostensibly include outstanding recreational opportunities and unique conditions of solitude. In actuality, however, the most prevalent criterion is often the absence of roads. These criteria are potentially dangerous for the following reasons.

First, recreation and solitude are relative and tenuous conditions at best. Clearly, extremely large uninhabited tracts of land would qualify and any large city would not. Would a grassland of 5,000 acres? If an old wagon road winds through it? Moreover, recreation and solitude can be found in settings like artificial climbing walls, isolation chambers, and virtual-reality simulators.

To destroy the possibility of an area being designated as wilderness, an opponent merely needs to show that the area fails to meet the wilderness designation standards. Prairies are poor candidates because they lack many attractive recreational features and are seldom isolated since distant towns and their noises, such as civil service sirens, can be seen and heard. Many prospective places will be disqualified by erstwhile native American habitations and forgotten logging roads. What's worse, many areas can be easily sabotaged, such as by building a radio tower nearby.

A second disturbing result, especially relevant to Australia and Africa, is an implication of the received view for the aboriginal humans who historically and perhaps still utilize a wilderness area. Because wilderness is, according to the received view, unoccupied by humans, humans residing in an area must be regarded as uncivilized barbarians (namely, a kind of wildlife) or removed from the area. Either alternative is unacceptable. Relegating aboriginals to the status of nonhumans was the feeling behind their attempted eradication in Australia. Removing entire peoples from their native environment has repeatedly been shown to be the equivalent of cultural genocide, a tragic case being the consignment of American Indians to so-called reservations.

What is needed is a concept of wilderness that includes certain human activities, especially those of indigenous peoples who have sustainably occupied the area for thousands of years, often longer even than some of the plant and animal species of the region, whose presence and evolution has been intertwined with the activities of the aboriginal humans.

Next, given the received concept, the potential to rejuvenate or resurrect wilderness is impossible. Britain can never have a wilderness. Contrary to the purist ideal, wilderness has actually always been a matter of degree. Due to the nonstatic quality of nature, specific areas can slide up or down a continuum according to their degree of "wildness." Consider, for example, the hauntingly beautiful island of Rhum. Owned by Scottish National Heritage, a conservation organization, Rhum is a mountainous island, roughly thirty-five square miles in size, off the northwest coast of Scotland. The island has had human inhabitants for thousands of years, and has been deforested. It has generally been trammeled until today, including having served as a hunting retreat for exotic red deer and being regularly overgrazed by sheep. According to the received view, Rhum can never be a wilderness. But, once the abusive practices are stopped and Rhum is allowed to go wild, it would seem that Rhum should at least be eligible for some kind of wilderness status, perhaps something like "in the process of becoming wilderness."

In this regard, I disagree with Leopold, who endorsed the purist perspective when he said that "the creation of new wilderness in the full sense of the word is impossible."[11] We know wilderness can be lost—a cedar bog can be turned into a shopping mall. But why can't it be gained? The purist notion begs the question against reclamation. Wilderness would better be conceived as a "process" and not so much as a "product." Instead of looking backward, we could look toward "future nature." When intrusive management regimes are halted, and evolutionary and ecological processes allowed to determine speciation and ecosystem destination, then an area would be in the process of wilderness.

CONCLUDING SUMMARY

The common, received concept of wilderness is that of a pristine place devoid of human habitation and influence. I have argued that this concept is unacceptable for philosophical, historical, scientific, and political reasons. The concept is an unrealistic ideal. Such remote places as the polar regions and the ocean depths may once have qualified—but only until roughly a hundred years ago. Today all of the earth's surface and even subsurface has been irretrievably altered by humans. According to ecological and evolutionary biology, nature is a process, always changing; but, in contrast, the received concept is static. Finally, the received concept has been used politically by the opponents of environmentalism. They argue that no area should be designated wilderness, because the ideal no longer obtains. The received concept also has unacceptable implications for the treatment of aboriginal peoples and their cultures.

Wilderness should be reconceived as a process, I have proposed. Accordingly, places could be reclaimed and resurrected, based on a standard of "wildness." When left alone, devoid of abusive human intrusions, any area could then be "in process of becoming wilderness."

Notes

1. Preparing an anthology on the concept of wilderness gave me the opportunity to explore my own views on this new debate. See J. Baird Callicott and Michael P. Nelson, eds., *The Great New Wilderness Debate* (Athens: University of Georgia Press, 1997).

2. Robert Marshall, "The Problem of Wilderness," *The Scientific Monthly* 30 (1930): 141.

3. *American Heritage Dictionary.*

4. Aldo Leopold, *A Sand County Almanac* (New York: Oxford University Press, 1949), 188.

5. Aldo Leopold, "The Wilderness and Its Place in Forest Recreation Policy," *Journal of Forestry* 19, no. 7 (1921): 719.

6. The Wilderness Act of 1964, Sec. 2(c).

7. The figure is taken from William Denevan, "The Pristine Myth: The Landscape of the Americas in 1492," *Annals of the Association of American Geographers* 82 (1992): 369-85; reprinted in Callicott and Nelson.

8. See, for example, Daniel B. Botkin's two books, *Discordant Harmonies: A New Ecology for the Twenty-First Century* (New York: Oxford University Press, 1990), and *Our National Heritage: The Lessons of Lewis and Clark* (New York: G. P. Putnam's Sons, 1995).

9. J. Baird Callicott, "A Critique of and an Alternative to the Wilderness Idea," *Wild Earth* 4, no. 4 (Winter 1994/95): 56.

10. Leopold, 1949, 199.

11. Ibid., 200.

For Further Discussion

1. What is your own understanding of wilderness? How has it changed after reading Nelson's essay?

2. Nelson suggests that "human activities, for better or worse, are *in principle* no less natural than the activities of beavers or pitcher-plants." Do you agree? Are some human activities "unnatural"? How? How could you distinguish human activities from the activities of other animals?

3. Nelson mentions several unacceptable practical implications of the muddled understanding of wilderness. Which of these do you think are most troubling? Why?

4. How would you manage areas already designated as wilderness areas? What human activities would you allow in these areas? Which not? Why?

The Trouble with Wilderness; or, Getting Back to the Wrong Nature

William Cronon

William Cronon opens his essay with an idea that he says may seem heretical to many environmentalists. He tells us that the time has come to rethink wilderness. Like Michael Nelson in the previous essay, Cronon believes that there are problems with our common and widely accepted understanding of wilderness. Cronon provides us with a historical review of various understandings of the wilderness and argues that, today, wilderness offers us the illusion that we can escape the cares and troubles of the world. This romantic and quasi-religious understanding of the wild serves as the unexamined foundation for much of modern environmentalism. On this view, wilderness is the standard against which the failings of the human world are measured. It is "the natural, unfallen antithesis of an unnatural civilization that has lost its soul." The trouble with such thinking, according to Cronon, is that it "expresses and reproduces the very values its devotees seek to reject." This view of wilderness reinforces a dualistic view of humans as entirely outside the natural world, and this dualism leaves us "little hope of discovering what an ethical, sustainable, honorable human place in nature might actually look like."

The time has come to rethink wilderness.

This will seem a heretical claim to many environmentalists, since the idea of wilderness has for decades been a fundamental tenet—indeed, a passion—of the environmental movement, especially in the United States. For many Americans wilderness stands as the last remaining place where civilization, that all too human disease, has not fully infected the earth. It is an island in the polluted sea of urban-industrial modernity, the one place we can turn for escape from our own too-muchness. Seen in this way, wilderness presents itself as the best antidote to our human selves, a refuge we must somehow recover if we hope to save the planet. As Henry David Thoreau once famously declared, "In Wildness is the preservation of the World."

But is it? The more one knows of its peculiar history, the more one realizes that wilderness is not quite what it seems. Far from being the one place on earth that stands apart from humanity, it is quite profoundly a human creation—indeed, the creation of very particular human cultures at very particular moments in human history. It is not a pristine sanctuary where the last remnant of an untouched, endangered, but still transcendent nature can for at least a little while longer be encountered without the contaminating taint of civilization. Instead, it is a product of that civilization, and could hardly be contaminated by the very stuff of which it is made. Wilderness hides its unnaturalness behind a mask that is all the more beguiling because it seems so natural. As we gaze into the mirror it holds up for us, we too easily imagine that what we behold is Nature when in fact we see the reflection of our own unexamined longings and desires. For this reason, we mistake ourselves when we suppose that wilderness can be the solution to our culture's problematic relationships with the nonhuman world, for wilderness is itself no small part of the problem.

To assert the unnaturalness of so natural a place will no doubt seem absurd or even perverse to many readers, so let me hasten to add that the nonhuman world we encounter in wilderness is far from being merely our own invention. I celebrate with others who love wilderness the beauty and power of the things it contains. Each of us who has spent time there can conjure images and sensations that seem all the more hauntingly real for having engraved themselves so indelibly on our memories. Such memories may be uniquely our own, but they are also familiar enough to be instantly recognizable to others. Remember this? The torrents of mist shoot out from the base of a great waterfall in the depths of a Sierra canyon, the tiny droplets cooling your face as you listen to the roar of the water and gaze up toward the sky through a rainbow that hovers just out of reach. Remember this too: looking out across a desert canyon in the evening air, the only sound a lone raven calling in the distance, the rock walls dropping away into a chasm so deep that its bottom all but vanishes as you squint into the amber light of the setting sun. And this: the moment beside the trail as you sit on a sandstone ledge, your boots damp with the morning dew while you take in the rich smell of the pines, and the small red fox—or maybe for you it was a raccoon or a coyote or a deer—that suddenly ambles across your path, stopping for a long moment to gaze in your direction with cautious indifference before continuing on its way. Remember the feelings of such moments, and you will know as well as I do that you were in the presence of something irreducibly nonhuman, something profoundly Other than yourself. Wilderness is made of that too.

And yet: what brought each of us to the places where such memories became possible is entirely a cultural invention. Go back 250 years in American and European history, and you do not find nearly so many people wandering around remote corners of the planet looking for what today we would call "the wilderness experience." As late as the eighteenth century, the most common usage of the word "wilderness" in the English language referred to landscapes that generally carried adjectives far different from the ones they attract today. To be a wilderness then was to be "deserted," "savage," "desolate," "barren"—in short, a "waste," the word's nearest synonym. Its connotations were anything but positive, and the emotion one was most likely to feel in its presence was "bewilderment"—or terror.

Many of the word's strongest associations then were biblical, for it is used over and over again in the King James Version to refer to places on the margins of civilization where it is all too easy to lose oneself in moral confusion and despair. The wilderness was where Moses had wandered with his people for forty years, and where they had nearly abandoned their God to worship a golden idol. "For Pharaoh will say of the Children of Israel," we read in Exodus, "They are entangled in the land, the wilderness hath shut them in." The wilderness was where Christ had struggled with the devil and endured his temptations: "And immediately the Spirit driveth him into the wilderness. And he was there in the wilderness for forty days tempted of Satan; and was with the wild beasts; and the angels ministered unto him." The "delicious Paradise" of John Milton's Eden was surrounded by "a steep wilderness, whose hairy sides / Access denied" to all who sought entry. When Adam and Eve were driven from that garden, the world they entered was a wilderness that only their labor and pain could redeem. Wilderness, in

short, was a place to which one came only against one's will, and always in fear and trembling. Whatever value it might have arose solely from the possibility that it might be "reclaimed" and turned toward human ends—planted as a garden, say, or a city upon a hill. In its raw state, it had little or nothing to offer civilized men and women.

But by the end of the nineteenth century, all this had changed. The wastelands that had once seemed worthless had for some people come to seem almost beyond price. That Thoreau in 1862 could declare wildness to be the preservation of the world suggests the sea change that was going on. Wilderness had once been the antithesis of all that was orderly and good—it had been the darkness, one might say, on the far side of the garden wall—and yet now it was frequently likened to Eden itself. When John Muir arrived in the Sierra Nevada in 1869, he would declare, "No description of Heaven that I have ever heard or read of seems half so fine." He was hardly alone in expressing such emotions. One by one, various corners of the American map came to be designated as sites whose wild beauty was so spectacular that a growing number of citizens had to visit and see them for themselves. Niagara Falls was the first to undergo this transformation, but it was soon followed by the Catskills, the Adirondacks, Yosemite, Yellowstone, and others. Yosemite was deeded by the U.S. government to the state of California in 1864 as the nation's first wildland park, and Yellowstone became the first true national park in 1872.

By the first decade of the twentieth century, in the single most famous episode in American conservation history, a national debate had exploded over whether the city of San Francisco should be permitted to augment its water supply by damming the Tuolumne River in Hetch Hetchy valley, well within the boundaries of Yosemite National Park. The dam was eventually built, but what today seems no less significant is that so many people fought to prevent its completion. Even as the fight was being lost, Hetch Hetchy became the battle cry of an emerging movement to preserve wilderness. Fifty years earlier, such opposition would have been unthinkable. Few would have questioned the merits of "reclaiming" a wasteland like this in order to put it to human use. Now the defenders of Hetch Hetchy attracted widespread national attention by portraying such an act not as improvement or progress but as desecration and vandalism. Lest one doubt that the old biblical metaphors had been turned completely on their heads, listen to John Muir attack the dam's defenders. "Their arguments," he wrote, "are curiously like those of the devil, devised for the destruction of the first garden—so much of the very best Eden fruit going to waste; so much of the best Tuolumne water and Tuolumne scenery going to waste." For Muir and the growing number of Americans who shared his views, Satan's home had become God's own temple.

The sources of this rather astonishing transformation were many, but for the purposes of this essay they can be gathered under two broad headings: the sublime and the frontier. Of the two, the sublime is the older and more pervasive cultural construct, being one of the most important expressions of that broad transatlantic movement we today label as romanticism; the frontier is more peculiarly American, though it too had its European antecedents and parallels. The two converged to remake wilderness in their own image, freighting it with moral values and cultural symbols that it carries to this day. Indeed, it is not too much to say that the modern environmental movement is itself a grandchild of romanticism and post-frontier ideology, which is why it is no accident that so much environmentalist discourse takes its bearings from the wilderness these intellectual movements helped create. Although wilderness may today seem to be just one environmental concern among many, it in fact serves as the foundation for a long list of other such concerns that on their face seem quite remote from it. That is why its influence is so pervasive and, potentially, so insidious.

To gain such remarkable influence, the concept of wilderness had to become loaded with some of the deepest core values of the culture that created and idealized it: it had to become sacred. This possibility had been present in wilderness even in the days when it had been a place of spiritual danger and moral temptation. If Satan was there, then so was Christ, who had found angels as well as wild beasts during His sojourn in the desert. In the wilderness the boundaries between human and nonhuman, between natural and supernatural, had always seemed less certain than elsewhere. This was why the early Christian saints and mystics had often emulated Christ's desert retreat as they sought to experience for themselves the visions and spiritual testing He had endured. One might meet devils and run the risk of losing one's soul in such a place, but one might also meet God. For some that possibility was worth almost any price.

By the eighteenth century this sense of the wilderness as a landscape where the supernatural lay just beneath the surface was expressed in the doctrine of the *sublime,* a word whose modern usage has been so watered down by commercial hype and tourist advertising that it retains only a dim echo of its former power. In the theories of Edmund Burke, Immanuel Kant, William Gilpin, and others, sublime landscapes were those rare places on earth where one had more chance than elsewhere to glimpse the face of God. Romantics had a clear notion of where one could be most sure of having this experience. Although God might, of course, choose to show Himself anywhere, He would most often be found in those vast, powerful landscapes where one could not help feeling insignificant and being reminded of one's own mortality. Where were these sublime places? The eighteenth-century catalog of their locations feels very familiar, for we still see and value landscapes as it taught us to do. God was on the mountaintop, in the chasm, in the waterfall, in the thundercloud, in the rainbow, in the sunset. One has only to think of the sites that Americans chose for their first national parks—Yellowstone, Yosemite, Grand Canyon, Rainier, Zion—to realize that virtually all of them fit one or more of these categories. Less sublime landscapes simply did not appear worthy of such protection; not until the 1940s, for instance, would the first swamp be honored, in Everglades National Park, and to this day there is no national park in the grasslands.

Among the best proofs that one had entered a sublime landscape was the emotion it evoked. For the early romantic writers and artists who first began to celebrate it, the sublime was far from being a pleasurable experience. The classic description is that of William Wordsworth as he recounted climbing the Alps and crossing the Simplon Pass in his autobiographical poem *The Prelude.* There, surrounded by crags and waterfalls, the poet felt himself literally to be in the presence of the divine—and experienced an emotion remarkably close to terror:

> The immeasurable height
> Of woods decaying, never to be decayed,
> The stationary blasts of waterfalls,
> And in the narrow rent at every turn
> Winds thwarting winds, bewildered and forlorn,
> The torrents shooting from the clear blue sky,
> The rocks that muttered close upon our ears,
> Black drizzling crags that spake by the way-side
> As if a voice were in them, the sick sight
> And giddy prospect of the raving stream,
> The unfettered clouds and region of the Heavens,

Tumult and peace, the darkness and the light—
Were all like workings of one mind, the features
Of the same face, blossoms upon one tree;
Characters of the great Apocalypse,
The types and symbols of Eternity,
Of first, and last, and midst, and without end.

This was no casual stroll in the mountains, no simple sojourn in the gentle lap of nonhuman nature. What Wordsworth described was nothing less than a religious experience, akin to that of the Old Testament prophets as they conversed with their wrathful God. The symbols he detected in this wilderness landscape were more supernatural than natural, and they inspired more awe and dismay than joy or pleasure. No mere mortal was meant to linger long in such a place, so it was with considerable relief that Wordsworth and his companion made their way back down from the peaks to the sheltering valleys.

Lest you suspect that this view of the sublime was limited to timid Europeans who lacked the American know-how for feeling at home in the wilderness, remember Henry David Thoreau's 1846 climb of Mount Katahdin, in Maine. Although Thoreau is regarded by many today as one of the great American celebrators of wilderness, his emotions about Katahdin were no less ambivalent than Wordsworth's about the Alps.

> It was vast, Titanic, and such as man never inhabits. Some part of the beholder, even some vital part, seems to escape through the loose grating of his ribs as he ascends. He is more lone than you can imagine. . . .Vast, Titanic, inhuman Nature has got him at disadvantage, caught him alone, and pilfers him of some of his divine faculty. She does not smile on him as in the plains. She seems to say sternly, why came ye here before your time? This ground is not prepared for you. Is it not enough that I smile in the valleys? I have never made this soil for thy feet, this air for thy breathing, these rocks for thy neighbors. I cannot pity nor fondle thee here, but forever relentlessly drive thee hence to where I *am* kind. Why seek me where I have not called thee, and then complain because you find me but a stepmother?

This is surely not the way a modern backpacker or nature lover would describe Maine's most famous mountain, but that is because Thoreau's description owes as much to Wordsworth and other romantic contemporaries as to the rocks and clouds of Katahdin itself. His words took the physical mountain on which he stood and transmuted it into an icon of the sublime: a symbol of God's presence on earth. The power and the glory of that icon were such that only a prophet might gaze on it for long. In effect, romantics like Thoreau joined Moses and the children of Israel in Exodus when "they looked toward the wilderness, and behold, the glory of the Lord appeared in the cloud."

But even as it came to embody the awesome power of the sublime, wilderness was also being tamed—not just by those who were building settlements in its midst but also by those who most celebrated its inhuman beauty. By the second half of the nineteenth century, the terrible awe that Wordsworth and Thoreau regarded as the appropriately pious stance to adopt in the presence of their mountaintop God was giving way to a much more comfortable, almost sentimental demeanor. As more and more tourists sought out the wilderness as a spectacle to be looked at and enjoyed for its great beauty, the sublime in effect became domesticated. The wilderness was still sacred, but the religious sentiments it evoked were more those of a pleasant parish church than those of a grand cathedral or a harsh desert retreat. The writer who best captures this late romantic sense of a domesticated sublime is undoubtedly John Muir, whose descriptions of Yosemite and the Sierra Nevada reflect none of the anxiety or terror one finds in earlier writers. Here he is, for instance, sketching on North Dome in Yosemite Valley:

> No pain here, no dull empty hours, no fear of the past, no fear of the future. These blessed mountains are so compactly filled with God's beauty, no petty personal hope or experience has room to be. Drinking this champagne water is pure pleasure, so is breathing the living air, and every movement of limbs is pleasure, while the body seems to feel beauty when exposed to it as it feels the campfire or sunshine, entering not by the eyes alone, but equally through all one's flesh like radiant heat, making a passionate ecstatic pleasure glow not explainable.

The emotions Muir describes in Yosemite could hardly be more different from Thoreau's on Katahdin or Wordsworth's on the Simplon Pass. Yet all three men are participating in the same cultural tradition and contributing to the same myth: the mountain as cathedral. The three may differ in the way they choose to express their piety—Wordsworth favoring an awe-filled bewilderment, Thoreau a stern loneliness, Muir a welcome ecstasy—but they agree completely about the church in which they prefer to worship. Muir's closing words on North Dome diverge from his older contemporaries only in mood, not in their ultimate content:

> Perched like a fly on this Yosemite dome, I gaze and sketch and bask, often-times settling down into dumb admiration without definite hope of ever learning much, yet with the longing, unresting effort that lies at the door of hope, humbly prostrate before the vast display of God's power, and eager to offer self-denial and renunciation with eternal toil to learn any lesson in the divine manuscript.

Muir's "divine manuscript" and Wordsworth's "Characters of the great Apocalypse" were in fact pages from the same holy book. The sublime wilderness had ceased to be a place of satanic temptation and become instead a sacred temple, much as it continues to be for those who love it today.

But the romantic sublime was not the only cultural movement that helped transform wilderness into a sacred American icon during the nineteenth century. No less important was the powerful romantic attraction of primitivism, dating back at least to Rousseau—the belief that the best antidote to the ills of an overly refined and civilized modern world was a return to simpler, more primitive living. In the United States, this was embodied most strikingly in the national myth of the frontier. The historian Frederick Jackson Turner wrote in 1893 the classic academic statement of this myth, but it has been part of American cultural traditions for well over a century. As Turner described the process, easterners and European immigrants, in moving to the wild unsettled lands of the frontier, shed the trappings of civilization, rediscovered their primitive racial energies, reinvented direct democratic institutions, and thereby reinfused themselves with a vigor, an independence, and a creativity that were the source of American democracy and national character. Seen in this way, wild country became a place not just of religious redemption but of national renewal, the quintessential location for experiencing what it meant to be an American.

One of Turner's most provocative claims was that by the 1890s the frontier was passing away. Never again would "such gifts of free land offer themselves" to the American people. "The frontier has gone," he declared, "and with its going has closed the first period of American history." Built into the frontier myth from its very beginning was the notion that this crucible of American identity was temporary and would pass away. Those who have celebrated the frontier have almost always looked backward as they did so, mourning an older, simpler, truer world that is about to disappear forever. That world and all of its attractions, Turner said, depended on free land—on wilderness. Thus, in the myth of the vanishing frontier lay the seeds of wilderness preservation in the United States, for if wild land had been so crucial in the making of the nation, then surely one must save its last remnants as monuments to the American past—and as an insurance policy to protect its future. It is no accident that the movement to set aside national parks and wilderness areas began to gain real momentum at precisely the time that laments about the passing frontier reached their peak. To protect wilderness was in a very real sense to protect the nation's most sacred myth of origin.

Among the core elements of the frontier myth was the powerful sense among certain groups of Americans that wilderness was the last bastion of rugged individualism. Turner tended to stress communitarian themes when writing frontier history, asserting that Americans in primitive conditions had been forced to band together with their neighbors to form communities and democratic institutions. For other writers, however, frontier democracy for communities was less compelling than frontier freedom for individuals. By fleeing to the outer margins of settled land and society—so the story ran—an individual could escape the confining strictures of civilized life. The mood among writers who celebrated frontier individualism was almost always nostalgic; they lamented not just a lost way of life but the passing of the heroic men who had embodied that life. Thus Owen Wister in the introduction to his classic 1902 novel *The Virginian* could write of "a vanished world" in which "the horseman, the cow-puncher, the last romantic figure upon our soil" rode only "in his historic yesterday" and would "never come again." For Wister, the cowboy was a man who gave his word and kept it ("Wall Street would have found him behind the times"), who did not talk lewdly to women ("Newport would have thought him old-fashioned"), who worked and played hard, and whose "ungoverned hours did not unman him." Theodore Roosevelt wrote with much the same nostalgic fervor about the "fine, manly qualities" of the "wild rough-rider of the plains." No one could be more heroically masculine, thought Roosevelt, or more at home in the western wilderness:

There he passes his days, there he does his life-work, there, when he meets death, he faces it as he has faced many other evils, with quiet, uncomplaining fortitude. Brave, hospitable, hardy, and adventurous, he is the grim pioneer of our race; he prepares the way for the civilization from before whose face he must himself disappear. Hard and dangerous though his existence is, it has yet a wild attraction that strongly draws to it his bold, free spirit.

This nostalgia for a passing frontier way of life inevitably implied ambivalence, if not downright hostility, toward modernity and all that it represented. If one saw the wild lands of the frontier as freer, truer, and more natural than other, more modern places, then one was also inclined to see the cities and factories of urban-industrial civilization as confining, false, and artificial. Owen Wister looked at the post-frontier "transition" that had followed "the horseman of the plains," and did not like what he saw: "a shapeless state, a condition of men and manners as unlovely as is that moment in the year when winter is gone and spring not come, and the face of Nature is ugly." In the eyes of writers who shared Wister's distaste for modernity, civilization contaminated its inhabitants and absorbed them into the faceless, collective, contemptible life of the crowd. For all of its troubles and dangers, and despite the fact that it must pass away, the frontier had been a better place. If civilization was to be redeemed, it would be by men like the Virginian who could retain their frontier virtues even as they made the transition to post-frontier life.

The mythic frontier individualist was almost always masculine in gender: here, in the wilderness, a man could be a real man, the rugged individual he was meant to be before civilization sapped his energy and threatened his masculinity. Wister's contemptuous remarks about Wall Street and Newport suggest what he and many others of his generation believed—that the comforts and seductions of civilized life were especially insidious for men, who all too easily became emasculated by the feminizing tendencies of civilization. More often than not, men who felt this way came, like Wister and Roosevelt, from elite class backgrounds. The curious result was that frontier nostalgia became an important vehicle for expressing a peculiarly bourgeois form of antimodernism. The very men who most benefited from urban-industrial capitalism were among those who believed they must escape its debilitating effects. If the frontier was passing, then men who had the means to do so should preserve for themselves some remnant of its wild landscape so that they might enjoy the regeneration and renewal that came from sleeping under the stars, participating in blood sports, and living off the land. The frontier might be gone, but the frontier experience could still be had if only wilderness were preserved.

Thus the decades following the Civil War saw more and more of the nation's wealthiest citizens seeking out wilderness for themselves. The elite passion for wild land took many forms: enormous estates in the Adirondacks and elsewhere (disingenuously called "camps" despite their many servants and amenities), cattle ranches for would-be rough riders on the Great Plains, guided big-game hunting trips in the Rockies, and luxurious resort hotels wherever railroads pushed their way into sublime landscapes. Wilderness suddenly emerged as the landscape of choice for elite tourists, who brought with them strikingly urban ideas of the countryside through which they traveled. For them, wild land was not a site for productive labor and not a permanent home; rather, it was a place of recreation. One went to the wilderness not as a producer but as a consumer, hiring guides and other backcountry residents who could serve as romantic surrogates for the rough riders and hunters of the frontier if one was willing to overlook their new status as employees and servants of the rich.

In just this way, wilderness came to embody the national frontier myth, standing for the wild freedom of America's past and seeming to represent a highly attractive natural alternative to the ugly artificiality of modern civilization. The irony, of course, was that in the process wilderness came to reflect the very civilization its devotees sought to escape. Ever since the nineteenth century, celebrating wilderness has been an activity mainly for well-to-do city folks. Country people generally know far too much about working the land to regard unworked land as their ideal. In contrast, elite urban tourists and wealthy sportsmen projected their leisure-time frontier fantasies onto the American landscape and so created wilderness in their own image.

There were other ironies as well. The movement to set aside national parks and wilderness areas followed hard on the heels of the final Indian wars, in which the prior human inhabitants of these areas were rounded up and moved onto reservations. The myth of the wilderness as "virgin," uninhabited land had always been especially cruel when seen from the perspective of the Indians who had once called that land home. Now they were forced to move elsewhere, with the result that tourists could safely enjoy the illusion that they were seeing their nation in its pristine, original state, in the new morning of God's own creation. Among the things that most marked the new national parks as reflecting a post-frontier consciousness was the relative absence of human violence within their boundaries. The actual frontier had often been a place of conflict, in which invaders and invaded fought for control of land and resources. Once set aside within the fixed and carefully policed boundaries of the modern bureaucratic state, the wilderness lost its savage image and became safe: a place more of reverie than of revulsion or fear. Meanwhile, it original inhabitants were kept out by dint of force, their earlier uses of the land redefined as inappropriate or even illegal. To this day, for instance, the Blackfeet continue to be accused of "poaching" on the lands of Glacier National Park that originally belonged to them and that were ceded by treaty only with the proviso that they be permitted to hunt there.

The removal of Indians to create an "uninhabited wilderness"—uninhabited as never before in the human history of the place—reminds us just how invented, just how constructed, the American wilderness really is. To return to my opening argument: there is nothing natural about the concept of wilderness. It is entirely a creation of the culture that holds it dear, a product of the very history it seeks to deny. Indeed, one of the most striking proofs of the cultural invention of wilderness is its thoroughgoing erasure of the history from which it sprang. In virtually all of its manifestations, wilderness represents a flight from history. Seen as the original garden, it is a place outside of time, from which human beings had to be ejected before the fallen world of history could properly begin. Seen as the frontier, it is a savage world at the dawn of civilization, whose transformation represents the very beginning of the national historical epic. Seen as the bold landscape of frontier heroism, it is the place of youth and childhood, into which men escape by abandoning their pasts and entering a world of freedom where the constraints of civilization fade into memory. Seen as the sacred sublime, it is the home of a God who transcends history by standing as the One who remains untouched and unchanged by time's arrow. No matter what the angle from which we regard it, wilderness offers us the illusion that we can escape the cares and troubles of the world in which our past has ensnared us.

This escape from history is one reason why the language we use to talk about wilderness is often perme-

ated with spiritual and religious values that reflect human ideals far more than the material world of physical nature. Wilderness fulfills the old romantic project of secularizing Judeo-Christian values so as to make a new cathedral not in some petty human building but in God's own creation, Nature itself. Many environmentalists who reject traditional notions of the Godhead and who regard themselves as agnostics or even atheists nonetheless express feelings tantamount to religious awe when in the presence of wilderness— a fact that testifies to the success of the romantic project. Those who have no difficulty seeing God as the expression of our human dreams and desires nonetheless have trouble recognizing that in a secular age Nature can offer precisely the same sort of mirror.

Thus it is that wilderness serves as the unexamined foundation on which so many of the quasi-religious values of modern environmentalism rest. The critique of modernity that is one of environmentalism's most important contributions to the moral and political discourse of our time more often than not appeals, explicitly or implicitly, to wilderness as the standard against which to measure the failings of our human world. Wilderness is the natural, unfallen antithesis of an unnatural civilization that has lost its soul. It is a place of freedom in which we can recover the true selves we have lost to the corrupting influences of our artificial lives. Most of all, it is the ultimate landscape of authenticity. Combining the sacred grandeur of the sublime with the primitive simplicity of the frontier, it is the place where we can see the world as it really is, and so know ourselves as we really are—or ought to be.

But the trouble with wilderness is that it quietly expresses and reproduces the very values its devotees seek to reject. The flight from history that is very nearly the core of wilderness represents the false hope of an escape from responsibility, the illusion that we can somehow wipe clean the slate of our past and return to the tabula rasa that supposedly existed before we began to leave our marks on the world. The dream of an unworked natural landscape is very much the fantasy of people who have never themselves had to work the land to make a living—urban folk for whom food comes from a supermarket or a restaurant instead of a field, and for whom the wooden houses in which they live and work apparently have no meaningful connection to the forests in which trees grow and die. Only people whose relation to the land was already alienated could hold up wilderness as a model for human life in nature, for the romantic ideology of wilderness leaves precisely nowhere for human beings actually to make their living from the land.

This, then, is the central paradox: wilderness embodies a dualistic vision in which the human is entirely outside the natural. If we allow ourselves to believe that nature, to be true, must also be wild, then our very presence in nature represents its fall. The place where we are is the place where nature is not. If this is so—if by definition wilderness leaves no place for human beings, save perhaps as contemplative sojourners enjoying their leisurely reverie in God's natural cathedral—then also by definition it can offer no solution to the environmental and other problems that confront us. To the extent that we celebrate wilderness as the measure with which we judge civilization, we reproduce the dualism that sets humanity and nature at opposite poles. We thereby leave ourselves little hope of discovering what an ethical, sustainable, *honorable* human place in nature might actually look like.

Worse: to the extent that we live in an urban-industrial civilization but at the same time pretend to ourselves that our *real* home is in the wilderness, to just that extent we give ourselves permission to evade responsibility for the lives we actually lead. We inhabit civilization while holding some part of ourselves—what we imagine to be the most precious part—aloof from its entanglements. We work our nine-to-five jobs in its institutions, we eat its food, we drive its cars (not least to reach the wilderness), we benefit from the intricate and all too invisible networks with which it shelters us, all the while pretending that these things are not an essential part of who we are. By imagining that our true home is in the wilderness, we forgive ourselves the homes we actually inhabit. In its flight from history, in its siren song of escape, in its reproduction of the dangerous dualism that sets human beings outside of nature—in all of these ways, wilderness poses a serious threat to responsible environmentalism at the end of the twentieth century.

By now I hope it is clear that my criticism in this essay is not directed at wild nature per se, or even at efforts to set aside large tracts of wild land, but rather at the specific habits of thinking that flow from this complex cultural construction called wilderness. It is not the things we label as wilderness that are the problem—for nonhuman nature and large tracts of the natural world *do* deserve protection—but rather what we ourselves mean when we use that label. Lest one doubt how pervasive these habits of thought actually are in contemporary environmentalism, let me list some of the places where wilderness serves as the ideological underpinning for environmental concerns that might otherwise seem quite remote from it. Defenders of biological diversity, for instance, although sometimes appealing to more utilitarian concerns, often point to "untouched" ecosystems as the best and richest repositories of the undiscovered species we must certainly try to protect. Although at first blush an apparently more "scientific" concept than wilderness, biological diversity in fact invokes many of the same sacred values, which is why organizations like the Nature Conservancy have been so quick to employ it as an alternative to the seemingly fuzzier and more problematic concept of wilderness. There is a paradox here, of course. To the extent that biological diversity (indeed, even wilderness itself) is likely to survive in the future only by the most vigilant and self-conscious management of the ecosystems that sustain it, the ideology of wilderness is potentially in direct conflict with the very thing it encourages us to protect.

The most striking instances of this have revolved around "endangered species," which serve as vulnerable symbols of biological diversity while at the same time standing as surrogates for wilderness itself. The terms of the Endangered Species Act in the United States have often meant that those hoping to defend pristine wilderness have had to rely on a single endangered species like the spotted owl to gain legal standing for their case—thereby making the full power of sacred land inhere in a single numinous organism whose habitat then becomes the object of intense debate about appropriate management and use. The ease with which anti-environmental forces like the wise-use movement have attacked such single-species preservation efforts suggests the vulnerability of strategies like these.

Perhaps partly because our own conflicts over such places and organisms have become so messy, the convergence of wilderness values with concerns about biological diversity and endangered species has helped produce a deep fascination for remote ecosystems, where it is easier to imagine that nature might somehow be "left alone" to flourish by its own pristine devices. The classic example is the tropical rain forest, which since the 1970s has become the most powerful modern icon of unfallen, sacred land—a veritable Garden of Eden—for many Americans and Europeans. And yet protecting the rain forest in the eyes of First World environmentalists all too often means protecting it from the people who live there. Those who seek to preserve such "wilderness" from the activities of native peoples run the risk of reproducing the same tragedy—being forceably removed from an ancient home—that befell American Indians. Third World countries face massive environmental problems and deep social conflicts, but these are not likely to be solved by a cultural myth that encourages us to "preserve" peopleless landscapes that have not existed in such places for millennia. At its worse, as environmentalists are beginning to realize, exporting American notions of wilderness in this way can become an unthinking and self-defeating form of cultural imperialism.

Perhaps the most suggestive example of the way that wilderness thinking can underpin other environmental concerns has emerged in the recent debate about "global change." In 1989 the journalist Bill McKibben published a book entitled *The End of Nature,* in which he argued that the prospect of global climate change as a result of unintentional human manipulation of the atmosphere means that nature as we once knew it no longer exists. Whereas earlier generations inhabited a natural world that remained more or less unaffected by their actions, our own generation is uniquely different. We and our children will henceforth live in a biosphere completely altered by our own activity, a planet in which the human and the natural can no longer be distinguished, because the one has overwhelmed the other. In McKibben's view, nature has died, and we are responsible for killing it. "The planet," he declares, "is utterly different now."

But such a perspective is possible only if we accept the wilderness premise that nature, to be natural, must also be pristine—remote from humanity and untouched by our common past. In fact, everything we know about environmental history suggests that people have been manipulating the natural world on various scales for as long as we have a record of their passing. Moreover, we have unassailable evidence that many of the environmental changes we now face also occurred quite apart from human intervention at one time or another in the earth's past. The point is not that our current problems are trivial, or that our devastating effects on the earth's ecosystems should be accepted as inevitable or "natural." It is rather that we seem unlikely to make much progress in solving these problems if we hold up to ourselves as the mirror of nature a wilderness we ourselves cannot inhabit.

To do so is merely to take to a logical extreme the paradox that was built into wilderness from the beginning: if nature dies because we enter it, then the only way to save nature is to kill ourselves. The absurdity of this proposition flows from the underlying dualism it expresses. Not only does it ascribe greater power to humanity than we in fact possess—physical and biological nature will surely survive in some form or another long after we ourselves have gone the way of all flesh—but in the end it offers us little more than a self-defeating counsel of despair. The tautology gives us no way out: if wild nature is the only thing worth saving, and if our mere presence destroys it, then the sole solution to our own unnaturalness, the only way to protect sacred wilderness from profane humanity, would seem to be suicide. It is not a proposition that seems likely to produce very positive or practical results.

And yet radical environmentalists and deep ecologists all too frequently come close to accepting this premise as a first principle. When they express, for instance, the popular notion that our environmental problems began with the invention of agriculture, they push the human fall from natural grace so far back into the past that all of civilized history becomes a tale of ecological declension. Earth First! founder Dave Foreman captures the familiar parable succinctly when he writes,

> Before agriculture was midwifed in the Middle East, humans were in the wilderness. We had no concept of "wilderness" because everything was wilderness and *we were a part of it.* But with irrigation ditches, crop surpluses, and permanent villages, we became *apart from* the natural world. . . . Between the wilderness that created us and the civilization created by us grew an ever-widening rift.

In this view the farm becomes the first and most important battlefield in the long war against wild nature, and all else follows in its wake. From such a starting place, it is hard not to reach the conclusion that the only way human beings can hope to live naturally on earth is to follow the hunter-gatherers back into a wilderness Eden and abandon virtually everything that civilization has given us. It may indeed turn out that civilization will end in ecological collapse or nuclear disaster, whereupon one might expect to find any human survivors returning to a way of life closer to that celebrated by Foreman and his followers. For most of us, though, such a debacle would be cause for regret, a sign that humanity had failed to fulfill its own promise and failed to honor its own highest values—including those of the deep ecologists.

In offering wilderness as the ultimate hunter-gatherer alternative to civilization, Foreman reproduces an extreme but still easily recognizable version of the myth of frontier primitivism. When he writes of his fellow Earth Firsters that "we believe we must return to being animal, to glorying in our sweat, hormones, tears, and blood" and that "we struggle against the modern compulsion to become dull, passionless androids," he is following in the footsteps of Owen Wister. Although his arguments give primacy to defending biodiversity and the autonomy of wild nature, his prose becomes most passionate when he speaks of preserving "the wilderness experience." His own ideal "Big Outside" bears an uncanny resemblance to that of the frontier myth: wide open spaces and virgin land with no trails, no signs, no facilities, no maps, no guides, no rescues, no modern equipment. Tellingly, it is a land where hardy travelers can support themselves by hunting with "primitive weapons (bow and arrow, atlatl, knife, sharp rock)." Foreman claims that "the primary value of wilderness is not as a proving ground for young Huck Finns and Annie Oakleys," but his heart is with Huck and Annie all the same. He admits that "preserving a quality wilderness experience for the human visitor, letting her or him flex Paleolithic muscles or seek visions, remains a tremendously important secondary purpose." Just so does Teddy Roosevelt's rough rider live on in the greener garb of a new age.

However much one may be attracted to such a vision, it entails problematic consequences. For one, it makes wilderness the locus for an epic struggle between malign civilization and benign nature, compared with which all other social, political, and moral concerns seem trivial. Foreman writes, "The preservation of wildness and native diversity is *the* most important issue. Issues directly affecting only humans pale in comparison." Presumably so do any environmental problems whose victims are mainly people, for such problems usually surface in landscapes that have already "fallen" and are no longer wild. This would seem to exclude from the radical environmentalist agenda problems of occupational health and safety in industrial settings, problems of toxic waste exposure on "unnatural" urban and agricultural sites, problems of poor children poisoned by lead exposure in the inner city, problems of famine and poverty and human suffering in the "overpopulated" places of the earth—problems, in short, of environmental justice. If we set too high a stock on wilderness, too many other corners of the earth become less than natural and too many other people become less than human, thereby giving us permission not to care much about their suffering or their fate.

It is not accident that these supposedly inconsequential environmental problems affect mainly poor people, for the long affiliation between wilderness and wealth means that the only poor people who count when wilderness is *the* issue are hunter-gatherers, who presumably do not consider themselves to be poor in the first place. The dualism at the heart of wilderness encourages its advocates to conceive of its protection as a crude conflict between the "human" and the "nonhuman"—or, more often, between those who value the nonhuman and those who do not. This in turn tempts one to ignore crucial differences *among* humans and the complex cultural and historical reasons why different peoples may feel very differently about the meaning of wilderness.

Why, for instance, is the "wilderness experience" so often conceived as a form of recreation best enjoyed by those whose class privileges give them the time and resources to leave their jobs behind and "get away from it all"? Why does the protection of wilderness so often seem to pit urban recreationists against rural people who actually earn their living from the land (excepting those who sell goods and services to the tourists themselves)? Why in the debates about pristine natural areas are "primitive" peoples idealized, even sentimentalized, until the moment they do something unprimitive, modern, and unnatural, and thereby fall from environmental grace? What are the consequences of a wilderness ideology that devalues productive labor and the very concrete knowledge that comes from working the land with one's own hands? All of these questions imply conflicts among different groups of people, conflicts that are obscured behind the deceptive clarity of "human" vs. "nonhuman." If in answering these knotty questions we resort to so simplistic an opposition, we are almost certain to ignore the very subtleties and complexities we need to understand.

But the most troubling cultural baggage that accompanies the celebration of wilderness has less to do with remote rain forests and peoples than with the ways we think about ourselves—we American environmentalists who quite rightly worry about the future of the earth and the threats we pose to the natural world. Idealizing a distant wilderness too often means not idealizing the environment in which we actually live, the landscape that for better or worse we call home. Most of our most serious environmental problems start right here, at home, and if we are to solve those problems, we need an environmental ethic that will tell us as much about *using* nature as about *not* using it. The wilderness dualism tends to cast any use as *ab*-use, and thereby denies us a middle ground in which responsible use and non-use might attain some kind of balanced, sustainable relationship. My own belief is that only by exploring this middle ground will we learn ways of imagining a better world for all of us: humans and nonhumans, rich people and poor, women and men, First Worlders and Third Worlders, white folks and people of color, consumers and producers—a world better for humanity in all of its diversity and for all the rest of nature too. The middle ground is where we actually live. It is where we—all of us, in our different places and ways—make our homes.

That is why, when I think of the times I myself have come closest to experiencing what I might call the sacred in nature, I often find myself remembering wild places much closer to home. I think, for instance, of a small pond near my house where water bubbles up from limestone springs to feed a series of pools that rarely freeze in winter and so play home to waterfowl that stay here for the protective warmth even on the coldest of winter days, gliding silently through steaming mists as the snow falls from gray February skies. I think of a November evening long ago when I found myself on a Wisconsin hilltop in rain and dense fog, only to have the setting sun break through the clouds to cast an otherworldly golden light on the misty farms and woodlands below, a scene so unexpected and joyous that I lingered past dusk so as not to miss any part of the gift that had come my way. And I think perhaps most especially of the blown-out, bankrupt farm in the sand country of central Wisconsin where Aldo Leopold and his family tried one of the first American experiments in ecological restoration, turning ravaged and infertile soil into carefully tended ground where the human and the nonhuman could exist side by side in relative harmony. What I celebrate about such places is not *just* their wildness, though that certainly is among their most important qualities; what I celebrate even more is that they remind us of the wildness in our own backyards, of the nature that is all around us if only we have eyes to see it.

Indeed, my principal objection to wilderness is that it may teach us to be dismissive or even contemptuous of such humble places and experiences. Without our

quite realizing it, wilderness tends to privilege some parts of nature at the expense of others. Most of us, I suspect, still follow the conventions of the romantic sublime in finding the mountaintop more glorious than the plains, the ancient forest nobler than the grasslands, the mighty canyon more inspiring than the humble marsh. Even John Muir, in arguing against those who sought to dam his beloved Hetch Hetchy valley in the Sierra Nevada, argued for alternative dam sites in the gentler valleys of the foothills—a preference that had nothing to do with nature and everything with the cultural traditions of the sublime. Just as problematically, our frontier traditions have encouraged Americans to define "true" wilderness as requiring very large tracts of roadless land—what Dave Foreman calls "The Big Outside." Leaving aside the legitimate empirical question in conservation biology of how large a tract of land must be before a given species can reproduce on it, the emphasis on big wilderness reflects a romantic frontier belief that one hasn't really gotten away from civilization unless one can go for days at a time without encountering another human being. By teaching us to fetishize sublime places and wide open country, these peculiarly American ways of thinking about wilderness encourage us to adopt too high a standard for what counts as "natural." If it isn't hundreds of square miles big, if it doesn't give us God's-eye views or grand vistas, if it doesn't permit us the illusion that we are alone on the planet, then it really isn't natural. It's too small, too plain, or too crowded to be *authentically* wild.

In critiquing wilderness as I have done in this essay, I'm forced to confront my own deep ambivalence about its meaning for modern environmentalism. On the one hand, one of my own most important environmental ethics is that people should always be conscious that they are part of the natural world, inextricably tied to the ecological systems that sustain their lives. Any way of looking at nature that encourages us to believe we are separate from nature—as wilderness tends to do—is likely to reinforce environmentally irresponsible behavior. On the other hand, I also think it no less crucial for us to recognize and honor nonhuman nature as a world we did not create, a world with its own independent, nonhuman reasons for being as it is. The autonomy of nonhuman nature seems to me an indispensable corrective to human arrogance. Any way of looking at nature that helps us remember—as wilderness also tends to do—that the interests of people are not necessarily identical to those of every other creature or of the earth itself is likely to foster *responsible* behavior. To the extent that wilderness has served as an important vehicle for articulating deep moral values regarding our obligations and responsibilities to the nonhuman world, I would not want to jettison the contributions it has made to our culture's ways of thinking about nature.

. . .

For Further Discussion

1. Compare Cronon's analysis of the idea of wilderness to Nelson's views. On what issues might they agree? Would they disagree?

2. Cronon describes two sources as being responsible for transforming our understanding of the wilderness as deserving preservation. One source is characterized as "sublime" and the other as "frontier." How do you understand these concepts? What does Cronon mean by sublime?

3. Cronon claims that "the trouble with wilderness is that it quietly expresses and reproduces the very values its devotees seek to reject." How do you interpret this statement? Do you agree with his analysis?

4. "When I think of the times I myself have come closest to experiencing what I might call the sacred in nature, I often find myself remembering wild places much closer to home." Can a small pond be thought of as a wilderness area? What differences exist between a small pond less than a mile from a neighborhood and a similar pond many miles from any human habitation?

The Wilderness Idea Reaffirmed

Holmes Rolston III

In this final selection of Chapter 11, Holmes Rolston comes to the defense of the concept of wilderness. Rolston argues that wild nature does differ significantly from human culture and that things of great value might be lost if we confuse this distinction. Rolston's essay was written in response to an article by philosopher J. Baird Callicott. Like Nelson and Cronon, Callicott had argued against the coherence of the received view of wilderness. Rolston finds numerous characteristics and values of wild nature that can be lost, or misunderstood, when the distinction between wild nature and human culture is blurred. He fears that the alternatives to the received view will allow for much greater human intervention in wild nature, including plans for managing wilderness areas. This, he believes, would be a mistake: "The values intrinsic to wilderness cannot, on pain of both logical and empirical contradiction, be 'improved' by deliberate human management."

INTRODUCTION

Revisiting the wilderness, Callicott (1991) is a doubtful guide; indeed he has gotten himself lost. That is a pity, because he is on the right track about sustainable development and I readily endorse his positive arguments for developing a culture more harmonious with nature. But these give no cause for being negative about wilderness.

The wilderness concept, we are told, is "inherently flawed," triply so. It metaphysically and unscientifically dichotomizes man and nature. It is ethnocentric, because it does not realize that practically all the world's ecosystems were modified by aboriginal peoples. It is static, ignoring change through time. In the flawed idea and ideal, wilderness respects wild communities where man is a visitor who does not remain. In the revisited idea(l), also Leopold's ideal, humans, themselves entirely natural, reside in and can and ought to improve wild nature.

HUMAN CULTURE
AND WILD NATURE

Wilderness valued without humans perpetuates a false dichotomy, Callicott maintains. Going back to Cartesian and Greek philosophy and Christian theology, such a contrast between humans and wild nature is a metaphysical confusion that leads us astray and also is unscientific. But this is not so. One hardly needs metaphysics or theology to realize that there are critical differences between wild nature and human culture. Humans now superimpose cultures on the wild nature out of which they once emerged. There is nothing unscientific or nonDarwinian about the claim that innovations in human culture make it radically different from wild nature.

Information in wild nature travels intergenerationally on genes; information in culture travels neurally as persons are educated into transmissible cultures. (Some higher animals learn limited behaviors from parents and conspecifics, but animals do not form transmissible cultures.) In nature, the coping skills are coded on chromosomes. In culture, the skills are coded in craftsman's traditions, religious rituals, or technology manuals. Information acquired during an organism's lifetime is not transmitted genetically; the essence of culture is acquired information transmitted to the next generation. Information transfer in culture can be several orders of magnitude faster and overlap genetic lines. I have but two children; copies of my books and my former students number in the thousands. A human being develops typically in some one or a few of ten thousand cultures, each heritage historically conditioned, perpetuated by language, conventionally established, using symbols with locally effective meanings. Animals are what they are genetically, instinctively, environmentally, without any options at all. Humans have myriads of lifestyle options, evidenced by their cultures: and each human makes daily decisions that affect his or her character. Little or nothing in wild nature approaches this.

The novelty is not simply that humans are more versatile in their spontaneous natural environments. Deliberately rebuilt environments replace spontaneous wild ones. Humans can therefore inhabit environments altogether different from the African savannas in which they once evolved. They insulate themselves from environmental extremes by their rebuilt habitations, with central heat from fossil fuel or by importing fresh groceries from a thousand miles away. In that sense, animals have freedom within ecosystems, but humans have freedom from ecosystems. Animals are adapted to their niches; humans adapt their ecosystems to their needs. The determinants of animal and plant behavior, much less the determinants of climate or nutrient recycling, are never anthropological, political, economic, technological, scientific, philosophical, ethical, or religious. Natural selection pressures are relaxed in culture; humans help each other out compassionately with medicine, charity, affirmative action, or headstart programs.

Humans act using large numbers of tools and things made with tools, extrasomatic artifacts. In all but the most primitive cultures, humans teach each other how to make clothes, thresh wheat, make fires, bake bread. Animals do not hold elections and plan their environmental affairs; they do not make bulldozers to cut down tropical rainforests. They do not fund development projects through the World Bank or contribute to funds to save the whales. They do not teach their religion to their children. They do not write articles revisiting and reaffirming the idea of wilderness. They do not get confused about whether their actions are natural or argue about whether they can improve nature.

If there is any metaphysical confusion in this debate, we locate it in the claim that "man is a natural, a wild, an evolving species, not essentially different in this respect from all the others" (p. 241). Poets like Gary Snyder perhaps are entitled to poetic license. But philosophers are not, especially when analyzing the concept of wildness. They cannot say that "the works of man, however precocious, are as natural as those of beavers," being "entirely natural," and then, hardly taking a breath, say that "the cultural component in human behavior is so greatly developed as to have become more a difference of kind than of degree" (p. 241). If this were only poetic philosophy it might be harmless, but proposed as policy, environmental professionals who operate with such contradictory philosophy will fail tragically.

"Anthropogenic changes imposed upon ecosystems are as natural as any other" (Callicott, 1990). Not so. Wilderness advocates know better; they do not gloss over these differences. They appreciate and criticize human affairs, with insight into their radically different characters. Accordingly, they insist that there are intrinsic wild values that are not human values. These ought to be preserved for whatever they can contribute to human values, and also because they are valuable on their own, in and of themselves. Just because the human presence is so radically different, humans ought to draw back and let nature be. Humans can and should see outside their own sector, their species self-interest, and affirm nonanthropogenic, noncultural values. Only humans have conscience enough to do this. That is not confused metaphysical dichotomy; it is axiological truth. To think that human culture is nothing but natural system is not discriminating enough. It risks reductionism and primitivism.

These contrasts between nature and culture were not always as bold as they now are. Once upon a time, culture evolved out of nature. The early hunter-gatherers had transmissible cultures but, sometimes, were not much different in their ecological effects from the wild predators and omnivores among whom they moved. In such cases, this was as much through lack of power to do otherwise as from conscious decision. A few such aboriginal peoples may remain.

But we Americans do not and cannot live in such a twilight society. Any society that we envision must be scientifically sophisticated, technologically advanced, globally oriented, as well as (we hope) just and charitable, caring for universal human rights and for biospheric values. This society will try to fit itself in intelligently with the ecosystemic processes on which it is superposed. It will, we plead, respect wildness. But none of these decisions shaping society are the processes of wild nature. There is no inherent flaw in our logic when we are discriminating about these radical discontinuities between culture and nature. The dichotomy charge is a half-truth, and, taken for the whole, becomes an untruth.

HUMANS IMPROVING WILD NATURE

Might a mature, humane civilization improve wild nature? Callicott thinks that it is a "fallacy" to think that "the best way to conserve nature is to protect it from human habitation and utilization" (p. 236). But, continuing the analysis, surely the fallacy is to think that a nature allegedly improved by humans is anymore real nature at all. The values intrinsic to wilderness cannot, on pain of both logical and empirical contradiction, be "improved" by deliberate human management, because deliberation is the antithesis of wildness. That is the sense in which civilization is the "antithesis" (p. 236) of wilderness, but there is nothing "amiss" in seeing an essential difference here. Animals take nature ready to hand, adapted to it by natural selection, fitted into their niches; humans rebuild their world through artifact and heritage, agriculture and culture, political and religious decisions.

On the meaning of "natural" at issue here, that of nature proceeding by evolutionary and ecological processes, any deliberated human agency, however well intended, is intention nevertheless and interrupts these spontaneous processes and is inevitably artificial, unnatural. (There is another meaning of "natural" by which even deliberated human actions break no laws of nature. Everything, better or worse, is natural in this sense, unless there is the supernatural.) The architectures of nature and of culture are different, and when culture seeks to improve nature, the man-

agement intent spoils the wildness. Wilderness management, in that sense, is a contradiction in terms— whatever may be added by way of management of humans who visit the wilderness, or of restorative practices, or monitoring, or other activities that environmental professionals must sometimes consider. A scientifically managed wilderness is conceptually as impossible as wildlife in a zoo.

To recommend that *Homo sapiens* "reestablish a positive symbiotic relationship with other species and a positive role in the unfolding of evolutionary processes" (p. 240) is, so far as wilderness preservation is involved, not just bad advice, it is impossible advice. The cultural processes by their very "nature" interrupt the evolutionary process; there is no symbiosis, there is antithesis. Culture is a post-evolutionary phase of our planetary history; it must be superposed on the nature it presupposes. To recommend, however, that we should build sustainable cultures that fit in with the continuing ecological processes is a first principle of intelligent action, and no wilderness advocate thinks otherwise.

If there are inherent conceptual flaws dogging this debate, we have located another: Callicott's allegedly "improved" nature. In such modified nature, the different historical genesis brings a radical change in value type. Every wilderness enthusiast knows the difference between a pine plantation in the Southeast and an old-growth grove in the Pacific Northwest. Even if the "improvement" is more or less harmonious with the ecosystem, it is fundamentally of a different order. Asian ring-tailed pheasants are rather well naturalized on the contemporary Iowa landscape. But they are there by human introduction, and they remain because farmers plow the fields, plant corn, and leave shelter in the fencerows. They are really as much like pets as like native wild species, because they are not really on their own.

BIODIVERSITY AND WILDERNESS

As an example of his recommended symbiosis where human culture enriches natural systems, Callicott cites a study (Nabhan et al., 1982) of two nearby communities, Quitovac (= Ki:towak) in Mexico, where sixty-five bird species were found, and Quitobaquito Springs (or A'al Waipia) in Organ Pipe National Monument, with only thirty-two species. His conclusion is that biodiversity is greater in such rural communities than in wild natural systems.

But this is an unusual case; the locale is desert, where water is the limiting factor. If you artificially water the desert, some things will come in that could not live there before. Similarly, if you heat up the tundra, where cold is the limiting factor. We will not be surprised if there are more birds around feeders offering food, water, and shelter than elsewhere. But bird feeders actually may not be increasing biodiversity. We will have to look more closely at what is meant by biodiversity and what is going on in the two communities.

A species count, uninterpreted, doesn't tell us much. In more sophisticated analyses, ecologists use up to a dozen and a half indices of diversity (Magurran, 1988; Pielou, 1975). These include within habitat diversity (alpha diversity), between habitat diversity (beta diversity) and regional diversity (gamma diversity). They include diversity of processes and heterogeneity of fauna and flora, and on and on. If all you do is count species, there are more animal species in the Denver zoo than in the rest of Colorado. Never mind that the processes of nature are entirely gone. Callicott knows that and wants ecosystem health as well as diversity.

Whether there is ecosystem health at Quitovac is less clear. Callicott thinks so, but Nabhan et al. (1982) are more circumspect. Though the bird species count was always higher at Quitovac, by a heterogeneity diversity index the avifauna at Quitovac has no advantage over Organ Pipe. (This asks what proportions of the birds are of what species, such as grackles, doves, English sparrows, pigeons.) They also find that Quitovac is "not nearly as diverse in mammals," that ever-present dogs, horses, and cattle, limit the presence of wild animals. Deer and javelina drink and browse frequently at Organ Pipe, seldom at Quitovac. Even rodents are more abundant at Organ Pipe.

They also found more plant diversity at Quitovac, one hundred and thirty-nine species there against eighty at Organ Pipe. It is hardly surprising that if you add some irrigated cultivated fields and orchards, new plant species will appear, and some insects will follow, and birds in turn follow the seeds and insects. They also note that seventeen of these plant species were planted intentionally and that of the fifty-nine species in fields and orchards, many were adventitious species, weeds of disturbed sites. Many were the Old World waifs that, like dandelions, have tagged along after civilization willy-nilly. Is this being offered as a wise symbiosis of nature and culture? Is that enhanced richness in biodiversity?

A species count, offered as evidence of biodiversity without further ado, assumes that if we have the species, we have what we want conserved. But we may have the parts, even extra, artificial parts, but no longer the composition of the former whole. Maybe Quitovac is about as much "ersatz world" (p. 243) as idyllic, humanized ecosystem health with optimized biodiversity. Even a new whole would not have the integrity of the once wild ecosystem. We can and ought to have rural nature, and we will be glad to have rural nature with a high bird count. But we can have a rural nature with a high species count and not have anywhere on the landscape the radical values of wild, pristine nature. That loss would not be compensated for by the stepped-up species count in agriculturally disturbed lands. In wilderness, we value the interactions as a fundamental component of biodiversity.

The predation pressures, for instance, are never the same on agricultural lands as they are on wildlands. Agriculture means an increase of disturbed soil, with most of these disturbances different in kind from those in wild nature. Different kinds of things grow in such soil, more r-selected species, fewer k-selected ones. Underground, the fungi and soil bacteria are different, so the decomposition regime is different, and that results in differences above ground. The energy flow and the nutrient cycling is different. It is often the case that the highest number of species are found in intermediately disturbed environments, but that considers species counts and alpha diversity alone. If all the environments are kept intermediately disturbed, we lose beta and gamma diversity. Indeed over the landscape as a whole, we lose even species counts, since in disturbed environments the sensitive species go extinct. We are not likely to retain the large carnivores.

Both these oases are water magnets for migrant birds. Quitovac, with its cultivated fields and orchards, draws more migrants into close proximity. Muddy shorelines attract some waders less frequent at Quitobaquito. All this tells us little about whether these migrating birds are safe in their wintering or breeding grounds. In fact, Central American agricultural development, destroying winter grounds, threatens many bird species. Quitovac may draw some breeding species that cannot survive at Quitobaquito or in the unwatered desert. But there are no bird species flourishing in Quitovac that were not flourishing already in their native habitats elsewhere. It is hard to think that much important bird conservation is going on there.

Quitobaquito Springs, far from being depauperate in birds, is one of the best-known sites for observing birds in that region of Arizona, and birders go there from all over the United States to see the migrants and to find the desert species. The oasis is but a small area. Organ Pipe Monument is designated to preserve many other kinds of habitats. Enlarging to consider beta and gamma diversity, the official Monument checklist contains 277 bird species, of which 63 are known to breed in various habitats there, and five more believed to breed as well (Groschupf et al., 1987). Only three are nonnative. Even if the diversity at Quitovac is greater, the diversity preserved by having both a rural area and a wilderness is higher than if we had two rural areas.

Also, whatever the possibilities, we do not want to forget the probabilities, which are that this (allegedly) idyllic picture will be upset by development pressures. Quitovac had been used for centuries, steadily but not intensively. When the comparisons here were made, only two or three dozen persons were using the area. The study concentrates on only a five hectare site, and the natives had only used ten percent of this for cultivated fields and orchards. Before the study could be completed, 125 hectares there were bulldozed to be used for intensive agriculture, including most of the study area, with disastrous results (Nabhan et al., 1982). There may be fewer species at Organ Pipe, but such a disaster is not likely to occur, owing to its sanctuary designation.

WILDERNESS AND CHANGE

Another alleged flaw in the concept of wilderness is that its advocates do not know the fourth dimension, time. That is a strange charge; my experience has been just the opposite. In wilderness, the day changes from dawn to dusk, the seasons pass, plants grow, animals are born, grow up, and age. Rivers flow, winds blow, even the rocks erode; change is pervasive. Indeed, wilderness is that environment in which one is most likely to experience geological time. Try a raft trip through the Grand Canyon.

On the scale of deep time, some processes continue on and on, so that the perennial givens—wind and rain, soil and photosynthesis, life and death and life renewed—can seem almost forever. Species survive for millions of years; individuals are ephemeral. Life persists in the midst of its perpetual perishing. Mountains are reliably there generation after generation. The water cycles back, always moving. In wilderness, time mixes with eternity; that is one reason we value it so highly.

Callicott writes as if wilderness advocates had studied ecology and never heard of evolution. But they know that evolution is the control of development by ecology, and what they value is precisely natural history. They do not object to natural changes. They may not even object to artificial changes in rural landscapes. But, since they know the difference between nature and culture, they know that cultural changes may be quite out of kilter with natural changes. Leopold uses the word "stability" when he is writing in the time frame of land-use planning. On that scale, nature typically does have a reliable stability, and farmers do well to figure in the perennial givens.

In an evolutionary time frame, Leopold knows that relative stability mixes with change. "Paleontology offers abundant evidence that wilderness maintained itself for immensely long periods; that its component species were rarely lost, and neither did they get out of hand; that weather and water built soil as fast or faster than it was carried away." That is why "wilderness . . . assumes unexpected importance as a laboratory for the study of land health" (Leopold, 1968, p. 196). Wilderness is the original sustainable development.

With natural processes, "protect" is perhaps a better word than either "preserve" or "conserve." Wilderness advocates do not seek to prevent natural change. There is nothing illusory, however, about appreciating today in wilderness processes that have a primeval character. There, the natural processes of 1992 do not differ much from those of 1492, half a millennium earlier. We may enjoy that perennial character, constancy in change, in contrast with the rapid pace of cultural changes, seldom as dramatic as those on the American landscape of the last few centuries.

A management program in the U.S. Forest Service seeks to evaluate the "limits of acceptable change." This emphasis worries about the rapid pace of cultural change as this contrasts with the natural pace on landscapes. Cordell and Reed (1990) are trying to decide the limits of acceptable humanly-introduced changes, artificial changes, since these are of such radically different kind and pace that they disrupt the processes of wilderness. They do not oppose natural changes. At this point, we have an example of how and why environmental professionals will make disastrous decisions, if confused by what is and is not natural. Callicott warns them that they do have to worry about "accelerating rates of environmental change" (p. 242). No one can begin to understand these rates of changes if the changes are thought of as being introduced by a species that is "entirely natural."

When we designate a desert wilderness in Nevada, there really isn't any problem deciding that mustangs are feral animals in contrast with desert bighorns, which are indigenous. There might have been ancestral horses in Paleolithic times in the American West, but they went extinct naturally. The present mustangs came from animals that the Europeans brought over in ships, originally from the plains of Siberia. Bighorns are what they are where they are by natural selection. Mustangs are not so. There is nothing conceptually problematic about that—unless one has never gotten clearly in mind the difference between nature and culture in the first place.

ABORIGINAL PEOPLES AND WILDERNESS

What of the argument that we cannot have any wilderness, because there is none to be had? This is a much stronger claim than that there is no real wilderness left on the American landscape after the European cultural invasion. Even the aboriginals had already extinguished wilderness. Now we have a somewhat different account of the human presence from that earlier advocated. The claim is no longer that the Indians were just another wild species, "entirely natural," but that they actively managed the landscape, so dramatically altering it that there was no wilderness even when Columbus arrived in 1492. It is ethnocentric to think otherwise. This is because we Caucasians exaggerate our own power to modify the landscape and diminish their power. This is a judgment based on prejudice, not on facts.

How much did the American Indians modify the landscape? That is an empirical question in anthropology and ecology. We do not disagree that where there was Indian culture, this altered the locales in which they resided, so that these locales were not wilderness in the pure sense. In that respect, Indian culture is not different in kind from the white man's culture. What we need to know is the degree. Had the Indians, when the white man arrived, already transformed the pre-Indian wilderness beyond the range of its spontaneous self-restoration?

Callicott concedes (pp. 241–242), rightly, that most of what has been presently designated as wilderness was infrequently used by the aborigines, since it is high, rough, or arid. We have no reason to think that in such areas the aboriginal modifications are irreversible. Were the more temperate regions modified so extensively and irreversibly that so little naturalness remains as to make wilderness designation an illusion? Callicott has "no doubt that most New World ecosys-

tems were in robust health" (p. 244). That suggests that they were not past self-regeneration.

The American Indians on forested lands had little agriculture; what agriculture they had tended to reset succession, and, when agriculture ceases, the subsequent forest regeneration will not be particularly unnatural. The Indian technology for larger landscape modification was bow and arrow, spear, and fire. The only one that extensively modifies landscapes is fire. Fire is—we have learned well by now—also quite natural. Fire suppression is unnatural, but no one argues that the Indians used that as a management tool, nor did they have much capacity for fire suppression. The argument is that they deliberately set fires. Does this make their fires radically different from natural fires? It does in terms of the source of ignition; the one is a result of environmental policy deliberation, the other of a lightning bolt.

But every student of fire behavior knows that on the scale of regional forest ecosystems, the source of ignition is not a particularly critical factor. The question is whether the forest is ready to burn, whether there is sufficient ground fuel to sustain the fire, whether the trees are diseased, how much duff there is, and so on. If conditions are not right, it will be difficult to get the fire going and it will burn out soon. If conditions are right, a human can start a regional fire this year. If some human does not, lightning will start it next year, or the year after that. On a typical summer day, the states of Arizona and New Mexico are each hit by several thousand bolts of lightning, mostly in the higher, forested regions. Doubtless the Indians started some fires too, but it is hard to think that their fires so dramatically and irreversibly altered the natural fire regime in the Southwest that meaningful wilderness designation is impossible today.

We do not want to be ethnocentric, but neither do we want to be naïve about the technological prowess of the American Indian cultures. They had no motors, indeed no wheels, no domestic animals, no horses (before the Spanish came), no beasts of burden. The Indians had a hard time getting so simple a thing as hot water. They had to heat stones and drop them in skins or tightly woven baskets. They lived on the landscape with foot and muscle, and in that sense, though they had complex cultures, they had culture with very reduced alterative power. Even in European cultures, in recent centuries the power of civilization to redo the world has accelerated logarithmically.

In Third World nations, perhaps areas that seem "natural" now are often the result of millennia of human modifications through fire, hunting, shifting cultivation, and selective planting and removal of species. This will have to be examined on a case-by-case basis, and we cannot prejudge the answers. We do not know yet how intensively the vast Brazilian rainforests were managed and whether no wilderness designation there is ecologically practicable, even if we desired it. Nor do wilderness enthusiasts advocate that such peoples be removed to accomplish this, where it is possible. What is protested is modern forms of development. Extractive reserves may be an answer, but extractive reserves for latex sold in world markets and manufactured into rubber products can hardly be considered aboriginal wisdom.

Sometimes we will have to make do with what wildness remains in the nooks and crannies of civilization. Meanwhile, where wilderness designation is possible and where there is an exploding population, what should we do? No one objects to trying to direct that explosion into more harmonious forms of human-nature encounter. But constraining an explosion takes some strong measures. One of these ought to be the designation of wilderness.

Perhaps the American Indians did not have enough contrast between their culture and the nature that surrounded them to produce the wilderness idea. It was not an idea that, within their limited power to remake nature, could occur to them. If you have only foot, muscle, bows, arrows, and fire, you do not think much about wilderness conservation. But we, in the twentieth century, do have the wilderness idea; it has crystallized with the possibility, indeed the impending threat, of destroying the last acre of primeval wilderness. It also has crystallized with our deepening scientific knowledge of how wild nature operates, of DNA, genes, and natural selection, and how dramatically different in kind, pace, and power the processes of culture can be. The Indians knew little of this: they lived still in an animistic, enchanted world.

And we need the wilderness idea desperately. When you have bulldozers that already have blacktopped more acreage than remains pristine, you can and ought to begin to think about wilderness. Such an idea, when it comes, is primitive in one sense: it preserves primeval nature, as much as it can. But it is morally advanced in another sense: it sees the intrinsic value of nature, apart from humans.

Ought implies can; the Indians could not, so they never thought much about the ought. We in the twentieth century can, and we must think about the ought. When we designate wilderness, we are not lapsing into some romantic atavism, reactionary and nostalgic to escape culture. We are breaking through culture to discover, nonanthropocentrically, that fauna and flora can count in their own right (an idea that Indians also might have shared). We realize that ecosystems sometimes can be so respected that humans only visit and do not remain (an idea that the Indians did not need or achieve). A "can" has appeared that has generated a new "ought."

Even some modern American Indians concur. In western Montana, the Salish and Kootenai tribes have set aside 93,000 acres of their reservation as the Mission Mountains Tribal Wilderness; in addition, they have designated the South Fork of the Jocko Primitive Area. In both areas, the Indian too is "a visitor who does not remain"; they want these areas "to be affected primarily by the forces of nature with the imprint of man's work substantially unnnoticeable" (*Tribal Wilderness Ordinance,* 1982). Indeed, in deference to the grizzly bears, in the summer season, the Indians do not permit any humans at all to visit 10,000 acres that are prime grizzly habitat. In both areas, they can claim even more restrictive environmental regulation on what people can do there than in the white man's wilderness. What, when, and how they hunt is an example.

Not a word of the above discussion disparages aboriginal Indian culture. To the contrary, that they survived with the bare skills they had is a credit to their endurance, courage, resolution, and wisdom. A wilderness enthusiast, if he or she has spent much time in the woods armed with only muscle and a few belongings in a backpack, is in an excellent position to appreciate the aboriginal skills.

SUSTAINABLE DEVELOPMENT AND WILDERNESS

Finding out how to remake civilization so that nature is conserved in the midst of sustainable development is indeed a more difficult and important task than saving wild remnants. Little wilderness can be safe unless the sustainable development problem is solved also. I can only endorse Callicott's desire to conserve nature in the midst of human culture. "Human economic activities should at least be compatible with the ecological health of the environments in which they occur" (p. 239). No party to the debate contests that. But this does not mean that wilderness ought not to be saved for what it is in itself.

"The farmer as a conservationist" is quite a good thing, and Leopold does well to hope that "land does well for its owner, and the owner does well by his land"; perhaps where a farmer begins, as did Leopold, with lands long abused, "both end up better by reason of their partnership." In that context, "conservation is a state of harmony between men and land" (p. 238). But none of that asks whether there also should be wilderness. Leopold tells us what he thinks about that after his trip to Germany. There was "something lacking. . . . I did not hope to find in Germany anything resembling the great 'wilderness areas' which we dream about and talk about." That was too much to hope; he could dream that only in America. But he did hope to find "a certain quality [—wildness—] which should be, but is not found" in the rural landscape, and, alas, not even that was there (Leopold, quoted by Callicott, p. 238). "In Europe, where wilderness now has retreated to the Carpathians and Siberia, every thinking conservationist bemoans its loss" (Leopold, [1949], 1968, p. 200). That loss would not be restored if every farmer were a restoration ecologist. All that Leopold says about sustainable development is true, but there is no implication that wilderness cannot or ought not to be saved. Affirming sustainable development is not to deny wilderness.

MONASTIC WILDERNESS AND CIVILIZED COMPLACENCY

Nor is affirming wilderness to deny sustainable development. Callicott alleges, "Implicit in the most passionate pleas for wilderness preservation is a complacency about what passes for civilization" (p. 236). Not so. I cannot name a single wilderness advocate who cherishes wilderness "as an alibi for the lack of private reform," any who "salve their consciences" by pointing to "the few odds and ends" of wilderness and thus "avoid facing up to the fact that the ways and means of industrial civilization lie at the root of the current global environmental crisis" (p. 239). The charge is flamboyant; the content runs hollow. Wilderness advocates want wilderness and they also want, passionately, to "re-envision civilization" (p. 236) so that it is in harmony with the nature that humans do modify and inhabit. There is no tension between these ideas in Leopold, nor in any of the other passionate advocates of wilderness that Callicott cites, nor in any with whom I am familiar.

The contrast of monastic sanctuaries with the wicked everyday world risks a flawed analogy. Unless we are careful, we will make a category mistake, be-

cause both monastery and lay world are in the domain of culture, while wilderness is a radically different domain. Monastery sets an ideal unattainable in the real civil world (if we must think of it that way) but both worlds are human, both moral. We are judging human behavior in both places, concerned with how far it can be godly. By contrast, the wilderness world is neither moral nor human; the values protected there are of a different order. We are judging evolutionary achievements and ecological stability, integrity, beauty—not censuring or praising human behavior.

Confusion about nature and culture is getting us into trouble again. We are only going to get confused if we think that the issue of whether there should be monasteries is conceptually parallel to the issue of whether there should be wilderness. The conservation of value in the one is by the cultural transmission of a social heritage, including a moral and religious heritage, to which the monastery was devoted. The conservation of value in the other is genetic, in genes subject to natural selection for survival value and adapted fit. There is something godly in the wilderness too, or at least a creativity that is religiously valuable, but the contrast between the righteous and the wicked is not helpful here. The sanctuary we want is a world untrammeled by man, a world left to its own autonomous creativity, not an island of saintliness in the midst of sinners.

We do not want the whole Earth without civilization, for we believe that humans belong on Earth; Earth is not whole without humans and their civilization, without the political animal building his *polis* (Socrates), without peoples inheriting their promised lands (as the Hebrews envisioned). Civilization is a broken affair, and in the long struggle to make and keep life human, moral, even godly, perhaps there should be islands, sanctuaries, of moral goodness within a civilization often sordid enough. But that is a different issue from whether, when we build our civilizations for better or worse, we also want to protect where and as we can those nonhuman values in wild nature that preceded and yet surround us. An Earth civilized on every acre would not be whole either, for a whole domain of value—wild spontaneous nature— would have vanished from this majestic home planet.

INTRINSIC WILDERNESS VALUES

I fear that we are seeing in Callicott's revisiting wilderness the outplay of a philosophy that does not think, fundamentally, that nature is of value in itself. Such a philosophy, though it may protest to the contrary, really cannot value nature for itself. All value in nature is by human projection; it is anthropogenic, generated by humans, though sometimes not anthropocentric, centered on humans. Callicott has made it clear that all so-called intrinsic value in nature is "grounded in human feelings" and "projected" onto the natural object that "excites" the value. "Intrinsic value ultimately depends upon human valuers." "Value depends upon human sentiments" (Callicott, 1984, p. 305).

He explains, "The source of all value is human consciousness, but it by no means follows that the locus of all value is consciousness itself. . . . An intrinsically valuable thing on this reading is valuable for its own sake, for itself, but it is not valuable in itself, i.e., completely independently of any consciousness, since no value can in principle . . . be altogether independent of a valuing consciousness. . . . Value is, as it were, projected onto natural objects or events by the subjective feelings of observers. If all consciousness were annihilated at a stroke, there would be no good and evil, no beauty and ugliness, no right and wrong; only impassive phenomena would remain." This, Callicott says, is a "truncated sense" of value where "'intrinsic value'
retains only half its traditional meaning" (Callicott, 1986, pp. 142–43, p. 156, and p. 143).

Talk about dichotomies! Only humans produce value; wild nature is valueless without humans. All it has without humans is the potential to be evaluated by humans, who, if and when they appear, may incline, sometimes, to value nature in noninstrumental ways. "Nonhuman species . . . may not be valuable in themselves, but they may certainly be valued for themselves. . . . Value is, to be sure, humanly conferred, but not necessarily homocentric" (Callicott, 1986, p. 160). The language of valuing nature for itself may be used, but it is misleading: value is always and only relational, with humans one of the relata. Nature in itself (a wilderness, for example) is without value. There is no genesis of wild value by nature on its own. Such a philosophy can value nature only in association with human habitation. But that—not some elitist wilderness conservation for spiritual meditation—is the view that many of us want to reject as "aristocratic bias and class privilege" (p. 237).

Sustainable development is, let's face it, irremediably anthropocentric. That is what we must have most places, and humans too have their worthy values. But must we have it everywhere? Must we have more of it and less wilderness? Maybe the value theory here is where the arrogance lies, not in some alleged ethno-centrism or misunderstood doctrine of the dominion of man.

A truncated value theory is giving us a truncated account of biodiversity. Callicott hardly wants wildernesses as "sanctuaries," only as "refugia" (p. 236). A refugia is a seedbed from which other areas get restocked. That is one good reason for wilderness conservation, but we do not want wilderness simply as a place from which the game on our rural lands can be restocked, or even, if we have a more ample vision of wildlife recreation, from which the wildlife that yet persists on the domesticated landscape can be resupplied steadily. Wildernesses are not hatcheries for rural or urban wildlife. Nor are they just "laboratories" (p. 238) for baseline data for sound scientific management. Nor are they raw materials on which we can work our symbiotic enhancements. Nor are they places that can excite us into projecting truncated values onto them. Some of these are sometimes good reasons for conserving wilderness. Leopold sums them up as "the cultural value of wilderness" (Leopold [1949] 1968, p. 200). But they are not the best reasons.

LEOPOLD AND WILDERNESS

Leopold pleads in the "Upshot," in his last book in the penultimate essay, entitled "Wilderness": "Wilderness was an adversary to the pioneer. But to the laborer in repose, able for the moment to cast a philosophical eye on his world, that same raw stuff is something to be loved and cherished, because it gives definition and meaning to his life" (Leopold, [1949] 1968, p. 188). He does not mean that wilderness is only a resource for personal development, though it is that. He means that we never know who we are or where we are until we know and respect our wild origins and our wild neighbors on this home planet. We never get our values straight until we value wilderness appropriately. The definition of the human kinds of values is incomplete until we have this larger vision of natural values.

Concluding his appeal for "raw wilderness" (p. 201), Leopold turns to the "Land Ethic," "The land ethic simply enlarges the boundaries of the community to include soils, waters, plants, and animals, or collectively: the land. . . . A land ethic of course cannot prevent the alteration, management, and use of these 'resources,' but it does affirm their right to continued existence, and, at least in spots, their continued existence in a natural state." We may certainly assert that the founder of the Wilderness Society believed that wilderness conservation is essential in this right to continued existence in a natural state.

"I am asserting that those who love the wilderness should not be wholly deprived of it, that while the reduction of wilderness has been a good thing, its extermination would be a very bad one, and that the conservation of wilderness is the most urgent and difficult of all the tasks that confront us" (Leopold, quoted in Meine, 1988, p. 245). We must take it as anomalous (else it would be amusing or even tragic) to see Leopold's principal philosophical interpreter, himself a foremost environmental philosopher who elsewhere has said many wise things, now trying to revisit the wilderness idea and deemphasize it in Leopold.

Just before Leopold plunges into his passionate plea for the land ethic, he calls for "wilderness-minded men scattered through all the conservation bureaus." "A militant minority of wilderness-minded citizens must be on watch throughout the nation, and available for action in a pinch" (Leopold [1949], 1968, p. 200). Alas! His trumpet call is replaced by an uncertain sound. Robert Marshall saluted Leopold as "The Commanding General of the Wilderness Battle" (cited in Meine, 1988, p. 248). How dismayed he would be by this dissension within his ranks.

On Earth, man is not a visitor who does not remain; this is our home planet and we belong here. Leopold speaks of man as both "plain citizen" and as "king." Humans too have an ecology, and we are permitted interference with, and rearrangement of, nature's spontaneous course; otherwise there is no culture. When we do this there ought to be some rational showing that the alteration is enriching, that natural values are sacrificed for greater cultural ones. We ought to make such development sustainable. But there are, and should be, places on Earth where the nonhuman community of life is untrammeled by man, where we only visit and spontaneous nature remains. If Callicott has his way, revisiting wilderness, there soon will be less and less wilderness to visit at all.

References

Callicott, J. B. 1984. Non-anthropocentric Value Theory and Environmental Ethics. *American Philosophical Quarterly* 21: 299–309.

Callicott, J. B. 1986. On the Intrinsic Value of Nonhuman Species. In *The Preservation of Species,* B. G. Norton, ed. Princeton University Press, Princeton, NJ.

Callicott, J. B. 1990. Standards of Conservation: Then and Now, *Conservation Biology* 4: 229–232.

Callicott, J. B. 1991. The Wilderness Idea Revisited: The Sustainable Development Alternative. *The Environmental Professional* 13: 235–247.

Cordell, H. K., and P. C. Reed 1990. Untrammeled by Man: Preserving Diversity through Wilderness. In *Preparing to Manage Wilderness in the 21st Century: Proceedings of the Conference.* Southeastern Forest Experiment Station, U.S. Department of Agriculture, Forest Service, Asheville, NC, pp. 30–33.

Groschupf, K., B. T. Brown, and R. R. Johnson. 1987. *A Checklist of the Birds of Organ Pipe Cactus National Monument.* Southwest Parks and Monument Association, Tucson, AZ.

Leopold, A. [1949] 1968. *A Sand County Almanac.* Oxford University Press, New York.

Leopold, A. S., S. A. Cain, C. M. Cottam, I. N. Gabrielson, and T. L. Kimball. 1963. Wildlife Management in the National Parks, Report of the Advisory Board on Wildlife Management. U.S. Government Printing Office, Washington, DC, March 4.

Meine, C. 1988. *Aldo Leopold: His Life and Work.* University of Wisconsin Press, Madison.

Magurran, A. E. 1988. *Ecological Diversity and Its Measurement.* Princeton University Press, Princeton, NJ.

Nabhan, G. P., A. M. Rea, K. L. Reichhardt, E. Mellink, and C. F. Hutchinson. 1982. Papago Influences on Habitat and Biotic Diversity: Quitovac Oasis Ethnoecology. *Journal of Ethnobiology* 2: 124–143.

Pielou, E. C. 1975. *Ecological Diversity.* John Wiley, New York. *Tribal Wilderness Ordinance of the Governing Body of the Confederated Salish and Kootenai Tribes.* 1982.

For Further Discussion

1. According to Rolston, what are the values of wilderness that would be lost if we are too quick to reject the received view that Nelson and Cronon question?

2. Rolston is skeptical of the human ability to improve wild nature. Do you agree? Do you think Nelson and Cronon would disagree?

3. Rolston suggests that there are important differences between the ways aboriginal peoples change their environment and the ways modern humans do. What are some of these differences, and do they make a difference ethically?

4. Rolston disagrees with Callicott, who claims that because intrinsic value depends on human valuers there is no intrinsic value to wild nature itself. With whom do you agree?

DISCUSSION AND STUDY QUESTIONS FOR CHAPTER 11

1. Why value the wilderness? What value is there in experiencing an area that is "untrammeled by man"? Is this merely a subjective preference? Would the experience change if you learned that a community had previously been established in this spot?

2. Is the wilderness experience an elitist and culturally bound concept? Would such a thing even make sense to indigenous peoples? Is there a difference between a wilderness area and land that is simply unpopulated?

3. Public policy decisions such as building dams and constructing roads into wilderness areas are often defended as providing services to meet public demand. But these decisions also create demand by making it easier to live in the desert or visit the wilderness area. Should government policy encourage people to visit wilderness areas? Can policy ever remain neutral in this respect?

4. How would you describe the difference between human culture and the wilderness? Is the wilderness a construct of human culture?

DISCUSSION CASES

Dams in Wild Areas: Hetch Hetchy and Glen Canyon

Two of the most famous environmental controversies in the United States concerned plans to construct dams in scenic wilderness areas. The first, at Hetch Hetchy Valley in the Yosemite National Park area, occurred during the first decades of the twentieth century, and in many ways it represented the beginning of environmentalism in the United States. The second concerned plans to dam the Colorado River at Glen Canyon, just upriver from the Grand Canyon.

Demand for water in the San Francisco Bay Area led to plans for a dam and reservoir in the Hetch Hetchy Valley. The controversy that resulted pitted two giants of the U.S. environmental movement against each other. Gifford Pinchot, the chief spokesman of the conservation movement, supported building the dam and John Muir, leader of the preservationists and founder of the Sierra Club, fought to protect this wilderness from development.

Conservationists pointed out that the growing population in the San Francisco area needed water, that spring melt-off from the Sierras was otherwise wasted, and that there were still plenty of wild areas left undeveloped. They believed that the aesthetic benefits that a few received from the wilderness

area should not outweigh the pressing needs of so many more. Preservationists sought to preserve the majestic area in its natural and pristine state.

Demand for water in the American Southwest also led to plans to build dams along the Colorado River. In the 1950s plans were announced to construct several dams along the Colorado, including one within Dinosaur National Monument and one at Glen Canyon. Led by Muir's Sierra Club, environmentalists fought against these projects; in the political debates that followed a compromise was reached to protect Dinosaur National Monument but to allow construction at Glen Canyon. The dam began operation in 1963.

Today, Lake Powell stretches 186 miles upriver from the Glen Canyon dam. The lake itself does not directly provide water to surrounding areas; rather, the dam acts as a valve, releasing water downstream at a rate that ensures continuous water access for people and agriculture in the more arid regions of the Southwest. The dam generates significant amounts of electricity for such cities as Phoenix and Las Vegas, and Lake Powell provides recreational opportunities to thousands of people each year.

Critics point out that the dam has had a major impact on the area's ecosystems. Several species of
(continued)

fish and plants have become extinct, and the physical features of the river, including areas within the Grand Canyon, have changed noticeably. Aside from the loss of Glen Canyon itself, over two hundred other neighboring canyons have been flooded. Sedimentation is also building up within Lake Powell, and some estimate that the intake pipes will be covered with sediment within a hundred years. So much water is diverted that the Colorado River no longer flows all the way to the Pacific. Some critics also point out that this damage was necessary only to feed the growing sprawl in desert areas like Phoenix and Las Vegas. In 1996, the Sierra Club's Board of Directors passed a resolution calling on the federal government to dismantle the Glen Canyon dam and restore the area to its natural state.

Boundary Waters Canoe Area

The Wilderness Act of 1964 defines the wilderness as those areas "where the earth and its community of life are untrammeled by man, where man himself is a visitor who does not remain." What happens when an area in which humans have lived and worked comes to be designated as a wilderness area?

The Boundary Waters Canoe Area contains thousands of lakes and over 1 million acres of land along the Canadian border of Minnesota. The site of fur trapping by French-Canadian voyageurs in the eighteenth century and significant logging and mining in the nineteenth century, by the middle of the twentieth century this area had become part of Superior National Forest. Designated as a "roadless primitive area" in the 1930s, its current name as the Boundary Waters Canoe Area was adopted in 1958 and its boundaries expanded to its current state with the BWCA Wilderness Bill of 1978. That law prohibited logging, mining, and motorized travel within the wilderness areas. The prohibition of motorboats was to be phased in over a twenty-year period.

Local residents have long opposed designating this as a wilderness area. Families have lived along some of these lakes for generations. Indeed, it was recognition of this fact that prevented all lakes from being designated as "canoe only" in the 1970s. Significant controversies continue as deadlines for removing motorized travel approach.

Some environmental groups argue that if this area is to be a wilderness area, all motorized travel and commercial activities must end and all permanent residences must be moved. They point out that there are over ten thousand other lakes within Minnesota that are open to motorboats. Local residents argue that they make a living with motorized travel. Many resorts, outfitters, fishing guides, and vacation homes would become worthless with a ban on motorboats. They point out that these activities existed in this area before the government designation as a wilderness area. Given the government's definition, this area is not now a wilderness area nor has it been so for hundreds of years.

Cars in Yosemite

During peak tourist season, many national parks are crowded with visitors. Traffic jams are common in such parks as Yosemite, the Grand Canyon, and Yellowstone. A recent U.S. Park Service proposal would establish policies to restrict access to the most crowded national park, Yosemite.

The plan was announced as a "vision for the 21st century . . . a comprehensive blueprint to reduce traffic, restore the natural resources within the valley and improve visitor facilities and services." The plan would ban most private cars within the park, rely on shuttle buses to move people within the park, increase trails for biking and hiking, and remove permanent structures such as some bridges, visitor centers, and ranger residences. Defenders of the plan claim that limiting access is the only way to preserve the park and provide a true wilderness experience. Critics charge that the plan would deny equal access to public lands and force citizens to conform to a particular conception of what a wilderness experience should be.

Both the Boundary Waters and Yosemite situations raise questions about the proper role of human management of wilderness areas. What implications for these cases do you draw from the essays by Nelson and Cronon? Is it worth designating either of these natural areas as wilderness? If so, what is your understanding of wilderness? How natural is a wilderness area if it is maintained only through human control? Of the values that Rolston mentions, which are honored and which are lost in such an area as Yosemite?

CHAPTER 12

Growth and Development

A brief consideration of present economic, ecological, and population realities suggests the ethical significance of dealing with population and economic growth and the deficiencies of any policy that does not place moral limits on growth. Consider three relevant facts. First, one-quarter of the world's population live in industrial countries and consume 80 percent of the world's goods. The remaining three-quarters of the population, many of whom live at or below minimal subsistence levels, consume only 20 percent of the world's production. Significant economic activity would be necessary even to begin to bring the living standards of the world's poorest citizens into line with the living standards of people in the industrialized world. One estimate holds that a fivefold increase in energy use and a five- to tenfold increase in economic activity would be required over the next fifty years to bring the standard of living for the present population of developing countries into line with that of those in the industrialized world.

Second, even conservative estimates suggest that during these fifty years world population will double, bringing the total world population to over 11 billion people. (During the mid-1990s estimates suggest worldwide population growth occurs at around 1.6 percent per year. This represents a slight decline from previous estimates and was greeted by some as reason for optimism. At this rate, population will double in about forty-three years!) Thus, economic activity needs to increase minimally by ten- to twentyfold to bring the standard of living of the actual world population in fifty years into line with that enjoyed by people in the industrialized present. And this might be an optimistic projection. Tremendous capital investment and accumulation would be necessary to convert all of these resources into final products, thus further increasing economic activity. Also, it is likely that the

resources needed for such growth would need to come from less accessible and lower-grade deposits because we've already taken the readily accessible and high-grade iron, coal, oil, timber, and so on.

Finally, we must recognize that the only source for this economic activity, ultimately, are the natural resources of the planet. The three standard factors of production—natural resources, capital, and labor—all derive from the productive capacity of the earth. In simple terms, raw material, energy, and food are the essential elements of all economic activity. To say that the productive capacity of the earth is already under significant stress is an understatement. For example, one estimate suggests that if the world's population in forty years consumed nonrenewable mineral and petroleum resources at current U.S. rates, these resources would last less than 10 years.

Thus, increasing population growth, increasing economic growth, and dwindling natural resources could be on a collision course with the well-being of billions of people and widespread environmental destruction is at stake.

Two general policy strategies seem open to us: control population growth and control either the type or rate of economic growth. These two options, of course, are not mutually exclusive. Presumably some combination of the two would be needed. Nevertheless, either strategy raises serious ethical questions.

Is there some ethically justifiable population goal? Some have suggested that the ecological concept of carrying capacity be applied to human population. This concept suggests that every ecosystem has an upper limit to what it can support. A population faces an impending catastrophe when it exceeds the carrying capacity of its ecosystem. The question is whether the earth itself can be considered a ecosystem that is threatened by human population growth.

But even if such a concept can be applied to human population, how would population growth be controlled? Who would make such decisions? On whom would the burden fall? How would the burden be distributed? Some argue that the population growth should be limited in those areas of the world where growing population already is creating health and environmental problems. Others argue that the industrialized countries of the northern hemisphere, with their high rate of resource use and economic activities, should first change their ways of life. On this view, the problems lie more with the distribution of economic and environmental burdens and benefits than they do with population size alone. This perspective emphasizes the need to change our approach to economic activity, controlling economic growth rather than controlling population growth.

But what would an economy that doesn't grow look like? Within economic theory, lack of growth is undesirable; it is equated with economic depression, unemployment, human suffering, and misery. Economic growth seems the sine qua non for human happiness. Without growth, increasing population will only face more unemployment, less food, fewer goods and services, more poverty and misery.

In recent decades, sustainable economics has arisen as an alternative to traditional growth-based market economies. Traditional economics addresses only questions of production and distribution. Resources ought to be allocated to those productive activities that most efficiently meet consumer demand. To these two fundamental questions, sustainable economics adds the question of scale: What is the optimal rate at which resources should move through the economy? Even in classically efficient economies, too many resources moving through the economy too quickly will result in social and ecological disaster. Instead, resources should move through at a rate that is sustainable.

Economist Herman Daly explains the difference in terms of a distinction between *growth* and *development*. Growth involves getting bigger in size by adding material. Development implies a qualitative improvement. There are, Daly reminds us, biophysical limits to growth. An economy cannot grow forever, and the concept of sustainable growth is an oxymoron. But economic development can be sustained indefinitely. Our goal should be a qualitative improvement in human life, not simply an economy that supplies more and more things.

But even the concept of sustainable development is not without controversy. A question that should be raised for any discussion of sustainability is: What do you want to sustain? Some critics charge that sustainability is attractive to wealthy countries only when it means sustaining the present way of life. But as long as industrialized countries maintain their way of life, according to these critics, worldwide sustainability can occur only by denying economic growth to poorer countries. On this view, sustainable development must include radical changes in the economic and consumption patterns among the industrial and postindustrial countries.

Lifeboat Ethics

Garrett Hardin

For many environmentalists, the image of Spaceship Earth is a helpful metaphor for understanding the present ecological situation. Popularized by economist Kenneth Boulding in the 1960s, this metaphor encourages us to conserve resources and eliminate pollution. After all, there are limits to growth within a spaceship.

In the following essay, Garrett Hardin argues that the spaceship metaphor should be replaced with the image of a lifeboat. Like us, people in a lifeboat are already in a crisis situation. Some, the comparatively rich, have reached a lifeboat that is relatively well-supplied. Others, the poor of the world, are on overcrowded lifeboats. What responsibilities do the people in the affluent world, those in the well-supplied lifeboats, have to those in the overcrowded lifeboats? The worst-case scenario would be one in which everyone drowns, in which our species does not survive. This will happen only if the well-off jeopardize their survival by coming to the aid of the poor. Lifeboat ethics demands that we not do this. We owe it to ourselves and we owe it to posterity.

No generation has viewed the problem of the survival of the human species as seriously as we have. Inevitably, we have entered this world of concern through the door of metaphor. Environmentalists have emphasized the image of the earth as a spaceship—Spaceship Earth. Kenneth Boulding is the principal architect of this metaphor. It is time, he says, that we replace the wasteful "cowboy economy" of the past with the frugal "spaceship economy" required for continued survival in the limited world we now see ours to be. The metaphor is notably useful in justifying pollution control measures.

Unfortunately, the image of a spaceship is also used to promote measures that are suicidal. One of these is a generous immigration policy, which is only a particular instance of a class of policies that are in error because they lead to the tragedy of the commons. These suicidal policies are attractive because they mesh with what we unthinkably take to be the ideals of "the best people." What is missing in the idealistic view is an insistence that rights and responsibilities must go together. The "generous" attitude of all too many people results in asserting inalienable rights while ignoring or denying matching responsibilities.

For the metaphor of a spaceship to be correct the aggregate of people on board would have to be under unitary sovereign control. A true ship always has a captain. It is conceivable that a ship could be run by a committee. But it could not possibly survive if its course were determined by bickering tribes that claimed rights without responsibilities.

What about Spaceship Earth? It certainly has no captain, and no executive committee. The United Nations is a toothless tiger, because the signatories of its charter wanted it that way. The spaceship metaphor is used only to justify spaceship demands on common resources without acknowledging corresponding spaceship responsibilities.

An understandable fear of decisive action leads people to embrace "incrementalism"—moving toward reform by tiny stages. As we shall see, this strategy is counterproductive in the area discussed here if it means accepting rights before responsibilities. Where human survival is at stake, the acceptance of responsibilities is a precondition to the acceptance of rights, if the two cannot be introduced simultaneously.

LIFEBOAT ETHICS

Before taking up certain substantive issues let us look at an alternative metaphor, that of a lifeboat. In developing some relevant examples the following numerical values are assumed. Approximately two-thirds of the world is desperately poor, and only one-third is comparatively rich. The people in poor countries have an average per capita GNP (Gross National Product) of about $200 per year; the rich, of about $3,000. (For the United States it is nearly $5,000 per year.) Metaphorically, each rich nation amounts to a lifeboat full of comparatively rich people. The poor of the world are in other, much more crowded lifeboats. Continuously, so to speak, the poor fall out of their lifeboats and swim for a while in the water outside, hoping to be admitted to a rich lifeboat, or in some other way to benefit from the "goodies" on board. What should the passengers on a rich lifeboat do? This is the central problem of "the ethics of a lifeboat."

First we must acknowledge that each lifeboat is effectively limited in capacity. The land of every nation has a limited carrying capacity. The exact limit is a matter for argument, but the energy crunch is convincing more people every day that we have already exceeded the carrying capacity of the land. We have been living on "capital"—stored petroleum and coal—and soon we must live on income alone.

Let us look at only one lifeboat—ours. The ethical problem is the same for all, and is as follows. Here we sit, say 50 people in a lifeboat. To be generous, let us assume our boat has a capacity of 10 more, making 60. (This, however, is to violate the engineering principle of the "safety factor." A new plant disease or a bad change in the weather may decimate our population if we don't preserve some excess capacity as a safety factor.)

The 50 of us in the lifeboat see 100 others swimming in the water outside, asking for admission to the boat, or for handouts. How shall we respond to their calls? There are several possibilities.

1. We may be tempted to try to live by the Christian ideal of being "our brother's keeper," or by the Marxian ideal of "from each according to his abilities, to each according to his needs." Since the needs of all are the same, we take all the needy into our boat, making a total of 150 in a boat with a capacity of 60. The boat is swamped, and everyone drowns. Complete justice, complete catastrophe.

2. Since the boat has an unused excess capacity of 10, we admit just 10 more to it. This has the disadvantage of getting rid of the safety factor, for which action we will sooner or later pay dearly. Moreover, *which* 10 do we let in? "First come, first served?" The best 10? The neediest 10? How do we *discriminate*? And what do we say to the 90 who are excluded?

3. Admit no more to the boat and preserve the small safety factor. Survival of the people in the lifeboat is then possible (though we shall have to be on our guard against boarding parties).

The last solution is abhorrent to many people. It is unjust, they say. Let us grant that it is.

"I feel guilty about my good luck," say some. The reply to this is simple: *Get out and yield your place to others.* Such a selfless action might satisfy the conscience of those who are addicted to guilt but it would not change the ethics of the lifeboat. The needy person to whom a guilt-addict yields his place will not himself feel guilty about his sudden good luck. (If he did he would not climb aboard.) The net result of conscience-stricken people relinquishing their unjustly held positions is the elimination of their kind of conscience from the lifeboat. The lifeboat, as it were, purifies itself of guilt. The ethics of the lifeboat persist, unchanged by such momentary aberrations.

This then is the basic metaphor within which we must work out our solutions. Let us enrich the image step by step with substantive additions from the real world.

REPRODUCTION

The harsh characteristics of lifeboat ethics are heightened by reproduction, particularly by reproductive differences. The people inside the lifeboats of the wealthy nations are doubling in numbers every 87 years; those outside are doubling every 35 years, on the average. And the relative difference in prosperity is becoming greater.

Let us, for a while, think primarily of the U.S. lifeboat. As of 1973 the United States had a population of 210 million people, who were increasing by 0.8% per year, that is, doubling in number every 87 years.

Although the citizens of rich nations are outnumbered two to one by the poor, let us imagine an equal number of poor people outside our lifeboat—a mere 210 million poor people reproducing at a quite differ-

ent rate. If we imagine these to be the combined populations of Colombia, Venezuela, Ecuador, Morocco, Thailand, Pakistan, and the Philippines, the average rate of increase of the people "outside" is 3.3% per year. The doubling time of this population is 21 years.

Suppose that all these countries, and the United States, agreed to live by the Marxian ideal, "to each according to his needs," the ideal of most Christians as well. Needs, of course, are determined by population size, which is affected by reproduction. Every nation regards its rate of reproduction as a sovereign right. If our lifeboat were big enough in the beginning it might be possible to live *for a while* by Christian-Marxian ideals. *Might.*

Initially, in the model given, the ratio of non-Americans to Americans would be one to one. But consider what the ratio would be 87 years later. By this time Americans would have doubled to a population of 420 million. The other group (doubling every 21 years) would now have swollen to 3,540 million. Each American would have more than eight people to share with. How could the lifeboat possibly keep afloat?

All this involves extrapolation of current trends into the future, and is consequently suspect. Trends may change. Granted: but the change will not necessarily be favorable. If—as seems likely—the rate of population increase falls faster in the ethnic group presently inside the lifeboat than it does among those now outside, the future will turn out to be even worse than mathematics predicts, and sharing will be even more suicidal.

RUIN IN THE COMMONS

The fundamental error of the sharing ethic is that it leads to the tragedy of the commons. Under a system of private property the man (or group of men) who own property recognize their responsibility to care for it, for if they don't they will eventually suffer. A farmer, for instance, if he is intelligent, will allow no more cattle in a pasture than its carrying capacity justifies. If he overloads the pasture, weeds take over, erosion sets in, and the owner loses in the long run.

But if a pasture is run as a commons open to all, the right of each to use it is not matched by an operational responsibility to take care of it. It is no use asking independent herdsmen in a commons to act responsibly, for they dare not. The considerate herdsman who refrains from overloading the commons suffers more than a selfish one who says his needs are greater. (As Leo Durocher says, "Nice guys finish last.") Christian-Marxian idealism is counterproductive. That it *sounds* nice is no excuse. With distribution systems, as with individual morality, good intentions are no substitute for good performance.

A social system is stable only if it is insensitive to errors. To the Christian-Marxian idealist a selfish person is a sort of "error." Prosperity in the system of the commons cannot survive errors. If *everyone* would only restrain himself, all would be well; but it takes *only one less than everyone* to ruin a system of voluntary restraint. In a crowded world of less than perfect human beings—and we will never know any other—mutual ruin is inevitable in the commons. This is the core of the tragedy of the commons.

WORLD FOOD BANKS

In the international arena we have recently heard a proposal to create a new commons, namely an international depository of food reserves to which nations will contribute according to their abilities, and from which nations may draw according to their needs. Nobel laureate Norman Borlaug has lent the prestige of his name to this proposal.

A world food bank appeals powerfully to our humanitarian impulses. We remember John Donne's celebrated line, "Any man's death diminishes me." But before we rush out to see for whom the bell tolls let us recognize where the greatest political push for international granaries comes from, lest we be disillusioned later. Our experience with Public Law 480 clearly reveals the answer. This was the law that moved billions of dollars worth of U.S. grain to food-short, population-long countries during the past two decades. When P.L. 480 first came into being, a headline in the business magazine *Forbes* revealed the power behind it: "Feeding the World's Hungry Millions: How it will mean billions for U.S. business."

And indeed it did. In the years 1960 and to 1970 a total of $7.9 billion was spent on the "Food for Peace" program, as P.L. 480 was called. During the years 1948 to 1970 an additional $49.9 billion were extracted from American taxpayers to pay for other economic aid programs, some of which went for food and food-producing machinery. (This figure does *not* include military aid.) That P.L. 480 was a give-away program was concealed. Recipient countries went through the motions of paying for P.L. 480 food—with IOU's. In December 1973 the charade was brought to an end as far as India was concerned when the United States "forgave" India's $3.2 billion debt. Public announcement of the cancellation of the debt was delayed for two months: one wonders why.

The search for a rational justification can be short-circuited by interjecting the word "emergency." Borlaug uses this word. We need to look sharply at it. What is an "emergency"? It is surely something like an accident, which is correctly defined as *an event that is certain to happen, though with a low frequency.* A well-run organization prepares for everything that is certain, including accidents and emergencies. It budgets for them. It saves for them. It expects them—and mature decision-makers do not waste time complaining about accidents when they occur.

What happens if some organizations budget for emergencies and others do not? If each organization is solely responsible for its own well-being, poorly managed ones will suffer. But they should be able to learn from experience. They have a chance to mend their ways and learn to budget for infrequent but certain emergencies. The weather, for instance, always varies and periodic crop failures are certain. A wise and competent government saves out of the production of the good years in anticipation of bad years that are sure to come. This is not a new idea. The Bible tells us that Joseph taught this policy to Pharaoh in Egypt more than 2,000 years ago. Yet it is literally true that the vast majority of the governments of the world today have no such policy. They lack either the wisdom or the competence, or both. Far more difficult than the transfer of wealth from one country to another is the transfer of wisdom between sovereign powers or between generations.

"But it isn't their fault! How can we blame the poor people who are caught in an emergency? Why must we punish them?" The concepts of blame and punishment are irrelevant. The question is, what are the operational consequences of establishing a world food bank? If it is open to every country every time a need develops, slovenly rulers will not be motivated to take Joseph's advice. Why should they? Others will bail them out whenever they are in trouble.

Some countries will make deposits in the world food bank and others will withdraw from it: there will be almost no overlap. Calling such a depository-transfer unit a "bank" is stretching the metaphor of *bank* beyond its elastic limits. The proposers, of course, never call attention to the metaphorical nature of the word they use.

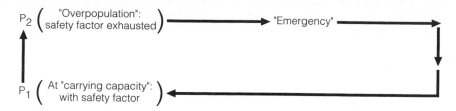

Figure 1. The population cycle of a nation that has no effective, conscious population control, and which receives no aid from the outside. P_2 is greater than P_1.

THE RATCHET EFFECT

An "international food bank" is really, then, not a true bank but a disguised one-way transfer device for moving wealth from rich countries to poor. In the absence of such a bank, in a world inhabited by individually responsible sovereign nations, the population of each nation would repeatedly go through a cycle of the sort shown in Figure 1. P_2 is greater than P_1, either in absolute numbers or because a deterioration of the food supply has removed the safety factor and produced a dangerously low ratio of resources to population. P_2 may be said to represent a state of overpopulation, which becomes obvious upon the appearance of an "accident," e.g., a crop failure. If the "emergency" is not met by outside help, the population drops back to the "normal" level—the "carrying capacity" of the environment—or even below. In the absence of population control by a sovereign, sooner or later the population grows to P_2 again and then the cycle repeats. The long-term population curve is an irregularly fluctuating one, equilibrating more or less about the carrying capacity.

A demographic cycle of this sort obviously involves great suffering in the restrictive phase, but such a cycle is normal to any independent country with inadequate population control. The third century theologian Tertullian expressed what must have been the recognition of many wise men when he wrote: "The scourges of pestilence, famine, wars, and earthquakes have come to be regarded as a blessing to overcrowded nations, since they serve to prune away the luxuriant growth of the human race."

Only under a strong and farsighted sovereign—which theoretically could be the people themselves, democratically organized—can a population equilibrate at some set point below the carrying capacity, thus avoiding the pains normally caused by periodic

and unavoidable disasters. For this happy state to be achieved it is necessary that those in power be able to contemplate with equanimity the "waste" of surplus food in times of bountiful harvests. It is essential that those in power resist the temptation to convert extra food into extra babies. On the public relations level it is necessary that the phrase "surplus food" be replaced by "safety factor."

But wise sovereigns seem not to exist in the poor world today. The most anguishing problems are created by poor countries that are governed by rulers insufficiently wise and powerful. If such countries can draw on a world food bank in times of "emergency," the population *cycle* of Figure 1 will be replaced by the population *escalator* of Figure 2. The input of food from a food bank acts as the pawl of a ratchet, preventing the population from retracing its steps to a lower level. Reproduction pushes the population upward, inputs from the world bank prevent its moving downward. Population size escalates, as does the absolute magnitude of "accidents" and "emergencies." The process is brought to an end only by the total collapse of the whole system, producing a catastrophe of scarcely imaginable proportions.

Such are the implications of the well-meant sharing of food in a world of irresponsible reproduction.

All this is terribly obvious once we are acutely aware of the pervasiveness and danger of the commons. But many people still lack this awareness and the euphoria of the "benign demographic transition" interferes with the realistic appraisal of pejoristic mechanisms. As concerns public policy, the deductions drawn from the benign demographic transition are these:

1. If the per capita GNP rises the birth rate will fall; hence, the rate of population increase will fall, ultimately producing ZPG (Zero Population Growth).

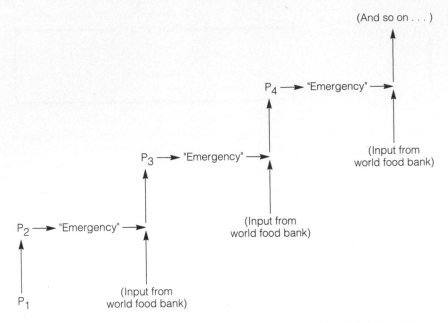

Figure 2. The population escalator. Note that input from a world food bank acts like the pawl of a ratchet, preventing the normal population cycle shown in Figure 1 from being completed. P_{n+1} is greater than P_n, and the absolute magnitude of the "emergencies" escalates. Ultimately the entire system crashes. The crash is not shown, and few can imagine it.

2. The long-term trend all over the world (including the poor countries) is of a rising per capita GNP (for which no limit is seen).

3. Therefore, all political interference in population matters is unnecessary; all we need to do is foster economic "development"—*note the metaphor*—and population problems will solve themselves.

Those who believe in the benign demographic transition dismiss the pejoristic mechanism of Figure 2 in the belief that each input of food from the world fosters development within a poor country thus resulting in a drop in the rate of population increase. Foreign aid has proceeded on this assumption for more than two decades. Unfortunately it has produced no indubitable instance of the asserted effect. It has, however, produced a library of excuses. The air is filled with plaintive calls for more massive foreign aid appropriations so that the hypothetical melioristic process can get started.

The doctrine of demographic laissez-faire implicit in the hypothesis of the benign demographic transi-

tion is immensely attractive. Unfortunately there is more evidence against the melioristic system than there is for it. On the historical side there are many counter-examples. The rise in per capita GNP in France and Ireland during the past century has been accompanied by a rise in population growth. In the 20 years following the Second World War the same positive correlation was noted almost everywhere in the world. Never in world history before 1950 did the worldwide population growth reach 1% per annum. Now the average population growth is over 2% and shows no signs of slackening.

On the theoretical side, the denial of the pejoristic scheme of Figure 2 probably springs from the hidden acceptance of the "cowboy economy" that Boulding castigated. Those who recognize the limitations of a spaceship, if they are unable to achieve population control at a safe and comfortable level, accept the necessity of the corrective feedback of the population cycle shown in Figure 1. No one who knew in his bones that he was living on a true spaceship would countenance political support of the population escalator shown in Figure 2.

ECO-DESTRUCTION VIA THE GREEN REVOLUTION

The demoralizing effect of charity on the recipient has long been known. "Give a man a fish and he will eat for a day: teach him how to fish and he will eat for the rest of his days." So runs an ancient Chinese proverb. Acting on this advice the Rockefeller and Ford Foundations have financed a multipronged program for improving agriculture in the hungry nations. The result, known as the "Green Revolution," has been quite remarkable. "Miracle wheat" and "miracle rice" are splendid technological achievements in the realm of plant genetics.

Whether or not the Green Revolution can increase food production is doubtful, but in any event not particularly important. What is missing in this great and well-meaning humanitarian effort is a firm grasp of fundamentals. Considering the importance of the Rockefeller Foundation in this effort it is ironic that the late Alan Gregg, a much-respected vice president of the Foundation, strongly expressed his doubts of the wisdom of all attempts to increase food production some two decades ago. (This was before Borlaug's work—supported by Rockefeller—had resulted in the development of "miracle wheat.") Gregg likened the growth and spreading of humanity over the surface of the earth to the metastasis of cancer in the human body, wryly remarking that "Cancerous growths demand food; but, as far as I know, they have never been cured by getting it."

"Man does not live by bread alone"—the scriptural statement has a rich meaning even in the material realm. Every human being born constitutes a draft on all aspects of the environment—food, air, water, unspoiled scenery, occasional and optional solitude, beaches, contact with wild animals, fishing, hunting—the list is long and incompletely known. Food can, perhaps, be significantly increased: but what about clean beaches, unspoiled forests, and solitude? If we satisfy the need for food in a growing population we necessarily decrease the supply of other goods, and thereby increase the difficulty of equitably allocating scarce goods.

The present population of India is 600 million, and it is increasing by 15 million per year. The environmental load of this population is already great. The forests of India are only a small fraction of what they were three centuries ago. Soil erosion, floods, and the psychological costs of crowding are serious. Every one of the net 15 million lives added each year stresses the Indian environment more severely. *Every life saved this year in a poor country diminishes the quality of life for subsequent generations.*

Observant critics have shown how much harm we wealthy nations have already done to poor nations through our well-intentioned but misguided attempts to help them. Particularly reprehensible is our failure to carry out postaudits of these attempts. Thus we have shielded our tender consciences from knowledge of the harm we have done. Must we Americans continue to fail to monitor the consequences of our external "do-gooding"? If, for instance, we thoughtlessly make it possible for the present 600 million Indians to swell to 1,200 millions by the year 2001—as their present growth rate promises—will posterity in India thank *us* for facilitating an even greater destruction of *their* environment? Are good intentions ever a sufficient excuse for bad consequences?

IMMIGRATION CREATES A COMMONS

I come now to the final example of a commons in action, one for which the public is least prepared for rational discussion. The topic is at present enveloped by a great silence which reminds me of a comment made by Sherlock Holmes in A. Conan Doyle's story, "Silver Blaze." Inspector Gregory had asked, "Is there any point to which you would wish to draw my attention?" To this Holmes responded:

"To the curious incident of the dog in the night-time."

"The dog did nothing in the night-time," said the Inspector.

"That was the curious incident," remarked Sherlock Holmes.

By asking himself what would repress the normal barking instinct of a watchdog Holmes realized that it must be the dog's recognition of his master as the criminal trespasser. In a similar way we should ask ourselves what repression keeps us from discussing something as important as immigration?

It cannot be that immigration is numerically of no consequence. Our government acknowledges a *net* flow of 400,000 a year. Hard data are understandably lacking on the extent of illegal entries, but a not implausible figure is 600,000 per year. The natural increase of the resident population is now about 1.7 million per year. This means that the yearly gain from immigration is at least 19%, and may be 37%, of the total increase. It is quite conceivable that educational campaigns like that of Zero Population Growth, Inc., coupled with adverse social and economic factors— inflation, housing shortage, depression, and loss of confidence in national leaders—may lower the fertility of American women to a point at which all of the yearly increase in population would be accounted for by immigration. Should we not at least ask if that is what we want? How curious it is that we so seldom discuss immigration these days!

Curious, but understandable—as one finds out the moment he publicly questions the wisdom of the status quo in immigration. He who does so is promptly charged with *isolationism, bigotry, prejudice, ethnocentrism, chauvinism,* and *selfishness.* These are hard accusations to bear. It is pleasanter to talk about other matters, leaving immigration policy to wallow in the cross-currents of special interests that take no account of the good of the whole—*or of the interests of posterity.*

We Americans have a bad conscience because of things we said in the past about immigrants. Two generations ago the popular press was rife with references to *Dagos, Wops, Pollacks, Japs, Chinks,* and *Krauts*—all pejorative terms which failed to acknowledge our indebtedness to Goya, Leonardo, Copernicus, Hiroshige, Confucius, and Bach. Because the implied inferiority of foreigners was *then* the justification for keeping them out, it is *now* thoughtlessly assumed that restrictive policies can only be based on the assumption of immigrant inferiority. *This is not so.*

Existing immigration laws exclude idiots and known criminals; future laws will almost certainly continue this policy. But should we also consider the quality of the average immigrant, as compared with the quality of the average resident? Perhaps we should, perhaps we shouldn't. (What is "quality" anyway?) But the quality issue is not our concern here.

From this point on, *it will be assumed that immigrants and native-born citizens are of exactly equal quality,* however quality may be defined. The focus is only on quantity. The conclusions reached depend on nothing else, so all charges of ethnocentrism are irrelevant.

World food banks move food to the people, thus facilitating the exhaustion of the environment of the poor. By contrast, unrestricted immigration moves people to the food, thus speeding up the destruction of the environment in rich countries. Why poor people should want to make this transfer is no mystery; but why should rich hosts encourage it? This transfer, like the reverse one, is supported by both selfish interests and humanitarian impulses.

The principal selfish interest in unimpeded immigration is easy to identify; it is the interest of the employers of cheap labor, particularly that needed for degrading jobs. We have been deceived about the forces of history by the lines of Emma Lazarus inscribed on the Statue of Liberty:

Give me your tired, your poor
Your huddled masses yearning to breathe free,
The wretched refuse of your teeming shore.
Send these, the homeless, tempest-tossed, to me:
I lift my lamp beside the golden door.

The image is one of an infinitely generous earth-mother, passively opening her arms to hordes of immigrants who come here on their own initiative. Such an image may have been adequate for the early days of colonization, but by the time these lines were written (1886) the force for immigration was largely manufactured inside our own borders by factory and mine owners who sought cheap labor not to be found among laborers already here. One group of foreigners after another was thus enticed into the United States to work at wretched jobs for wretched wages.

At present, it is largely the Mexicans who are being so exploited. It is particularly to the advantage of certain employers that there be many illegal immigrants. Illegal immigrant workers dare not complain about their working conditions for fear of being repatriated. Their presence reduced the bargaining power of all Mexican-American laborers. Cesar Chavez has repeatedly pleaded with congressional committees to close the doors to more Mexicans so that those here can negotiate effectively for higher wages and decent working conditions. Chavez understands the ethics of a lifeboat.

The interests of the employers of cheap labor are well served by the silence of the intelligentsia of the country. WASPS—White Anglo-Saxon Protestants— are particularly reluctant to call for a closing of the doors to immigration for fear of being called ethnocentric bigots. It was, therefore, an occasion of pure delight for this particular WASP to be present at a meeting when the points he would like to have made were made better by a non-WASP speaking to other non-WASPS. It was in Hawaii, and most of the people in the room were second-level Hawaiian officials of Japanese ancestry. All Hawaiians are keenly aware of the limits of their environment, and the speaker had asked how it might be practically and constitutionally possible to close the doors to more immigrants to the islands. (To Hawaiians, immigrants from the other 49 states are as much of a threat as those from other nations. There is only so much room in the islands, and the islanders know it. Sophistical arguments that imply otherwise do not impress them.)

Yet the Japanese-Americans of Hawaii have active ties with the land of their origin. This point was raised by a Japanese-American member of the audience who asked the Japanese-American speaker: "But how can we shut the doors now? We have many friends and relations in Japan that we'd like to bring to Hawaii some day so that they can enjoy this beautiful land."

The speaker smiled sympathetically and responded slowly: "Yes, but we have children now and someday we'll have grandchildren. We can bring more people here from Japan only by giving away some of the land that we hope to pass on to our grandchildren some day. What right do we have to do that?"

To be generous with one's own possessions is one thing; to be generous with posterity's is quite another. This, I think, is the point that must be gotten across to those who would, from a commendable love of distributive justice, institute a ruinous system of the commons, either in the form of a world food bank or that of unrestricted immigration. Since every speaker is a member of some ethnic group it is always possible to charge him with ethnocentrism. But even after purging an argument of ethnocentrism the rejection of the commons is still valid and necessary if we are to save at least some parts of the world from environmental ruin. Is it not desirable that at least some of the grandchildren of people now living should have a decent place in which to live?

Plainly many new problems will arise when we consciously face the immigration question and seek rational answers. No workable answers can be found if we ignore population problems. And—if the argument of this essay is correct—so long as there is no true world government to control reproduction everywhere it is impossible to survive in dignity if we are to be guided by Spaceship ethics. Without a world government that is sovereign in reproductive matters mankind lives, in fact, on a number of sovereign lifeboats. For the foreseeable future survival demands that we govern our actions by the ethics of a lifeboat. Posterity will be ill served if we do not.

For Further Discussion

1. In your own opinion, is a lifeboat or a spaceship a better metaphor for the present environment situation? What are the strengths and weaknesses of each metaphor?

2. Hardin suggests that the Christian ideal of being our brother's keeper would result in a catastrophe in a lifeboat. Do you agree with his understanding of this ethical ideal?

3. What reasons are offered for rejecting the idea of a world food bank? Are you convinced by them?

4. Do you agree with Hardin that the need for cheap labor is a major reason for allowing continuous and unimpeded immigration? Who benefits from immigration policy? Who is burdened by it?

5. People do not ordinarily end up in a lifeboat until a larger ship sinks. Why is our larger ship sinking? Is it worth asking who is responsible for the disaster? Should innocent passengers receive priority for getting into the lifeboat?

Fables About Population and Food

Paul Ehrlich
Anne Ehrlich

As we noticed in Chapter 3, not everyone agrees that we face serious environmental and population problems. Many deny that population growth, economic growth, and the supply of natural resources are on a collision course. In this essay, which is Chapter 5 from their book Betrayal of Science and Reason, *Paul and Anne Ehrlich respond to these more optimistic scenarios. The Ehrlichs refute ten separate claims, what they identify as fables, made by those who deny the reality of a population and food crisis. In reading this essay, it would be helpful to return to Chapter 3 to review the selections by Simon, Easterbrook, and Bast, Hill, and Rue, all of whom are discussed in the following reading.*

Ever since Reverend Thomas Malthus at the end of the eighteenth century warned about the dangers of overpopulation, analysts have been concerned about maintaining a balance between human numbers and the human food supply. That concern remains valid today in a world where a tenth of the population goes to bed hungry each night and millions die every year from hunger-related causes. Few subjects have been closer to our hearts and minds for the past three decades than the race between population growth and increasing food production. That race was a major focus of Paul's first popular book, *The Population Bomb,* published in 1968; it was also the principal focus of our 1995 book, written with our colleague Gretchen Daily, *The Stork and the Plow.*

The world food situation has been a favorite arena for brownlash writers and spokespeople who deny, often vehemently, that a growing population might someday run into absolute food shortages. The essence of their argument takes two forms: population growth is not a problem and (for some of them) is even virtually an unmitigated blessing; and food production can be increased more or less forever without constraint. Some of the more extreme holders of the latter view still occasionally quote an old and long-discredited estimate publicized by Catholic bishops several years ago that theoretically 40 billion people could be fed on Earth. Needless to say, these groups aren't fond of our positions on these matters, and the brownlash attacks our analyses regularly.

Let's take a look at some of the brownlash claims about population and food. Here and throughout the rest of this book, we summarize or directly quote brownlash statements in boldface, then present what we believe to be the consensus or majority view of environmental scientists. We begin with one of the most extreme brownlash claims.

- "We now have in our hands—in our libraries, really—the technology to feed, clothe, and supply energy to an ever-growing population for the next 7 billion years." (Julian Simon, 1994)

Does Julian Simon really mean to suggest the world's population can continue to grow for billions of years at the rate it was growing when he wrote that statement? The world population was growing in 1994 at the rate of about 1.6 percent per year, which corresponds to a doubling time of about 43 years. A bit of arithmetic reveals that such a population growth rate could not persist even for hundreds of years, let alone

millions of years. To suggest that an "ever-growing" population can persist for billions of years is, of course, ridiculous.

Consider how long it would take for the 1994 world population of 5.6 billion to increase to a size where there were *ten human beings for each square meter* of ice-free land on the planet. At the 1994 growth rate, it would take only 18 doublings to bring the population to that point, and the population was then doubling every 43 years. Thus the required doublings would take only $18 \times 43 = 774$ years—somewhat short of 7 billion. After 1900 years at this growth rate, the mass of the human population would be equal to the mass of the Earth; after 6000 years, the mass of the human population would equal the mass of the universe.

Of course, Professor Simon may have had a somewhat lower population growth rate in mind than that of 1994, the year he made his remarkable statement. But if the growth rate were *one million times* smaller than the actual 1994 value—that is, if it were only an infinitesimal 0.0000016 percent per year—Earth's population would still reach a mass exceeding that of the universe before the end of the 7-billion-year period Simon mentioned. Such is the power of exponential growth. Simon's statement is nonsense, pure and simple. But it is only an extreme version of claims of other technological optimists who, living on a planet rife with human hunger, glibly assert that many multiples of the present population can be fed.

- There is no overpopulation today.

This is a popular theme with many brownlash writers. To understand how fallacious this statement is requires recognizing that overpopulation can be reached very quickly by exponentially growing populations in situations of seeming abundance. There is overpopulation when organisms (people in this case) become so numerous that they degrade the ability of the environment to support their kind of animal in the future. The number of people Earth can support *in the long term* (without degrading the environment)—giving existing socioeconomic systems, consumption patterns, and technological capabilities—is called the *human carrying capacity* of the planet at that time. And carrying capacity can be exceeded without causing immediate effects obvious to the untutored observer. "Overshoots" commonly occur in nature with all kinds of organisms. A population has an "outbreak," grows far beyond its carrying capacity, consumes its resources (for animals, usually food), and "crashes" to a size far below the previous carrying capacity.

The surprise element in overshoots and the finite nature of all habitats (including Mother Earth) to support organisms is recognized in the saying, "A long history of exponential growth does not imply a long future of such growth." This can be illustrated simply. Suppose you have an aquarium that has a carrying capacity of 1000 guppies—more than that will start to exhaust the oxygen supply. Suppose the tank is stocked with a pair of adult guppies, and that their numbers grow exponentially with a doubling time of one month. After eight doublings, taking eight months, the guppies reach half the tank's carrying capacity: $2 \rightarrow 4 \rightarrow 8 \rightarrow 16 \rightarrow 32 \rightarrow 64 \rightarrow 128 \rightarrow 256 \rightarrow 512$ guppies. For this whole period, the population is safely far below the tank's carrying capacity. Then, in the ninth month, a further doubling to 1024 suddenly results in an overshoot of carrying capacity—with the last 100 guppies added in less than five days. After 35 weeks of apparent prosperity, overshoot occurs within a week. At first, there are no obvious symptoms, but gradually more and more of the fishes are gasping at the surface, and then they begin to die.

Overshoots can occur in human populations, too. Humanity has already overshot Earth's carrying capacity by a simple measure: no nation is supporting its present population on *income*—that is, the sustainable flow of renewable resources. Instead, key "renewable" resources, the natural *capital* of humanity, are being used so rapidly that they have become effectively non-renewable. *Homo sapiens* is collectively acting like a person who happily writes ever larger checks without considering what's happening to the balance of the account.

Warning signs that the human enterprise is nearing the end of exponential growth include declines in the amount or availability of good farmland, soil, fresh water, and biodiversity, all of which are crucial elements of natural capital essential for sustaining humanity and especially for sustaining agricultural production. A more fundamental but indirect indicator of how close humanity is to its limits is that it is already consuming, co-opting, or destroying some 40 percent of the terrestrial food supply of all animals (not just human beings).

Deep, rich agricultural soils are being eroded away in many areas at rates of inches per decade; soils are normally formed at rates of inches per millennium. Accumulations of "fossil" fresh water, stored underground over thousands of years during glacial periods,

are being mined as if they were metals—and often for low-value uses such as irrigating forage crops. Water from aquifers that are recharged at rates measured in inches per year is being pumped out at rates measured in feet per year—and the freshwater-holding capacity of the aquifers is being compromised in the process. Becoming dependent on such largely irreplaceable sources of water—especially for non-essential purposes such as irrigating feed crops in arid regions—is both shortsighted and risky.

The worst form of capital depletion is biological. Microorganisms, plants, and other animals are being exterminated at a rate unprecedented in 65 million years—on the order of 10,000 times faster than the stock can be replaced. These are the working parts of our global life-support system; if we destroy them, the price will be a catastrophic decline in carrying capacity. Natural ecosystems and the species they contain maintain biodiversity and the production of ecosystem goods such as forest products and food from the sea, the harvesting and trading of which are familiar and important parts of the human economy.

Ecosystems also provide essential life-support functions of cleansing, recycling, and renewal, upon which the economy is utterly dependent. From the perspective of agriculture alone, these ecosystem services are vitally important. Among them are amelioration of climate and weather; generation and maintenance of soils and soil fertility; recycling of nutrients; running of the hydrological cycle, which supplies rainfall and irrigation water; control of more than 95 percent of the potential pests of crops; and maintenance of a vast natural library of biodiversity. That library is the source of the innumerable potential and actual ecosystem goods such as medicines and genetic material essential for development of crop varieties resistant to pests and diseases and able to cope with changing conditions, such as adverse weather or increasing soil salinity.

The depletion of environmental capital by people or other animals is a sure sign of overpopulation. But the carrying capacity of an environment is not fixed. Through evolution, organisms can become more efficient in the ways they exploit their environments and thus expand their carrying capacity. In non-human animals, this evolution is mostly genetic and requires numerous generations. Since the animals' environment is also perpetually changing (for example, plants being devoured by a herbivore are coevolving with the herbivore in response to its attacks), it is usually impossible to track changes in carrying capacity.

But human beings are specialists in cultural evolution, which can proceed much more rapidly than can genetic evolution. Through ingenuity and invention, it is possible to enlarge human carrying capacity—as indeed has happened in the past. Today, widespread behavioral changes—such as becoming vegetarian—potentially could increase Earth's carrying capacity for human beings in a short time as well. Assuming full cooperation in the needed changes, it might be possible to support 6 billion people indefinitely (that is, to end human overpopulation, if there were no further population growth). But we doubt that most people in today's rich nations would willingly embrace the changes in lifestyle necessary to increase global carrying capacity. How many Americans would be willing to adjust their lifestyles radically to live, say, like the Chinese, so that more Dutch or Australians or Mexicans could be supported? How many Chinese would give up their dreams of American-style affluence for the same reason? Such lifestyle changes certainly seem unlikely to us, since most current trends among those who can afford it are toward more affluence and consumption, which tend to *decrease* carrying capacity and intensify the degree of overpopulation.

Consider the rise in consumption of animal products that almost always accompanies rising affluence. That behavioral change, now seen around the world, is contributing to the human overshoot. The feeding base for humanity is the 1.8 billion metric tons of cereal grains produced each year by the world's farmers, which amounts to roughly half of all the food produced. Of the grain harvest, about one-third is fed to livestock to produce meat and dairy foods. Unfortunately, anywhere from 60 to 90 percent of the calories fed to animals are lost in the process of supplying the animals' needs first before passing on the energy to people. At least three times as many people could be sustained by directly eating the grain as by eating products from grain-fed animals.

Unfortunately, when nations industrialize and attain higher incomes, a strong trend usually develops toward consumption of more animal products. This trend is spectacularly under way in China today, where the rapid switch of a billion-plus people from almost entirely vegetarian diets to diets based more on meat could have profound repercussions on the world food supply. Although China has been essentially self-sufficient in food production for decades, the country's increasing consumption of animal foods may mean that massive amounts of grain will need to be imported for livestock feed. This would intensify the pressures on the world's agricultural resources, thereby further increasing global overpopulation.

- One needn't worry about population growth in the United States, because it's still nowhere near as densely populated as the Netherlands. (Malcolm Forbes, 1989)

The idea that the number of people per square mile is a key determinant of population pressure is as widespread and persistent as it is wrong—Paul and physicist John Holdren (now at the Kennedy School of Government at Harvard) long ago named it the "Netherlands fallacy." Nicholas Eberstadt, in his contribution to the Competitive Enterprise Institute's book *The True State of the Planet,* wrote: "What are the criteria by which to judge a country 'overpopulated'? Population density is one possibility that comes to mind." He then proceeded to extend the Netherlands fallacy to Bermuda and Monaco.

The fascination with how many people can be crowded into how much land is common to many brownlash writings. In *Eco-Scam,* Ronald Bailey repeats the tired Netherlands fallacy and quotes Eberstadt to the effect that "There is absolutely no content to the notion of overpopulation." In *Apocalypse Not,* Ben Bolch and Harold Lyons point out correctly that if the 1990 world population were placed in Texas, less than half of 1 percent of Earth's land surface, "each person would have an area equal to the floor space of a typical U.S. home." They also say: "Anyone who has looked out an airplane window while traveling across the country knows how empty the United States really is."

Our response is perfectly straightforward. First, the key issue in judging overpopulation is not how many people can fit in any given space but whether the population's requirements for food, water, other resources, and ecosystem services can be met on a sustainable basis. Most of the "empty" land in the United States either grows the food essential to the well-being of Americans and much of the world (as in Iowa) or supplies us with forestry products (northern Maine), or, lacking water, good soil, and a suitable climate (as in much of Nevada), it is land that cannot directly contribute much to the support of civilization.

The key point here is that the Netherlands, Bermuda, and Monaco (and Singapore, Hong Kong, São Paulo, Mexico City, Tokyo, London, and New York) *can be crowded with people only because the rest of the world is not.* The Netherlands, for example, imports large amounts of food and extracts from other parts of the world much of the energy and virtually all the materials it requires. It uses an estimated seventeen times more land for food and energy than exists within its borders.

• "Predictions of a 'population explosion' . . . were wrong because they were based on projections of past trends." (Joseph Bast, Peter Hill, and Richard Rue, 1994) "Nor does [Ehrlich] acknowledge that predicted population growth has not exploded, as he had predicted [sic]." (Dixy Lee Ray and Lou Guzzo, 1993)

When Paul wrote *The Population Bomb* in 1968, there were 3.5 billion people on Earth. Two years later we published a book in which we cited a *projection* of the United Nations, which "predicted" there would be 5.65 billion people in 1995. In 1995 the actual population size turned out to be 5.70 billion, just a little higher than our "prediction." Although our critics continually claim otherwise, we have never "predicted" a future population size or rate of growth but instead have depended entirely on (and have always cited) the work of professional demographers, primarily those at the United Nations and the Population Reference Bureau.

We're perpetually (and correctly) cited as population alarmists, and while we deserve partial credit for sounding the alarm, we can't take credit for the accuracy of the projections we cite. Those projections are made by demographers, who build on history and existing population structures and say what will happen if certain trends continue. The projections are based on reasonable assumptions about birth, death, and migration rates but over time those assumptions may well be violated, moving a population's trajectory away from the projection. Yet given all the variables, the demographers do remarkably well. For example, in 1977, we and John Holdren reprinted demographer Tomas Frejka's high and low projections for 2000— 6.67 and 5.92 billion, respectively. At the moment, Frejka's projections, made in 1974, seem pretty close to the mark: the world's population in 2000 seems likely to fall somewhere between 6.1 and 6.2 billion.

Whether our "predictions" of a population explosion were wrong is not a matter of projection but a matter of history. In the nearly thirty years since the *Bomb* was written, some 2.3 billion people have been added to Earth's population—more than existed when we were born. That's equivalent to the *addition* each year of roughly the present population of Germany. We've already seen more than a doubling of America's population in our lifetime, and the most recent U.S. population projections indicate continued growth well beyond 2050. If we are lucky enough to live to the turn of the century, we'll have seen a *tripling* of the number of human beings with which we share Earth. If that isn't a population explosion, we don't know what is!

• Enviros hate people.

In *A Moment on the Earth,* Gregg Easterbrook correctly quotes something we have been saying in speeches for decades—that no one has ever suggested a sane reason for having more than 135 million Americans. We chose that number because one can make the argument (although it's not a very good one) that a population of that size was necessary to put together a military force large enough to win World War II. That was the largest land war ever likely to be fought, one requiring a lot of "cannon fodder." But Easterbrook came up with a counterargument: a sane reason for wanting more than 135 million Americans is that "people like to be alive."

One can imagine some ripostes to this statement, such as "Do people hate not being born?" or "Would people like to be alive if they had to live like chickens in factory farms?" But ripostes are unnecessary. As we often point out, the best way to maximize the number of Americans (or Chinese, or Nigerians) who live isn't accomplished by cramming into the United States (or China, or Nigeria) as many people as possible in the next few decades until those nations self-destruct. Rather, the way to maximize the number of lives lived is to have permanently sustainable populations in those nations (and on this planet) for tens of thousands, perhaps millions, of years.

As for hating people, virtually everyone we know who is concerned about the population problem is gregarious and likes people—likes them enough to think beyond their brute numbers at any given time and care deeply about the conditions in which they live. Interestingly, many individuals who claim to love people and want the population to keep growing shield themselves from the growing numbers in their own vicinity. For instance, the Vatican relentlessly pushes for unlimited immigration into the United States and other countries while also opposing most forms of birth control. Yet the Vatican City doesn't admit immigrants or refugees, even though it has one-fourth the population density of New York City.

Like the Vatican, many brownlash writers who express a supposed devotion to people are carrying on an old tradition of the upper classes and the Vatican (*not* of Catholics in general): promoting reproduction among the poor. They do this by downplaying the seriousness of the population problem and opposing programs to give people control over their reproductive activities. Not surprisingly, their views are often supported by agribusiness and big corporations, which naturally are still interested in abundant cheap labor, as evidenced by their lobbying efforts against immigration reform. Overall, it seems to us that "enviros" care a lot more about people than do the promoters of the brownlash.

- Modern medicine has eliminated one big threat connected with overpopulation: the rampant spread of infectious diseases.

Brownlash writers generally overlook the deterioration of the epidemiological environment, which is quite directly related to population size as well as to poverty and environmental deterioration. In *A Moment on the Earth,* Gregg Easterbrook does not discuss what the epidemiological risk means for people—especially poor people—even though the hazards are widely discussed in the literature he attacks. Indeed, in an earlier piece criticizing us, he ridiculed (and quoted out of context) Paul's 1968 statement in *The Population Bomb* that it was "not inconceivable" that a novel viral disease could kill 500 million people.

Yet scientists who grapple with outbreaks of infectious disease care deeply. It is instructive to note that the World Health Organization estimates that as many as 40 million people could be infected with HIV (the virus that causes AIDS) by the year 2000; virtually all infected people can be expected to die of the disease unless some other cause of death intervenes or a miracle cure is found in time. Consider the words spoken in 1989 by Joshua Lederberg, microbiologist at Rocke-feller University and recipient of a Nobel Prize for his work on the sharing of genetic material by bacteria: "Nature is not benign. . . . The survival of the human species is *not* a preordained evolutionary program. Abundant sources of genetic variation exist for viruses to learn new tricks, not necessarily confined to what happens routinely or even frequently."

For a number of population-related reasons, human beings are more vulnerable than ever to epidemics. For one thing, the increase in human numbers alone makes us more susceptible. Some infectious agents can persist only in populations above a certain size (the measles virus, for example, requires 50,000 or more). Thus an ever-increasing population opens the door to a new range of pathogens. Many human groups are coming into more frequent contact with wild animal populations, and other animals are the source of most epidemic diseases. Recently, exposure to infected wild animals has led to transfers of viruses such as those that cause AIDS and various hemorrhagic fevers (among them the Marburg, Ebola, Hanta, and Machupo viruses). Increased contact by larger groups of people, frequently associated with the clearing of forests, raises the chances both of a transfer from a wild animal into the human population and of the new disease causing an epidemic. Larger human populations also lead to urbanization, so infected and susceptible people are jammed together in cities, where sexual promiscuity and drug use are often rampant. Cities have rightly been called "ecosystem[s] that can amplify infectious diseases" or, more bluntly, "graveyards of mankind."

To make matters worse, international air travel creates a mechanism for the rapid spread around the globe of infectious agents (bacteria, viruses, fungi, etc.), including drug-resistant strains of bacteria, vectors (organisms such as mosquitoes that move infectious agents from one individual to another), and animal reservoirs (animals that play host to diseases that also infect human beings). Furthermore, human-induced global climate change threatens to increase the range of the tropical vectors of serious diseases such as malaria and dengue fever. And finally, but far from exhaustively, a combination of widespread malnutrition, exposure to immune-suppressing chemical pollutants, and HIV infection has produced a human population with a very large subpopulation of immune-compromised individuals; as easy targets for pathogens, they increase humanity's vulnerability to epidemics. In rich nations such as the United States, infectious diseases are mostly relatively minor causes of death, but there is no guarantee that situation will continue.

- Those who are concerned about world hunger are simply wrong.

The Population Bomb starts with these words: "The battle to feed all of humanity is over. In the 1970s the world will undergo famines—hundreds of millions of people are going to starve to death in spite of any crash programs embarked upon now." In fact, some 250 million people, about as many as now live in the United States, have perished for lack of food since those words were written. Roughly 100 to 140 million of them died in the 1970s. But because the famines of the 1970s were less catastrophic than we predicted, we've probably been criticized more for the "battle is over" statement than any other. Our response is that partly because of *The Population Bomb* and similar warnings by food specialists, global programs to improve food distribution were initiated and emergency systems set up to feed famine victims. Those programs were largely successful in heading off the massive famines we had thought would occur during that decade.

Typical of the criticism directed toward us is an assertion in a brownlash book issued by the Heartland Institute, which claims that "None of [the Ehrlichs'] predictions has come true, or is ever likely to come true." Lawyer Michael Fumento, author of *Science Under Siege,* also quotes our prediction of famine in his book, adding that it was "off by hundreds of millions." But there were substantial food shortages and some acute famines in sub-Saharan Africa in the early 1970s, which recurred throughout the 1980s. Even in the 1990s, the Refugee Policy Group estimated that roughly 150,000 to 200,000 people have starved in acute famines each year. This estimate may be too high, but even if it is double the actual figure, it represents a tragic failure for our civilization.

The principal problem, of course, is not acute famines; it is chronic undernutrition of huge numbers of extremely poor people. Overall, since *The Population Bomb* was published, roughly 8 to 10 million people (mostly young children) have died each year from hunger and hunger-related diseases, according to studies by the World Bank and other international agencies. And such numbers may well be underestimates. First of all, governments don't like to admit they can't feed their people; and second, starvation compromises the immune system, so often the proximate cause of death—the final blow—is not starvation per se but disease.

Today some 700 to 800 million people, perhaps even as many as a billion, don't get enough food to support normal daily activities. Even if the actual number of hungry people were only half as high, it would still indicate a level of human suffering that doesn't match the rosy views of the brownlash. The vast extent of chronic hunger, mainly in developing countries, reflects extreme poverty; hundreds of millions of poor people, largely in rural areas, simply cannot afford to buy sufficient food and lack the means to grow enough for themselves.

Acute famines are a different matter. Although crop failures and production shortfalls continue to occur frequently, modern communications and distribution mechanisms have enabled national and international agencies to prevent large-scale famines by rushing emergency food supplies to the hungry from areas of surplus production. Only when such assistance has been blocked by local wars or politics, as in the Sudan, or, earlier, in Ethiopia, have acute famines resulted. Of course, the severity of the famines and the consequences suffered by the people in these situations are intensified when they occur, as is usually the case, in societies that are already chronically underfed.

In short, we believe the statement that the battle to feed all of humanity is over was correct; we still think it unlikely that large-scale hunger will be banished. Should all human beings become well fed in the future, we shall be delighted to be proven wrong.

• Feeding the world's population is a problem of distribution, not supply. "Famine is a thing of the past for most of the world's people." (Dennis Avery, 1995)

These are two of the most common assertions about food supplies made by the brownlash. There is some truth to the first statement. If there were no maldistribution, if everyone shared equally, and if no grain were fed to animals, all of humanity could be adequately nourished today. Unfortunately, such claims are irrelevant. Although people in developed countries could eat lower on the food chain, it is as unrealistic to think we will all suddenly become vegetarian saints as it is to think we will suddenly trade in our cars for bicycles or go to bed at sunset to save energy.

Overpopulation and carrying capacity are calculated on the basis of animals—and people—as they are, not as we might wish them to be. Human carrying capacity is the *long-term* ability of an area to support human beings. When people are living on natural capital rather than what might be called "natural interest"—sustainable resource flows based on natural capital—then, by definition, carrying capacity is exceeded and the area is overpopulated. The circumstance of overpopulation sometimes can be remedied by changing patterns of behavior without changing the numbers of people—for instance, by adopting vegetarian diets or better soil husbandry practices. Overpopulation exists whenever people trying to produce food allow soil to erode faster than new soil can be generated, or drain aquifers faster than they can be recharged, or exterminate populations and species that are working parts of the ecosystems that support agriculture and fisheries faster than recolonization and speciation can reestablish them. Today overpopulation prevails worldwide.

But suppose everyone *were* willing to eat a largely vegetarian diet today, with only a small supplement from fish and range-fed animals. Suppose, in addition, that people were also willing to share food absolutely equitably, varying only according to the caloric and other nutritional needs of individuals of different ages, sizes, and levels of activity. Today's harvests could feed about 7 billion such vegetarians (assuming that crops to feed humans were planted in place of those now grown to feed livestock). Since our population size is nearly 6 billion already, that is hardly a comforting number.

For comparison, assume that everyone in the world switched to the equivalent of an average Latin American diet, in which some 15 percent of calories are of animal origin (as opposed to 30 percent or more in rich countries). Under such a scenario, only about 4 billion would have enough food; almost 2 billion people—a third of today's population—would get no food at all. In contrast, if everyone were to eat something resembling a North American or northern European diet, only half of today's population could be fed. The other half would just be out of luck.

But humanity shows precious little sign of turning into a species of vegetarians. Indeed, if recent world trends are any indicator, equal sharing is likely to become less, not more, common. So is it sensible to describe today's situation as "just" being caused by maldistribution? We don't think so. Wouldn't it be more sensible to work toward a world in which *everyone* could have a healthy, satisfying diet, regardless of food preferences? Those who say we shouldn't worry about the food situation because it's just a matter of sharing are being wildly overoptimistic about human social behavior.

The statement that famines are history was made by Dennis Avery of the Hudson Institute, a prominent technological optimist about agricultural prospects. We agree with him that most people may be safe from famine at the moment because improved distribution and emergency food programs can head off famine when local food shortages occur. But that says nothing about the future. So Avery's statement cannot be said with assurance to be either true or false. We think his confidence is unwarranted, considering that food production is being undermined by the depletion of natural capital—that is, by erosion and paving of farmland, draining of aquifers, loss of biodiversity, and destruction of marine fisheries. Per-capita yields from the sea are already dropping precipitously, and the prospects of turning that trend around are not encouraging. Can the harvests from land be far behind? We still fear that the apprehensions of environmental scientists will be justified and that Avery's assurance is misplaced, for reasons that will become obvious in what follows.

- Even though millions of people are still inadequately nourished, advances in agriculture will eliminate the remaining pockets of hunger early in the next century.

The argument that technology will save us is a frequent theme of the brownlash, here applied to agriculture. The claim is rooted in past technological successes but is usually made without considering the totally unprecedented nature of today's situation. Technological optimism is nowhere more rampant than in connection with increasing food production.

In addition, agricultural optimists often exaggerate past successes by choosing a time scale for comparison that is congenial to the notion that hunger can be easily eliminated. Thus they compare total food production in 1950 directly with that of the present, calling attention to the great increase in output over the intervening period. This is proof, they suggest, that hunger will be easily eliminated. But simply considering the increase over the entire period obscures the much less favorable trend that started about a decade ago. . . .

The optimists also ignore or understate the depletion of natural capital, biophysical limits beyond which yields simply cannot be increased, and other factors that make a repetition of the 1950–1985 food production surge very unlikely. They do not point out, moreover, that the institutions and infrastructure needed to translate technological developments into greater agricultural productivity are largely absent in food-short (i.e., less developed) regions and that financial support for those institutions has diminished in recent years. Furthermore, they either neglect to mention or dismiss the potential impacts of global change on food production. And, finally, they seem unaware of the principle advanced by the eminent demographer Nathan Keyfitz that "bad policies are widespread and persistent."

Knowledgeable scientists are greatly concerned that these constraints on food production may soon result in serious food shortages. As Mahabub Hossain, head of the Social Sciences Division of the International Rice Research Institute (the organization that created the green revolution in rice, the grain that sustains more people than any other), stated in 1994:

> The race to avoid a collision between population growth and rice production in Asia goes on, amid worrying signs that gains of the recent past may be lost over the next few decades. . . . If [current] trends continue, demand for rice in many parts of Asia will outstrip supply within a few years.

Since Hossain's comment appeared in the *International Herald Tribune,* the development of a new, higher-yielding variety of rice was announced, which is expected to increase rice production roughly 10 percent after full field testing and deployment. This was good news following a period when agronomists feared they had encountered a "yield cap" on rice—that is, a biophysical limit beyond which increased seed production by a rice plant would be impossible. But this latest breakthrough is no guarantee that others will follow; on the contrary, there's every reason to think the limits aren't far off. Moreover, a 10 percent increase in rice production may sound impressive, but set against growing human numbers in Asia (where 90 percent of the world's rice is consumed), that increase would be barely sufficient to support five years of population growth.

We suspect agricultural optimism can be traced mostly to a great faith in the potential of science to pull technological rabbits out of a hat. An example of that optimistic attitude appeared recently in one of the world's best business journals, *The Economist.* The article discussed the probable impact in China of both population growth and increasing per-capita demand for grain (accompanied by losses of farmland). After describing a pessimistic assessment by agricultural policy analyst Lester R. Brown, president of the Worldwatch Institute, and several Chinese scholars, the last paragraph exemplified this blind faith in technology: "Optimists counter the gloomy Malthusians by pointing to as yet unimagined scientific breakthroughs that will boost crop yields around the world. Economists add the simple point that China could then easily export other goods to pay for its imported grain."

The first statement, converted to a sports analogy, is like claiming that "unimagined breakthroughs in training will allow a runner to run a two-minute mile." The second sentence simply ignores the economics of supply and demand. If China's demand skyrockets, grain prices will be driven up, and global access of people to food will quite likely *decrease.* The agricultural scientists we know don't look to "unimagined breakthroughs" to solve their problems; that would hardly be a prudent strategy. Rather, they are deeply concerned about keeping humanity nourished. As for China's grain imports, what other countries produce enough surplus grain to fill the gap for over a billion consumers? China's need for imported grain by 2025 could exceed all the grain traded today on the world market.

Analyses of food production trends over the past few decades suggest that there is indeed cause to worry about maintaining food supplies. It is certainly true that the most important indicator of human nutrition, world grain production, has roughly tripled since 1950. What the food optimists overlook is that the rate of increase has markedly slowed since the 1950s and 1960s. More sobering is that since 1985, grain production increases have failed to keep up with population growth, even though population growth itself has slackened; and since 1990 there has been no increase in absolute terms, causing severe shrinkage in grain reserves by 1996.

There are many reasons for this change. It's true, as Avery points out, that some of the slowdown can be attributed to changes in agricultural policies and land use in rich countries—enacted in part to reduce surpluses. But far more relevant to future food production are tightening constraints such as degradation and losses of land, limited water supplies, and biophysical barriers to increased yields, all of which are increasingly evident.

The agricultural achievements of the past half-century did not come easily; they took a great deal of effort, cooperation, and investment. Even before the great post–World War II surge in agricultural productivity, the tools needed to do the job were in hand: fertilizer-sensitive crops, synthetic fertilizers, and irrigation technologies. Effecting such far-reaching changes required decades of lead time, and the result is now clear, in terms of both impressive success and growing constraints.

The green revolution has already been put in place in most suitable areas of the world, and most of the expected yield gains have been achieved. Farmers now are seeing diminishing returns from fertilizer applications on some major crops over much of the world, and the biological potential for genetically increased yields in some crops may now be approaching the maximum. Even if some unanticipated breakthrough were to be made, it would take many years, if not decades, to develop and deploy new crop varieties—years during which demand would continue rising as the population expands.

A global study by the United Nations Environment Programme found that significant degradation, ranging from slight to severe, has occurred on vegetated land on every continent just since 1945. This deterioration is the result of human activities—chiefly overcultivation, overgrazing, and deforestation. Soils in many areas have deteriorated beyond the ability of fertilizers to mask the impacts of soil erosion. A subsequent study estimated that the capacity of all the world's productive land to supply food and other direct benefits has declined by about 10 percent. And the degradation continues.

Land is not only being rapidly degraded; it is also increasingly being diverted to uses other than food production. Most commonly, land is taken over simply for living space. As the population expands, more people need room for housing, stores, offices, industry, roads, and other infrastructure. Millions of acres of prime farmland are being lost to urbanization each year.

Impending severe water shortages also threaten world food production. Providing a dependable and abundant supply of water is essential for achieving the high yields of modern grain varieties; indeed, the great increase in grain production of past decades can be ascribed largely to the more than doubling of the amount of irrigated land since 1950. But today, most of the readily available sources of irrigation water have already been tapped, and the trend is beginning to reverse. More and more land is being taken out of production because of the rising costs of pumping water from depleted aquifers or because of the classic problems of irrigation: salting up or waterlogging of soils. In addition, urban demands for water often outbid farmers in water-short areas. By 1980, irrigated acreage was no longer expanding faster than the population, and the day when it ceases to increase at all may not be far off.

More ominous, perhaps, is the possible impact of global climate change on agricultural ecosystems. Rapid climate change could deliver the coup de grace to humanity's chances of even restricting hunger to the present levels. Frightening indeed is the possibility that long-standing climatic patterns could be disrupted by more frequent or severe floods or droughts, farming areas could experience too much or too little rainfall or temperatures higher or lower than those at which crops currently planted thrive, and so on. Such changes could be exceedingly disruptive to farming. They would necessitate adjusting to a new regime, during which time food production would likely drop. And since the climate is unlikely simply to shift to a new, stable regime overnight, farmers may suffer a protracted period of grappling with the vicissitudes of an unpredictable climate.

Climatologists on the International Panel on Climate Change think that climate-related agricultural problems will be most severe in low-latitude developing nations, where agriculture is less adaptable than it is in richer countries like the United States. We suspect the reverse is true, for several reasons. First, modern intensive agriculture as practiced in rich countries centers on monocultures—large-scale plantings of a single genetic strain of a crop, usually a high-yielding variety. These strains are often more sensitive to adverse weather than are traditional crop strains. Second, the most agriculturally productive regions are in temperate zone areas where exceptionally good soils and climate coincide. Climate change, however, may decouple the favorable climate and the good soils, leaving farmers struggling to maintain food production with new handicaps—either poorer soils or a less benign climate.

In contrast, traditional agriculture in developing nations depends less on monocultures and frequently involves planting several crops together to provide some insurance against failures. We would put our bets on the flexibility of traditional farmers. They not only might adapt more readily to changing conditions than can operators of large industrial farms, they also don't depend on recommendations of an agricultural administration to change crops or planting times. Nor are they constrained by government subsidies and restrictions on their crops. Furthermore, most global warming models project relatively little change in tropical climates from greenhouse gas buildups; more dramatic changes are likely in the temperate zones. In all food-growing regions, global climate change is expected to manifest itself in the short to medium term principally as abnormally severe storms, floods, droughts, and other disruptive weather factors, all of which are potentially catastrophic for agriculture.

The impressive array of tools developed in the 1930s and 1940s to expand food production were deployed around the world in the 1960s and 1970s and did their job well. A new kit of tools is required to carry us into the future; yet no such kit appears to be on the horizon, although some help will be provided by genetic engineering. Meanwhile, the world's farmers must persist in their monumental struggle to increase production in the face of deteriorating natural capital and faltering ecosystem services (such as natural pest control being disrupted by misuse of pesticides), and they must win the battle year after year without doing irreparable damage to those services. Continued success might be assured, but only if (among other things) the size of the human population can soon be stabilized and a gradual decline initiated. Yet even the most optimistic assumptions for success in reducing population growth suggest that at least 2 to 3 billion more people will need to be fed within a few decades.

A potentially powerful approach to ending and eventually reversing population growth more rapidly would be greatly increasing socioeconomic equity in opportunity between the sexes and among families, regions and nations—as we have discussed elsewhere. To a large degree, past successes in raising crop yields have come from pushing back biophysical frontiers. Now humanity is faced with pushing back socioeconomic frontiers. That may be the toughest task of all. Confronted with such an enormous challenge, it is difficult to be complacent.

• There is no need to worry about any population-related problem. The human mind is the "ultimate resource," and growing populations will always be able to solve their resource problems. We need more people in order to have more geniuses; it would be immoral to keep them from being born.

How typical of the brownlash to define away population problems as self-solving while arguing that limiting the number of people alive at any given time—something virtually all societies have tried to do to one degree or another—is immoral! As Julian Simon put it, "What business do I have trying to help arrange it that fewer human beings will be born, each one of whom might be a Mozart or a Michelangelo or an Einstein—or simply a joy to his or her family or community, and a person who will enjoy life?" The obvious response might be: What business does anyone have trying to help arrange it that more human beings

will be born, each one of whom might be a Judas, an Attila the Hun, or a Hitler—or simply a burden to his or her family and community and a person who will live a life that is nasty, brutish and short?

Of course, having additional people to work on problems does not necessarily lead to solutions. Consider what happened to the people of Easter Island. This triangular sixty-four-square-mile Pacific island, isolated some 2000 miles west of Chile, was colonized by Polynesians some 1500 years ago. When the first colonists landed, they found an island covered by a tall subtropical forest, with the commonest tree a palm that grew to some eighty feet in height and had a trunk up to six feet in diameter. Yet when the first European visitor, Dutch explorer Jacob Roggeveen, arrived on Easter Sunday, 1722, he found a grassy wasteland inhabited by perhaps 2000 people.

By then the island lacked firewood and had no mammals (not even bats), land birds, reptiles, or snails. The only domestic animals raised by the Easter Islanders were chickens. While their ancestors had hunted porpoises from seagoing canoes crafted from the trunks of the now-extinct palms, the islanders of Roggeveen's day had to make do with small, flimsy, leaky canoes put together from small planks.

During the time of the carving of the some two hundred giant stone statues that once lined the island's coast (ca. A.D. 1200–1500), archaeologists estimate that the island's population had exploded to between 7000 and 20,000 people. But all those minds couldn't solve the island's resource problems. The fate of the population has been reconstructed from pollen records, archaeological evidence, and legends. The large trees were harvested more rapidly than they could regenerate, particularly when rats (which arrived with the Polynesians) ate the seedlings. Once the trees were gone, there was no way to build canoes for porpoise hunting, and there were no materials for making ropes and rollers to drag statues to their places of erection. Without the forest to absorb and meter out rainfall, streams and springs dried up, unprotected soil eroded away, and crop yields dropped. Rival clans warred over resources, and famine struck the once-rich island.

Unlike most premodern peoples, the Easter Islanders apparently didn't limit their fertility. Instead, as food supplies became short they switched to cannibalism, which turned out to be an effective—if not very attractive—method of population control. A common curse became "The flesh of your mother sticks between my teeth."

Long ago, Robert Louis Stevenson commented on overpopulation in the South Pacific:

> Over the whole extent of the South Seas, from one tropic to another, we find traces of a bygone state of over-population, when the resources of even a tropic soil were taxed, and even the improvident Polynesian trembled for the future. . . . we may suppose, more soberly, a people of sea-rovers, emigrants from a crowded country, to strike upon and settle island after island, and as time went on to multiply exceedingly in their new seats. In either case the end must be the same; soon or late it must grow apparent that the crew are too numerous, and that famine is at hand.

Ecologist Jared Diamond of the University of California, Los Angeles, drew similar conclusions but added a modern parallel: "Any islander who tried to warn about the dangers of progressive deforestation would have been overridden by vested interests of carvers, bureaucrats, and chiefs, whose jobs depended on continued deforestation. Our Pacific Northwest loggers are only the latest in a long line of loggers to cry, 'Jobs over trees!'"

As Stevenson implied, Polynesia historically generated a lot of environmental refugees. So, obviously, did many ancient societies that suffered ecological collapse. Today such refugees remain a well-known problem related to imbalances between human numbers and resources, a problem that is totally ignored by the brownlash. Of the more than 50 million refugees counted in 1995, at least 25 million were environmental refugees. Ecologists Norman Myers and Jennifer Kent recently wrote:

> The total may well double by the year 2010 if not before, as increasing numbers of impoverished people press ever harder on over-loaded environments. Their numbers seem likely to grow still more rapidly if predictions of global warming are borne out, whereupon sea-level rise and flooding of many coastal communities, plus agricultural dislocations through droughts and disruption of monsoon and other rainfall systems, could eventually cause as many as 200 million people to be put at risk of displacement.

But can't today's population, with its knowledge of the histories of past civilizations and billions of working brains keep us from repeating the gigantic mistakes of the past? Surely we can avoid the fate of the Easter Islanders, the Henderson Islanders (a population that died out completely on an island of the Pitcairn group in the South Pacific), the classic Maya, the early inhabitants of the Tigris and Euphrates Valleys, the Greeks, the Anasazi (Native Americans who built the vast pueblos of Chaco Canyon), and others who destroyed the environmental supports of their societies!

We wish the answer were yes. But the billions of human brains we have today are not stopping civilization from destroying its natural capital even faster than the Easter Islanders, Sumerians, or Anasazi destroyed theirs. Instead, an overpopulated global society is inflicting enormous, multifaceted damage on its life-support systems and simultaneously generated climate change—a situation eerily similar to the one analysts think may have faced the classic Maya civilization just before its collapse. As archaeologist Jeremy Sabloff remarked:

> [R]esearch [on climate change and the Mayan collapse] also raises an anthropological question that is relevant not only to the ancient Maya but to the contemporary world was well, and is certainly deserving of continued attention—namely, how severe do internal stresses in a civilization have to become before relatively minor climate shifts can trigger widespread cultural collapse?

There is no evidence that larger numbers of human beings translates into more talented or more humane individuals. Of course, there is a certain threshold—a community of 50 souls cannot have 100 geniuses. But arguably, Athens with perhaps 50,000 people in the year 425 B.C., had more geniuses than does San Francisco today, with fifteen times the number of citizens.

Of course, environmental rather than genetic differences determine what proportion of a population will display genius. Thus the San Francisco Bay Area has a far greater reputation for science, art, and education than, say, Kinshasa, Zaire, or Dallas–Fort Worth, although all three metropolitan areas have populations of the same magnitude. The point, of course, is that genius, like most human characteristics, results from an interaction between a genetic heritage and an environment—and there is no shred of evidence that any *group* of human beings has a greater genetic potential to produce geniuses than any other. The Bay Area's advantage can be traced to its environment: among other things, the area has long supported education and the arts; it can claim some of the best universities and colleges in the world; and (not unconnected to the others) it hosts a concentration of forward-looking and high-tech industries. But as the population of the Bay Area keeps growing and California's financial situation becomes more precarious, school systems have gone into decline and many valuable programs are being cut back. It is quite possible that Dallas–Fort Worth will soon surpass the San Francisco Bay Area as a cultural center—more evidence that it was not California genes but the intellectual and physical environment of the Bay Area that made it a center of excellence.

Whatever the causal relations, contrary to Julian Simon's expectations, as California's population has soared, its quality of life and intellectual potential have fallen. A state school system that ranked as one of the best in the nation now ranks thirty-eighth in expenditures per child and essentially ties with Utah for having the largest classes. The University of California, Berkeley, has lost much of the financial support it needs to remain one of the world's best institutions of higher education, a development that may negatively influence California's economy and quality of life for generations.

In China's imperial past, the Chinese ruled a large part of East Asia while Europeans were living in caves and painting themselves blue. That makes it crystal clear that China has no lack of genes to produce talented administrators, scientists, artists, and others. Although there are many, many more Chinese today, and many of them are also highly talented, they live in a political milieu that often suppresses rather than promotes intellectual expression. Art and science also thrived in ancient India and still thrive in India today despite the handicap of poverty; Arabs many centuries ago invented the number system that made modern science possible. And who could doubt the genius of the people who produced Benin bronzes or the great ruined cities of Zimbabwe? African genius (aided by English incompetence) allowed the Zulus to destroy a British army at Islawanda in southern Africa more than a century ago. Extraordinarily smart administrators clearly lived in Mayan, Aztec, and Inca civilizations; and Native Americans on the Great Plains could boast the finest light cavalry in the world a few centuries after the Spanish reintroduced the horse into North America. Germany in World War II, despite its abominable and ignorant Nazi overlords, nearly beat a

combination of nations with many times its population by being more clever militarily.

The point is that virtually all human individuals have enormous potential, and as soon as there are hundreds or thousands of them together, the appearance of genius of one sort or another is almost assured. Once there are tens of millions, the success of a society cannot be seriously constrained by a lack of brains, only by the environments in which those brains must develop and operate. Put another way, it's very hard to become the next Mozart if one is starving to death on the outskirts of Port-au-Prince! Having more people today is not the solution for generating more geniuses. Creating environments in which the inherent talents of people now disadvantaged—by race or gender discrimination, poverty, malnutrition, or whatever—can be fully expressed is. Indeed, there is every reason to believe that having *fewer* people would make it easier to create those environments. Thus Simon seems to have the genius-population relationship exactly backward: smaller, smoothly-functioning, nurturing societies are far more apt to give rise to geniuses than are large, debt-ridden, disintegrating, and inequitable societies.

If human resources are not necessarily increased by population growth, supplies of other resources—non-living and living—certainly are not enlarged by the explosion of human numbers. . . .

For Further Discussion

1. Review the essay by Julian Simon in Chapter 3. Do you think that the Ehrlichs have been fair in their criticism of his views? How might Simon respond to their claims?

2. After reading this essay, return to the essays in Chapter 3. Where are your own views located along the continuum from Simon to Easterbrook to Meadows?

3. The Ehrlichs quote Gregg Easterbrook as offering as a reason for increased population the fact that people like to be alive. Can people not yet alive like being alive? Can we deny this enjoyment to someone who is not born?

4. Who benefits from an increasing world population? Who bears the burden? Who benefits from zero population growth? Who would bear the burdens?

The Myth of Catching-up Development

Maria Mies

Economic development or, perhaps more accurately, eco-
nomic growth, is the policy solution to overpopulation and
ecological destruction most commonly offered by industri-
alized northern hemisphere countries. The goal, on this
model, is for the undeveloped, nonindustrial, and poor
countries of the world to catch up with and share in the
affluent good life of those northern societies.

In this selection, Maria Mies critically examines this
catch-up model of development. Mies suggests that this
model is a myth; nowhere has it attained the goals cited.
She believes that this policy is both impossible and undesir-
able for biophysical, ethical, cultural, ecological, and eco-
nomic reasons.

Virtually all development strategies are based on the
explicit or implicit assumption that the model of "the
good life" is that prevailing in the affluent societies of
the North: the USA, Europe and Japan. The question
of how the poor in the North, those in the countries of
the South, and peasants and women worldwide may
attain this "good life" is usually answered in terms of
what, since Rostow, can be called the "catching-up development" path. This means that by following the
same path of industrialization, technological progress
and capital accumulation taken by Europe and the
USA and Japan the same goal can be reached. These
affluent countries and classes, the dominant sex—the
men—the dominant urban centres and lifestyles are
then perceived as the realized utopia of liberalism, a
utopia still to be attained by those who apparently still
lag behind. Undoubtedly the industrialized countries'
affluence is the source of great fascination to all who
are unable to share in it. The so-called "socialist" countries' explicit aim was to catch up, and even to overtake capitalism. After the breakdown of socialism in
Easter Europe, particularly East Germany, the aim is
now to quickly catch up with the lifestyle of the so-
called market economies, the prototype of which is
seen in the USA or West Germany.

A brief look at the history of the underdeveloped
countries and regions of the South but also at present

day East Europe and East Germany can teach us that
this catching-up development path is a myth: nowhere
has it led to the desired goal.

This myth is based on an evolutionary, linear understanding of history. In this concept of history the peak
of the evolution has already been reached by some,
namely, men generally, white men in particular, industrial countries, urbanites. The "others"—women,
brown and black people, "underdeveloped" countries,
peasants—will also reach this peak with a little more
effort, more education, more "development." Technological progress is seen as the driving force of this evolutionary process. It is usually ignored that, even in
the early 1970s, the catching-up development theory
was criticized by a number of writers. Andre Gunder
Frank,[1] Samir Amin,[2] Johan Galtung,[3] and many others have shown that the poverty of the underdeveloped nations is not as a result of "natural" lagging
behind but the direct consequence of the overdevel-
opment of the rich industrial countries who exploit
the so-called periphery in Africa, South America and
Asia. In the course of this colonial history, which
continues today, these areas were progressively underdeveloped and made dependent on the so-called me-
tropolis. The relationship between these overdeveloped centres or metropoles and the underdeveloped
peripheries is a colonial one. Today, a similar colonial
relationship exists between Man and Nature, between
men and women, between urban and rural areas. We
have called these the colonies of White Man. In order
to maintain such relationships force and violence are
always essential.[4]

But the emotional and cognitive acceptance of the
colonized is also necessary to stabilize such relationships. This means that not only the colonizers but also
the colonized must accept the lifestyle of "those on top"
as the only model of the good life. This process of acceptance of the values, lifestyle and standard of living
of "those on top" is invariably accompanied by a devaluation of one's own: one's own culture, work, technology, lifestyle and often also philosophy of life and
social institutions. In the beginning this devaluation is

often violently enforced by the colonizers and then re-inforced by propaganda, educational programmes, a change of laws, and economic dependency, for example, through the debt trap. Finally, this devaluation is often accepted and internalized by the colonized as the "natural" state of affairs. One of the most difficult problems for the colonized (countries, women, peasants) is to develop their own identity after a process of formal decolonization—identity no longer based on the model of the colonizer as the image of the true human being; a problem addressed by Fanon,[5] Memmi,[6] Freire,[7] and Blaise.[8] To survive, wrote Memmi, the colonized must oppress the colonization. But to become a true human being he/she, him/herself, must oppress the colonized which, within themselves, they have become.[9] This means that he/she must overcome the fascination exerted by the colonizer and his lifestyle and re-evaluate what he/she is and does.

To promote the elimination of the colonizers from within the colonized, it is useful to look more closely at the catching-up development myth.

It may be argued that those who have so far paid the price for development also look up to those at the top as their model of the future, as their concrete utopia; that this is a kind of universal law. But if we also consider the price nature had to pay for this model, a price that now increasingly affects people in the affluent societies too, it may be asked why do not these people question this myth? Because even in the North, the paradigm of unlimited growth of science and technology, goods and services—of capital—and GNP have led to an increasing deterioration in the environment, and subsequently the quality of life.

DIVIDE AND RULE: MODERN INDUSTRIAL SOCIETY'S SECRET

Most people in the affluent societies live in a kind of schizophrenic or "double-think" state. They are aware of the disasters of Bhopal and Chernobyl, of the "greenhouse" effect, the destruction of the ozone layer, the gradual poisoning of ground-water, rivers and seas by fertilizers, pesticides, herbicides, as well as industrial waste, and that they themselves increasingly suffer the effects of air pollution, allergies, stress and noise, and the health risks due to industrially produced food. They also know that responsibility for these negative impacts on their quality of life lies in their own lifestyles and an economic system based on

constant growth. And yet (except for very few) they fail to act on this knowledge by modifying their lifestyles.

One reason for this collective schizophrenia is the North's stubborn hope, even belief, that they can have their cake and eat it: ever more products from the chemical industry *and* clean air and water; more and more cars and no "greenhouse" effect; an ever increasing output of commodities, more fast- and processed-foods, more fancy packaging, more exotic, imported food *and* enjoy good health and solve the waste problem.

Most people expect science and technology to provide a solution to these dilemmas, rather than taking steps to limit their own consumption and production patterns. It is not yet fully realized that a high material living standard militates against a genuinely good quality of life, especially if problems of ecological destruction are clearly understood.

The belief, however, that a high material living standard is tantamount to a good or high quality of life is the ideological support essential to uphold and legitimize the constant growth and accumulation model of modern industrial society. Unless the masses of people accept this the system cannot last and function. This equation is the real ideological-political hegemony that overlies everyday life. No political party in the industrialized countries of the North dares question this schizophrenic equation, because they fear it would affect their election prospects.

We have already shown that this double-think is based on assumptions that there are no limits to our planet's resources, no limits to technological progress, no limits to space, to growth. But as, in fact, we inhabit a limited world, this limitlessness is mythical and can be upheld only by colonial divisions: between centres and peripheries, men and women, urban and rural areas, modern industrial societies of the North and "backward," "traditional," "underdeveloped" societies of the South. The relationship between these parts is hierarchical not egalitarian, and characterized by exploitation, oppression and dominance.

The economic reason for these colonial structures is, above all, the *externalization of costs*[10] from the space and time horizon of those who profit from these divisions. The economic, social and ecological costs of constant growth in the industrialized countries have been and are shifted to the colonized countries of the South, to those countries' environment and their peoples. Only by dividing the international workforce into workers in the colonized peripheries and workers in the industrialized centres and by maintaining these relations of dominance even after formal decolonization, is it possible for industrial countries' workers to be paid wages ten times and more higher than those paid to workers in the South.

Much of the social costs of the reproduction of the labour force within industrial societies is externalized *within* those societies themselves. This is facilitated through the patriarchal-capitalist sexual division of labour whereby women's household labour is defined as non-productive or as non-work and hence not remunerated. Women are defined as housewives and their work is omitted from GNP calculations. Women can therefore be called the internal colony of this system.

The ecological costs of the industrial production of chemical fertilizers, pesticides, atomic energy, and of cars and other commodities, and the waste and damage for which they are responsible during both the production and the consumption process, are being inflicted on nature. They manifest themselves as air-, water-, soil-pollution and poisoning that will not only affect the present, but all future generations. This applies particularly to the long-term effects of modern high technology: atomic industry, genetic engineering, computer technology and their synergic effects which nobody can either predict or control. Thus, both nature and the future have been colonized for the short-term profit motives of affluent societies and classes.

The relationship between colonized and colonizer is based not on any measure of partnership but rather on the latter's coercion and violence in its dealings with the former. This relationship is in fact the secret of unlimited growth in the centres of accumulation. If externalization of all the costs of industrial production were not possible, if they had to be borne by the industrialized countries themselves, that is if they were internalized, an immediate end to unlimited growth would be inevitable.

CATCHING-UP IMPOSSIBLE AND UNDESIRABLE

The logic of this accumulation model, based on exploitation and colonizing divisions, implies that anything like "catching-up development" is impossible for the colonies, for all colonies. This is because just as one colony may, after much effort, attain what was considered the ultimate in "development," the industrial centres themselves have already "progressed" to a

yet more "modern" stage of development; "development" here meaning technological progress. What today was the TV is tomorrow the colour TV, the day after the computer, then the ever more modern version of the "computer generation" and even later artificial intelligence machines and so forth.[11] This catching-up policy of the colonies is therefore always a lost game. Because the very progress of the colonizers is based on the existence and the exploitation of those colonies.

These implications are usually ignored when development strategies are discussed. The aim, it is usually stated, is not a reduction in the industrialized societies' living standards but rather that all the "underdeveloped" should be enabled to attain the same level of affluence as in those societies. This sounds fine and corresponds to the values of the bourgeois revolutions: equality for all! But that such a demand is not only a logical, but also a material impossibility is ignored. The impossibility of this demand is obvious if one considers the ecological consequences of the universalization of the prevailing production system and lifestyle in the North's affluent industrial societies to everyone now living and for some further 30 years on this planet. If, for example, we note that the six per cent of the world's population who live in the USA annually consume 30 per cent of all the fossil energy produced, then, obviously, it is impossible for the rest of the world's population, of which about 80 per cent live in the poor countries of the South, to consume energy on the same scale.[12]

According to Trainer, those living in the USA, Europe and Japan, consume three-quarters of the world's energy production. "If present world energy production were to be shared equally, Americans would have to get by on only one-fifth of the per capita amount they presently consume."[13] Or, put differently, world population may be estimated at eleven billion people after the year 2050; if of these eleven billion people the per capita energy consumption was similar to that of Americans in the mid-1970s, conventional oil resources would be exhausted in 34–74 years;[14] similar estimations are made for other resources.

But even if the world's resource base was unlimited it can be estimated that it would be around 500 years before the poor countries reached the living standard prevailing in the industrialized North; and then only if these countries abandoned the model of permanent economic growth, which constitutes the core of their economic philosophy. It is impossible for the South to "catch-up" with this model, not only because of the limits and inequitable consumption of the resource base, but above all, because this growth model is based on a colonial world order in which the gap between the two poles is increasing, especially as far as economic development is concerned.

These examples show that catching-up development is not possible for all. In my opinion, the powers that dominate today's world economy are aware of this, the managers of the transnational corporations, the World Bank, the IMF, the banks and governments of the club of the rich countries; and in fact they do not really want this universalization, because it would end *their* growth model. Tacitly, they accept that the colonial structure of the so-called market economy is maintained worldwide. This structure, however, is masked by such euphemisms as "North-South relations," "sustainable development," "threshold-countries" and so on which suggest that all poor countries can and will reach the same living standard as that of the affluent countries.

Yet, if one tries to disregard considerations of equity and of ecological concerns it may be asked if this model of the good life, pursued by the societies in the North, this paradigm of "catching-up development" has at least made people in the North happy. Has it fulfilled its promises there? Has it at least made women and children there more equal, more free, more happy? Has their quality of life improved while the GDP grew?

We read daily about an increase in homelessness and of poverty, particularly of women and children,[15] of rising criminality in the big cities, of growing drug and other addictions, including the addiction to shopping. Depression and suicides are on the increase in many of the affluent societies, and direct violence against women and children seems to be growing—both public and domestic violence as well as sexual abuse; the media are full of reports of all forms of violence. Additionally, the urban centres are suffocating from motor vehicle exhaust emissions; there is barely any open space left in which to walk and breathe, the cities and highways are choked with cars. Whenever possible people try to escape from these urban centres to seek relief in the countryside or in the poor South. If, as is commonly asserted, city-dwellers' quality of life is so high, why do they not spend their vacations in the cities?

It has been found that in the USA today the quality of life is lower than it was ten years ago. There seems to be an inverse relationship between GDP and the quality of life; the more GDP grows, the more the quality of life deteriorates.[16] For example: growing market forces have led to the fact that food, which so far was still prepared in the home is now increasingly bought from fast-food restaurants; preparing food has become a service, a commodity. If more and more people buy this commodity the GDP grows. But what also grows at the same time is the erosion of community, the isolation and loneliness of individuals, the indifference and atomization of the society. As Polanyi remarked, market forces destroy communities.[17] Here, too, the processes are characterized by polarizations: the higher the GDP the lower the quality of life.

But "catching-up development" not only entails immaterial psychic and social costs and risks, which beset even the privileged in the rich countries and classes. With the growing number of ecological catastrophes—some man-made like the Gulf War or Chernobyl—material life also deteriorates in the rich centres of the world. The affluent society is one society which in the midst of plenty of commodities lacks the fundamental necessities of life: clean air, pure water, healthy food, space, time and quiet. What was experienced by mothers of small children after Chernobyl is now experienced by mothers in Kuwait. All the money of oil-rich Kuwait cannot buy people sunlight, fresh air, or pure water. This scarcity of basic common necessities for survival affects the poor and the rich, but with greater impact on the poor.

In short, the prevailing world market system, oriented towards unending growth and profit, cannot be maintained unless it can exploit external and internal colonies: nature, women and other people, but it also needs people as consumers who never say: "IT IS ENOUGH." The consumer model of the rich countries is not generalizable worldwide, neither is it desirable for the minority of the world's population who live in the affluent societies. Moreover, it will lead increasingly to wars to secure ever-scarcer resources; the Gulf War was in large part about the control of oil resources in that region. If we want to avoid such wars in the future the only alternative is a deliberate and drastic change in lifestyle, a reduction of consumption and a radical change in the North's consumer patterns and a decisive and broad-based movement towards energy conservation. . . .

These facts are widely known, but the myth of catching-up development is still largely the basis of development policies of the governments of the North and the South, as well as the ex-socialist countries. A TV discussion[18] in which three heads of state participated—Robert Mugabe of Zimbabwe, Vaclav Havel of the CSFR, and Richard von Weizsacker, President of the then FRG—is a clear illustration of this. The discussion took place after a showing of the film *The March,* which depicted millions of starving Africans trying to enter rich Europe. The President of the FRG said quite clearly that the consumption patterns of the 20 per cent of the world's population who live in the affluent societies of the industrialized North are using 80 per cent of the world's resources, and that these consumption patterns would, in the long run, destroy the natural foundations of life—worldwide. When, however, he was asked, if it was not then correct to criticize and relinquish the North's consumption patterns and to warn the South against imitating the North he replied that it would be wrong to preach to people about reducing consumption. Moreover, people in the South had the right to the same living standard as those in the North. The only solution was to distribute more of "our" wealth, through development aid, to the poor in the South, to enable them to "catch-up." He did not mention that this wealth originated as a result of the North's plundering of the colonies, as has been noted.

The President of socialist Zimbabwe was even more explicit. He said that people in the South wanted as many cars, refrigerators, TV sets, computers, videos and the same standard of living as the people in the North; that this was the aim of his politics of development. Neither he nor von Weizsacker asked whether this policy of universalizing the North's consumption patterns through a catching-up strategy was materially feasible. They also failed to question the ecological consequences of such a policy. As elected heads of state they dared not tell the truth, namely that the lifestyle of the rich in the North cannot be universalized, and that it should be ended in these countries in order to uphold the values of an egalitarian world.

Despite these insights, however, the catching-up development myth remains intact in the erstwhile socialist countries of the East. Development in East Germany, Poland and the ex-Soviet Union clearly demonstrate the resilience of this myth; but also the disaster that follows when the true nature of the "free" market economy becomes apparent. People in East Germany, the erstwhile GDR, were anxious to participate in the consumer model of capitalist FRG and, by voting for the destruction of their own state and the unification of Germany, hoped to become "equal." Political democracy, they were told, was the key to affluence. But they now realize, that in spite of political democracy and that they live in the same nation state as the West Germans, they are *de facto* treated as a cheap labour pool or a colony for West German capital, which is interested in expanding its market to the East but hesitates to invest there because the unification of Germany means that the East German workers will demand the same wages as their counterparts in West Germany. Where, then, is the incentive to go East? Less than a year after the unification, people in East Germany were already disappointed and depressed: unemployment had risen rapidly; the economy had virtually broken down; but no benefits had accrued from the new market system. According to the politicians, however, a period of common effort will be rewarded by catching-up with the West Germans. And, inevitably, the women in East Germany are worst affected by these processes. They who formerly had a participation rate of 90 per cent in the labour force are the first to lose their jobs, and more rapidly than men; they form the bulk of the unemployed. Simultaneously, they are losing whatever benefits the socialist state had provided for them: creches, a liberal abortion law, job security as mothers, time off for child-care, and so on.

But due to their disappointment with the socialist system people do not, yet, understand that this is the normal functioning of capitalism; that it needs colonies for its expansionism, that even democracy and formal equality do not result automatically in an equal standard of living or equal economic rewards.

In East Germany, the anger and the disappointment about what people call their betrayal by West German politicians, particularly Chancellor Kohl, has been converted into hostility towards other minorities, ethnic and racial minorities, foreign workers, other East Europeans, all of whom wanted to enter the "European House" and sit at the table of the rich.

In other parts of the world the collapse of the catching-up development myth leads to waves of fundamentalism and nationalism directed against religious, ethnic, racial, "others" within and outside their own territory. The main target of both nationalism and fundamentalism, and communalism, is women, because religious, ethnic and cultural identity are always based on a patriarchy, a patriarchal image of women, or rather control over "our" women, which, as we know from many examples, almost always amounts to more violence against women, more inequality for women.[19] Moreover, the collapse of the myth of catching-up development results in a further militarization of men. Practically all the new nationalisms and fundamentalisms have led to virtual civil war in which young, militarized men play the key role. As unacceptable as equals by the rich men's club and unable to share their lifestyle they can only show their manhood—as it is understood in a patriarchal world—by shouldering a machine-gun.

The myth of catching-up development, therefore, eventually leads to further destruction of the environment, further exploitation of the "Third World," further violence against women and further militarization of men.

DOES CATCHING-UP DEVELOPMENT LIBERATE WOMEN?

So far we have looked at the ecological cost effects of the catching-up strategy for the countries of the South. This strategy has been pursued, virtually since the Enlightenment and the bourgeois revolutions, as well as in the various movements for emancipation from oppression and exploitation: the working-class movement, the national liberation movements, and the women's movement. For women living in the industrialized countries catching-up development meant and continues to mean the hope that the patriarchal man-woman relationship will be abolished by a policy of equal rights for women. This policy is at present pursued by demands for positive discrimination for women, special quotas or reservations for women in political bodies, and in the labour market. Several state governments in Germany have issued special promotion programmes for women. Efforts are made to draw women into those sectors of the economy that formerly were exclusively men's domains, such as the new high-tech industries. Women's resistance to these technologies is seen as a handicap for their liberation, because technology as such is considered as men's area of power and therefore one that women must invade if they are to be "equal." All these efforts and initiatives at the political level add up to the strategy of women catching-up with men. This equalization policy is usually promulgated by the political parties in power or formally in opposition; it is shared by many in the women's movement, conversely, it is also opposed by many women. They see that there is a wide gap between the rhetoric and the actual performance of the political and economic system, which continues to marginalize women. What is more important, this strategy of catching-up with the men means that men generally, and white men in influential positions, are seen as the model to which women must aspire. The implications of this strategy are that the structure of the world economy remains stable, that nature and external colonies continue to be exploited, and that to maintain this structure militarism is necessary as a final resort.

For affluent societies' middle-class women this catching-up policy presupposes that they will get a share of the White Man's loot. Since the Age of Enlightenment and the colonization of the world the White Man's concept of emancipation, of freedom and equality is based on dominance over nature, and other peoples and territories. The division between nature and culture, or civilization, is integral to this understanding. From the early women's movement up to the present, a large section of women have accepted the strategy of catching-up with men as the main path to emancipation. This implied that women must overcome within themselves what had been defined as "nature," because, in this discourse, women were put on the side of nature, whereas men were seen as the representatives of culture. Theoreticians of the women's movement, such as Simone de Beauvoir[20] and Shulamith Firestone,[21] made this culture-nature divide the core of their theory of emancipation. Today this dichotomy again turns up in the discourse on reproductive engineering and gene technology. . . .

But more specifically let us ask why, for women, the catching-up development path even in the affluent societies of the industrialized North, is and will remain an illusion.

1. The promises of freedom, equality, self-determination of the individual, the great values of the French Revolution, proclaimed as universal rights and hence also meant for women, are betrayed for many women because all these rights depend on the possession of property, and of money. Freedom is the freedom of those who possess money. Equality is the equality of money. Self-determination is the freedom of choice in the supermarket. This freedom, equality, self-determination is always dependent on those who control the money/property. And in the industrialized societies and nations they are mostly the husbands or the capitalists' state. This at least is the relationship between men and women that is protected by law; the man as breadwinner, the woman as housewife.[22]

Self-determination and freedom are *de facto* limited for women, not only because they themselves are treated as commodities but also because, even if they possess money, they have no say in what is to be offered as commodities on the market. Their own desires and needs are constantly manipulated by those whose aim is to sell more and more goods. Ultimately, women are also persuaded that they want what the market offers.

2. This freedom, equality and self-determination, which depend on the possession of money, on purchasing power, cannot be extended to all women in the world. In Europe or the USA the system may be able to fulfil some of women's demand for equity with men, as far as income and jobs are concerned (or wages for housework, or a guaranteed minimum income), but only as long as it can continue the unrestricted exploitation of women as producers and consumers in the colonies. It cannot guarantee to *all* women worldwide the same standard of living as that of middle-class women in the USA or Europe. Only while women in Asia, Africa or Latin America can be forced to work for much lower wages than those in the affluent societies—and this is made possible through the debt trap—can enough capital be accumulated in the rich countries so that even unemployed women are guaranteed a minimum income; but all unemployed women in the world cannot expect this. Within a world system based on exploitation, "some are more equal than others."

3. This, however, also means that with such a structure there is no real material base for international women's solidarity. Because the core of individual freedom, equality, self-determination, linked to money and property, is the *self-interest of the individual* and not altruism or solidarity; these interests will always compete with the self-interests of others. Within an exploitative structure interests will necessarily be antagonistic. It may be in the interest of Third World women, working in the garment industry for export, to get higher wages, or even wages equivalent to those paid in the industrialized countries; but if they actually received these wages then the working-class woman in the North could hardly afford to buy those garments, or buy as many of them as she does now. In her interest the price of these garments must remain low. Hence the interests of these two sets of women who are linked through the world market are antagonistic. If we do not want to abandon the aim of international solidarity and equality we must abandon the materialistic and self-centred approach to fighting only for our own interests. The interests' approach must be replaced by an ethical one.

4. To apply the principle of self-interest to the ecological problem leads to intensified ecological degradation and destruction in other parts of the world. This became evident after Chernobyl, when many women in Germany, desperate to know what to feed to their babies demanded the importation of unpolluted food from the Third World. One example of this is the poisoning of mothers' milk in the affluent countries by DDT and other toxic substances as a result of the heavy use of fertilizers, pesticides and insecticides in industrialized agriculture. Rachel Carson had already warned that poisoning the soil would eventually have its effect on people's food, particularly mothers' milk;[23] now that this has happened many women in the North are alarmed. Some time ago a woman phoned me and said that in Germany it was no longer safe to breastfeed a baby for longer than three months; mothers' milk was poisoned. As a solution she suggested starting a project in South India for the production of safe and wholesome baby food. There, on the dry and arid Deccan Plateau, a special millet grows, called *ragi*. It needs little water and no fertilizer and is poor people's cheap subsistence food. This millet contains all the nutrients an infant needs. The woman suggested that ragi should be processed and canned as baby food and exported to Germany. This, she said, would solve the problem of desperate mothers whose breast milk is poisoned and give the poor in South India a new source of money income. It would contribute to their development!

I tried to explain that if ragi, the subsistence food of the poor, entered the world market and became an export commodity it would no longer be available for the poor; its price would soar and that, provided the project worked, pesticides and other chemicals would soon be used to produce more ragi for the market in the North. But ragi production, she answered, would have to be controlled by people who would guarantee it was not polluted. This amounts to a new version of eco-colonialism. When I asked her, why as an alternative, she would not rather campaign in Germany for a change in the industrialized agriculture, for a ban on the use of pesticides, she said that this would take too much time, that the poisoning of mothers' milk was an emergency situation. In her anxiety and concerned only with the interests of mothers in Germany she was willing to sacrifice the interests of poor women in South India. Or rather she thought that these conflicting interests could be made compatible by an exchange of money. She did not realize that this money would never suffice to buy the same healthy food for South Indian women's infants that they now had free of cost.

This example clearly shows that the myth of catching-up development, based on the belief of the miraculous workings of the market, particularly the world market, in fact leads to antagonistic interests even of mothers, who want only to give their infants unpolluted food.

Notes

1. Frank, A. G., *World Accumulation 1492-1789.* Macmillan, New York, 1978.
2. Amin, S., *Accumulation on a World Scale: A Critique of the Theory of Underdevelopment.* Monthly Review Press, New York, 1974.
3. Galtung, J., Eine Strukturelle Theorie des Imperialismus, in D. Senghaas (ed.) *Imperialismus und strukturelle Gewalt. Analysen über abhängige Reproduktion.* Suhrkamp, Frankfurt, 1972.
4. Mies, M., *Patriarchy and Accumulation on a World Scale, Women in the International Division of Labour.* Zed Books, London, 1989.
5. Fanon, F., *Peau Noire, Masques Blancs.* Edition du Seuil, Paris, 1952; English version: *Black Skin, White Masks.* Paladin, London, 1970.
6. Memmi, A., *Portrait du Colonise.* Edition Payot, Paris, 1973.
7. Freire, P., *Pedagogy of the Oppressed.* Penguin Books, Harmondsworth, 1970.
8. Blaise, S., *Le Rapt des Origines. ou: Le Meurtre de la Mere.* Maison des Femmes, Paris, 1988.
9. Memmi, op. cit., quoted in Blaise (1988) p. 74.
10. Kapp, W. K., Social Costs of Business Enterprise: Asia Publishing House, Bombay, 1963.
11. Ullrich, O., *Weltniveau. In der Sackgasse des Industriesystems.* Rotbuchverlag, Berlin, 1979, p. 108.
12. See *The Global 2000 Report to the President* US Foreign Ministry (ed.) Washington, Appendix, 1980, p. 59.
13. Trainer, F. E., *Developed to Death. Rethinking World Development.* Green Print, London, 1989.
14. Ibid., p. 61.
15. Sheldon, Danzinger and Stern.
16. Trainer, op. cit., p. 130.
17. Polanyi, K., *The Great Transformation,* Suhrkamp, Frankfurt, 1978.
18. This discussion took place under the title: "Die Zukunft gemeinsam meistern" on 22 May 1990 in Norddeutscher Rundfunk (NDR). It was produced by Rolf Seelmann-Eggebert.
19. Chhachhi, A. "Forced Identities: The State, Communalism, Fundamentalism and Women in India," in Kandiyoti, D. (ed.) *Women, Islam and the State.* University of California Press, 1991.
20. de Beauvoir, S., *The Other Sex.* Alfred A. Knopf Inc., New York, 1952.
21. Firestone, S., *The Dialectic of Sex.* William Morrow & Co., New York, 1970.
22. Mies, M., op. cit., 1989.
23. Carson, R. *Silent Spring.* Fawcett Publications, Greenwich, 1962. Hynes, P. H. *The Recurring Silent Spring.* Pergamon Press, New York, 1989.

For Further Discussion

1. Mies poses two challenges to the model of catching-up development: It is a myth in the sense that it has never occurred anywhere, and it is undesirable even if possible. What are the differences between these two claims?

2. Mies suggests that past economic development has been closely connected to colonialism that, she claims, continues today. How would you characterize a colonial relationship? Where, if at all, do you see such relationships in today's world?

3. What is your own understanding of economic development? What features characterize a developed economy? What is affluence?

4. How might Mies answer those who defend the free market as the best means for setting public policy?

5. In Mies's opinion, who bears the greatest burden for economic development?

Sustainable Growth: An Impossibility Theorem

Herman E. Daly

Economist Herman Daly challenges the assumption and value of economic growth. A well-known defender of sustainable economics, Daly argues that the very concept of sustainable growth is an oxymoron. Economic growth cannot continue indefinitely. Growth may be impossible to sustain even into the short-term future. This essay outlines Daly's criticism of economic growth and sketches some policy implications of the alternative model. The crucial concept of a sustainable economy is optimal scale, an idea that Daly defines in the following essay.

Impossibility statements are the very foundation of science. It is impossible to: travel faster than the speed of light; create or destroy matter-energy; build a perpetual motion machine, etc. By respecting impossibility theorems we avoid wasting resources on projects that are bound to fail. Therefore economists should be very interested in impossibility theorems, especially the one to be demonstrated here, namely that it is impossible for the world economy to grow its way out of poverty and environmental degradation. In other words, sustainable growth is impossible.

In its physical dimensions the economy is an open subsystem of the earth ecosystem, which is finite, nongrowing, and materially closed. As the economic subsystem grows it incorporates an ever greater proportion of the total ecosystem into itself and must reach a limit at 100 percent, if not before. Therefore its growth is not sustainable. The term "sustainable growth" when applied to the economy is a bad oxymoron—self-contradictory as prose, and unevocative as poetry.

CHALLENGING THE ECONOMIC OXYMORON

Economists will complain that growth in GNP is a mixture of quantitative and qualitative increase and therefore not strictly subject to physical laws. They have a point. Precisely because quantitative and qualitative change are very different it is best to keep them separate and call them by the different names already provided in the dictionary. To grow means "to increase naturally in size by the addition of material through assimilation or accretion." To develop means "to expand or realize the potentialities of; to bring gradually to a fuller, greater, or better state." When something grows it gets bigger. When something develops it gets different. The earth ecosystem develops (evolves), but does not grow. Its subsystem, the economy, must eventually stop growing, but can continue to develop. The term "sustainable development" therefore makes sense for the economy, but only if it is understood as "development without growth"—i.e., qualitative improvement of a physical economic base that is maintained in a steady state by a throughput of matter-energy that is within the regenerative and assimilative capacities of the ecosystem. Currently the term "sustainable development" is used as a synonym for the oxymoronic "sustainable growth." It must be saved from this perdition.

Politically it is very difficult to admit that growth, with its almost religious connotations of ultimate goodness, must be limited. But it is precisely the non-sustainability of growth that gives urgency to the concept of sustainable development. The earth will not tolerate the doubling of even one grain of wheat 64 times, yet in the past two centuries we have developed a culture dependent on exponential growth for its economic stability (Hubbert, 1976). Sustainable development is a cultural adaptation made by society as it becomes aware of the emerging necessity of nongrowth. Even "green growth" is not sustainable. There is a limit to the population of trees the earth can support, just as there is a limit to the populations of humans and of automobiles. To delude ourselves into believing that growth is still possible and desirable if only we label it "sustainable" or color it "green" will just delay the inevitable transition and make it more painful.

LIMITS TO GROWTH?

If the economy cannot grow forever then by how much can it grow? Can it grow by enough to give everyone in the world today a standard of per capita resource use equal to that of the average American? That would turn out to be a factor of seven,[1] a figure that is neatly bracketed by the Brundtland Commission's call (Brundtland et al., 1987) for the expansion of the world economy by a factor of five to ten. The problem is that even expansion by a factor of four is impossible if Vitousek et al. (1986, pp. 368–373) are correct in their calculation that the human economy currently preempts one-fourth of the global net primary product of photosynthesis (NPP). We cannot go beyond 100 percent, and it is unlikely that we will increase NPP since the historical tendency up to now is for economic growth to reduce global photosynthesis. Since land-based ecosystems are the more relevant, and we preempt 40 percent of land-based NPP, even the factor of four is an overestimate. Also, reaching 100 percent is unrealistic since we are incapable of bringing under direct human management all the species that make up the ecosystems upon which we depend. Furthermore it is ridiculous to urge the preservation of biodiversity without being willing to halt the economic growth that requires human takeover of places in the sun occupied by other species.

If growth up to the factor of five to ten recommended by the Brundtland Commission is impossible, then what about just sustaining the present scale—i.e., zero net growth? Every day we read about stress-induced feedbacks from the ecosystem to the economy, such as greenhouse buildup, ozone layer depletion, acid rain, etc., which constitute evidence that even the present scale is unsustainable. How then can people keep on talking about "sustainable growth" when: (a) the present scale of the economy shows clear signs of unsustainability, (b) multiplying that scale by a factor of five to ten as recommended by the Brundtland Commission would move us from unsustainability to imminent collapse, and (c) the concept itself is logically self-contradictory in a finite, nongrowing ecosystem? Yet sustainable growth is the buzz word of our time. Occasionally it becomes truly ludicrous, as when writers gravely speak of "sustainable growth in the rate of increase of economic activity." Not only must we grow forever, we must accelerate forever! This is hollow political verbiage, totally disconnected from logical and physical first principles.

ALLEVIATING POVERTY, NOT ANGELIZING GNP

The important question is the one that the Brundtland Commission leads up to, but does not really face: How far can we alleviate poverty by development without growth? I suspect that the answer will be a significant amount, but less than half. One reason for this belief is that if the five- to tenfold expansion is really going to be for the sake of the poor, then it will have to consist of things needed by the poor—food, clothing, shelter—not information services. Basic goods have an irreducible physical dimension and their expansion will require growth rather than development, although development via improved efficiency will help. In other words, the reduction in resource content per dollar of GNP observed in some rich countries in recent years cannot be heralded as severing the link between economic expansion and the environment, as some have claimed. Angelized GNP will not feed the poor. Sustainable development must be development without growth—but with population control and wealth redistribution—if it is to be a serious attack on poverty.

In the minds of many people, growth has become synonymous with increase in wealth. They say that we must have growth to be rich enough to afford the cost of cleaning up and curing poverty. That all problems are easier to solve if we are richer is not in dispute.

What is at issue is whether growth at the present margin really makes us richer. There is evidence that in the US it now makes us poorer by increasing costs faster than it increases benefits (Daly and Cobb, 1989, appendix). In other words we appear to have grown beyond the optimal scale.

DEFINING THE OPTIMAL SCALE

The concept of an optimal scale of the aggregate economy relative to the ecosystem is totally absent from current macroeconomic theory. The aggregate economy is assumed to grow forever. Microeconomics, which is almost entirely devoted to establishing the optimal scale of each microlevel activity by equating costs and benefits at the margin, has neglected to inquire if there is not also an optimal scale for the aggregate of all micro activities. A given scale (the product of population times per capita resource use) constitutes a given throughput of resources and thus a given load on the environment, and can consist of many people each consuming little, or fewer people each consuming correspondingly more.

An economy in sustainable development adapts and improves in knowledge, organization, technical efficiency, and wisdom; and it does this without assimilating or accreting, beyond some point, an ever greater percentage of the matter-energy of the ecosystem into itself, but rather stops at a scale at which the remaining ecosystem (the environment) can continue to function and renew itself year after year. The nongrowing economy is not static—it is being continually maintained and renewed as a steady-state subsystem of the environment.

What policies are implied by the goal of sustainable development, as here defined? Both optimists and pessimists should be able to agree on the following policy for the US (sustainable development should begin with the industrialized countries). Strive to hold throughput constant at present levels (or reduced truly sustainable levels) by taxing resource extraction, especially energy, very heavily. Seek to raise most public revenue from such resource severance taxes, and compensate (achieve revenue neutrality) by reducing the income tax, especially on the lower end of the income distribution, perhaps even financing a negative income tax at the very low end. Optimists who believe that resource efficiency can increase by a factor of ten should welcome this policy, which raises resource prices considerably and would give powerful incentive to just those technological advances in which they have so much faith. Pessimists who lack that technological faith will nevertheless be happy to see restrictions placed on the size of the already unsustainable throughput. The pessimists are protected against their worst fears; the optimists are encouraged to pursue their fondest dreams. If the pessimists are proven wrong and the enormous increase in efficiency actually happens, then they cannot complain. They got what they most wanted, plus an unexpected bonus. The optimists, for their part, can hardly object to a policy that not only allows but gives a strong incentive to the very technical progress on which their optimism is based. If they are proved wrong at least they should be glad that the throughput-induced taxes are harder to avoid than income taxes and do not reduce incentives to work.

At the project level there are some additional policy guidelines for sustainable development. Renewable resources should be exploited in a manner such that:

1. harvesting rates do not exceed regeneration rates; and

2. waste emissions do not exceed the renewable assimilative capacity of the local environment.

BALANCING NONRENEWABLE AND RENEWABLE RESOURCES

Nonrenewable resources should be depleted at a rate equal to the rate of creation of renewable substitutes. Projects based on exploitation of nonrenewable resources should be paired with projects that develop renewable substitutes. The net rents from the nonrenewable extraction should be separated into an income component and a capital liquidation component. The capital component would be invested each year in building up a renewable substitute. The separation is made such that by the time the nonrenewable is exhausted, the substitute renewable asset will have been built up by investment and natural growth to the point where its sustainable yield is equal to the income component. The income component will have thereby become perpetual, thus justifying the name "income," which is by definition the maximum available for consumption while maintaining capital intact. It has been shown (El Serafy, 1989, pp. 10–18) how this division of rents into capital and income depends upon: (1) the

discount rate (rate of growth of the renewable substitute); and (2) the life expectancy of the nonrenewable resource (reserves divided by annual depletion). The faster the biological growth of the renewable substitute and the longer the life expectancy of the nonrenewable, the greater will be the income component and the less the capital set-aside. "Substitute" here should be interpreted broadly to include any systemic adaptation that allows the economy to adjust the depletion of the nonrenewable resource in a way that maintains future income at a given level (e.g., recycling in the case of minerals). Rates of return for the paired projects should be calculated on the basis of their income component only.

However, before these operational steps toward sustainable development can get a fair hearing, we must first take the conceptual and political step of abandoning the thought-stopping slogan of "sustainable growth."

Note

1. Consider the following back-of-the-envelope calculation, based on the crude estimate that the US currently uses 1/3 of annual world resource flows (derived from National Commission on Materials Policy, 1973). Let R be current world resource consumption. Then $R/3$ is current US resource consumption, and $R/3$ divided by 250 million is present per capita US resource consumption. Current world per capita resource consumption would be R divided by 5.3 billion. For future world per capita resource consumption to equal present US per capita consumption, assuming constant population, R must increase by some multiple, call it M. Then M times R divided by 5.3 billion must equal $R/3$ divided by 250 million. Solving for M gives 7. World resource flows must increase sevenfold if all people are to consume resources at the present US average.

But even the sevenfold increase is a gross underestimate of the increase in environmental impact, for two reasons. First, because the calculation is in terms of current flows only with no allowance for the increase in accumulated stocks of capital goods necessary to process and transform the greater flow of resources into final products. Some notion of the magnitude of the extra stocks needed comes from Harrison Brown's estimate that the "standing crop" of industrial metals already embodied in the existing stock of artifacts in the ten richest nations would require more than 60 years' production of these metals at 1970 rates. Second, because the sevenfold increase of net, usable minerals and energy will require a much greater increase in gross resource flows, since we must mine ever less accessible deposits and lower grade ores. It is the gross flow that provokes environmental impact.

References

Brundtland, G. H., et al. 1987. *Our Common Future: Report of the World Commission on Environment and Development.* Oxford: Oxford University Press.

Daly, H. E., and J. B. Cobb, Jr. 1989. *For the Common Good: Redirecting the Economy toward Community, the Environment and a Sustainable Future.* Boston: Beacon Press.

El Serafy, S. 1989. "The Proper Calculation of Income from Depletable Natural Resources." In Y. J. Ahmad, S. El Serafy, and E. Lutz, eds., *Environmental Accounting for Sustainable Development,* a UNEP–World Bank Symposium. Washington, D.C.: The World Bank.

Hubbert, M. King. 1976. "Exponential Growth as a Transient Phenomenon in Human History." In Margaret A. Storm, ed., *Societal Issues: Scientific Viewpoints.* New York: American Institute of Physics.

National Commission on Materials Policy. 1973. *Material Needs and the Environment Today and Tomorrow.* Washington, D.C.: US Government Printing Office.

Vitousek, Peter M., Paul R. Ehrlich, Anne H. Ehrlich, and Pamela A. Matson. 1986. "Human Appropriation of the Products of Photosynthesis." *BioScience* 34. (6 May).

For Further Discussion

1. How does Daly distinguish between economic growth and economic development? Can you think of other examples of things that can both grow and develop?

2. Why does Daly believe that "sustainable growth" is an oxymoron? Are you convinced by his analysis?

3. Using Daly's understanding of growth, respond to Julian Simon's claim (Chapter 3) that natural resources are infinite. How might Simon respond to Daly?

4. William Baxter (Chapter 8) speaks of optimal pollution and the optimal satisfaction of consumer demand. Herman Daly speaks here of optimal scale. What's the difference? What is being optimized in each case?

5. "Sustainability" is a concept that has received a great deal of attention among environmentalists. What do you understand to be the goal of sustainable economics?

Strategies for a Sustainable Food Supply

Lester W. Milbrath

Environmentalists must do more than simply criticize the economic growth model of development. Beyond speaking in general terms of sustainability, defenders of this approach owe us some specific details about what a sustainable economy might look like.

In this selection from his book Envisioning a Sustainable Society, *Lester Milbrath outlines several strategies that a sustainable food supply should pursue. Milbrath describes some traditional farming techniques, still being practiced throughout much of the world, that could contribute to increased food supplies. He also argues that decreasing agricultural reliance on fossil fuels and major changes in land-use planning would be advisable. Milbrath offers a cautious endorsement of biotechnology and bioengineering technologies to support increased and ecologi-*

cally suitable crops. Finally, he recommends reconceptualizing agriculture in terms more sensitive to ecological realities.

The idea of a sustainable agriculture was discussed thoroughly at several recent conferences; the following definitions came out of those deliberations: "Sustainable agriculture does not deplete soil or people. . . . Sustainable agriculture protects soil and water and promotes the health of people and rural communities. . . . Sustainable agriculture needs to be ecologically sound, economically viable, socially just, and humane." (Quoted in *The Land Report,* Spring 1986, p. 8)

These definitions are stated more as goals than as programs, but they can guide programmatic choices. It is clear from our previous discussion that limiting human population and easing away from our dependence on fossil energy will be essential. What else can we do to maintain an adequate supply of food in our sustainable society? Actually, much can be done but it will require hard work.

LESSONS FROM THE PAST

We could study cultivation that has been persistently successful for thousands of years. Wendell Berry did just that when he visited Peruvian peasants high in the Andes Mountains. They had practiced agriculture in that region for 4,000 years. He summed up the themes of Andean agriculture this way: frugality, care, security in diversity, ecological sensitivity, and correctness of scale.

China combines the vastness of America with traditional agricultural practices that have been worked out by trial and error for thousands of years—as in the Andes. Much of China is mountainous or desert: the cultivable land constitutes only 11 percent to 14 percent of the total, according to various best estimates. One-half of that land is cultivable only because of irrigation. In the United States, by contrast, 21 percent is now cropland and, if we took "China-like" measures to use land, that figure could rise to 30 percent. The ratio of total population to cultivable land is *seven times as dense* in China as it is in the United States. Furthermore, the Chinese use very little fossil energy (mainly for chemical fertilizer and some pesticides). How do they manage to feed their 1,000,000,000 people?

Mainly, they follow the same themes as the Andes farmer: plots are small, they are frugally and carefully tended, they rotate crops, they recycle human and animal wastes back to the fields, they do not waste anything, and they tune their usage to climate, topography, and soil. No land is planted to lawns. No habitat is saved for wildlife in farming areas (they do maintain wildlife habitat in preserves in the mountains). They grow two or three crops per year, by rotation, on each plot that is suitable. The revolutionary communist government tried to improve production by importing methods from the Soviet Union: large fields with big machinery, organized into communes with an emphasis on grain. The reforms failed, all have been abandoned; the Chinese have returned to traditional methods, including allowing peasants to sell their produce in free markets. Agricultural productivity has risen and is now sufficient to feed their people.

In 1985, the Chinese Academy of Sciences invited a "Pollution Control Delegation," of which I was a member, to visit China. We visited a pig farm on the outskirts of Guangzhou where the pigs are fed with food wastes from the city. Some of the pig feces go into underground digesters to produce methane that is used for cooking and to run a small generator for electric lights. Some of the feces go into a fish pond to nourish fish that are periodically harvested for food. The remainder of the feces are spread on the fields where vegetables are grown. China leads the world in developing technology for biogas generation and in methods for maximizing yields from aquaculture. Food supply worldwide could be greatly increased if intensive cultivation of the land were practiced as in China. We should keep in mind, however, that this strategy takes over the entire ecosystem to serve humans; the needs of other species are given little weight.

FOOD GROWN IN WATER

What are the prospects and limitations of using fishing, aquaculture, and mariculture to produce food? The deep sea fishing industry has now developed such technical proficiency and capital equipment that it overfishes, reduces reproductive capability of fish populations, and forces fish stocks into decline. This industry illustrates the "tragedy of the commons." It was the victim of linear thinking aimed only at increasing immediate yield and ignored the maxim: "We can never do merely one thing." Fishing yields, worldwide, are down so drastically that we can expect less rather than more food from that sector.

In some places, the trend toward fish farming is growing. Salmon farmers in Norway raise young salmon in cohorts—each age group within a separate large round net shaped like a bowl. They are tended with plenty of food just as cattle are fattened in feedlots in the United States. Because the food for the fish comes from plants grown on land, it is not clear that there is any gain in total food availability from fish farming. Eating salmon is perceived as superior to eating grain, just as eating beef or pork is perceived as superior to eating corn; therefore, the market is good and the enterprises are profitable. Their whole investment was threatened by red-tide algae in May 1988. They towed their nets deep into the fjord and managed to escape the poisonous red-tide. Now the fjords are becoming so eutrophicated that sport fishing and other water recreation is losing its attraction.

Growing fish for food rather than sport in freshwater ponds and lakes is not practiced much outside Asia. Additional food could be obtained from aquaculture in many areas that are not well-suited for farming by impounding streams on steep slopes or digging ponds in marshes. Much additional research is needed to guard against ecosystem degradation from these human intrusions into nature.

Food from algae can also be grown in ponds. Normally, we think of algae as a smelly unpleasant nuisance. Actually, many marine animals live off these tiny plants that use sunlight and nutrients as they grow in water; their energy is passed up to predators at the top of the food chain, like ourselves. Certain kinds of algae, Spirulina for example, have a protein content as high as 66 percent and can be eaten directly as food by humans, as is done by some African tribes. Entrepreneurs are beginning to grow algae commercially to be used as a food supplement for cattle, hogs, and chickens; they have not yet made it sufficiently palatable that people in the West are willing to consume it directly. Spirulina algae reproduce quickly with high yields in very hot weather. An almost continuous input stream of nutrients and almost continuous harvesting must be programmed into the process to make it work effectively. At this early stage of development, algae growing apparently can be as productive, and possibly more productive, as growing plants in soil.

BECOMING VEGETARIANS

Feeding plants to animals and then eating the animals is one of the oldest ways of acquiring food that humans have used. It is also a much less efficient use of plant nutrients than eating the plants or seeds directly. A person receiving nutrients from eating meat requires six to eight times as much plant nutrients as would be required if the plants were consumed directly. Most world hunger could be alleviated if food grains were diverted from feeding animals and distributed instead to needy people in other lands. This simple fact has been understood for a long time; yet, this strategy has never been implemented. The reason is economics. The poor people in far away lands have little money to buy grain. The much richer people in developed countries are willing to pay a premium to be able to eat meat instead of grain. Some nutritionists argue that meat and fish are needed for a well-balanced human diet; others claim that a perfectly adequate diet can be derived from eating only plants. Certainly, the vegetarian option could someday be implemented but people would have to learn and value new dietary patterns.

Grasses and leaves cannot be digested by humans; therefore, it would make sense to continue to graze cattle, sheep, and goats on land that cannot be cultivated for grains—herds of chickens, ducks, or geese also can be grazed. We could, of course, leave such lands to wild animals if we could disabuse ourselves of the notion that such lands are unproductively "wasted."

WEANING OURSELVES FROM FOSSIL ENERGY

As fossil energy becomes scarcer and scarcer, costs are likely to rise so high that energy intensive agriculture will no longer be feasible. How do we shift away from that style of agriculture? As we contemplate the restructuring of agriculture, we should keep in mind there is only one imperative value: the life and health of the world's ecosystems. For them to be kept viable, healthy topsoil must be regenerated and maintained.

A second reason for turning away from energy intensive agriculture is that chemical inputs become less and less effective for their intended purpose but more destructive to the ecosystem. As noted, heavy inputs of nitrogen fertilizer kill energy-fixing bacteria; soils then require increasing amounts of fertilizer to maintain productivity. Insects and weeds develop immunity to pesticides and larger doses or different chemicals are required to kill them. These chemicals sink into the soil and form a barrier that roots cannot penetrate; when dry spells come along their shallow roots cause the plants to wilt. Experiments show that plants on nearby plots that have not been treated with chemicals grow deeper roots enabling them to withstand dry spells better. More and more farmers are discovering that chemicals are destroying the ecological vitality of their farms.

Some farmers are turning to regenerative agriculture; between 30,000 and 40,000 organic farmers are found throughout the United States. Time and thoughtful husbandry are needed to restore ecological vitality to a farm that has been heavily treated with chemicals. A number of organizations and research institutes are developing the appropriate knowledge to help guide farmers through this transition.

Wendell Berry favors a return to traditional agriculture with smaller farms, smaller machinery, fewer chemical inputs, diverse crops, and an emphasis on regenerating the topsoil. He studied and wrote about Amish rural culture. The Amish have deliberately separated themselves from the cultural influences of modern North America. They have flourished by maintaining traditional farming methods with few fossil energy inputs. While the rest of American agriculture was transforming to larger and larger corporate-style farms, and more and more farmers were going broke and leaving the land, the Amish have doubled their population in twenty years. They have taken over hundreds of run-down farms and transformed them into viable homesteads, mainly by dint of hard work and sensible husbandry. Six to eight Amish families can support themselves on the same amount of land (about 500 acres) that an energy-intensive farm requires to support one family. Berry's description of Amish farming was similar to the style of farming that I grew up with as a boy on a 150-acre farm in Minnesota more than fifty years ago. Much of the relevant knowledge is still available; reverting to it would not be that difficult.

A major reason Berry likes smaller, less energy-intensive farms is that they nourish rural community life.

> To farmers who give priority to the maintenance of their community, the economy of scale (that is, the economy of *large* scale, of "growth") can make no sense, for it requires the ruination and displacement of neighbors. A farm cannot be increased except by the decrease of a neighborhood. What the interest of the community proposes is invariably an economy of *proper* scale. (W. Berry, 1981, p. 261)

> The culture that sustains agriculture and that it sustains must form its consciousness and its aspiration upon the correct metaphor or the Wheel of Life. The appropriate agricultural technology would therefore be diverse; it would aspire to diversity; it would enable the diversification of economies, methods, and species to conform to the diverse kinds of land. It would always use plants and animals together. It would be as attentive to decay as to growth, to maintenance as to production. It would return all wastes to the soil, control erosion, and conserve water. To enable care and devotion and to safeguard the local communities and cultures of agriculture, it would use the land in small holdings. It would aspire to make each farm so far as possible the source of its

own operating energy, by the use of human energy, work animals, methane, wind or water or solar power. The mechanical aspect of technology would serve to harness or enhance the energy available on the farm. It would not be permitted to replace such energies with imported fuels, to replace people, or to replace or reduce human skills. (W. Berry, 1977, pp. 89–90)

The February 1986 issue of *Mother Earth News* headlined a story about a farmer in southern Minnesota who had lost his farm to the bank because he could not meet payments on his loan; he was left with only five acres. He decided to farm those five acres intensively, but organically, concentrating on producing a variety of fresh vegetables. He used only small and appropriate technology to lower capital and operating costs. He developed his markets and emphasized high-quality fresh produce. In 1985, he derived $27,000 net income from those five organic acres. His model has proved so successful that Minnesota has invested several hundred thousand dollars to help him carry his model to another 100 displaced farmers, hoping that they too can become productive once more on small acreage. This "small is beautiful" success story illustrates how we could wean ourselves from our dependence on fossil energy while still maintaining an agriculture that would be highly productive of nutritious food.

LAND USE PLANNING AND CONTROLS

Land, and the uses people put it to, is so fundamental to society that every society has found it necessary to control the use of land. The more crowded our planetary surface becomes, the greater the necessity for land-use planning and controls. When the Europeans colonized North America, they found such a plenitude of land that planning seemed unnecessary and controls were kept to a minimum. Many Americans have now come to believe that people should be free to do anything they wish with their land. That assumption really is not true, because no society can allow people to do *anything* they choose with their land, but the myth stands in the way of sensible planning and controls.

If we are to preserve farmland from being "developed" into building lots, roads, parking lots, and so on, we must plan the use of land and employ legal strategies to ensure that good land keeps producing food. Some countries utilize very tight controls. In Norway, for example, the national Ministry of Agriculture must grant permission to take good farm land out of production. In China, title to all land—and therefore ultimate control—is held by the government. Such tight controls would be unpalatable in North America, but other incentives are being used to keep farms producing food. For example, New York and several other states have passed Agricultural District Laws enabling the farms on the periphery of a metropolitan area to form a district and be recognized as a legal entity. The farmers in the district pledge not to sell their land for development for a certain period of time (for example, seven years); in return, the local and county governments assess and tax their land for its agricultural value, not its development value. Such weak controls may not be adequate to the need. When the need to protect prime land becomes powerfully manifest, we can be sure that government will have no other choice than to plan and control rigorously.

A BIOTECHNICAL SOLUTION?

Because they live and die within a single season, most cultivated annual crops must be more enterprising than perennials and produce myriad seeds to ensure survival. Humans, desiring lots of food from seeds, propagate the more productive annuals, but in doing so they must use till agriculture with its attendant problems of erosion and loss of topsoil. Wes Jackson of the Land Institute wants to find and propagate perennials that will produce as many seeds as annuals. Because they would not need to be planted each year, we could dispense with till agriculture and rebuild the topsoil.

Jackson's ideal is to propagate a group of high-yielding, seed-producing, herbacious perennials in a polyculture. His dream system would reorient agriculture toward the ways of nature; it would emphasize diversity; it would rebuild and maintain the topsoil; it would use natural controls for weeds and pests; and it would diminish fossil energy inputs. Alas, nature has not visibly and abundantly produced the kinds of plants he seeks. He hopes to assist nature by finding special mutant varieties that can be crossbred with currently used varieties to develop new ones coming closer to the high-yielding perennial ideal (a perennial corn has been found, for example). Even though he is cautious about gene splicing, he sees that new technology as possibly the only way to develop species

with the qualities he seeks. If perennials could successfully be propagated to produce high yields of seeds, it would stimulate a revolution in agriculture— promising a sustainable food supply not only for humans but many other species as well in a diversified ecosystem that draws from and nourishes a healthy topsoil.

BIOENGINEERING

Bioengineering is a special case of biotechnology that is so important it has become a new concept in our lexicon; it goes by such other phrases as *recombinant DNA, genetic engineering,* and *gene splicing.* Scientists discovered the DNA structure in the 1960's. They have since learned how to "splice genes," in effect, remanufacturing biological structures to emphasize characteristics more to the liking of humans; this procedure is called *recombinant DNA.* Genetic structures are extremely complicated. A tiny microorganism contains 3,000 to 5,000 genes; more complex organisms have approximately 50,000 genes. Humans are estimated to have about 500,000 genes. Scientists are slowly linking specific genes to specific traits in organisms. By adding, subtracting, or substituting genes they can engineer organisms to have desired traits or perform desired functions.

This new line of inquiry has spurred a vastly expanded biological research effort. Many firms see the possibility to make a lot of money by inventing, patenting, and selling new chemicals, cures, plants and animals. They have hired scientists and established laboratories specifically for this purpose. They have sent gene prospectors to the far corners of the globe looking for genes with special characteristics that can be used in their recombinant research. Less developed countries have begun to worry about their genetic treasure being stolen to make more wealth in developed countries and are beginning talks to establish worldwide controls on gene trafficking.

As is usual with powerful new knowledge, some people make strong claims for the benefits, while others make dire predictions of calamity if something goes wrong; both may be correct. The potential for changing things is enormous, hence, the list of potential benefits is lengthy, with no end of new ideas in sight. Examples of immediate prospects that would expand food production are: A bovine growth hormone that could increase milk production by 30 percent per cow; an ice-minus bacteria that could reduce frost damage; plants engineered to resist herbicides that would kill neighboring weeds.

> Genetic engineering signals the most radical change in our relationship with the natural world since the dawn of the age of pyrotechnology. Up until the past decade, we were still wholly dependent on the gift of fire bestowed on us by Prometheus. . . . With recombinant DNA technology it is now possible to snip, insert, stitch, edit, program and produce new combinations of living things just as our ancestors were able to heat, burn, melt and solder together various inert materials creating new shapes, combinations and forms. The transition from the Age of Pyrotechnology to the age of Biotechnology is the most important and disturbing technological change in recorded history. (Rifkin, 1985, p. 41)

While few benefits have yet been produced, genetic engineers hope to create: microbes that eat pollution, that leach useful minerals out of ore, that accelerate the production of cheap energy through biomass conversion. They hope to develop plants that are resistant to certain diseases or perennial plants that could produce grain yields as high as annual plants. They hope to design new animals that produce leaner meat as well as animals that convert their food to meat (or milk or eggs) more efficiently. Recently, a newspaper carried a story about a bioengineer who is trying to produce a cow that will give only low fat milk.

These are not hybrids but animals that reproduce just as natural animals; they are new beings. The U.S. Patent Office recently decided that new creatures are patentable. Bioengineers even hope to use cloning to reproduce millions of identical copies of desirable biostructures. Genetic surgery is also contemplated for humans. Current efforts focus on correcting gene abnormalities for victims of crippling hereditary diseases such as sickle cell anemia. If this line of research is successful, however, we are likely to see efforts to correct the gene structures of people who tend to gain weight too easily, who are too short or too tall, and so on. Is there no end? Should there be?

> We are in the early dawn hours of a new epoch in history, one in which we become the sovereigns over our own biological destiny. Though reluctant to predict a timetable for the conversion of our species from alchemist to algenist, those involved in the biological sciences are confident that they have at last opened the door onto a new horizon, one in which the biology of the planet will be remodeled, this time in our own image. Our generation, they say, stands at the crossroads of this new journey, one whose final consequences won't be fully grasped for centuries to come. . . .
>
> The question of whether we should embark on a long journey in which we become the architects of life is, along with the nuclear issue, the most important ever to face the human family. The ecological and environmental questions raised by such a prospect are mind boggling: the economic, political and ethical questions that accompany this new technology are without parallel. (Rifken, 1985, pp. 43–45)

This line of research is very powerful, with the potential to change the life of every living creature; it puts humans very close to the position of playing God. Those who urge us to go slow, or desist altogether, point out that we cannot possibly know the ramifications of what we are doing. Some have suggested that a virus or microbe created in a laboratory could accidentally find its way into the environment. Because it would have no natural enemies, and other organisms would not have developed natural defenses, it could create an epidemic. Obviously, this is a powerful agent of social change. For example, when European colonists came to New England, they displaced Native Americans not so much by force of arms as by infecting them with smallpox, to which most of the colonists were already immune. In some villages, 90 percent of the native inhabitants succumbed to smallpox; their culture was devastated and their power to resist the Europeans was almost nonexistent. National defense establishments, or terrorists, probably are trying to create new critters to wield more effective biological warfare against their enemies.

History tells us what happens to an ecosystem when new species are introduced from another continent. They usually fit in well with those already present, but sometimes the consequences have been nearly catastrophic. North Americans wish the following species had never been introduced: Gypsy moth, Kudzu vine, Dutch elm disease, chestnut blight, starlings, Mediterranean fruit fly. Natural evolution in Australia had never produced mammals; they were introduced by humans from other continents. When rabbits were introduced, they had no natural enemies and reproduced at epidemic rates, nearly stripping the land of vegetation. The epidemic was slowed only when humans introduced a virus that specifically targeted rabbits. Now the rabbits have developed immunity to the virus and are once more reproducing in epidemic numbers. Foreseeing deleterious consequences is always difficult; but doubly so for species with which we have had little experience.

Some of the longer-range effects of genetic engineering are more subtle and have the potential to profoundly change society. Genetic researchers intend to bring more and more of the biosphere under human control; it is the ultimate expression, to date, of human domination of nature. Increasing numbers of wild animals and plants will be crowded from their niches, their life support appropriated for human purposes. Unless humans learn to restrain their growth in numbers and use of resources, we will turn the biosphere into an overwhelmingly human-dominated entity. Our special sense of the wonder and magic of nature will be diminished; our view of it will be more akin to the way we relate to machines.

Genetic engineering clearly has the potential to produce our food more efficiently. Keep in mind we are expanding biological efficiency but not life support systems; limits of water, nutrients, and energy will still limit productivity. The socioeconomic-political consequences are difficult to foresee but they will be great. For example, a 30 percent increase in milk productivity from new cows raised with a bovine growth hormone could produce such a glut of milk as to drive thousands of dairy farmers, already on the brink of insolvency, to leave their farms. Moving into this new era could destroy our rural way of life. Predictably, we will not confront that choice directly. Our linear mode of thinking will accept each increment of technological change for its visible benefit and will ignore the hidden cost until it becomes painfully manifest many years later.

This new technology cannot be stopped; the genie cannot be stuffed back in the bottle; our only hope is to find effective controls. Scientists and public officials in the United States have recognized the potential dangers of genetic engineering and have instituted an elaborate set of reviews before a new organism can be cleared for release. Most bioengineers proclaim confidence in the safety of these procedures, but here are some cautions concerning them: (1) The proposed new entity should be considered "guilty" until proven innocent; exhaustive testing must precede development—even if it adds delays and increases development costs; (2) we must carefully license who can do this kind of research; it is not appropriate for high school biology classes; and (3) we must move quickly

to worldwide controls; it will do little good to have effective controls in one's own country if lax controls in another allow an epidemic to start.

We should be as foresighted as possible about the long-range consequences of these new technological developments. We should consider them just as carefully as we consider new legislation, for they will shape our lives just as permanently as legislation—perhaps more so. It is easier to repeal a law than to repeal a technology or a new creature. We should require assessments of their environmental, social, and value impacts before allowing deployment. We urgently need to learn how to do that.

RECONCEPTUALIZING AGRICULTURE

Agriculture traditionally has been considered an economic activity to which economic analysis applies. A new wave of thinking views agriculture in a much broader perspective. Dahlberg reports on a conference that considered new directions for agriculture; it pointed to four major but often neglected elements: (1) inadequate attention to externalities—social, environmental, and health effects; (2) inadequate attention to societal goals, ethics, and values; (3) local and regional concerns and developments that do not take into account the larger national setting; and (4) the way all of these considerations are integral parts of the larger global setting.

In a similar vein, Lowrance, Hendrix, and Odum identify four levels of sustainability that must be considered if we are to have a sustainable agriculture: (1) *agronomic sustainability:* the ability of a tract of land to maintain acceptable levels of production over a long period of time; (2) *microeconomic sustainability:* the ability of a farm, as the basic economic unit, to stay in business; (3) *ecological sustainability:* maintenance of life-support systems of larger-scale landscape units (forests, river basins) far into the future; and, (4) *macroeconomic sustainability:* monetary and fiscal policy at the national and international level. Policies could be aimed at any level but need to consider implications at other levels if they are to be successful.

Agroecology is the name given to a new scientific subdiscipline concerned with sustainable agriculture. Altieri defines it as being concerned with the optimization and stability of the agroecosystem as a whole. Altieri believes the transition to a sustainable agriculture can be accelerated by (1) focusing research attention on long-range problems; (2) integrating agricultural planning with an ecological perspective for all land use; (3) encouraging local producer–consumer cooperatives to set production goals; (4) organizing small farmers into a strong political constituency; (5) subordinating agricultural resource interests to broader political and economic interests; and, (6) consumers becoming more effective in challenging agriculture research that ignores nutrition and environmental issues.

These are but three examples of the kind of rethinking that is needed, and now being carried out, if we are to develop a sustainable agriculture.

CONCLUSION

A sustainable society can flourish only if it is founded on a sustainable agriculture. An agriculture that is trying to support more people than its carrying capacity is not sustainable. An agriculture that wastes and loses its topsoil is not sustainable. An agriculture dependent that wastes and loses its topsoil is not sustainable. An agriculture that poisons soils, waters, and the air is not sustainable. Modern agriculture does all those things and must be turned around if it is ever to become sustainable.

We need a regenerative agriculture that works with nature rather than against it. Regenerative agriculture will help the soil to renew its vitality. It will propagate plant communities that are diverse and will serve other animals in addition to humans. Such an agriculture will need many hours of human labor in loving nurturance of natural elements and systems. We can expect a surplus of labor in the future, therefore, many people could turn to doing their own work nurturing plants and animals on the land. These husbandmen of regenerative agriculture will have to understand ecological principles and pursue strategies that enhance the vitality and sustainability of ecosystems. We will have to re-vision man as a member of that system, not its master. Our very lives and those of our children will depend on how well we understand what we are doing and on how faithful our stewardship is as we nurture the land and its creatures.

For Further Discussion

1. How do you compare the definitions of sustainable agriculture offered by Milbrath with the model of farming discussed by Donald Worster and Wes Jackson in Chapter 10?

2. Milbrath's vision of a sustainable future involves both high-tech and low-tech activities. In this way, his views might appear compatible with both Julian Simon (Chapter 3) and Wes Jackson (Chapter 10). In what ways might they be incompatible?

3. Milbrath claims that a sustainable society can flourish only if it is founded on a sustainable agriculture. Do you agree?

4. Which of Milbrath's suggestions might be easily adapted to in both industrialized and nonindustrialized countries? Which might not?

DISCUSSION AND STUDY QUESTIONS FOR CHAPTER 12

1. In economic terms, what is the difference between income and capital? Does it make sense to think of natural resources in either term?

2. During the 1990s, the growth rate of the world population has been estimated to be between 1.7 and 2 percent annually. At a 2-percent growth rate, the world's population of approximately 6 billion people would double in 34 years; at the 1.7-percent rate, in 100 years the world's population would be more than 25 billion people. Other estimates suggest that the 25 percent of the world's population living in industrialized countries consume 80 percent of the world's resources. Which do you think is the more pressing ethical issue? Why?

3. Would the world be a better place, economically and ethically, if all countries lived the way North Americans live? When we speak of sustainable development, what do we wish to sustain?

4. Population control policies range from persuasion and economic incentives to coercion and force. Are policies other than persuasion ever justified? Should women bear greater responsibilities for population control than men? Should men have any voice in such policies?

DISCUSSION CASES

Poverty and Subsistence Farming

What makes someone poor? What makes a country impoverished or underdeveloped? Compare the relative wealth and poverty of two families.

The first family is a farm family—perhaps an Amish family in Pennsylvania or a family living in the northern region of India. This family lives and works on a small farm, grows much of their own food, lives without electricity in a small and simple house, relies on animals for heavy work, and travels either by foot or by horse. Most waking hours are spent in activities to support life on the farm. They have modest wants and, as a result, have most of what they desire.

The second family lives in the suburbs of a large U.S. city; both parents work steady jobs, and they live in a large home equipped with many electronic and consumer goods. The family has a large mortgage on their home, significant debt on various credit cards, loans for two cars, and college tuition. Their income is sufficient to make payments on their debts, and they never go without a meal. Now, imagine that this family's TV set is black and white rather than color, that they have neither a home computer nor a stereo, that they buy their clothes at a second-hand store, and their cars are older used models. They lack many things that they want.

Use these two fictional families to discuss the nature of poverty and the goal of economic activity, both for individual families and for societies. Is it true that the more you have the happier you are? What implications does your answer have for your own understanding of development, growth, affluence? Which of these families live a lifestyle that you would describe as sustainable?

What would happen to the cultures of the Amish or Indian families if they pursued a lifestyle similar to the average American?

Licensing Parents

Does every person have a right to be a parent? Does every person have a right to have a child? Or, given the potential harms to innocent people, should a license be required before having a child? What responsibilities do parents have to their children?

Consider the following proposal. For reasons of public safety, society requires a license for many activities—from driving cars, to building houses, to practicing law and medicine. Given the potential harms that face a child and given the increasing threats associated with population growth, let us require that people obtain a license before having children. The licenses would permit one child per person; they could be traded so people who desire large families could purchase a license from someone who is childless, and the licenses could involve an evaluation to assess the parents' qualifications. By enforcing such a program, population growth could be stabilized and children could have a better chance for a decent life.

What, if anything, could be wrong with such a proposal? Could there ever come a time when such a policy is justified? What implications do your answers to these questions have on your views on population control and growth? What implications do your answers have for your reaction to the essays by Hardin and the Ehrlichs? Is licensing parents preferable to facing a lifeboat situation?

In Chapter 2 we read a selection from the Book of Genesis in which humans are told to "be fruitful and multiply." How do you interpret that statement? Does this mean that humans have a responsibility to have children? A right to have children? What responsibilities do people have to their children? Do we have comparable responsibility to our great-grandchildren?

Overpopulation or Overconsumption?*

The 1990 annual report of the United Nations Fund for Population Activity (*The State of World Population*, 1990) offered the following observations on

---

*This case was adapted from an example used by Maria Mies and Vandana Shiva, "People or Population: Towards a New Ecology of Reproduction," chapter 19 of their book *Ecofeminism* (London: Zed Books, 1993).

(*continued*)

population growth and consumption: "For any given type of technology, for any given level of consumption or waste, for any given level of poverty or inequality, the more people there are the greater the impact on the environment." The report continued, "By far the largest share of the resources used, and waste created, is currently the responsibility of the 'top billion' people, those living in the industrialized countries. These are the countries overwhelmingly responsible for the damage to the ozone layer and acidification, as well as for roughly two-thirds of global warming."

These quotes seem to suggest that both population and consumption are responsible for ecological destruction. The report goes on to consider a specific issue: "As incomes grow, lifestyles and technologies will come to resemble those of Europe, North America, and Japan. . . . There will be an increase in car ownership. Since 1950 the human population has doubled, but the car population has increased seven times. The world car fleet is projected to grow from the present 400 million to 700 million over the next twenty years—twice as fast as the human population."

What policy implications do you draw from this? Should developing countries limit their population growth? Should they limit their reliance on cars? What responsibilities should industrialized countries shoulder?

Maximum Wage

Despite periodic attempts to limit increases, there is widespread agreement on the legitimacy of a minimum wage. But should there also be a maximum wage?

In an economic theory that assumes infinite growth, the pursuit of an ever-increasing income is acceptable, if not desirable. Assuming that income is related to desert and desert is related to produc-tivity—that is, assuming that income was fairly earned—then limiting income is counterproductive. The old adage that "a rising tide lifts all boats" hints at the justification for unlimited income. As long as some of this income is diverted through taxation to provide a safety net for the poor and disadvantaged, there ought to be no limit on personal income. Even John Rawls's liberal theory of justice would accept the legitimacy of this scenario.

But what if growth is not infinite? What if the availability of natural resources creates real limits on economic activity? In a finite world, every increase in the amount of resources going to one person decreases the amount of resources available to others. In a finite world, inequality of wealth and resources has one of two implications: either the worst-off must accept their status and live with the inequality or steps must be taken to close the gap. The infinite growth model acknowledges that this inequality is undesirable, if not unfair, but seeks to close the gap exclusively by raising the income level of the poor. In a finite world, this only places a greater strain on the available resources and results in widespread continual environmental destruction. An alternative is to establish a maximum wage.

Some proposals assess the maximum wage at a factor of ten times the minimum wage; others suggest a factor of twenty or twenty-five. Minnesota Congressman Martin Sabo has proposed an Income Equity Act, which would limit tax deductions for firms whose highest-paid workers earn more than twenty-five times the wages of their lowest-paid full-time workers. Thus, a company whose lowest-paid full-time employee earns $20,000 a year would lose some tax breaks if its CEO earns more than $500,000. In 1995, *Business Week* magazine estimated the average pay for U.S. chief executives was 135 times the salary of the average factory worker.

CHAPTER 13

Environmental Justice

Chapter 1 defined ethics as involving two fundamental questions: How should I live as an individual? and How should we live as a society? Traditionally, justice has been a central norm of social institutions. Living together generates benefits and creates burdens that exist only within a society. Social justice establishes the norms for distributing these social benefits and burdens.

How should the benefits and burdens of social living be distributed? What would be an unjust distribution of benefits and burdens? Within Western philosophy, answers to these questions often begin with Aristotle's assertion that justice requires treating equals equally. This formal principle of justice requires that each individual be treated as an equal unless relevant differences exist that would justify differential treatment.

For example, justice would demand that teachers treat each student with equal respect and consideration. Does this mean that every student should receive the same grades? Presumably not, because individual classroom achievement is a difference that is relevant to receiving grades—some students receive As while others receive Cs, and so forth. As long as each student's work is judged according to the same standards and given the same consideration, justice seems to allow, if not demand, different grades. Similarly, employees deserve equal treatment, but a more productive or harder-working employee seems to deserve higher wages. But what do citizens deserve from their government? What do people deserve from their society?

According to most modern political theories, each citizen deserves a strong presumption of equality. There can be reasons for treating citizens differently, as when a government jails a criminal or supplies income to the elderly, but the presumption rests on equal treatment. Thus, on this model, government

policies of segregating races or denying voting rights to women were obviously unjust. The differences that exist between whites and African Americans, between men and women, do not justify different classes of citizenship.

No doubt the clearest example of unjust discrimination occurs when an individual or public body deliberately establishes a policy that denies equal treatment. An employer who refuses to hire Jewish workers or a legislature passing a law prohibiting African Americans from entering a state university would be plain examples of unjust treatment. But often injustice is more subtle. Sometimes very unequal benefits and burdens result from public policies that are not established consciously to promote inequality. When this occurs, closer examination is necessary.

Unequal distribution of benefits and burdens is not in itself unjust, as we have seen. Randomness alone can account for a wide distribution of any social benefit or burden. But when the inequality is correlated with variables like race, gender, or social class, then the inequality is suspect. For example, it might turn out that unemployment rates are higher in the Northeast than they are in the Southwest. Given the economic variability of business cycles, this might not involve any justice issue. But suppose that we discovered that, across various geographic areas and diverse industries, unemployment rates among African American citizens were three and four times higher than unemployment rates among whites. Given the history of racial discrimination in this country, that situation raises serious questions of social justice. Even without evidence of deliberate discrimination, a society should be troubled by such widely unequal practices.

Environmental justice explores the distribution of environmental benefits and burdens, with particular attention to how environmental burdens are distributed across racial, ethnic, and class lines. According to many observers, environmental burdens—such as pollution, toxic waste sites, landfills, and polluting industries—are more likely to be located where there are high concentrations of poor people and people of color. Further, worldwide distribution of environmental burdens seem to fall more often on women, the poor, indigenous people, and minorities. Such observations raise questions of environmental racism and sexism. Further, as was described in Chapter 12, environmental benefits and burdens are distributed unjustly on the international level as well. Poor countries become sites for toxic waste and polluting industries. Rain forest and wilderness areas within poor countries are developed to produce goods for export, typically at very low prices, to wealthier countries.

Thus, across the world environmental issues are closely tied to questions of social justice. Indeed, it may be impossible to successfully address one without also resolving the other.

Justice and Environmental Decision Making

Robert D. Bullard

Sociologist Robert Bullard is a leading researcher on questions of environmental racism and environmental justice. Bullard has found numerous instances in which hazardous and polluting industries and land uses are disproportionately located within poor and minority communities. He has also found evidence that when government agencies do address environmental cleanup or when they do prosecute environmental crime, they are more likely to concentrate their efforts on protecting white and wealthy communities. In this selection, Bullard outlines some of his findings and reviews the inequities involved. Bullard goes on to sketch several principles of environmental justice that he believes should be adopted by those who make environmental policy decisions.

Despite the recent attempts by federal agencies to reduce environmental and health threats in the United States, inequities persist. If a community is poor or inhabited largely by people of color, there is a good chance that it receives less protection than a community that is affluent or white. This situation is a result of the country's environmental policies, most of which "distribute the costs in a regressive pattern while providing disproportionate benefits for the educated and wealthy." Even the Environmental Protection Agency (EPA) was not designed to address environmental policies and practices that result in unfair outcomes. The agency has yet to conduct a single piece of disparate

impact research using primary data. In fact, the current environmental protection paradigm has institutionalized unequal enforcement, traded human health for profit, placed the burden of proof on the "victims" rather than on the polluting industry, legitimated human exposure to harmful substances, promoted "risky" technologies such as incinerators, exploited the vulnerability of economically and politically disenfranchised communities, subsidized ecological destruction, created an industry around risk assessment, delayed cleanup actions, and failed to develop pollution prevention as the overarching and dominant strategy. As a result, low-income and minority communities continue to bear greater health and environmental burdens, while the more affluent and white communities receive the bulk of the benefits.

The geographic distribution of both minorities and the poor has been found to be highly correlated to the distribution of air pollution, municipal landfills and incinerators, abandoned toxic waste dumps, lead poisoning in children, and contaminated fish consumption. Virtually all studies of exposure to outdoor air pollution have found significant differences in exposure by income and race. Moreover, the race correlation is even stronger than the class correlation. The National Wildlife Federation recently reviewed some sixty-four studies of environmental disparities; in all but one, disparities were found by either race or income, and disparities by race were more numerous than those by income. When race and income were compared for significance, race proved to be the more important factor in twenty-two out of thirty tests. And researchers at Argonne National Laboratory recently found that

in 1990, 437 of the 3,109 counties and independent cities failed to meet at least one of the EPA ambient air quality standards. . . . 57 percent of whites, 65 percent of African-Americans, and 80 percent of Hispanics live in 437 counties with substandard air quality. Out of the whole population, a total of 33 percent of whites, 50 percent of African-Americans, and 60 percent of Hispanics live in the 136 counties in which two or more air pollutants exceed standards. The percentage living in the 29 counties designated as nonattainment areas for three or more pollutants are 12 percent of whites, 20 percent of African-Americans, and 31 percent of Hispanics.

The public health community has very little information on the magnitude of many air pollution-related health problems. For example, scientists are at a loss to explain the rising number of deaths from asthma in recent years. However, it is known that persons suffering from asthma are particularly sensitive to the effects of carbon monoxide, sulfur dioxide, particulate matter, ozone, and oxides of nitrogen.

Current environmental decision making operates at the juncture of science, technology, economics, politics, special interests, and ethics and mirrors the larger social milieu where discrimination is institutionalized. Unequal environmental protection undermines three basic types of equity: procedural, geographic, and social.

PROCEDURAL EQUITY

Procedural equity refers to fairness—that is, to the extent that governing rules, regulations, evaluation criteria, and enforcement are applied in a nondiscriminatory way. Unequal protection results from nonscientific and undemocratic decisions, such as exclusionary practices, conflicts of interest, public hearings held in remote locations and at inconvenient times, and use of only English to communicate with and conduct hearings for non-English-speaking communities.

A 1992 study by staff writers from the *National Law Journal* uncovered glaring inequities in the way EPA enforces its Superfund laws: "There is a racial divide in the way the U.S. government cleans up toxic waste sites and punishes polluters. White communities see faster action, better results and stiffer penalties than communities where blacks, Hispanics and other minorities live. This unequal protection often occurs whether the community is wealthy or poor."

After examining census data, civil court dockets, and EPA's own record of performance at 1,177 Superfund toxic waste sites, the authors of the *National Law Journal* reported the following:

- Penalties applied under hazardous waste laws at sites having the greatest white population were 500 percent higher than penalties at sites with the greatest minority population. Penalties averaged out at $335,566 at sites in white areas but just $55,318 at sites in minority areas.

- The disparity in penalties applied under the toxic waste law correlates with race alone, not income. The average penalty in areas with the lowest median income is $113,491—3 percent more than the average penalty in areas with the highest median income.

- For all the federal environmental laws aimed at protecting citizens from air, water, and waste pollution, penalties for noncompliance were 46 percent higher in white communities than in minority communities.

- Under the Superfund cleanup program, abandoned hazardous waste sites in minority areas take 20 percent longer to be placed on the National Priority List than do those in white areas.

- In more than half of the ten autonomous regions that administer EPA programs around the country, action on cleanup at Superfund sites begins from 12 to 42 percent later at minority sites than at white sites.

- For minority sites, EPA chooses "containment," the capping or walling off of a hazardous waste dump site, 7 percent more frequently than the cleanup method preferred under the law: permanent "treatment" to eliminate the waste or rid it of its toxins. For white sites, EPA orders permanent treatment 22 percent more often than containment.

These findings suggest that unequal environmental protection is placing communities of color at risk. The *National Law Journal* study supplements the findings of several earlier studies and reinforces what grassroots activists have been saying all along: Not only are people of color differentially affected by industrial pollution but also they can expect different treatment from the government.

GEOGRAPHIC EQUITY

Geographic equity refers to the location and spatial configuration of communities and their proximity to environmental hazards and locally unwanted land uses (LULUs), such as landfills, incinerators, sewage treatment plants, lead smelters, refineries, and other noxious facilities. Hazardous waste incinerators are not randomly scattered across the landscape. Communities with hazardous waste incinerators generally have large minority populations, low incomes, and low property values.

A 1990 Greenpeace report (*Playing with Fire*) found that communities with existing incinerators have 89 percent more people of color than the national average; communities where incinerators are proposed for construction have minority populations that are 60 percent higher than the national average; the average income in communities with existing incinerators is 15 percent lower than the national average; property values in communities that host incinerators are 38 percent lower than the national average; and average property values are 35 percent lower in communities where incinerators have been proposed.

The industrial encroachment into Chicago's Southside neighborhoods is a classic example of geographic inequity. Chicago is the nation's third largest city and one of the most racially segregated cities in the country. More than 92 percent of the city's 1.1 million African American residents live in racially segregated areas. The Altgeld Gardens housing project, located on the city's southeast side, is one of these segregated enclaves. The neighborhood is home to 150,000 residents, of whom 70 percent are African American and 11 percent are Latino.

Altgeld Gardens is encircled by municipal and hazardous waste landfills, toxic waste incinerators, grain elevators, sewage treatment facilities, smelters, steel mills, and a host of other polluting industries. Because of its location, the area has been dubbed a "toxic doughnut" by Hazel Johnson, a community organizer in the neighborhood. There are 50 active or closed commercial hazardous waste landfills; 100 factories, including 7 chemical plants and 5 steel mills; and 103 abandoned toxic waste dumps.

Currently, health and risk assessment data collected by the state of Illinois and the EPA for facility permitting have failed to take into account the cumulative and synergistic effects of having so many "layers" of poison in one community. Altgeld Gardens residents wonder when the government will declare a moratorium on permitting any new noxious facilities in their neighborhood and when the existing problems will be cleaned up. All of the polluting industries imperil the health of nearby residents and should be factored into future facility-permitting decisions.

In the Los Angeles air basin, 71 percent of African Americans and 50 percent of Latinos live in areas with the most polluted air, whereas only 34 percent of whites live in highly polluted areas. The "dirtiest" zip code in California (90058) is sandwiched between South-Central Los Angeles and East Los Angeles. The one-square-mile area is saturated with abandoned toxic waste sites, freeways, smokestacks, and wastewater pipes from polluting industries. Some eighteen industrial firms in 1989 discharged more than 33 million pounds of waste chemicals into the environment.

Unequal protection may result from land-use decisions that determine the location of residential amenities and disamenities. Unincorporated communities of poor African Americans suffer a "triple" vulnerability to noxious facility siting. For example, Wallace, Louisiana, a small unincorporated African American community located on the Mississippi River, was re-zoned from residential to industrial use by the mostly white officials of St. John the Baptist Parish to allow construction of a Formosa Plastics Corporation plant. The company's plants have been major sources of pollution in Baton Rouge, Louisiana; Point Comfort, Texas; Delaware City, Delaware; and its home country of Taiwan. Wallace residents have filed a lawsuit challenging the rezoning action as racially motivated.

Environmental justice advocates have sought to persuade federal, state, and local governments to adopt policies that address distributive impacts, concentration, enforcement, and compliance concerns. Some states have tried to use a "fair share" approach to come closer to geographic equity. In 1990, New York City adopted a fair share legislative model designed to ensure that every borough and every community within each borough bears its fair share of noxious facilities. Public hearings have begun to address risk burdens in New York City's boroughs.

Testimony at a hearing on environmental disparities in the Bronx points to concerns raised by African
Americans and Puerto Ricans who see their neighborhoods threatened by garbage transfer stations, salvage
yards, and recycling centers.

On the Hunts Point peninsula alone there are at
least thirty private transfer stations, a large-scale
Department of Environmental Protection (DEP)
sewage treatment plant and a sludge dewatering
facility, two Department of Sanitation (DOS) marine transfer stations, a city-wide privately regulated medical waste incinerator, a proposed DOS
resource recovery facility and three proposed DEP
sludge processing facilities. That all of the facilities
listed above are located immediately adjacent to the
Hunts Point Food Center, the biggest wholesale
food and meat distribution facility of its kind in the
United States, and the largest source of employment in the South Bronx, is disconcerting. A policy
whereby low-income and minority communities
have become the "dumping grounds" for unwanted
land uses, works to create an environment of disin-
centives to community-based development initiatives. It also undermines existing businesses.

Some communities form a special case for environmental justice. For example, Native American reservations are geographic entities but are also quasi-
sovereign
nations. Because of less-stringent environmental regulations than those at the state and federal levels, Native
American reservations from New York to California
have become prime targets for risky technologies. Indian natives do not fall under state jurisdiction. Similarly, reservations have been described as the "lands
the feds forgot." More than one hundred industries,
ranging from solid waste landfills to hazardous waste
incinerators and nuclear waste storage facilities, have
targeted reservations.

SOCIAL EQUITY

Social equity refers to the role of sociological factors,
such as race, ethnicity, class, culture, lifestyles, and
political power, in environmental decision making.
Poor people and people of color often work in the
most dangerous jobs and live in the most polluted
neighborhoods, and their children are exposed to all
kinds of environmental toxins on the playgrounds and
in their homes and schools.

Some government actions have created and exacerbated environmental inequity. More stringent environmental regulations have driven noxious facilities
to follow the path of least resistance toward poor,
overburdened communities. Governments have even
funded studies that justify targeting economically disenfranchised communities for noxious facilities. Cer-
rell Associates, Inc., a Los Angeles-based consulting
firm, advised the state of California on facility siting
and concluded that "ideally . . . officials and companies should look for lower socioeconomic neighborhoods that are also in a heavy industrial area with
little, if any, commercial activity."

The first state-of-the-art solid waste incinerator
slated to be built in Los Angeles was proposed for the
south-central Los Angeles neighborhood. The city-
sponsored project was defeated by local residents. The
two permits granted by the California Department of
Health Services for state-of-the-art toxic waste incinerators were proposed for mostly Latino communities:
Vernon, near East Los Angeles, and Kettleman City,
a farm-worker community in the agriculturally rich
Central Valley. Kettleman City has 1,200 residents of
which 95 percent are Latino. It is home to the largest
hazardous waste incinerator west of the Mississippi
River. The Vernon proposal was defeated, but the
Kettleman City proposal is still pending.

PRINCIPLES OF
ENVIRONMENTAL JUSTICE

To end unequal environmental protection, governments should adopt five principles of environmental
justice: guaranteeing the right to environmental protection, preventing harm before it occurs, shifting the
burden of proof to the polluters, obviating proof of intent to discriminate, and redressing existing inequities.

The Right to Protection

Every individual has a right to be protected from environmental degradation. Protecting this right will require enacting a federal "fair environmental protection
act." The act could be modeled after the various federal civil rights acts that have promoted nondiscrimination—with the ultimate goal of achieving "zero tolerance"—in such areas as housing, education, and
employment. The act ought to address both the intended and unintended effects of public policies and

industrial practices that have a disparate impact on racial and ethnic minorities and other vulnerable groups. The precedents for this framework are the Civil Rights Act of 1964, which attempted to address both *de jure* and *de facto* school segregation, the Fair Housing Act of 1968, the same act as amended in 1988, and the Voting Rights Act of 1965.

For the first time in the agency's twenty-three-year history, EPA's Office of Civil Rights has begun investigating charges of environmental discrimination under Title VI of the 1964 Civil Rights Act. The cases involve waste facility siting disputes in Michigan, Alabama, Mississippi, and Louisiana. Similarly, in September 1993, the U.S. Civil Rights Commission issued a report entitled *The Battle for Environmental Justice in Louisiana: Government, Industry, and the People.* This report confirmed what most people who live in "Cancer Alley"—the 85-mile stretch along the Mississippi River from Baton Rouge to New Orleans—already knew: African American communities along the Mississippi River bear disproportionate health burdens from industrial pollution.

A number of bills have been introduced into Congress that address some aspect of environmental justice:

- The Environmental Justice Act of 1993 (H.R. 2105) would provide the federal government with the statistical documentation and ranking of the top one hundred "environmental high impact areas" that warrant attention.

- The Environmental Equal Rights Act of 1993 (H.R. 1924) seeks to amend the Solid Waste Act and would prevent waste facilities from being sited in "environmentally disadvantaged communities."

- The Environmental Health Equity Information Act of 1993 (H.R. 1925) seeks to amend the Comprehensive Environmental Response, Compensation, and Liability Act of 1990 (CERCLA) to require the Agency for Toxic Substances and Disease Registry to collect and maintain information on the race, age, gender, ethnic origin, income level, and educational level of persons living in communities adjacent to toxic substance contamination.

- The Waste Export and Import Prohibition Act (H.R. 3706) banned waste exports as of 1 July 1994 to countries that are not members of the Organization for Economic Cooperation and Development (OECD); the bill would also ban waste exports to and imports from OECD countries as of 1 January 1999.

The states are also beginning to address environmental justice concerns. Arkansas and Louisiana were the first two to enact environmental justice laws. Virginia has passed a legislative resolution on environmental justice. California, Georgia, New York, North Carolina, and South Carolina have pending legislation to address environmental disparities.

Environmental justice groups have succeeded in getting President Clinton to act on the problem of unequal environmental protection, an issue that has been buried for more than three decades. On 11 February 1994, Clinton signed an executive order entitled "Federal Actions to Address Environmental Justice in Minority Populations and Low-Income Populations." This new executive order reinforces what has been law since the passage of the 1964 Civil Rights Act, which prohibits discriminatory practices in programs receiving federal financial assistance.

The executive order also refocuses attention on the National Environmental Policy Act of 1970 (NEPA), which established national policy goals for the protection, maintenance, and enhancement of the environment. The express goal of NEPA is to ensure for all U.S. citizens a safe, healthful, productive, and aesthetically and culturally pleasing environment. NEPA requires federal agencies to prepare detailed statements on the environmental effects of proposed federal actions significantly affecting the quality of human health. Environmental impact statements prepared under NEPA have routinely downplayed the social impacts of federal projects on racial and ethnic minorities and low-income groups.

Under the new executive order, federal agencies and other institutions that receive federal monies have a year to implement an environmental justice strategy. For these strategies to be effective, agencies must move away from the "DAD" (decide, announce, and defend) modus operandi. EPA cannot address all of the environmental injustices alone but must work in concert with other stakeholders, such as state and local governments and private industry. A new interagency approach might include the following:

- Grassroots environmental justice groups and their networks must become full partners, not silent or junior partners, in planning the implementation of the new executive order.

- An advisory commission should include representatives of environmental justice, civil rights, legal, labor, and public health groups, as well as the relevant governmental agencies, to advise on the implementation of the executive order.

- State and regional education, training, and outreach forums and workshops on implementing the executive order should be organized.

- The executive order should become part of the agenda of national conferences and meetings of elected officials, civil rights and environmental groups, public health and medical groups, educators, and other professional organizations.

The executive order comes at an important juncture in this nation's history: Few communities are willing to welcome LULUs or to become dumping grounds for other people's garbage, toxic waste, or industrial pollution. In the real world, however, if a community happens to be poor or inhabited by persons of color, it is likely to suffer from a "double whammy" of unequal protection and elevated health threats. This is unjust and illegal.

The civil rights and environmental laws of the land must be enforced even if it means the loss of a few jobs. This argument was a sound one in the 1860s, when the Thirteenth Amendment to the Constitution, which freed the slaves in the United States, was passed over the opposition of proslavery advocates who posited that the new law would create unemployment (slaves had a zero unemployment rate), drive up wages, and inflict undue hardship on the plantation economy.

Prevention of Harm

Prevention, the elimination of the threat before harm occurs, should be the preferred strategy of governments. For example, to solve the lead problem, the primary focus should be shifted from treating children who have been poisoned to eliminating the threat by removing lead from houses.

Overwhelming scientific evidence exists on the ill effects of lead on the human body. However, very little action has been taken to rid the nation's housing of lead even though lead poisoning is a preventable disease tagged the "number one environmental health threat to children."

Lead began to be phased out of gasoline in the 1970s. It is ironic that the "regulations were initially developed to protect the newly developed catalytic converter in automobiles, a pollution-control device that happens to be rendered inoperative by lead, rather than to safeguard human health." In 1971, a child was not considered "at risk" unless he or she had 40 micrograms of lead per deciliter of blood (μg/dl). Since that time, the amount of lead that is considered safe has

continually dropped. In 1991, the U.S. Public Health Service changed the official definition of an unsafe level to 10 μg/dl. Even at that level, a child's IQ can be slightly diminished and physical growth stunted.

Lead poisoning is correlated with both income and race. In 1988, the Agency for Toxic Substances and Disease Registry found that among families earning less than $6,000, 68 percent of African American children had lead poisoning, as opposed to 36 percent of white children. In families with incomes exceeding $15,000, more than 38 percent of African American children suffered from lead poisoning, compared with 12 percent of white children. Thus, even when differences in income are taken into account, middle-class African American children are three times more likely to be poisoned with lead than are their middle-class white counterparts.

A 1990 report by the Environmental Defense Fund estimated that under the 1991 standard of 10 μg/dl, 96 percent of African American children and 80 percent of white children of poor families who live in inner cities have unsafe amounts of lead in their blood—amounts sufficient to reduce IQ somewhat, harm hearing, reduce the ability to concentrate, and stunt physical growth. Even in families with annual incomes greater than $15,000, 85 percent of urban African American children have unsafe lead levels, compared to 47 percent of white children.

In the spring of 1991, the Bush administration announced an ambitious program to reduce lead exposure of children, including widespread testing of homes, certification of those who remove lead from homes, and medical treatment for affected children. Six months later, the Centers for Disease Control announced that the administration "does not see this as a necessary federal role to legislate or regulate the cleanup of lead poisoning, to require that homes be tested, to require home owners to disclose results once they are known, or to establish standards for those who test or clean up lead hazards."

According to the *New York Times*, the National Association of Realtors pressured President Bush to drop his lead initiative because it feared that forcing home owners to eliminate lead hazards would add from $5,000 to $10,000 to the price of those homes, further harming a real estate market already devastated by the aftershocks of Reaganomics. The public debate has pitted real estate and housing interests against public health interests. Right now, the housing interests appear to be winning.

For more than two decades, Congress and the nation's medical and public health establishments have waffled, procrastinated, and shuffled papers while the lead problem steadily grows worse. During the years of President Reagan's "benign neglect," funding dropped very low. Even in the best years, when funding has risen to as much as $50 million per year, it has never reached levels that would make a real dent in the problem.

Much could be done to protect at-risk populations if the current laws were enforced. For example, a lead smelter operated for fifty years in a predominately African American West Dallas neighborhood, where it caused extreme health problems for nearby residents. Dallas officials were informed as early as 1972 that lead from three lead smelters was finding its way into the bloodstreams of children who lived in two mostly African American and Latino neighborhoods: West Dallas and East Oak Cliff.

Living near the RSR and Dixie Metals smelters was associated with a 36 percent increase in childhood blood lead levels. The city was urged to restrict the emissions of lead into the atmosphere and to undertake a large screening program to determine the extent of the public health problem. The city failed to take immediate action to protect the residents who lived near the smelters.

In 1980, EPA, informed about possible health risks associated with the Dallas lead smelters, commissioned another lead-screening study. This study confirmed what was already known a decade earlier: Children living near the Dallas smelters were likely to have greater lead concentrations in their blood than children who did not live near the smelters.

The city only took action after the local newspapers published a series of headline-grabbing stories in 1983 on the "potentially dangerous" lead levels discovered by EPA researchers in 1981. The articles triggered widespread concern, public outrage, several class-action lawsuits, and legal action by the Texas attorney general.

Although EPA was alarmed with a wealth of scientific data on the West Dallas lead problem, the agency chose to play politics with the community by scrapping a voluntary plan offered by RSR to clean up the "hot spots" in the neighborhood. John Hernandez, EPA's deputy administrator, blocked the cleanup and called for yet another round of tests to be designed by the Centers for Disease Control with EPA and the Dallas Health Department. The results of the new study were released in February 1983. This study again established the smelter as the source of elevated lead levels in West Dallas children. Hernandez's delay of cleanup actions in West Dallas was tantamount to waiting for a body count.

After years of delay, the West Dallas plaintiffs negotiated an out-of-court settlement worth more than $45 million. The lawsuit was settled in June 1983 as RSR agreed to pay for cleaning up the soil in West Dallas, a blood-testing program for children and pregnant women, and the installation of new antipollution equipment. The settlement was made on behalf of 370 children—almost all of whom were poor black residents of the West Dallas public housing project—and forty property owners. The agreement was one of the largest community lead-contamination settlements ever awarded in the United States. The settlement, however, did not require the smelter to close. Moreover, the pollution equipment for the smelter was never installed.

In May 1984, however, the Dallas Board of Adjustments, a city agency responsible for monitoring land-use violations, asked the city attorney to close the smelter permanently for violating the city's zoning code. The lead smelter had operated in the mostly African American West Dallas neighborhood for fifty years without having the necessary use permits. Just four months later, the West Dallas smelter was permanently closed. After repeated health citations, fines, and citizens' complaints against the smelter, one has to question the city's lax enforcement of health and land-use regulations in African American and Latino neighborhoods.

The smelter is now closed. Although an initial cleanup was carried out in 1984, the lead problem has not gone away. On 31 December 1991, EPA crews began a cleanup of the West Dallas neighborhood. It is estimated that the crews will remove between 30,000 and 40,000 cubic yards of lead-contaminated soil from several West Dallas sites, including school property and about 140 private homes. The project will cost EPA from $3 million to $4 million. The lead content of the soil collected from dump sites in the neighborhood ranged from 8,060 to 21,000 parts per million. Under federal standards, levels of 500 to 1,000 parts per million are considered hazardous. In April 1993, the entire West Dallas neighborhood was declared a Superfund site.

There have been a few other signs related to the lead issue that suggest a consensus on environmental justice is growing among coalitions of environmental, social justice, and civil libertarian groups. The Natural Resources Defense Council, the National Association for the Advancement of Colored People Legal Defense and Education Fund, the American Civil Liberties Union, and the Legal Aid Society of Alameda County joined forces and won an out-of-court settlement worth between $15 million and $20 million for a blood-testing program in California. The lawsuit (*Matthews v. Coye*) arose because the state of California was not performing the federally mandated testing of some 557,000 poor children who receive Medicaid. This historic agreement will likely trigger similar actions in other states that have failed to perform federally mandated screening.

Lead screening is important but it is not the solution. New government-mandated lead abatement initiatives are needed. The nation needs a "Lead Superfund" clean-up program. Public health should not be sacrificed even in a sluggish housing market. Surely, if termite inspections (required in both booming and sluggish housing markets) can be mandated to protect individual home investment, a lead-free home can be mandated to protect human health. Ultimately, the lead debate—public health (who is affected) versus property rights (who pays for cleanup)—is a value conflict that will not be resolved by the scientific community.

Shift the Burden of Proof

Under the current system, individuals who challenge polluters must prove that they have been harmed, discriminated against, or disproportionately affected. Few poor or minority communities have the resources to hire the lawyers, expert witnesses, and doctors needed to sustain such a challenge. Thus, the burden of proof must be shifted to the polluters who do harm, discriminate, or do not give equal protection to minorities and other overburdened classes.

Environmental justice would require the entities that are applying for operating permits for landfills, incinerators, smelters, refineries, and chemical plants, for example, to prove that their operations are not harmful to human health, will not disproportionately affect minorities or the poor, and are nondiscriminatory.

A case in point is Louisiana Energy Services' proposal to build the nation's first privately owned uranium enrichment plant. The proposed plant would handle about 17 percent of the estimated U.S. requirement for enrichment services in the year 2000. Clearly, the burden of proof should be on Louisiana Energy Services, the state government, and the Nuclear Regulatory Commission to demonstrate that local residents' rights would not be violated in permitting the plant. At present, the burden of proof is on local residents to demonstrate that their health would be endangered and their community adversely affected by the plant.

According to the Nuclear Regulatory Commission's 1993 draft environmental impact statement, the proposed site for the facility is Claiborne Parish, Louisiana, which has a per capita income of only $5,800 per year—just 45 percent of the national average. The enrichment plant would be just one-quarter mile from the almost wholly African American community of Center Springs, founded in 1910, and one and one-quarter miles from Forest Grove, which was founded by freed slaves. However, the draft statement describes the socioeconomic and community characteristics of Homer, a town that is five miles from the proposed site and whose population is more than 50 percent white, rather than those of Center Springs or Forest Grove. As far as the draft is concerned, the communities of Center Springs and Forest Grove do not exist; they are invisible.

The racial composition of Claiborne Parish is 53.43 percent white, 46.09 percent African American, 0.16 percent American Indian, 0.07 percent Asian, 0.23 percent Hispanic, and 0.01 percent "other." Thus, the parish's percentage population of African Americans is nearly four times greater than that of the nation and nearly two and one-half times greater than that of Louisiana. (African Americans composed 12 percent of the U.S. population and 29 percent of Louisiana's population in 1990.)

Clearly, Claiborne Parish's current residents would receive fewer of the plant's potential benefits—high-paying jobs, home construction, and an increased tax base—than would those who moved into the area or commuted to it to work at the facility. An increasing number of migrants will take jobs at the higher end of the skill and pay scale. These workers are expected to buy homes outside of the parish. Residents of Claiborne Parish, on the other hand, are likely to get the jobs at the low end of the skill and pay scale.

Ultimately, the plant's social costs would be borne by nearby residents, while the benefits would be more dispersed. The potential social costs include increased noise and traffic, threats to public safety and to mental and physical health, and LULUs.

The case of Richmond, California, provides more evidence of the need to shift the burden of proof. A 1989 study, *Richmond at Risk,* found that the African American residents of this city bear the brunt of toxic releases in Contra Costa County and the San Francisco Bay area. At least, thirty-eight industrial sites in and around the city store up to ninety-four million pounds of forty-five different chemicals, including ammonia, chlorine, hydrogen fluoride, and nitric acid. However, the burden of proof is on Richmond residents to show that they are harmed by nearby toxic releases.

On 26 July 1993, sulfur trioxide escaped from the General Chemical plant in Richmond, where people of color make up a majority of the residents. More than twenty thousand citizens were sent to the hospital. A September 1993 report by the Bay Area Air Quality Management District confirmed that "the operation was conducted in a negligent manner without due regard to the potential consequences of a miscalculation or equipment malfunction, and without required permits from the District."

When Richmond residents protested the planned expansion of a Chevron refinery, they were asked to prove that they had been harmed by Chevron's operation. Recently, public pressure has induced Chevron to set aside $4.2 million to establish a new health clinic and help the surrounding community.

A third case involves conditions surrounding the 1,900 *maquiladoras,* assembly plants operated by U.S., Japanese, and other countries' companies along the 2,000-mile U.S.–Mexican border. A 1983 agreement between the United States and Mexico requires U.S. companies in Mexico to export their waste products to the United States, and plants must notify EPA when they are doing so. However, a 1986 survey of 772 *maquiladoras* revealed that only twenty of the plants informed EPA when they were exporting waste to the United States, even though 86 percent of the plants used toxic chemicals in their manufacturing processes. And in 1989, only ten waste-shipment notices were filed with EPA.

Much of the waste from the *maquiladoras* is illegally dumped in sewers, ditches, and the desert. All along the Rio Grande, plants dump toxic wastes into the river, from which 95 percent of the region's residents get their drinking water. In the border cities of Brownsville, Texas, and Matamoros, Mexico, the rate of anencephaly—being born without a brain—is four times the U.S. national average. Affected families have filed lawsuits against eighty-eight of the area's one hundred *maquiladoras* for exposing the community to xylene, a cleaning solvent that can cause brain hemorrhages and lung and kidney damage. However, as usual, the burden of proof rests with the victims. Unfortunately, Mexico's environmental regulatory agency is understaffed and ill equipped to enforce the country's environmental laws adequately.

Obviate Proof of Intent

Laws must allow disparate impact and statistical weight—as opposed to "intent"—to infer discrimina-

tion because proving intentional or purposeful discrimination in a court of law is next to impossible. The first lawsuit to charge environmental discrimination in the placement of a waste facility, *Bean v. Southwestern Waste,* was filed in 1979. The case involved residents of Houston's Northwood Manor, an urban, middle-class neighborhood of homeowners, and Browning-Ferris Industries, a private disposal company based in Houston.

More than 83 percent of the residents in the subdivision owned their single-family, detached homes. Thus, the Northwood Manor neighborhood was an unlikely candidate for a municipal landfill except that, in 1978, it was more than 82 percent black. An earlier attempt had been made to locate a municipal landfill in the same general area in 1970, when the subdivision and local school district had a majority white population. The 1970 landfill proposal was killed by the Harris County Board of Supervisors as being an incompatible land use; the site was deemed to be too close to a residential area and a neighborhood school. In 1978, however, the controversial sanitary landfill was built only 1,400 feet from a high school, football stadium, track field, and the North Forest Independent School District's administration building. Because Houston has been and continues to be highly segregated, few Houstonians are unaware of where the African American neighborhoods end and the white ones begin. In 1970, for example, more than 90 percent of the city's African American residents lived in mostly black areas. By 1980, 82 percent of Houston's African American population lived in mostly black areas.

Houston is the only major U.S. city without zoning. In 1992, the city council voted to institute zoning, but the measure was defeated at the polls in 1993. The city's African American neighborhoods have paid a high price for the city's unrestrained growth and lack of a zoning policy. Black Houston was allowed to become the dumping ground for the city's garbage. In every case, the racial composition of Houston's African American neighborhoods had been established before the waste facilities were sited.

From the early 1920s through the late 1970s, all five of the city-owned sanitary landfills and six out of eight of Houston's municipal solid-waste incinerators were located in mostly African American neighborhoods. The other two incinerator sites were located in a Latino neighborhood and a white neighborhood. One of the oldest waste sites in Houston was located in Freedman's Town, an African American neighborhood settled by former slaves in the 1860s. The site has been built over with a charity hospital and a low-income public housing project.

Private industry took its lead from the siting pattern established by the city government. From 1970 to 1978, three of the four privately owned landfills used to dispose of Houston's garbage were located in mostly African American neighborhoods. The fourth privately owned landfill, which was sited in 1971, was located in the mostly white Chattwood subdivision. A residential part, or "buffer zone," separates the white neighborhood from the landfill. Both government and industry responded to white neighborhood associations and their NIMBY (not in my backyard) organizations by siting LULUs according to the PIBBY (place in blacks backyards) strategy.

The statistical evidence in *Bean v. Southwestern Waste* overwhelmingly supported the disproportionate impact argument. Overall, fourteen of the seventeen (82 percent) solid-waste facilities used to dispose of Houston's garbage were located in mostly African American neighborhoods. Considering that Houston's African American residents comprised only 28 percent of the city's total population, they clearly were forced to bear a disproportionate burden of the city's solid-waste facilities. However, the federal judge ruled against the plaintiffs on the grounds that "purposeful discrimination" was not demonstrated.

Although the Northwood Manor residents lost their lawsuit, they did influence the way the Houston city government and the state of Texas addressed race and waste facility siting. Acting under intense pressure from the African American community, the Houston city council passed a resolution in 1980 that prohibited city-owned trucks from dumping at the controversial landfill. In 1981, the Houston city council passed an ordinance restricting the construction of solid-waste disposal sites near public facilities such as schools. And the Texas Department of Health updated its requirements of landfill permit applicants to include detailed land-use, economic, and sociodemographic data on areas where they proposed to site landfills. Black Houstonians had sent a clear signal to the Texas Department of Health, the city of Houston, and private disposal companies that they would fight any future attempts to place waste disposal facilities in their neighborhoods.

Since *Bean v. Southwestern Waste,* not a single landfill or incinerator has been sited in an African American neighborhood in Houston. Not until nearly a decade after that suit did environmental discrimination resur-

face in the courts. A number of recent cases have challenged siting decisions using the environmental discrimination argument: *East Bibb Twiggs Neighborhood Association v. Macon-Bibb County Planning and Zoning Commission* (1989), *Bordeaux Action Committee v. Metro Government of Nashville* (1990), *R.I.S.E. v. Kay* (1991), and *El Pueblo para El Aire y Agua Limpio v. County of Kings* (1991). Unfortunately, these legal challenges are also confronted with the test of demonstrating "purposeful" discrimination.

Redress Inequities

Disproportionate impacts must be redressed by targeting action and resources. Resources should be spent where environmental and health problems are greatest, as determined by some ranking scheme—but one not limited to risk assessment. The EPA already has geographic targeting that involves selecting a physical area, often a naturally defined area such as a watershed; assessing the condition of the natural resources and range of environmental threats, including risks to public health; formulating and implementing integrated, holistic strategies for restoring or protecting living resources and their habitats within that area; and evaluating the progress of those strategies toward their objectives.

Relying solely on proof of a cause-and-effect relationship as defined by traditional epidemiology disguises the exploitative way the polluting industries have operated in some communities and condones a passive acceptance of the status quo. Because it is difficult to establish causation, polluting industries have the upper hand. They can always hide behind "science" and demand "proof" that their activities are harmful to humans or the environment.

A 1992 EPA report, *Securing Our Legacy,* described the agency's geographic initiatives as "protecting what we love." The strategy emphasized "pollution prevention, multimedia enforcement, research into causes and cures of environmental stress, stopping habitat loss, education, and constituency building." Examples of geographic initiatives under way include the Chesapeake Bay, Great Lakes, Gulf of Mexico, and Mexican Border programs.

Such targeting should channel resources to the hot spots, communities that are burdened with more than their fair share of environmental problems. For example, EPA's Region VI has developed geographic information systems and comparative risk methodologies to evaluate environmental equity concerns in the region. The methodology combines susceptibility factors, such as age, pregnancy, race, income, preexisting disease, and lifestyle, with chemical release data from the Toxic Release inventory and monitoring information; state health department vital statistics data; and geographic and demographic data—especially from areas around hazardous waste sites—for its regional equity assessment.

Region VI's 1992 Gulf Coast Toxics Initiatives project is an outgrowth of its equity assessment. The project targets facilities on the Texas and Louisiana coast, a "sensitive . . . eco-region where most of the releases in the five-state region occur." Inspectors will spend 38 percent of their time in this "multimedia enforcement effort." It is not clear how this percentage was determined, but, for the project to move beyond the "first-step" phase and begin addressing real inequities, most of its resources (not just inspectors) must be channeled to the areas where most of the problems occur.

A 1993 EPA study of Toxic Release Inventory data from Louisiana's petrochemical corridor found that "populations within two miles of facilities releasing 90% of total industrial corridor air releases feature a higher proportion of minorities than the state average; facilities releasing 88% have a higher proportion than the Industrial Corridor parishes' average."

To no one's surprise, communities in Corpus Christi, neighborhoods that run along the Houston Ship Channel and petrochemical corridor, and many unincorporated communities along the 85-mile stretch of the Mississippi River from Baton Rouge to New Orleans ranked at or near the top in terms of pollution discharges in EPA Region VI's Gulf Coast Toxics Initiatives equity assessment. It is very likely that similar rankings would be achieved using the environmental justice framework. However, the question that remains is one of resource allocation—the level of resources that Region VI will channel into solving the pollution problem in communities that have a disproportionately large share of poor people, working-class people, and people of color.

Health concerns raised by Louisiana's residents and grassroots activists in such communities as Alsen, St. Gabriel, Geismer, Morrisonville, and Lions—all of which are located in close proximity to polluting industries—have not been adequately addressed by local parish supervisors, state environmental and health officials, or the federal and regional offices of EPA.

A few contaminated African American communities in southeast Louisiana have been bought out or are in the process of being bought out by industries under their "good neighbor" programs. Moving people away from the health threat is only a partial solution, however, as long as damage to the environment continues. For example, Dow Chemical, the state's largest chemical plant, is buying out residents of mostly African American Morrisonville. The communities of Sun Rise and Reveilletown, which were founded by freed slaves, have already been bought out.

Many of the community buyout settlements are sealed. The secret nature of the agreements limits public scrutiny, community comparisons, and disclosure of harm or potential harm. Few of the recent settlement agreements allow for health monitoring or surveillance of affected residents once they are dispersed. Some settlements have even required the "victims" to sign waivers that preclude them from bringing any further lawsuits against the polluting industry.

A FRAMEWORK FOR ENVIRONMENTAL JUSTICE

The solution to unequal protection lies in the realm of environmental justice for all people. No community—rich or poor, black or white—should be allowed to become a "sacrifice zone." The lessons from the civil rights struggles around housing, employment, education, and public accommodations over the past four decades suggest that environmental justice requires a legislative foundation. It is not enough to demonstrate the existence of unjust and unfair conditions: the practices that cause the conditions must be made illegal.

The five principles already described—the right to protection, prevention of harm, shifting the burden of proof, obviating proof of intent to discriminate, and targeting resources to redress inequities—constitute a framework for environmental justice. The framework incorporates a legislative strategy, modeled after landmark civil rights mandates, that would make environmental discrimination illegal and costly.

Although enforcing current laws in a nondiscriminatory way would help, a new legislative initiative is needed. Unequal protection must be attacked via a federal "fair environmental protection act" that redefines protection as a right rather than a privilege. Legislative initiatives must also be directed at states because many of the decisions and problems lie with state actions.

Noxious facility siting and cleanup decisions involve very little science and a lot of politics. Institutional discrimination exists in every social arena, including environmental decision making. Burdens and benefits are not randomly distributed. Reliance solely on "objective" science for environmental decision making—in a world shaped largely by power politics and special interests—often masks institutional racism. For example, the assignment of "acceptable" risk and use of "averages" often results from value judgments that serve to legitimate existing inequities. A national environmental justice framework that incorporates the five principles presented above is needed to begin addressing environmental inequities that result from procedural, geographic, and societal imbalances.

The antidiscrimination and enforcement measures called for here are not more regressive than the initiatives undertaken to eliminate slavery and segregation in the United States. Opponents argued at the time that such actions would hurt the slaves by creating unemployment and destroying black institutions, such as businesses and schools. Similar arguments were made in opposition to sanctions against the racist system of apartheid in South Africa. But people of color who live in environmental "sacrifice zones"—from migrant farm workers who are exposed to deadly pesticides to the parents of inner-city children threatened by lead

poisoning—will welcome any new approaches that will reduce environmental disparities and eliminate the threats to their families' health.

For Further Discussion

1. In your experience, where are the most polluted sites located? Are you familiar with any toxic waste sites located in wealthy neighborhoods? What is the most polluted neighborhood in your own hometown? How do you think this situation came about? Does it matter if the results were unintentional?

2. Think of two or three industries that pose significant health threats to their workers and neighbors. Where are these industries located? In which countries? In which geographic regions?

3. What do you consider the most important environmental problem? Are there any benefits associated with this problem? Who received the benefits? Who bears the burden most directly?

4. Bullard speaks of shifting the burden of proof from those who claim to be affected by pollution to those who pollute. Do you agree with this? What are the costs and benefits of either alternative?

Just Garbage

Peter S. Wenz

Philosopher Peter Wenz examines the environmental racism that seems evident in much of the United States. Wenz analyzes the arguments that might be offered to defend or rationalize these apparent facts. The most common defense claims that economic factors, not racial discrimination, account for the unequal distribution of environmental burdens. Thus, even if inequalities are found, they can be explained by extraneous factors having nothing to do with race. Although these inequalities are regrettable, they are not unjust or immoral.

Wenz examines this argument and finds it seriously flawed. He then reviews other possible rationales based on several theories of justice and determines that they are all wanting. Wenz concludes with a proposal of his own that,

he believes, is more consistent with the principle of commensurate benefits and burdens.

Environmental racism is evident in practices that expose racial minorities in the United States, and people of color around the world, to disproportionate shares of environmental hazards.[1] These include toxic chemicals in factories, toxic herbicides and pesticides in agriculture, radiation from uranium mining, lead from paint on older buildings, toxic wastes illegally dumped, and toxic wastes legally stored. In this chapter, which concentrates on issues of toxic waste, both illegally dumped and legally stored, I will examine the justness of current practices as well as the arguments commonly given in their defense. I will then propose an alternative practice that is consistent with prevailing principles of justice.

A DEFENSE OF CURRENT PRACTICES

Defenders often claim that because economic, not racial, considerations account for disproportionate impacts of nonwhites, current practices are neither racist nor morally objectionable. Their reasoning recalls the Doctrine of Double Effect. According to that doctrine, an effect whose production is usually blameworthy becomes blameless when it is incidental to, although predictably conjoined with, the production of another effect whose production is morally justified. The classic case concerns a pregnant woman with uterine cancer. A common, acceptable treatment for uterine cancer is hysterectomy. This will predictably end the pregnancy, as would an abortion. However, Roman Catholic scholars who usually consider abortion blameworthy consider it blameless in this context because it is merely incidental to hysterectomy, which is morally justified to treat uterine cancer. The hysterectomy would be performed in the absence of pregnancy, so the abortion effect is produced neither as an end-in-itself, nor as a means to reach the desired end, which is the cure of cancer.

Defenders of practices that disproportionately disadvantage nonwhites seem to claim, in keeping with the Doctrine of Double Effect, that racial effects are blameless because they are sought neither as ends-in-themselves nor as means to reach a desired goal. They are merely predictable side effects of economic and political practices that disproportionately expose poor people to toxic substances. The argument is that burial of toxic wastes, and other locally undesirable land uses (LULUs), lower property values. People who can afford to move elsewhere do so. They are replaced by buyers (or renters) who are predominately poor and cannot afford housing in more desirable areas. Law professor Vicki Been puts it this way: "As long as the market allows the existing distribution of wealth to allocate goods and services, it would be surprising indeed if, over the long run, LULUs did not impose a disproportionate burden upon the poor." People of color are disproportionately burdened due primarily to poverty, not racism.[2] This defense against charges of racism is important in the American context because racial discrimination is illegal in the United States in circumstances where economic discrimination is permitted.[3] Thus, legal remedies to disproportionate exposure of nonwhites to toxic wastes are available if racism is the cause, but not if people of color are exposed merely because they are poor.

There is strong evidence against claims of racial neutrality. Professor Been acknowledges that even if there is no racism in the process of siting LULUs, racism plays at least some part in the disproportionate exposure of African Americans to them. She cites evidence that "racial discrimination in the sale and rental of housing relegates people of color (especially African Americans) to the least desirable neighborhoods, regardless of their income level."[4]

Without acknowledging for a moment, then, that racism plays no part in the disproportionate exposure of nonwhites to toxic waste, I will ignore this issue to display a weakness in the argument that justice is served when economic discrimination alone is influential. I claim that even if the only discrimination is economic, justice requires redress and significant alteration of current practices. Recourse to the Doctrine of Double Effect presupposes that the primary effect, with which a second effect is incidentally conjoined, is morally justifiable. In the classic case, abortion is justified only because hysterectomy is justified as treatment for uterine cancer. I argue that disproportionate impacts on poor people violate principles of distributive justice, and so are not morally justifiable in the first place. Thus, current practices disproportionately exposing nonwhites to toxic substances are not justifiable even if incidental to the exposure of poor people.

Alternate practices that comply with acceptable principles of distributive justice are suggested below. They would largely solve problems of environmental racism (disproportionate impacts on nonwhites) while ameliorating the injustice of disproportionately exposing poor people to toxic hazards. They would also discourage production of toxic substances, thereby reducing humanity's negative impact on the environment.

THE PRINCIPLE OF COMMENSURATE BURDENS AND BENEFIT

We usually assume that, other things being equal, those who derive benefits should sustain commensurate burdens. We typically associate the burden of work with the benefit of receiving money, and the burdens of monetary payment and tort liability with the benefits of ownership.

There are many exceptions. For example, people can inherit money without working, and be given ownership without purchase. Another exception, which dissociates the benefit of ownership from the burden of tort liability, is the use of tax money to protect the public from hazards associated with private property, as in Superfund legislation. Again, the benefit of money is dissociated from the burden of work when governments support people who are unemployed.

The fact that these exceptions require justification, however, indicates an abiding assumption that people who derive benefits should shoulder commensurate burdens. The ability to inherit without work is justified as a benefit owed to those who wish to bequeath their wealth (which someone in the line of inheritance is assumed to have shouldered burdens to acquire). The same reasoning applies to gifts.

Using tax money (public money) to protect the public from dangerous private property is justified as encouraging private industry and commerce, which are supposed to increase public wealth. The system also protects victims in case private owners become bankrupt as, for example, in Times Beach, Missouri, where the government bought homes made worthless due to dioxin pollution. The company responsible for the pollution was bankrupt.

Tax money is used to help people who are out of work to help them find a job, improve their credentials, or feed their children. This promotes economic growth and equal opportunity. These exceptions prove the rule by the fact that justification for any deviation from the commensuration of benefits and burdens is considered necessary.

Further indication of an abiding belief that benefits and burdens should be commensurate is grumbling that, for example, many professional athletes and corporate executives are overpaid. Although the athletes and executives shoulder the burden of work, the complaint is that their benefits are disproportionate to their burdens. People on welfare are sometimes criticized for receiving even modest amounts of taxpayer money without shouldering the burdens of work, hence recurrent calls for "welfare reform." Even though these calls are often justified as means to reducing government budget deficits, the moral issue is more basic than the economic. Welfare expenditures are minor compared to other programs, and alternatives that require poor people to work are often more expensive than welfare as we know it.

The principle of commensuration between benefits and burdens is not the only moral principle governing distributive justice, and may not be the most important, but it is basic. Practices can be justified by showing them to conform, all things considered, to this principle. Thus, there is no move to "reform" the receipt of moderate pay for ordinary work, because it exemplifies the principle. On the other hand, practices that do not conform are liable to attack and require alternate justification, as we have seen in the cases of inheritance, gifts, Superfund legislation, and welfare.

Applying the principle of commensuration between burdens and benefits to the issue at hand yields the following: In the absence of countervailing considerations, the burdens of ill health associated with toxic hazards should be related to benefits derived from processes and products that create these hazards.

TOXIC HAZARDS AND CONSUMERISM

In order to assess, in light of the principle of commensuration between benefits and burdens, the justice of current distributions of toxic hazards, the benefits of their generation must be considered. Toxic wastes result from many manufacturing processes, including those for a host of common items and materials, such as paint, solvents, plastics, and most petrochemical-based materials. These materials surround us in the paint on our homes, in our refrigerator containers, in our clothing, in our plumbing, in our garbage pails, and elsewhere.

Toxins are released into the environment in greater quantities now than ever before because we now have a consumer-oriented society where the acquisition, use, and disposal of individually owned items is greatly desired. We associate the numerical dollar value of the items at our disposal with our "standard of living," and assume that a higher standard is conducive to, if not identical with, a better life. So toxic wastes needing disposal are produced as by-products of the general pursuit of what our society defines as valuable, that is, the consumption of material goods.

Our economy requires increasing consumer demand to keep people working (to produce what is demanded). This is why there is concern each Christmas season, for example, that shoppers may not buy enough. If demand is insufficient, people may be put out of work. Demand must increase, not merely hold steady, because commercial competition improves labor efficiency in manufacture (and now in the service sector as well), so fewer workers can produce desired items. More items must be desired to forestall labor efficiency-induced unemployment, which is grave in a society where people depend primarily on wages to secure life's necessities.

Demand is kept high largely by convincing people that their lives require improvement, which consumer purchases will effect. When improvements are seen as needed, not merely desired, people purchase more readily. So our culture encourages economic expansion by blurring the distinction between wants and needs.

One way the distinction is blurred is through promotion of worry. If one feels insecure without the desired item or service, and so worries about life without it, then its provision is easily seen as a need. Commercials, and other shapers of social expectations, keep people worried by adjusting downward toward the trivial what people are expected to worry about. People worry about the provision of food, clothing, and housing without much inducement. When these basic needs are satisfied, however, attention shifts to indoor plumbing, for example, then to stylish indoor plumbing. The process continues with need for a second or third bathroom, a kitchen disposal, and a refrigerator attached to the plumbing so that ice is made automatically in the freezer, and cold water can be obtained without even opening the refrigerator door. The same kind of progression results in cars with CD players, cellular phones, and automatic readouts of average fuel consumption per mile.

Abraham Maslow was not accurately describing people in our society when he claimed that after physiological, safety, love, and (self-) esteem needs are met, people work toward self-actualization, becoming increasingly their own unique selves by fully developing their talents. Maslow's Hierarchy of Needs describes people in our society less than Wenz's Lowerarchy of Worry. When one source of worry is put to rest by an appropriate purchase, some matter less inherently or obviously worrisome takes its place as the focus of concern. Such worry-substitution must be amenable to indefinite repetition in order to motivate purchases needed to keep the economy growing without inherent limit. If commercial society is supported by consumer demand, it is worry all the way down. Toxic wastes are produced in this context.

People tend to worry about ill health and early death without much inducement. These concerns are heightened in a society dependent upon the production of worry, so expenditure on health care consumes an increasing percentage of the gross domestic product. As knowledge of health impairment due to toxic substances increases, people are decreasingly tolerant of risks associated with their proximity. Thus, the same mindset of worry that elicits production that generates toxic wastes, exacerbates reaction to their proximity. The result is a desire for their placement elsewhere, hence the NIMBY syndrome—Not In My Back Yard. On this account, NIMBYism is not aberrantly selfish behavior, but integral to the cultural value system required for great volumes of toxic waste to be generated in the first place.

Combined with the principle of Commensurate Burdens and Benefits, that value system indicates who should suffer the burden of proximity to toxic wastes. Other things being equal, those who benefit most from the production of waste should shoulder the greatest share of burdens associated with its disposal. In our society, consumption of goods is valued highly and constitutes the principal benefit associated with the generation of toxic wastes. Such consumption is generally correlated with income and wealth. So other things being equal, justice requires that people's proximity to toxic wastes be related positively to their income and wealth. This is exactly opposite to the predominant tendency in our society, where poor people are more proximate to toxic wastes dumped illegally and stored legally.

REJECTED THEORIES OF JUSTICE

Proponents of some theories of distributive justice may claim that current practices are justified. In this section I will explore such claims.

A widely held view of justice is that all people deserve to have their interests given equal weight. John Rawls's popular thought experiment in which people choose principles of justice while ignorant of their personal identities dramatizes the importance of equal consideration of interests. Even selfish people behind the "veil of ignorance" in Rawls's "original position" would choose to accord equal consideration to everyone's interests because, they reason, they may themselves be the victims of any inequality. Equal consideration is a basic moral premise lacking serious challenge in our culture, so it is presupposed in what follows. Disagreement centers on application of the principle.

Libertarianism

Libertarians claim that each individual has an equal right to be free of interference from other people. All burdens imposed by other people are unjustified unless part of, or consequent upon, agreement by the party being burdened. So no individual who has not consented should be burdened by burial of toxic wastes (or the emission of air pollutants, or the use of agricultural pesticides, etc.) that may increase risks of disease, disablement, or death. Discussing the effects of air pollution, libertarian Murray Rothbard writes, "The remedy is simply to enjoin anyone from injecting pollutants into the air, and thereby invading the rights of persons and property. Period."[5] Libertarians John Hospers and Tibor R. Machan seem to endorse Rothbard's position.[6]

The problem is that implementation of this theory is impractical and unjust in the context of our civilization. Industrial life as we know it inevitably includes production of pollutants and toxic substances that threaten human life and health. It is impractical to secure the agreement of every individual to the placement, whether on land, in the air, or in water, of every chemical that may adversely affect the life or health of the individuals in question. After being duly informed of the hazard, someone potentially affected is bound to object, making the placement illegitimate by libertarian criteria.

In effect, libertarians give veto power to each individual over the continuation of industrial society. This seems a poor way to accord equal consideration to everyone's interests because the interest in physical safety of any one individual is allowed to override all other interests of all other individuals in the continuation of modern life. Whether or not such life is worth pursuing, it seems unjust to put the decision for everyone in the hands of any one person.

Utilitarianism

Utilitarians consider the interests of all individuals equally, and advocate pursuing courses of action that promise to produce results containing the greatest (net) sum of good. However, irrespective of how "good" is defined, problems with utilitarian accounts of justice are many and notorious.

Utilitarianism suffers in part because its direct interest is exclusively in the sum total of good, and in the future. Since the sum of good is all that counts in utilitarianism, there is no guarantee that the good of some will not be sacrificed for the greater good of others. Famous people could receive (justifiably according to utilitarians) particularly harsh sentences for criminal activity to effect general deterrence. Even when fame results from honest pursuits, a famous

felon's sentence is likely to attract more attention than sentences in other cases of similar criminal activity. Because potential criminals are more likely to respond to sentences in such cases, harsh punishment is justified for utilitarian reasons on grounds that are unrelated to the crime.

Utilitarianism suffers in cases like this not only from its exclusive attention to the sum total of good, but also from its exclusive preoccupation with future consequences, which makes the relevance of past conduct indirect. This affects not only retribution, but also reciprocity and gratitude, which utilitarians endorse only to produce the greatest sum of future benefits. The direct relevance of past agreements and benefits, which common sense assumes, disappears in utilitarianism. So does direct application of the principle of Commensurate Burdens and Benefits.

The merits of the utilitarian rejection of common sense morality need not be assessed, however, because utilitarianism seems impossible to put into practice. Utilitarian support for any particular conclusion is undermined by the inability of anyone actually to perform the kinds of calculations that utilitarians profess to use. Whether the good is identified with happiness or preference-satisfaction, the two leading contenders at the moment, utilitarians announce the conclusions of their calculations without ever being able to show the calculation itself.

When I was in school, math teachers suspected that students who could never show their work were copying answers from other students. I suspect similarly that utilitarians, whose "calculations" often support conclusions that others reach by recourse to principles of gratitude, retributive justice, commensuration between burdens and benefits, and so forth, reach conclusions on grounds of intuitions influenced predominantly by these very principles.

Utilitarians may claim that, contrary to superficial appearances, these principles are themselves supported by utilitarian calculations. But, again, no one has produced a relevant calculation. Some principles seem *prima facie* opposed to utilitarianism, such as the one prescribing special solicitude of parents for their own children. It would seem that in cold climates more good would be produced if people bought winter coats for needy children, instead of special dress coats and ski attire for their own children. But utilitarians defend the principle of special parental concern. They declare this principle consistent with utilitarianism by appeal to entirely untested, unsubstantiated assumptions about counterfactuals. It is a kind of "Just So" story that explains how good is maximized by adherence to current standards. There is no calculation at all.

Another indication that utilitarians cannot perform the calculations they profess to rely upon concerns principles whose worth are in genuine dispute. Utilitarians offer no calculations that help to settle the matter. For example, many people wonder today whether or not patriotism is a worthy moral principle. Detailed utilitarian calculations play no part in the discussion.

These are some of the reasons why utilitarianism provides no help to those deciding whether or not disproportionate exposure of poor people to toxic wastes is just.

Free Market Approach

Toxic wastes, a burden, could be placed where residents accept them in return for monetary payment, a benefit. Since market transactions often satisfactorily commensurate burdens and benefits, this approach may seem to honor the principle of commensuration between burdens and benefits.

Unlike many market transactions, however, whole communities, acting as corporate bodies, would have to contract with those seeking to bury wastes. Otherwise, any single individual in the community could veto the transaction, resulting in the impasse attending libertarian approaches.[7] Communities could receive money to improve such public facilities as schools, parks, and hospitals, in addition to obtaining tax revenues and jobs that result ordinarily from business expansion.

The major problem with this free market approach is that it fails to accord equal consideration to everyone's interests. Where basic or vital goods and services are at issue, we usually think equal consideration of interests requires ameliorating inequalities of distribution that markets tend to produce. For example, one reason, although not the only reason, for public education is to provide every child with the basic intellectual tools necessary for success in our society. A purely free market approach, by contrast, would result in excellent education for children of wealthy parents and little or no education for children of the nation's poorest residents. Opportunities for children of poor parents would be so inferior that we would say the children's interests had not been given equal consideration.

The reasoning is similar where vital goods are concerned. The United States has the Medicaid program for poor people to supplement market transactions in health care precisely because equal consideration of interests requires that everyone be given access to health care. The 1994 health care debate in the United States was, ostensibly, about how to achieve universal coverage, not about whether or not justice required such coverage. With the exception of South Africa, every other industrialized country already has universal coverage for health care. Where vital needs are concerned, markets are supplemented or avoided in order to give equal consideration to everyone's interests.

Another example concerns military service in time of war. The United States employed conscription during the Civil War, both world wars, the Korean War, and the war in Vietnam. When the national interest requires placing many people in mortal danger, it is considered just that exposure be largely unrelated to income and market transactions.

The United States does not currently provide genuine equality in education or health care, nor did universal conscription (of males) put all men at equal risk in time of war. In all three areas, advantage accrues to those with greater income and wealth. (During the Civil War, paying for a substitute was legal in many cases.) Imperfection in practice, however, should not obscure general agreement in theory that justice requires equal consideration of interests, and that such equal consideration requires rejecting purely free market approaches where basic or vital needs are concerned.

Toxic substances affect basic and vital interests. Lead, arsenic, and cadmium in the vicinity of children's homes can result in mental retardation of the children.[8] Navaho teens exposed to radiation from uranium mine tailings have seventeen times the national average of reproductive organ cancer.[9] Environmental Protection Agency (EPA) officials estimate that toxic air pollution in areas of South Chicago increases cancer risks one hundred to one thousand times.[10] Pollution from Otis Air Force base in Massachusetts is associated with alarming increases in cancer rates.[11] Non-Hodgkin's Lymphoma is related to living near stone, clay, and glass industry facilities, and leukemia is related to living near chemical and petroleum plants.[12] In general, cancer rates are higher in the United States near industries that use toxic substances and discard them nearby.[13]

In sum, the placement of toxic wastes affects basic and vital interests just as do education, health care, and wartime military service. Exemption from market decisions is required to avoid unjust impositions on the poor, and to respect people's interests equally. A child dying of cancer receives little benefit from the community's new swimming pool.

Cost-Benefit Analysis (CBA)

CBA is an economist's version of utilitarianism, where the sum to be maximized is society's wealth, as measured in monetary units, instead of happiness or preference satisfaction. Society's wealth is computed by noting (and estimating where necessary) what people are willing to pay for goods and services. The more people are willing to pay for what exists in society, the better off society is, according to CBA.

CBA will characteristically require placement of toxic wastes near poor people. Such placement usually lowers land values (what people are willing to pay for property). Land that is already cheap, where poor people live, will not lose as much value as land that is currently expensive, where wealthier people live, so a smaller loss of social wealth attends placement of toxic wastes near poor people. This is just the opposite of what the Principle of Commensurate Burdens and Benefits requires.

The use of CBA also violates equal consideration of interests, operating much like free market approaches. Where a vital concern is at issue, equal consideration of interests requires that people be considered irrespective of income. The placement of toxic wastes affects vital interests. Yet CBA would have poor people exposed disproportionately to such wastes.[14]

In sum, libertarianism, utilitarianism, free market distribution, and cost-benefit analysis are inadequate principles and methodologies to guide the just distribution of toxic wastes.

LULU POINTS

An approach that avoids these difficulties assigns points to different types of locally undesirable land uses (LULUs) and requires that all communities earn LULU points.[15] In keeping with the Principle of Commensurate Benefits and Burdens, wealthy communities would be required to earn more LULU points than poorer ones. Communities would be identified by currently existing political divisions, such as villages, towns, city wards, cities, and counties.

Toxic waste dumps are only one kind of LULU. Others include prisons, half-way houses, municipal waste sites, low-income housing, and power plants,

whether nuclear or coal fired. A large deposit of extremely toxic waste, for example, may be assigned twenty points when properly buried but fifty points when illegally dumped. A much smaller deposit of properly buried toxic waste may be assigned only ten points, as may a coal-fired power plant. A nuclear power plant may be assigned twenty-five points, while municipal waste sites are only five points, and one hundred units of low-income housing are eight points.

These numbers are only speculations. Points would be assigned by considering probable effects of different LULUs on basic needs, and responses to questionnaires investigating people's levels of discomfort with LULUs of various sorts. Once numbers are assigned, the total number of LULU points to be distributed in a given time period could be calculated by considering planned development and needs for prisons, power plants, low-income housing, and so on. One could also calculate points for a community's already existing LULUs. Communities could then be required to host LULUs in proportion to their income or wealth, with new allocation of LULUs (and associated points) correcting for currently existing deviations from the rules of proportionality.

Wherever significant differences of wealth or income exist between two areas, these areas should be considered part of different communities if there is any political division between them. Thus, a county with rich and poor areas would not be considered a single community for purposes of locating LULUs. Instead, villages or towns may be so considered. A city with rich and poor areas may similarly be reduced to its wards. The purpose of segregating areas of different income or wealth from one another is to permit the imposition of greater LULU burdens on wealthier communities. When wealthy and poor areas are considered as one larger community, there is the danger that the community will earn its LULU points by placing hazardous waste near its poorer members. This possibility is reduced when only relatively wealthy people live in a smaller community that must earn LULU points.

PRACTICAL IMPLICATIONS

Political strategy is beyond the scope of this . . . , so I will refrain from commenting on problems and prospects for securing passage and implementation of the foregoing proposal. I maintain that the proposal is just. In a society where injustice is common, it is no surprise that proposals for rectifications meet stiff resistance.

Were the LULU points proposal implemented, environmental racism would be reduced enormously. To the extent that poor people exposed to environmental hazards are members of racial minorities, relieving the poor of disproportionate exposure would also relieve people of color.

This is not to say that environmental racism would be ended completely. Implementation of the proposal requires judgment in particular cases. Until racism is itself ended, such judgment will predictably be exercised at times to the disadvantage of minority populations. However, because most people of color currently burdened by environmental racism are relatively poor, implementing the proposal would remove 80 to 90 percent of the effects of environmental racism. While efforts to end racism at all levels should continue, reducing the burdens of racism is generally advantageous to people of color. Such reductions are especially worthy when integral to policies that improve distributive justice generally.

Besides improving distributive justice and reducing the burdens of environmental racism, implementing the LULU points proposal would benefit life on earth generally by reducing the generation of toxic hazards. When people of wealth, who exercise control of manufacturing processes, marketing campaigns, and media coverage, are themselves threatened disproportionately by toxic hazards, the culture will evolve quickly to find their production largely unnecessary. It will be discovered, for example, that many plastic items can be made of wood, just as it was discovered in the late 1980s that the production of many ozone-destroying chemicals is unnecessary. Similarly, necessity being the mother of invention, it was discovered during World War II that many women could work in factories. When certain interests are threatened, the impossible does not even take longer.

The above approach to environmental injustice should, of course, be applied internationally and intranationally within all countries. The same considerations of justice condemn universally, all other things being equal, exposing poor people to vital dangers whose generation predominantly benefits the rich. This implies that rich countries should not ship their toxic wastes to poor countries. Since many poorer countries, such as those in Africa, are inhabited primarily by nonwhites, prohibiting shipments of toxic wastes to them would reduce significantly worldwide environmental racism. A prohibition on such shipments would also discourage production of dangerous wastes, as it would require people in rich countries to live with whatever dangers they create. If the principle of LULU points were applied in all countries, including poor ones, elites in those countries would lose interest in earning foreign currency credits through importation of waste, as they would be disproportionately exposed to imported toxins.

In sum, we could reduce environmental injustice considerably through a general program of distributive justice concerning environmental hazards. Pollution would not thereby be eliminated, since to live is to pollute. But such a program would motivate significant reduction in the generation of toxic wastes, and help the poor, especially people of color, as well as the environment.

Notes

1. See the introduction to this volume for studies indicating the disproportionate burden of toxic wastes on people of color.
2. Vicki Been, "Market Forces, Not Racist Practices, May Affect the Siting of Locally Undesirable Land Uses," in *At Issue: Environmental Justice,* ed. by Jonathan Petrikin (San Diego, Calif.: Greenhaven Press, 1995), 41.
3. See *San Antonio Independent School District v. Rodriguez,* 411 R.S. 1 (1973) and *Village of Arlington Heights v. Metropolitan Housing Development Corporation,* 429 U.S. 252 (1977).
4. Been, 41.
5. Murray Rothbard, "The Great Ecology Issue," *The Individualist* 21, no. 2 (February 1970): 5.
6. See Peter S. Wenz, *Environmental Justice* (Albany, N.Y.: State University of New York Press, 1988), 65–67 and associated endnotes.
7. Christopher Boerner and Thomas Lambert, "Environmental Justice Can Be Achieved Through Negotiated Compensation," in *At Issue: Environmental Justice.*
8. F. Diaz-Barriga et al., "Arsenic and Cadmium Exposure in Children Living Near to Both Zinc and Copper Smelters," summarized in *Archives of Environmental Health* 46, no. 2 (March/April 1991): 119.
9. Dick Russell, "Environmental Racism," *Amicus Journal* (Spring 1989): 22–32, 34.
10. Marianne Lavelle, "The Minorities Equation," *National Law Journal* 21 (September 1992): 3.
11. Christopher Hallowell, "Water Crisis on the Cape," *Audubon* (July/August 1991): 65–74, especially 66 and 70.
12. Athena Linos et al., "Leukemia and Non-Hodgkin's Lymphoma and Residential Proximity to Industrial Plants," *Archives of Environmental Health* 46, no. 2 (March/April 1991): 70–74.

13. L. W. Pickle et al., *Atlas of Cancer Mortality among Whites: 1950–1980,* HHS publication # (NIH) 87–2900 (Washington, D.C.: U.S. Department of Health and Human Services, Government Printing Office: 1987).

14. Wenz, 216–18.

15. The idea of LULU points comes to me from Frank J. Popper, "LULUs and Their Blockage," in *Confronting Regional Challenges: Approaches to LULUs, Growth, and Other Vexing Governance Problems,* ed. by Joseph DiMento and Le Roy Graymer (Los Angeles, Calif.: Lincoln Institute of Land Policy, 1991), 13–27, especially 24.

For Further Discussion

1. If, as Wenz suggests, cost should not determine where toxic sites are located, what should the criterion be?

2. Wenz suggests that in general those who derive benefits from public policy should be the same ones who sustain the burdens associated with that policy. If you apply that principle to such social practices as nuclear power, industrial pollution, wilderness preservation, and the growth of agribusiness, what is the result?

3. What does Wenz mean by LULU points? Do you find his proposal reasonable? Practical?

4. Can you develop a response to Wenz's rejection of free market theories of justice?

5. Would it be just for an impoverished community to accept toxic waste for pay? Should poor communities be free to accept a disproportionate burden for compensation?

Indigenous Environmental Perspectives: A North American Primer

Winona LaDuke

Winona LaDuke is a Native American writer and activist. In this essay, she offers a Native American perspective on a variety of environmental issues, with particular attention to the implications of environmental policy for questions of social justice. From uranium mining and processing to toxic waste disposal to stripmining, she offers evidence that native peoples bear a disproportionate burden of environmental destruction. Despite these burdens, native peoples reap few if any benefits. LaDuke reminds us that for many generations, numerous native cultures maintained lifestyles that today would be characterized as sustainable development. Unfortunately, these cultures have systematically been devastated by the growth of colonialism and industrial economies.

Pimaatisiiwin, or the "good life," is the basic objective of the Anishinabeg and Cree people who have historically, and to this day, occupied a great portion of the north-central region of the North American continent. An alternative interpretation of the word is "continuous birth."

This is how we traditionally understand the world. Two tenets are essential to this concept: cyclical thinking and reciprocal relations and responsibilities to the Earth and Creation. Cyclical thinking, common to most Indigenous or landbased cultures and value systems is an understanding that the world—time, and all parts of the natural order including the moon, the tides, women, lives, seasons, or age—flows in cycles. Within this understanding is a clear sense of birth and rebirth, and a knowledge that what one does today will affect one in the future—on the return. A second concept, reciprocal relations, defines the responsibilities and ways of relating between humans and the ecosystem. Simply stated, the "resources" of the economic system, whether wild rice or deer, are recognized as animate, and as such, gifts from the Creator. Within that context, one could not take life without a reciprocal offering, usually tobacco, or some other recognition of the reliance of the Anishinabeg on the Creator. There must always be this reciprocity. Additionally assumed in the "code of ethics" is an understanding that "you take only what you need, and you leave the rest."

Implicit in the concept of the Pimaatisiiwin is a continuous inhabitation of place, an intimate understanding of the relationship between humans and the ecosystem, and the need to maintain that balance. These values, and basic tenets of culture made it possible for the Cree, Ojibway, and many other Indigenous peoples to maintain economic, political, religious, and other institutions for generations in a manner which would today be characterized as "sustainable development."[1]

A MODEL

By its very nature, "development" or concomitantly, an "economic system" based on these ascribed Indigenous values must be decentralized, self-reliant, and very closely based on the carrying capacity of that ecosystem. The nature of northern economies has been a diversified mix of hunting, harvesting, and gardening, all utilizing a balance of human intervention or care, and in keeping with the religious and cultural systems' reliance upon the wealth and generosity of nature. Since by their very nature, Indigenous cultures are not in an adversarial relationship with nature, this reliance is recognized as correct and positive.

The Anishinabeg or Ojibway nation, as an example, encompasses people and land within four Canadian provinces and five American states. This nation has shared common culture, history, governance, language, and land base—the five indicators, according to international law, of the existence of a Nation of people. This nation historically and presently functions within a decentralized economic and political system, with much of the governance left to local bands (villages or counties), through clan and extended family systems. The vast natural wealth of this region and the resource management systems of the Anishinabeg enabled people to prosper for many generations. In one study of Anishinabeg harvesting technologies and systems, a scientist noted

> Economically, these family territories were regulated in a very wise and interesting manner. The game was kept account of very closely, so that the proprietors knew how abundant each kind of animal was, and hence could regulate killing and not deplete the stock. Beaver were made the object of the most careful "farming," the numbers of occupants, old and young to each cabin being kept

count of. . . . The killing of game was regulated by each family. . . .[2]

Anishinabeg and other resource management systems contained techniques for sustained yield. Although such systems show a high degree of unification of conception and execution, there has only been limited imitation by the scientific profession.[3]

This system enabled the economy to prosper. Conceptually the system provided for both domestic production and production for exchange or export. Hence, whether the resource was wild rice or white fish, the extended family as a production unit harvested within a social and resource management code which insured sustained yield. In addition, it is important to note that previous North American Indigenous populations were substantially greater than today, indicating that the carrying capacity for an intact or managed ecosystem is substantially higher than under present circumstances. While this conceptual discussion is a meager overview of one Indigenous economic system, it is offered as a framework for understanding the larger context of Indigenous values, economic systems, and their relevance in the present discussions of "sustainable development."

COLONIALISM AND UNDERDEVELOPMENT

The holocaust of the Americas is unmatched on a world scale, and its aftermath caused the disruption necessary to unseat many of our Indigenous economic and governmental systems. There can be no accurate estimate of the number of people killed since the invasion, but one conservative estimate provides for 112,554,000 Indigenous people in the western hemisphere in 1492, and an estimated 28,264,000 in 1980. Needless to say, this is a significant depopulation.[4] This intentional and unintentional genocide facilitated a subsequent process of colonialism which served to establish a new set of relations between Indigenous nations and colonial or "settler" nations in the Americas.

Three basic concepts encapsulate the set of relations. First, colonialism has been extended through a set of "center-periphery relations" in which the center has expanded through (1) the cultural practice of the spread of Christianity and later western science and other forms of western thought, (2) the socioeconomic practice of capitalism, and (3) the military-political practice of colonialism.[5]

This has resulted in a set of relations between Indigenous economies and peoples and the North American colonial economy which are characterized by dependency and underdevelopment. "Underdevelopment," or more accurately "underdeveloping" (since it is an ongoing practice) is "the process by which the economy both loses wealth and undergoes the structural transformation which accentuates and institutionalizes this process. . . ."[6]

This process, underway for at least the past two hundred years, is characterized by appropriation of land and resources from Indigenous nations, for the purpose of "development" of the American and Canadian economies, and subsequently, the "underdevelopment" of Indigenous economies. The resulting loss of wealth (closely related to loss of control over traditional territories) has created a situation where most Indigenous nations are forced to live in circumstances of material poverty. It is no coincidence that Native Americans and Native Hawaiians (as well as First Nations in Canada) are the poorest people in the United States. Subsequently, Indigenous peoples are subjected to an array of socioeconomic and health problems which are a direct consequence of poverty.[7]

In this process of colonialism, and eventually marginalization, Indigenous nations become peripheral to the colonial economy, and eventually involved in a set of relations characterized by "dependency." Theotonio Dos Santos states,

> By dependence we mean a situation in which the economy of certain countries is conditioned by the development and expansion of another economy to which the former is subjected. . . .[8]

These circumstances, and indeed, the forced underdevelopment of sustainable Indigenous economic systems for the purpose of colonial exploitation of land and resources, are an essential backdrop to any discussion of present environmental circumstances in the North American community, and indeed, a discussion of "sustainable development" in a North American context. Perhaps most alarming is the understanding that even today, this process continues since a vast portion of the remaining natural resources in the North American continent still underlie Native lands, or, as in the case of the disposal of toxic wastes on Indian reservations, the residual structures of colonialism make Native communities focal points for the excrement of industrial society.

INDIGENOUS NATIONS TODAY

At the outset, it is useful to note that there are over 5,000 nations in the world today, and just over 170 states. "Nations" are defined under international law as those in possession of a common language, land-base, history, culture and territory.[9] North America is similarly comprised of a series of nations, known as "First Nations" in Canada, and, with few exceptions, denigrated in the United States with the term "tribes." Demographically, Indigenous nations represent the majority population north of the 55th parallel in Canada (the 50th parallel in the eastern provinces), and occupy approximately two-thirds of the Canadian landmass.

Although the United States has ten times the population, Indian people do not represent the majority, except in few cases, particularly the "four corners" region of the United States, or the intersection of Arizona, Utah, New Mexico, and Colorado, where Ute, Apache, Navajo, and Pueblo people reside. Inside our reservations however (approximately four percent of our original land base in the United States), Indian people remain the majority population.

In our territories and our communities, a mix of old and new co-exist, sometimes in relative harmony, and at other times, in a violent disruption of the way of life. In terms of economic and land tenure systems (the material basis for relating to the ecosystem), most Indigenous communities are a melange of colonial and traditional structures and systems. While American or Canadian laws may restrict and allocate resources and land on reservations (or aboriginal territory), Indigenous practice of "usufruct rights" is still maintained, and with it traditional economic and regulatory institutions like the trapline, "rice boss," and family hunting, grazing (for those peoples who have livestock), or harvesting territories.

These subsistence lifestyles continue to provide a significant source of wealth for domestic economies on the reservation—whether for nutritional consumption or for household use, as in the case of firewood. They also, in many cases, provide the essential ingredients of foreign exchange—wild rice, furs, or woven rugs and silverwork. These economic and land tenure systems (specific to each region) are largely "invisible" to American and Canadian government agencies, economic analysts who consistently point to Native "unemployment," with no recognition of the traditional economy.

In many northern communities, over seventy-five percent of local food, and a significant amount of income is garnered from this traditional economic system. In other cases, for instance, on the Northern Cheyenne reservation in Montana, over ninety percent of the land is held by Cheyenne, and is utilized primarily for ranching. Although not formal "wage work" in the industrial system, these land-based economies are essential to our communities. The lack of recognition for Indigenous economic systems, though long entrenched in the North American colonial view of Native peoples, is particularly frustrating in terms of the present debate over development options.

Resource extraction plans or energy mega-projects proposed for Indigenous lands do not consider the significance of these economic systems, nor their value for the future. A direct consequence is that environmentally destructive development programs ensue, many times foreclosing the opportunity to continue the lower scale, intergenerational economic practices which had been underway in the Native community. For many Indigenous peoples, the reality is that, as sociologist Ivan Illich has noted, "the practice of development is in fact a war on subsistence."

The following segment of this paper includes an overview of North American Indigenous environmental issues, in the format of generalized discussions and case studies. The paper is far from exhaustive, but is presented with the intention of providing information on the environmental crises pending or present in our communities. It is the belief of many Native people that due to our historic and present relations with the United States and Canadian governments, we may be "the miner's canary," or a microcosm of the larger environmental crisis facing the continent. In a final segment of this paper, we return to the discussion of sustainable development, and offer, once again, some present documentation of the practice of "pimaatisi-iwin," interpreted as good life or continuous birth, within the context of our Indigenous economic and value systems.

URANIUM MINING

Uranium mining and milling are the most significant sources of radiation exposure to the public of the entire nuclear fuel cycle far surpassing nuclear reactors and nuclear waste disposal.

—Victor Gillinsky, U.S. Nuclear
Regulatory Commission, 1978

Perhaps the solution to the radon emission problem is to zone the land into uranium mining and milling districts so as to forbid human habitation.

—Los Alamos Scientific Laboratory,
February 1978

The production of uranium or yellowcake from uranium ore usually requires the discharge of significant amounts of water and the disposal of significant portions of radioactive material. Uranium mill tailings, the solid wastes from the uranium milling stage of the cycle, contain eighty-five percent of the original radioactivity in the uranium ore. One of these products, Radium 226, remains radioactive for at least 16,000 years.

In 1975, 100% of all federally produced uranium came from Indian reservations. That same year there were 380 uranium leases on Indian lands, as compared to four on public and acquired lands. In 1979, there were 368 operating uranium mines in the United States. Worldwide, it is estimated that seventy percent of uranium resources are contained on Indigenous lands.[10]

NAVAJO NATION

Spurred by the advice of the Bureau of Indian Affairs and promises of jobs and royalties, the Navajo Tribal Council approved a mineral agreement with the Kerr McGee Corporation. In return for access to uranium deposits and a means to fulfill risk-free contracts with the U.S. Atomic Energy Commission, Kerr McGee employed 100 Navajo men as uranium miners in the underground mines.

Wages for the non-union miners were low—$1.60 per hour, or approximately two thirds of off-reservation wages. In addition, regulation and worker safety enforcement was exceedingly lax. In 1952, the mine inspector found that ventilation units were not in operation. In 1954, the inspector found that the fan was operating only during the first half of the shift. When he returned in 1955, the blower ran out of gas during an inspection. One report from 1959 noted that radiation levels at the operations were ninety times above tolerable levels.

Seventeen years later most of the readily retrievable ore had been exhausted, and the company began to phase out the mines. By 1975, eighteen of the miners who had worked in the Kerr McGee mines had died of lung cancer and twenty-one more were feared to be dying. By 1980, thirty-eight had died and ninety-five more had contracted respiratory ailments and cancers. The incidence of skin and bladder cancer, birth defects, leukemia and other diseases associated with uranium mining also accelerated.

In its departure from the Shiprock area of Navajo, Kerr McGee abandoned approximately seventy-one acres worth of uranium mill tailings on the banks of the San Juan River, the only major waterway in the arid region. As a result, radioactive contamination spread downstream. Southeast of the facility, the Churchrock uranium mine discharged 80,000 gallons of radioactive water from the mine shaft (in "dewatering") annually into the local water supply.

In July of 1979, the largest radioactive spill in United States history occurred at the United Nuclear uranium mill near Churchrock on the Navajo reservation. The uranium mill tailings dam at the site broke under pressure and 100 million gallons of sludge flooded the Rio Puerco River. Although the company had known of the cracks in the dam for two months prior to the incident, no repairs had been made. The water supply of 1,700 Navajo people was irretrievably contaminated, and subsequently over 1,000 sheep and cattle ingested radioactive water.

By 1980, forty-two operating uranium mines, ten uranium mills, five coal fired power plants and four coal stripmines (spanning 20–40,000 acres each) were in the vicinity of the Navajo reservation. Approximately fifteen new uranium mining operations were under construction on the reservation itself. Although eighty-five percent of Navajo households had no electricity, each year, the Navajo nation exported enough energy resources to fuel the needs of the state of New Mexico for thirty-two years.

The birth defect rate in the Shiprock Indian Health Service area is two to eight times higher than the national average, according to a study supported by the March of Dimes Research Support Grant #15-8, and undertaken by Lora Magnum Shields, a professor at Navajo Community College in Shiprock and Alan B. Goodman, Arizona Department of Health.[11]

LAGUNA PUEBLO

Approximately fifty miles to the east of the Navajo reservation lies Laguna Pueblo, until 1982 the site of the largest uranium strip mine in the world. The Anaconda Jackpile mine comprised 7,000 acres of the reservation, operating from 1952 to 1971, when the eco-

nomically retrievable ore was exhausted. An "Indian preference" clause in hiring ensured the employment of Laguna workers, and by 1979, 650 persons were employed at the mine, with this reservation reflecting some of the highest per capita income in the region. The significance of this employment, as indicated in other health and economic statistics, had a mixed impact on the local community.

Prior to 1952, the Rio Paguate coursed through an agricultural and ranching valley that provided food for the Pueblo. Rio Paguate now runs through the remnants of the stripmine, emerging on the other side a fluorescent green in color. In 1973, the Environmental Protection Agency discovered that Anaconda had contaminated the Laguna water with radiation.

In 1975, the EPA returned to find widespread groundwater contamination in the Grants Mineral Belt. And in 1978, the EPA came back again, this time to inform Laguna that the water was contaminated and to inform the people that the Tribal Council building, the Paguate Community Center, and the newly constructed housing were all radioactive. In addition, Anaconda was reprimanded for having used low-grade uranium ore to repair the road system on the reservation.[12]

PINE RIDGE RESERVATION

On June 11, 1962, 200 tons of uranium mill tailings from the uranium mill in Edgemont, South Dakota, washed into the Cheyenne River and traveled into the Angostora Reservoir. The Cheyenne River flows from this reservoir down through the hills and across the reservation. The Cheyenne passes within several hundred feet of the Red Shirt Table. These tailings are a part of an estimated seven and a half million tons of radioactive material, abandoned from the uranium mill at the Edgemont mine.

Water samples taken from the Cheyenne River and from a subsurface well on the Redshirt Table revealed a gross alpha radioactivity level of nineteen and fifteen picocuries per liter respectively. Federal safety regulations state that a reading greater than five picocuries per liter is considered dangerous to life. In June of 1980, the Indian Health Service revealed water test results for the Pine Ridge reservation community of Slim Buttes (adjacent to the same area) to indicate gross alpha radiation levels at three times the federal safety maximum. A June 10, 1980 report of the Office of Environmental Health, Bureau of Indian Affairs,

Aberdeen indicated that the gross alpha radiation reading for the Slim Buttes water sample was fifty picocuries per liter. A water sample taken from Cherry Creek, on the Cheyenne River Reservation to the north acted as a control sample. It contained 1.9 picocuries per liter, one tenth of that on the Red Shirt Table.

A preliminary study of 1979 reported that fourteen women, or thirty-eight percent of the pregnant women on the Pine Ridge reservation miscarried, according to records at the Public Health Service Hospital in Pine Ridge. Most miscarriages were before the fifth month of pregnancy, and in many cases there was excessive hemorrhaging. Of the children who were born, some sixty to seventy percent suffered from breathing complications as a result of undeveloped lungs and/or jaundice. Some were born with such birth defects as cleft palate and club foot. Subsequent information secured under Freedom of Information Act requests from the Indian Health Service verified the data. Between 1971 and 1979, 314 babies had been born with birth defects, in a total Indian population of under 20,000.[13]

CANADIAN URANIUM MINING

We have always been here, and we will stay here. The mines will come and go, they won't have to live with their consequences. We will . . .

—Dene elder, Wollaston Lake

Previous uranium mining in the north of Canada including Port Hope, has left over 222,000 cubic meters of radioactive waste. Other dumps, including those in the villages of Port Granby and Welcome, contain a further 573,000 cubic meters of toxic radioactive waste. By 1985, over 120 million tons of low level radioactive waste was abandoned near now defunct uranium mines. This amount represents enough material to cover the Trans-Canada Highway two meters deep from Halifax to Vancouver. Present production of uranium waste from the province of Saskatchewan alone occurs at the rate of over one million tons annually.

Uranium mining began at Elliot Lake, Ontario, in 1958, adjacent to and upstream from the Serpent River reserve of Anishinabeg. The uranium mining has continued until recently, but now most of the mine shafts are being closed, as uranium mining investment has shifted to the open pit mines of northern Saskatchewan.

During the mine operations at Elliot Lake, a significant number of miners were exposed to high levels of radiation. According to a report commissioned by Member of Parliament Steven Lewis (and former Ambassador to the United Nations), "in no individual year between 1959 and 1974 inclusive . . . did the average underground dust counts for the uranium mines of Elliot Lake fall below the recommended limits. In another instance, workers at the thorium separation plant operated by Rio Algom (in Serpent River) until the 1970s were exposed to up to forty times the radiation level recommended by the International Commission for Radiological Protection."

Between 1955 and 1977, eighty-one of the 956 deaths of Ontario uranium miners were from lung cancers. This figure is almost twice that anticipated. Another inquiry into the health of uranium miners revealed that there are ninety-three persons suffering from silicosis, ascribed to the Eliot Lake operations by the end of 1974. By early 1975, nearly 500 miners had lung disabilities, wholly or partly ascribed to dust exposure in the mines and mills. Over the next five years, more than 500 new cases had developed.

The Elliot Lake uranium mines produced over 100 million tons of uranium wastes. Most of these were left abandoned by the roadsides and mine sites, where, according to one observer ". . . local residents continue to pick blueberries within a forty foot high wall of uranium tailings. . . ."

In 1978, the Elliot Lake uranium operations continued to spew out 14,000 tons of solid and liquid effluent daily. Most of that effluent was discharged, untreated, into the Serpent River basin. By 1980, dumping of tailings was so extensive that liquid wastes from the mines comprised between one half and two thirds of the total flow. In 1976, the International Joint Commission on the Great Lakes identified the outflow of the Serpent River into Lake Huron as the greatest single source of Radium 226 and thorium isotopes into the freshwater. Perhaps more significant was a 1976 report by the Ontario Ministry of the Environment, which concluded that eighteen lakes in the Serpent River system had been contaminated as a result of uranium mining to the extent that they were unfit for human use and all fish life had been destroyed.

The most significant longterm impact of the Elliot Lake operations will be borne by the Serpent River and other Anishinabeg bands downstream from the operation. A study carried out by the Toronto Jesuit Center for the Environmental Impact Protection Program between 1982–4 documented the following:

• Twice as many young adults (people under the age of thirty-six) reported chronic disease at Serpent

River than at two adjacent reserves. The other two reserves had no direct uranium impacts.

- The Serpent River band reported the largest proportion of participants of all ages with chronic disease.

- Pregnancies ending prematurely with fetal death were more prevalent at Serpent River.

In males over 45, "ill health" is reported by more men with exposure to the plant or uranium mine (seventy-five percent) as compared to forty-three percent in men who did not work at the facilities. Additional birth defects were reported in children of men who worked in the uranium related facilities.[14]

By 1986, uranium production in Saskatchewan had doubled, up to over $923 million in exports annually to American utilities. From previous exposure to radiation (resulting from uranium mining over the past twenty years in the north of Saskatchewan), it is estimated that Native people already have eighty times as much radiation in their bodies as residents of the south. The contamination of the north will only become worse with increased uranium mining.[15]

Uranium mining is projected to expand 100% in Canada in the next few years. Five proposed projects are now awaiting environmental impact assessment. (Three of them will be using existing EIS, with two new programs.) The mines include Dominique-Janine Extension (AMOK-a French Company), South McMahon Lake Project (Midwest Joint Venture), McClean Lake Project (Minatco Ltd.), MacArthur River and Cigar Lake mines. Two of the existing three uranium mines, Cluff Lake and Rabbit Lake/Collins Bay are currently expanding, after exhausting original ore bodies. Most of the exploration and test mine work is occurring just west of Wollaston Lake. The Canadian Crown Corporation, Cameco remains the key owner in most of the mining ventures with 20% ownership of Cluff Lake, 66.6% ownership of Key Lake and Rabbit Lake and 48.75% ownership of Cigar Lake.

In November of 1989, a two million liter spill of radioactive water occurred at Rabbit Lake uranium mine. The spill was not reported by the company to the community, but was seen by the community people, who requested to be informed on the contents of the spill from the facility. Almost half of the spill ended up in Wollaston Lake adjacent to a Dene community. The following spring, Cameco pleaded guilty to negligence in the spill, and paid a $50,000 fine.[16]

Evidence of the possible impact on future generations from longterm exposure to low levels of radiation was recently released by a British scientific study.

This evidence may have some bearing on uranium mining communities, where longterm health studies have not been undertaken. The study found that exposure of male workers to consistent levels of radiation may cause a mutation in sperm resulting in higher rates of leukemia in their offspring. According to a study by the Medical Research Council of Southampton University (Great Britain), workers exposed to radiation may father children with an increased risk of leukemia. After examining a host of variables, the study team found that children of fathers who worked at the Sellafield Nuclear power plant had a two and one-half times higher risk of contracting leukemia. Fathers who had received the highest dosages over their working life stood six to eight times as high a chance of producing a child with leukemia. Overall, the study found fifty-two cases of childhood leukemia in the health district, with the town of Sellafield exhibiting a ten-fold excess over average figures.

KERR MCGEE SEQUOYAH FUELS FACILITY, CHEROKEE NATION, OKLAHOMA

In 1968, when Kerr McGee began building a uranium processing plant in our community, the people were happy at the thought of employment. Kerr McGee assured the safety of the plant, and now seventeen years later, we have a situation that the majority of us would have never imagined. Most of the waste has been stored in plastic lined ponds, one of which has been leaking since 1974.

—Jessie Deer in Water, Vian, Oklahoma

Sequoyah Fuels Corporation (Kerr McGee) operates a uranium fuels processing plant in Gore, Oklahoma. The plant is within the borders of the Cherokee nation, home to a resident population of over 100,000 Cherokees. The plant converts yellowcake (U 308) into uranium hexaflouride using a "wet process." The products of the plant are trucked onto nearby Interstate 40 for delivery to more than fifty customers, including twenty-five nuclear power plants, seven nations, and the Department of Energy.

The process generates two main streams of liquid wastes, the fluoride stream and the nitrate stream. The fluoride stream is treated, and then discharged into the Illinois River under a National Pollutant Discharge Elimination System permit issued by the state. The nitrate stream is processed and discharged into a series of sludge ponds, most of which are at their capacity.

Some of the wastes are processed into a "byproduct" known as raffinate. Essentially a toxic sludge, the material contains radioactive elements like radium 26, thorium 230, and uranium as well as a host of other toxic and heavy metals. According to Kerr McGee's own data, the raffinate it sprayed in 1982 contained 178,000 percent more molybdenum than the maximum allowable concentration for irrigation water. Each year, the liquid (now in the form of raffinate) is used as a fertilizer on various portions of the site and other lands owned by the corporation. In total, over 10,000 acres of land are further exposed to radiation by the use of raffinate fertilizer. Over 11.8 million gallons of the fertilizer were used in 1986 alone.[17] One of these sites is the Rabbit Hill Farms, where in 1987, 11,000 bales of hay from the farm were donated to Navajo sheepherders during a hard winter. "Although it hasn't been a banner year for the company, Kerr McGee decided to help because of our longstanding relationship with the Navajos because of mining and oil and gas leases in the area," the company explained.[18]

There have been a number of spills at the site including an overflow of a settling basin in the spring of 1972, 1,450 pounds of uranium hexafluoride spilled into a surface stream in December of 1978, a major spill in December of 1980, and ongoing leaks at the bottom of raffinate pond number two, which has been leaching continuously into the groundwater for ten years. Dr. Richard Hayes Phillips, of the University of Oregon conducted some research into the effluent discharge of the facility. His findings included the above statistics, and documentation of concentrations of uranium, radium, and thorium in the surface effluent stream which have been measured at 21.3, 2,387 and 5.15 times higher (respectively) than permissible levels.

Within a ten mile radius of the plant, over 200 cancers and birth defects have been recorded. In the town of Vian, population 1,500, 124 persons, or eight percent of the population, had cancer.[19]

COAL STRIPMINING

In 1976, four out of the ten largest coal stripmines in the country were on Indian lands.[20] Today, the circumstances are very much the same, with over one-third of all western low-sulfur strippable coal reserves underlying Indian lands. The majority of the remaining resources are adjacent to the reservations. These statistics are particularly stark in light of a present move to develop low-sulfur coal resources as an alternative to more polluting coal supplies. The North Cheyenne reservation has been at the center of this conflict for almost two decades.

NORTH CHEYENNE COAL

The North Cheyenne reservation lies at the very center of the country's largest deposit of coal. The reservation itself has billions of tons of strippable coal, but the Cheyenne have vigorously opposed this exploitation for over thirty years. Multinational energy corporations working with the approval of the federal and state governments are surrounding the Cheyenne reservation with coal stripmines, railroads, electric generating plants, and transmission lines. Indeed, the largest coal stripmine in the United States is fifteen miles from the reservation, and the four coal generating plants at Colstrip loom just off the reservation border.

The 500,000 acre reservation now sits adjacent to the Powder River Coal lease. In 1982, the Secretary of the Interior sold federal public coal for pennies a ton along the entire eastern boundary or the reservation. This was the largest federal coal sale in the history of the United States. It stretches from the Wyoming border, running along the major water source, the Tongue River. Five to seven new coal stripmines are planned proximate to the reservation.

The environmental and social impacts of the coal stripmining are devastating. First of all, the present mining area is included in a vast region known for meager rainfall and limited reclamation potential. According to a 1973 study by the National Academy of Sciences, "those areas receiving less than seven inches of rainfall, reclamation should not be attempted at all, and instead those lands should be designated as a National Sacrifice Area." The stripmining process divests the land of much of the aquifer system, disrupts groundwater systems, and contaminates a good portion of the remaining groundwater.

Centralized electrical generation causes relatively pristine regions, like the North Cheyenne and the adjacent Crow reservation to bear the burden of reduced air quality, while the end user of electricity is relatively free from the pollution of production. The North Cheyenne, for a number of years, have sought to keep their air quality at a premium by designating their airspace as a "Class One" air quality standard, as recognized by the Environmental Protection Agency. Unfortunately, airborne contaminants from both the mining and the power plant process cause adverse health conditions for the Cheyenne people, including higher incidence of respiratory disease, and lower birth weights.[21]

NAVAJO NATION: BIG MOUNTAIN

My Mother Earth has been totally hurt. The Peabody Coal mine is going on and also uranium has been mined. We have been badly hurt and she has been hurt. Our ancestor told us that the land has guts just like a human. Heart, liver, lung. All of these things she has inside of her. And now all of these things are in critical condition.

—Roberta Blackgoat, Dine Elder-Big Mountain

Similar circumstances occur in the region of the Navajo nation, where at least five coal-fired power plants are located on or adjacent to the reservation. One of those plants, the Four Corners Power Plant was the only manmade object seen by Gemini Two astronauts from outer space. These power plants are fueled by coal from mines such as Black Mesa, where after two decades of stripmining by Peabody Coal and General Electric, the groundwater is contaminated and the water table lowered. This has caused severe hardship to Navajo and Hopi people who live in the area. Perhaps the most significant impact, however, is the forced relocation of over ten thousand Navajo people (an estimated 2,553 families) from the area over the Black Mesa Coal Field. The field contains over twenty-two billion tons of coal, and is presently being mined, with new power plants and a new coal slurry pipeline proposed.

The relocation is legislated under federal law—the Navajo–Hopi Indian Relocation Act, which partitions into two equal parts 1.8 million acres of land formerly held in common by the two peoples. The removal of people from the area makes mining possible, causing their imminent cultural and psychological destruction. There is no word for relocation in Navajo, to move away simply means to disappear.[22]

GWICHIN NATION

Gwichin territory spans the United States–Canadian border in what is known as the Yukon and Alaska. They have continuously inhabited that region for per-

haps 30,000 years and retain a way of life based on the land, primarily the porcupine caribou herd, which numbers around 170,000. The health of the herd is essential to Gwichin survival since every year they may harvest up to 10,000 animals just for domestic consumption.

In light of the Persian Gulf war and insatiable demands for oil, pressure has been building to open up the Arctic National Wildlife Refuge to oil exploitation. The Refuge, presently referred to as America's Serengetti is huge and contains vast concentrations of wildlife. The nineteen million acre refuge hosts perhaps the largest complex Arctic ecosystem, as yet unaltered by industrialization. It is also the only coastline in Alaska still off limits to oil leasing, but it makes up only about 105 miles out of 1,200 miles of coastline.

Prudoe Bay and the Beaufort Sea lie just north of the Refuge, and through the Trans Alaska Pipeline system to Valdez, Alaska, presently supply around two and two tenths million barrels a day. This represents about a quarter of the United States' domestically produced crude oil, and one eighth of total daily consumption. It is estimated that by the year 2000, this oil will be depleted to the point where the pipeline will be operating at twenty-five percent capacity.

Oil in the Gwichin territory of the ANWR represents possibly less than 200 days of American oil needs. If the oil is exploited it will devastate the calving grounds of the porcupine caribou herd, and cause widespread desolation in the animals and the Gwichin. In the fall of 1991, legislation to open the refuge, pending in the American Congress, was narrowly defeated.[23]

HOBBEMA: THE IMPACT OF OIL EXPLOITATION

The Cree people near Hobbema, Alberta, have come under great stress as a direct result of oil exploitation in their territory. Four bands live adjacent to Hobbema— the Samson, Ermineskin, Louis Bull, and Montana bands—with a total population of around 6,500. Oil was discovered in their territory in the 1940s, but it was not until the 1970s that oil royalties began to flood into their coffers. By 1983, at the peak of the oil boom, the four bands were receiving 185 million dollars annually in royalties. Oil royalties gave the average Hobbema family 3,000 or more in monthly payments.

Social upheaval was a direct result of the rapid transition from a land based economy into a cash economy. A century ago, French sociologist Emil Durk-heim found a strong connection between suicide and collective crisis. In an 1897 study of suicides in Italy, he charted the relationship between large-scale in-dustrial growth, economic prosperity, and suicides. Incomes skyrocketed by thirty five percent between 1873 and 1889, suicides jumped thirty-six percent be-tween 1871 and 1877 and another eight percent be-tween 1877 and 1889. Similar statistics were reported in other countries undergoing rapid industrialism.

A study of the oil-rich bands of Alberta was com-missioned by the Department of Indigenous Affairs in 1984. The study confirmed a similar syndrome—sud-den wealth was causing profound social disruption in the bands. The unexpected influx of money led to al-coholism, drug addiction, and suicides. "When we have no money, we had a lot of family unity," recalls Theresa Bull, vice chairman of the Hobbema Health board. "Then we had all this money and people could buy anything they wanted. It replaced the old val-ues. . . . It doesn't bring happiness. It put more value on materialistic possessions. The family and the value of spirituality got lost. . . ."

The town of Hobbema had one of the highest sui-cide rates in North America from 1980 to 1987. From 1985 to 1987, there was a violent death almost every week in Hobbema, and the suicide rate for young men was eighty-three times the national average. There are as many as three hundred suicide attempts by Hob-bema Indians every year. The oil money "stripped them of self respect and dignity," said a social worker. Researchers believe that the true rate for Indian sui-cides nationally (in Canada) is twelve times the na-tional average.[24]

MERCURY CONTAMINATION AT GRASSY NARROWS

In 1956, the first massive outbreak of mercury poison-ing occurred in Minimata Japan. By 1968, the Mini-mata disaster and the nature of mercury poisoning were well documented. No less than 183 medical pa-pers have been published on the subject. By that year, the number of deaths from the poisoning in Japan had approached 100, with several thousand maimed.

In 1960, four years after the Minimata disaster, Dryden Pulp and Paper began to contaminate the Wabigoon River with suspended solids. In March of 1962, Reed Paper opened its Dryden chlor alkali plant

which used mercury to bleach paper products, and then released it into the river to form toxic methyl mercury. An estimated twenty pounds of mercury per day were released into the river. On September 8, 1975, the Ontario Minister of Health publicly admit-ted that twenty to thirty of the Native people living on the Grassy Narrows reserve had shown symptoms of Minimata disease. He also publicly admitted that they may have underestimated the severity of the problem. Tests were reported where Native people had as high as 358 parts per billion mercury in their blood, but at that time did not show signs of mercury poisoning. Mercury poisoning, however, accumulates in the body over time.

Commercial fishing was closed down, causing a to-tal disruption in the economy. An attempt at getting unpolluted fish from a lake near the White Dog re-serve failed when non-Indian lodge owners in the vicinity successfully appropriated the lakes' fish for tourism. With commercial fishing banned since 1970, unemployment on the reserves rose over eighty per-cent. Between 1969 and 1974 welfare tripled on the White Dog Reserve. At Grassy Narrows, it nearly quadrupled from 29,000 dollars to 122,000 dollars. During that same period of time (between January 1970 and June 1972), 200 Native people in the Ke-nora area died violent deaths. Since in its early stages mercury poisoning can lead to highly destructive be-havior, sometimes falsely associated with alcoholism, this report should have been enough to initiate a full-scale study of mercury poisoning among the Native people.

Mercury discharge continued, virtually unabated until 1970, when more than 20,000 pounds of mer-cury had been dumped into the English-Wabigoon River system (with 30,000 pounds more unaccounted for). Mercury is in the English River system, and un-like Minimata Japan, which has the advantage of ocean currents, it will take an estimated sixty to one hundred years to clean itself out. There is no evidence that Dry-den is even going to clean up the mercury pool below the plant, which is still being flushed into the river system.[25]

ALBERTA TIMBER SALES

The province of Alberta has dealt away timberlands almost the size of Great Britain. This new land rush was completed in December of 1988, with the primary beneficiary being the Japanese multinational Diashowa. Diashowa just completed construction of a pulp mill ten kilometers north of the village of Peace River, and has plans for building two more.

The Alberta government granted Diashowa a twenty year lease to 25,000 square kilometers adjacent to the Peace River, and an additional 15,000 square kilometers plus money for roads, rail lines and a bridge. The company also purchased the rights to log the Wood Buffalo National Park, the last great stand of old-growth spruce in Alberta. The lease expires in 2002, and the mill will pump 5,000 tons of chlorinated organic compounds into the Peace River annually. The land leased to Diashowa overlaps with the traditional lands of the Lubicon Lake First Nation.

The Lubicon Cree have been opposing any development in their territory since they were invaded by oil companies in the late 1970s. By 1982, there were over 400 oil wells in their territory, and traditional hunting and trapping trails were turned into company roads. The ability of the people to sustain themselves from the land decreased significantly. Trapping incomes were devastated, and welfare soared from ten percent in 1980 to ninety-five percent in 1983. While the people suffered, an estimated one million dollars worth of oil was extracted daily from the land.[26]

EMISSIONS DISCHARGE

Formaldehyde is a water soluble gas that is known to irritate the eyes, respiratory tract, and skin. It is also a suspected carcinogen, and has been linked to cancer of the skin, lungs, and nasal passages. Airborne formaldehyde can cause ocular damage and allergic dermatitis. Young children and the elderly are especially vulnerable to the effects of this toxin.

At the rate of between twenty-two and thirty-two pounds per hour, the Potlatch Timber Corporation near Cook, Minnesota is emitting formaldehyde into the air over the Nett Lake reservation. At the conservative rate of twenty-two pounds per hour, this figure represents 164,208 pounds annually of emissions.[27]

HYDRO-ELECTRIC EXPLORATION

James Bay, at the base of Hudson Bay, is the largest drainage system on the North American continent. Virtually every major river in the heartland ends up there. This makes the bay a rich ecosystem teeming with wildlife, the staging ground for migratory birds, and a feeding area for the largest migratory herd of mammals on the continent—the George's River Caribou herd. Approximately 35,000 Cree, Innu, Inuit, and Ojibway people live within the region and are dependent upon the ecosystem. The way of life is land-based subsistence—hunting, harvesting, and tourism economy in which at least fifty percent of the food and income for the region originates.

The James Bay I project, introduced in 1972, was intended to produce 10,000 megawatts of electricity by putting eleven and one-half square kilometers of land under water and behind dams. The project concentrated along the East Main and Rupert River and ruined the ecology of some 176,000 square kilometers, an area about two-thirds the size of West Germany. The Native people of the area did not hear of the project until planning was well underway.

Following years of futile litigation, 400 kilometers of paved road, three power stations, and five reservoirs were built. Four major rivers were destroyed and five 735 KV powerlines cut a swath through the wilderness. The environmental impact is enormous. Mercury levels at the reservoirs are six times safe levels, and some two-thirds of the people downstream from the reservoirs have mercury contamination in their bodies, some at thirty times the allowable level. Vast amounts of hunting and trapping territory have been devastated, causing economic and social dislocation from loss of food and cultural activities.

If the project continues, Phase Two will be even more devastating. The area to be impacted is the size of New England, or 356,000 square miles. These projects, according to Jan Bayea of the National Audobon Society will mean that ". . . in fifty years, this entire ecosystem will be lost."[28]

TOXIC WASTE DUMPS

When Waste Tech wanted to build an incinerator and dump on our land, they said they would give us thousands of dollars and a nice two-story house. But I thought about the land and how we rely on it—this dump would poison the water and the land. It's not just temporary, my children and grandchildren will have to live on this land forever. Don't listen to these thieves that want our land—we need to protect Mother Earth.

> —Jane Yazzie, one of the organizers of Citizens
> Against Ruining our Environment, which
> helped defeat a toxic waste dump proposal at
> Dilcon on the Navajo reservation

Most insidious is the recent set of proposals to dispose of toxic wastes in Native communities. Largely a result of stiff and successful opposition in urban areas, toxic waste operators have increasingly looked to reservations and so-called third world countries as possible disposal sites. The "labor pool" and "underdeveloped economies" of many reservations provide an apparent mandate for development. Additionally, the sovereign status of Indian nations, exempt from state and local laws, in which there is minimal federal regulation of waste dumping, has provided an additional incentive.

In the past few years, over forty-five Indian communities have been approached by waste companies offering multi-million dollar contracts in exchange for the right to dump or incinerate on Indian lands. East of San Diego alone, over eighteen rancherias (small Indian communities) have been approached as possible dumping grounds for garbage and toxic wastes. One community that has already been impacted by a toxic waste dump is the Akwesasne Mohawk reservation.

AKWESASNE

The Akwesasne Mohawk reservation spans the United States/Canadian border, and is home to approximately 8,000 Mohawk people. Through the center of their territory is the St. Lawrence River, the waterway of their people. For generations, the Mohawk have relied upon the river for fish, food, and transportation. Today, the river is full of poison.

In Canada, the Akwesasne reserve has been singled out from sixty-three Native communities located in the Great Lakes basin as the most contaminated. On the American side in 1983, the Environmental Protection Agency designated the area as one of the top "Superfund sites" in the United States.

The General Motors Massena Central Foundry is possibly the most significant PCB dump site in North America. The chemical is known to cause brain, nerve, liver and skin disorders in humans and cancer and reproductive disorders in laboratory animals. Five PCB saturated lagoons, and a number of sludge pits dot GM's 258 acre property, a site adjacent to the reserva-

tion. Contaminant Cove, at the conflux of the Grasse and St. Lawrence Rivers, is perhaps the worst. According to the Environmental Protection Agency, fifty parts per million PCBs is classified as hazardous waste. Sludge and vegetation at the bottom of Contaminant Cove have been documented at 3,000 parts per million. A male snapping turtle was located with 3,067 parts per million of PCBs in its body.[29]

At present, a project known as the Akwesasne Mothers Milk Project is undertaking a study of breastmilk, fetal cord, and urine samples of Mohawk mothers on the reservation. A total of 168 women are participating in the study. The primary organizer of the project, Katsi Cook Barreiro, is a practicing midwife who has delivered many children on the reservation. "I've got myself four one hundredths parts per million of mirex (a flame retardant) and eighty-four one-thousandths parts per million PCBs in my body," she explains. "This means that there may be a potential exposure of our future generations. The analysis of Mohawk mother's milk shows that our bodies are, in fact, a part of the landfill."[30]

NUCLEAR WASTE CONTAMINATION

The Hanford nuclear reservation is well within the treaty area of the Yakima Indian Nation on the Columbia River. The nuclear site contains 570 square miles of land, and a significant portion of it is contaminated with radiation. In August of 1973, over 115,000 gallons of liquid high-level radioactive waste seeped into the ground from a leaking storage tank. The waste contained cesium 137, strontium 90, and plutonium. Other leaks from August of 1958 to June two decades later included over 422,000 gallons reported on the site.

Soil at the site is so contaminated that much has been removed as high-level radioactive waste. The Department of Energy changed concentrations from ten nanocuries per gram of soil permitting a rise to 100 nanocuries per gram of soil allowing for plutonium contamination to rise indefinitely at the site by changing the definition of high-level waste.

Hanford produces a dry fallout of small respirable dust particles that are contaminated with plutonium. This airborne dust is released from smokestacks at the site, and is not contained by the site boundaries.

A significant portion of these wastes are contaminating the air and water in the region of the Hanford reservation, an area in which approximately twenty different Indigenous peoples live.[31]

INUIT BREASTMILK

Studies of Inuit breastmilk in the Hudson Bay region of northern Canada indicate that Inuit women have levels of PCB contamination higher than those recorded anywhere else in the world. The maximum PCB concentration considered "safe" by the Canadian government is one and five tenths parts per million. A Laval University Study, conducted by Dr. Eric Dewailly, in 1988, discovered much higher levels in Inuit women. Dewailly's study of twenty-four samples (one third of all nursing mothers in that year), recorded an average concentration of three and fifty-nine hundredths parts per million of PCBs in breastmilk. Some samples were recorded at fourteen and seven tenths parts per million. The average concentration of PCBs in breastmilk in Quebec is five tenths parts per million.

According to an Inuit spokesperson, Mary Kaye May of the Kativik Regional Health Council, the findings brought "fear and great sadness" to the village. It is assumed that the higher levels are attributed to the Inuit diet of fish and marine mammals (at least nine meals per month) which are known to be concentrating PCBs in the food chain. The PCBs have appeared in the Arctic food chain in recent years, largely attributed to atmospheric distribution of heavy metals, and toxins from southern industries, and from abandoned military, radar, and communications installations utilizing PCBs.

While the toxic contamination of infants is of great concern to the Inuit, there are, for all major purposes, no alternatives. "If the women stop breastfeeding," Mary Kaye May continues, "and with the cost of baby formula at seventeen dollars a can, we will face a frightening number of cases of infant malnutrition."

Related contamination includes radioactive cesium, DDT, toxophene, and other pesticides. Many of them have been banned in North America, but are still used in many developing countries. Dr. Lyle Lockhart (Canadian government Fisheries and Oceans Department, Winnipeg, Manitoba) indicates that "toxins will distill off the warm land and plant surfaces in many of these countries and circulate in the air possibly for years, and gradually condense and accumulate in colder regions like the Arctic and Antarctic, which are becoming the world's dump for these things." As an example,

toxophene, a pesticide most commonly used in cotton fields was discovered in a study by Lockhart to be located in the livers of two freshwater fish taken in the Mackenzie River of the Northwest Territories. The severity of the problem is indicated in another study of the polar bear which, because of its high level on the food chain, could be forced into extinction by the year 2006 due to sterility caused by PCB contamination.[32]

ACID RAIN CONTAMINATION

The Anishinabeg treaty areas of northern Michigan, Wisconsin, and Minnesota are impacted by airborne contamination from acid rain at an alarming level. The traditional Anishinabeg diet consists, to a great extent of wild rice, fish, deer, waterfowl, fruits, and herbs. The major protein sources comprising the traditional diet are foods most impacted by acid rain.

Acid pollution is accumulating in the northern lakes, and during spring run-off causes toxic shock to the northern ecosystem, eliminating millions of fish eggs, young fry, molting crayfish, and other animals on the food chain. The secondary impact is the leaching of heavy metals, including mercury into the food system as a result of the acidification process. Presently a number of lakes in the northern area have fish consumption advisories due to mercury contamination. Since 1970, Minnesota has documented a five percent per year increase of mercury in fish tissue. This contamination is attributed to atmospheric deposits originating from coal fired power plants and garbage incinerators. At the present rate of increase, by the year 2015 mercury concentrations in fish tissue will be dangerously high.[33]

MILITARY OCCUPATION

"The militarization, that's what you have to fight," explains Francesca Snow, an Innu woman who has been working actively to stop the siting of a NATO base in her homeland. "They will destroy the land. They will destroy the animals, and they will destroy our life." Father Alexis Jouveneau, a priest who has lived with the Innu for many years, issued a warning to the Canadian government about the NATO base, "You are destroying not only their lifestyle, you are destroying their whole life so that you may proceed with military exercises. At that point, you might as well build a psychiatric clinic right here, and it will soon be over-filled." "If the military goes ahead," says Innu elder Antoine Malec, "You will not see us cry. We will not cry, but our hearts will bleed. . . ."

THE WESTERN SHOSHONE NATION

All nuclear weapon states explode their bombs on unconsenting nations. No nuclear state tests bombs on its own lands and people. The United States doesn't set off nuclear weapons in Santa Barbara or Washington D.C. It bombs the Western Shoshone Nation.

Since 1963, the United States has exploded 651 nuclear weapons and "devices" on Newe Sogobia, the Western Shoshone nation. Because they cause destruction, the 670 nuclear explosions in Newe Sogobia have been classified by the Western Shoshone National Council as bombs rather than tests.

In 1863, the representatives of the United States and Western Shoshone nation signed the Treaty of Ruby Valley. The United States proposed the treaty in order to end Shoshone armed defense of Sogobia, acquire gold from the territory, and establish safe routes to California. The United States Senate ratified the treaty in 1866 and President Grant confirmed it in 1869. The treaty is still in effect.

The nation of Newe Sogobia has an area of 43,000 square miles, about the size of Honduras, and is bounded by western Nevada, southern Idaho and southeastern California. To maintain control over the area, the United States has usurped almost ninety percent of Shoshone land and resources and placed them under the Departments of Interior, Energy, Defense, Transportation, and other agencies. The Western Shoshone, however continue their occupancy of their traditional lands. They have maintained this, and resisted government efforts to terminate their title through the Court of Claims, by continuing their opposition to a twenty-six million dollar proposal to "settle" their claims and compensate them for their land.

In a recent effort to secure occupancy of the land, the Bureau of Land Management has been attempting to remove the livestock of the Dann family, one of the leading families in challenging the occupation.[34]

HAWAII

Hawaii was the last frontier in an era of United States expansion. It was designated as a state in 1959, eighteen years after Pearl Harbor. Over 180,000 Native Hawaiian people have traditional rights to most of the Islands, and today maintain a precarious land-based existence in the face of increasing tourism and military occupation.

Hawaii is the most militarized state in the country, serving as the center for the Pentagon's Pacific Command. It serves as the headquarters for military activities which purport to control more than half the earth's surface—from the west coast of North America to the east coast of Africa, from the Antarctic to the Arctic.

There are more than 100 military installations in the Hawaiian islands and fully ten percent of the state and twenty-five percent of Oahu are under direct federal control. Hawaii is the loading and reloading base for the Pacific fleet. In 1972, Oahu alone was the storage site for some 32,000 nuclear weapons. One island of the Hawaiian archipelago, Kaho'olawe was a bombing site for almost five decades for Pacific rim countries.

PIMAATISIIWIN: THE GOOD LIFE, OR CONTINUOUS BIRTH

Indigenous peoples remain on the front lines of the North American struggle to protect our environment. We understand very clearly that our lives, and those of our future generations are totally dependent on our ability to resist colonialism and industrialization in our lands, and to rebuild our communities.

There are cases where Indigenous people have successfully defended their homelands. The Innu people in the area referred to as Labrador successfully defeated the siting of a NATO base in their territory, after a decade of litigation, rallies, and civil disobedience.

The Northern Cheyenne, who now face a new series of coal developments, have successfully defeated and opposed coal strip mining on their reservation for almost three decades.

The Gwichin people and a coalition of environmental groups successfully averted the opening of the Arctic National Wildlife Refuge in the fall of 1991, and once again have saved their land, their people and the porcupine caribou herd.

The White Earth Anishinabeg from northern Minnesota successfully repudiated the siting of a nuclear waste repository on the reservation. The nuclear waste dump was proposed for the headwaters of the Mississippi River.

The Native Hawaiian people, after four decades of struggle, stopped the bombing of their sacred land,

Kaho'olawe, by the American military. It was, until then, the only national historic landmark regularly shelled with artillery.

The Navajo people, Kaibab Paiute, Oglala Lakota and others successfully opposed the siting of hazardous waste incinerators and dumps on their reservation lands, in spite of millions of dollars of jobs and "benefits" promised to the people.

The Cree of the Moose River have thus far stopped the proposal for twelve dams in the River basin, and now the government says it will not go ahead with any dam construction without Native consent.

These, and many others, are examples of Native people engaging in a successful resistance to the destruction of their land and lives. On the other side, there are also many examples of communities rebuilding traditional economies, social structures and ways of life.

On the White Earth reservation in northern Minnesota, a group of craftswomen and wild rice harvesters banded together to protect the wild rice crop, and to seek a decent price for their products. The wild ricers succeeded in designating a number of reservation lakes as "organic," to preserve the water quality in the lakes. The wild rice is harvested for home use, and is now sold at a fair market value, with more money coming to the reservation. Similarly, the Ikwe Marketing Collective teaches crafts and traditional arts for the purpose of maintaining the traditions, and also offering an opportunity to sell these products at a fair market value to non-Indians.

Near Leupp on the Navajo reservation, a number of families have banded together to develop a community farm. The gardening project, called Navajo Family Farms, is similar to Navajo agriculture of a century ago, and seeks to provide food to the local families and surplus for sale at a decent price to the local communities. The project utilizes "drip irrigation" methods, and is intended to be within the balance of the ecosystem.

The Chippewa of Northern Wisconsin, embroiled for a decade in a confrontational relationship with many racist anti-Indian groups, have proposed an ecological co-management agreement which will protect the treaty area they have legal rights to in northern Wisconsin, and provide a model for sustainable development merging Indigenous and non-Native values and land use.

In upstate New York, a group of Mohawk families are trying to reestablish traditional agriculture systems in an Indian Corn Project, sponsored in part by Cornell University. The project seeks to develop nutritional alternatives, and at the same time provide a tangible example of cultural revitalization for the Native community.

Notes

1. For discussion see Colin Scott, "Knowledge Construction Among Cree Hunters: Metaphors and Literal Understanding." *Société des Americanistes,* JSA 1989, LXXV:193–208.
2. F. G. Speck, "Family Hunting Territories and Social Life," *American Anthropologist,* Spring 1915.
3. Peter J. Usher, "Property Rights: The Basis of Wildlife Management," in National and Regional Interests in the North: Third National Workshop on People, Resources and the Environment North of 60, Canadian Arctic Resources Committee, Ottawa, Ontario, 1984.
4. Robert Venables, "The Cost of Columbus—Was There a Holocaust?" in *View from the Shore: American Indian Perspectives on the Quincentary,* Cornell University: Akwe:kon Press, Fall 1990.
5. John Galtung, et al., eds., *Self Reliance: A Strategy for Development* (London: Bogle-L'Ouverture Publications, Ltd, 1980), 20.
6. Samir Amin, "Unequal Development: An Essay on the Social Formations of Peripheral Capitalism," (New York: Monthly Review Press, 1976).
7. U.S. Congress, American Indian Policy Review Commission, Final Report, Washington DC, 1977: 90.
8. Theotonio Dos Santos, "The Structure of Dependence" in *Readings in United States Imperialism,* K. T. Finn and Donald C. Hodges, ed. (New York: Monthly Review Press, 1971).
9. Jason Clay, "What's a Nation?" *Mother Jones,* November–December 1990.
10. OECD Nuclear Energy Agency Joint Report with the International Atomic Energy Agency, 1988, and Winona LaDuke, "Native America: The Economics of Radioactive Colonialism," In *Review of Radical Political Economics,* Volume XV, Number 3, Fall 1985.
11. Christopher McLeod, Randy Hayes and Glenn Switkes, in "Four Corners A National Sacrifice Area?" *Workbook,* Volume X, No. 2, (Albuquerque, NM: Southwest Research and Information Center, April–June 1985).
12. Rich Nafziger "Indian Uranium: Profits and Perils," Americans for Indian Opportunity, Red Paper, Albuquerque, 1976.
13. Women of All Red Nations, Preliminary Health Report, March 1980. Also Rocky Mountain Greenpeace Press Release, March 10, 1980, Denver, Colorado and Attorney Frances Wise, cited in interview with Winona LaDuke, *Science for the People,* September/October 1982.

14. Roger Moody, CIMRA, Unpublished research on Rio Tinto Zinc, and in *Plunder,* (London, England: April–June 1985, 1990).

15. Greenpeace Canada, Press Release, November 2, 1988.

16. Jamie Kneen, Hatchet Lake Band Environmental Officer, correspondence, January 1991, Wollaston Lake, Saskatchewan.

17. Joel Bleifuss, "Kerr McGee Lays Waste to Eastern Oklahoma," in *In These Times,* August 19–September 1, 1987.

18. Native Americans for a Clean Environment, brochure, 1988.

19. Native Americans for a Clean Environment, Interview with Jessie Deer in Water, September 1988.

20. Council on Economic Priorities, 1979, New York, NY.

21. Native Action Funding Proposal, 1991, Lame Deer, Montana.

22. Cate Gillis, "Big Mountain Weavers Collective," in *Indigenous Woman,* Fall 1991.

23. Winona LaDuke, "Norma Kassi: The Gwichin, the Porcupine Caribou Herd and Big Oil," in *Indigenous Woman,* Spring 1991.

24. Geoffrey York, *The Dispossessed* (Toronto, Ontario: Lester, Orpen and Dennys, Ltd., 1989), 88–95.

25. Jim Harding, "Mercury Poisoning," *Canadian Dimension,* Spring 1988.

26. Peter Hamel, *Quaker Newsletter,* 1990.

27. *Minnesota North Newsletter,* Orr Minnesota, Spring 1987.

28. Andre Picard, "James Bay II," in *The Amicus Journal,* Fall 1990.

29. Akwesasne Mothers Milk Project Report for the Global Assembly: Women and the Environment, Miami, Florida, November 1991.

30. Winona LaDuke, Lea Foushee, "Nuclear Native America," Research paper of Indigenous Women's Network, 1988.

31. *Toronto Globe and Mail,* February 7, 1989, and *Ottawa Citizen,* March 15, 1989.

32. Lea Foushee, "Acid Rain Research Paper," Indigenous Women's Network, testimony before the International Council of Indigenous Women, Samiland, Norway, August 1990.

33. Bernard Nietschman, "Nuclearization of the Western Shoshone Nation," in International Working Group on Indigenous Affairs Documents, Number 62, January 1989, Copenhagen, Denmark.

34. Mililani Traskookok, testimony presented at the National Minority Leadership Summit, Washington DC, October 23, 1991.

For Further Discussion

1. LaDuke says that cyclical thinking and reciprocal relationships are each essential to her culture's understanding of the world. What exactly does she mean by these two concepts? How do they differ, if at all, from your own thinking or from the thinking of your own culture?

2. LaDuke suggests that the lifestyle of many indigenous peoples would today be called "sustainable development." Do you agree?

3. Did any part of LaDuke's essay surprise you? Did any part anger you? If so, why?

4. Like Maria Mies (Chapter 12), LaDuke speaks of colonialism as a partial explanation for environmental injustice. From the Native American perspective, is the United States a colonial power? From your own perspective, is this true?

5. In your opinion, what explains the environmental adversity faced by so many indigenous peoples? Is it an accident? Is it any different from what other peoples face?

DISCUSSION AND STUDY QUESTIONS FOR CHAPTER 13

1. Should industries, local communities, and countries be forced to keep the toxic waste they generate? Should nuclear waste be kept on the site of the nuclear plant? Should people be allowed to pay others to bear this risk for them?

2. Should every state be required to bear its fair share of toxic sites, or should such sites be located in the most remote areas of the country?

3. Opposition to undesirable land uses, such as toxic waste storage, is often characterized as the NIMBY ("not in my back yard") attitude. Can you offer criteria for distinguishing justified from unjustified cases of NIMBY?

4. In economic terms, costs are defined as opportunities forgone. The more that you must give up for a choice, the more costly it is. Could the costs of pollution be different for two different communities? How so? Could cost–benefit analysis be unfair?

DISCUSSION CASES

Locating Nuclear Plants

Decisions on where to build nuclear plants have been made traditionally on economic grounds, using a type of cost–benefit analysis. Assuming the desire to build a plant, policymakers would consider the costs and benefits of alternative sites. The plants could be located within heavily populated urban areas or in rural areas some distance away. The benefits of building within the urban area would include being close to the energy demand, but the costs would involve higher land prices, higher labor costs, and, if any accident would occur, greater risks to a greater number of people. As a result of such seemingly neutral calculations, most nuclear plants within the United States are located outside of large urban areas. Are such decisions fair to the people who live in these areas? How many different nuclear plants are you aware of? Where are they located? How many are located within metropolitan areas, and how many are located in more rural areas just outside of an urban center? What do you think explains the location of nuclear plants?

Most nuclear waste produced at nuclear plants is stored on site. Did you know that? Until a permanent central site is located and built, nuclear waste will continue to be stored locally. Is this wise? Fair? Should the people who benefit from the electricity generated from these plants bear the burden of storing the waste? Or is nuclear power a general social benefit whose burden should be distributed widely?

Voluntary Consent and Toxic Waste

Society faces numerous decisions concerning storage of toxic waste, including the waste generated by nuclear plants. In most cases, nuclear waste presently is stored on-site until a national nuclear waste storage facility can be located and built. This means that, in effect, every nuclear power plant is also a nuclear waste storage site.

Most would agree that a question of justice is raised when citizens uninvolved in the generation of this waste are coerced into accepting the risks of storing it. But, suppose that those responsible for the waste—for example, consumers who use energy generated by the nuclear plant acting through their public utility company—pay other people to bear this risk?

In some cases, native people, often acting through tribal leaders, have expressed a willingness to accept toxic waste for storage on their tribal lands. They are, of course, paid for this land and for this service. The sites are often on remote areas of reservation land, and the revenue might represent a substantial increase in the tribe's income; it might also be the only external source of income for the people.

Similar situations exist in less industrialized countries. Toxic waste from the industrialized world is often shipped to other countries that are willing to accept the waste for a price. Are such actions fair? Can voluntary decisions, in the face of few alternatives and pressing economic need, ever be coercive? How do you think Wenz and LaDuke might respond to these situations? Is voluntary consent sufficient to justify disproportionate burdens?

Structural Injustice

Injustice can occur as the result of intentional decisions to place an unfair and undeserved burden on the least advantaged members of a society. But similar results can occur as the unintentional consequences of seemingly neutral policy-making methods. Such injustices are called structural or institutional injustices, and these can raise important ethical questions.

Consider the decision to build a trash-to-steam plant or to construct a landfill. Economic logic would seem to suggest that these facilities should be located on least valuable land in the least attractive sections of a town. Unfortunately, such areas

(continued)

are often also the neighborhoods of the poorest residents. Could economic logic be biased and unjust?

Consider the following quote attributed to Lawrence Summers, chief economist of the World Bank: "The measurement of costs of health-impairing pollution depends on the foregone earnings from increased morbidity and mortality. From this point of view a given amount of health-impairing pollution should be done in the country with the lowest cost, which will be the country with the lowest wages. I think the economic logic behind dumping a load of toxic waste in the lowest-wage country is impeccable and we should face up to that." How would a just society distribute its "health-impairing" pollution?

_____

Summers is quoted by Laura Westra and Peter Wenz, *Faces of Environmental Racism* (Totowa, NJ: Roman Littlefield, 1995), p. xvi.

CHAPTER 14

International Relations and the Environment

Environmentalists often have ambivalent opinions about international relations. On one hand, environmental issues and controversies seldom respect national borders, and satisfactory solutions require international cooperation. Such issues as global warming, ozone depletion, air and water pollution, ocean fishery depletion, and species preservation affect numerous nations and require international attention. To address these issues, environmentalists encourage international cooperation. On the other hand, many regard too much international cooperation, particularly on issues of trade and economic development, as threatening environmental regulation and progress. Economic pressure and international debt provide incentives for countries to exploit their resources at unsustainable rates. Free trade appears to give an economic advantage to industries that face few or no environmental regulations in their home country. In short, as long as environmental regulation imposes unequal costs within a global marketplace, there will be incentives for all countries to lower environmental standards.

In 1987 twenty-four countries signed the Montreal Protocol, an international agreement that limited and eventually will eliminate the production and use of chlorofluorocarbons (CFCs) and other ozone-depleting chemicals. Many see this agreement as a model of international cooperation on environmental matters. In the decade following the original signing, the number of countries agreeing to the Protocol rose to 163. On several occasions during that decade, regulations were strengthened with little opposition. Industrialized countries have generally met their goals for reducing use of CFCs, and developing countries have made significant strides in switching to alternatives. Evidence suggests that, at least at lower atmospheric levels, the amount of ozone-depleting chemicals is in decline. Although some problems remain—CFC

smuggling continues and some countries are less diligent at policing the agreement—there is reason for cautious optimism on this issue.

Since the time of the Montreal Protocol, a number of other international agreements have been pursued. Most notably, perhaps, the United Nations Conference on Environment and Development (the Earth Summit) held in Rio de Janeiro during June 1992 provided the forum for addressing a wide variety of international environmental concerns. As this book goes to press, over 160 nations are again meeting in Kyoto, Japan, to sign an international treaty addressing global warming and climatic change.

But international agreement and cooperation is difficult to attain. In Chapter 1 we suggested that public policy involves three fundamental questions: What are our goals? How can we attain them? Why are they valued? The variety of answers to these questions is greatly magnified on the international scene. Simply put, there is typically much greater diversity of goals, policies, and values across nations than there is within one nation. Thus, forging agreements among nations is not only a more challenging task politically, but it also makes evaluating international issues a more challenging task ethically.

One major ethical issue requiring international cooperation on environmental matters concerns the distribution of benefits and burdens. As Chapter 13 pointed out, environmental justice requires that the benefits and burdens be distributed fairly. From the perspective of poorer countries especially, environmental goals may seem at best a costly luxury or at worst an unfair burden. The very name of the Rio conference, "Environment *and* Development," reflects the concern of many countries that environmental protection not come at the expense of economic development. One implicit theme of this conference pitted the industrialized northern countries against the developing southern countries. What many in the north saw as environmental issues—preservation of rain forests and biodiversity, global warming—many in the south interpreted as unfair barriers to economic development. According to some, it was as if the industrialized north was saying, "Now that *we've* attained a comfortable lifestyle by exploiting natural resources throughout the world, we think that *you* should preserve the remaining natural resources by not seeking a similar lifestyle." A key ethical question to keep in mind, especially when considering environmental issues on the international level, is: Who benefits and who pays for this policy?

A second area of concern for environmentalists is international trade policy. In recent decades several international trade agreements have caught the attention of the environmental community. Policy debates over trade agreements such as GATT (General Agreement on Tariffs and Trade) and NAFTA (North American Free Trade Agreement) have often focused on environmental concerns. Many environmentalists charge that these international agreements work to undermine environmental regulation and policy.

Defenders of free trade argue, on familiar economic grounds, that free trade between countries, like a free market within countries, promotes economic growth. When countries open their borders to free trade, economic forces such as competition, specialization, and the law of supply and demand will ensure the optimal satisfaction of consumer demand. In short, on a global level, more people will get more of what they desire. Just as free markets work toward this goal within countries, international free trade works toward this goal globally. Barriers to trade, such as tariffs and import quotas, frustrate market forces and prevent global economic growth. Isolationist policy—the political name for such barriers—can only have a detrimental effect on the world economy, according to free trade defenders.

In recent years, a coalition of labor groups and environmentalists have challenged the economic policy of free trade. These people point out that among the barriers to trade are such social policies as child labor and minimum wage laws and environmental protection. Consider how environmental regulation within a country can affect international trade.

A country with strict environmental regulation of pollution (or child labor laws, for that matter) creates an incentive for industry to relocate to countries without such regulation. It is cheaper for the industry to produce goods in a country that does not place restrictions on air and water pollution or waste disposal. Companies that remain within the original country will therefore be at a comparative disadvantage when competing with the companies that exported its production. The net effect is that there is an incentive for the original country to lower its environmental standards.

Or consider a country that places environmental restrictions on the goods sold within its borders. A country that places fuel-efficiency requirements on automobiles sold within its borders raises the price of automobiles for its citizens. Such rules also create an incentive for domestic automobile producers to relocate production and jobs out of the country or it places its domestic producers at a worldwide competitive disadvantage.

During the late 1980s, several trade disputes between the United States and Canada highlighted these issues. As discussed in the reading by Tom Athanasiou, the United States had, on health and environmental grounds, banned the import, production, and use of asbestos. The government of Quebec, where asbestos is a major industry, argued that the asbestos ban violated the free trade agreement between the two countries. The ban violated the ability of Canadian companies to do business in the United States. Also during the late 1980s Canada challenged a U.S. law requiring a portion of the materials used to manufacture newsprint to come from recycled sources. These environmental regulations were seen as nontariff barriers to trade because Canadian companies lacked access to recycled materials and were therefore at a competitive disadvantage when selling to U.S. consumers.

Although many debates over trade involve arcane discussions of economics and finance, part of this dispute is ethical and philosophical. Both sides claim that their goal is fair trade between countries. Fairness requires that business, industry, and countries compete on a level playing field. Defenders of free trade argue that the most efficient way to attain this level and fair competition is to remove all barriers to trade and allow a free and open competition. Environmentalists and other critics argue that the playing field should be at a level of equally demanding environmental protection. Until all countries guarantee environmental protection, restricting trade might be the best way to guarantee fair and equal opportunity to industries that operate within environmental law.

Few environmental issues can be adequately addressed within the borders of any particular country. Environmentalists need to be familiar with a range of issues involved in international relations and international trade if they hope to have a voice in some of the most important environmental problems of the twenty-first century.

The Rio Declaration on Environment and Development

United Nations

During the first two weeks of June 1992, the United Nations sponsored an international conference in Rio de Janeiro, Brazil. This meeting occurred on the twentieth anniversary of a 1972 meeting in Stockholm, Sweden, which was the first major international conference that addressed environmental concerns. The Rio conference, commonly called the Earth Summit, explicitly focused on balancing environmental protection and economic development. The phrase "sustainable development," an idea traceable to the Stockholm meeting, was a central item on the agenda at the Earth Summit.

The central idea of sustainable development is to allow for continued economic development, but only in ways that use natural resources at rates that can be continued into the indefinite future. The goal is to meet the needs of the present without jeopardizing the needs of the future. To many critics, sustainable development is little more than a smoke screen to hide behind as environmental destruction continues. They charge that the principles and agreements coming out of the Earth Summit do little to restrict economic activity in any environmentally significant way.

As you read through this Rio Declaration, the official proclamation that was agreed upon in Rio, ask yourself if the goals of economic development and environmental protection are equally balanced or if one or the other is given priority. You should also seek to identify the ethical values and assumptions of this document.

PREAMBLE

The United Nations Conference on Environment and Development,

Having met at Rio de Janeiro from 3 to 14 June 1992,

Reaffirming the Declaration of the United Nations Conference on the Human Environment, adopted at Stockholm on 16 June 1972, and seeking to build upon it,

With the goal of establishing a new and equitable global partnership through the creation of new levels of cooperation among States, key sectors of societies and people,

Working towards international agreements which respect the interests of all and protect the integrity of the global environmental and developmental system,

Recognizing the integral and interdependent nature of the Earth, our home,

Proclaims that:

PRINCIPLE 1

Human beings are at the centre of concerns for sustainable development. They are entitled to a healthy and productive life in harmony with nature.

PRINCIPLE 2

States have, in accordance with the Charter of the United Nations and the principles of international law, the sovereign right to exploit their own resources pursuant to their own environmental and developmental policies, and the responsibility to ensure that activities within their jurisdiction or control do not cause damage to the environment of other States or of areas beyond the limits of national jurisdiction.

PRINCIPLE 3

The right to development must be fulfilled so as to equitably meet developmental and environmental needs of present and future generations.

PRINCIPLE 4

In order to achieve sustainable development, environmental protection shall constitute an integral part of the development process and cannot be considered in isolation from it.

PRINCIPLE 5

All States and all people shall cooperate in the essential task of eradicating poverty as an indispensable requirement for sustainable development, in order to decrease the disparities in standards of living and better meet the needs of the majority of the people of the world.

PRINCIPLE 6

The special situation and needs of developing countries, particularly the least developed and those most environmentally vulnerable, shall be given special priority. International actions in the field of environment and development should also address the interests and needs of all countries.

PRINCIPLE 7

States shall cooperate in a spirit of global partnership to conserve, protect and restore the health and integrity of the Earth's ecosystem. In view of the different contributions to global environmental degradation, States have common but differentiated responsibilities. The developed countries acknowledge the responsibility that they bear in the international pursuit of sustainable development in view of the pressures their societies place on the global environment and of the technologies and financial resources they command.

PRINCIPLE 8

To achieve sustainable development and a higher quality of life for all people, States should reduce and eliminate unsustainable patterns of production and consumption and promote appropriate demographic policies.

PRINCIPLE 9

States should cooperate to strengthen endogenous capacity-building for sustainable development by improving scientific understanding through exchanges of scientific and technological knowledge, and by enhancing the development, adaptation, diffusion and transfer of technologies, including new and innovative technologies.

PRINCIPLE 10

Environmental issues are best handled with the participation of all concerned citizens, at the relevant level. At the national level, each individual shall have appropriate access to information concerning the environment that is held by public authorities, including information on hazardous materials and activities in their communities, and the opportunity to participate in decision-making processes. States shall facilitate and encourage public awareness and participation by making information widely available. Effective access to judicial and administrative proceedings, including redress and remedy, shall be provided.

PRINCIPLE 11

States shall enact effective environmental legislation. Environmental standards, management objectives and priorities should reflect the environmental and developmental context to which they apply. Standards applied by some countries may be inappropriate and of unwarranted economic and social cost to other countries, in particular developing countries.

PRINCIPLE 12

States should cooperate to promote a supportive and open international economic system that would lead to economic growth and sustainable development in all countries, to better address the problems of environmental degradation. Trade policy measures for environmental purposes should not constitute a means of arbitrary or unjustifiable discrimination or a disguised restriction on international trade. Unilateral actions to deal with environmental challenges outside the jurisdiction of the importing country should be avoided. Environmental measures addressing transboundary or global environmental problems should, as far as possible, be based on an international consensus.

PRINCIPLE 13

States shall develop national law regarding liability and compensation for the victims of pollution and other environmental damage. States shall also cooperate in an expeditious and more determined manner to develop further international law regarding liability and compensation for adverse effects of environmental damage caused by activities within their jurisdiction or control to areas beyond their jurisdiction.

PRINCIPLE 14

States should effectively cooperate to discourage or prevent the relocation and transfer to other States of any activities and substances that cause severe environmental degradation or are found to be harmful to human health.

PRINCIPLE 15

In order to protect the environment, the precautionary approach shall be widely applied by States according to their capabilities. Where there are threats of serious or irreversible damage, lack of full scientific certainty shall not be used as a reason for postponing cost-effective measures to prevent environmental degradation.

PRINCIPLE 16

National authorities should endeavour to promote the internalization of environmental costs and the use of economic instruments, taking into account the approach that the polluter should, in principle, bear the cost of pollution, with due regard to the public interest and without distorting international trade and investment.

PRINCIPLE 17

Environmental impact assessment, as a national instrument, shall be undertaken for proposed activities that are likely to have a significant adverse impact on the environment and are subject to a decision of a competent national authority.

PRINCIPLE 18

States shall immediately notify other States of any natural disasters or other emergencies that are likely to produce sudden harmful effects on the environment of those States. Every effort shall be made by the international community to help States so afflicted.

PRINCIPLE 19

States shall provide prior and timely notification and relevant information to potentially affected States on activities that may have a significant adverse transboundary environmental effect and shall consult with those States at an early stage and in good faith.

PRINCIPLE 20

Women have a vital role in environmental management and development. Their full participation is therefore essential to achieve sustainable development.

PRINCIPLE 21

The creativity, ideals and courage of the youth of the world should be mobilized to forge a global partnership in order to achieve sustainable development and ensure a better future for all.

PRINCIPLE 22

Indigenous people and their communities, and other local communities, have a vital role in environmental management and development because of their knowledge and traditional practices. States should recognize and duly support their identity, culture and interests and enable their effective participation in the achievement of sustainable development.

PRINCIPLE 23

The environment and natural resources of people under oppression, domination and occupation shall be protected.

PRINCIPLE 24

Warfare is inherently destructive of sustainable development. States shall therefore respect international law providing protection for the environment in times of armed conflict and cooperate in its further development, as necessary.

PRINCIPLE 25

Peace, development and environmental protection are interdependent and indivisible.

PRINCIPLE 26

States shall resolve all their environmental disputes peacefully and by appropriate means in accordance with the Charter of the United Nations.

PRINCIPLE 27

States and people shall cooperate in good faith and in a spirit of partnership in the fulfilment of the principles embodied in this Declaration and in the further development of international law in the field of sustainable development.

For Further Discussion

1. The Rio Declaration claims that human beings are at the center of concerns for sustainable development. Who could disagree with this statement?

2. In your opinion, is this document more of an economic or an environmental statement? Which specific passages support your opinion?

3. This declaration asserts that states have the sovereign right to exploit their own resources pursuant to their own environmental and developmental policies. Does this mean that there is no single correct environmental perspective? Are you surprised to find a document written by a group of nations that makes such a claim about the right of sovereign nations?

4. Does the concept of sustainable development mean the same for the United States as it would mean for a nonindustrialized country such as Indonesia or Brazil?

5. Are the first two principles of this declaration consistent with or in tension with the principles that follow?

Agenda 21: Sustainable Living

United Nations

Leaders from over 170 countries met in Rio for the Earth Summit. The Rio Declaration offers a general statement of the principles and values agreed on by these countries. Agenda 21 is a more detailed plan of action also adopted by these countries during the Earth Summit. Agenda 21 presents a vision of the future that is economically, environmentally, and ethically sound. This action plan was intended to carry out the promised balance between economic development and environmental protection.

This reading from Agenda 21 presents the goal of sustainable living patterns. The document acknowledges that "economic growth is essential to meet basic human needs and achieve acceptable levels of personal well-being." The goal is to achieve this growth in ways that do not destroy the environment. Thus, the key is to find appropriate patterns of growth and certain kinds of consumption. Population growth, particularly in developing countries, is recognized as heightening the urgent need for economic growth while posing even greater environmental challenges. Again, as you read this document you should look for the implicit ethical and environmental values and assumptions.

The nations of the world have begun to realize that the Earth's carrying capacity is finite, and that global consumption, production and demographic patterns will have to be made sustainable if future generations are to live healthy, prosperous and satisfying lives. Achieving sustainable living for all requires an environmentally responsible global approach to modify these unsustainable patterns, involving efficiency and waste minimization changes in production processes, less wasteful consumption, reducing demographic pressures and ensuring access to health care.

More than 1 billion people on our planet are poor, malnourished and diseased—a certain indication of today's disparate and unsustainable patterns of production and consumption. Over 800 million people go hungry each day. Many among these are children. Nearly one and a half billion people are denied access

to primary health care and are threatened by a host of diseases, most of which are easily avoidable. In light of present demographic pressures, meeting the needs of all the world's inhabitants will be an ever greater challenge. Alleviating poverty is a moral imperative and a *sine qua non* when addressing issues of sustainable development.

This century has seen a massive increase in the world's production and consumption, particularly in developed countries. Although stimulating economic growth in the short term, Governments now recognize that this globally unsustainable use of the Earth's resources has degraded the environment and generated unmanageable amounts of waste and pollution.

Natural resources and living standards are often the casualties of population growth. The impact of population on environment and development issues should be further analyzed. Human vulnerability in sensitive areas should be determined. It is proposed that population factors be thoroughly researched and incorporated into national planning, policy and decision making.

Providing primary health care to everyone is a key aspect of alleviating poverty. Standards of health care, for those who receive indifferent or middling services, and specialized health care for environmentally related problems should be increased. Access to affordable health care, and facilities that communities can maintain on their own, are important factors.

In the decade of the 1990s, it is important that the world community make a social transition to poverty alleviation and sustainable consumption patterns. A world where poverty is endemic would always be susceptible to ecological and human crises. An improvement in living standards and development progress would also contribute toward the demographic transition to stable populations. Global cooperation and a vigorous and mutually beneficial partnership are essential to achieving the fundamental goal of improving human welfare throughout the world.

COMBATING POVERTY

Of the nearly 4.2 billion people in the developing world, some 25 per cent live in conditions of crippling poverty, without adequate food, basic education and health care, and deprived, in many cases, of their very cultural identity. Even in industrialized societies there are as many as 100 million poor people, many of whom are homeless and unemployed but have, at the very least, some access to social security benefits and health care.

Poverty in rural areas compels people to cultivate marginal lands, which results in soil erosion, depletion of shallow water resources and, consequently, lower crop yields, and tightens further the noose of poverty. Trying to survive on a daily basis, the poor have often little choice but to continue to overexploit their resources, which in turn reduces the changes of their offspring ever breaking free of the cycle of poverty and an unsustainable environment. Of the estimated 1 billion poor people living in developing regions, some 450 million live in low-potential agricultural areas. A similar number live in areas that are ecologically vulnerable, especially to soil erosion, land degradation, floods and other disasters; about 100 million others dwell in urban slums.

The poor are often the victims of environmental stresses caused by the actions of the rich. They are forced to live in areas more vulnerable to natural disasters, industrial hazards and water and air pollution. The growing migration from rural poverty to urban squalor is also of increasing concern, as this concentration of poverty leads to the breakdown of urban services and social systems, resulting in increased crime and a destabilizing socio-political environment. The burden of poverty and deprivation also falls far more heavily on women and children and certain ethnic and minority groups. These target groups require specific measures in poverty alleviation programmes.

An environmental policy that focuses mainly on the conservation and protection of resources, without due regard to the livelihoods of those who depend on them, would not only have an adverse impact on poverty but would also be unsuccessful. On the other hand, development policies that focus mainly on increasing production without concerning themselves with the long-term potential of the resources on which production is based would, sooner or later, run into problems of declining productivity and consequently

greater poverty. An effective and sustainable development strategy to tackle the problems of poverty and environmental degradation should focus simultaneously on resources, production and people.

Governments, at the national and international level, need to provide the means and the commitment to combat poverty in all countries. A multifaceted strategy targeting the causes of poverty and focusing on vulnerable groups and fragile, low-potential ecosystems would require considerable funding and manpower resources. It should feature imaginative programmes to generate employment opportunities, as well as remunerative activities, and provide essential social services such as health, nutrition, education and safe water. The participation of local communities in the planning, formulation and implementation of such programmes is essential. Promoting the empowerment and improvement of the socio-economic situation of vulnerable target groups, such as women, children, indigenous and minority communities, landless households, refugees and migrants, would also require specific priority measures.

Enabling the poor to achieve sustainable livelihoods should provide an integrating factor that allows policies to address issues of development, sustainable resource management and poverty eradication simultaneously. This would entail improving the means for information-gathering on poverty target groups and areas in order to facilitate the design of focused programmes. The programmes should work to equitably increase resource productivity; develop infrastructure, marketing systems, technologies and credit systems; and implement mechanisms for public participation to support and widen sustainable development options for resource-poor households.

Combating poverty involves promoting sustainable livelihoods. Programmes should cover a wide range of sectoral interventions and should be geographically and ecologically specific, aimed at specific vulnerable target groups and take into consideration the physical characteristics of the ecosystem concerned. They should include immediate steps to alleviate extreme poverty as well as long-term strategies for sustainable national socio-economic development to eliminate mass poverty and reduce inequalities.

A focal point in the United Nations system should be established to facilitate the exchange of information and the formulation and implementation of replicable pilot projects. In degraded and ecologically vulnerable areas, intensive efforts should be made to implement integrated policies and programmes for rehabilitating resource management and redressing poverty. This should involve development activities of lasting value, such as food-for-work programmes to build infrastructure. A review of the progress made in eradicating poverty through the implementation of the above activities should be given high priority in the follow-up of Agenda 21.

CHANGING CONSUMPTION PATTERNS

Economic growth is essential to meet basic human needs and achieve acceptable levels of personal well-being. Today's modern industrial economy has, however, led to the unprecedented use of energy and raw materials and the generation of wastes. Industrialized countries consume most of the world's energy, and many other resources, and far outstrip the consumption in developing countries.

Present levels of certain kinds of consumption, such as energy resources, in industrialized countries are already giving rise to serious environmental problems and are unlikely to be sustainable over the longer term. In addition, growing economies, incomes and population in the developing world seem destined to push human activities well beyond sustainable levels if similar consumption patterns take hold there. At the same time, consumption remains important as a driving force for development and for the creation of income and export markets needed to promote world-wide growth and prosperity.

This calls for a practical strategy to bring about a fundamental transition from the wasteful consumption patterns of the past to new consumption patterns based on efficiency and a concern for the future. In this respect, a new awareness and understanding, as well as a partnership between nations—industrial and developing—are crucial.

Since the mid-20th century, world energy production has risen some twentyfold; industrial output for the corresponding period has gone up some fourfold. Further, world population has doubled since then from 2.5 billion to 5.4 billion. This rapid growth in production has, for some, resulted in increased consumption and higher living standards. Most of this growth, however, has occurred in the developed countries, which, although comprising only one fifth of the total world population, account for some four fifths of the consumption of fossil fuels and other resources. Even with regard to the world's basic food commodities, such as cereals, meat and milk, industrialized countries consume between 48 and 72 per cent. Continuing these consumption levels in industrialized countries, while adopting them in developing countries, would not only be unsustainable but also gravely threaten the Earth's ecology. The world community needs to assess and curtail all wasteful and inefficient consumption patterns, particularly those that would seriously and irreparably damage our common environment.

Altering consumption patterns is one of the greatest challenges in the quest for environmentally sound and sustainable development, given the depth to which they are rooted in the basic values and lifestyles of industrial societies and emulated throughout much of the rest of the world. It requires the combined efforts of Governments, individuals and industry in a gradual process to examine new concepts of growth and prosperity which rely less on the flow of energy and natural resource materials through the economy and which take into account the availability and true value of natural resource capital.

The Governments and industries of many industrialized countries are intensifying their efforts to reduce the amount of resources used in the generation of economic growth, recognizing that this can both reduce environmental stress and contribute to greater economic and industrial competitiveness. Some progress has been achieved in making energy and raw material use more efficient through the adoption of environmentally sound technologies and recycling. These advances should be continued and generalized, and the experiences passed on to developing countries by helping them acquire these technologies, and develop others suited to their particular circumstances.

While the need to examine the role of consumption in dealing with environment and development is widely recognized, it is not yet matched by an understanding of the nature of the issue or how to address it. Governments and regional and international economic and environmental organizations, as well as private research and policy institutes, should make concerted efforts to compile basic data on consumption patterns and analyze the relationship between production, consumption, technological adaptation and innovation, economic growth and development, and population dynamics. Further, they should assess how modern economies can grow and prosper while reducing energy, material use and production of harmful wastes.

The costs and consequences of wasteful and environmentally damaging consumption are generally not borne fully by the producers and consumers who cause them. It is essential to design and implement market signals and sound environmental pricing that explicitly take into account the costs and consequences of consumption and waste generation. These may comprise environmental charges and taxes, deposit refund systems, emissions standards and charges and "polluter pays" principles. Governments should consider the adoption of such measures on a widespread and effective scale.

Information on the real environmental costs and consequences of various production processes and services can be critical in influencing and enabling consumers to make environmentally conscious product choices. These changes in demand will, in turn, also induce industry to adapt. Information dissemination and transparency are critical. Governments should review and improve the environmental impact of their procurement policies and encourage industry to introduce the environmental labelling of products, including energy, so that consumers can be clearly informed of the consequences of their consumption and behaviour.

The replication throughout the developing world of the present consumption patterns of industrialized countries is simply not a viable option, since this would place a tremendous stress on the Earth's environment. At the same time, levels of consumption in developing countries must increase in order to improve living standards. This has to be achieved through improving efficiencies, reducing wasteful consumption and the related waste generation. Without responsible attitudes and rational measures, both the consumption needs and living environment for future generations will be compromised.

DEMOGRAPHIC DYNAMICS AND SUSTAINABILITY

The world's population, in mid-1991, reached 5.4 billion, of which 77 per cent lived in developing countries and the rest in industrially advanced countries. In the 1960s, global population grew at about 2.1 per cent annually. This has now declined to some 1.7 per cent a year. The number of people added to the total each year, now amounting to 92 million, is higher than ever before. It is projected that world population will reach some 6.3 billion people in the year 2000 and 8.5 billion in the year 2025.

Over 90 per cent of the population increase today occurs in developing countries, having risen from 1.7 billion in 1950 to 4.2 billion in 1991, and expected to soar to nearly 5 billion by the year 2000.

By the turn of this century, some 2 billion people in the developing world will live in urban areas: over 40 per cent of the people in Africa and Asia—excluding Japan—and 76 per cent of those in Latin America will be urbanized. Of the world's 20 largest cities, 17 will be in the developing countries. At present, over 40 per cent of the urban population in the developing world lives in squalor, without access to essential services such as health care. Coping with these projected urban populations in the future poses major challenges to sustainable development.

Rapidly increasing demands for natural resources, employment, education and social services will make it difficult to protect natural resources and improve living standards. The migration of large numbers of people within countries and across national boundaries will, more than likely, continue to increase, driven by a combination of factors, including population growth, concentration of wealth and land, poverty and economic polarization.

Developmental and environmental conservation plans have generally recognized population variables as critical factors which influence consumption patterns, production, lifestyles and long-term sustainability. But more attention will have to be given to these issues in general policy formulation and the design of development plans. All countries will have to improve their capacities to assess the environmental and developmental implications of their population patterns, and to formulate and implement appropriate policies and action programmes. These policies should be designed to cope with the inevitable increase in population numbers, while at the same time incorporating measures to bring about the demographic transition.

Population programmes are at the interface between people and their environment. Local-level sustainability requires a new action framework that examines population dynamics in conjunction with other factors such as the social dimensions of gender, access to resources, livelihoods and the structure of authority. Population programmes should empower people, and be consistent with socio-economic and environmental planning for sustainability. Integrated population/environment programmes should closely correlate action on demographic variables with resources management activities and development goals.

A major priority is *developing and disseminating knowledge concerning the links between demographic trends and factors and sustainable development*. Population dynamics must be incorporated in the global analysis and research of environment and development issues. Governments should work to develop a better understanding of the relationships between human populations, technology, cultural behaviour, natural capital and life-support systems. In order to determine the priorities for action at global and regional levels, human vulnerability in sensitive areas and centres of population must be carefully assessed.

Another priority is *formulating integrated national policies for environment and development, taking into account demographic trends and factors*. National population issues must be integrated into the national planning, policy and decision-making process as part of national development and conservation plans. Policies and programmes should combine environment and development issues with population concerns to foster a holistic view of sustainable development that holds as its goals the alleviation of poverty, secure livelihoods, integrated health care, the reduction of maternal and infant mortality, education and services for the responsible planning of family size, the improvement of the status and income of women, the fulfilment of women's personal aspirations and individual and community empowerment. These policies and programmes should be people-centred, working to increase the quality and improve the capacity of human resources for environmental conservation.

Lastly, *implementing integrated environment and development programmes at the local level, taking into account demographic trends and factors* is also a critical priority. These programmes should work to improve the quality of life, ensure the sustainable use of natural resources and enhance environmental quality.

These priorities will require the strengthening of research activities that integrate population and environment issues, such as integrated demographic analysis, which embraces a broad social science perspective of environmental changes, and models methodologies and socio-demographic information in a suitable format for interfacing with physical and biological data. They will also require increased awareness of the need to sustain the world's resources through the rationalization of resource use, consumption patterns and population. Better "population literacy"—an awareness of the population/environment interactions taking place at local and national levels—should be developed among decision-makers, parliamentarians, journalists, teachers and students, civil and religious authorities and the general public.

To strengthen institutional capacity, Governments should promote the collaborative exchange of information between research institutions and international, regional and national agencies involved in population programmes. Governments should enhance national capacities to deal with integrated population/environment/development issues by strengthening population committees and commissions, population planning units, and advisory committees on population to enable them to elaborate population policies consistent with national strategies for sustainable development.

For Further Discussion

1. What would you describe as the guiding values of this document? Can you cite specific passages where value claims are made explicit?

2. What does this document say about the relationship among poverty, population, and environmental destruction? Must economic justice and equality be achieved as part of environmental protection?

3. This reading tells us that economic growth is essential for meeting basic human needs and achieving acceptable levels of personal well-being. Do you agree?

4. Do you agree with the statement that altering consumption patterns, especially in the industrialized world, is one of the greatest challenges to the goal of sustainable development? Can you cite some specific examples of such changes?

5. Many U.N. documents of the environment speak of "sustainable development." What do you think should be sustained? Is this phrase just an empty promise that lacks meaning when considered in detail?

Biodiversity, Ethics, and International Law

Farhana Yamin

The 1992 Convention on Biological Diversity (the Biodiversity Convention) is an international agreement addressing loss of biodiversity and setting goals for conserving biodiversity. Whereas the Rio Declaration and Agenda 21 were clearly anthropocentric, the Convention on Biological Diversity acknowledges greater ethical standing for non-human life. In this selection, Farhana Yamin examines the ethnical and international dimensions of biodiversity conservation. She suggests that ethical approaches that are more ecocentric will help in understanding and implementing the goals of the Biodiversity Convention. She also points out that principles of international distributive justice must play a central role in the protection of global biodiversity.

INTRODUCTION

International efforts to conserve biological diversity (biodiversity) raise two fundamental types of ethical or value-centred questions. First, what obligations does humankind, as a whole, owe to the natural world; and second, how should the benefits and burdens of conservation efforts be apportioned?[1] The 1992 Convention on Biological Diversity (the Biodiversity Convention) provides a legal, institutional and normative framework for international efforts to conserve biodiversity for its 118 parties.[2] The provisions of the Convention address both kinds of ethical issues. Examination of ethical issues has been relatively neglected by prevailing international relations literature, which has instead given greater emphasis to studying negotiating strategy, power politics, effectiveness of institutions or instrumental strategies for reaching given goals.[3] This emphasis is due in part to the strong "realist" tradition in international relations which denies the relevance of moral and ethical considerations to the behaviour of nation-states.[4] It is also linked to positivist jurisprudence which regards questions concerning the application and observance of law as separate and distinct from morality.[5]

This article sets out the problem of diminishing biodiversity. It then examines the ethical dimensions of biodiversity conservation which challenge some of the central themes of jurisprudence concerning the relationships between morality, rights and justice. The article argues that implementation of the Biodiversity Convention—the international community's response to tackling the loss of biodiversity—will raise fundamental questions about our ethical goals and values and, ultimately, our national and international political and institutional arrangements. In particular, the article suggests that by shedding light on the difficult question of burden-sharing, the Convention makes an important practical contribution to the subject of international justice which has been described as "the great philosophic issue of the 1990s."[6]

THE PROBLEM OF DIMINISHING BIODIVERSITY

What Is Biodiversity?

Biodiversity refers to the biological diversity which exists on Earth. It includes genetic diversity (the variation of genes within a species); species diversity (the variation of species within a region); and ecosystem diversity (the variety of ecosystems within a region). The Earth's genes, species and ecosystems have evolved over 3,000 million years.[7] While the total number of species is not known, biologists estimate that there are between 5 million and 30 million species, with a likely figure of about 10 million.[8]

This diversity of life is not evenly distributed across the globe. Generally speaking, temperate regions tend to have less diversity than tropical regions. Closed tropical forests, for example, are estimated to contain at least 50 per cent and perhaps 90 per cent of the world's species, even though they cover only 9 per cent of the Earth's land surface.[9] The following countries have been identified as "megadiversity countries," as between them they are estimated to hold up to 70 per cent of the world's species diversity: Mexico, Colombia, Ecuador, Peru, Brazil, Zaire, Madagascar, China, India, Malaysia, Indonesia and Australia. With the exception of Australia, these are developing countries, many of which are considered to be the poorest in economic terms in the world.[10]

The Importance of Biodiversity

The loss of biodiversity has profound implications for human welfare and for the planet. Biological resources are renewable. Forests, fisheries, wildlife and crops reproduce without human intervention. Natural habitats provide human beings with the means for survival, supplying food (meat, nuts, fruits, vegetables), fodder, firewood, construction materials, medicinal plants and wild genes for domestic plants and animals.[11] Furthermore, the highly diverse natural ecosystems which support this wealth of species also provide important ecological services, including maintaining hydrological cycles, regulating climate, contributing to the process of soil formation and maturation, storing and cycling essential nutrients, absorbing and breaking down pollutants and providing sites for tourism, recreation, and research and education.[12] The importance of biodiversity is becoming more significant as biotechnology techniques involving recombinant DNA, tissue culture, cell fusion, fermentation, and enzyme technology develop and increase our ability to manipulate forms of life. While it is difficult to quantify the actual and potential value of biodiversity, it is clear that biodiversity conservation is essential for human existence at many different levels.

Rapid, Mass Extinction

Yet the existence of this diversity is being put at risk by our actions through habitat destruction, pollution, over-exploitation, or competition with species introduced by humans.[13] Of these, extinction caused by direct activities (including hunting and collection) and indirect activities (including habitat degradation, destruction or modification from agricultural, industrial and other activities) are the most significant.[14]

Species extinction is a natural evolutionary process and habitat modification has taken place wherever human beings have altered their natural environment.[15] It is the unprecedented scale and speed of extinction caused by humankind which is of international concern. For many developed countries, such concern

might have come too late as much of the biodiversity found within their jurisdictions has already disappeared. The United States, for example, has lost all of its grasslands and savannah. Germany and the Netherlands both lost over 50 per cent of their wetlands between 1950 and 1980.[16] Similar developments are occurring at an alarming rate in developing countries, where the world's remaining biodiversity is concentrated. While it is impossible to quantify future rates of loss with certainty, all estimates are disturbing. The World Resources Institute has estimated that up to 15 per cent of the Earth's species could be lost by 2020.[17] Endemic species (those that are limited to a certain area) and tropical species (which tend to have smaller, localized populations) are at greater risk of extinction.[18] Harvard biologist Edward O. Wilson has estimated that at a minimum, 50,000 invertebrate species per year, almost 140 a day, are doomed to extinction by the destruction of their tropical rainforest habitats.[19]

Precaution and Practicability

If we wish to maintain biodiversity, conservation must be undertaken in a coherent, coordinated and comprehensive manner. The concept of biodiversity is itself ambiguous providing little guidance as to what proxy (index) we are to conserve or maximize. Furthermore, our lack of knowledge and our limited understanding of ecosystems means we cannot be certain of the consequences of our actions.[20] We must therefore act on the basis of uncertainty. For this reason, ecologists have long advocated implementing the precautionary principle. This states that where there is a threat of actual or potential irreversible damage, lack of full scientific certainty should not be used as a reason for postponing measures to avoid or minimize the threat.[21] The need for additional research is obvious; otherwise "there is always the danger . . . of winning the innings—establishing a comprehensive system of reserves through political action—but losing the game for lack of ecological understanding."[22]

It is important to note, however, that the fact that some habitats have been disturbed, either in whole or in part, does not mean that we cannot reinstate them or minimize the effects of current activities. This point was stressed by the Association of Conservation Biology workshop, which noted that "it is important to understand that protecting biological diversity, as a practical matter, is independent of the pursuit of the Holy Grail of 'pristine.' Just because a system is not pristine does not mean that it has no value for conservation. The task of conservation is not to preserve some ideal, pristine nature. Rather, its task is to protect diversity."[23]

Acting on the precautionary principle or regarding biodiversity conservation as a "practical matter" begs discussion of the underlying rationale for biodiversity conservation. This is because urging precaution and practicability does not get us very far in deciding how much biodiversity extinction we should tolerate and who or what in nature is to be exploited and suppressed. The answers to these questions are rooted in different moral and ethical assumptions about the nature of humankind's relationship to the natural world. This relationship is socially and culturally dependent. Accordingly, reasons for conserving biodiversity differ across cultures and societies.

THE NATURE OF HUMAN OBLIGATIONS TO CONSERVE BIODIVERSITY

Broadly speaking, we might group ethical justifications into two categories, which provide different ethical guidance about how much biodiversity we should seek to conserve and at what cost to humanity. The first consists of ecocentric reasons, which tend to stress moral imperatives to respect the intrinsic value, integrity and interdependence of all forms of life. The second consists of anthropocentric reasons which stress the value of biodiversity to human beings as an economic resource or as less overtly instrumental educational, cultural, recreational and aesthetic values. This distinction follows that drawn by Arne Naess in 1973 between deep and shallow ecology, which has to do with the "difference between a shallow concern at 'pollution and resource depletion' for the deleterious effects this might have on human life, and the deep concern—for its own sake—for ecological principles such as complexity, diversity and symbiosis."[24]

Anthropogenic Versus Ecocentric Approaches

Conservation has a long history of being justified on anthropocentric grounds. The doctrine of Christian stewardship and other early writing had, according to Keith Thomas, the implicit underlying philosophy that "the earth was a larder created for the use of mankind" and was "justified by reference to Genesis, which set out the charter for human dominion over the inferior creatures, and by the easy teleology of Aristotle, who had taught that plants existed for the sake of animals and animals for the sake of man."[25] In this view, conservation is merely an exercise of self-restraint designed to maximize human welfare over the long-term. The anthropogenic approach dominated the governmental proceedings of the 1992 UN Conference on Environment and Development held in Rio de Janeiro. These proceedings, which had a profound impact on the negotiations of the Biodiversity Convention, culminated in adoption of Principle 1 of the 1992 Rio Declaration on Environment and Development which provides that: "Human beings are at the centre of concerns for sustainable development. They are entitled to a healthy and productive life in harmony with nature."[26]

Conservation based on "ecocentric" grounds tempers the concept of absolute dominion by emphasizing humankind's guardianship role and respect for and appreciation of God's creations. Aldo Leopold, for example, has pointed out that: "It just occurs to me . . . that God started his show a good many million years before he had any men for audience—a sad waste of both actors and music—and in answer to both, that it is just barely possible that God himself likes to hear birds sing and see flowers grow."[27] A recent exponent of similar views is Vice-President Al Gore, who has stressed that the concept of "dominion does not mean that the earth belongs to humankind; on the contrary, whatever is done to the earth must be done with an awareness that it belongs to God."[28]

Conservation stressing the interdependency of all forms of life can also be regarded as ecocentric in that it accords human beings no privileged position in the natural world and the distinction between humankind and nature, living and non-living, as meaningless. Paiakan, the Kayapo leader in Brazil, is an eloquent exponent of such approaches. He points out: "The forest is one big thing: it has people, animals, and plants. There is no point saving the animals if the forest is burned down; there is no point in saving the forest if the people and animals who live in it are killed or driven away. The groups trying to save the race of animals cannot win if the people trying to save the forest lose."[29]

Such indigenous belief systems and the world's five major religions, Buddhism, Christianity, Hinduism, Islam and Judaism are enormously influential, and in many cases are the original sources of some of the best as well as the worst concepts now found in ecological philosophy.[30] What these traditions share, and what secular conservationists resist, is regarding our duties to the natural world as one element of an overarching ethical framework which stresses the concepts of good and bad and right and wrong.[31]

More recent Western philosophic tradition has explored ecocentric and anthropocentric rationales for biodiversity conservation in the context of discussions about (1) what moral rights the natural world should enjoy, (2) the extent of correlative duties this imposes on humankind, and (3) the kind of political, legal and institutional structures in which such rights should be respected. By raising these questions, biodiversity conservation has connected environmental issues to the central themes of jurisprudence—the relationship between morality, rights and law. Furthermore, it is becoming clear that ecocentric responses to these questions demand a fundamental reappraisal of the place of non-humans in ethical discourse, a reconsideration of the relationship between humans and nature, and ultimately a review of the appropriate political, legal and institutional arrangements at all levels of national and international society.[32] While the speed of any resulting changes is likely to be incremental, the magnitude of their implications for our notions of society and the international order could be revolutionary. For these reasons, even seasoned practitioners of realpolitik may be interested in the emerging theories.

Perhaps the most accessible contemporary reference point for exploring these issues is Christopher Stone's "Should trees have standing? Towards legal rights for 'natural objects,'" in which he proposed that the United States do the unthinkable—"give legal rights to forests, oceans, rivers and other so-called 'natural objects' in the environment—indeed, to the natural environment as a whole."[33] The intention behind the conferral of legal rights would be to ensure that harm to the environment receives proper consideration in all aspects of decision-making, and also to suggest that law can "contribute to a change in popular consciousness" by facilitating a shift in popular moral attitudes towards an ecocentric approach to conservation, so that we do not continually ask: "What is in it for us?"[34] Professor Stone's suggestions have inspired a generation of ethical philosophers, environmentalists, activists, lawyers and economists but not without controversy. The following by no means exhaustive summary of views illustrates some aspects of the lively debate about two separate but related questions: which natural elements ought we to consider morally significant, and to which should we accord legal rights?

Professor Stone has argued that if conservation is our goal it may be unnecessary to insist on moral rights, as it may make more sense "to talk of our duties to animals, rather than of animal rights."[35] The view is shared by John Passmore, who has written: "The idea of 'rights' is simply not applicable to what is non-human. . . . It is one thing to say that it is wrong to treat animals cruelly, quite another to say that animals have rights."[36] Humphrey Primatt and Andrew Linzey are more sympathetic to the notion of rights, arguing that human beings have no right to treat animals unmercifully since, like humans, they can feel pain and so we should regard their treatment as morally significant.[37] Other theorists offer more radical suggestions. Joel Feinberg, for example, has written that plants "are not 'mere things'; they are vital objects with inherited biological propensities determining their natural growth . . . so that certain conditions are 'good' and 'bad' for plants, thereby suggesting that plants, unlike rocks, are capable of having a 'good.'"[38] As a moral foundation for rights, this argument is rejected by others, including J. Baird Callicott, a long-standing "ecocentrist," who writes: "A more or less reasonable case might be made for rights for some animals, but when we come to plants, soils and waters, the frontier between plausibility and absurdity appears to have been crossed."[39] These views raise complex ontological issues about what principles we should use to divide the natural world into things which count as worthy of our moral consideration and things which do not.

Further difficult issues arise if either animate beings or inanimate objects (or both) are accorded rights or moral consideration, concerning how we should balance these different rights or considerations when they conflict. Existing legal and political theories, be they utilitarian, rights-based or communitarian, find it difficult enough to justify and balance the rights possessed by humans. Justifying and balancing how rights or moral considerations between humans and other elements of nature are accorded rights, and as between these elements, will provide a further considerable challenge.

Finally, discussion of rights focuses attention on another element of jurisprudential debate, namely the correlativity between rights and duties. A body of theorists has argued that a right must carry with it a correlative duty (otherwise it cannot be protected and amount to a right).[40] Thus, if a duty to conserve the environment exists, we might need to consider on whom it rests, how it can be given effect in the structure of our society and how it can be enforced. In the biodiversity context, does it rest, for example, on all humankind as a corollary of the fact that the Biodiversity Convention designates biodiversity a "common concern of humankind?" Or does the duty rest on states in whose territory the biodiversity is found, as a consequence of the fact that the Convention explicitly provides that states are "responsible for conserving their biological diversity and for using their biological resources in a sustainable manner"?

Practical Implications

Do these utopian moral and legal visions serve any practical purpose? In particular, can they play any part in the effective implementation of law already in place, as is the case with the Biodiversity Convention? In *Law's empire,* Ronald Dworkin argues that philosophic argument offers a distinct yet complementary role in the larger politics of law.[41] Philosophers can "offer large programs that can, if they take hold in lawyers' imagination, make its [law's] progress more deliberate and reflective." He argues that philosophers' dreams advance competing visions of law's future from among which the judiciary and legislature choose. In this way, utopian thought not only enriches our understanding of existing law, but also provides reasons for changing its base, thus provoking significant further change.

The discussion above illuminates some of the visions that may be relevant to the explication, implementation, and future development of the Biodiversity Convention. The degree to which any one of these visions contributes to these tasks depends on its intellectual cogency and persuasiveness in underpinning responsible human behaviour. Three concepts embedded in the Biodiversity Convention that might benefit from further philosophic consideration are the "sovereign responsibility for conservation and sustainable use for present and future generations," "fair and equitable benefit sharing" and the concept of "differentiated responsibilities." These are discussed more fully below.

EQUITABLE BURDEN-SHARING

Conceptually, the cluster of ethical issues concerning international burden-sharing—who should undertake conservation activities, what compensation (if any)

they should receive and on what basis—are distinct from ethical questions concerning why humankind should undertake conservation at all. These burden-sharing issues raise fundamental questions about international justice, because addressing them requires confronting the unequal distribution of power, wealth and resource endowments between countries, between present and future generations, and within countries. The following discussion is intended to shed light on how countries have approached these questions and the obligations and practical arrangements they have devised to establish a just distribution of burden efforts under the Biodiversity Convention.[42]

Burden-Sharing Between Countries

Any discussion of international burden-sharing issues must confront two important facts: first, that biodiversity is not evenly distributed across the globe; and second, that countries are not equal in terms of wealth, power and resource endowments, which means that they have different developmental needs and environment priorities.

The unequal distribution of biodiversity springs from the fact that, generally speaking, temperate regions tend to have less biodiversity than areas in or near the tropics. As discussed above, many developed countries have already lost much of their biodiversity. Much of the world's current biodiversity is found, and is being lost, in developing countries. Policies to address biodiversity loss therefore impose a larger and more direct burden on these countries' economic and developmental policies than on those of developed countries. This is because most biodiversity conservation measures have an adverse short-term impact on key industries, such as forestry and agriculture, which developing countries are compelled to expand to meet the immediate needs of their growing populations and to service their international debt obligations.[43]

The conservation/development and developed/developing country conflicts are further exacerbated by the fact that while developing countries bear much of the cost of biodiversity conversation, they are rarely able to realize biodiversity's full economic value.[44] Developed country industries such as the biotechnology, pharmaceutical, chemical, agricultural and cosmetics industries, all of which depend on open access to developing countries' genetic diversity, can appropriate the full economic value of this resource, and often the knowledge of local communities, without making any payment to these communities and the states in which the resources are found.[45] Developing countries are particularly aggrieved when the monopoly profits, products, technologies and knowledge of such industries are protected through intellectual property rights.[46]

The vast differences in wealth and power, coupled with the history of Southern colonialism by Northern countries, fuel mistrust and suspicion about the motives of developed countries leading inevitably to friction in international negotiations.[47] Ashish Kothari, outlining an Indian perspective, writes:

> In a large number of Southern countries the seeds of this [biodiversity] destruction were laid during the colonial era; in India, for instance, large scale commercial forestry started in British colonial times. Neo-colonial exploitation continues today . . . Adverse terms of trade, protectonist policies of the North, dumping of hazardous and environmentally destructive technologies and materials by the North into the South, and a host of other factors continue to cause severe and widespread biodiversity destruction.[48]

Similarly, Vandana Shiva argues that there are two primary causes for the large-scale destruction of biodiversity, both attributable in some shape or form to the North.[49] The first concerns habitat destruction due to internationally financed mega projects such as dam and highway construction and mining operations in biodiversity-rich forests. The second concerns the technological and economic push to replace diversity with homogeneity in forestry, agriculture, fisheries and animal husbandry through the use of modern agriculture methods relying on uniformity and monocultures.[50] These sorts of factors justify, according to Parvez Hassan, the claim that developed countries should adjust their consumption patterns and trading policies and should pay their "ecological debt" to developing countries. The payment of this debt is regarded as part of the "quest for equity" as this "would, of necessity, make allowances for past policies of exploitation." Hassan argues that "no aspect of the international effort to protect climate and forests and to preserve biodiversity is more important than the financing of such effort" because "governments shackled with the unbearable burden of ever-mounting foreign debts will not find the resources to meet their obligations."[51] In his view, global cooperation should

strive to achieve "a just world order, an order which is based on and recognizes equity as the dominant principle governing relations between states."

Practical Implications

What practical consequences flow from these ethical analyses? The identification of the North's past exploitation and present patterns of production and consumption as the root causes of biodiversity destruction shifts ethical responsibility to developed countries. This provides a legitimate ethical basis for developing countries to demand, as of right, that developed countries:

- stop exploitation of developing countries' natural resources;
- make amends for previous exploitation;
- take the lead in stopping future destruction by changing their consumption patterns and policies;
- provide financial assistance and technology transfer to developing countries to enable the latter to fulfil international obligations accepted by them in the interests of the global environment, as their own limited resources are needed to address other legitimate priorities, such as poverty eradication, health, social development and education.

Those involved in international environmental negotiations will be familiar with the kinds of claims being put forward by the Group of 77 (G77), collectively representing more than one hundred developing countries who between them account for 70 per cent of the world's population, but only 30 per cent of its income. The emphasis on the moral nature of these claims, and in particular the use of the language of equity and rights, enables G77 countries not only to retain their political pride but also to resist pressure from developed countries to impose conditionalities on transfer of financial and other assistance, such as land reform, population policy and human rights violations.[52]

Have these claims influenced the negotiations of the Biodiversity Convention and its burden-sharing approach?[53] Article 1 of the Convention states that:

> The objectives of this Convention, to be pursued in accordance with its relevant provisions, are the conservation of biological diversity, the sustainable use of its components and the fair and

equitable sharing of the benefits arising out of the utilizations of genetic resources, including by appropriate access to genetic resources and by appropriate transfer of relevant technologies, taking into account all rights over those resources and technologies, and by appropriate funding.

Incorporating equitable benefit-sharing as regards genetic resources, technology transfer and funding into the objective of the Convention reflects G77 insistence that conservation of biodiversity be linked to wider economic and developmental, essentially anthropogenic, concerns. For many developed countries such an incorporation goes way beyond what they originally intended a conservation treaty to be about.[54] The incorporation of these wider concerns in the Convention's objective bolsters developing countries' ethical stance that conservation and development should be balanced against each other and that developing countries' interests and circumstances should be taken into account in international partnerships to protect the environment.

It could be persuasively argued that the addition of these wider economic and development objectives roots the Convention in an anthropocentric approach. Many of the Convention's provisions, for example, are predicated on an anthropocentric, utilitarian approach to biodiversity conservation. The preamble of the Convention states, for example, that biodiversity is a "common concern of humankind." Preparatory drafts of the Convention had referred to biodiversity as a "common heritage of mankind."[55] However, developing-country concerns that the common ownership aspects might impinge on the principle of national sovereignty over natural resources led to the abandonment of the common heritage principle and the incorporation, for the very first time in the operative text of a treaty, of Principle 21 of the Stockholm Declaration, which provides that "States have . . . the sovereign right to exploit their own resources pursuant to their own environmental policies, and the responsibility to ensure that activities within their national jurisdiction or control do not cause damage to the environment of other States or of areas beyond the limits of national jurisdiction." The principle of sovereignty over natural resources forms an essential bedrock of the Convention. Its inclusion is intended to send an important political signal that it is sovereign states, not the international community, that are in charge of resources located within national boundaries. The inclusion of Principle 21 is also intended to

support states' having "the authority to determine access to genetic resources," which must be on mutually agreed terms, subject to the principle of prior informed consent and in accordance with national legislation.[56] This contrasts with the provisions of the 1983 FAO International Undertaking on Plant Genetic Resources which characterized plant genetic resources as a "heritage of mankind" and stipulated that they "should be available without restriction" and "free of charge, on the basis of mutual exchange or mutually agreed terms."[57]

In addition to the objectives stated in Article 1, the need for "equitable sharing" or "fair and equitable" sharing of benefits is mentioned in the Biodiversity Convention at least four times. These references, particularly in relation to biotechnology, regarded by many countries as a key industry of the future, again reflect developing countries' concerns to link conservation and development goals and to avoid exploitation of their natural resources by the North on unfavourable terms. The Biodiversity Convention contains a number of substantive provisions that give practical effect to these equity concerns, including the following:

- provision of new and additional financial assistance to developing countries, over and above ODA, to cover "agreed full incremental costs" of implementing the Convention and enabling compliance;[58]

- transfer of environmentally safe technology, including biotechnology and technologies covered by intellectual property rights, on "fair and most favourable" terms, facilitated by the financial mechanism if necessary;[59]

- obligations for developed and developing countries to share equitably benefits arising from utilization of the knowledge, innovations and practices of indigenous and local communities with the communities concerned;[60]

- obligations to advance priority access to developing-country parties and to share equitably the benefits and the results of research and development arising from the commercial or other utilization of genetic resources, particularly with developing-country parties providing access to such genetic resources;[61]

- obligations to advance priority access to developing countries and to share equitably the results and benefits arising from biotechnologies, based on genetic resources, particularly with developing-country parties providing such genetic resources;[62]

- acknowledgment that developing-country parties' abilities to comply are conditional upon the "effective implementation" of developed country parties' financial cooperation and technology transfer and that "economic and social development and the eradication of poverty are the first and overriding priorities of the developing countries;[63]

- in funding and technology transfer, special consideration for least-developed countries,[64] and consideration to special conditions resulting from the dependence on distribution and location of biodiversity within developing-country parties, especially small island states and those that are environmentally vulnerable.[65]

Thus, while the Convention does not include the concept of "ecological debt" for past exploitation, which was fiercely resisted by developed countries, it does take on board many of the anthropogenic ethical arguments presented by developing countries in championing their version of an equitable international order.

Four considerations temper optimism about the extent to which these provisions will actually result in biodiversity conservation and a more equitable international order. First, parties to the Convention have considerable leeway in interpreting and implementing many of its provisions because the extent of their obligations is subject to the caveats usually found in international law, such as "as far as possible and as appropriate," "in accordance with particular circumstances and conditions" or "subject to national legislation." Second, the concept of "new and additional" financial assistance from developed to developing countries through the Convention's interim financial mechanism, the Global Environment Facility (GEF), has resulted only in modest assistance so far, and this situation looks likely to continue.[66] Third, some of the more controversial provisions, such as those relating to access to and transfer of technology, including biotechnologies, are ambiguously drafted. This leaves unclear the extent to which such provisions will actually alter the implementation of rights and obligations relating to, for example, intellectual property and free trade, which are governed by mainstream international economic law, in particular the GATT and the new Trade Related Aspects of Intellectual Property Agreements.[67] Finally, it is not clear whether providing

access to genetic resources will generate commensurate benefits and whether the Convention's "fair and equitable" provisions will succeed in channeling these benefits to developing countries, local communities and indigenous peoples.[68]

The ambiguities and shortcomings of the Convention in these respects might certainly benefit from a philosophical programme offering explication and choices for future interpretation and development by negotiators.

Burden-Sharing Between Present and Future Generations

Discussion of burden-sharing issues at the international level tends to focus on the unequal economic needs and capabilities of the present-day generation (intragenerational equity) rather than on possible economic and ecological imbalances between present and future generations (intergenerational equity).[69] This fact highlights the problem that while the international community finds it relatively easy to acknowledge its moral obligation to consider the needs of future generations, it finds it more difficult to decide how to give this moral concern concrete legal and institutional form.[70]

The Convention's designation of biological diversity as a "common concern of mankind" and the need to conserve it for the benefit of present and future generations illustrates the ease with which the international community accepts this moral imperative. The Convention does go beyond this platitude by defining sustainable use for the first time as "use of components of biological diversity in a way and at a rate that does not lead to the long-term decline of biological diversity, thereby maintaining its potential to meet the needs and aspirations of present and future generations."[71] By requiring parties to base conservation on sustainable use, the Convention obliges parties to balance the needs and aspirations of present and future generations. This is an important legal precedent.

The Convention fails, however, to provide guidance on how this should be done. Previous drafts of the Convention had, for example, included provisions to establish "global lists of biographic areas of particular importance for the conservation of biological diversity, and of species threatened with extinction on a global scale."[72] Once listed, these areas would have had official priority in conservation activities, making monitoring and financing of areas and species specifically conserved for future generations more consistent, more transparent and, above all, secure from future encroachment. Notwithstanding the fact that areas would only have been listed with the voluntary consent of parties, the global list concept was abandoned precisely because many countries felt it might foreclose their choice of future development options.

This illustrates the difficulty of allowing present generations to play a guardianship role in the absence of well-developed legal concepts and institutional mechanisms to champion the interests of future generations. Establishing these mechanisms through, for example, bestowing rights on the natural world, as suggested by Professor Stone,[73] or establishing an international ombudsman along the lines suggested by Edith Brown Weiss,[74] will prove difficult in the face of short-term national, anthropocentrically based interests. This merely illustrates the urgent need for cogent, persistent expositions of the merits of such suggestions to our decision-makers.

Burden-Sharing Within Countries

The state-based nature of international society tends to focus on issues concerning the distribution of burdens between countries, and, more recently, across generations. Yet implementation of international environmental commitments affects the livelihood and well-being of particular individuals, regions, and economic and social groups, and of course, biodiversity itself. A particularly wide range of actors is affected by the Biodiversity Convention because its scope—covering all global living natural resources—affects many economic sectors and social groups ranging from transnational agriculture, pharmaceutical enterprises and the biotechnology industry to farmers, foresters, indigenous groups and local communities.

While the provisions of the Convention are addressed to states, it is widely recognized that national implementation of these commitments will only be successful if the resources of private actors are used in accordance with the Convention's provisions. This is particularly true in the case of provisions concerning access to in situ genetic resources, which are effectively controlled by rural, local and indigenous communities, and access to the benefits of research and technology transfer, which are effectively in the hands of private companies.

Traditionally, the way in which each state implements its international obligations and apportions burdens between these groups has been primarily a do-

mestic matter for that state. Increased international environmental regulation—necessitated by the fact that environmental issues transcend national boundaries, and the fact that they are increasingly bound up with economic issues—challenges the way in which states approach the distribution of burdens at the national level. International cooperation is essential because uncoordinated, unilateral policies might not only be ineffective at solving the problem but might also place citizens and businesses of one state at an economic disadvantage if other states do not follow suit.

These limitations make clear that if they are to make sense and provide any practical guidance to international policy-makers faced with implementation choices, discussions of equity and distributive justice must encompass both a national and an international component, and not "stop at national boundaries." Yet the predominant accounts of distributive justice do precisely this. The most influential theoretical work, John Rawls' *Theory of justice,* is intended to apply to questions of justice within states, not between them.[75] Meanwhile, the conception of a just international order presented by Southern advocates operates only on the international plane between developed and developing countries, drawing a thick veil over distributional questions at the national level, as these might impinge on sovereignty.

CONCLUSION

The Conference of the Parties to the Biodiversity Convention must keep implementation of the Convention under constant review.[76] This review process will invariably require an assessment of the adequacy of parties' current commitments and implementation efforts in meeting the objectives of the Convention, which in turn requires revisiting questions about the underlying rationale for biodiversity conservation and an equitable allocation of additional burdens. It is difficult to see how these issues could be addressed without reference to underlying moral values governing choices about the kinds of burdens that *ought* to be undertaken by different countries and individuals given their respective contributions to biodiversity loss and their capacity to address this. Will the lack of any coherent account of why we should undertake biodiversity conservation and how we ought to distribute the burdens of global environmental protection, nationally and internationally, impede this review process?

Certainly it is true to say, as Professor Stone does, that treaty-makers are not academics and that the absence of a cogent moral framework should not be considered problematic provided negotiators can forge compromises they perceive to be just and practical.[77] The development of a new theory or theories of distributive justice that encompass the distribution of goods, opportunities and other resources not only between humans, but also between humans and non-human elements of nature, as well as between present and future generations, might show such negotiators the best route for a better global future. A thorough appreciation of the underlying moral and ethical issues involved in developing such theories might, *en route* to that future, actually generate consensus in the international context and also provide a much-needed ethical framework guiding how, as individuals, we ought to live our lives. Such theories, and our choice of ethical principles, will certainly have weighty practical implications in the biodiversity context and in the context of other international issues involving global partnerships.

Notes

1. For a general discussion of these questions in the broader context of global environmental issues, see Christopher D. Stone, *The gnat is older than man: global environment and the human agenda* (Princeton, NJ: Princeton University Press, 1993).
2. Convention on Biological Diversity, *International Legal Materials (ILM)*, 21, 1992, p. 848.
3. Ken Booth, "Human wrongs and international relations," *International Affairs* 72: 1, January 1995.
4. Exemplified by e.g. Hume, Hobbes and Hans Morgenthau. See Marshall Cohen, "Moral scepticism and international relations," in Charles R. Beitz, ed., *International ethics* (Princeton, NJ: Princeton University Press, 1985).
5. For an introduction to this discourse, see N. E. Simmonds, *Central issues in jurisprudence, justice, law and rights* (London: Sweet and Maxwell, 1986).
6. Stone, *The gnat is older than man*, p. 54.
7. Jeffrey A. McNeely, "Critical issues in the implementation of the Convention on Biological Diversity," in Anatole Krattiger *et al.*, eds, *Widening perspectives on biodiversity* (IUCN and the International Academy of the Environment, 1994), p. 7.
8. World Resources Institute (WRI), *A guide to the global environment 1992–1993* (Oxford: Oxford University Press, 1992), p. 127.
9. Ibid., p. 130.
10. World Conservation Monitoring Centre, *Global biodiversity: status of the Earth's living resources* (London: Chapman and Hall), p. 154.
11. Jeff McNeely, "Economic incentives for conserving biological diversity," in Simone Bilderbeek, ed., *Biodiversity and international law* (Amsterdam: IOS Press, 1992), p. 49.
12. Ibid. The question of valuing biodiversity is explored more fully in World Conservation Monitoring Centre, *Global biodiversity*.
13. Michael E. Soule and Kathryn A. Kohm, *Research priorities for conservation biology* (Washington DC: Island Press, 1988), p. 7.
14. For a detailed account of how modern agricultural practices destroy biodiversity, see R. Velleve, *Saving the seed: genetic diversity and European agriculture* (London: Earthscan, 1992).
15. See T. Swanson, *The economics of a Biodiversity Convention*, CSERGE Discussion Paper GEC 92–108.
16. Soule and Kohm, *Research priorities for conservation biology*, p. 7.
17. WRI, *Guide to the global environment 1992–1993*, p. 127.
18. Ibid., p. 130.
19. John C. Ryan, "Conserving biological diversity," in Lester R. Brown *et al.*, eds., *State of the world 1992*, (London: W. W. Norton, 1992), p. 8.
20. Current estimates, for example, suggest that taxonomists have recorded only 1.4 million of the estimated 5–30 million species on the earth.
21. The Preamble to the Biodiversity Convention incorporates a qualified version of the principle, noting "that where there is a threat of *significant* reduction or loss of biological diversity, lack of full scientific certainty should not be used as a reason for postponing measures to avoid or minimize such a threat." For a fuller discussion of the evolution and significance of the precautionary principle, see Timothy O'Riordan and James Cameron, *Interpreting the precautionary principle* (London: Earthscan, 1994).
22. Conclusion of a 1988 workshop of the Association of Conservation Biology, cited in Soule and Kohm, *Research priorities for conservation biology*, p. 1.
23. Soule and Kohm, *Research priorities for conservation biology*, p. 2.
24. Arne Naess, "The shallow and the deep, long-range ecology movement: a summary," *Inquiry* 16, 1973, succinctly explained by Andrew Dobson, *Green political thought* (London: Routledge, 1992), p. 47.
25. Keith Thomas, "Introduction," in C. C. W. Taylor, ed., *Ethics and the environment*, proceedings of a conference held at Corpus Christi College, Oxford, 20–21 September 1991, p. 6.
26. UN Doc.A/ConF.151/6/Rev.1, 13 June 1992.
27. Aldo Leopold, "Conservation as a moral issue," in Donald Scherer and Thomas Attig, eds, *Ethics and the*

environment (Englewood Cliffs, NJ: Prentice Hall, 1983), p. 9.

28. Al Gore, *Earth in the balance* (Boston: Houghton-Mifflin, 1992), p. 244, quoted in Stone, *The gnat is older than man,* p. 238.

29. Paiakan continues by stressing the interdependencies in political context by pointing out: "the people trying to save the Indians cannot win if either of the others lose; the Indians cannot win without the support of either of these groups; but the groups cannot win without the help of the Indians, who know the forest and the animals and can tell what is happening to them. No one of us is strong enough to win alone; together we can be strong enough to win." Quoted in Susanna Hecht and Alexander Cockburn, *The fate of the forest: developers, destroyers and defenders of the Amazon* (New York: Harper Perennial, 1990), p. 217.

30. M. Palmer, *Genesis or nemesis: belief, meaning and ecology* (London: Dryad Press, 1988), p. 113.

31. For a brief statement from the world's five major religions on their attitudes to the environment, see World Wide Fund for Nature, "The Assissi Declarations," in *Environmental Law and Policy,* 17: 1, 1987.

32. Robert Elliot, in the introduction to Robert Elliot, ed., *Environmental ethics* (Oxford: Oxford University Press, 1995).

33. Christopher Stone, "Should trees have standing? Towards legal rights for natural objects," *Southern California Law Review* 45, 1972, p. 456.

34. Ibid., p. 491.

35. Stone, *The gnat is older than man,* p. 274.

36. John Passmore, *Man's responsibility for nature* (New York: Scribner's, 1974).

37. Andrew Linzey, *Animal theology* (London: SCM Press, 1994), p. 20.

38. Cited in Kenneth E. Goodpaster, "On being morally considerable," in Scherer and Attig, eds, *Ethics and the environment,* p. 36.

39. J. Baird Callicott, "Animal liberation: a triangular affair," in Scherer and Attig, eds, *Ethics and the environment,* p. 56.

40. As, for example, Hohfeld has argued. For a detailed discussion of this issue, see Lord Lloyd of Hampstead and M. D. Freeman, *Lloyd's introduction to jurisprudence* (London: Stevens, 1985).

41. Ronald Dworkin, *Law's empire* (London: Fontana, 1986), p. 408.

42. For a background to the Biodiversity Convention negotiations and an explanation of its key provisions, see Clare Shine and Palitha Kohona, "The Convention on Biological Diversity: Bridging the Gap Between Conservation and Development," *Review of European Community and International Environmental Law (RECIEL)* 1: 3, 1992, p. 278.

43. Ibid., pp. 278–88.

44. See Pat Mooney, "Genetic resources in the international commons," *RECIEL* 2: 2, 1993.

45. Ibid.

46. See Farhana Yamin and Darrell Posey, "Indigenous peoples, biodiversity and intellectual property rights," *RECIEL* 2: 2, 1993.

47. Parvez Hassan, "Moving towards a just international environmental law," in Simone Bilderbeek, ed., *Biodiversity and international law* (Amsterdam: IOS Press, 1992).

48. Ashish Kothari, "Beyond the Biodiversity Convention: a view from India," *Biopolicy International* 13 (Nairobi: ACTS Press, 1993), p. 4.

49. Vandana Shiva, *Monocultures of the mind: perspectives on biodiversity and biotechnology* (London: Zed Books; Penang, Malaysia: Third World Network, 1993), p. 68.

50. Ibid.

51. Hassan, "Moving towards a just international environmental law," p. 75.

52. Stone, *The gnat is older than man,* pp. 245, 246.

53. For a brief introduction to the Biodiversity Convention negotiations, see Shine and Kohona, "The Convention on Biological Diversity," p. 278.

54. Ibid.

55. Ibid.

56. Biodiversity Convention, Art. 15(1), (4), (5).

57. The International Undertaking on Plant Genetic Resources, FAO Resolution 8/83, adopted November 1983. FAO and the parties to the Biodiversity Convention are currently examining how the Undertaking can be revised to ensure consistency with the Convention.

58. Biodiversity Convention, Art. 20(2).

59. Biodiversity Convention, Art. 20.

60. Biodiversity Convention, Art. 8 and 12th preambular para. on indigenous and local communities.

61. Biodiversity Convention, Art. 15.

62. Biodiversity Convention, Art. 19.

63. Biodiversity Convention, Art. 20(4).

64. Biodiversity Convention, Art. 20(5).

65. Biodiversity Convention, Art. 20 (6), (7).

66. So far only $2 billion has been pledged to the GEF for the period 1994–7 to fund four global environmental problems: climate change, ozone depletion, international watercourses and biodiversity.

67. The United States refused to sign the Convention in 1992 because it found the Convention's provisions on financial resources and intellectual property ambiguous. See United States, Declaration made at the UNEP Conference for the Adoption of the Agreed Text of the Convention on Biological Diversity, 22 May 1992. *ILM* 21, 1992, p. 848. The United States has now signed the Convention.

68. WRI, *Biodiversity prospecting: using genetic resources for sustainable development* (Washington, DC: WRI, 1993).

69. See Edith Brown Weiss, In *Fairness to future generations: international law, common patrimony and international equity* (New York: Dobbs Ferry, 1989) and Grundling. "Our responsibility to future generations," *American Journal of International Law* 84, 1990, p. 190, which also contains a number of other useful articles on intergenerational equity.

70. See Patrici Birnié and Alan Boyle, *International law and the environment* (Oxford: Oxford University Press, 1992), p. 212, who provide a limited number of examples in which the intergenerational concept has been implemented in international law.

71. Biodiversity Convention, Art. 2; see also preambular paras 5, 23.

72. Shine and Kohona, "The Convention on Biological Diversity," p. 278.

73. Stone, "Should trees have standing?"

74. Weiss, *In fairness to future generations.*

75. For a more detailed treatment of the possible application of Rawls' insights into the relationship between states, see Stone, *The gnat is older than man,* pp. 252–66.

76. Biodiversity Convention, Art. 23.

77. Stone, *The gnat is older than man,* p. 265.

For Further Discussion

1. Why preserve biodiversity? How many different reasons are identified in this essay? Do you disagree with any? Is anything missing?

2. Yamin cites the legal philosopher Ronald Dworkin in support of the practical relevance of philosophic and utopian thinking. How does Dworkin understand this practical role for philosophy?

3. Do countries in regions that lack a great amount of biodiversity have less responsibility for preserving biodiversity than countries with great amounts? Does a country's economic status matter?

4. Yamin talks about "fair and equitable" sharing of benefits and burdens. How do you understand these concepts?

The Second Coming of "Free Trade"

Tom Athanasiou

Journalist Tom Athanasiou suggests that the Rio Conference and other international environmental agreements accomplish very little of environmental or social worth. His book Divided Planet, *from which this selection is taken, argues that the roots of the environmental crisis lie in the global divide between rich and poor nations. He agrees with many at Rio that environmental problems must be addressed within a context of economic and social justice. However, he does not believe that the development strategies as practiced by the industrialized countries offer a just or adequate solution to either environmental or social ills.*

In this selection, Athanasiou offers a critique of trade agreements such as GATT and NAFTA. He warns that without radical social change, trade policies will benefit only the powerful elites of both the North and South, leaving environmental destruction and continued poverty in their wake. Defenders of free trade argue that open markets promote "harmonization," a closing of the gap between rich and poor. Athanasiou fears that this amounts to a "harmonization down" rather than a "harmonization up" by pushing all countries toward the lowest common environmental standards.

During the 1980s the world's elite politicians set out on a program of trade liberalization that was unprecedented in its ambition. Negotiations over the European Common Market, the North American Free Trade Agreement (NAFTA), and the Asia Pacific Economic Cooperation Forum (APEC) defined three vast regional trade blocs, overarched by the General Agreement on Tariffs and Trade (GATT), the mother of all trade treaties. By the end of 1994, when GATT and its new World Trade Organization (WTO) became an institutional fact, nations everywhere had been wired into a single hair-trigger planetary economy.

Recall the 1980s. In the United States, Ronald Reagan was president and the economy seemed to be booming along. Corporate lobbyists, especially U.S. agricultural lobbyists, set out an ambitious trade agenda based on economic dogma and the short-term self-interests of largely stateless transnational corporations. "Free trade" became the defining political orthodoxy, taking over the role that anticommunism had played during the 1950s. Few people foresaw the debate that lay ahead. The idea that "trade" had something basic to do with "environment" would have sent the average economic bureaucrat into a state of irritated incomprehension. Trade negotiations were still the private pursuits of cloistered corporate and diplomatic specialists who all spoke the same, narrow, technical language. The enemy of "free trade" was "protectionism," pure and simple. "Fair trade" was a term that hadn't been heard much for the better part of a century.

Those days are over. Hardly a decade into what U.S. economic historian Doug Dowd calls "the second coming of free trade," NAFTA, GATT's WTO, the Common Market, APEC, and the ever-threatening trade war between the United States and Japan are constant front-page news. Everywhere trade is the focus of incessant heated conversation, and it's not hard to see why. Unemployment and economic insecurity are high and rising, and trade, a fundamental mechanism of global economic integration, is always implicated. Together with automation, and global financial markets, and all the other devices of the new globalization, "trade" comes to the losers in garments far more somber than those it wears to greet the rich and even the comfortable.

Life in the North is cheap, but life in the South is cheaper still. In Mexico, the Third World country that most Americans know best, the "economic openness" of the 1980s set off a dramatic increase in malnutrition and in economic desperation of all kinds. It brought, too, a sharp increase in the number of millionaires and even billionaires, even as between 1982 and 1990, real Mexican wages dropped by over half. In the wake of 1995's financial collapse and Mexico's reborn commitment to "economic discipline," they do not promise to rebound. Meanwhile, once-lovely border rivers, now the unregarded sewers of unregulated export-processing zones, have grown so profoundly toxic that in some cases their waters cannot even be safely approached, let alone entered or drunk.

Trade pacts come and go, but the larger pattern of events does not often change. There are striking, damning similarities between the 1840s world of free trade and potato famine and our 1990s world of stateless capital, breakneck technological revolution, anxious but self-satisfied elites, globally integrated production and marketing, and ecological deterioration. The parallel was borne out by the 1984–85 Ethiopian famine, which killed a million people, but did not interrupt the export of green beans to England, as well as by the civil war and famine in Somalia in 1992–93. Somalia is a clear marker of the new times. The first famine broadcast live on satellite TV, it was as well a tale of trade—the arms trade, the oil trade (four oil companies have leased drilling rights to two-thirds of Somali territory), and the seafood trade. Even at the height of the UN-U.S. joint pacification and relief operation, Operation Restore Hope, fleets of fishing trawlers (Taiwanese, Korean, Spanish, Greek, and Italian) were working the undefended Somali fishing grounds, removing far more protein than was entering the country in "aid."

After the potato famine, a great wave of Irish headed to America to start new lives. Today no country would accept so many refugees, for today immigration is not so free. Today it is goods and investment capital that can ignore national boundaries. There is no longer a New World, and the old is enclosed by GATT, the master free-trade deal that governs over 120 countries and more than 90 percent of all international trade.

GATT is an institution far too few of us understand in any detail. Aside from a few brief moments—first the early 1990s, when the ratification or death of GATT's "Uruguay Round" negotiations was imminent and the related maneuvering took on the air of a high-stakes horserace, and then again in late 1994, when the U.S. Republican establishment pretended to oppose GATT's final passage through Congress—newspapers rarely discuss it, and then only in the business pages. Most people find it only another incomprehensible acronym, another boring detail of a winner-take-all economy no one expects them to understand. And yet GATT is anything but boring. The World Trade Organization, GATT's newly created implementing body, could easily evolve to be the most powerful institution since the World Bank and the IMF, and certainly it will have a profound effect on the shape of global environmental treaties. Maurice Strong, secretary-general of the Earth Summit, repeatedly warned negotiators that their final treaties would have to be "GATT legal." It is a phrase (often softened to "GATT consistent") well known to politicians, executives, and, recently, green activists.

Strong's meaning was clear. The Earth Summit treaties would not be allowed to contradict the doctrine of free trade as codified in secretly negotiated trade agreements (many key GATT documents were, and still are, classified) interpreted by an unelected cabal of unknown, Geneva-based bureaucrats. GATT was not even nominally a democratic institution, whereas the Summit was organized under the auspices of the United Nations and was being hyped around the world as a new beginning for both mankind and nature. But none of this mattered much. It was GATT, not the Summit, that embodied the emerging future.

Until recently, environmentalists seldom gave much thought to geopolitics, and those few who did were almost always of the World Resources Institute school of "global environmental governance." In this elite circle, it has long been taken as an unchallenged truth that we all, whether officers of planetary corporations or residents of impoverished Brazilian favelas, have the same ultimate interests in enlightened planning and rational global management. Today this is no longer a near-universal illusion, and the trade debate is one of the reasons why—it has made it impossible to deny that this is a world of winners and losers.

The character of the late 1980s should be recalled. The Bush years, following upon the Reagan "boom," saw the bill come due. The United States had become the world's largest debtor nation, and the Bush administration, ideologically zealous in a fashion Lord Trevelyan would have easily understood, took the opportunity to prescribe for the country a mild dose of the bitter medicine so well known in the South as structural adjustment. Bush's men fell to the job with gusto, setting out anew on the path that Reagan had cleared—to cripple regulatory agencies like the Environmental Protection Agency and the Occupational Safety and Health Administration; to privatize national forests, mines, and oil fields; to promote "free market reforms" in agriculture, education, and low-income housing; to cut social programs from health care to alternative energy development; and, of course, to pursue aggressively both the GATT and NAFTA agreements.

The trade deals that crystallized in the 1980s were of a piece with a larger agenda, and to understand that agenda is to understand the challenge of ecology. As Mark Ritchie, director of the Institute of Agriculture and Trade Policy and, according to no less a source than the *Wall Street Journal*, the man who jump-started the anti-GATT campaign in the United States, put it, "Since many of the citizens active in the trade debate

started out as committed environmentalists, this debate has generated a pool of people with a serious understanding of the links between ecology and economy—the fundamental knowledge needed to prosper in the next century. Thank you, Ron and George."

As GATT and its cousins—from NAFTA to APEC—have become better known, "trade" has lost its aura of unchallenged promise and prosperity. *The Economist,* for over a century the ideological flagship of free trade, notes that "if liberal capitalism is to continue advancing it must overcome the entrenched skepticism of the people that it benefits." This is surely true—which is just why that skepticism must be understood. *The Economist's* editors cannot print "fair trade" without adding "whatever that means," and have written of the "sinister alliance of environmentalists and economic isolationists" that threatens their chosen creed. But this is only smoke and self-delusion. The actual cause of the skepticism is an economy now mutating in strange and frightening ways. "Free trade," long an incantation designed to conjure images of affluence and "growth," has come for many to denote instead an "upturn" for the few, and for themselves only pain. "Fair trade," for its part, names the almost unnamable, unbearably difficult project of reforming the economy.

HARMONIC CONVERGENCE

Since its establishment after World War II, GATT has evolved through a series of negotiating sessions called rounds. Initially, its business was the elimination of tariffs, and by the mid-1970s this goal had been largely achieved. GATT's Uruguay Round (1986–93) did not focus on tariffs, subsidies, quotas, or any of the usual fronts in the ongoing low-intensity conflict that is international trade. Instead, the Uruguay negotiations broadened GATT's agenda to include investments, services, and intellectual property rights, and drew attention to the at first obscure goal of eliminating "technical" or "nontariff" barriers to trade. These bland and abstract terms name the real trade controversy, or would, if only more people understood that in the minds of free-trade true believers, ecological protection and human rights are just such barriers.

GATT looms large on the trade and environment landscape, but it is hardly the whole of it. We come to it with lessons learned elsewhere, lessons that, in North America, were first taught by the 1988 U.S.-Canadian Free Trade Agreement (FTA). It was the FTA that first awakened U.S. and Canadian environmentalists, health and safety advocates, and labor activists to the fact that their most crucial laws and regulations, precious fruits of decades of hard-fought campaigns, were actually "import restrictions" and other forms of "protectionism." In 1989 Quebec, on behalf of its huge asbestos industry, invoked the FTA to attack a U.S. ban on the import, production, and use of asbestos. At about the same time, urged on by its pulp and paper industry, Canada challenged U.S. laws requiring that a portion of the fiber used in the manufacture of newsprint be recycled; weak though they are, these laws remain, according to a Canadian industrial group, "disguised non-tariff barriers to trade because Canada does not have the supply needed of recycled fiber to maintain market share in the U.S." In the United States, too, industrial interests sprang into action, invoking the FTA to attack Canadian acid-rain reduction programs, energy export rules, reforestation and fish conservation policies, pesticide laws, and even automobile fuel-efficiency standards as "nontariff trade barriers."

The pesticide episode illustrated with particular clarity the deregulatory dangers hidden in the free-trade agenda. It is not simply that corporations sought to open Canada's markets for their products by attacking its pesticide laws. It is that they made this attack by challenging the legal foundation of these laws. Unlike U.S. rulings, which require only that the benefits of use be "shown" to outweigh the risks, Canada's laws went so far as to require that a pesticide be *shown to be safe* before it could be approved. Six out of seven pesticide products legal in the United States could not be used in Canada, a situation viewed by transnational agrochemical corporations as an unfair and even irrational restraint on trade. In the end, Canada defused the dispute by agreeing to work toward "equivalence" with U.S. laws. The larger matter, however, remains. It is "harmonization," trade jargon that means nothing less than the global leveling of divergent national rules about exactly what can be traded, and under what social and environmental conditions.

"Harmonization," pursued intelligently, could be a boon to both humanity and the earth. The question, to use the jargon, is whether we will get "harmonization up" or "harmonization down"—that is, whether trade agreements will treat international standards as "floors" that nations are free to build upon as their people wish, or "ceilings" that define maximum standards that no country will be free to exceed. It's a simple question to pose, but in a free-trade world that pits every nation and community against every other, a world in which ecological rapine and human slavery are both means of lowering prices on merciless global markets, it is anything but simple to answer.

We cannot know where the trade debate will take us, though we can easily understand what is at stake. Harmonization, as brought to us by Uruguay GATT and NAFTA, is, in the words of Lori Wallach of Public Citizen, a key U.S.-based fair-trade organization, "a one-way ratched down." The weight of the evidence argues that she is right. GATT demands that American food safety laws be watered down to conform to rules set by a highly secretive, industry-dominated, Rome-based organization named Codex Alimentarius (literally, "the law of food"), which consistently sets standards far laxer than those in force in the United States and Europe. Under Codex, peaches sold in the United States would be allowed to contain fifty times as much DDT as they do today. NAFTA says that the United States will not have to lower its food safety standards *if it can be proven that those laws are necessary in the first place,* a test that precious few laws would pass.

This is just the beginning of the threat. There is a fundamental conflict between the logic of social and environmental protection and the demands of deregulated international trade. The roster of U.S. laws that face challenges as restraints on trade includes the Marine Mammals Protection Act, the Endangered Species Act, the Resource Conversion and Recovery Act, the Pelly Amendment (a key enforcement tool for whaling and other treaties), various food safety laws, Corporate Average Fuel Economy (CAFE) standards, and even (according to a European Commission study) the Nuclear Non-Proliferation Act. All these laws are easily interpreted, under GATT rules, as nontariff trade barriers. Internationally, matters are even worse. At least seventeen environmental treaties involve limitations on trade and are highly vulnerable to challenge under GATT. They include not only the Convention on International Trade in Endangered Species and the Basel Convention on hazardous waste export, which have as their explicit *purpose* the restriction of trade, but also the London Dumping Convention, the Migratory Bird Treaty, the Montreal Protocol on Substances That Deplete the Ozone Layer, the International Convention on the Regulation of Whaling, the International Agreement for the Conversation of Fishes, and a variety of UN environmental protocols.

The inclusion of the ozone treaty on this list should make the issues crystal clear. The Montreal Protocol is constantly cited as the strongest basis for reasoned optimism about the future of global environmental governance. And although the Montreal Protocol is flawed, it is a model of serious intent in comparison with most other environmental treaties, for it explicitly allows nations to block imports from countries that violate its terms. Judged by GATT rules, it would be illegal, though it is unlikely to be challenged. The Protocol is too popular, and such a step would loudly advertise both GATT's antienvironmental bias and the public secret of most other environmental treaties— that they are not only indifferently enforced but also largely unenforceable. Steven Shrybman, a counsel to the Canadian Environmental Law Association and a leading fair-trade theorist, put the point succinctly: "When countries are serious about enforcing environmental measures, they include trade sanctions among the enforcement measures." The question is, why is the Montreal Protocol almost alone among environmental treaties in containing such sanctions?

There are myriad issues here, but the most difficult of them is "national sovereignty." GATT, as an instrument of economic globalization, ultimately implies a deregulation of trade so complete that it would circumscribe the ability of even *strong* governments to control health and environmental standards within their own borders. The pattern was visible as early as 1981, when a landmark Danish law mandating the use of returnable bottles was overturned as a barrier to the European Common Market. By the early 1990s, GATT had already been used to challenge European rules against hormone-tainted beef, to force Austria to abandon plans to introduce a 70 percent tax on tropical timber, and to overturn Thai smoking restrictions in the interests of U.S. tobacco companies. In this last and particularly notorious case, GATT ruled that no country could use trade restrictions to enforce public health laws *unless it could demonstrate that they were the least trade-restrictive means to that end.*

Conflicts over sovereignty are crucial to the politics of the new world, which is just why *The Economist* derides concern for sovereignty as "economic isolationism." The real issue, however, is not isolationism but democracy, not nationalism but a globalization that is swamping nation-states and putting nothing in their place but global corporations and unrelated international markets. Fair traders, to be sure, have often been less than coherent on these points, but the debates they provoked in battles against NAFTA and GATT at least made the stakes clear. In 1993, for example, Public Citizen, joined by the Sierra Club and Friends of the Earth, filed suit to protest NAFTA's summary enactment in terms that stressed not vague nationalism

but the specific matter of national legislative sovereignty. Their argument was simple—since NAFTA would affect the U.S. environment, it was properly subject to the terms of the National Environmental Policy Act, and required a U.S. environmental impact statement before it could legally go into effect.

The lawsuit failed, but it marked the frontiers of a new territory. Environmentalists worry about the efficacy of their hard-won laws, as food safety advocates worry about theirs, as just about everyone worries about security and jobs. We are all, today, subject to the logic of almost unrelated global economic processes that are far more powerful than the laws of individual nations. It is a sign of the times that concern about national sovereignty, long a hot button in the South, has become an issue in the United States as well. The North no longer stands above the global fray, and judging by the strength of 1993's anti-NAFTA coalition, a lot of people know it.

Environmental standards, and the fight to maintain and even raise them, is only one aspect of a larger argument over the terms of globalization. These standards, along with health standards and labor rights, are part of the living history of the United States and Europe, a history that, from a hard-line free trader's point of view, has an inconveniently democratic character. GATT review panels seek to iron out the resulting frictions, usually by lowering our "unreasonably high" standards to the levels enjoyed throughout most of the world.

Free trade inevitably pits the relatively strong health, environmental, welfare, and labor rules of the North against the far laxer standards that can always be found somewhere else—in Asia, the former East bloc, Latin America, or Africa. Moreover, it *must* do so, and precisely because it proposes to throw the poor together with the rich, the weak together with the strong, in one fantastic, chaotic, unregulated, planetary marketplace.

The sultans of globalization argue that a harmonized "level playing field" will be fair by definition. "Free trade," in the textbook-perfect world of economics, is said to improve life for all, even the lowliest, by removing economic "inefficiencies" and freeing the economy to approach its optimum performance. These, though, are illusions possible only to those who can ignore the population's increasingly sharp division into rich and poor, into a well-equipped and adaptable minority habituated to consumption and affluence and a hard-pressed majority still reeling from the destruction of their traditional cultures.

The richest can harmonize up, if they must. The rest lack both the ability and the means to raise social and environmental standards, and could not in any case do so without risking the small "comparative advantage" that world markets allow them, an advantage that turns on low wages and ecological looting. In a world thus partitioned, "free trade" can only be a hoax, as David Ricardo, the nineteenth-century economist who invented the idea of comparative advantage, would almost certainly have admitted. Ricardo believed that

> the fancied or real insecurity of capital, when not under the immediate control of its owner, together with the natural disinclination which every man has to quit the country of his birth and connections, and intrust himself, with all his habits fixed, to a strange government and new laws, check the emigration of capital. These feelings, which I should be sorry to see weakened, induce most men of property to be satisfied with a low rate of profits in their own country, rather than seek a more advantageous employment for their wealth in foreign nations.

Adam Smith himself, when discussing his famous "invisible hand," assumed it to be in the interest of capitalists to invest at home:

> By preferring the support of domestic to that of foreign industry, he intends only to his own security; and by directing that industry in such a manner as its produce may be of greatest value, he intends only his own gain, and he is in this, as in many other cases, led by an invisible hand to promote an end which was no part of his intention.

Today's world, Ricardo and Smith would find, is one in which such homey nationalism is visibly absurd; it is a world in which capital is free to roam the planet in search of investment opportunities, in which goods are free to travel in search of better markets and higher prices, in which only labor is enjoined, by law and fortified frontier, against easy mobility. Chinese manufacturers may dump cheap clothes and shoddy artifacts on the U.S. markets, but this does not mean the Chinese people are free to follow. Not unless they can shell out $30,000 to risk their lives crossing the Pacific on a tramp steamer.

For Further Discussion

1. How would you distinguish between free trade and fair trade? Is fair trade indistinguishable from protectionism?

2. Has Athanasiou made a convincing case for the ethical and environmental problems of free trade?

3. What is meant by "technical or nontariff barriers to trade"? Why do some defenders of free trade believe that ecological and human rights principles constitute such barriers to trade?

4. What does Athanasiou mean by "harmonization up" and "harmonization down"?

5. How might a defender of free trade respond to Athanasiou's claim that "free trade inevitably pits the relatively strong health, environmental, welfare, and labor rules of the North against the far laxer standards that can always be found somewhere else"?

Trade and the Environment

Daniel Finn

Daniel Finn is an economist and theologian. This selection is Chapter 7 from his book Just Trading, *which is an analysis of trade issues from the ethical perspective of a Christian theologian. In this reading, Finn addresses the environmental implications of international trade and assesses claims that trade causes significant environmental destruction.*

Finn begins his book with several fundamental ethical commitments. First, he believes that a theological and ethical evaluation of issues such as trade is both possible and necessary, these are not simply technical, scientific issues best left to experts. As a Christian, he also is committed to a "preferential option for the poor." Economic policies ought to be judged by how they serve the most disadvan-

taged. He understands the world as God's creation and recognizes a duty to respect the integrity of nature. Finally, he acknowledges an ethical commitment to the equality of all people.

In light of these values, Finn offers his assessment of the environmental impact of international trade. Critics of trade cite five major environmental objections: trade contributes to unsustainable depletion of resources, it increases pollution, it provides incentives to lower global environmental standards, it undermines domestic environmental protection, and it allows more powerful countries to impose their own standards on the less powerful.

After a careful analysis, Finn concludes that trade can be a helpful and necessary tool for attaining a just and environmentally healthy world. But international trade must be regulated by ethical considerations and not simply left to the market.

Christian faith requires a commitment to sustainability and to the integrity of God's creation. As a result, an assessment of the severity and urgency of anticipated environmental damage stands as a crucial background conviction in the debate over international trade. In an ideal world, controversies about environmental problems might be resolved before a view of international trade were constructed. It is no surprise that controversies around trade abound when some commentators anticipate cataclysmic problems or attribute a moral status to the biosphere (or even "biotic rights"), while others anticipate smooth adjustments or ascribe no more than an instrumental value to the nonhuman natural world.

Many critics of increased international trade see environmental problems as not only vastly serious but immensely urgent, due to both the severity of the coming damage and the irreversibility of the natural and social processes involved. John Cobb has employed as a symbol for our environmental situation today the image of a train hurtling down a hillside toward an anticipated but nonexistent bridge over a deep gorge. With this sort of assessment of environmental problems, it is understandable that growth-producing international trade makes little sense. Instead, Daly and Cobb call for steady-state "development," but not growth.

Some trade proponents on the right openly deride environmentalists for extremism and faulty science, and they accuse the environmental movement of covertly statist goals aimed at political control of the economy. More frequently, economists and other proponents of increased trade acknowledge the existence of environmental problems caused by economic growth, but tend to view those problems as less severe than do critics of trade. Of course, economists differ significantly among themselves, but it seems clear that most share a position fundamentally opposite that of Daly and Cobb. To some extent this may be attributable to a history of cataclysmic predictions within economics that have proved to be unfounded. Thomas Malthus was neither the first nor the last economist to predict dire consequences, only to be proved wrong because of later, unanticipated adjustments in the system.

FIVE ENVIRONMENTAL CHALLENGES TO TRADE

Those opposing increased trade out of concern for the environment tend to object on the basis of five distinct mechanisms of interaction between trade and damage to the ecosphere.

The first objection is that trade contributes to an unsustainable depletion of the planet's natural resources. The second is that trade accelerates the unsustainable use of the earth's absorptive capacity. Each of these is a particular form of the more general objection that trade promotes economic growth, which is the primary cause of environmental destruction.

The third sort of relation between increased trade and more responsible environmental policies arises from the competitive advantage of firms located in nations with weak environmental laws. Many critics of trade have predicted that curbing import restrictions in industrialized nations will simply encourage large corporations to move their manufacturing operations overseas to escape the costs of mandated pollution abatement which they currently face.

The fourth objection to increased trade focuses on the environmental effects of international agreements designed to increase trade (e.g., the GATT/WTO, the Canada-U.S. Free Trade Agreement, NAFTA, etc.). From this perspective, trade is not simply an activity between nations; it has become an international legal force which threatens national sovereignty. Individual nations have surrendered authority to international trade organizations, which then have the power to overturn national environmental laws.

The fifth point of contention between advocates and opponents of increased trade concerns the unilateral use of restrictions on imports as a lever to get other nations to improve their own environmental regulations.

These five points of tension between trade and the environment provide a helpful typology for the remainder of this chapter.

1. USING UP THE EARTH'S ASSETS: TRADE AND RESOURCE DEPLETION

Each year the nations of the world export to one another total merchandise trade in excess of three and a half trillion dollars ($3,500,000,000,000), a number so large that it overwhelms even and active imagination. Not only is the dollar value immense, but the amount of materials shipped internationally is equally stunning. Careful record is kept only on the mass of materials loaded and unloaded in maritime shipments, but . . . mountains or, in some cases, lakes of raw materials and processed goods are transported by ships between nations, more than four billion metric tons each year.

Although much of that amount includes manufactured products and renewable primary products such as food and timber, a large portion consists of nonrenewable natural resources that are consumed by production processes and people far distant from the point of origin. The scale on which human activity is transforming the planet is astounding. Is this a problem? if so, how big?

Depleting Nonrenewable Resources

Environmental concerns about resource depletion first came to general awareness in 1972 with the publication of *The Limits to Growth,* a report of the Club of Rome. The report was pathbreaking, though because of its novelty it contained a number of methodological limitations. Most important was the tendency to extrapolate contemporary trends into the future without considering the effects of price changes on patterns of consumption. For example, the study reported that the price of mercury had gone up 500 percent in the previous twenty years and the price of lead 300 percent in the previous thirty. In summary, the report asserted,

> Given present resource consumption rates and the projected increase in these rates, the great majority of the currently important nonrenewable resources will be extremely costly one hundred years from now.

This sort of prediction took more concrete and visible form in a celebrated wager between environmentalist and population specialist Paul Ehrlich and futurist Julian Simon. In 1980 the two agreed to a ten-year bet on which direction the real (i.e., inflation-adjusted) price of raw materials would move. The two drew up a futures contract requiring Simon to sell to Ehrlich ten years later the same quantities of five metals (copper, chrome, nickel, tin, and tungsten) which could be purchased for $1,000 in 1980. If the prices rose over that ten-year period, Simon would pay the difference; if they fell. Ehrlich would ante up.

Economists have generally refrained from Simon's optimism about a drop in resource prices. Yet for another reason, they almost always meet predictions about "running out" of natural resources with skepticism. Economic theory predicts that resource depletion will never occur completely (because that last unit will be too expensive to extract) and will occur very slowly, far more slowly than simply extrapolation of current trends would indicate. The reason for this is that as the price of resource A rises (due to higher extraction costs as the most convenient sources are used up first), users of A will have an incentive to switch to a "backstop technology," an alternative resource B which *had* been more expensive but which is now cheaper than A. The process is an iterative one, in that this increased demand for resource B tends to raise its price and, in turn, some users of resource B may find it economically attractive to substitute yet another resource C in place of B. There is a good bit of technological optimism in this presumption, for it assumes that as the price of any one resource rises there *will be* other resources that are close technological substitutes for it.

Just how much of this substitution can occur? This is a critically important issue but no one really knows the answer. One's position on the long-term result of this iterative substitution process stands as a part of a background commitment concerning technology generally: Some are optimistic, some pessimistic.

Even if we don't know how much substitution *can* occur in the future, how much has already occurred? Although economic theory encourages a reliance on higher resource prices to trigger this iterative substitution, it is one of the ironies of the twentieth century that this process has not much been in play in recent decades. No poll was taken in the early 1970s, but it is likely that, along with the Club of Rome and Paul Ehrlich, a large majority of people concerned about the environment would have predicted sizable increases in the price of metals and minerals over the subsequent twenty years. However, . . . real raw commodity prices for metals and minerals, as well as for petroleum and raw agricultural products, were actually lower in 1992 than they were in 1975.

. . . Incidentally, the record shows that in October of 1990, when the ten-year wager was up, Paul Ehrlich sent a check for $576.07 to Julian Simon.

Of all this data, the information about the price of metals and minerals is the most critical. Part of the drop in prices comes from more efficient production processes. Just as important, there has actually been an increase in the reserves of nearly all such resources in the meantime. It remains true that such natural resources are in a finite and "fixed" supply on the planet; this is a geologic fact. From the point of human activity, known reserves (those physical supplies which are economically feasible to extract) depend on the effort of humans to find them and to devise more effective ways to extract them.

Optimists on the right have been excessively exuberant in their conclusions. Ronald Bailey has summarized the depletion issue in glowing terms:

> There are no permanent resource shortages— future food supplies are ample, world population will level off before overcrowding becomes a problem and pollution can be controlled at modest cost.

Those less optimistic about long-term shortages of natural resources warn that the impact of humanity on the resource base is approaching, if it has not already reached, a maximum sustainable rate. Herman Daly argues that we now live in a "full world," quite unlike the empty world of the past where natural resources were considered to be "free" goods. Daly warns that economists and many others have overlooked the complementarity between human capital and natural capital (that the productivity of the former depends on the presence of the latter) and instead have stressed their substitutability (which is true to a degree but not without limit). The value of human capital is in many situations threatened by a diminishing flow of natural capital (for example, in fishing and forestry in some parts of the world), and this leads to unsustainability because,

> In the era of full-world economics, this threat is real and is met by liquidating stocks of natural capital to temporarily keep up the flow of natural resources that support the value of man-made capital.

Glib optimism in the face of broad scientific uncertainties is not available to Christians committed to respect for the integrity of creation. Nonetheless, trends in resource prices do make a difference. Although no one doubts that there is a finite amount of all these materials in the earth, in a recent edition of its environmental sourcebook, the World Resources Institute has in effect downplayed the significance of natural resource depletion in the face of these changes: "Evidence suggests however that the world is not yet running out of most nonrenewable resources and is not likely to, at least in the next few decades." In addition, when that process of "running out" begins to raise prices, the Institute argues,

> New technology is increasingly making possible substitutes for many traditional natural resource-based materials. Technology development is also yielding to more efficient means of providing light, motive power, and other energy-related services. Such changes are paving the way to economies less dependent on natural resources. When shortages do emerge, experience and economic theory suggest that prices will rise, accelerating technological change and substitution.

An investigation and ultimate evaluation of this debate cannot be undertaken here, but it is clear that the uncertainties about the future call for a morally induced caution in public policy.

One of the most important changes required in such public policy is ending subsidies for the production and consumption of nonrenewable natural resources. Examples are numerous. Oil companies receive indirect tax subsidies through depletion allowances and, as we will see in more detail presently, consumers of petroleum products in North America pay far less for the externalities they generate than consumers in the rest of the industrialized world. We have earlier noted the shift of a sizable portion of U.S. dairy production to semi-arid regions of California where water is priced far below its governmentally subsidized cost of production. Such unsustainable stress on water resources has led, among other tragedies, to the demise of the Colorado River long before it flows to the sea; except in years of unusually high precipitation, it is simply used up. Similarly, in many regions of the world, irrigation projects are depleting irreplenishable aquifers. In the Ludhiana District of the Punjab in India, the water table fell nearly a meter during the 1980s, the result of government subsidies that had reduced the cost of irrigation to the farmer. Both irrigation and the logging of forests are important to the economic development of the Third World, but subsidies in the use of natural resources are short-sighted and counterproductive. They give producers and consumers false signals, implying that resources cost less than is both economically and morally true.

The nonrenewable resource perhaps most important economically is the supply of carbon-based fuels, particularly natural gas and petroleum (which make up about half the world's international maritime trade). As these two energy sources are depleted, there are three alternative backstop technologies available. The first is coal, abundant in supply but threatening even more global warming per unit of energy than do petroleum and natural gas (a problem to which we will soon turn). The second is nuclear power, an energy source growing in world importance but whose long-term waste storage problems remain an unresolved threat (and entail a background judgment concerning which huge disparities of opinion exist). The third is really a cluster of renewable energy sources including energy generated from wind, from solar photovoltaic cells, and from geothermal or solar heat collection. As we shall see in the discussion of pollution problems, a carbon tax designed to internalize the social pollution costs of commercial energy sources would render these renewable source of energy more commercially competitive, given technological improvements in recent decades.

Depleting Renewable Resources: Forests

. . . The one type of raw commodity whose price *did* rise between 1975 and 1992 was timber, a thoroughly renewable resource. Combined with the loss of millions of acres of forest through land use changes (e.g., the burning of vast areas of the Amazon rain forests), consumer demand for wood presents a serious environmental threat to the sustainability of forests in many parts of the world.

Trade has indeed played a role in this process. . . .

Yet . . . with the critically important exception of Malaysia, the leading exporters of roundwood in the world are more advanced economies. Forests in the industrialized world certainly need careful environmental attention, and public policy must be developed to preserve critical portions of the now mostly logged "ancient" forests. Still, a recent study done by the FAO (the United Nations Food and Agriculture Organization) and data gathered by the World Resource Institute suggest that the total forested area of the indus-

trialized nations of the temperate zone (Europe, the Soviet Union, Canada, the United States, and Japan) actually increased between 1980 and 1990. The primary problem of unsustainable forest use in the world occurs in the developing world, primarily with tropical and subtropical forests.

. . . Malaysia stands out as a sore thumb among the world's leading exporters of wood. In many ways, Malaysia's case is an extreme one. Numerous reports and studies have been done both within Malaysia and abroad documenting the devastation of forests, particularly in the provinces of Borneo, and the government corruption, collusion, and flouting of both true national interest and international law that characterize the high-profit mining of Malaysia's forests. A number of steps have been taken internationally to rein in such practices, in particular, the International Tropical Timber Agreement (ITTA). Entering into force in 1985, the ITTA has been approved by most of the world's leading importers and exporters of tropical forest products, including Malaysia. Its fundamental aim is to ensure that by the year 2000 all tropical timber sold in international trade will be taken from forests under sustainable management.

Although bad public policy and venal private interests play important parts in the world's current unsustainable use of its forests, the most important force in this process is far more ordinary, though sometimes overlooked by environmentalists. . . . In those parts of the world where forests are most severely threatened—Asia, South America, and Africa—the vast majority of wood harvested becomes fuel, almost exclusively for domestic use.

Much world attention has rightly been focused on the loss of rain forest in recent decades, but other types of forests have seen even more rapid losses worldwide. Consider the example of Sudan, home to more of the world's "very dry" forests than any other nation. It annually produces more than twenty-four million cubic meters of wood, exports none of it, and consumes 90 percent of it as fuelwood or charcoal, resulting in a loss of 81,000 hectares (more than 300 square miles) of very dry forests each year.

This is not an isolated case. About half the people on the planet use wood, charcoal, or a biomass substitute (such as crop residues or dung) to cook all or some of their meals. The stoves traditionally used are terribly inefficient, consuming six or seven times more energy than modern stoves using commercial fuels. The gas range in the typical kitchen of the industrialized world is a sizable capital investment but is not simply a great convenience; it is more frugal with the earth's energy resources than traditional cookstoves.

We should be careful to note that such comparisons are in no way an indictment of the poor of the world for their necessity. Their behavior is perfectly reasonable and morally justifiable. The point here is that an environmentally superior alternative can be made available, even within the budgets of the vast majority of the poor.

Considerable effort is being exerted to introduce more energy-efficient biomass stoves in the developing world. Improved design reduces indoor smoke and dramatically cuts fuel consumption, making these newer stoves an attractive option. Consumers can often pay for the stove with the fuel saved during a few months' use. Convincing large numbers of people to change even to lower-cost technologies is difficult; traditional practices change only slowly. Nonetheless, considerable success has been achieved where the introduction of improved stoves has been undertaken with proper reliance on local producers and the reactions of local consumers. Reducing household biomass smoke reduces disease; cutting fuel consumption by up to 50 percent means a significant rise in economic welfare: a dramatic reduction in the time necessary to collect daily fuelwood in rural areas and substantial increases in effective annual income for urban dwellers who must purchase their charcoal or fuelwood.

The unsustainable use of the world's forests is a critical concern for a Christian ethics committed to the principle of respect for the integrity of God's creation. Trade plays a role in that unsustainability, and existing treaties on illegal timber trade need to be enforced and strengthened. Nonetheless, in addition to the threat to forests from land use changes, it is the consumption of wood for fuel that most threatens the world's forests. In fact, the availability of commercial fuels, often provided through international trade, will be an important step on the way to dependence on renewable energy sources that can sustain the annual energy demand of the increasing prosperity of developing nations.

Biodiversity

Threats to global biodiversity from indiscriminate economic growth are severe. How to value losses in the variety of species and in the genetic variety within individual species stands as an important background assumption that cannot be examined here. Although the Christian's commitment to the integrity of God's creation is the driving force behind concern about biodiversity, a careful specification of precisely what losses in biodiversity are morally acceptable is a matter of great debate. Far less debatable is the conviction that humankind has been all too casual in mining the biological resources of the world and that strong efforts need to be made to reverse the trend.

In many ways, the interplay of science and biodiversity remains at a critical threshold. We still suffer from large deficiencies in knowledge about the range and character of uncounted species on the earth. In addition to the value any species possesses in itself as a part of God's creation, the very diversity of species hosted by, for example, tropical ecosystems represents a critically important source of genetic material in an era of genetic manipulation. Though for a long time drug companies and other users of such genetic material took it for granted as a part of "the global commons," more recent agreements between pharmaceutical firms and peoples indigenous to those ecosystems have begun to recognize and repay the debt these firms and their customers owe. More public pressure to continue this trend toward strengthened international agreement for the equitable treatment of indigenous peoples and host Third World countries is needed.

The single most important threat to biodiversity has been the loss of habitat, in nearly every type of ecosystem on the earth. International trade plays a role in this process, and international agreements about a host of issues affecting pollution, overexploitation of plant and animal species, agriculture and forestry, and the loss of habitat are critical. The Convention on International Trade in Endangered Species of Wild Fauna and Flora (CITES) has been in force for some twenty years. It has helped curtail inappropriate trade in endangered species, even though stronger compliance by signatory nations is still required. The Global Environment Facility, a joint program of the U.N. and the World Bank, has made protecting biodiversity a high priority, and though successes have been limited, such cooperative international efforts show promise.

Looking back on this cluster of issues related to trade and resource depletion, it becomes clear that, like purely domestic economic activity, international commerce requires a framework of rules. In most cases these must be rooted both in national legislation and in international agreement. Because of the importance of background assumptions concerning technological optimism in evaluating the depletion of nonrenewable resources and of the value of species diversity in assessing prospects for biodiversity, participants to these debates have unusually wide differences to negotiate in structuring such agreements. Significant progress has been made in the daunting task of establishing a scientifically rigorous definition for global biogeophysical sustainability, but far more work has yet to be done.

2. OVERTAXING THE EARTH'S ABSORPTIVE CAPACITY: TRADE AND POLLUTION

The second reason trade is often accused of threatening the environment is the concern that more economic growth will further overwhelm the earth's capacity to absorb the wastes which humankind annually generates. Two central distinctions are critically important. The first concerns the difference between primarily "local" environmental problems (such as air and water quality) and the more clearly "global" environmental problems (such as greenhouse warming and upper-atmosphere ozone depletion). The second distinction is between increased international trade and increased economic growth. Economic growth without increased trade can have impacts on the environment discernibly different from growth accompanied by increased trade.

Trade and Primarily Local Environmental Problems

At an elementary level it is obvious that environmental problems grow as economies shift from being predominantly agricultural to predominantly industrial. Atmospheric emissions from motor vehicles and factories, to take simply one sort of pollution, cause deteriorating health conditions. If the environment were the only value to which Christian ethics needed to attend, international trade might appropriately be condemned simply because it leads to greater economic growth and, therefore, to more pressure on the planet's absorptive capacity. However, Christians must also attend to other values, like the need for an increase in economic welfare for the world's poor; moreover, Christian ethics must take a more careful look at the relation between increasing environmental problems and increasing prosperity.

The impact of international trade on local pollution problems is notoriously difficult to assess against the background of pollution problems arising from local economic activity alone. Nonetheless, it is clear that firms involved in international trade, like those which operate only locally, regularly exert political power to resist local or national regulations aimed at reducing pollution, because of the increased costs of production they entail. We postpone until later in this chapter the treatment of one critically important question here: to what extent "dirty" industries tend to migrate from nations with strict environmental laws to others where environmental regulation is more permissive.

Of all the forms of international trade that may cause greater pollution, the international waste trade stands out. Solid and liquid wastes from production and consumption are ultimately deposited back into the biosphere in one particular location or another. These range from very low-risk wastes that can be composted or landfilled to high-risk wastes that are highly toxic, persistent, mobile, and bioaccumulative. Examples of the latter include cyanide wastes, chlorinated solvent wastes from the degreasing of metals, and dioxin-based wastes.

Concerned about environmental damage, the industrialized nations of the world have imposed stricter regulations on the disposal of toxic wastes, greatly raising the cost of disposition. This has led to the rise of hazardous waste disposal firms and an international trade in toxic wastes. In principle there is good reason to allow the transport of toxic wastes across national boundaries. Particularly for hazardous wastes generated in developing nations, transport to other nations with more adequate disposal facilities makes both economic and moral sense. The problem, however, is that far greater quantities of hazardous wastes have flowed in the other direction, not because of adequate disposal potential but rather because the governments of developing nations may be willing to provide disposal sites for wastes in return for money, even though this may not serve the long-term interests of the nation. For example, the African nation of Guinea-Bissau negotiated—and later under public outcry rescinded—a $120 million-per-year deal to store industrial wastes from the north. Such amounts of money are simply part of the cost of doing business to large industrial firms, but this sum represented nearly 75 percent of Guinea-Bissau's annual gross domestic product. The prospects for graft and other forms of governmental corruption in the presence of large amounts of money will predictably threaten democratic accountabilities. However, the issue is even more complicated than that.

To understand the decision from the point of view of developing nations, it might be helpful to consider the possibility that an imaginary and vastly wealthy trading partner offered a very lucrative hazardous waste contract to the government of Canada or the United States. What would occur if the Canadian government received an offer for, say, $300 billion, or if the U.S. government received an offer of $3 trillion? These dollar amounts, of course, are completely unrealistic. Nonetheless, the scale of these offers is the same as that sometimes presented to small Third World countries, and it helps to recognize that more than a few legislators and citizens in even the wealthy industrialized nations would likely advocate accepting such a hazardous waste contract even if that meant rendering some portion of the country permanently uninhabitable.

The Basel Convention on the control of transboundary movements of hazardous wastes and their disposal is a critically important development. Although representatives of nearly all the industrialized nations signed the convention, a number of these nations have been slow to ratify and move toward strengthening the convention. The agreement prohibits the export of wastes for disposal in Antarctica and requires the acceptance in writing by the nation of import for any transboundary hazardous waste shipments. This attempts to prevent bribery of local government officials in contravention of national laws. Because national governments themselves can be corrupted or simply shortsighted, waste exports are permitted under the convention only if the exporting nation does "not have the technical capacity" or "suitable disposal sites," and only if the country of import has appropriate facilities for environmentally appropriate treatment and disposal. Although it is in principle possible for a developing nation to be technically ready to accept, treat, and store the results of hazardous waste, the risk of corruption or myopic self-interest of the current generation is so high that a Christian moral commitment to the integrity of creation would seem to require that each nation agree to treat and dispose of its own hazardous wastes as the usual procedure. Although such an agreement is "paternalistic" and rejects the economies of scale that international trade promises, the risks arising from the vast disparities of wealth between nations render it prudent.

Shifting attention to more "ordinary" kinds of pollution, we should recognize that it is both empirically and ethically incorrect simply to indict trade and economic growth because of local pollution problems. As one researcher has put it, "The air is harder to breathe, the water is dirtier and the sanitation and health conditions are poorer in Calcutta, Lagos, or Mexico City, than in New York, Tokyo, and London." . . .

Although the development of cleaner technologies has been essential to stemming air and water pollution problems in wealthier nations, those technologies have been necessitated by environmental regulations, which do indeed "retard economic growth" because they increase the costs of production. They are, in effect, a national decision to "spend" a portion of national income to offset the environmentally damaging effects of economic growth. In the United States, this amounts to more than 2 percent of GDP. What conservative critics of environmentalists so often overlook is that most environmental regulations were and often still are opposed at nearly every turn by the corporations affected (and at times by their workers).

To a large extent, lower-income nations spend a smaller portion of their national income to alleviate pollution problems because there are other more pressing needs to be addressed first. A contributing factor in this slower response to pollution problems in developing nations is clearly the resistance of powerful economic interests to regulations that increase costs and lower profits, just as it is in the industrialized world. Though this problem is more severe in the developing world, many reputable Third World policymakers have described the decision as concerning what a nation judges it "can afford."

. . .

Trade and Global Pollution

Although even local pollution problems have global effects eventually, it is helpful to distinguish them from pollution threats that are predominantly global.

. . .

There is no doubt that economic growth brings about more greenhouse gas emissions. However, . . . it is not simply overall economic growth alone that contributes to such pollution. Nations that produce and export petroleum (for example, Qatar, Gabon, and the United Arab Emirates) and nations where national policy keeps energy costs low (for example, the U.S.A. and Canada) tend to rank higher on the list.

Do nations that engage in more international trade emit more greenhouse gases because of it? Environmental critics of trade have argued that if nations became more economically self-sufficient, goods would not have to be shipped so far, cutting energy consumption and greenhouse emissions. Thus, it is helpful to ask how the actual transportation of goods in international trade contributes to the buildup of greenhouse gases. Unfortunately, no one can answer this question. Much international trade (for example, most trade between European nations or between Canada and the U.S.) takes place by rail or truck, and current national mileage figures for aggregate trucking or railroad hauling do not accurately distinguish between domestic and international destinations. Nonetheless, international accounting for CO_2 emissions does keep track of maritime "bunker fuels," those consumed by ships in international transport. These represent about 1.5 percent of CO_2 emissions from all industrial processes. This is not a small number in absolute terms (more than 250 million metric tons of CO_2 emissions each year!), but is certainly small relative to the other sources of CO_2 emitted from industrial processes. In addition, about half the tonnage transported by ship in international trade consists of petroleum or petroleum products, pointing the finger toward domestic consumption of commercial energy.

. . .

A small but significant proportion of anthropogenic CO_2 emissions arises from the burning of biomass fuels, mostly firewood, predominantly in the developing world. The policy options we saw earlier concerning forest sustainability apply here as well. Increasing the efficiencies of biomass stoves is a first step; this reduces CO_2 emissions while improving economic welfare. International assistance is important in that process, and international trade will likely prove essential in the move from biomass to cleaner-burning fuels, particularly renewables, as Third World incomes rise.

The primary threat of greenhouse global warming arises from the use of commercial energy—throughout the world, but especially in industrialized nations. Commercial energy consumption per person is by far the highest there. . . .

. . . Most of the difference in energy efficiencies among industrialized nations is directly attributable to differences in the prices users of energy must pay. Although the data sets for many energy types are incomplete, we can get an idea of international differences in energy prices by examining the data for gasoline. . . .

. . . The gasoline prices consumers face in Japan and several European nations are two to three times higher than those faced by Canadian and U.S. consumers. Although every nation's taxes are higher on gasoline than on other forms of energy—to help pay for the costs of highways—similar ratios exist for other types of energy.

. . . Economic theory predicts that consumers facing higher prices for a good will consume less of that good. The data . . . illustrate this strong inverse corre-

lation between energy prices and energy consumption. This provides the most persuasive explanation for energy use patterns in North America. Because United States and Canadian energy users—both individuals and businesses—face significantly lower energy prices than their peers in other industrialized nations, they are less frugal in their use of energy. Of course, the explanation for lower energy supply costs *and* lower energy taxes in Canada and the United States is that both have large domestic energy stocks, politically powerful energy companies, a heavy dependence on the automobile, and a citizenry culturally accustomed to all of the above. It is, for example, far easier in a nation like Japan to build a national consensus supporting higher energy taxes and better energy efficiency when practically all energy is imported and very few citizens have vested interests in energy firms.

The way to reduce energy consumption is to raise energy prices. This will no doubt occur in the coming decades as petroleum stocks decline, but a prudent concern about global warming requires a quicker response. Because any tax on basic commodities such as energy causes serious hardship for the poor, a concern for justice requires offsetting subsidies for low-income people, though these should be general income supplements and not directly related to energy consumption rates, so the poor also experience the incentive to conserve energy.

Different forms of energy release different amounts of carbon, so the single best way to reduce CO_2 emissions is to raise the price of commercial energy in proportion to the amount of CO_2 each form of energy emits: a "carbon tax." . . . A \$100 tax for every metric ton of carbon emitted in the United States or Canada would raise oil prices about 39 percent, natural gas prices by 46 percent, and coal prices by 187 percent. These figures, and those for a \$300 tax, come from a study by the International Energy Agency designed to predict the effect of carbon taxes on energy consumption in the year 2010. For simplicity, they presume a currently untaxed market for these three products.

. . .

In sum, it would seem from the perspective of both Christian ethics and secular policy analysis that too strong a focus on the detrimental aspects of international trade in the pollution problems of the world actually draws attention away from steps that are far more likely to reduce environmental problems. Even ending all trade tomorrow morning would not make much difference in the local or global pollution problems we currently face. Local problems will continue to require the expenditure of local resources (usually mandated through government). Although the industrialized world has far to go in the control of local pollution problems, experience to date indicates that such strategies are indeed effective. Global problems require global solutions. International agreements to reduce chlorofluorocarbons have been effective, though far more needs to be done to provide incentives to developing nations to adopt ozone-friendly technologies. The threat of global warming, in spite of its uncertainties, calls for an international agreement on a sizable carbon tax to reduce energy consumption, to channel it away from the most carbon-intensive fossil fuels, and, simultaneously, to channel it away from carbon fuels as a group toward alternative, renewable energy sources. For both CFCs and greenhouse gases, continuing international trade relations may prove essential to the dissemination of cleaner technologies in the developing world.

3. TRADE AND THE COMPETITIVE DISADVANTAGE OF ENVIRONMENTAL LAWS

Following the concerns that increased trade accelerates the depletion of resources and the pollution of the biosphere is a third issue: Trade transforms local environmental laws into an international competitive disadvantage because of lax environmental laws in other nations. Paul Hawken, for example, predicted that

> the inevitable result of the present GATT treaty is that the rewards of international trade would go to the cheapest producer, not the most responsible producer. . . . International economic advantage would go to the companies that were best able to externalize environmental and social costs; companies that internalized those costs and took full responsibility for their environmental impact would be placed at a disadvantage.

We saw in the foregoing section that many local pollution problems—in particular air and water quality—actually tend to improve as national incomes rise because nations begin to spend more of that income to alleviate environmental problems once more basic needs are met. This "expenditure" usually comes through government regulations that raise the cost of production in industries that cause pollution. Thus both standard economic theory and everyday common sense predict that firms in nations with high pollution abatement costs will be tempted to move production facilities abroad if more permissive environmental regulations are available there.

For example, numerous reports have documented the environmental degradation brought about by the hundreds of maquiladora plants built in the 1980s in a "free trade zone" along Mexico's northern border. Toxic wastes have been openly dumped into the air and water. There seems to be agreement, even among the critics of increased trade, that Mexico's environmental law is not the primary problem here; enforcement is. Nonetheless, whether it results from weak environmental laws or lack of enforcement, a cost advantage available to firms that migrate would be an incentive for them to do so. The question, of course, is *how prevalent* is the flight of dirty industries from First World nations to the developing world? How *strong* is the tendency to migrate?

Consider the United States. Pollution abatement costs brought about by federal regulations have been substantial, with estimates for 1988 accounting them to be in excess of $14 billion for all U.S. industry. Still, that amount is being paid out of the total value of U.S. industrial production for that year of about $2.6 trillion, resulting in perhaps a .5 percent cost for pollution abatement on the average. Even "dirty" industries (those with the highest pollution abatement costs as a percent of total costs) still have relatively low abatement costs, ranging from a high of 3.2 percent of total costs for cement production to 2.2 percent for industrial chemicals to 1.8 percent for iron and steel foundries. Even if such industries were assured of zero pollution abatement costs for the long run in developing nations (an unlikely prospect), such cost differentials of 2 or 3 percent are not of a magnitude that ordinarily leads to plant relocation, though it would make it more likely for firms at the margin. In addition, of course, firms *might* estimate that pollution abatement costs in the industrialized world will rise faster than in the developing world and thus might expect the cost differential to widen, though this, too, is unlikely given the increase of environmental concern worldwide.

A number of researchers have undertaken empirical investigations to detect any migration of dirty industries toward the developing world. Consider one recent study in Latin America. There has indeed been an increase in dirty industry production and in the trade of dirty industry products in Third World nations, but it remains unclear whether any sizable portion of that comes from plant relocation. The reason for this is that developing nations predictably expand their industrial sectors, including "dirty" industrial sectors, as development proceeds. In any case, there seems to be a clear difference in the choice of "clean" or polluting technologies among developing nations. . . .

The study found significant differences between open economies (those more oriented to international trade) and closed economies. It also distinguished between low-, middle-, and high-income nations and between nations that were experiencing slow and fast economic growth during the 1980s. . . . Lower-income nations had higher growths in toxic intensity during the decade than higher-income nations, but there was a dramatic difference in nations that were more open to international trade. For open economies, faster economic growth moved in tandem with *lower* rates of growth in toxic intensity. For closed economies faster economic growth entailed *higher* growth rates of toxic intensity regardless of the initial level of income. Why?

The primary explanation for this difference seems to be that open economies generally, and fast-growth open economies in particular, benefit from faster transfers of clean production technologies from the developed world. It seems that because the newest and most economically efficient production technologies were developed to include cleaner processes (mandated in the industrialized nations), the relatively small costs of pollution abatement included in the new technologies often lead transnational firms producing in both places to invest in the cleaner technologies when producing in the developing world. In contrast, where trade is curtailed and investment decisions are made by local firms subject to more lax environmental law, plant expansion is more frequently a simple replication of older, "dirty" technologies. There are "pollution havens" in the Third World, but there is reason to believe that they occur primarily in closed economies.

A 1995 survey of existing empirical research on the topic concludes that "overall, the evidence of industrial flight to developing countries is weak, at best." Such studies are not definitive. Nor do they contradict the fundamental presumption of trade critics *and* mainstream economists that firms in dirty industries will make relocation decisions to further their own interests, not to reduce global pollution. They do show, however, that such theory-based presumptions identify only one force among many which shape the effects of trade on environmental problems.

4. THE RULES OF TRADE AND NATIONAL SOVEREIGNTY

The fourth point of conflict between environmental values and trade concerns the authority of some trade agreements to overthrow national environmental laws if they conflict with the principles of trade. Thus, for example, environmentalists were irate that U.S. laws (requiring tuna sold in the United States to be caught using environmentally superior nets, to prevent the death of dolphins) could be overthrown by the GATT after an appeal from Mexico (whose fishing fleets chose to use the more "economical" but environmentally more destructive fishing nets for their tuna catch). Similarly, many Canadians resented the threat to Canadian provincial decisions (to invest tax dollars in reforestation) after a U.S. challenge under the 1989 Canada–U.S. Free Trade Agreement. What appears to Mexican fishers as an unfair trade restriction and to U.S. loggers as an unfair production subsidy appears to others as sorely needed government decisions to offset environmental damage caused by business as usual. The nations of Europe have found a similar problem with the trade rules of the European Union. In the current international scheme of things, trade agreements have more legal clout than environmental agreements, and this leads many environmentalists to become critics of continuing efforts to increase trade. Though each of these agreements is worthy of analysis, it will be helpful to focus on the most critical one, the World Trade Organization, the successor to the General Agreement on Tariffs and Trade (the GATT).

If it were not for its narrow focus on trade and growth, the WTO might universally be recognized as a much-needed communal structure to prevent individual nations from choosing communally destructive policies of protectionism in a kind of "prisoner's dilemma" situation. (If my nation imposes tariffs and yours does not retaliate, I will be better off. However, since we both know that, we may both impose tariffs and both be worse off than if neither had.) Still, it is morally irresponsible to grant authority over the rules of trade to the WTO when only one goal (economic growth) of the many held by member nations is incorporated into its mission. Both religious ethics and secular policy analysis require a change.

There have been a number of good analyses of controversial GATT decisions, in particular the tuna-dolphin case, where the GATT ruled that a U.S. law discriminating against tuna caught with dolphin-endangering nets contravened the GATT agreement and had to be rescinded. The central issue in the case from the perspective of the GATT was that this law entailed a restriction not on the kind of *product* to be allowed into the nation but on the sort of *process* used prior to bringing the product to the border for sale. There were other issues involved, but the most controversial was that the GATT historically allowed nations to ban only particular products, and then, only as long as the ban applied equally to those inside and outside the country. The underlying reason why *process* restrictions were outlawed under the GATT was that it is all too easy for a nation to decide that the dominant process within its own borders is preferred and to erect protectionist barriers against processes used predominantly in other nations.

In the years since the founding of the GATT, nations throughout the world have come to understand the importance of the environmental effects of alternative production processes. Protectionism is no longer the dominant reason that a nation importing a product would care about the process that produced it. In fact, a "product life cycle" analysis is necessary to render production and consumption environmentally responsible—from the extractions of raw natural resources to the wastes released during and at the end of the product's life. As Daniel Esty has phrased it, "This differentiation between products and production processes cannot be sustained in an ecologically interdependent world." Some argue that the WTO's own internal logic requires such an extension: The WTO now holds antitrust as a goal in its prevention of "dumping" and supports human rights in at least a limited way by treating differently products made by prison labor. Thus it could consistently hold environmental quality as a similar goal. Others have proposed steps such as opening WTO sessions to public scrutiny (its dispute resolution panels are notorious for their secrecy) and involving nongovernmental organizations in WTO deliberations as a means of moving toward a "greener" WTO.

There is, of course, another possibility besides transforming the World Trade Organization. The nations of the world could establish a separate and parallel international umbrella agreement, a "Global Environmental Organization," for all environmental issues. This organization could bear an authority parallel to the WTO, and these two international treaty bodies would work out the inevitable conflicts between trade and the environment by negotiation. However, there seem to be, at least in the near term, insurmountable political difficulties in establishing so strong an international organization for the environment. As a result, efforts must continue to transform WTO standards to include environmental sustainability as one of its fundamental commitments.

Those who decide ultimately to oppose the rules of trade of the WTO or other international organizations rather than attempt to improve those rules, tend to take this stance after concluding that such an attempt to change the rules will bear little fruit. Although such a stance is morally defensible, it is at least as morally ambiguous as a commitment to changing the rules. It seems quite unlikely that the environmental problems before us can be solved without international cooperation through binding agreements. Any selective use of the sovereignty argument against rules we do not like will quite likely come back to haunt us when others choose selectively to flout the rules we have worked hard to establish. In fact, a similar form of "unilateralism" (as it has come to be called in the literature) makes up the fifth environmental argument for restricting trade.

5. IMPORT RESTRICTIONS AS INTERNATIONAL ENVIRONMENTAL LEVERAGE

The last of the five forms of conflict between those who would oppose increased trade and those who would encourage it concerns the active use of trade restrictions as leverage to induce other nations to change their environmental policies. A single nation or group of nations might unilaterally place environmental conditions on imports in an attempt to address ecological problems. To cite simply one example, the United States threatened Japan with restrictions on its animal product exports to the U.S. if Japan did not stop trade in products made from the shell of the rare hawksbill turtle.

At first hearing, it may seem contradictory for environmentalists to propose such a tactic and at the same time complain about the power of the WTO to overthrow national environmental legislation. However the issue here is different, particularly from a moral perspective. Rather than appealing to an international legal authority with the power to *force* changes, proponents of this approach argue that they are simply stating the conditions under which their own nation will engage in trade, leaving other nations with a free choice: They can either accept those conditions and trade with us, or reject them and find somewhere else to sell their products.

For example, Denmark instituted regulations on containers for mineral waters, soft drinks, and beer, requiring that they be reusable within a system of deposit and return. Because this in effect required glass containers, foreign firms supplying those beverages to Denmark objected. The extra weight (compared with plastic or aluminum containers) significantly increased transportation costs for beverages imported into Denmark, and mandatory reuse required an expensive shipping of empty bottles back across international borders, eliminating the simpler recycling of aluminum or plastic containers at Danish recycling centers. This issue was litigated through the European Commission and not the GATT, but the arguments would have been similar. The Danes defended their law as necessary to minimize packaging waste. Outsiders argued that the laws gave Danish beverage firms an artificial advantage by raising production costs for foreign firms. A similar issue arose under the Canada–U.S. Free Trade Agreement when American beer producers objected to a 10¢ levy on all metallic beer cans sold in the province of Ontario.

Critics of such import restrictions respond in three different ways. The first is that although this process may seem fair enough between nations that have roughly the same economic power, it is all too easy for larger and economically more successful nations to, in effect, "force" economically less influential nations into compliance. In such cases, the distinction between a legal requirement (which many environmentalists oppose as a loss of sovereignty) and a "free" economic choice is simplistic.

The second objection is that this use of trade restrictions to accomplish nontrade goals would open Pandora's box. For every nation that may take this approach for a laudable environmental goal, there may well be a number of others that threaten trade restrictions to further protectionist or geopolitical objectives far less benign. As a result, proponents of trade usually want to "depoliticize" trade decisions. More accurately phrased, they want to insulate trade policy from other political or economic goals of the nation.

The third response by critics against the use of import restrictions to further environmental goals is that they usually don't work. Relations among nations are enough like those among individuals that the use of threats and unilateral mandates to accomplish even morally appropriate goals usually brings about a resentment that threatens not only the goal at hand but a number of others on which nations must cooperate. As Robert Repetto of the World Resources Institute put it,

> Even if legal under international rules, these policies are problematic because they rely on one welfare-reducing measure (trade restrictions) to discourage another (non-cooperation in environmental protection). . . . The use of trade *concessions* to elicit international environmental *cooperation* is an approach much more likely to generate economic and environmental gains and an overall improvement in welfare.

From the perspective of Christian ethics as well as secular policy analysis, the issue here concerns the relative advantages and disadvantages of any nation's claiming the right (and thus allowing others the right) to act unilaterally in trade issues that affect the environment. It seems clear that the best solution would be an international agreement covering such issues, and the nation's effort to achieve such a multilateral agreement might be a "moral test" of the sincerity with which the environmental regulation was initially enacted. For example, it would be obviously duplicitous if Ontario wanted to penalize U.S. brewers for exporting cans, but defended the right of Canadian brewers to export metal cans to the United States without penalty. While helpful, this test does not guarantee moral rectitude, since a nation might decide that its protectionist self-interest (rather than an environmental value) is best served by an internationally recognized trade restriction. Because of the potential for insincere environmental concern to cloak underlying protectionist self-interest, international trade agreements should include in their dispute-resolution processes a judgment about the ultimate social and scientific legitimacy of environmental claims.

Nonetheless, it is undeniable that the environmental health of the planet will depend on more ecologically benign production and distribution processes, and, as we have already seen, future trade agreements will need to provide for life-cycle regulation of production. Because progress on environmental issues will require strengthening both international trust and the institutions on which such trust is founded, unilateral actions—even for good causes—should be a last resort, not a first step.

CONCLUSION

The Christian commitment to respecting the integrity of God's creation presents severe challenges to believers today, living within an economic system that will, if unchecked, profligately abuse that gift. In an economy based primarily on markets, the individual decisions of a multitude of economic actors, and in particular the highly influential decisions of the most powerful, will tend to ignore the long-term environmental costs of their actions. Only a community decision through government can channel those forces toward less environmentally destructive outcomes. At the same time, the widespread environmental problems throughout the former Soviet bloc demonstrate that markets are not the only source of threats to the ecosphere.

In this chapter, we have addressed environmental concerns about international trade. The two most fundamental charges against trade are, in effect, charges against economic growth: that it leads to the depletion of the earth's storehouse of resources and simultaneously overwhelms the earth's capacity to absorb wastes. Implicit here, as we have seen, are a number of "background" issues especially concerning the likelihood and severity of future environmental problems. Those more convinced that the likelihood is high and the severity extreme will be more likely to insist on restrictions on economic growth, and on trade. Nonetheless, it would seem that there are more effective and politically more feasible options for reducing each of these long-term problems than trying to do so by restricting trade itself.

Local pollution problems have been mitigated significantly through concerted political decisions to do so, even though much work remains to be done even in wealthy nations such as the United States and Canada. Worldwide pollution problems are critically important. Significant strides in reducing ozone depletion have been made through the Montreal Protocol and related agreements. Trade and increased aid look to be absolutely essential in persuading the developing world to adopt ozone-friendly technologies since CFC refrigeration processes are more cheaply available. Global warming remains a tremendous uncertainty and threat for the world. Nonetheless, restricting trade in an effort to reduce CO_2 emissions will be less effective, politically more difficult, and more likely to stunt the aspirations of Third World people in poverty than would be alternative strategies. These in-

clude increasing energy efficiency (particularly in the United States and Canada) through tax policies as well as working toward improving the sustainability of Third World forests through enforcement of timber trade treaties and subsidies for forest preservation and for the dissemination of environmentally friendly technologies such as improved biomass stoves. The single most important step to be taken will be internalizing the social costs of commercial energy through some form of carbon tax. Not only would a substantial carbon tax slow the growth of consumption of carbon-based fuels through higher prices, but perhaps even more important, those higher prices would make backstop renewable technologies more and more economically feasible. While a "soft landing" in the ultimate switch to renewable energy sources is by no means ensured, it is a definite possibility.

This chapter also dealt with three additional points of conflict between advocates of trade and environmentalists. Common sense and economic theory would predict that tighter environmental laws in some nations would create an incentive for firms to move to other countries with looser regulations. However, the empirical record indicates that this cost advantage is small in comparison with other costs of production and that in fact there has been little capital migration due to governmentally imposed pollution abatement costs.

Ultimately, of course, international agreements on pollution standards and other environmental concerns represent the best hope for an ecologically responsible economy worldwide. Thus Christians and others concerned about the environment ought to be more cautious in their assertions of national sovereignty as a defense against rules of trade that currently have too much international legal force. . . . The moral danger in the sovereignty argument should be recognized: It undercuts the ultimately necessary deference of local, case-by-case decisions to an international body with the capacity to engender worldwide compliance with higher environmental standards.

All this should not be taken to mean that nothing but good things comes from international trade. It remains true that powerful firms engaged in trade will attempt to subvert efforts to improve the environment because these raise their costs of production. Hard political work will continue to be needed to bring about such improvements. In addition, trade by definition increases the geographic separation between environmental costs caused in production and the economic benefits enjoyed in consumption. Inevitably this causes a greater psychic distance between consumers and the environmental problems their consumption entails, a highly unfortunate result from the perspective of Christian ethics. Even further, as Herman Daly has argued, to the extent that trade allows industrialized nations to exceed their domestic limits on resource generation or waste absorption (by depending on other nations for these), the gradual tightening of biophysical environmental constraints will be felt more simultaneously around the planet than would otherwise have been the case. The result, of course, will be less opportunity for one nation to learn from another's experience with environmental problems and less time to alter behavior when those constraints begin to bind more tightly.

Such dangers are real and too often overlooked by glib advocates of "free trade" on the right. Nonetheless, the moral risks identified in these concerns, though different in scale, seem analogous to those faced domestically by Christians who have a good job, own a car, use a dishwasher, or take a summer vacation at some distance from their home. The environmental threats arising from such "nonexcessive" activities are real and illustrate an important tension implicit in Christian faith today. On the one hand, Christians remain convinced that God sees the material world as good and intends a fulfilling life for everyone. On the other hand, even a moderate modern lifestyle harms the environment—threatening the integrity of God's creation—and of itself does little to address the grinding poverty of so many of our fellow inhabitants on the globe. There is great moral risk in such a lifestyle, as there is in international trade. It is only with hard work to offset those threats to the integrity of creation and to better the prospects of the world's poor that Christians can (conditionally at best) justify such a lifestyle and the patterns of exchange that support it.

In addition to this integration of moral ambiguity within our view of the world, however, we must also acknowledge the positive effects of trade in the transfer of environmentally necessary technologies and in its potential to increase the economic well-being of the poor of the developing world. This recognition has led many in the environmental movement to part company with those who would reduce international trade because of its threat to the environment. A number of environmental organizations have come to share the view articulated in the Worldwatch Institute's *1993 State of the World* report.

Trade can be a tool for shaping a world that is ecologically sustainable and socially just. . . . Now policy makers must get on with the task of determining how the rules of trade can be revised to help achieve it.

For Further Discussion

1. Daniel Finn mentions the five distinct challenges to free trade raised by environmentalists. Summarize these in your own terms. Have you previously thought about any of these?

2. Why does economic theory predict that resource depletion will never be complete and that depletion occurs very slowly, far more slowly than simply extrapolation of current trends would indicate? Do you think that Finn agrees with Julian Simon (Chapter 3) that natural resources are infinite?

3. Why does Finn reject "glib optimism" in the face of broad scientific uncertainties about the future availability of resources? Do you agree with him?

4. Finn tells us that in the parts of the world where forests are most severely threatened, the vast majority of the wood harvested is used as fuel for home cooking and heating. Does this affect your own views on deforestation? Why or why not?

5. Finn is convinced that higher energy prices would result in lower energy consumption and thus less pollution as a result of fossil fuel use. The United States and Canada have significantly lower energy prices that other industrialized countries. Should the United States and Canada increase the taxes on gasoline? Would you support such a decision as a citizen? Would you willingly pay more as a consumer?

6. How might Finn answer Athanasiou's criticism of free trade?

DISCUSSION AND STUDY QUESTIONS FOR CHAPTER 14

1. Enforcement is a major challenge to any international agreement. Military action is an unattractive option. Trade sanctions can harm both sides, and often cause harm to poor and innocent citizens. That generally leaves persuasion and domestic political pressure as the most likely enforcement mechanisms for international agreements. What policy would you recommend for enforcing environmental agreements?

2. Do you believe that industrialized countries have a greater responsibility for implementing environmental policy than developing countries?

3. Do poorer, developing countries have as much of a right to exploit their domestic resources as did the industrialized countries in years past? Has the change in environmental consciousness created greater responsibilities today than existed a century ago?

4. What principles should govern management of migratory species? Should countries in which breeding grounds are located have priority of ownership?

5. Would you be willing to pay more for imported products as a means to encourage exporting countries to meet environmental standards? Should the United States create a system of environmental protection tariffs?

DISCUSSION CASES

Salmon Fishing

Few countries have had relations as amicable as the United States and Canada, but recent disputes over fishing rights have strained these relations considerably. Some go so far as to describe the dispute as a fishing war. During the summer of 1997, Canadian salmon fishermen blockaded and held captive for three days a U.S. ferry with 385 passengers on board in retaliation for alleged overfishing by U.S. salmon fishermen.

In 1985 the United States and Canada signed a treaty regulating salmon fishing along the Pacific Coast from Oregon to Alaska. The need for such a treaty stemmed from two facts: Salmon migrate from the waters of one country through the international waters of the Pacific Ocean and into the waters of the other country, and salmon numbers are in significant decline. Salmon fishing is a major industry in Alaska, Washington, and British Columbia. Declining salmon populations have been traced to a number of causes. Development, and especially construction of dams, along the banks of rivers, which the salmon use as spawning grounds, is a significant factor. Soil run-off and pollution, in both rivers and the ocean, is also a factor. But many view overfishing as the major cause.

The migratory patterns of the salmon do not respect national borders. The fish spawn far inland in rivers in both countries. Young fish migrate to the ocean where they may live for many years before returning to their original river system to spawn and die. Overfishing in the ocean lowers the numbers of fish returning upriver and significantly reduces the fish harvest during spawning runs. Overfishing during these spawning runs reduces the population of new fish significantly. The 1985 Pacific Salmon Treaty committed each country to manage their fisheries to ensure that "each party receives benefits equivalent to the production of salmon originating in its waters."

Commercial salmon catch has declined sharply along many rivers, and the Canadian fishermen attribute the decline to U.S. fishing practices. Canadian fishermen accuse Alaskan and Washington State fishermen of overfishing salmon in the Pacific. The U.S. fishermen claim that salmon out in the ocean should be available to anyone.

Many animal species—fish, whales, birds—migrate across national borders. On what grounds should rights to migratory animals be allocated?

Commercial and Subsistence Whaling

The International Whaling Commission was established in 1946 to "provide for the proper conservation of whale stocks and thus make possible the orderly development of the whaling industry." Like many local fish and game agencies, the mission of

(*continue*)

the IWC was to manage whaling on conservation principles. Properly regulated, whaling could continue indefinitely. By the mid-1980s, it was clear that the IWC had failed in this mission, the population of many whale species—including the blue, right, humpback, sperm, and fin whales—had been devastated. Oceangoing factory ships, most from Japan, Russia, and Norway, had pushed these species to near extinction. In 1985, the IWC established a moratorium on commercial whaling to give whale species time to recover.

The Norwegians disagreed with the initial moratorium and chose not to participate; they continued to hunt whales, especially the minke whales off their coast. The Japanese claimed the right to continue to kill up to 300 minke whales each year for what they described as scientific research. The Icelanders argued that the IWC had ignored their pressing economic concerns and resigned from the IWC in 1992. IWC countries may impose trade sanctions on countries that disregard the moratorium. Private groups such as Greenpeace have called for boycotts of Norwegian and Japanese products, but such steps have had little effect. Few other legal actions are possible to enforce the moratorium. In November 1997, a whaling ship owned by one of Norway's best-known whalers was sunk at its dock. Press reports suggested that the Sea Shepherds, an anti-whaling group known to support acts of eco-sabotage, were responsible for the sinking.

Despite the moratorium, the IWC has allowed continued subsistence whaling by indigenous peoples. This policy is defended on both ethical grounds (respecting local cultural practices) and conservation grounds (catching a limited number of whales). The amount of such whaling is restricted, and the allowable uses of the whale products are limited. With the legal support of the indigenous peoples' national governments, the IWC requires that the whale products be used solely for local consumption by local people.

During its meeting in October 1997, the IWC granted the Makah Indians of Washington State a quota of harvesting four gray whales per year. The Makah people have not whaled in over seventy years. Critics charge that Japanese whalers hope to use quotas granted to indigenous peoples as a means for avoiding the international moratorium.

Many indigenous peoples themselves claim that the restrictions placed on their use of whales are unfair and oppressive. If they are allowed to hunt whales for subsistence, should they be allowed to sell those whales to others in order to earn cash that can then be used to buy the goods and services they need?

Polluted Borders

Nogales, Arizona, is a border town located 60 miles south of Tucson and adjacent to Nogales, Mexico. During the time when the North American Free Trade Agreement (NAFTA) between Canada, the United States, and Mexico was being approved, attention was focused on Nogales as an example of what could result from free trade.

Nogales can be thought of as a single large city, mostly located in the Mexican state of Sonora with a smaller section lying across the U.S. border in Arizona. Since 1967, about the time that air and water pollution laws were first being considered in the United States, more than ninety primarily U.S.-owned factories have located on the Mexican side of Nogales. Many of these factories produce plastic and electronic products that generate significant toxic waste. By being located on one side of the city rather than the other, these factories can avoid a variety of U.S. environmental, health, and employment laws. There are, of course, costs associated with locating in Mexico rather than in the United States, many of which are removed with the passage of NAFTA.

Many citizens of Nogales, Arizona, claim that health problems—including a reported rise in cancer and lupus rates and an increase in birth defects—are caused by chemical pollution coming from the Mexican side of the border. The Nogales Wash, a small tributary of the Santa Cruz River that flows through town, is so polluted that it actually caught fire in May 1991. Air pollution in Nogales exceeds pollution levels of Phoenix, a city with a population fifty times its size.

Health and pollution data from the Mexican side of the border is less available. The population on the Sonoran side has grown tremendously in recent decades as thousands of poor people have moved into the area in search of jobs. Many of these people live in shantytowns, called *colonias,* without

(continue)

running water and sewage systems. Lacking education and job security, few workers on the Mexican side are likely to report adverse health effects to authorities. Faced with strong economic pressures, Mexican authorities have little incentive to investigate or police pollution.

Ozone Depletion and Greenhouse Gases

Many people regard the Montreal Protocol, the 1987 agreement that limited the production and use of ozone-depleting chemicals, as a model of international cooperation. Industrialized countries have met their timetables, developing countries have reduced their reliance on such chemicals, and, in the time since its original passage, most of the world's countries have signed on to the agreement. Early evidence suggests that the levels of ozone-depleting gases in the atmosphere have been declining.

Environmentalists have hoped that the Montreal Protocol could be used as a prototype for international agreements limiting the production of greenhouse gases—mostly carbon dioxide and other gases that can trap solar energy within the atmosphere and that have been identified as a possible cause of global warming. Such optimism has turned out to be misplaced.

In 1992, the United States and most other countries signed a treaty committing industrialized countries to reduce their emissions of carbon dioxide and other greenhouse gases back to 1990 levels by the year 2000. In December 1997, 160 nations met in Kyoto, Japan, to negotiate an international treaty to limit greenhouse gases to prevent global warming. In late October 1997, President Clinton presented the plan that the United States would take to Kyoto. Described by the administration as far-reaching and meaningful, the U.S. plan would not meet the target of 1990-level emissions until sometime after 2010. Since 1990, the United States has increased in greenhouse gas emissions by 8 percent. A Department of Energy report in 1997 showed that greenhouse emissions increased by 3.4 percent in 1996 alone.

Why is it more difficult to reach agreement on greenhouse gases than it was on ozone-depleting gases? Some point out that the threat to human health from ozone depletion was regarded as more direct than the potential threat posed by global warming. Others suggest that the connection of CFCs to ozone depletion is scientifically clearer and more direct than the connection between greenhouse gases and global warming. It is also true that ozone-depleting chemicals represented a minor part of the world economy, and there were alternative technologies available. Carbon dioxide, the major greenhouse gas, is the by-product of burning fossil fuels. Any significant reduction in carbon dioxide would require major economic changes worldwide. Without alternatives readily available, few countries are willing to take drastic steps to reduce carbon emissions.

In response to the Clinton announcement, Meg McDonald, the Australian Ambassador for the Environment, stated, "We think it is better to do what is realistic than to have unrealistic targets which are never reached." Critics charge that without such targets, industry will have little incentive to develop alternative technologies.

PART IV *Philosophy and Theory of Environmental Ethics*

In Chapter 1, we spoke of different levels of ethics. At the first level, ethics refers to those beliefs, values, and attitudes that do, in fact, guide behavior. In this sense, ethics is not far removed from the original Greek meaning of *ethos:* customs and social conventions. We spoke also of normative ethics as the first level of stepping back from the customary and appealing to various norms to explain and justify ethical behavior. At a still more abstract level, philosophical ethics attempts to synthesize ethical norms and beliefs into a comprehensive and systematic theory. In the following four chapters, we consider whether it is possible to articulate a systematic and comprehensive environmental theory or environmental philosophy.

Philosophers have long attempted to provide comprehensive ethical theories. From at least as far back as Plato, a major goal of philosophical ethics has been to integrate an account of the good life, of how we should live, with more general accounts of human nature (metaphysics) and rationality (epistemology). To many, philosophical ethics would be incomplete if it did not situate ethics within a broader philosophical worldview.

But why the need for a systematic theory? This challenge is the same whether asked of ethics in general or environmental ethics in particular. In the view of many philosophers, unless we can situate our positions regarding various environmental issues within a unified and coherent framework, we have not answered the most fundamental of all philosophical questions: Why? As you have worked through this book, you have probably found yourself inclined one way or another on each issue. I would hope that this inclination has been increasingly the result of reasoned analysis rather than mere subjective preference. But, has this reasoning developed out of a systematic framework or not? Are there reasons why you hold one particular set of positions rather than another? Are your views on moral standing for animals consistent with your views on pollution? Are such issues at all related, or are your views simply a set of independently arrived at conclusions? If they are related, how are they related? If they are independent, why did you end up with these conclusions rather than others? If you answer that they are not related, this suggests that your ethical views are arbitrary, and that is worrisome. Lack of a unifying framework seems to suggest that things could have turned out differently, and this suggests that there is no underlying reason or objectivity to your views.

Consider as an example the voting record of a member of Congress. Imagine reviewing this record and listing all of the separate votes on hundreds of different issues. Would you be able to find a set of principles or values—a political philosophy—that unifies these different votes? Imagine that you couldn't. No doubt, you would be troubled by this. Shouldn't you be able to discern an underlying perspective that explains each vote? Shouldn't that perspective be consistent across each issue? If so, then we would be on our way to finding a political philosophy. If not, we might reasonably question this legislator's integrity or rationality.

An individual's position on the various policy issues that we have examined is like the voting record. The broad philosophical perspectives examined in the following four chapters can be understood as reflections on the underlying political philosophy. As we shall see, not everyone agrees that such a unifying theory can or should be found.

Chapter 15 introduces three philosophical perspectives that have received significant attention among environmental philosophers. Deep ecology, social ecology, and ecofeminism are often characterized as radical ecophilosophies because each calls attention to deeper social and cultural roots of environmental problems. Stepping back from environmental policy issues, these philosophical theories all suggest that environmental controversies are symptomatic of broader questions and controversies. The theories differ in their analysis of these underlying causes.

Chapter 16 reflects on environmental policy from a political rather than philosophical perspective. Perhaps more in line with the voting record analogy used above and using the more common terms of conservative and liberal politics, this chapter examines environmentalism from the political perspective to consider if there is or can be an underlying political philosophy of environmentalism.

Chapter 17 steps back from environmental policy in yet another way to ask about the cultural underpinnings and implications of environmentalism. Readers of this book are from the United States and Canada almost exclusively. Even recognizing minority voices and minority cultures within these countries, the fact remains that these two countries share a dominant cultural perspective. Along with some previous readings that point out the limitations inherent in that perspective, Chapter 17 provides occasions for more systematic reflections on the cultural assumptions of much environmental policy.

Finally, Chapter 18 asks us to step back one more time. If environmental philosophers and theorists step back from policy questions to look for a comprehensive and unified theory, environmental pragmatists step back even further. Environmental pragmatists challenge us to consider if theory is necessary, desirable, or even possible. With environmental pragmatism, we have come full circle. Theory and practice, philosophy and policy, continue to pull us in different directions. Although each is necessary and worthy, reconciliation is perhaps the ultimate philosophical challenge.

CHAPTER 15

Deep Ecology, Social Ecology, and Ecofeminism

The phrase "deep ecology" was first used by philosopher Arne Naess in his 1973 article, "The Shallow and the Deep, Long-Range Ecology Movements." Naess described the shallow ecological approach, the approach that was most common in the early 1970s, as committed "to the fight against pollution and resource depletion." The shallow approach is clearly anthropocentric, focusing on environmental issues that threaten human well-being. Perhaps more precisely, issues such as pollution and resource depletion threaten the lifestyles of many humans. Deep ecology, on the other hand, examines the underlying causes of environmental destruction. The three philosophical approaches considered in this chapter share the assumption that there is a deeper, more fundamental explanation of environmental destruction.

As discussed previously, an environmental philosophy seeks to articulate a perspective that unifies the wide range of environmental issues. Philosophers who study deep ecology, social ecology, and ecofeminism agree that there is a way to unify these diverse issues: Environmental and ecological troubles are manifestations of broader cultural and social problems. Just as a sneeze and a cough are symptomatic of an underlying medical ailment, pollution and resource depletion are symptoms of an underlying cultural ailment. Unless we understand and address the roots of our environmental crisis, we have little hope of any long-term and satisfactory solution.

Although followers of these three environmental philosophies agree that there is an underlying factor that explains the variety of environmental problems, they disagree on what that underlying factor is. Deep ecologists attribute the environmental crisis to a general worldview. Social ecologists and ecofeminists blame instead social practices of domination and control; social ecologists claim the culprit is general patterns of hier-archy and dominance, whereas ecofeminists attribute environmental problems more specifically to gender-based hierarchies and oppression.

Deep ecologists believe that the dominant worldview of technocratic-industrial societies is responsible for most environmental destruction. This worldview regards humans as separate from and superior to the natural world. It justifies an attitude not only of superiority but of dominance as well. Deep ecology is an attempt to articulate a comprehensive religious and philosophical alternative to this worldview of dominance and separateness. Deep ecologists often address questions of metaphysics and epistemology in their search for an alternative worldview.

Social ecologists and ecofeminists share some of these beliefs. They, too, believe that patterns of domination and control must be addressed as part of an environmental philosophy. But unlike the deep ecologists, they tend to be more focused on social and political issues. Social ecologists have been particularly concerned with distancing themselves from the more abstract musings of deep ecology. Both social ecologists and ecofeminists are troubled by misanthropic (anti-human) tendencies in the writings of some self-described deep ecologists.

Social ecology sees environmental destruction as part of a more general pattern of social domination. Particular social arrangements tend to reinforce attitudes that justify the domination of others. Until and unless we address these social problems—racism, classism, sexism, for example—we will fail to resolve the many environmental problems that plague us. For social ecologists, environmental problems are fundamentally problems of social justice.

Ecofeminists sympathize with these conclusions. In general, however, they identify the oppression of women as a major form of social domination. It is important to recognize that there are a variety of approaches that can be identified as feminist and therefore a variety of approaches that might be ecofeminist. Some ecofeminists conclude that the oppression of women is the single most important form of social domination; others hold that women's oppression is but one among many oppressive practices. Some argue that a distinctive "women's way" of knowing or valuing should replace the dominant masculine epistemology

and ethics; others reject such dualistic ways of thinking altogether. The unifying theme of ecofeminism, however, is that there are important connections between the domination of nature and the domination of women. These connections are ethical, political, epistemological, symbolic, and practical. The social problems associated with environmentalism and justice for women should be addressed simultaneously.

Deep Ecology

Bill Devall
George Sessions

Sociologist Bill Devall and philosopher George Sessions provide an overview of deep ecology in this first selection of Chapter 15. They seek an alternative to the "dominant worldview of technocratic-industrial societies." That worldview sees humans as separate from and superior to nature. Instead, Devall and Sessions explain two fundamental norms of deep ecology: self-realization and biocentric equality. Self-realization is a concept familiar to a long-standing tradition in ethics: Human beings have a potential not fully realized in our day-to-day lives. The ethical life should be spent attempting to realize or fulfill that potential. ("Be all that you can be" reminds the U.S. Army.) Deep ecologists suggest that a more ecologically informed understanding of the self provides a deeper interpretation of the goal of self-realization. Likewise, biocentric equality is a more complete understanding of the traditional ethical commitment to equality. From the point of view of the ecosphere, all organisms are equal. These fundamental norms underlie the more practical principles of deep ecology. Devall and Sessions outline eight such general principles in the following reading, which is from their book Deep Ecology.

The term *deep ecology* was coined by Arne Naess in his 1973 article, "The Shallow and the Deep, Long-Range Ecology Movements." Naess was attempting to describe the deeper, more spiritual approach to Nature exemplified in the writings of Aldo Leopold and Rachel

Carson. He thought that this deeper approach resulted from a more sensitive openness to ourselves and non-human life around us. The essence of deep ecology is to keep asking more searching questions about human life, society, and Nature as in the Western philosophical tradition of Socrates. As examples of this deep questioning, Naess points out "that we ask why and how, where others do not. For instance, ecology as a science does not ask what kind of a society would be the best for maintaining a particular ecosystem—that is considered a question for value theory, for politics, for ethics." Thus deep ecology goes beyond the so-called factual scientific level to the level of self and Earth wisdom.

Deep ecology goes beyond a limited piecemeal shallow approach to environmental problems and attempts to articulate a comprehensive religious and philosophical worldview. The foundations of deep ecology are the basic intuitions and experiencing of ourselves and Nature which comprise ecological consciousness. Certain outlooks on politics and public policy flow naturally from this consciousness. And in the context of this book, we discuss the minority tradition as the type of community most conducive both to cultivating ecological consciousness and to asking the basic questions of values and ethics addressed in these pages.

Many of these questions are perennial philosophical and religious questions faced by humans in all cultures over the ages. What does it mean to be a unique human individual? How can the individual self maintain and increase its uniqueness while also being an inseparable aspect of the whole system wherein there are no sharp breaks between self and the *other*? An ecological perspective, in this deeper sense, results in what Theodore Roszak calls "an awakening of wholes greater than the sum of their parts. In spirit, the discipline is contemplative and therapeutic."

Ecological consciousness and deep ecology are in sharp contrast with the dominant worldview of technocratic-industrial societies which regard humans as isolated and fundamentally separate from the rest of Nature, as superior to, and in charge of, the rest of creation. But the view of humans as separate and superior to the rest of Nature is only part of larger cultural patterns. For thousands of years, Western culture has become increasingly obsessed with the idea of *dominance:* with dominance of humans over nonhuman Nature, masculine over the feminine, wealthy and powerful over the poor, with the dominance of the West over non-Western cultures. Deep ecological consciousness allows us to see through these erroneous and dangerous illusions.

For deep ecology, the study of our place in the Earth household includes the study of ourselves as part of the organic whole. Going beyond a narrowly materialist scientific understanding of reality, the spiritual and the material aspects of reality fuse together. While the leading intellectuals of the dominant worldview have tended to view religion as "just superstition," and have looked upon ancient spiritual practice and enlightenment, such as found in Zen Buddhism, as essentially subjective, the search for deep ecological consciousness is the search for a more objective consciousness and state of being through an active deep questioning and meditative process and way of life.

Many people have asked these deeper questions and cultivated ecological consciousness within the context of different spiritual traditions—Christianity, Taoism, Buddhism, and Native American rituals, for example. While differing greatly in other regards, many in these traditions agree with the basic principles of deep ecology.

Warwick Fox, an Australian philosopher, has succinctly expressed the central intuition of deep ecology: "It is the idea that we can make no firm ontological divide in the field of existence: That there is no bifurcation in reality between the human and the non-human realms . . . to the extent that we perceive boundaries, we fall short of deep ecological consciousness."

From this most basic insight or characteristic of deep ecological consciousness, Arne Naess has developed two *ultimate norms* or intuitions which are themselves not derivable from other principles or intuitions. They are arrived at by the deep questioning process and reveal the importance of moving to the philosophical and religious level of wisdom. They cannot be validated, of course, by the methodology of modern science based on its usual mechanistic assumptions and its very narrow definition of data. These ultimate norms are *self-realization* and *biocentric equality.*

I. SELF-REALIZATION

In keeping with the spiritual traditions of many of the world's religions, the deep ecology norm of self-realization goes beyond the modern Western *self* which is defined as an isolated ego striving primarily for hedonistic gratification or for a narrow sense of individual salvation in this life or the next. This socially programmed sense of the narrow self or social self dislocates us, and leaves us prey to whatever fad or fashion is prevalent in our society or social reference group. We are thus robbed of beginning the search for our unique spiritual/biological personhood. Spiritual growth, or unfolding, begins when we cease to understand or see ourselves as isolated and narrow competing egos and begin to identify with other humans from our family and friends to, eventually, our species. But the deep ecology sense of self requires a further maturity and growth, an identification which goes beyond humanity to include the nonhuman world. We must see beyond our narrow contemporary cultural assumptions and values, and the conventional wisdom of our time and place, and this is best achieved by the meditative deep questioning process. Only in this way can we hope to attain full mature personhood and uniqueness.

A nurturing nondominating society can help in the "real work" of becoming a whole person. The "real work" can be summarized symbolically as the realization of "self-in-Self" where "Self" stands for organic wholeness. This process of the full unfolding of the self can also be summarized by the phrase, "No one is saved until we are all saved," where the phrase "one" includes not only me, an individual human, but all humans, whales, grizzly bears, whole rain forest ecosystems, mountains and rivers, the tiniest microbes in the soil, and so on.

II. BIOCENTRIC EQUALITY

The intuition of biocentric equality is that all things in the biosphere have an equal right to live and blossom and to reach their own individual forms of unfolding and self-realization within the larger Self-realization. This basic intuition is that all organisms and entities in the ecosphere, as parts of the interrelated whole, are equal in intrinsic worth. Naess suggests that biocentric equality as an intuition is true in principle, although in the process of living, all species use each other as food, shelter, etc. Mutual predation is a biological fact of life, and many of the world's religions have struggled with the spiritual implications of this. Some animal liberationists who attempt to side-step this problem by advocating vegetarianism are forced to say that the entire plant kingdom including rain forests have no right to their own existence. This evasion flies in the face of the basic intuition of equality. Aldo Leopold expressed this intuition when he said humans are "plain citizens" of the biotic community, not lord and master over all other species.

Biocentric equality is intimately related to the all-inclusive Self-realization in the sense that if we harm the rest of Nature then we are harming ourselves. There are no boundaries and everything is interrelated. But insofar as we perceive things as individual organisms or entities, the insight draws us to respect all human and nonhuman individuals in their own right as parts of the whole without feeling the need to set up hierarchies of species with humans at the top.

The practical implications of this intuition or norm suggest that we should live with minimum rather than maximum impact on other species and on the Earth in general. Thus we see another aspect of our guiding principle: "simple in means, rich in ends." . . .

A fuller discussion of the biocentric norm as it unfolds itself in practice begins with the realization that we, as individual humans, and as communities of humans, have vital needs which go beyond such basics as food, water, and shelter to include love, play, creative expression, intimate relationships with a particular landscape (or Nature taken in its entirety) as well as intimate relationships with other humans, and the vital need for spiritual growth, for becoming a mature human being.

Our vital material needs are probably more simple than many realize. In technocratic-industrial societies there is overwhelming propaganda and advertising which encourages false needs and destructive desires designed to foster increased production and consumption of goods. Most of this actually diverts us from facing reality in an objective way and from beginning the "real work" of spiritual growth and maturity.

Many people who do not see themselves as supporters of deep ecology nevertheless recognize an overriding vital human need for a healthy and high-quality natural environment for humans, if not for all life, with minimum intrusion of toxic waste, nuclear radiation from human enterprises, minimum acid rain and smog, and enough free flowing wilderness so humans can get in touch with their sources, the natural rhythms and the flow of time and place.

Drawing from the minority tradition and from the wisdom of many who have offered the insight of interconnectedness, we recognize that deep ecologists can offer suggestions for gaining maturity and encouraging the processes of harmony with Nature, but that there is no grand solution which is guaranteed to save us from ourselves.

The ultimate norms of deep ecology suggest a view of the nature of reality and our place as an individual (many in the one) in the larger scheme of things. They cannot be fully grasped intellectually but are ultimately experiential. . . .

As a brief summary of our position thus far, [the] figure summarizes the contrast between the dominant worldview and deep ecology.

Dominant Worldview	Deep Ecology
Dominance over Nature	Harmony with Nature
Natural environment as resource for humans	All nature has intrinsic worth/biospecies equality
Material/economic growth for growing human population	Elegantly simple material needs (material goals serving the larger goal of self-realization)
Belief in ample resource reserves	Earth "supplies" limited
High technological progress and solutions	Appropriate technology; nondominating science
Consumerism	Doing with enough/recycling
National/centralized community	Minority tradition/bioregion

III. BASIC PRINCIPLES OF DEEP ECOLOGY

In April 1984, during the advent of spring and John Muir's birthday, George Sessions and Arne Naess summarized fifteen years of thinking on the principles of deep ecology while camping in Death Valley, California. In this great and special place, they articulated these principles in a literal, somewhat neutral way, hoping that they would be understood and accepted by persons coming from different philosophical and religious positions.

Readers are encouraged to elaborate their own versions of deep ecology, clarify key concepts and think through the consequences of acting from these principles.

Basic Principles

1. The well-being and flourishing of human and nonhuman Life on Earth have value in themselves (synonyms: intrinsic value, inherent value). These values are independent of the usefulness of the nonhuman world for human purposes.

2. Richness and diversity of life forms contribute to the realization of these values and are also values in themselves.

3. Humans have no right to reduce this richness and diversity except to satisfy *vital* needs.

4. The flourishing of human life and cultures is compatible with a substantial decrease of the human population. The flourishing of nonhuman life requires such a decrease.

5. Present human interference with the nonhuman world is excessive, and the situation is rapidly worsening.

6. Policies must therefore be changed. These policies affect basic economic, technological, and ideological structures. The resulting state of affairs will be deeply different from the present.

7. The ideological change is mainly that of appreciating *life quality* (dwelling in situations of inherent value) rather than adhering to an increasingly higher standard of living. There will be a profound awareness of the difference between big and great.

8. Those who subscribe to the foregoing points have an obligation directly or indirectly to try to implement the necessary changes.

Naess and Sessions Provide Comments on the Basic Principles

RE (1). This formulation refers to the biosphere, or more accurately, to the ecosphere as a whole. This includes individuals, species, populations, habitat, as well as human and nonhuman cultures. From our current knowledge of all-pervasive intimate relationships, this implies a fundamental deep concern and respect. Ecological processes of the planet should, on the whole, remain intact. "The world environment should remain 'natural'" (Gary Snyder).

The term "life" is used here in a more comprehensive nontechnical way to refer also to what biologists classify as "nonliving"; rivers (watersheds), landscapes, ecosystems. For supporters of deep ecology, slogans such as "Let the river live" illustrate this broader usage so common in most cultures.

Inherent value as used in (1) is common in deep ecology literature. ("The presence of inherent value in a natural object is independent of any awareness, interest, or appreciation of it by a conscious being.")

RE (2). More technically, this is a formulation concerning diversity and complexity. From an ecological standpoint, complexity and symbiosis are conditions for maximizing diversity. So-called simple, lower, or primitive species of plants and animals contribute essentially to the richness and diversity of life. They have value in themselves and are not merely steps toward the so-called higher or rational life forms. The second principle presupposes that life itself, as a process over evolutionary time, implies an increase of diversity and richness. The refusal to acknowledge that some life forms have greater or lesser intrinsic value than others (see points 1 and 2) runs counter to the formulations of some ecological philosophers and New Age writers.

Complexity, as referred to here, is different from complication. Urban life may be more complicated than life in a natural setting without being more complex in the sense of multifaceted quality.

RE (3). The term "vital need" is left deliberately vague to allow for considerable latitude in judgment. Differences in climate and related factors, together with differences in the structures of societies as they now exist, need to be considered (for some Eskimos, snow-mobiles are necessary today to satisfy vital needs).

People in the materially richest countries cannot be expected to reduce their excessive interference with the nonhuman world to a moderate level overnight. The stabilization and reduction of the human population will take time. Interim strategies need to be developed. But this in no way excuses the present complacency—the extreme seriousness of our current situation must first be realized. But the longer we wait the more drastic will be the measures needed. Until deep changes are made, substantial decreases in richness and diversity are liable to occur: the rate of extinction of species will be ten to one hundred times greater than any other period of earth history.

RE (4). The United Nations Fund for Population Activities in their State of World Population Report (1984) said that high human population growth rates (over 2.0 percent per annum) in many developing countries "were diminishing the quality of life for many millions of people." During the decade 1974–1984, the world population grew by nearly 800 million—more than the size of India. "And we will be adding about one Bangladesh (population 93 million) per annum between now and the year 2000."

The report noted that "The growth rate of the human population has declined for the first time in human history. But at the same time, the number of people being added to the human population is bigger than at any time in history because the population base is larger."

Most of the nations in the developing world (including India and China) have as their official government policy the goal of reducing the rate of human population increase, but there are debates over the types of measures to take (contraception, abortion, etc.) consistent with human rights and feasibility.

The report concludes that if all governments set specific population targets as public policy to help alleviate poverty and advance the quality of life, the current situation could be improved.

As many ecologists have pointed out, it is also absolutely crucial to curb population growth in the so-called developed (i.e., overdeveloped) industrial societies. Given the tremendous rate of consumption and waste production of individuals in these societies, they represent a much greater threat and impact on the biosphere per capita than individuals in Second and Third World countries.

RE (5). This formulation is mild. For a realistic assessment of the situation, see the unabbreviated version of the I.U.C.N.'s *World Conservation Strategy*. There are other works to be highly recommended, such as Gerald Barney's *Global 2000 Report to the President of the United States*.

The slogan of "noninterference" does not imply that humans should not modify some ecosystems as do other species. Humans have modified the earth and will probably continue to do so. At issue is the nature and extent of such interference.

The fight to preserve and extend areas of wilderness or near-wilderness should continue and should focus on the general ecological functions of these areas (one such function: large wilderness areas are required in the biosphere to allow for continued evolutionary speciation of animals and plants). Most present designated wilderness areas and game preserves are not large enough to allow for such speciation.

RE (6). Economic growth as conceived and implemented today by the industrial states is incompatible with (1)–(5). There is only a faint resemblance between ideal sustainable forms of economic growth and present policies of the industrial societies. And "sustainable" still means "sustainable in relation to humans."

Present ideology tends to value things because they are scarce and because they have a commodity value. There is prestige in vast consumption and waste (to mention only several relevant factors).

Whereas "self-determination," "local community," and "think globally, act locally," will remain key terms in the ecology of human societies, nevertheless the implementation of deep changes requires increasingly global action—action across borders.

Governments in Third World countries (with the exception of Costa Rica and a few others) are uninterested in deep ecological issues. When the governments of industrial societies try to promote ecological measures through Third World governments, practically nothing is accomplished (e.g., with problems of desertification). Given this situation, support for global action through nongovernmental international organizations becomes increasingly important. Many of these organizations are able to act globally "from grassroots to grassroots," thus avoiding negative governmental interference.

Cultural diversity today requires advanced technology, that is, techniques that advance the basic goals of each culture. So-called soft, intermediate, and alternative technologies are steps in this direction.

RE (7). Some economists criticize the term "quality of life" because it is supposed to be vague. But on closer inspection, what they consider to be vague is actually the nonquantitative nature of the term. One cannot quantify adequately what is important for the quality of life as discussed here, and there is no need to do so.

RE (8). There is ample room for different opinions about priorities: what should be done first, what next? What is most urgent? What is clearly necessary as opposed to what is highly desirable but not absolutely pressing?

Interview with Arne Naess

The following excerpts are from an interview with Arne Naess conducted at the Zen Center of Los Angeles in April 1982. It was originally published as an interview in *Ten Directions*. In the interview, Naess further discusses the major perspective of deep ecology. . . .

The essence of deep ecology is to ask deeper questions. The adjective "deep" stresses that we ask why and how, where others do not. For instance, ecology as a science does not ask what kind of a society would be the best for maintaining a particular ecosystem—that is considered a question for value theory, for politics, for ethics. As long as ecologists keep narrowly to their science, they do not ask such questions. What we need today is a tremendous expansion of ecological thinking in what I call ecosophy. *Sophy* comes from the Greek term *sophia,* "wisdom," which relates to ethics, norms, rules, and practice. Ecosophy, or deep ecology, then, involves a shift from science to wisdom.

For example, we need to ask questions like, Why do we think that economic growth and high levels of consumption are so important? The conventional answer would be to point to the economic consequences of not having economic growth. But in deep ecology, we ask whether the present society fulfills basic human needs like love and security and access to nature, and, in so doing, we question our society's underlying assumptions. We ask which society, which education, which form of religion, is beneficial for all life on the planet as a whole, and then we ask further what we need to do in order to make the necessary changes. We are not limited to a scientific approach; we have an obligation to verbalize a total view.

Of course, total views may differ. Buddhism, for example, provides a fitting background or context for deep ecology, certain Christian groups have formed platforms of action in favor of deep ecology, and I myself have worked out my own philosophy, which I call ecosophy. In general, however, people do not question deeply enough to explicate or make clear a total view. If they did, most would agree with saving the planet from the destruction that's in progress. A total view, such as deep ecology, can provide a single motivating force for all the activities and movements aimed at saving the planet from human exploitation and domination.

. . . It's easier for deep ecologists than for others because we have certain fundamental values, a fundamental view of what's meaningful in life, what's worth maintaining, which makes it completely clear that we're opposed to further development for the sake of increased domination and an increased standard of living. The material standard of living should be drastically reduced and the quality of life, in the sense of basic satisfaction in the depths of one's heart or soul, should be maintained or increased. This view is intuitive, as are all important views, in the sense that it can't be proven. As Aristotle said, it shows a lack of education to try to prove everything, because you have to have a starting point. You can't prove the methodology of science, you can't prove logic, because logic presupposes fundamental premises.

All the sciences are fragmentary and incomplete in relation to basic rules and norms, so it's very shallow to think that science can solve our problems. Without basic norms, there is no science.

. . . People can then oppose nuclear power without having to read thick books and without knowing the myriad facts that are used in newspapers and periodicals. And they must also find others who feel the same and form circles of friends who give one another confidence and support in living in a way that the majority find ridiculous, naive, stupid and simplistic. But in order to do that, one must already have enough self-confidence to follow one's intuition—a quality very much lacking in broad sections of the populace. Most people follow the trends and advertisements and become philosophical and ethical cripples.

There is a basic intuition in deep ecology that we have no right to destroy other living beings without sufficient reason. Another norm is that, with maturity, human beings will experience joy when other life forms experience joy and sorrow when other life forms experience sorrow. Not only will we feel sad when our brother or a dog or a cat feels sad, but we will grieve when living beings, including landscapes, are destroyed. In our civilization, we have vast means of destruction at our disposal but extremely little maturity in our feelings. Only a very narrow range of feelings have interested most human beings until now.

For deep ecology, there is a core democracy in the biosphere. . . . In deep ecology, we have the goal not only of stabilizing human population but also of reducing it to a sustainable minimum without revolution or dictatorship. I should think we must have no more than 100 million people if we are to have the variety of cultures we had one hundred years ago. Because we need the conservation of human cultures, just as we need the conservation of animal species.

. . . Self-realization is the realization of the potentialities of life. Organisms that differ from each other in three ways give us less diversity than organisms that differ from each other in one hundred ways. Therefore, the self-realization we experience when we identify with the universe is heightened by an increase in the number of ways in which individuals, societies, and even species and life forms realize themselves. The greater the diversity, then, the greater the self-realization. This seeming duality between individuals and the totality is encompassed by what I call the Self and the Chinese call the Tao. Most people in deep ecology have had the feeling—usually, but not always, in nature—that they are connected with something greater than their ego, greater than their name, their family, their special attributes as an individual—a feeling that is often called oceanic because many have it on the ocean. Without that identification, one is not easily drawn to become involved in deep ecology. . . .

. . . Insofar as these deep feelings are religious, deep ecology has a religious component, and those people who have done the most to make societies aware of the destructive way in which we live in relation to natural settings have had such religious feelings. Rachel Carson, for example, says that we *cannot* do what we do, we have no religious or ethical justification for behaving as we do toward nature. . . . She is saying that we are simply not permitted to behave in that way. Some will say that nature is not man's property, it's the property of God; others will say it in other ways. The main point is that deep ecology has a religious component, fundamental intuitions that everyone must cultivate if he or she is to have a life based on values and not function like a computer.

. . . To maximize self-realization—and I don't mean self as ego but self in a broader sense—we need maximum diversity and maximum symbiosis. . . . Diversity, then, is a fundamental norm and a common delight. As deep ecologists, we take a natural delight in diversity, as long as it does not include crude, intrusive forms, like Nazi culture, that are destructive to others.

For Further Discussion

1. How would you explain the difference between deep ecology and shallow ecology to a friend? Do you think that the typical environmentalist is deep or shallow?

2. How would you react if someone characterized your environmental views as shallow?

3. Could you come up with a list of five or ten characteristics of the dominant worldview that Devall and Sessions discuss?

4. Devall and Sessions quote the Australian philosopher Warwick Fox as suggesting that "we can make no firm ontological divide in the field of existence." What do you think he means by this? Do you agree?

5. Some critics claim the deep ecologists are misanthropic (anti-human). Do you find any evidence of this in this essay?

6. Do you disagree with any of the basic norms outlined by Devall and Sessions?

Social Ecology Versus Deep Ecology

Murray Bookchin

Murray Bookchin, a social and political theorist, has written about social and ecological issues for decades; he is the person most often associated with social ecology. Social ecologists focus on more specific social causes of the environmental crisis than do deep ecologists. In the following essay, Bookchin describes the differences between social ecology and deep ecology. Encouraging us to "look beyond the spiritual ecobabble" of some deep ecologists, Bookchin emphasizes the particular social context in which environmental destruction occurs. Bookchin likens the deep ecologists to "deep" Malthusians, a reference to Thomas Malthus (1766–1854). Bookchin sees Malthus as "an apologist for the misery of the Industrial Revolution," blaming poverty, starvation, and unhappiness on population growth alone. But such a view fails to make important distinctions among people. Not all people suffer from overpopulation; not everyone shares the misery equally. By blaming environmental destruction indiscriminately on humans, deep ecology fails to acknowledge that the benefits and burdens of industrial and consumerist society are not equally distributed. Bookchin believes that human decisions and human actions have allowed the domination of both nature and the poor and powerless. Human decisions and human actions must take responsibility for solving these injustices as well.

BEYOND "ENVIRONMENTALISM"

The environmental movement has travelled a long way beyond those annual "Earth Day" festivals when millions of school kids were ritualistically mobilized to clean up streets and their parents were scolded by Arthur Godfrey, Barry Commoner, and Paul Ehrlich. The movement has gone beyond a naive belief that patchwork reforms and solemn vows by EPA bureaucrats will seriously arrest the insane pace at which we are tearing down the planet.

This shopworn "Earth Day" approach toward "engineering" nature so that we can ravage the Earth with minimal effects on ourselves—an approach that

I called "environmentalism"—has shown signs of giving way to a more searching and radical mentality. Today, the new word in vogue is "ecology"—be it "deep ecology," "human ecology," "biocentric ecology," "anti-humanist ecology," or, to use a term uniquely rich in meaning, "*social ecology.*"

Happily, the new relevance of the word "ecology" reveals a growing dissatisfaction with attempts to use our vast ecological problems for cheaply spectacular and politically manipulative ends. Our forests disappear due to mindless cutting and increasing acid rain; the ozone layer thins out from widespread use of fluorocarbons; toxic dumps multiply all over the planet; highly dangerous, often radioactive pollutants enter into our air, water, and food chains. These innumerable hazards threaten the integrity of life itself, raising far more basic issues than can be resolved by "Earth Day" cleanups and faint-hearted changes in environmental laws.

For good reason, more and more people are trying to go beyond the vapid "environmentalism" of the early 1970s and toward an *ecological* approach: one that is rooted in an ecological philosophy, ethics, sensibility, image of nature, and, ultimately, an ecological movement that will transform our domineering market society into a nonhierarchical cooperative one that will live in harmony with nature, because its members live in harmony with each other. They are beginning to sense that there is a tie-in between the way people deal with each other as social beings—men with women, old with young, rich with poor, white with people of color, first world with third, elites with "masses"—and the way they deal with nature.

The questions that now face us are: what do we really mean by an *ecological* approach? What is a *coherent* ecological philosophy, ethics, and movement? How can the answers to these questions and many others *fit together* so that they form a meaningful and creative whole? If we are not to repeat all the mistakes of the early seventies with their hoopla about "population control," their latent anti-feminism, elitism, arrogance, and ugly authoritarian tendencies, so we must honestly and seriously appraise the new tendencies that today go under the name of one or another form of "ecology."

TWO CONFLICTING TENDENCIES

Let us agree from the outset that the word "ecology" is no magic term that unlocks the real secret of our abuse of nature. It is a word that can be as easily abused, distorted, and tainted as words like "democracy" and "freedom." Nor does the word "ecology" put us all—whoever "we" may be—in the same boat against environmentalists who are simply trying to make a rotten society work by dressing it in green leaves and colorful flowers, while ignoring the deep-seated *roots* of our ecological problems.

It is time to face the fact that there are differences within the so-called "ecology movement" of the present time that are as serious as those between the "environmentalism" and "ecologism" of the early seventies. There are barely disguised racists, survivalists, macho Daniel Boones, and outright social reactionaries who use the word "ecology" to express their views, just as there are deeply concerned naturalists, communitarians, social radicals, and feminists who use the word "ecology" to express theirs.

The differences between these two tendencies in the so-called "ecology movement" consists not only in quarrels over theory, sensibility, and ethics. They have far-reaching *practical* and *political* consequences on the way we view nature, "humanity," and ecology. Most significantly, they concern how we propose to *change* society and by what *means*.

The greatest differences that are emerging within the so-called "ecology movement" of our day are between a vague, formless, often self-contradictory ideology called "deep ecology" and a socially oriented body of ideas best termed "social ecology." Deep ecology has parachuted into our midst quite recently from the Sunbelt's bizarre mix of Hollywood and Disneyland, spiced with homilies from Taoism, Buddhism, spiritualism, reborn Christianity, and, in some cases, eco-fascism. Social ecology, on the other hand, draws its inspiration from such radical decentralist thinkers as Peter Kropotkin, William Morris, and Paul Goodman, among many others who have challenged society's vast hierarchical, sexist, class-ruled, statist, and militaristic apparatus.

Bluntly speaking, deep ecology, despite all its social rhetoric, has no real sense that our ecological problems have their roots in society and in social problems. It preaches a gospel of a kind of "original sin" that accuses a vague species called "humanity"—as though people of color were equatable with whites, women with men, the third world with the first, the poor with the rich, and the exploited with their exploiters. This vague, undifferentiated humanity is seen as an ugly "anthropocentric" thing—presumably a malignant product of natural evolution—that is "overpopulating" the planet, "devouring" its resources, destroying its wildlife and the biosphere. It assumes that some vague domain called "nature" stands opposed to a constellation of non-natural things called "human beings," with their "technology," "minds," "society," and so on. Formulated largely by privileged white male academics, deep ecology has brought sincere naturalists like Paul Shepard into the same company with patently anti-humanist and macho mountain-men like David Foreman, who writes in *Earth First!*—a Tucson-based journal that styles itself as the voice of a wilderness-oriented movement of the same name—that "humanity" is a cancer in the world of life.

It is easy to forget that this same kind of crude eco-brutalism led Hitler to fashion theories of blood and soil that led to the transport of millions of people to murder camps like Auschwitz. The same eco-brutalism now reappears a half-century later among self-professed deep ecologists who believe that famines are nature's "population control" and immigration into the US should be restricted in order to preserve "our" ecological resources.

Simply Living, an Australian periodical, published this sort of eco-brutalism as part of a laudatory interview of David Foreman by Professor Bill Devall, co-author of *Deep Ecology,* the manifesto of the deep ecology movement. Foreman, who exuberantly expressed his commitment to deep ecology, frankly informs Devall that

> When I tell people how the worst thing we could do in Ethiopia is to give aid—the best thing would be to just let nature seek its own balance, to let the people there just starve—they think this is monstrous. . . . Likewise, letting the USA be an overflow valve for problems in Latin America is not solving a thing. It's just putting more pressure on the resources we have in the USA.

One could reasonably ask what it means for "nature to seek its own balance" in a part of the world where agribusiness, colonialism, and exploitation have ravaged a once culturally and ecologically stable area like East Africa. And who is this all-American "our" that owns the "resources we have in the USA"? Is it the ordinary people who are driven by sheer need to cut timber, mine ores, operate nuclear power plants? Or are they the giant corporations that are not only wrecking the good old USA, but have produced the main problems in Latin America that are sending Indian folk across the Rio Grande? As an ex-Washington lobbyist and political huckster, David Foreman need not be expected to answer these subtle questions in a radical way. But what is truly surprising is the reaction—more precisely, the *lack* of any reaction—which marked Professor Devall's behavior. Indeed, the interview was notable for his almost reverential introduction and description of Foreman.

WHAT IS "DEEP ECOLOGY"?

Deep ecology is enough of a "black hole" of half-digested and ill-formed ideas that a man like Foreman can easily express utterly vicious notions and still sound like a fiery pro-ecology radical. The very words "deep ecology" clue us into the fact that we are not dealing with a body of clear ideas, but with an ideological toxic dump. Does it make sense, for example, to counterpost "deep ecology" with "superficial ecology" as though the word "ecology" were applicable to *everything* that involves environmental issues? Does it not completely degrade the rich meaning of the word "ecology" to append words like "shallow" and "deep" to it? Arne Naess, the pontiff of deep ecology—who, together with George Sessions and Bill Devall, inflicted this vocabulary upon us—have taken a pregnant word—ecology—and stripped it of any inner meaning and integrity by designating the most pedestrian environmentalists as "ecologists," albeit "shallow" ones, in contrast to their notion of "deep."

This is not an example of mere wordplay. It tells us something about the mindset that exists among these "deep" thinkers. To parody the word "shallow" and "deep ecology" is to show not only the absurdity of this terminology but to reveal the superficiality of its inventors. In fact, this kind of absurdity tells us more than we realize about the confusion Naess-Sessions-Devall, not to mention eco-brutalists like Foreman, have introduced into the current ecology movement. Indeed, this trio relies very heavily on the ease with which people forget the history of the ecology movement, the way in which the wheel is reinvented every few years by newly arrived individuals who, well-meaning as they may be, often accept a crude version of highly developed ideas that appeared earlier in a richer context and tradition of ideas. At worst, they shatter such contexts and traditions, picking out tasty pieces that become utterly distorted in a new, utterly alien framework. No regard is paid by such "deep thinkers" to the fact that *the new context in which an idea is placed may utterly change the meaning of the idea itself.* German "National Socialism" was militantly "anti-capitalist." But its "anti-capitalism" was placed in a strongly racist, imperialist, and seemingly "naturalist" context which extolled wilderness, a crude biologism, and anti-rationalism—features one finds in latent or explicit form in Sessions' and Devall's *Deep Ecology.*[1]

Neither Naess, Sessions, nor Devall have written a single line about decentralization, a nonhierarchical society, democracy, small-scale communities, local autonomy, mutual aid, communalism, and tolerance that was not already conceived in painstaking detail and brilliant contextualization by Peter Kropotkin a century ago. But what the boys from Ecotopia do is to totally recontextualize the framework of these ideas, bringing in personalities and notions that basically change their racial libertarian thrust. *Deep Ecology* mingles Woody Guthrie, a Communist Party centralist who no more believed in decentralization than Stalin, with Paul Goodman, an anarchist who would have been mortified to be placed in the same tradition with Guthrie. In philosophy, the book also intermingles Spinoza, a Jew in spirit if not in religious commitment, with Heidegger, a former member of the Nazi party in spirit as well as ideological affiliation—all in the name of a vague word called "process philosophy." Almost opportunistic in their use of catch-words and what Orwell called "double-speak," "process philosophy" makes it possible for Sessions-Devall to add Alfred North Whitehead to their list of ideological ancestors because he called his ideas "processual."

One could go on indefinitely describing this sloppy admixture of "ancestors," philosophical traditions, social pedigrees, and religions that often have nothing in common with each other and, properly conceived, are commonly in sharp opposition with each other. Thus, a reactionary like Thomas Malthus and the tradition he spawned is celebrated with the same enthusiasm in *Deep Ecology* as Henry Thoreau, a radical libertarian who fostered a highly humanistic tradition. Eclecticism would be too mild a word for this kind of hodgepodge, one that seems shrewdly calculated to embrace everyone under the rubric of deep ecology who is prepared to reduce ecology to religion rather than a systematic and critical body of ideas. This kind of "ecological" thinking surfaces in an appendix to the Devall-Sessions book, called *Ecosophy T,* by Arne Naess, who regales us with flow diagrams and corporate-type tables of organization that have more in common with logical positivist forms of exposition (Naess, in fact, was an acolyte of this school of thought for years) than anything that could be truly called organic philosophy.

If we look beyond the spiritual eco-babble, and examine the *context* in which demands like decentralization, small-scale communities, local autonomy, mutual aid, communalism, and tolerance are placed, the blurred images that Sessions and Devall create come into clearer focus. These demands are not intrinsically ecological or emancipatory. Few societies were more decentralized than European feudalism, which was structured around small-scale communities, mutual aid, and the communal use of land. Local autonomy was highly prized, and autarchy formed the economic key to feudal communities. Yet few societies were more hierarchical. The manorial economy of the Middle Ages placed a high premium on autarchy or "self-sufficiency" and spirituality. Yet oppression was often intolerable and the great mass of people who belonged to that society lived in utter subjugation by their "betters" and the nobility.

If "nature-worship," with its bouquet of wood sprites, animistic fetishes, fertility rites and other such ceremonies, paves the way to an ecological sensibility and society, then it would be hard to understand how ancient Egypt, with its animal deities and all-presiding goddesses, managed to become one of the most hierarchical and oppressive societies in the ancient world. The Nile River, which provided the "life-giving" waters of the valley, was used in a highly ecological manner. Yet the entire society was structured around the oppression of millions of serfs by opulent nobles, such that one wonders how notions of spirituality can be given priority over the need for a critical evaluation of social structures.

Even if one grants the need for a new sensibility and outlook—a point that has been made repeatedly in the literature of social ecology—one can look behind even this limited context of deep ecology to a still broader context. The love affair of deep ecology with Malthusian doctrines, a spirituality that emphasizes self-effacement, a flirtation with a *super*naturalism that stands in flat contradiction to the refreshing naturalism that ecology has introduced into social theory, a crude positivism in the spirit of Naess—all work against a truly organic dialectic so needed to understand *development*. We shall see that all the bumper-sticker demands like decentralization, small-scale communities, local autonomy, mutual aid, communalism, tolerance, and even an avowed opposition to hierarchy, go awry when we place them in the larger context of anti-humanism and "biocentrism" that mark the authentic ideological infrastructure of deep ecology.

THE ART OF EVADING SOCIETY

The seeming ideological "tolerance" and pluralism which deep ecology celebrates has a sinister function of its own. It not only reduces richly nuanced ideas and conflicting traditions to their lowest common denominator; it legitimates extremely primitivistic and reactionary notions in the company of authentically radical contexts and traditions.

Deep ecology reduces people from social beings to a simple species—to zoological entities that are interchangeable with bears, bisons, deer, or, for that matter, fruit flies and microbes. The fact that people can consciously change themselves and society, indeed enhance that natural world in a free ecological society, is dismissed as "humanism." Deep ecology essentially ignores the social nature of humanity and the social origins of the ecological crises.

This "zoologization" of human beings and of society yields sinister results. The role of capitalism with its competitive "grow or die" market economy—an economy that would devour the biosphere whether there were 10 billion people on the planet or 10 million—is simply vaporized into a vapid spiritualism. Taoist and Buddhist pieties replace the need for social and economic analysis, and self-indulgent encounter groups replace the need for political organization and action. Above all, deep ecologists explain the destruction of human beings in terms of the same "natural laws" that are said to govern the population vicissitudes of lemmings. The fact that major reductions of populations would not diminish levels of production and the destruction of the biosphere in a capitalist economy totally eludes Devall, Sessions, and their followers.

In failing to emphasize the unique characteristics of human societies and to give full due to the self-reflective role of human consciousness, deep ecologists essentially evade the *social* roots of the ecological crisis. Deep ecology contains no history of the emergence of society out of nature, a crucial development that brings social theory into organic contact with ecological theory. It presents no explanation of—indeed, it reveals no interest in—the emergence of hierarchy out of society, of classes out of hierarchy, of the state out of classes—in short, the highly graded social as well as ideological developments which are at the roots of the ecological problem.

Instead, we not only lose sight of the social differences that fragment "humanity" into a host of human beings—men and women, ethnic groups, oppressors and oppressed—we lose sight of the individual self in an unending flow of eco-babble that preaches the "realization of self-in-Self where the 'Self' stands for organic wholeness." More of the same cosmic eco-babble appears when we are informed that the "phrase 'one' includes not only men, an individual human, but all humans, grizzly bears, whole rain forest ecosystems, mountains and rivers, the tiniest microbes in the soil, and so on."

ON SELFHOOD AND VIRUSES

Such flippant abstractions of human individuality are extremely dangerous. Historically, a "Self" that absorbs all real existential selves has been used from time immemorial to absorb individual uniqueness and freedom into a supreme "Individual" who heads the state, churches of various sorts, adoring congregations, and spellbound constituencies. The purpose is the same, no matter how much such a "self" is dressed up in ecological, naturalistic, and "biocentric" attributes. The Paleolithic shaman, in reindeer skins and horns, is the predecessor of the Pharaoh, the Buddha, and, in more recent times, of Hitler, Stalin, and Mussolini.

That the egotistical, greedy, and soloist bourgeois "self" has always been a repellent being goes without saying, and deep ecology as put forth by Devall and Sessions makes the most of it. But is there not a free, independently minded, ecologically concerned, idealistic self with a unique personality that can think of itself as different from "whales, grizzly bears, whole rain forest ecosystems (no less!), mountains and rivers, the tiniest microbes in the soil, and so on"? Is it not indispensable, in fact, for the individual self to disengage itself from a Pharonic "Self," discover its own capacities and uniqueness, and acquire a sense of per-

sonality, of self-control and self-direction—all traits indispensable for the achievement of *freedom?* Here, one can imagine Heidegger grimacing with satisfaction at the sight of this self-effacing and passive personality so yielding that it can easily be shaped, distorted, and manipulated by a new "ecological" state machinery with a supreme "Self" at its head. And this all in the name of a "biocentric equality" that is slowly reworked as it has been so often in history, into a social hierarchy. From Shaman to Monarch, from Priest or Priestess to Dictator, our warped social development has been marked by "nature worshippers" and their ritual Supreme Ones who produced unfinished individuals at best or deindividuated the "self-in-Self" at worst, often in the name of the "Great Connected Whole" (to use *exactly* the language of the Chinese ruling classes who kept their peasantry in abject servitude, as Leon E. Stover points out in his *The Cultural Ecology of Chinese Civilization*).

What makes this eco-babble especially dangerous today is that we are already living in a period of massive de-individuation. This is not because deep ecology or Taoism is making any serious inroads into our own cultural ecology, but because the mass media, the commodity culture, and a market society are "reconnecting" us into an increasingly depersonalized "whole" whose essence is passivity and a chronic vulnerability to economic and political manipulation. It is not an excess of "selfhood" from which we are suffering, but rather the surrender of personality to the security and control of corporations, centralized government, and the military. If "selfhood" is identified with a grasping, "anthropocentric," and devouring personality, these traits are to be found not so much among ordinary people, who basically sense they have no control over their destinies, but among the giant corporations and state leaders who are not only plundering the planet, but also robbing from women, people of color, and the underprivileged. It is not deindividuation that the oppressed of the world require, but *re*individuation that will transform them into active agents in the task of remaking society and arresting the growing totalitarianism that threatens to homogenize us all into a Western version of the "Great Connected Whole."

We are also confronted with the delicious "and so on" that follows the "tiniest microbes in the soil" with which our deep ecologists identify the "Self." Taking their argument to its logical extreme, one might ask: why stop with the "tiniest microbes in the soil" and ignore the leprosy microbe, the viruses that give us smallpox, polio, and, more recently, AIDS? Are they, too, not part of "all organisms and entities in the ecosphere . . . of the interrelated whole . . . equal in intrinsic worth. . . ," as Devall and Sessions remind us in their effluvium of eco-babble? Naess, Devall, and Sessions rescue themselves by introducing a number of highly debatable qualifiers:

> The slogan of "noninterference" does not imply that humans should not modify some ecosystems as do other species. Humans have modified the Earth and will probably continue to do so. At issue is the nature and extent of such interference.

One does not leave the muck of deep ecology without having mud all over one's feet. Exactly *who* is to decide the "nature" of human "interference" in nature and the "extent" to which it can be done? What are "some" of the ecosystems we can modify and which ones are not subject to human "interference"? Here, again, we encounter the key problem that deep ecology poses for serious, ecologically concerned people: the *social* bases of our ecological problems and the role of the human species in the evolutionary scheme of things.

Implicit in deep ecology is the notion that a "Humanity" exists that accurses the natural world; that individual selfhood must be transformed into a cosmic "Selfhood" that essentially transcends the person and his or her uniqueness. Even nature is not spared from a kind of static, prepositional logic that is cultivated by the logical positivists. "Nature," in deep ecology and David Foreman's interpretation of it, becomes a kind of scenic view, a spectacle to be admired around the campfire. It is not viewed as an *evolutionary* development that is cumulative and *includes* the human species.

The problems deep ecology and biocentricity raise have not gone unnoticed in the more thoughtful press in England. During a discussion of "biocentric ethics" in *The New Scientist* 69 (1976), for example, Bernard Dixon observed that no "logical line can be drawn" between the conservation of whales, gentians, and flamingoes on the one hand and the extinction of pathogenic microbes like the smallpox virus. At which point David Ehrenfeld, in his *Arrogance of Humanism*,[2]—a work that is so selective and tendentious in its use of quotations that it should validly be renamed "The Arrogance of Ignorance"—cutely observes that the smallpox virus is "an endangered species." One wonders what to do about the AIDS virus if a vaccine or therapy should threaten its "survival"? Further,

given the passion for perpetuating the "ecosystem" of every species, one wonders how smallpox and AIDS viruses should be preserved? In test tubes? Laboratory cultures? Or, to be truly "ecological," in their "native habitat," the human body? In which case, idealistic acolytes of deep ecology should be invited to offer their own bloodstreams in the interests of "biocentric equality." Certainly, "if nature should be permitted to take its course"—as Foreman advises for Ethiopians and Indian peasants—plagues, famines, suffering, wars, and perhaps even lethal asteroids of the kind that exterminated the great reptiles of the Mesozoic should not be kept from defacing the purity of "first nature" by the intervention of human ingenuity and—yes!—*technology.* With so much absurdity to unscramble, one can indeed get heady, almost dizzy, with a sense of polemical intoxication.

At root, the eclecticism which turns deep ecology into a goulash of notions and moods is insufferably reformist and surprisingly environmentalist—all its condemnations of "superficial ecology" aside. Are you, perhaps, a mild-mannered liberal? Then do not fear: Devall and Sessions give a patronizing nod to "reform legislation," "coalitions," "protests," the "women's movement" (this earns all of ten lines in their "Minority Tradition and Direct Action" essay), "working in the Christian tradition," "questioning technology" (a hammering remark, if there ever was one), "working in Green politics" (which faction, the "fundies" of the "realos"?). In short, everything can be expected in so "cosmic" a philosophy. Anything seems to pass through deep ecology's donut hole: anarchism at one extreme and eco-fascism at the other. Like the fast food emporiums that make up our culture, deep ecology is the fast food of quasi-radical environmentalists.

Despite its pretense of "radicality," deep ecology is more "New Age" and "Aquarian" than the environmentalist movements it denounces under those names. Indeed, the extent to which deep ecology accommodates itself to some of the worst features of the "dominant view" it professes to reject is seen with extraordinary clarity in one of its most fundamental and repeatedly asserted demands—namely, that the world's population must be drastically reduced, according to one of its devotees, to 500 million. If deep ecologists have even the faintest knowledge of the "population theorists" Devall and Sessions invoke with admiration—notably, Thomas Malthus, William Vogt, and Paul Ehrlich—then they would be obliged to add: by measures that are virtually eco-fascist. This specter clearly looms before us in Devall's and Sessions' sinister remark: ". . . the longer we wait [for population control], the more drastic will be the measures needed."

THE "DEEP" MALTHUSIANS

Devall and Sessions often write with smug assurance on issues they know virtually nothing about. This is most notably the case in the so-called "population debate," a debate that has raged for over two hundred years and more and involves explosive political and social issues that have pitted the most reactionary elements in English and American society against authentic radicals. In fact, the eco-babble which Devall and Sessions dump on us in only two paragraphs would require a full-sized volume of careful analysis to unravel.

Devall and Sessions hail Thomas Malthus (1766–1854) as a prophet whose warning "that human population growth would exponentially outstrip food production . . . was ignored by the rising tide of industrial/technological optimism." First of all, Thomas Malthus was not a prophet; he was an apologist for the misery that the Industrial Revolution was inflicting on the English peasantry and working classes. His utterly fallacious argument that population increases exponentially while food supplies increase arithmetically was not ignored by England's ruling classes; it was taken to heart and even incorporated into social Darwinism as an explanation of why oppression was a necessary feature of society and why the rich, the white imperialists, and the privileged were the "fittest" who were equipped to "survive"—needless to say, at the expense of the impoverished many. Written and directed in great part as an attack upon the liberatory vision of William Godwin, Malthus' mean-spirited *Essay on the Principle of Population* tried to demonstrate that hunger, poverty, disease, and premature death are *inevitable* precisely because population and food supply increase at different rates. Hence war, famines, and plagues (Malthus later added "moral restraint") were necessary to keep population down—needless to say, among the "lower orders of society," whom he singles out as the chief offenders of his inexorable population "laws."[3] Malthus, in effect, became the ideologue par excellence for the land-grabbing English nobility in its effort to dispossess the peasantry of their traditional common lands and for the English capitalists to work children, women, and men to death in the newly emergent "industrial/technological" factory system.

Malthusianism contributed in great part to that meanness of spirit that Charles Dickens captured in his famous novels, *Oliver Twist* and *Hard Times*. The doctrine, its author, and its overstuffed wealthy beneficiaries were bitterly fought by the great English anarchist, William Godwin, the pioneering socialist, Robert Owen, and the emerging Chartist movement of English workers in the early 19th century. However, Malthusianism was naively picked up by Charles Darwin to explain his theory of "natural selection." It then became the bedrock theory for the new *social* Darwinism, so very much in vogue in the late nineteenth and early twentieth centuries, which saw society as a "jungle" in which only the "fit" (usually, the rich and white) could "survive" at the expense of the "unfit" (usually, the poor and people of color). Malthus, in effect, had provided an ideology that justified class domination, racism, the degradation of women, and, ultimately, British imperialism.

Malthusianism was not only revived in Hitler's Third Reich; it also reemerged in the late 1940s, following the discoveries of antibiotics to control infectious diseases. Riding on the tide of the new Pax Americana after World War II, William F. Vogt and a whole bouquet of neo-Malthusians were to challenge the use of the new antibiotic discoveries to control disease and prevent death—as usual, mainly in Asia, Africa, and Latin America. Again, a new "population debate" erupted, with the Rockefeller interests and large corporate sharks aligning themselves with the neo-Malthusians, and caring people of every sort aligning themselves with third world theorists like Josua de Castro, who wrote damning, highly informed critiques of this new version of misanthropy.

Zero Population Growth fanatics in the early seventies literally polluted the environmental movement with demands for a government bureau to "control" population, advancing the infamous "triage" ethic, according to which various "underdeveloped" countries would be granted or refused aid on the basis of their compliance to population control measures. In *Food First,* Francis Moore Lappé and Joseph Collins have done a superb job in showing how hunger has its origins not in "natural" shortages of food or population growth, but in social and cultural dislocations. (It is notable that Devall and Sessions do *not* list this excellent book in their bibliography.) The book has to be read to understand the reactionary implications of deep ecology's demographic positions.

Demography is a highly ambiguous and ideologically charged social discipline that cannot be reduced to a mere numbers game in biological reproduction. Human beings are not fruit flies (the species which the neo-Malthusians love to cite). Their reproductive behavior is profoundly conditioned by cultural values, standards of living, social traditions, gender relations, religious beliefs, socio-political conflicts, and various socio-political expectations. Smash up a stable, precapitalist culture and throw its people off the land into city slums, and, due to demoralization, population may soar rather than decline. As Gandhi told the British, imperialism left India's wretched poor and homeless with little more in life than the immediate gratification provided by sex and an understandably numbed sense of personal, much less social, responsibility. Reduce women to mere reproductive factories and population rates will explode.

Conversely, provide people with decent lives, education, a sense of creative meaning in life, and, above all, expand the role of women in society—and population growth begins to stabilize and population rates even reverse their direction. Nothing more clearly reveals deep ecology's crude, often reactionary, and certainly superficial ideological framework—all its decentralist, antihierarchical, and "radical" rhetoric aside—than its suffocating "biological" treatment of the population issue and its inclusion of Malthus, Vogt, and Ehrlich in its firmament of prophets.

Not surprisingly, the *Earth First!* newsletter, whose editor professes to be an enthusiastic deep ecologist, carried an article titled "Population and AIDS" which advanced the obscene argument that AIDS is desirable as a means of population control. This was no spoof. It was earnestly argued and carefully reasoned in a Paleolithic sort of way. Not only will AIDS claim large numbers of lives, asserts the author (who hides under the pseudonym of "Miss Ann Thropy," a form of black humor that could also pass as an example of macho-male arrogance), but it "may cause a breakdown in technology (read: human food supply) and its export which could also decrease human population." These people feed on human disasters, suffering, and misery, preferably in third world countries where AIDS is by far a more monstrous problem than elsewhere.

We have little reason to doubt that this mentality is perfectly consistent with the "more drastic . . . measures" Devall and Sessions believe we will have to explore. Nor is it inconsistent with Malthus and Vogt that we should make no effort to find a cure for this disease which may do so much to depopulate the world. "Biocentric democracy," I assume, should call for nothing less than a "hands-off" policy on the AIDS virus and perhaps equally lethal pathogens that appear in the human species.

WHAT IS SOCIAL ECOLOGY?

Social ecology is neither "deep," "tall," "fat," nor "thick." It is *social*. It does not fall back on incantations, sutras, flow diagrams or spiritual vagaries. It is avowedly *rational*. It does not try to regale metaphorical forms of spiritual mechanism and crude biologism with Taoist, Buddhist, Christian, or shamanistic ecobabble. It is a coherent form of *naturalism* that looks to *evolution* and the *biosphere,* not to deities in the sky or under the earth for quasi-religious and supernaturalistic explanations of natural and social phenomena.

Philosophically, social ecology stems from a solid organismic tradition in Western philosophy, beginning with Heraclitus, the near-evolutionary dialectic of Aristotle and Hegel, and the critical approach of the famous Frankfurt School—particularly its devastating critique of logical positivism (which surfaces in Naess repeatedly) and the primitivistic mysticism of Heidegger (which pops up all over the place in deep ecology's literature).

Socially, it is revolutionary, not merely "radical." It critically unmasks the entire evolution of hierarchy in all its forms, including neo-Malthusian elitism, the eco-brutalism of David Foreman, the anti-humanism of David Ehrenfeld and "Miss Ann Thropy," and the latent racism, first-world arrogance, and Yuppie nihilism of postmodernistic spiritualism. It is rooted in the profound eco-anarchistic analyses of Peter Kropotkin, the radical economic insights of Karl Marx, the emancipatory promise of the revolutionary Enlightenment as articulated by the great encyclopedist, Denis Diderot, the *Enrages* of the French Revolution, the revolutionary feminist ideals of Louise Michel and Emma Goldman, the communitarian visions of Paul Goodman and E. A. Gutkind, and the various eco-revolutionary manifestoes of the early 1960s.

Politically, it is green—radically green. It takes its stand with the left-wing tendencies in the German Greens and extra-parliamentary street movements of European cities; with the American radical ecofeminist movement; with the demands for a new politics based on citizens' initiatives, neighborhood assemblies, and New England's tradition of town-meetings; with non-aligned anti-imperialist movements at home and abroad; with the struggle by people of color for complete freedom from the domination of privileged whites and from the superpowers.

Morally, it is *humanistic* in the high Renaissance meaning of the term, not the degraded meaning of "humanism" that has been imparted to the world by David Foreman, David Ehrenfeld, and a salad of academic deep ecologists. Humanism from its inception has meant a shift in vision from the skies to the earth, from superstition to reason, from deities to people— who are no less products of natural evolution than grizzly bears and whales. Social ecology accepts neither a "biocentricity" that essentially denies or degrades the uniqueness of human beings, human subjectivity, rationality, aesthetic sensibility, and the ethical potentiality of humanity, nor an "anthropocentricity" that confers on the privileged few the right to plunder the world of life, including human life. Indeed, it opposes "centricity" of *any* kind as a new word for hierarchy and domination—be it that of nature by a mystical "Man" or the domination of people by an equally mystical "Nature." It firmly denies that nature is a static, scenic view which Mountain Men like a Foreman survey from a peak in Nevada or a picture window that spoiled yuppies view from their ticky-tacky country homes. To social ecology, nature *is* natural *evolution,* not a cosmic arrangement of beings frozen in a moment of eternity to be abjectly revered, adored, and worshipped like Gods and Goddesses in a realm of "*super*nature." Natural evolution is nature in the very real sense that it is composed of atoms, molecules that have evolved into amino acids, proteins, unicellular organisms, genetic codes, invertebrates and vertebrates, amphibia, reptiles, mammals, primates, and human beings—all, in a cumulative thrust toward ever-greater complexity, ever-greater subjectivity, and finally, an ever-greater capacity for conceptual thought, symbolic communication, and self-consciousness.

This marvel we call "Nature" has produced a marvel we call homo sapiens—"thinking man"—and, more significantly for the development of society, "thinking woman," whose primeval domestic domain provided the arena for the origins of a caring society, human empathy, love, and idealistic commitment. The human species, in effect, is no less a product of natural evolution and differentiation than blue-green algae. To degrade the human species in the name of "anti-humanism," to deny people their uniqueness as thinking beings with an unprecedented gift for conceptual thought, is to deny the rich fecundity of natural evolution itself. To separate human beings and society from nature is to dualize and truncate nature itself,

to diminish the meaning and thrust of natural evolution in the name of a "biocentricity" that spends more time disporting itself with mantras, deities, and supernature than with the realities of the biosphere and the role of society in ecological problems.

Accordingly, social ecology does not try to hide its critical and reconstructive thrust in metaphors. It calls "technological/industrial" society *capitalism*—a word which places the onus for our ecological problems on the *living* sources and *social* relationships that produce them, not on a cutesy "Third Wave" abstraction which buries these sources in technics, a technical "mentality," or perhaps the technicians who work on machines. It sees the domination of women not simply as a "spiritual" problem that can be resolved by rituals, incantations, and shamannesses, important as ritual may be in solidarizing women into a unique community of people, but in the long, highly graded, and subtly nuanced development of hierarchy, which long preceded the development of classes. Nor does it ignore class, ethnic differences, imperialism, and oppression by creating a grab-bag called "Humanity" that is placed in opposition to a mystified "Nature," divested of all development.

All of which brings us as social ecologists to an issue that seems to be totally alien to the crude concerns of deep ecology: natural evolution has conferred on human beings the capacity to form a "second" or cultural nature out of "first" or primeval nature. Natural evolution has not only provided humans with the *ability,* but also the *necessity* to be purposive interveners into "first nature," to consciously *change* "first nature" by means of a highly institutionalized form of community we call "society." It is not alien to natural evolution that a species called human beings have emerged over the billions of years who are capable of thinking in a sophisticated way. Nor is it alien for human beings to develop a highly sophisticated form of symbolic communication which a new kind of community— institutionalized, guided by thought rather than by instinct alone, and ever-changing—has emerged called "society."

Taken together, all of these human traits—intellectual, communicative, and social—have not only emerged from natural evolution and are inherently human; they can also be placed at the *service* of natural evolution to consciously increase biotic diversity, diminish suffering, foster the further evolution of new and ecologically valuable life-forms, reduce the impact of disastrous accidents or the harsh effects of mere change.

Whether this species, gifted by the creativity of natural evolution, can play the role of a nature rendered self-conscious or cut against the grain of natural evolution by simplifying the biosphere, polluting it, and undermining the cumulative results of organic evolution is above all a *social* problem. The primary question ecology faces today is whether an ecologically oriented society can be created out of the present anti-ecological one.

Unless there is a resolute attempt to fully anchor ecological dislocations in social dislocations; to challenge the vested corporate and political interests we should properly call *capitalism;* to analyze, explore, and attack hierarchy as a *reality,* not only as a sensibility; to recognize the material needs of the poor and of third world people; to function politically, and not simply as a religious cult; to give the human species and mind their due in natural evolution, rather than regard them as "cancers" in the biosphere; to examine economies as well as "souls," and freedom instead of scholastic arguments about the "rights" of pathogenic viruses—unless, in short, North American Greens and the ecology movement shift their focus toward a *social ecology* and let deep ecology sink into the pit it has created for us, the ecology movement will become another ugly wart on the skin of society.

What we must do, today, is return to *nature,* conceived in all its fecundity, richness of potentialities, and subjectivity—not to *super*nature with its shamans, priests, priestesses, and fanciful deities that are merely anthropomorphic extensions and distortions of the "Human" as all-embracing divinities. And what we must "enchant" is not only an abstract image of "Nature" *that often reflects our own systems of power, hierarchy, and domination*—but rather human beings, the human mind, and the human spirit.

Notes

1. Unless otherwise indicated, all future references and quotes come from Bill Devall and George Sessions, *Deep Ecology* (Layton, UT: Gibbs M. Smith, 1985), a book which has essentially become the bible of the "movement" that bears its name.
2. David Ehrenfeld, *The Arrogance of Humanism* (New York: The Modern Library, 1978) pp. 207–211.
3. Chapter five of his *Essay,* which, for all its "concern" over the misery of the "lower classes," inveighs against the poor laws and argues that the "pressures of distress on this part of the community is an evil so deeply seated that no human ingenuity can reach it." Thomas Malthus, *On Population* (New York: The Modern Library), p. 34.

For Further Discussion

1. Bookchin suggests that the present ecological movement is beginning to sense that there is a tie-in between the way people deal with one another and the way they deal with nature. Can you think of any specific examples that would support this tie-in?

2. Bookchin describes deep ecology as a "vague, formless, often self-contradictory ideology." Review the essay by Devall and Sessions. How would you respond to Bookchin on their behalf?

3. Why is Bookchin dismayed by proposals to let nature seek its own balance?

4. Bookchin criticizes deep ecology for reducing humans to zoological entities that are interchangeable with bears, bison, and other animals. Among the many essays you've read, which authors would agree with this criticism? What reasons lie behind Bookchin's criticism? Of those who would agree with his conclusion, who might disagree with his reasons?

5. How would you explain the basic ideas of social ecology to a friend? Would you consider yourself a social ecologist?

The Power and the Promise of Ecological Feminism

Karen J. Warren

Philosopher Karen Warren argues that ecological feminism can provide a framework for understanding both feminism and environmental ethics. The key to understanding the connections between feminist concerns and ecological concerns lies in the conceptual framework that legitimizes hierarchies and domination. A conceptual framework is a set of basic beliefs, values, attitudes, and assumptions that determine how one understands the world. Certain of these frameworks explain and justify oppression; some frameworks ("patriarchal" ones) explain and justify the subordination of women by men. Ecofeminists claim that within Western culture, women have been closely identified with nature. Accordingly, patterns of thinking that explain and justify the subordination of women also explain and justify the subordination of the natural world. To break this cycle, we must break the conceptual framework that legitimizes all forms of domination. Warren ends her reading by describing the use of first-person narratives as an alternative methodological approach for exploring ethical and environmental concerns.

INTRODUCTION

Ecological feminism (ecofeminism) has begun to receive a fair amount of attention lately as an alternative feminism and environmental ethic.[1] Since Francoise d'Eaubonne introduced the term *ecofeminisme* in 1974 to bring attention to women's potential for bringing about an ecological revolution,[2] the term has been used in a variety of ways. As I use the term in this paper, ecological feminism is the position that there are important connections—historical, experiential, symbolic, theoretical—between the domination of women and the domination of nature, an understanding of which is crucial to both feminism and environmental ethics. I argue that the promise and power of ecological feminism is that *it provides a distinctive framework both for reconceiving feminism and for developing an environmental ethic which takes seriously connections between the domination of women and the domination of*

nature. I do so by discussing the nature of a feminist ethic and the ways in which ecofeminism provides a feminist and environmental ethic. I conclude that any feminist theory *and* any environmental ethic which fails to take seriously the twin and interconnected dominations of women and nature is at best incomplete and at worst simply inadequate.

FEMINISM, ECOLOGICAL FEMINISM, AND CONCEPTUAL FRAMEWORKS

Whatever else it is, feminism is at least the movement to end sexist oppression. It involves the elimination of any and all factors that contribute to the continued and systematic domination or subordination of women. While feminists disagree about the nature of and solutions to the subordination of women, all feminists agree that sexist oppression exists, is wrong, and must be abolished.

A "feminist issue" is any issue that contributes in some way to understanding the oppression of women. Equal rights, comparable pay for comparable work, and food production are feminist issues wherever and whenever an understanding of them contributes to an understanding of the continued exploitation or subjugation of women. Carrying water and searching for firewood are feminist issues wherever and whenever women's primary responsibility for these tasks contributes to their lack of full participation in decision making, income producing, or high status positions engaged in by men. What counts as a feminist issue, then, depends largely on context, particularly the historical and material conditions of women's lives.

Environmental degradation and exploitation are feminist issues because an understanding of them contributes to an understanding of the oppression of women. In India, for example, both deforestation and reforestation through the introduction of a monoculture species tree (e.g., eucalyptus) intended for commercial production are feminist issues because the loss of indigenous forests and multiple species of trees has drastically affected rural Indian women's ability to maintain a subsistence household. Indigenous forests provide a variety of trees for food, fuel, fodder, household utensils, dyes, medicines, and income-generating uses, while monoculture-species forests do not.[3] Although I do not argue for this claim here, a look at the global impact of environmental degradation on women's lives suggests important respects in which environmental degradation is a feminist issue.

Feminist philosophers claim that some of the most important feminist issues are *conceptual* ones: these issues concern how one conceptualizes such mainstay philosophical notions as reason and rationality, ethics, and what it is to be human. Ecofeminists extend this feminist philosophical concern to nature. They argue that, ultimately, some of the most important connections between the domination of women and the domination of nature are conceptual. To see this, consider the nature of conceptual frameworks.

A *conceptual framework* is a set of *basic* beliefs, values, attitudes, and assumptions which shape and reflect how one views oneself and one's world. It is a socially constructed lens through which we perceive ourselves and others. It is affected by such factors as gender, race, class, age, affectional orientation, nationality, and religious background.

Some conceptual frameworks are oppressive. An *oppressive conceptual framework* is one that explains, justifies, and maintains relationships of domination and subordination. When an oppressive conceptual framework is *patriarchal,* it explains, justifies, and maintains the subordination of women by men.

I have argued elsewhere that there are three significant features of oppressive conceptual frameworks: (1) value-hierarchical thinking, i.e., "up-down" thinking which places higher value, status, or prestige on what is "up" rather than on what is "down"; (2) value dualisms, i.e., disjunctive pairs in which the disjuncts are seen as oppositional (rather than as complementary) and exclusive (rather than as inclusive), and which place higher value (status, prestige) on one disjunct rather than the other (e.g., dualisms which give higher value or status to that which has historically been identified as "mind," "reason," and "male" than to that which has historically been identified as "body," "emotion," and "female"); and (3) logic of domination, i.e., a structure of argumentation which leads to a justification of subordination.[4]

The third feature of oppressive conceptual frameworks is the most significant. A logic of domination is

not *just* a logical structure. It also involves a substantive value system, since an ethical premise is needed to permit or sanction the "just" subordination of that which is subordinate. This justification typically is given on grounds of some alleged characteristic (e.g., rationality) which the dominant (e.g., men) have and the subordinate (e.g., women) lack.

Contrary to what many feminists and ecofeminists have said or suggested, there may be nothing *inherently* problematic about "hierarchical thinking" or even "value-hierarchical thinking" in contexts other than contexts of oppression. Hierarchical thinking is important in daily living for classifying data, comparing information, and organizing material. Taxonomies (e.g., plant taxonomies) and biological nomenclature seem to require *some* form of "hierarchical thinking." Even "value-hierarchical thinking" may be quite acceptable in certain contexts. (The same may be said of "value dualisms" in nonoppressive contexts.) For example, suppose it is true that what is unique about humans is our conscious capacity to radically reshape our social environments (or "societies"), as Murray Bookchin suggests.[5] Then one could truthfully say that humans are better equipped to radically reshape their environments than are rocks or plants—a "value-hierarchical" way of speaking.

The problem is not simply *that* value-hierarchical thinking and value dualisms are used, but *the way* in which each has been used in *oppressive conceptual frameworks* to establish inferiority and to justify subordination.[6] It is the logic of domination, *coupled with* value-hierarchical thinking and value dualisms, which "justifies" subordination. What is explanatorily basic, then, about the nature of oppressive conceptual frameworks is the logic of domination.

For ecofeminism, that a logic of domination is explanatorily basic is important for at least three reasons. First, without a logic of domination, a description of similarities and differences would be just that—a description of similarities and differences. Consider the claim, "Humans are different from plants and rocks in that humans can (and plants and rocks cannot) consciously and radically reshape the communities in which they live; humans are similar to plants and rocks in that they are both members of an ecological community." Even if humans are "better" than plants and rocks with respect to the conscious ability of humans to radically transform communities, one does not *thereby* get any *morally* relevant distinction between humans and nonhumans, or an argument for the domination of plants and rocks by humans. To get *those* conclusions one needs to add at least two powerful assumptions, viz., (A2) and (A4) in argument A below:

(A1) Humans do, and plants and rocks do not, have the capacity to consciously and radically change the community in which they live.

rrA2) Whatever has the capacity to consciously and radically change the community in which it lives is morally superior to whatever lacks this capacity.

(A3) Thus, humans are morally superior to plants and rocks.

(A4) For any X and Y, if X is morally superior to Y, then X is morally justified in subordinating Y.

(A5) Thus, humans are morally justified in subordinating plants and rocks.

Without the two assumptions that *humans are morally superior* to (at least some) nonhumans, (A2), and that *superiority justifies subordination,* (A4), all one has is some difference between humans and some nonhumans. This is true *even if* that difference is given in terms of superiority. Thus, it is the logic of domination, (A4), which is the bottom line in ecofeminist discussions of oppression.

Second, ecofeminists argue that, at least in Western societies, the oppressive conceptual framework which sanctions the twin dominations of women and nature is a patriarchal one characterized by all three features of an oppressive conceptual framework. Many ecofeminists claim that, historically, within at least the dominant Western culture, a patriarchal conceptual framework has sanctioned the following argument B:

(B1) Women are identified with nature and the realm of the physical; men are identified with the "human" and the realm of the mental.

(B2) Whatever is identified with nature and the realm of the physical is inferior to ("below") whatever is identified with the "human" and the realm of the mental; or, conversely, the latter is superior to ("above") the former.

(B3) Thus, women are inferior to ("below") men; or, conversely, men are superior to ("above") women.

(B4) For any X and Y, if X is superior to Y, then X is justified in subordinating Y.

(B5) Thus, men are justified in subordinating women.

If sound, argument B establishes *patriarchy,* i.e., the conclusion given at (B5) that the systematic domination of women by men is justified. But according to ecofeminists, (B5) is justified by just those three features of an oppressive conceptual framework identified earlier: value-hierarchical thinking, the assumption at (B2); value dualisms, the assumed dualism of the mental and the physical at (B1) and the assumed inferiority of the physical vis-à-vis the mental at (B2); and a logic of domination, the assumption at (B4), the same as the previous premise (A4). Hence, according to ecofeminists, insofar as an oppressive patriarchal conceptual framework has functioned historically (within at least dominant Western culture) to sanction the twin dominations of women and nature (argument B), both argument B and the patriarchal conceptual framework, from whence it comes, ought to be rejected.

Of course, the preceding does not identify which premises of B are false. What is the status of premises (B1) and (B2)? Most, if not all, feminists claim that (B1), and many ecofeminists claim that (B2), have been assumed or asserted within the dominant Western philosophical and intellectual tradition.[7] As such, these feminists assert, as a matter of historical fact, that the dominant Western philosophical tradition has assumed the truth of (B1) and (B2). Ecofeminists, however, either deny (B2) or do not affirm (B2). Furthermore, because some ecofeminists are anxious to deny any ahistorical identification of women with nature, some ecofeminists deny (B1) when (B1) is used to support anything other than a strictly historical claim about what has been asserted or assumed to be true within patriarchal culture—e.g., when (B1) is used to assert that women properly are identified with the realm of nature and the physical.[8] Thus, from an ecofeminist perspective, (B1) and (B2) are properly viewed as problematic though historically sanctioned claims: they are problematic precisely because of the way they have functioned historically in a patriarchal conceptual framework and culture to sanction the dominations of women and nature.

What *all* ecofeminists agree about, then, is the way in which *the logic of domination* has functioned historically within patriarchy to sustain and justify the twin dominations of women and nature.[9] Since *all* feminists (and not just ecofeminists) oppose patriarchy, the conclusion given at (B5), all feminists (including ecofeminists) must oppose at least the logic of domination, premise (B4), on which argument B rests—whatever the truth-value status of (B1) and (B2) *outside of* a patriarchal context.

That *all* feminists must oppose the logic of domination shows the breath and depth of the ecofeminist critique of B: it is a critique not only of the three assumptions on which this argument for the domination of women and nature rests, viz., the assumptions at (B1), (B2), and (B4); it is also a critique of patriarchal conceptual frameworks generally, i.e., of those oppressive conceptual frameworks which put men "up" and women "down," allege some way in which women are morally inferior to men, and use that alleged difference to justify the subordination of women by men. Therefore, ecofeminism is necessary to *any* feminist critique of patriarchy, and, hence, necessary to feminism (a point I discuss again later).

Third, ecofeminism clarifies why the logic of domination, and any conceptual framework which gives rise to it, must be abolished in order both to make possible a meaningful notion of difference which does not breed domination and to prevent feminism from becoming a "support" movement based primarily on shared experiences. In contemporary society, there is no one "woman's voice," no *woman* (or *human*) *simpliciter*: every woman (or human) is a woman (or human) of some race, class, age, affectional orientation, marital status, regional or national background, and so forth. Because there are no "monolithic experiences" that all women share, feminism must be a "solidarity movement" based on shared beliefs and interests rather than a "unity in sameness" movement based on shared experiences and shared victimization.[10] In the words of Maria Lugones, "Unity—not to be confused with solidarity—is understood as conceptually tied to domination."[11]

Ecofeminists insist that the sort of logic of domination used to justify the domination of humans by gender, racial or ethnic, or class status is also used to justify the domination of nature. Because eliminating a logic of domination is part of a feminist critique—whether a critique of patriarchy, white supremacist culture, or imperialism—ecofeminists insist that *naturism* is properly viewed as an integral part of any feminist solidarity movement to end sexist oppression and the logic of domination which conceptually grounds it.

ECOFEMINISM RECONCEIVES FEMINISM

The discussion so far has focused on some of the oppressive conceptual features of patriarchy. As I use the phrase, the "logic of traditional feminism" refers to the location of the conceptual roots of sexist oppression, at least in Western societies, in an oppressive patriarchal conceptual framework characterized by a logic of domination. Insofar as other systems of oppression (e.g., racism, classism, ageism, heterosexism) are also conceptually maintained by a logic of domination, appeal to the logic of traditional feminism ultimately locates the basic conceptual interconnections among *all* systems of oppression in the logic of domination. It thereby explains at a *conceptual* level why the eradication of sexist oppression requires the eradication of the other forms of oppression.[12] It is by clarifying this conceptual connection between systems of oppression that a movement to end sexist oppression—traditionally the special turf of feminist theory and practice—leads to a reconceiving of feminism as *a movement to end all forms of oppression.*

Suppose one agrees that the logic of traditional feminism requires the expansion of feminism to include other social systems of domination (e.g., racism and classism). What warrants the inclusion of nature in these "social systems of domination"? Why must the logic of traditional feminism include the abolition of "naturism" (i.e., the domination or oppression of nonhuman nature) among the "isms" feminism must confront? The conceptual justification of expanding feminism to include ecofeminism is twofold. One basis has already been suggested: by showing that the conceptual connections between the dual dominations of women and nature are located in an oppressive and, at least in Western societies, patriarchal conceptual framework characterized by a logic of domination, ecofeminism explains how and why feminism, conceived as a movement to end sexist oppression, must be expanded and reconceived as also a movement to end naturism. This is made explicit by the following argument C:

(C1) Feminism is a movement to end sexism.

(C2) But Sexism is conceptually linked with naturism (through an oppressive conceptual framework characterized by a logic of domination).

(C3) Thus, Feminism is (also) a movement to end naturism.

Because, ultimately, these connections between sexism and naturism are conceptual—embedded in an oppressive conceptual framework—the logic of traditional feminism leads to the embrace of ecological feminism.[13]

The other justification for reconceiving feminism to include ecofeminism has to do with the concepts of gender and nature. Just as conceptions of gender are socially constructed, so are conceptions of nature. Of course, the claim that women and nature are social constructions does not require anyone to deny that there are actual humans and actual trees, rivers, and plants. It simply implies that *how* women and nature are conceived is a matter of historical and social reality. These conceptions vary cross-culturally and by historical time period. As a result, any discussion of the "oppression or domination of nature" involves reference to historically specific forms of social domination of nonhuman nature by humans, just as discussion of the "domination of women" refers to historically specific forms of social domination of women by men. Although I do not argue for it here, an ecofeminist defense of the historical connections between the dominations of women and of nature, claims (B1) and (B2) in argument B, involves showing that within patriarchy the feminization of nature and the naturalization of women have been crucial to the historically successful subordinations of both.[14]

If ecofeminism promises to reconceive traditional feminism in ways which include naturism as a legitimate feminist issue, does ecofeminism also promise to reconceive environmental ethics in ways which are feminist? I think so. This is the subject of the remainder of the paper.

CLIMBING FROM ECOFEMINISM TO ENVIRONMENTAL ETHICS

Many feminists and some environmental ethicists have begun to explore the use of first-person narrative as a way of raising philosophically germane issues in ethics often lost or underplayed in mainstream philosophical ethics. Why is this so? What is it about narrative which makes it a significant resource for theory and practice in feminism and environmental ethics? Even if appeal to first-person narrative is a helpful literary device for describing ineffable experience or a legitimate social science methodology for documenting personal and social history, how is first-person narrative a valuable vehicle of argumentation for ethical decision making and theory building? One fruitful way to begin answering these questions is to ask them of a particular first-person narrative.

Consider the following first-person narrative about rock climbing:

For my very first rock climbing experience, I chose a somewhat private spot, away from other climbers

and on-lookers. After studying "the chimney," I focused all my energy on making it to the top. I climbed with intense determination, using whatever strength and skills I had to accomplish this challenging feat. By midway I was exhausted and anxious. I couldn't see what to do next—where to put my hands or feet. Growing increasingly more weary as I clung somewhat desperately to the rock, I made a move. It didn't work. I fell. There I was, dangling midair above the rocky ground below, frightened but terribly relieved that the belay rope had held me. I know I was safe. I took a look up at the climb that remained. I was determined to make it to the top. With renewed confidence and concentration, I finished the climb to the top.

On my second day of climbing, I rappelled down about 200 feet from the top of the Palisades at Lake Superior to just a few feet above the water level. I could see no one—not my belayer, not the other climbers, no one. I unhooked slowly from the rappel rope and took a deep cleansing breath. I looked all around me—really looked—and listened. I heard a cacophony of voices—birds, trickles of water on the rock before me, waves lapping against the rocks below. I closed my eyes and began to feel the rock with my hands—the cracks and crannies, the raised lichen and mosses, the almost imperceptible nubs that might provide a resting place for my fingers and toes when I began to climb. At that moment I was bathed in serenity. I began to talk to the rock in an almost inaudible, child-like way, as if the rock were my friend. I felt an overwhelming sense of gratitude for what it offered me—a chance to know myself and the rock differently, to appreciate unforeseen miracles like the tiny flowers growing in the even tinier cracks in the rock's surface, and to come to know a sense of *being in relationship* with the natural environment. It felt as if the rock and I were silent conversational partners in a longstanding friendship. I realized then that I had come to care about this cliff which was so different from me, so unmovable and invincible, independent and seemingly indifferent to my presence. I wanted to be with the rock as I climbed. Gone was the determination to conquer the rock, to forcefully impose my will on it; I wanted simply to work respectfully with the rock as I climbed. And as I climbed, that is what I felt. I felt myself *caring* for this rock and feeling thankful that climbing provided the opportunity for me to know it and myself in this new way.

There are at least four reasons why use of such a first-person narrative is important to feminism and environmental ethics. First, such a narrative gives voice to a felt sensitivity often lacking in traditional analytical ethical discourse, viz., a sensitivity to conceiving of oneself as fundamentally "in relationship with" others, including the nonhuman environment. It is a modality which *takes relationships themselves seriously*. It thereby stands in contrast to a strictly reductionist modality that takes relationships seriously only or primarily because of the nature of the *relators* or parties to those relationships (e.g., relators conceived as moral agents, right holders, interest carriers, or sentient beings). In the rock-climbing narrative above, it is the climber's relationship with the rock she climbs which takes on special significance—which it itself a locus of value—in addition to whatever moral status or moral considerability she or the rock or any other parties to the relationship may also have.[15]

Second, such a first-person narrative gives expression to a variety of ethical attitudes and behaviors often overlooked or underplayed in mainstream Western ethics, e.g., the difference in attitudes and behaviors toward a rock when one is "making it to the top" and when one thinks of oneself as "friends with" or "caring about" the rock one climbs.[16] These different attitudes and behaviors suggest an ethically germane contrast between two different types of relationship humans or climbers may have toward a rock: an imposed conqueror-type relationship, and an emergent caring-type relationship. This contrast grows out of, and is faithful to, felt, lived experience.

The difference between conquering and caring attitudes and behaviors in relation to the natural environment provides a third reason why the use of first-person narrative is important to feminism and environmental ethics: it provides a way of conceiving of ethics and ethical meaning as *emerging out of* particular situations moral agents find themselves in, rather than as being *imposed on* those situations (e.g., as a derivation or instantiation of some predetermined abstract principle or rule). This emergent feature of narrative centralizes the importance of *voice*. When a multiplicity of cross-cultural *voices* are centralized, narrative is able to give expression to a range of attitudes, values, beliefs, and behaviors which may be overlooked or silenced by imposed ethical meaning and theory. As a reflection of and on felt, lived experiences, the use of narrative in ethics provides a stance from which ethical discourse can be held accountable to the historical, material, and social realities in which moral subjects find themselves.

Lastly, and for our purposes perhaps most importantly, the use of narrative has argumentative significance. Jim Cheney calls attention to this feature of narrative when he claims, "To contextualize ethical deliberation is, in some sense, to provide a narrative or story, from which the solution to the ethical dilemma emerges as the fitting conclusion."[17] Narrative has argumentative force by suggesting *what counts* as an appropriate conclusion to an ethical situation. One ethical conclusion suggested by the climbing narrative is that what counts as a proper ethical attitude toward mountains and rocks is an attitude of respect and care (whatever that turns out to be or involve), not one of domination and conquest.

In an essay entitled "In and Out of Harm's Way: Arrogance and Love," feminist philosopher Marilyn Frye distinguishes between "arrogant" and "loving" perception as one way of getting at this difference in the ethical attitudes of care and conquest.[18] Frye writes:

> The loving eye is a contrary of the arrogant eye.
>
> The loving eye knows the independence of the other. It is the eye of a seer who knows that nature is indifferent. It is the eye of one who knows that to know the seen, one must consult something other than one's own will and interests and fears and imagination. One must look at the thing. One must look and listen and check and question.
>
> The loving eye is one that pays a certain sort of attention. This attention can require a discipline but *not* a self-denial. The discipline is one of self-knowledge, knowledge of the scope and boundary of the self. . . . In particular, it is a matter of being able to tell one's own interests from those of others and of knowing where one's self leaves off and another begins. . . .
>
> The loving eye does not make the object of perception into something edible, does not try to assimilate it, does not reduce it to the size of the seer's desire, fear and imagination, and hence does not have to simplify. It knows the complexity of the other as something which will forever present new things to be known. The science of the loving eye would favor The Complexity Theory of Truth [in contrast to The Simplicity Theory of Truth] and presuppose The Endless Interestingness of the Universe.[19]

According to Frye, the loving eye is not an invasive, coercive eye which annexes others to itself, but one which "knows the complexity of the other as something which will forever present new things to be known."

When one climbs a rock as a conqueror, one climbs with an arrogant eye. When one climbs with a loving eye, one constantly "must look and listen and check and question." One recognizes the rock as something very different, something perhaps totally indifferent to one's own presence, and finds in that difference joyous occasion for celebration. One knows "the boundary of the self," where the self—the "I," the climber—leaves off and the rock begins. There is no fusion of two into one, but a complement of two entities *acknowledged* as separate, different, independent, *yet in relationship;* they are in relationship *if only* because the loving eye is perceiving it, responding to it, noticing it, attending to it.

An ecofeminist perspective about both women and nature involves this shift in attitude from "arrogant perception" to "loving perception" of the nonhuman world. Arrogant perception of nonhumans by humans presupposes and maintains *sameness* in such a way that it expands the moral community to those beings who are thought to resemble (be like, similar to, or the same as) humans in some morally significant way. Any environmental movement or ethic based on arrogant perception builds a moral hierarchy of beings and assumes some common denominator of moral considerability in virtue of which like beings deserve similar treatment or moral consideration and unlike beings do not. Such environmental ethics are or generate a "unity in sameness." In contrast, "loving perception" presupposes and maintains *difference*—a distinction between the self and other, between human and at least some nonhumans—in such a way that perception of the other as other *is* an expression of love for one who/which is recognized at the outset as independent, dissimilar, different. As Maria Lugones says, in loving perception, "Love is seen not as fusion and erasure of difference but as incompatible with them."[20] "Unity in sameness" alone is an *erasure of difference.*

"Loving perception" of the nonhuman natural world is an attempt to understand what it means *for humans* to care about the nonhuman world, a world *acknowledged* as being independent, different, perhaps even indifferent to humans. Humans *are* different from rocks in important ways, even if they are also both members of some ecological community. A moral community based on loving perception of oneself *in relationship with* a rock, or with the natural environment as a whole, is one which acknowledges and respects difference, whatever "sameness" also exists.[21] The limits of loving perception are determined only by the limits of one's (e.g., a person's, a community's) ability to respond lovingly (or with appropriate care, trust, or friendship)—whether it is to other humans or to the nonhuman world and elements of it.[22]

If what I have said so far is correct, then there are very different ways to climb a mountain and *how* one climbs it and *how* one narrates the experience of climbing it matter ethically. If one climbs with "arrogant perception," with an attitude of "conquer and control," one keeps intact the very sorts of thinking that characterize a logic of domination and an oppressive conceptual framework. Since the oppressive conceptual framework which sanctions the domination of nature is a patriarchal one, one also thereby keeps intact, even if unwittingly, a patriarchal conceptual framework. Because the dismantling of patriarchal conceptual frameworks is a feminist issue, *how* one climbs a mountain and *how* one narrates—or tells the story—about the experience of climbing also are *feminist issues.* In this way, ecofeminism makes visible why, at a conceptual level, environmental ethics is a feminist issue. I turn now to a consideration of ecofeminism as a distinctively feminist and environmental ethic.

ECOFEMINISM AS A FEMINIST AND ENVIRONMENTAL ETHIC

A feminist ethic involves a twofold commitment to critique male bias in ethics wherever it occurs, and to develop ethics which are not male-biased. Sometimes this involves articulation of values (e.g., values of care, appropriate trust, kinship, friendship) often lost or underplayed in mainstream ethics.[23] Sometimes it involves engaging in theory building by pioneering in new directions or by revamping old theories in gender sensitive ways. What makes the critiques of old theories or conceptualizations of new ones "feminist" is that they emerge out of sex-gender analyses and reflect whatever those analyses reveal about gendered experience and gendered social reality.

As I conceive feminist ethics in the pre-feminist present, it rejects attempts to conceive of ethical theory in terms of necessary and sufficient conditions, because it assumes that there is no essence (in the sense of some transhistorical, universal, absolute abstraction) of feminist ethics. While attempts to formulate joint necessary and sufficient conditions of a feminist ethic are unfruitful, nonetheless, there are some necessary conditions, what I prefer to call "boundary conditions," of a feminist ethic. These boundary conditions clarify some of the minimal conditions of a feminist ethic without suggesting that feminist ethics has some ahistorical essence. They are like the boundaries of a quilt or collage. They delimit the territory of the piece without dictating what the interior, the design, the actual pattern of the piece looks like. Because the actual design of the quilt emerges from the multiplicity of voices of women in a cross-cultural context, the design will change over time. It is not something static.

What are some of the boundary conditions of a feminist ethic? First, nothing can become part of a feminist ethic—can be part of the quilt—that promotes sexism, racism, classism, or any other "isms" of social domination. Of course, people may disagree about what counts as a sexist act, racist attitude, classist behavior. What counts as sexism, racism, or classism may vary cross-culturally. Still, because a feminist ethic aims at eliminating sexism and sexist bias, and (as I have already shown) sexism is intimately connected in conceptualization and in practice to racism, classism, and naturism, a feminist ethic must be anti-sexist, anti-racist, anti-classist, anti-naturist and opposed to any "ism" which presupposes or advances a logic of domination.

Second, a feminist ethic is a *contextualist* ethic. A contextualist ethic is one which sees ethical discourse and practice as emerging from the voices of people located in different historical circumstances. A contextualist ethic is properly viewed as a *collage* or *mosaic,* a *tapestry* of voices that emerges out of felt experiences. Like any collage or mosaic, the point is not to have *one picture* based on a unity of voices, but a *pattern* which emerges out of the very different voices of people located in different circumstances. When a contextualist ethic is *feminist,* it gives central place to the voices of women.

Third, since a feminist ethic gives central significance to the diversity of women's voices, a feminist ethic must be structurally pluralistic rather than unitary or reductionistic. It rejects the assumption that there is "one voice" in terms of which ethical values, beliefs, attitudes, and conduct can be assessed.

Fourth, a feminist ethic reconceives ethical theory as theory in process which will change over time. Like all theory, a feminist ethic is based on some generalizations.[24] Nevertheless, the generalizations associated with it are themselves a pattern of voices within which the different voices emerging out of concrete and alternative descriptions of ethical situations have meaning. The coherence of a feminist theory so conceived is given within a historical and conceptual context, i.e., within a set of historical, socioeconomic circumstances (including circumstances of race, class, age, and affectional orientation) and within a set of basic beliefs, values, attitudes, and assumptions about the world.

Fifth, because a feminist ethic is contextualist, structurally pluralistic, and "in-process," one way to evaluate the claims of a feminist ethic is in terms of their *inclusiveness:* those claims (voices, patterns of voices) are morally and epistemologically favored (preferred, better, less partial, less biased) which are more inclusive of the felt experiences and perspectives of oppressed persons. The condition of inclusiveness requires and ensures that the diverse voices of women (as oppressed persons) will be given legitimacy in ethical theory building. It thereby helps to minimize empirical bias, e.g., bias rising from faulty or false generalizations based on stereotyping, too small a sample size, or a skewed sample. It does so by ensuring that any generalizations which are made about ethics and ethical decision making include—indeed cohere with—the patterned voices of women.[25]

Sixth, a feminist ethic makes no attempt to provide an "objective" point of view, since it assumes that in contemporary culture there really is no such point of view. As such, it does not claim to be "unbiased" in the sense of "value-neutral" or "objective." However, it does assume that whatever bias it has as an ethic centralizing the voices of oppressed persons is a *better bias*—"better" because it is more inclusive and therefore less partial—than those which exclude those voices.[26]

Seventh, a feminist ethic provides a central place for values typically unnoticed, underplayed, or misrepresented in traditional ethics, e.g., values of care, love, friendship, and appropriate trust.[27] Again, it need not do this at the exclusion of considerations of rights, rules, or utility. There may be many contexts in which talk of rights or of utility is useful or appropriate. For instance, in contracts or property relationships, talk of rights may be useful and appropriate. In deciding what is cost-effective or advantageous to the most people, talk of utility may be useful and appropriate. In a feminist *qua* contextualist ethic, whether or not such talk is useful or appropriate depends on the context; *other values* (e.g., values of care, trust, friendship) are *not* viewed as reducible to or captured solely in terms of such talk.[28]

Eighth, a feminist ethic also involves a reconception of what it is to be human and what it is for humans to engage in ethical decision making, since it rejects as either meaningless or currently untenable any gender-free or gender-neutral description of humans, ethics, and ethical decision making. It thereby rejects what Alison Jaggar calls "abstract individualism," i.e., the position that it is possible to identify a human essence or human nature that exists independently of any particular historical context.[29] Humans and human moral conduct are properly understood essentially (and not merely accidentally) in terms of networks or webs of historical and concrete relationships.

All the props are now in place for seeing how ecofeminism provides the framework for a distinctively feminist and environmental ethic. It is a feminism that critiques male bias wherever it occurs in ethics (including environmental ethics) and aims at providing an ethic (including an environmental ethic) which is not male biased—and it does so in a way that satisfies the preliminary boundary conditions of a feminist ethic.

First, ecofeminism is quintessentially anti-naturist. Its anti-naturism consists in the rejection of any way of thinking about or acting toward nonhuman nature that reflects a logic, values, or attitude of domination. Its anti-naturist, anti-sexist, anti-racist, anti-classist (and so forth, for all other "isms" of social domination) stance forms the outer boundary of the quilt: nothing gets on the quilt which is naturist, sexist, racist, classist, and so forth.

Second, ecofeminism is a contextualist ethic. It involves a shift *from* a conception of ethics as primarily a matter of rights, rules, or principles predetermined and applied in specific cases to entities viewed as competitors in the contest of moral standing, *to* a conception of ethics as growing out of what Jim Cheney calls "defining relationships," i.e., relationships conceived in some sense as defining who one is.[30] As a contextualist ethic, it is not that rights, or rules, or principles are *not* relevant or important. Clearly they are in certain contexts and for certain purposes.[31] It is just that what *makes* them relevant or important is that those to whom they apply are entities *in relationship with* others.

Ecofeminism also involves an ethical shift *from* granting moral consideration to nonhumans *exclusively* on the grounds of some similarity they share with humans (e.g., rationality, interests, moral agency, sentiency, right-holder status) *to* "a highly contextual account to see clearly what a human being is and what the nonhuman world might be, morally speaking, *for* human beings."[32] For an ecofeminist, *how* a moral agent is in relationship to another becomes of central significance, not simply *that* a moral agent is a moral agent or is bound by rights, duties, virtue, or utility to act in a certain way.

Third, ecofeminism is structurally pluralistic in that it presupposes and maintains difference—difference among humans as well as between humans and at least some elements of nonhuman nature. Thus, while ecofeminism denies the "nature/culture" split, it affirms that humans are both members of an ecological community (in some respects) and different from it (in other respects). Ecofeminism's attention to relationships and community is not, therefore, an erasure of difference but a respectful acknowledgement of it.

Fourth, ecofeminism reconceives theory as theory in process. It focuses on patterns of meaning which emerge, for instance, from the storytelling and first-person narratives of women (and others) who deplore the twin dominations of women and nature. The use of narrative is one way to ensure that the content of the ethic—the pattern of the quilt—may/will change over time, as the historical and material realities of women's lives change and as more is learned about women-nature connections and the destruction of the nonhuman world.[33]

Fifth, ecofeminism is inclusivist. It emerges from the voices of women who experience the harmful domination of nature and the way that domination is tied to their domination as women. It emerges from listening to the voices of indigenous peoples such as Native Americans who have been dislocated from their land and have witnessed the attendant undermining of such values as appropriate reciprocity, sharing, and kinship that characterize traditional Indian culture. It emerges from listening to voices of those who, like Nathan Hare, critique traditional approaches to environmental ethics as white and bourgeois, and as failing to address issues of "black ecology" and the "ecology" of the inner city and urban spaces.[34] It also emerges out of the voices of Chipko women who see the destruction of "earth, soil, and water" as intimately connected with their own inability to survive economically.[35] With its emphasis on inclusivity and difference, ecofeminism provides a framework for recognizing that what counts as ecology and what counts as appropriate conduct toward both human and nonhuman environments is largely a matter of context.

Sixth, as a feminism, ecofeminism makes no attempt to provide an "objective" point of view. It is a social ecology. It recognizes the twin dominations of women and nature as social problems rooted both in very concrete, historical, socioeconomic circumstances and in oppressive patriarchal conceptual frameworks which maintain and sanction these circumstances.

Seventh, ecofeminism makes a central place for values of care, love, friendship, trust, and appropriate reciprocity—values that presuppose that our relationships to others are central to our understanding of who we are.[36] It thereby gives voice to the sensitivity that in climbing a mountain, one is doing something in relationship with an "other," an "other" whom one can come to care about and treat respectfully.

Lastly, an ecofeminist ethic involves a reconception of what it means to be human, and in what human ethical behavior consists. Ecofeminism denies abstract individualism. Humans are who we are in large part by virtue of the historical and social contexts and the relationships we are in, including our relationships with nonhuman nature. Relationships are not something extrinsic to who we are, not an "add on" feature of human nature; they play an essential role in shaping what it is to be human. Relationships of humans to the nonhuman environment are, in part, constitutive of what it is to be a human.

By making visible the interconnections among the dominations of women and nature, ecofeminism shows that both are feminist issues and that explicit acknowledgment of both is vital to any responsible environmental ethic. Feminism *must* embrace ecological feminism if it is to end the domination of women because the domination of women is tied conceptually and historically to the domination of nature.

A responsible environmental ethic also *must* embrace feminism. Otherwise, even the seemingly most revolutionary, liberational, and holistic ecological ethic will fail to take seriously the interconnected dominations of nature and women that are so much a part of the historical legacy and conceptual framework that sanctions the exploitation of nonhuman nature. Failure to make visible these interconnected, twin dominations results in an inaccurate account of how it is that nature has been and continues to be dominated and exploited and produces an environmental ethic that lacks the depth necessary to be truly *inclusive* of the realities of persons who at least in dominant Western culture have been intimately tied with that exploitation, viz., women. Whatever else can be said in favor of such holistic ethics, a failure to make visible ecofeminist insights into the common denominators of the twin oppressions of women and nature is to perpetuate, rather than overcome, the source of that oppression.

This last point deserves further attention. It may be objected that as long as the end result is "the same"—the development of an environmental ethic which does not emerge out of or reinforce an oppressive conceptual framework—it does not matter whether that ethic (or the ethic endorsed in getting there) is feminist or not. Hence, it simply is *not* the case that any adequate environmental ethic must be feminist. My argument, in contrast, has been that it *does* matter, and for three important reasons. First, there is the scholarly issue of accurately representing historical reality, and that, ecofeminists claim, requires acknowledging the historical feminization of nature and naturalization of women as part of the exploitation of nature. Second, I have shown that the conceptual connections between the domination of women and the domination of nature are located in an oppressive and, at least in Western societies, patriarchal conceptual framework characterized by a logic of domination. Thus, I have shown that failure to notice the nature of this connection leaves at best an incomplete, inaccurate, and partial account of what is required of a conceptually adequate environmental ethic. An ethic which *does not* acknowledge this is simply *not* the same as one that does, whatever else the similarities between them. Third, the claim that, in contemporary culture, one can have an adequate environmental ethic which is *not* feminist assumes that, in contemporary culture, the label *feminist* does not add anything crucial to the nature or description of environmental ethics. I have shown that at least in contemporary culture this is false, for the word *feminist* currently helps to clarify just *how* the domination of nature is conceptually linked to patriarchy and, hence, how the liberation of nature is conceptually linked to the termination of patriarchy. Thus, because it has critical bite in contemporary culture, it serves as an important reminder that in contemporary sex-gendered, raced, classed, and naturist culture, an unlabeled position functions as a privileged and "unmarked" position. That is, without the addition of the word *feminist,* one presents environmental ethics as if it has no bias, including male-gender bias, which is just what ecofeminists deny: failure to notice the connections between the twin oppressions of women and nature is male-gender bias.

One of the goals of feminism is the eradication of all oppressive sex-gender (and related race, class, age, affectional preference) categories and the creation of a world in which *difference does not breed domination*— say, the world of 4001. If in 4001 an "adequate environmental ethic" is a "feminist environmental ethic," the word *feminist* may then be redundant and unnecessary. However, this is *not* 4001, and in terms of the current historical and conceptual reality the dominations of nature and of women are intimately connected. Failure to notice or make visible that connection in 1990 perpetuates the mistaken (and privileged) view that "environmental ethics" is *not* a feminist issue, and that *feminist* adds nothing to environmental ethics.[37]

CONCLUSION

I have argued in this paper that ecofeminism provides a framework for a distinctively feminist and environmental ethic. Ecofeminism grows out of the felt and theorized about connections between the domination of women and the domination of nature. As a contextualist ethic, ecofeminism refocuses environmental ethics on what nature might mean, morally speaking, *for* humans, and on how the relational attitudes of humans to others—humans as well as nonhumans—sculpt both what it is to be human and the nature and ground of human responsibilities to the nonhuman environment. Part of what this refocusing does is to take seriously the voices of women and other oppressed persons in the construction of that ethic.

A Sioux elder once told me a story about his son. He sent his seven-year-old son to live with the child's grandparents on a Sioux reservation so that he could "learn the Indian ways." Part of what the grandparents taught the son was how to hunt the four leggeds of the forest. As I heard the story, the boy was taught, "to shoot your four-legged brother in his hind area, slowing it down but not killing it. Then, take the four legged's head in your hands, and look into his eyes. The eyes are where all the suffering is. Look into your brother's eyes and feel his pain. Then, take your knife and cut the four-legged under his chin, here, on his neck, so that he dies quickly. And as you do, ask your brother, the four-legged, for forgiveness for what you do. Offer also a prayer of thanks to your four-legged kin for offering his body to you just now, when you need food to eat and clothing to wear. And promise the four-legged that you will put yourself back into the earth when you die, to become nourishment for the earth, and for the sister flowers, and for the brother deer. It is appropriate that you should offer this blessing for the four-legged and, in due time, reciprocate in turn with your body in this way, as the four-legged gives life to you for your survival." As I reflect upon that story, I am struck by the power of the environmental ethic that grows out of and takes seriously narrative, context, and such values and relational attitudes as care, loving perception, and appropriate reciprocity, and doing what is appropriate in a given situation—however that notion of appropri-

ateness eventually gets filled out. I am also struck by what one is able to see, once one begins to explore some of the historical and conceptual connections between the dominations of women and of nature. A *re-conceiving* and *re-visioning* of both feminism and environmental ethics, is, I think, the power and promise of ecofeminism.

Notes

1. Explicit ecological feminist literature includes works from a variety of scholarly perspectives and sources. Some of these works are Leonie Caldecott and Stephanie Leland, eds., *Reclaim the Earth: Women Speak Out for Life on Earth* (London: The Women's Press, 1983); Jim Cheney, "Eco-Feminism and Deep Ecology," *Environmental Ethics* 9 (1987): 115–45; Andrée Collard with Joyce Contrucci, *Rape of the Wild: Man's Violence Against Animals and the Earth* (Bloomington: Indiana University Press, 1988); Katherine Davies, "Historical Associations: Women and the Natural World," *Women & Environments* 9, no. 2 (Spring 1987): 4–6; Sharon Doubiago, "Deeper Than Deep Ecology: Men Must Become Feminists," in *The New Catalyst Quarterly,* no. 10 (Winter 1987/88): 10–11; Brian Easlea, *Science and Sexual Oppression: Patriarchy's Confrontation with Women and Nature* (London: Weidenfeld & Nicholson, 1981); Elizabeth Dodson Gray, *Green Paradise Lost* (Wellesley, MA: Roundtable Press, 1979); Susan Griffin, *Women and Nature: The Roaring Inside Her* (San Francisco: Harper and Row, 1978); Joan L. Griscom, "On Healing the Nature/History Split in Feminist Thought," in *Heresies #13: Feminism and Ecology* 4, no. 1 (1981): 4–9; Ynestra King, "The Ecology of Feminism and the Feminism of Ecology," in *Healing Our Wounds: The Power of Ecological Feminism,* ed. Judith Plant (Boston: New Society Publishers, 1989), pp. 18–28; "The Eco-feminist Imperative," in *Reclaim the Earth,* ed. Caldecott and Leland (London: The Women's Press, 1983), pp. 12–16; "Feminism and the Revolt of Nature," in *Heresies #13: Feminism and Ecology* 4, no. 1 (1981), 12–16, and "What Is Ecofeminism?" *The Nation,* 12 December 1987; Marti Kheel, "Animal Liberation Is a Feminist Issue," *The New Catalyst Quarterly,* no. 10 (Winter 1987–88): 8–9; Carolyn Merchant, *The Death of Nature: Women, Ecology and the Scientific Revolution* (San Francisco: Harper and Row, 1980); Patrick Murphy, ed., "Feminism, Ecology, and the Future of the Humanities," special issue of *Studies in the Humanities* 15, no. 2 (December 1988); Abby Peterson and Carolyn Merchant, "Peace with the Earth: Women and the Environmental Movement in Sweden," *Women's Studies International Forum* 9, no. 5–6

(1986): 465–79; Judith Plant, "Searching for Common Ground: Ecofeminism and Bioregionalism," in *The New Catalyst Quarterly,* no. 10 (Winter 1987/88): 6–7; Judith Plant, ed., *Healing Our Wounds: The Power of Ecological Feminism,* (Boston: New Society Publishers, 1989); Val Plumwood, "Ecofeminism: An Overview and Discussion of Positions and Arguments," *Australasian Journal of Philosophy,* Supplement to vol. 64 (June 1986): 120–37; Rosemary Radford Ruether, *New Woman/New Earth: Sexist Ideologies & Human Liberation* (New York: Seabury Press, 1975); Kirkpatrick Sale, "Ecofeminism—A New Perspective," *The Nation,* 26 September 1987, 302–05; Ariel Kay Salleh, "Deeper Than Deep Ecology: The Eco-Feminist Connection," *Environmental Ethics* 6 (1984): 339–45, and "Epistemology and the Metaphors of Production: An Eco-Feminist Reading of Critical Theory," in *Studies in the Humanities* 15 (1988): 130–39; Vandana Shiva, *Staying Alive: Women, Ecology and Development* (London: Zed Books, 1988); Charlene Spretnak, "Ecofeminism: Our Roots and Flowering," *The Elmswood Newsletter,* Winter Solstice 1988; Karen J. Warren, "Feminism and Ecology: Making Connections," *Environmental Ethics* 9 (1987): 3–21; "Toward an Ecofeminist Ethic," *Studies in the Humanities* 15 (1988): 140–156; Mariam Wyman, "Explorations of Ecofeminism," *Women & Environments* (Spring 1987): 6–7; Iris Young, "'Feminism and Ecology' and 'Women and Life on Earth: Eco-Feminism in the 80s'," *Environmental Ethics* 5 (1983): 173–80; Michael Zimmerman, "Feminism, Deep Ecology, and Environmental Ethics," *Environmental Ethics* 9 (1987): 21–44.

2. Francoise d'Eaubonne, *Le Feminisme ou la Mort* (Paris: Pierre Horay, 1974), pp. 213-52.

3. I discuss this in my paper, "Toward an Ecofeminist Ethic."

4. The account offered here is a revision of the account given earlier in my paper "Feminism and Ecology: Making Connections." I have changed the account to be about "oppressive" rather than strictly "patriarchal" conceptual frameworks in order to leave open the possibility that there may be some patriarchal conceptual frameworks (e.g., in non-Western cultures) which are *not* properly characterized as based on value dualisms.

5. Murray Bookchin, "Social Ecology Versus 'Deep Ecology'," in *Green Perspectives: Newsletter of the Green Program Project,* no. 4–5 (Summer 1987): 9.

6. It may be that in contemporary Western society, which is so thoroughly structured by categories of gender, race, class, age, and affectional orientation, that there simply is no meaningful notion of "value-hierarchical thinking" which does not function in an oppressive context. For purposes of this paper, I leave that question open.

7. Many feminists who argue for the historical point that claims (B1) and (B2) have been asserted or assumed to be true within the dominant Western philosophical tradition do so by discussion of that tradition's conceptions of reason, rationality, and science. For a sampling of the sorts of claims made within that context, see "Reason, Rationality, and Gender," ed. Nancy Tuana and Karen J. Warren, a special issue of the American Philosophical Association's *Newsletter on Feminism and Philosophy* 88, no. 2 (March 1989): 17–71. Ecofeminists who claim that (B2) has been assumed to be true within the dominant Western philosophical tradition include Gray, *Green Paradise Lost;* Griffin, *Woman and Nature: The Roaring Inside Her;* Merchant, *The Death of Nature;* Ruether, *New Woman/New Earth.* For a discussion of some of these ecofeminist historical accounts, see Plumwood, "Ecofeminism." While I agree that the historical connections between the domination of women and the domination of nature is a crucial one, I do not argue for that claim here.

8. Ecofeminists who deny (B1) when (B1) is offered as anything other than a true, descriptive, historical claim about patriarchal culture often do so on grounds that an objectionable sort of biological determinism, or at least harmful female sex-gender stereotypes, underlie (B1). For a discussion of this "split" among those ecofeminists ("nature feminists") who assert and those ecofeminists ("social feminists") who deny (B1) as anything other than a true historical claim about how women are described in patriarchal culture, see Griscom, "On Healing the Nature/History Split."

9. I make no attempt here to defend the historically sanctioned truth of these premises.

10. See, e.g., bell hooks, *Feminist Theory: From Margin to Center* (Boston: South End Press, 1984), pp. 51–52.

11. Maria Lugones, "Playfulness, 'World-Travelling,' and Loving Perception," *Hypatia* 2, no. 2 (Summer 1987): 3.

12. At an *experiential* level, some women are "women of color," poor, old, lesbian, Jewish, and physically challenged. Thus, if feminism is going to liberate these women, it also needs to end the racism, classism, heterosexism, anti-Semitism, and discrimination against the handicapped that is constitutive of their oppression as black, or Latina, or poor, or older, or lesbian, or Jewish, or physically challenged women.

13. This same sort of reasoning shows that feminism is also a movement to end racism, classism, ageism, heterosexism and other "isms" which are based in oppressive conceptual frameworks characterized by a logic of domination. However, there is an important caveat: ecofeminism is *not* compatible with all feminisms and all environmentalisms. For a discus-

sion of this point, see my article, "Feminism and Ecology: Making Connections." What it is compatible with is the minimal condition characterization of feminism as a movement to end sexism that is accepted by all contemporary feminisms (liberal, traditional Marxist, radical, socialist, Blacks and non-Western).

14. See, e.g., Gray, *Green Paradise Lost;* Griffin, *Women and Nature;* Merchant, *The Death of Nature;* and Ruether, *New Woman/New Earth.*

15. Suppose, as I think is the case, that a necessary condition for the existence of a moral relationship is that at least one party to the relationship is a moral being (leaving open for our purposes what counts as a "moral being"). If this is so, then the *Mona Lisa* cannot properly be said to have or stand in a moral relationship with the wall on which she hangs, and a wolf cannot have or properly be said to have or stand in a moral relationship with a moose. Such a necessary-condition account leaves open the question whether *both* parties to the relationship must be moral beings. My point here is simply that however one resolves *that* question, recognition of the relationships themselves as a locus of value is a recognition of a source of value that is different from and not reducible to the values of the "moral beings" in those relationships.

16. It is interesting to note that the image of being friends with the Earth is one which cytogeneticist Barbara McClintock uses when she describes the importance of having "a feeling for the organism," "listening to the material [in this case the corn plant]," in one's work as a scientist. See Evelyn Fox Keller, "Women, Science, and Popular Mythology," in *Machina Ex Dea: Feminist Perspectives on Technology,* ed. Joan Rothschild (New York: Pergamon Press, 1983), and Evelyn Fox Keller, *A Feeling for the Organism: The Life and Work of Barbara McClintock* (San Francisco: W. H. Freeman, 1983).

17. Cheney, "Eco-Feminism and Deep Ecology," 144.

18. Marilyn Frye, "In and Out of Harm's Way: Arrogance and Love," *The Politics of Reality* (Trumansburg, New York: The Crossing Press, 1983), pp. 66–72.

19. Ibid., pp. 75–76.

20. Maria Lugones, "Playfulness," p. 3.

21. Cheney makes a similar point in "Eco-Feminism and Deep Ecology," p. 140.

22. Ibid., p. 138.

23. This account of a feminist ethic draws on my paper "Toward an Ecofeminist Ethic."

24. Marilyn Frye makes this point in her illuminating paper, "The Possibility of Feminist Theory," read at the American Philosophical Association Central Division Meetings in Chicago, 29 April–1 May 1986. My discussion of feminist theory is inspired largely by that paper and by Kathryn Addelson's paper

"Moral Revolution," in *Women and Values: Reading in Recent Feminist Philosophy,* ed. Marilyn Pearsall (Belmont, CA: Wadsworth Publishing Co., 1986) pp. 291–309.

25. Notice that the standard of inclusiveness does not exclude the voices of men. It is just that those voices must cohere with the voices of women.

26. For a more in-depth discussion of the notions of impartiality and bias, see my paper, "Critical Thinking and Feminism," *Informal Logic* 10, no. 1 (Winter 1988): 31–44.

27. The burgeoning literature on these values is noteworthy. See, e.g., Carol Gilligan, *In a Different Voice: Psychological Theories and Women's Development* (Cambridge: Harvard University Press, 1982); *Mapping the Moral Domain: A Contribution of Women's Thinking to Psychological Theory and Education,* ed. Carol Gilligan, Janie Victoria Ward, and Jill McLean Taylor, with Betty Bardige (Cambridge: Harvard University Press, 1988); Nel Noddings, *Caring: A Feminine Approach to Ethics and Moral Education* (Berkeley: University of California Press, 1984); Maria Lugones and Elizabeth V. Spelman, "Have We Got a Theory for You! Feminist Theory, Cultural Imperialism, and the Women's Voice," *Women's Studies International Forum* 6 (1983): 573–81; Maria Lugones, "Playfulness"; Annette C. Baier, "What Do Women Want in a Moral Theory?" *Nous* 19 (1985): 53–63.

28. Jim Cheney would claim that our fundamental relationships to one another as moral agents are not as moral agents to rights-holders, and that whatever rights a person properly may be said to have are relationally defined rights, not rights possessed by atomistic individuals conceived as Robinson Crusoes who do not exist essentially in relation to others. On this view, even rights talk itself is properly conceived as growing out of a relational ethic, not vice versa.

29. Alison Jaggar, *Feminist Politics and Human Nature* (Totowa, NJ: Rowman and Allanheld, 1980), pp. 42–44.

30. Henry West has pointed out that the expression "defining relations" is ambiguous. According to West, "the 'defining' as Cheney uses it is an adjective, not a principle—it is not that ethics defines relationships; it is that ethics grows out of conceiving of the relationships that one is in as defining what the individual is."

31. For example, in relationships involving contracts or promises, those relationships might be correctly described as that of moral agent to rights holders. In relationships involving mere property, those relationships might be correctly described as that of moral agent to objects having only instrumental value, "relationships of instrumentality." In comments on an earlier draft of this paper, West suggested that possessive individualism, for instance, might be recast in

such a way that an individual is defined by his or her property relationships.

32. Cheney, "Eco-Feminism and Deep Ecology," p. 144.

33. One might object that such permission for change opens the door for environmental exploitation. This is not the case. An ecofeminist ethic is anti-naturist. Hence, the unjust domination and exploitation of nature is a "boundary condition" of the ethic; no such actions are sanctioned or justified on ecofeminist grounds. What it *does* leave open is some leeway about what counts as domination and exploitation. This, I think, is a strength of the ethic, not a weakness, since it acknowledges that *that* issue cannot be resolved in any practical way in the abstract, independent of a historical and social context.

34. Nathan Hare, "Black Ecology," in *Environmental Ethics,* ed. K. S. Shrader-Frechette (Pacific Grove, CA: Boxwood Press, 1981), pp. 229–36.

35. For an ecofeminist discussion of the Chipko movement, see my "Toward an Ecofeminist Ethic," and Shiva's *Staying Alive.*

36. See Cheney, "Eco-Feminism and Deep Ecology," p. 122.

37. I offer the same sort of reply to critics of ecofeminism such as Warwick Fox who suggest that for the sort of ecofeminism I defend, the word *feminist* does not add anything significant to environmental ethics and, consequently, that an ecofeminist like myself might as well call herself a deep ecologist. He asks: "Why doesn't she just call it [i.e., Warren's vision of a transformative feminism] deep ecology? Why specifically attach the label *feminist* to it . . . ?" (Warwick Fox, "The Deep Ecology-Ecofeminism Debate and Its Parallels," *Environmental Ethics* 11, no. 1 [1989]: 14, n.22). Whatever the important similarities between deep ecology and ecofeminism (or, specifically, my version of ecofeminism)—and, indeed, there are many—it is precisely my point here that the word *feminist* does add something significant to the conception of environmental ethics, and that any environmental ethic (including deep ecology) that fails to make explicit the different kinds of interconnections among the domination of nature and the domination of women will be, from a feminist (and ecofeminist) perspective such as mine, inadequate.

For Further Discussion

1. Why does Warren claim that environmental degradation and exploitation are feminist issues? Do you agree?

2. Warren speaks of the "historical, experiential, symbolic, theoretical" connections between the domination of women and the domination of nature. Can you think of examples for each type of connection?

3. What does Warren mean by the "logic of domination"?

4. Warren suggests that various value dualisms and value hierarchies are used to justify oppressive domination. Can you think of concepts that might be paired with the following words to suggest both value dualism and value hierarchies: *mind, reason, humans, objectivity, man?*

5. Why do you think that Warren uses the first-person narrative about rock climbing in her essay?

6. What was your reaction to the story offered about the Sioux elder? What ethical implications does Warren draw from this story?

DISCUSSION AND STUDY QUESTIONS FOR CHAPTER 15

1. Deep ecologists claim that we cannot make firm ontological distinctions among individuals. Is this true? From where you are right now, how many individuals can you see? Whom or what did you count as an individual and why? Are cells individuals? Molecules? Ecosystems?

2. Is humanity in general at fault for the ecological crisis, or are specific humans more responsible than others? Are those with power and money more responsible than those without power and money?

3. Are women more often and more closely associated with nature than men? Develop a list of characteristics typically attributed to nature. Do the same for women and for men. Are there parallels?

CHAPTER 16

Political Theory and the Environment

Throughout this textbook, our understanding of ethics has included both individual and social components. In this chapter, we turn explicitly to the social and political side of that distinction. We examine some of the most fundamental questions of political theory: How should we live together in community? What values should guide our political institutions and decisions?

One of the oldest challenges of political theory concerns the justification of government to make decisions based on a particular understanding of what is good. As mentioned in Chapter 1, Plato's political philosophy argued that society should be governed by those people (Plato called them philosopher-kings) who know what is good. Knowledge of the good justifies political authority. Kings, to the extent that they are philosophers, have legitimate political authority only to the degree that they know what is good for society. Political decisions must be guided by this vision of the good.

Chapter 1 also described Plato's critique of democracy. Because most people lack knowledge of the good, there is little chance for a democratic decision to attain the good. Most people make decisions based on mere opinions of what is good, on what they think is good, but not on what they know to be good. In more familiar terms, most people make decisions based on what they want, and what they want is not always what is good for them. Democracy, with its commitment to individual freedom, allows people to pursue their wants even at the expense of what is good.

The modern political theory of liberal democracy rejects all of this. Democratic theory justifies political authority by appeal to the consent of the governed, rather than appeal to knowledge of the good. Modern political theory is liberal in the sense that it places a high ethical value on individual freedom or liberty.

Modern political theory raises two major objections to the classical political theory of Plato. First, there are real questions about the possibility of ever really knowing what is good. This objection is raised on both practical and philosophical grounds. History shows that most political leaders—from ancient kings to modern dictators—who claim to know what is good for the people—more often than not were corrupt and despotic. Even if in theory it were possible to know what is good, in practice this approach has proved unjust and oppressive. Philosophically, modern political theory rejects our ability to ever know what is good. On this view, the good and all other value questions are matters of subjective opinion rather than objective fact. Because these value questions are a matter of opinion, political decisions ought to give every opinion equal weight.

Modern political theory also rejects the classical model on grounds that it denies individuals the moral respect that they deserve. Justice, on the modern view, demands that the interests of each individual be given equal moral consideration. Again, as we saw in Chapter 1, the dignity owed to each individual is usually derived from the inherent worth attributed to each self-governing person. Rather than concern for what is good, modern political theory emphasizes such procedural issues as rights, fairness, equality, liberty. These values, rather than any particular understanding of the good, protect the dignity of each individual.

Environmental concerns raise interesting challenges for political theory. At the level of party politics, environmentalists are typically identified as liberals and Democrats. At the level of political philosophy, however, it is more difficult to integrate environmental concerns into liberal democratic institutions. The two major decision-making avenues of liberal democracies are the voting booth and free markets. Each promotes and protects the right of individuals to pursue their own understanding of what is good. But neither guarantees environmentally benign results. In fact, as we have seen, many environmentalists fault the political and economic systems of market-based democracies for allowing environmental destruction.

Although some environmental concerns do raise procedural questions of fairness and equality (recall Chapter 13), many more raise questions of the good and the valuable. But what happens when democratic

decision procedures do not promote environmental values and environmental outcomes?

The selections in this chapter consider where environmental concerns should be located along the political spectrum. They suggest that the common identification of environmentalism with liberalism and anti-environmentalism with conservatism is misguided. The authors of these readings suggest the need to rethink several key concepts of political theory in light of environmental challenges.

Traditionalist Conservatism and Environmental Ethics

John R. E. Bliese

Traditional conservatism can be traced to the classical political tradition of Plato and Aristotle. As the name suggests, conservatives would argue that not every change is worth making, that not every change is progress. There are some things that are good and ought to be conserved, even if people don't want to conserve them or people don't understand why.

In the present political climate, conservatives are often identified by their commitment to a small range of social issues such as traditional gender roles for men and women and traditional models of family life. In this selection, John Bliese argues that the philosophical foundations of traditional conservatism could accommodate a strong environmental position. Focusing on Anglo-American conservatism as it has developed since World War II, Bliese finds a solid foundation for a conservative environmentalism.

INTRODUCTION

Environmentalists are usually thought of as liberals, and environmentalism is often thought to be a liberal political position. However, this perception is in large part merely an historical accident. Environmentalism does not have any single political home. In fact, the philosophical foundations of conservatism also support environmental protection and resource conservation.

The term *conservative* is one of the most abused in our political vocabulary. Here I am concerned only with the conservative intellectual movement in America and England that began shortly after World War II. Within that movement there are two primary schools of thought: libertarian, free market advocates who trace their lineage back to Adam Smith, and traditionalists who are the intellectual heirs of Edmund Burke.

. . .

At the heart of traditionalist philosophy are several crucial principles that have profound implications for environmentalism, and our three authors are noteworthy among leading theoreticians of the traditionalist position for explicitly applying them to environmental issues. The principles to be discussed here are:

1. Rejection of materialism
2. Rejection of ideology
3. Piety toward nature
4. Social contract across generations
5. Prudence, the prime political virtue

1. *Rejection of materialism.* I start with a negative statement as principle number one because it is perhaps the most prominent aspect of all traditionalist writings: traditionalists are not materialists. According to Rossiter, the conservative "places moral above material values and ends. . . . The state of culture, learning, law, charity, and morality are of more concern to

the Conservative than the annual output of steel and aluminum." Kirk insists that the conservative rejects the "reduction of human striving to material production and consumption."

Weaver contends that civilization is not to be equated with material success. "There is no correlation between the degree of comfort enjoyed and the achievement of a civilization. On the contrary, absorption in ease is one of the most reliable signs of present or impending decay." A person's character develops in response to inner and outer checks on appetites and desires, so what is really needed is not more consumer goods but training that includes "the disciplines of rigor and denial." In stark contrast to the "spoiled child psychology" of modern America, "self-denial [is] the greatest lesson to be learned." Weaver was a student of the culture of the Old South, and while he well realized its faults, he thought it admirable in one overriding respect: it was *the last non-materialist civilization in the Western World*" and it stands as a challenge: "to save the human spirit by re-creating a non-materialist society." Whether his historical judgment is accurate or not, the conservative principle that he advocates is clear.

For the traditionalist conservative, because production and consumption are very low level concerns, economics does not provide the standard for life, or even for government. Edmund Burke, after all, put economists in the same category with "sophisters."

The traditionalists are especially critical of modern industrial capitalism. "They view unregulated capitalism as an integral part of the materialism and hedonism they so dislike in the modern Western world; they spurn the civilization that has grown up around the capitalist order. They believe that unregulated capitalism has placed great burdens on the way of life they think necessary for the development of human freedom along virtuous lines." Rossiter notes their "outspoken distaste for the excesses, vulgarities, and dislocations of the industrial way of life." Kirk contends that "industrialism was a harder knock to conservatism than the books of the French egalitarians," for it "turned the world inside out," replacing personal loyalties with financial relationships, turning the wealthy from being patrons to pursuing mere aggrandizement, and destroying community.

Industrial capitalism has destroyed the world that the traditionalist conservative prefers, but Kirk is realistic: ". . . once in an industrial society, we cannot get out of it without starving half the world's population.

What we can do, nevertheless, is to humanize the industrial system, so far as lies in our power," and "to stand guard against over-industrialism, a real menace nowadays."

The rejection of materialism can have wide-ranging implications for environmentalism. For example, at the most basic level, a traditionalist is not willing to sacrifice environmental quality merely to get more consumer goods. At the macro level, environmentalists often see the crux of many problems in the current "worship" (for that is what it must be called) of economic growth. Ecologists would perhaps universally agree that unlimited growth on a finite planet is a recipe for disaster. Traditionalist conservatives would also reject the policy of growth as an end in itself because it is a manifestation of materialism pure and simple. Where Edward Abbey calls the desire for unlimited growth "the ideology of the cancer cell," Gray calls it "the most vulgar ideal ever put before suffering humankind." If some restrictions on growth are needed to preserve environmental quality, the traditionalist is quite willing to accept them.

2. *Rejection of ideology.* A second negative principle may also be found throughout the writings of the traditionalists: a rejection of all ideology, and a corresponding humility in the face of the limits of human reason. Traditionalists reject all attempts to refashion society after abstract notions. Society is infinitely complex, always a historical creation, and therefore is not malleable for the creation of utopias, as ideologists believe. John Gray rejects rationalism in politics, because it

> . . . represents political reasoning as an application of first principles of justice or rights, rather than as circumstantial reasoning aiming at the achievement of a *modus vivendi*. It supposes that the functions and limits of state activity can be specified, once and for all, by a theory, instead of varying with the history, traditions, and circumstances that peoples and their governments inherit. . . . It thereby contributes to the corruption of political practice by ideological hubris, which is one of the most distinctive vices of an age wedded to political religions.

Similarly, Gerhart Niemeyer condemns "the ideological mind building thought systems around utopian fantasies with which to manipulate human beings into false hopes."

Imposing abstract ideologies on a society invariably produces disaster. Burke saw that the attempt to re-mold France produced not a utopia but a revolution, vast amounts of human suffering, and eventually tyranny. The scholars who revived traditionalism after World War II had the example of Stalin's Russia as a stark warning of how much suffering an abstract ideology, attempting to create a new type of human being, can bring upon a vast nation.

The other side of this coin, of course, is an insistence that conservatism is non-ideological. Niemeyer contends that "no conservative ideology can be dreamed into existence." Gray claims that "it is by returning to the homely truths of traditional conservatism that we are best protected from the illusions of ideology."

This rejection of ideology (as they define it) has some important implications for environmental policies. It provides a basis from within conservatism for limiting the excesses of free market thinking, where that has itself become an ideology, i.e., an abstract theory to be imposed on society, promising utopia. Much of Gray's *Beyond the New Right* is devoted to precisely this type of a critique, including the very long chapter on environmental policies. It should be valuable to environmentalists, especially in America today, to find such a critique firmly within the conservative intellectual movement.

To this point, I have been dealing with things that the traditionalist opposes. I now turn to a consideration of principles that the traditionalist advocates. Some of the major virtues for conservatism have profound implications that would lead the traditionalist conservative to be an environmentalist.

3. *Piety toward nature*. I begin with what Weaver calls "a crowning concept which governs [a person's] attitude toward the totality of the world": the capstone virtue, piety. Weaver defines piety as "a discipline of the will through respect. It admits the right to exist of things larger than the ego, of things different from the ego." In order to bring harmony back into our world, we need to regard three things "with the spirit of piety": nature, other people, and the past. It is the first of these, of course, that is most important here.

Weaver explains the proper, pious attitude toward nature in several of his works. The point is a crucial one, so we need to consider it, in his own words, at some length. He contends that

> . . . the attitude toward nature . . . is a matter . . . basic to one's outlook or philosophy of life . . . [A

pious person] tends to look upon nature as something which is given and something which is finally inscrutable. This is equivalent to saying that he looks upon it as the creation of a Creator. There follows from this attitude an important deduction, which is that man has a duty of veneration toward nature and the natural. Nature is not something to be fought, conquered and changed according to any human whims. To some extent, of course, it has to be used. But what man should seek in regard to nature is not a complete dominion but a *modus vivendi*—that is, a manner of living together, a coming to terms with something that was here before our time and will be here after it. The important corollary of this doctrine, it seems to me, is that man is not the lord of creation, with an omnipotent will, but a part of creation, with limitations, who ought to observe a decent humility in the face of the inscrutable.

In his only autobiographical essay, Weaver traces the development of his thought toward conservatism: "I found myself in decreasing sympathy with those social and political doctrines erected upon the concept of a man-dominated universe." "Creation," he concludes, "is not ours and [it] exhibits the marks of a creative power that we do not begin to possess."

In one of his last works, he returns to this theme: "Civilized man carries a sense of restraint into his behavior both toward nature and his fellow beings. The first of these is piety; the second ethics. Piety comes to us as a warning voice that we must think as mortals, that it is not for us either to know all or to control all. It is a recognition of our own limitations and a cheerful acceptance of the contingency of nature, which gives us the productive virtue of humility. . . . [Nature] is the matrix of our being." "Piety respects the mystery of nature," and an "acceptance of nature, with an awareness of the persistence of tragedy, is the first element of spirituality."

A similar conclusion may be found in an analysis of Christian society by T. S. Eliot. "We may say that religion . . . implies a life in conformity with nature. It may be observed that the natural life and the supernatural life have a conformity to each other which neither has with the mechanistic life . . . a wrong attitude towards nature implies, somewhere, a wrong attitude toward God, and . . . the consequence is an inevitable doom."

The dominant attitude toward nature, traditionalist conservatives believe, is completely wrong. Weaver sums up "the offense of modern man" by saying that "he is impious," and this impiety is nothing less than a "sin." Concerning the modern attitude toward the natural world, he writes:

> For centuries now we have been told that our happiness requires an unrelenting assault upon [nature]: dominion, conquest, triumph—all these names have been used as if it were a military campaign. Somehow the notion has been loosed that nature is hostile to man or that her ways are offensive or slovenly, so that every step of progress is measured by how far we have altered these. Nothing short of a recovery of the ancient virtue of *pietas* can absolve man from this sin. . . . [The attempt to win] an unconditional victory over nature . . . is impious, for . . . it violates the belief that creation or nature is fundamentally good, that the ultimate reason for its laws is a mystery, and that acts of defiance such as are daily celebrated by the newspapers are subversive of cosmos. Obviously a degree of humility is required to accept this view.

Weaver maintained this position throughout his life. In *Visions of Order,* he again rejects the notion of conquering nature. "If nature is something ordained by a creator, one does not speak of 'conquering' it. The creation of a benevolent creator is something good, and conquest implies enmity and aggression." And he again denounces industrialism for "constantly making war upon nature, disfiguring and violating her."

Kirk also condemns the consequences of impiety toward nature. "The modern spectacle of vanished forests and eroded lands, wasted petroleum and ruthless mining . . . is evidence of what an age without veneration does to itself and its successors." Those words were originally written in the 1950s. How much worse must a traditionalist believe we are today, considering, for instance, the national forests "hammered" by clearcutting, the "cow bombed" grazing lands of the west, the lunar landscapes created by heap leach mining, or the fragile deserts torn up by ORVs. Those are not just the results of poor policies; they are impieties toward nature which, as the traditionalist insists, are nothing less than sins—a conclusion with which environmentalists would certainly concur.

4. *Social contract across generations.* A fundamental tenet of traditionalist conservatism is that society is intergenerational, "joined in perpetuity by a moral bond among the dead, the living, and those yet to be born—

the community of souls." Of all of Burke's magnificent prose, the passage perhaps most often quoted by conservatives elaborates this principle: "Society is indeed a contract. . . . It is a partnership in all science; a partnership in all art; a partnership in every virtue, and in all perfection. As the ends of such a partnership cannot be obtained in many generations, it becomes a partnership not only between those who are living, but between those who are living, those who are dead, and those who are to be born."

In this partnership, we who are now alive have an obligation to posterity. Our fundamental role toward future generations is one of stewardship: "The spirit of trusteeship—the sense of receiving a precious heritage and handing it on intact and perhaps even slightly strengthened—pervades Conservatism."

The conservative principle of trusteeship for future generations has obvious links with environmentalism. Indeed, as Carl Pope, Executive Director of the Sierra Club, states, "If there is anything that has distinguished the environmental movement during the past 100 years, it has been our insistence that we not plan for a one-generation society, that the future matters."

5. *Prudence, the prime political virtue.* According to Burke, "prudence is not only first in rank of the virtues political and moral, but she is the director of all the others." Rossiter explains that prudence "represents a cluster of urges—toward caution, deliberation, and discretion, toward moderation and calculation, toward old ways and good form—which gives every other standard virtue a special look when displayed by a true Conservative."

One basic reason that the traditionalist insists on prudence in altering or "reforming" society is because of its complexity. In Burke's words, "the nature of man is intricate: the objects of society are of the greatest possible complexity," so complex that individual human reason is not able to grasp them. Our culture has been created over many generations, with contributions from countless minds and hearts. Therefore, says the traditionalist, we need to be very careful when trying to change society on the basis of some new notion an individual has dreamed up. We need to remember that the individual is foolish, but the species is wise.

Such reasoning, of course, applies *a fortiori* to the environment. As Forest Service Chief Jack Ward Thomas is fond of quoting: "ecosystems are not only more complex than we think: they are more complex than we *can* think." If we need to be very careful, humble and prudent in introducing social and political changes and reforms, we need to be much more so in changing the environment.

IMPLICATIONS FOR POLICY

The political philosophy of traditionalist conservatism, we have seen, contains several important principles that support environmental protection and resource conservation. While this is not the place for a detailed analysis of specific policies, we may consider briefly a few general implications of the traditionalist perspective for environmental policy.

1. *Conservation of natural resources and sustainable development.* The conservative notions of stewardship, of obligation for the future, of holding our world and our civilization in trusteeship, have some important implications for environmental policies. Both environmentalists and traditionalist conservatives reject many current practices as "unconscionable effort[s] to transfer . . . costs and risks from the present to future generations." Thus, T. S. Eliot objects "to the exhaustion of natural resources." He contends that "a good deal of our material progress is a progress for which succeeding generations may have to pay dearly." He illustrates this contention by pointing to soil erosion, a result of "the exploitation of the earth, on a vast scale for two generations, for commercial profit: immediate benefits leading to dearth and desert."

Kirk likewise objects to squandering natural resources that future generations will need. "In America especially, we live beyond our means by consuming the portion of posterity, insatiably devouring minerals and forests and the very soil, lowering the water-table, to gratify the appetites of the present tenants of the country." He recommends "turning away from the furious depletion of natural resources, we ought to employ our techniques of efficiency in the interest of posterity, voluntarily conserving our land and our minerals and our forest and our water and our old towns and our countryside for the future partners in our contract of eternal society."

The current concept of "sustainable development" is an eminently conservative idea. Traditionalists would favor measures to ensure that renewable resources are used no faster than they can be replenished and that nonrenewable resources are used very carefully and ef-

ficiently, ideally no faster than substitutes can be found and introduced. An economy designed on these lines would meet our obligations to future generations.

Two of the most daunting problems that we face—global warming and the collapse of biodiversity—call for the prime conservative virtue of prudence above all. In both cases, we are conducting uncontrolled and irreversible experiments with the entire planet, surely the height of imprudence.

2. *Climate change.* Our massive use of fossil fuels has significantly changed the composition of the atmosphere by increasing the amount of greenhouse gases. The ultimate consequences are not at all clear, but they could be devastating. It may take another decade or more for scientists to study the question.

Thus, what policies should we pursue now? Prudence would surely dictate that we not continue blindly with the status quo. We should cut our energy use dramatically, in order to reduce the addition of greenhouse gases until we have greater knowledge about their ultimate effects. Nor would this require drastic, imprudent policies. Quite the contrary, it could be done rather easily because we now use energy very inefficiently. Major improvements in efficiency would be worth making even in the absence of concern over climate change.

3. *Declining biodiversity: the need for prudent and pious action.* The incredible diversity of life is one of the most outstanding features of our marvelous, and so far as we know, unique planet. These myriads of life forms have evolved over hundreds of millions of years. In the natural course of events, species are constantly becoming extinct and other species evolving. But we are now causing extinctions at rates hundreds or thousands of times higher than normal. We are causing the loss of untold numbers of species without knowing what potential value they might be to us—as say, medicines or food—without knowing anything about their roles in their ecosystems and without knowing at what point whole ecosystems will simply collapse, as one after another of their important elements are destroyed. This loss surely is the height of impiety, but it is also enormously imprudent. In the often quoted words of biologist Edward O. Wilson, "this is the folly our descendants are least likely to forgive us." A traditionalist conservative would appreciate the homely advice of Aldo Leopold, that it is the first rule of the tinkerer, when you take something apart, to save all the pieces.

Moreover, traditionalist conservatism, with its veneration for religion, would find support for preserving biodiversity in the *Bible*. The story of Noah is a relevant allegory for our time: God may have given humans "dominion" over the Earth, but He also ordered Noah to save *all* the creatures, not just the charismatic megafauna. A traditionalist would see no reason to believe that God expects less of us today.

To be prudent and pious, we should do everything we can to preserve the diversity of life on Earth. The key is to preserve sufficient habitat intact and unpolluted. If that means some restrictions on "development," so be it. For the traditionalist, piety and prudence in preserving God's creatures would be far more important than a few more golf courses or shopping malls.

4. *Costs of environmental protection.* The business community often objects to pollution abatement measures because they increase costs of production. The traditionalist would here agree with the libertarian that precisely those costs *ought* to be borne by the polluter. But the traditionalist would go even farther and add that there is a basic duty, a moral responsibility on the part of industry, and everyone else for that matter, not to pollute the environment. No one has the right to hurt his or her neighbors or to damage creation just to save a little money—a position shared with environmentalists.

5. *Role of government and restrictions on the market.* Traditionalist conservatives apply their rejection of ideology to free market theory as well, for it has too often become the ideology of the libertarian. Traditionalists, to be sure, accept the importance of a free market as a means of ensuring freedom. But the traditionalist conservative "would . . . resist making a fetish of market institutions" and "would also be ready to curb them when their workings are demonstrably harmful." The traditionalist is perfectly willing to grant government a role, especially in solving environmental problems. John Gray contends that "market institutions, although they are indispensably necessary, are insufficient as guarantors of the integrity of the environment, human as well as natural. They must be supplemented by governmental activity when . . . private investment cannot by itself sustain the public environment of the common life."

When confronting specific environmental problems, Kirk in similar fashion looks to the government for solutions. In a newspaper column about the damages caused by pesticides, Kirk concludes that "the public needs to bring pressure upon state authorities, everywhere, to take measures against polluters of every sort." In another column he laments the loss of "farmlands, woodlands and wetlands" to urban sprawl, industrial expansion and tourist development. Arguing that "restraining action has become necessary," he praises a policy adopted by the government of Vermont to "restrain thoughtless development by bulldozer and speculator."

Traditionalist principles would call for limits on some specific systemic elements in our economy that can be particularly damaging to the environment. For example, a traditionalist conservative would want to put checks on the market in those areas where it can be used to "justify" harvesting naturally renewable resources to depletion. Our economic system can lead a company to view a resource as simply an investment, and if an alternative investment offers higher returns, the company can find it "economical" to deplete the resource and put its money elsewhere. With equally damaging results, "discount rates" can be used to justify immediate consumption and destruction of natural resources, leaving nothing to the future. A traditionalist would want to device checks on such workings of the market, where they violate our moral obligations of piety and stewardship.

6. *Preservation of wilderness.* There are many other areas in which traditionalist conservative ideas coincide with and reinforce those of environmentalists. It may suffice here to mention just a couple that relate to preservation of natural areas.

Now that our wilderness is virtually all gone, we need to preserve what little is left for a wide variety of reasons, one of which is for its heritage value. Aldo Leopold claims that "many of the attributes most distinctive of America and Americans are [due to] the impress of the wilderness and the life that accompanied it." Because traditionalists are interested in preserving those things that have shaped what is best in the American character, their preservation is the basis for an argument for wilderness preservation that they would support.

Environmentalists believe that natural areas should be preserved for their aesthetic and spiritual values. We need periodically to get back to nature for spiritual renewal, to recover from the stresses of modernity and our hectic urban lives. With his belief that nature is to be venerated, it is not surprising to find a similar sentiment expressed by Weaver: "What humane spirit, after reading a newspaper or attending a popular motion picture or listening to the farrago of nonsense on a radio program, has not found relief in fixing his gaze upon some characteristic bit of nature? It is escape from the sickly metaphysical dream. Out of the surfeit of falsity born of technology and commercialism we rejoice in returning to primary data." And he wrote these remarks before the age of television.

The political philosophy of traditionalist conservatism, we have seen, contains several important principles that support conservation of natural resources and protection of the environment. They establish a basis for limiting some of the excesses of rampant industrialism and thoughtless "development." And they offer, from within the conservative movement, grounds for curbing some of the excesses that result from a doctrinaire or ideological application of free market theory. In the political climate of America in the mid-1990s, environmentalists might find it very useful to discover these intellectual resources from impeccably conservative sources.

For Further Discussion

1. Prior to reading this essay, how would you have defined a conservative? Have your views changed as a result of this essay?

2. Which other authors in this book might agree with traditional conservative views on the environment? Which might disagree?

3. How are the concepts of conservative and conservationist similar? How are they different?

4. How do you understand the phrase "piety toward nature"?

5. How might a conservative react to Herman Daly's distinction (Chapter 4) between economic growth and economic development? Can you cite specific passages from this essay to support your views?

Green Democracy: The Search for an Ethical Solution

Mike Mills

"Green" political thought is concerned with the outcomes of political decision making: creating sustainable communities, preserving sensitive natural areas, reducing toxic pollution, and so on. However, this outcomes approach to political decision making often conflicts with the procedural values of democracy: participation, equality, fairness. In short, environmental goods are not always protected by democratic institutions and procedures. Environmentalists, therefore, seem caught between two repugnant alternatives: jeopardize their environmental goals or be willing to compromise on democratic procedures.

In this selection, Mike Mills seeks to integrate environmental goals with democratic procedures. After rejecting two branches of green political theory (ecoauthoritarianism and ecoradicalism), he attempts to synthesize a more procedural environmentalism into the framework of democratic theory.

Green democracy has, perhaps surprisingly, now become controversial. When such controversies appear the tendency is to search for principles or values that, when consistently applied, may properly guide our behavior and our thoughts. This . . . is no exception and seeks to look for those principles within green political theory and environmental ethics itself. By looking, very briefly, at two branches of green political theory which do, indeed, compete in their use of democracy (ecoauthoritarianism and what I will call ecoradicalism), I will suggest that both suffer by failing adequately to consider two things. First, and most importantly, both advance policy prescriptions without purposefully expanding the moral community to which that policy should be addressed. My argument will be that however we characterise what greens believe or what they want to do, the question of the moral community—its expansion and the implications of its expansion—is logically prior to all others. Second, I will argue that green political theory (in both its ecoauthoritarian and its ecoradical sense) has been perhaps too concerned with outcome, and could risk being more concerned with process. I hope to make it clear that this will have a number of distinct advantages over taking a more consequentialist line. First it will help protect against autocracy; second, it is much easier to reconcile with a fundamentally green (ecocentric) ethic; and third, it is very useful in helping to promote the idea that democracy of various forms and at various levels is, indeed, absolutely central to green political theory.

ECOAUTHORITARIANISM AND ECORADICALISM

Ecoauthoritarianism

Whether it argues the perils of over-population (Hardin 1968) or of scarcity more generally (Ophuls 1977; Ophuls and Boyan 1992) the thrust of ecoauthoritarianism appears to remain the same. There are ecological imperatives which have to be addressed and the political organisation necessary to resolve them may not be particularly democratic. Primarily, this disorientation towards democracy is a function of two things; first, a strongly Hobbesian conception of human nature which implies that individuals will not, of their own free will, make selfless, co-operative personal choices; and second, an apparently overwhelming orientation towards ends-based policies, structures and institutions, rather than means-based ones. In other words, ecoauthoritarians take a very consequentialist moral line in which stark and mutually exclusive choices exist based upon either avoiding the ecological crisis through authoritarian measures, or suffering it. In Hardin's case he argues for "mutual coercion mutually agreed upon" (1968); in Ophuls' case he argues for some kind of competent aristocracy to distribute scarce resources. Yet, a generous interpretation of Ophuls' model of scarcity politics would lead us to conclude that although he sees democracy as constrained at higher levels—perhaps national or regional—he is very keen on democracy lower down because he sees that type of freedom (at the micro-level) as differentiating his eco-society from other authoritarian regimes. Although this does not rescue the theory as a whole it should sensitise us to the fact that different types of democracy can exist at different levels.

The major problem with taking this sort of line is the internal inconsistency of the arguments themselves. To my mind, any theory which accepts a Hobbesian version of human nature and then argues that political power should be wielded (by Hobbesians) without cursory political checks and balances (or the equiva-

lent) is asking for trouble. In other words, the very arguments given by ecoauthoritarians to justify the circumvention of democracy will do quite as well in arguing the complete opposite—the greater the problem, the more severe the risk, the more pressing the imperative, the more necessary it becomes that democracy becomes extended, entrenched and practised. As Sagoff has argued:

> Democracy, as everyone knows, is susceptible to abuse and has all kinds of problems, but I know of no other mechanism for making policy decisions that has this ethical underpinning (citizens having the opportunity to present their views to legislators). (Sagoff 1988: 115)

The counter-argument to this given by Saward (directed against the ecoradicals, 1993) is that even if democracy is desirable (whether for intrinsic or instrumental reasons) it is simply incompatible with other green goals. Although I cannot resolve this question fully now, this point depends entirely on what these goals are. If green political theory is goal or ends oriented (as the ecoauthoritarians are) then this argument has some validity although it will depend on the extent to which we accept that democracy can exist within ecological constraints. However, it is not only possible, but also perfectly reasonable, to argue that green political theory should be more process oriented, in which case democracy becomes not only compatible with, but also essential to, green political theory.

Without labouring the critique, let me just make one more point, the full importance of which will also become more evident as we go on. For ecoauthoritarians the ecological crisis presents us with the need to manage the finite resources of the planet effectively to trust in political or technocratic decisions that will, presumably, secure and distribute goods and control the ecological effects of their consumption. I would argue, however, that our ability to manage on the scale envisaged by ecoauthoritarians must be limited. In this respect then, it may be better to ensure that whatever solution we find to the ecological crisis (and to the centrality of democracy to green political theory), it had better be a process-oriented one, that is, one which minimises the need for management on the basis of inadequate information.

Ecoradicalism

Ecoradicals, on the other hand, are a much messier proposition. My central point will simply be that ecoradicals (if we take ecoradicals to belong in part to green parties and radical green movements) could afford to be more concerned with political processes (as opposed to political ends) and this in turn would help to resolve problems of green democracy. I will not be able to resolve these arguments, however, until later on. Goodin (1992) suggests that while there is a great deal of policy prescription within green programmes (as you would expect) there is actually very little on institutional change and on political processes. He notes that "it is ecological values that form the focus of the green programme" (1992: 183) and in this respect, green politics and political theory posits a largely holistic view of both problems and their solutions. In this, Goodin suggests that it is the nondiscriminatory nature of green values and their push for diversity which lead them "positively [to] embrace pluralism" (1992: 199) and "cherish diversity in its social every bit as much as in its biological form" (1992: 199).

There is, apparently, some gap between principles (values) on the one hand and policies on the other because there does not seem to be much on the question of policy or political processes to join the two, even at a theoretical level. In the economic programme there is an onus on individuals to act, as Goodin notices (and as rallying calls like "think globally, act locally" imply), and for policy and lifestyle to become one and the same thing in places. In economic relations, as in politics, the green message appears to be the same and appears, still, to imply the type of political (democratic) relations necessary for individuals to take control.

Dobson (1990, 1993) sees decentralisation as the central green prescription and sees the guiding principles of ecologism as subsumed under the broad headings of limits to growth (which implies interdependence, finite resources, the paucity of technological solutions) and ecologism's commitment to nonanthropocentric principles and policies. As far as the principle goes, Dobson, quite rightly, says that: "much of ecologism's momentum is controversially engaged in widening the community of rights holders to animals, trees, plants and even inanimate nature" (1993: 223). This provides us with our first tentative link between democracy on the one hand and philosophy on the other.

Ecoradicals, at a philosophical level, though, fall into a variety of camps. Eckersley, for example, distinguishes three varieties of ecocentrism—autopoietic intrinsic value; transpersonal (deep) ecology; and ecofeminist (1992: 60-74). Dobson's (1989) major distinction is between those who advocate the "state of

being" (deep ecology) approach to ethics and those who see nature as having intrinsic value. Both appear to agree that the transpersonal/deep ecology approach (which sees our perceptions, attitudes and behaviour toward the environment as changing with changes in consciousness, experience and intuition) is problematic as the basis of a political theory largely because it is itself non-axiological (Eckersley 1992; Dobson 1989). While Mathews (1988) is I believe, quite right in arguing, that a commitment to deep ecology is not a fatalistic, do-nothing choice, nevertheless, I will proceed on the basis that if green political theory is ethically informed, then it is primarily some version of intrinsic value rather than deep ecology that is likely to be most useful.[1]

The criticisms levelled at ecoradicals focus on the internal consistency of their ideas. One observation made by Saward (1993) concerns the relationship between direct democracy and other green goals such as intrinsic value and holism. Here, he makes the point, rightly, that it is difficult to work to imperatives on the one hand and have a fairly arbitrary decision-making process such as democracy on the other hand. The fact that the green movement does have a perception of a Good Life (Dobson 1990: 14) would seem to reinforce the point that the ecoradicals have a problem—how can you go for the Good Life when your democracy may not take you there? Given, in addition, that there clearly is a bottom line green commitment to political liberalism as seen in "diversity" and "direct democracy" for example, then Sagoff's point is relevant:

> Liberal political theory cannot commit a democracy beforehand to adopt any general rule or principle that answers the moral questions that confront it; if political theory could do this, it would become autocratic and inconsistent with democracy. (Sagoff 1988: 162)

Of course, we are not dealing with an exclusively liberal political theory, but green theory does have liberal political elements which make the principle the same in both theories. In fact liberal political theory does commit us to certain rules and principles which are thought to be, in some sense, prior to the democratic process itself. Mostly, these revolve around who or what might be considered as worthy of participation in such a process—who decides over these moral questions? Liberal political theory does, then, have something to say about the nature of the democratic process, it is simply reticent about policy outcomes. Indeed, although Paehlke (1988) has quite rightly argued that democracy can be enhanced by environmental policy, nevertheless, it is also true to say that the ecoradicals have few safeguards against the triumph of one laudable principle, say, sustainability, over another, decentralisation for example. Dobson observes that:

> It has been suggested that ecologism's commitment to principles such as liberty and democracy is compromised by apparently laying such great emphasis on the *ends* rather than the *means* of political association. (Dobson 1993: 234)

I will be arguing that both of these points are correct, and that they are related. In other words, philosophically, greens have been concerned to expand the community of rights holders (or something similar) and they have also been concerned with ends rather than means. A commitment to some form of liberalism may help in counteracting some forms of strongly prescriptive ideology but it is likely that democracy will suffer unless other safeguards are built in.

Of all the difficulties with the ecoradical's view of democracy, that is by far the most difficult to surmount. Yet in a broader philosophical sense, ecoradicals are far from as consequentialist as their programmes suggest: they lean quite strongly towards doing what is "right," as much as they do towards what is ultimately "good." Certainly in terms of their views on the political system, we can see fairly clearly that their support for participatory democracy and decentralisation denotes a view of how people can develop, grow and take control which is independent of what outcomes that might entail.

Ecoauthoritarians and ecoradicals do display some of the same problems, but in different degrees. Primarily, we can see that both do not necessarily protect (or even advance) a position on democracy which is defensible against strongly asserted and pressing political, economic and social goals. While in the case of ecoauthoritarians I would argue that they are perhaps a little more democratic than we sometimes give them credit for, nevertheless, there is little room theoretically for a consistent commitment to democracy. Ecoradicals, on the other hand, do emphasise democracy but fail to reconcile it with broader imperatives. Such a reconciliation is possible if we change the emphasis of their thinking away from goals and direct it more towards political mechanisms.

Both approaches, to my mind at least, have not taken on board the consequences of their very starting point sufficiently. Before all else, for both, comes the assertion that the ecological crisis is primarily a crisis of our ethical system. Although we may perceive our current problems as those of over-population, resource depletion, and food scarcity, these are first and foremost symptoms of an ethical crisis. Environmental philosophers have been criticised for not providing the types of guidelines to political action which political theorists prefer (see, for example, Dobson 1989) but it has to be said that there is plenty to be getting on with if an ethically based green political theory is what theorists are after.

THE EXPANSION OF THE MORAL COMMUNITY

If we consider what greens argue is distinctive about their ideology, their political theory and their practical concerns, it is invariably the case that these can be reduced to a concern to expand the moral community.[2] Eckersley (1992) argues that the fundamental characteristic of green political theory is the fact that it is ecocentric and this ecocentricity is logically prior to all other political considerations:

> A non-anthropocentric perspective is one that ensures that the interests of non-human species and ecological communities . . . are not ignored in human decision making simply because they are not human or because they are not of instrumental value to humans. (Eckersley 1992: 57)

By accepting that humans are not the only ones with values or the only measure of value, we accept that our moral community expands because it is now necessary to accommodate others within our ethical choices. Eckersley, here, has made two interrelated points—one is that the non-human world should be given consideration and the second is that such consideration can be ensured only when a non-anthropocentric perspective is taken. This is a position I would support, indeed, my argument makes it essential that a non-anthropocentric approach is adopted—without this there is only a contingent expansion of the moral community, and if that happens, then democracy cannot be secured either. Goodin (1992) continues this line of argument when he says that it is "naturalness" that greens value. If it is naturalness that has value, then presumably natural things should be given moral consideration. Dobson too (1990), as we saw earlier, took as a fundamental axiom of ecologism the idea that we should look toward a biocentric, or ecocentric (non-anthropocentric), basis for our political theory.

In terms of theory, we can see our ethical responsibilities shifting away from a largely human-centred approach to be ecocentric. These have been the primary terms of the debate and it is perhaps because of this that the political consequences of expanding the moral community—which I take to be a corollary of switching to an essentially ecocentric ethic—have not received the attention they might deserve. The 1992 General Election Manifesto of the British Green Party, for example, can illustrate this point quite well, not because there is no evidence of a shift, but rather because it surfaces in a rather *ad hoc* way and is not, to any large extent, directed at political variables.

From what we have said so far, we would expect to find the core axiology—that the Green Party was ecocentric and this entailed greater moral considerability of non-human species—represented somewhere within the manifesto itself, or at least an indication of such. And, indeed, it is perfectly possible to find such indications, but they tend to come in very specific forms. Largely, these are either in the forms of policy prescriptions (for example, rapidly phasing out factory farming or the greater protection of the soil) or in terms of changing the basis upon which future decisions will be made:

> [the Green Party] would revolutionize the system of national accounts by rigorously identifying real costs and real benefits in our industrial society. In so doing we would attribute equal value to the natural capital on which we depend (topsoil, water, clean air, fossil fuels etc.) as we do to the financial capital which greases the wheels of the world economy. (Green Party 1992: 11)

To the extent that decision making would change (in terms of values and outputs at least) then clearly this change is in line with a broadening of the moral community. However, green political theory might suggest that concomitant changes might also occur in the political system more broadly—for example, in the attribution of rights to non-humans on the basis of their having value. This type of analysis is not in the manifesto, the overtly political provisions are aimed largely, although not exclusively, at resolving problems of the British political system. Without wishing to labour the point, the critique of Saward (1993) and the warning of Sagoff (1988) still remain. If there is to be an essentially goal-oriented policy-making process (and one based on ecological imperatives) then the Green Party

may have problems with democracy. The commitment that the manifesto shows to *ends* might need to be matched by a similar commitment to the reform of the *process* of politics in a non-anthropocentric way which would ensure that green goals did not take precedence over due process.[3]

A more consistent application of the holistic principle, and the greater incorporation of non-human interests into the political features of the state would make green positions defensible against the accusation that highly deterministic, imperative-driven, consequentialistic policy runs the risk of forsaking democracy. If we accept that value exists non-anthropocentrically, then it clearly is the case that what political movements (and philosophies) do should reflect this change and should find more things morally considerable. We can see in these programmes the obvious wish to expand the moral community but this is invariably associated with ends-based policies, rather than means-based processes.

So, greens propose the expansion of the moral community for very particular reasons (non-anthropocentric) and although such a position (which will be expanded below) does exist theoretically, in practice there are two problems. The first is that programmes are not necessarily informed by ecocentrism and holism and the second is that only certain formulations of this ethic will actually serve the dual function of both securing the democratic basis of green politics and the ethical basis that greens want. Importantly, greens appear to prefer, as you might expect, to argue their case in terms of policy ends, rather than in terms of the more deontological (axiological) premises of the policy process.

ACHIEVING THE EXPANSION OF THE MORAL COMMUNITY

The expansion of the moral community can be achieved in a number of different ways depending, to a great extent, upon how far we want to go and what arguments we are going to use along the way. I have already argued that the type of expansion associated with transpersonal/deep ecology is not one that I am going to pursue here but, rather, will follow Dobson and Eckersley in suggesting that it is difficult to formulate politically.[4] I am further restricted by the critique of green politics offered by Saward and the observations of Sagoff, both of which suggest that too great an emphasis on the ends of policy rather than the means will undermine the democratic basis of green politics. This inclines me to believe that it is best to avoid ethical arguments that are entirely consequentialist. Lastly, I have argued that I am concerned with an ecocentric ethic, one which finds value in the non-human world and leads to the expansion of the moral community.

In expanding the moral community the most obvious first step would be to include animals or sentient beings. For Regan (1984), it is the ability to be "the experiencing subject of life" which denotes whether we get "rights" or not. Once sentience is established Regan argues that all sentient beings are equal in having value and that it is only just that claims which these beings make upon us morally should be seen as valid (i.e. they have rights). For Regan then our moral community expands to incorporate sentient beings, although these rights do not have to be the same rights as those held by humans (because in some cases, such as the vote, it would be silly) and to establish a moral right is not the same as establishing a legal right (the classic case is that of keeping a promise).

So, the boundaries are pretty clear here, and Regan is looking very much to replicate for animals the existing ethic for humans. In this sense, it is not an "environmental ethic" (Rolston 1987) because it does not look to the environment for the source of its value. Neither, then, is it holistic because Regan is concerned with individual sentient beings, not with species, nor with the systems which sustain the animals and to which the animals contribute. Consequently, we could not use this as the only basis from which to expand the moral community, primarily because it could not hope to resolve all of the ecological problems we might be interested in.

Another possibility is the "reverence for life" literature which draws the line of moral considerability differently. Here the fundamental axiom, according to Goodpaster (1983), is not whether something is sentient but whether it has life or not: "Nothing short of the condition of *being alive* seems to me to be a plausible and non-arbitrary criteria (for moral considerability)" (1983: 31). This distinguishes Goodpaster both from those who take a more holistic approach (and would value the systems of those who lived above the individuals themselves) and from animal rights/liberation authors who do construct distinctions (which Goodpaster appears to believe are "arbitrary") between living things, usually on the basis of sentience, the ability to suffer (Singer 1975) or having inherent value which it would be unjust to ignore (Regan 1984). Reverence for life theorists do not, therefore, distinguish between plants and animals as far as moral considerability goes,

although this does not mean, of course, that each are as morally *significant*.

Those who are holistically minded have criticised the reverence for life approach as an ethical system because it takes a conventional view of ethics (that it should be concerned with discrete individuals) and extends it into the non-human world (Callicott 1983: 301)—in this respect the argument is very similar to those heard against animal rights. Presumably the complaint is that no psychological or perceptual change need accompany this ethical change. More important, though, is the argument that reverence for life is difficult if not impossible to live by because we would not be able to do anything any harm (Callicott 1983: 301). Interestingly, then, reverence for life (or life-principle) theorists provide an ethic which is ecocentric but which is not holistic. Again, perhaps, the predicament is that holism is making very particular demands of green theorists, and I will return to this below.

An environmental ethic which is going to be of any use is going to have to allow us to value systems as well as individuals—otherwise, as Sylvan (1984) pointed out, we may save the tree and spoil the forest. Equally though, we cannot risk having an ethic which does not protect individuals otherwise we are all dispensable in the scheme of things. Similarly, this ethic (which will allow us to expand our moral community) must find the source of worth or value of that which will be morally considerable, in nature, rather than in any instrumental value for people.

There are other approaches to the study of environmental ethics which can broaden our understanding of issues such as "value." Rolston (1975), for example, has argued very convincingly that it is possible to draw objective values from nature itself and this has made it possible for him to view ethics at a systemic level. However, he has also argued (1987) that we can find not only instrumental and intrinsic value, but systemic value as well in nature (this resides in the productive processes which generate intrinsic value). Consequently, it is possible to see both individuals and systems as the bearers or objects of value and our failure to appreciate this is simply a failure to discover it. If we take this line we are committing ourselves to a view of the moral community which, because it truly draws value from nature, includes the whole of nature. It does, therefore, have a very expansive view of our responsibilities.

A second holistic theorist is Baird Callicott (1983), who is often seen as someone who draws heavily on the work of Aldo Leopold and his land ethic (Johnson 1984: 353). He combines Leopold's views on biotic communities (that the "good" of these communities is the "ultimate measure of moral value," Baird Callicott, quoted in Moline 1986: 101) with Hume's ideas on the motivation of humans (passion, emotion, feeling or sentiment) and Darwin's concept of moral evolution to argue for social sympathy towards both members of communities, and society itself. It is the fact that these sentiments are excited by objects external to us which makes the ethic non-anthropocentric. Equally, Baird Callicott argues that this appreciation is not simply of the system but can be of the individuals as well, so it is not exclusively system oriented, or individualistic. He provides a very good basis for the belief that both systems and individuals deserve our moral consideration and should be members of our moral community. Interestingly, too, he rests his argument upon a defense of affection and sentiment which differentiates his theory from that of, say, Rolston.

Lastly, Attfield (1983) argues from an unorthodox "holistic" perspective. He is more of the persuasion that in protecting the individual we imply some extended considerability to those things—habitats, systems or whatever—which support those individuals but he is very much against any extended holistic view. His argument is that we get no sense of obligation necessarily from establishing interdependence beyond "strengthen[ing] the argument from human and animal interests for the preservation of the systems on which they depend" (Attfield 1983: 203). Here, we do have a non-anthropocentrism to the extent that animals are included in Attfield's moral community, but he has not attributed any intrinsic value to, for example, plants and rivers.

Although I do not really need to defend or promote one of these arguments over and above the others, I will make just one point. Green political theory, by virtue of it being "green," emphasises holism, that is, it emphasises the value of the wholes of which individuals are a part, rather than the individuals themselves. In this respect, theories which are not properly holistic (Attfield's, for example) cannot be said to be green in any meaningful sense.[5]

PROCESS AND DEMOCRACY

Let me begin by briefly returning to the argument that greens should be seen as more process oriented than is the case at the moment.[6] Greens want to secure, for example, sustainability and through this the circumstances of future generations. It is perfectly reasonable

to see this as a goal towards which any green polity should work. Following the arguments of Saward and Sagoff, making such an "end" integral to the form of a green democracy would undermine democracy itself. Now if, as I argued earlier, greens are concerned with expanding the moral community, this need not be a problem—it has been perceived as such only because the implications of such an expansion have not been taken to their logical conclusion. If we ensure that those future generations, non-human species, and eco-systems are afforded the political consideration that we might expect for a member of the moral community then we do not have to prescribe as many of the policy outcomes in advance (i.e. we don't have to be as ends oriented). We simply have to construct our political institutions (which would include rules, structures, basic laws) in a way which guarantees that the political process will be "considerate" of all those interests which are represented. If we take for example the four basic principles of the German greens (ecology, social responsibility, grassroots democracy and non-violence: Spretnak and Capra 1985) it is perfectly possible to see all of these as principles which are as much guides to the way the political system might operate, as they are ends which the system must achieve.

An anthropocentric polity, such as we have at the moment, displays certain characteristics which may give us clues as to how we may proceed to reconstruct democracy on the basis of green principles. Primary amongst these is that the polity is constructed for, by and of people—in other words there is some congruence between the nature of the political and the moral communities. This is not to say they are, or have to be, identical, but rather that there is a relationship between what, or who, is thought morally considerable, and what, or who, is politically represented. Presumably, then, a green polity would want to do the same thing—it would want to make the nature of its (expanded) moral community more congruent with the political community. To my mind, if we are following both the logic of the philosophical basis of green political theory and the idea that moral and political communities are similar, then we end up, broadly, with four central political areas which would need to be changed to accommodate the new moral community: standing, quality of democracy, decision making and political representation.

Standing

Under such arguments, which are largely based upon liberal conceptions of citizenship and the rule of law, it would be possible to allow various nonhuman entities (perhaps, a river, marsh, brook, beach, national monument, commons, tree, species)[7] to have action taken on their behalf against those who injure them. Although the "liberalness" of this type of approach has been criticised because it reinforces current perceptions of individualism in law and in nature I would argue, first, that such redress *should* be available no matter what type of political theory we construct (because it would be arrogant to believe our theory to be so good that redress would never be necessary),[8] and, second, that it would be wrong to see this as an approach which could only reinforce individualistic stereotypes of our place in nature. Indeed, it is perfectly possible to argue that the idea of a whole "system" being damaged and complaining about it could do a great deal to change cultural perceptions of ecosystems.

There is no reason why it would be only individual members of sentient species that would have standing—it is perfectly possible to have many non-individualistic aspects of ecosystems given legal and moral standing. It may be difficult to do this in a strictly "holistic" way but on some (limited) versions of holism such an approach could be helpful if standing was not restricted to individual examples of species or ecosystems. I think the obvious qualification to make is not that liberal-legal solutions are unnecessary but, rather, that they are insufficient. Their problem, from a green point of view, is that all other aspects of the political, economic and social system remain unchanged and hence moral considerability, significance and community membership becomes far too contingent upon purely legal processes.

We could make a similar case on the basis of attributing rights to the non-human world. These may be specific rights, say, to thrive, exist without threat of damage, and may exist independently of standing in other areas which may offer more general protection. If we are to take the idea of rights to their logical conclusion we could, for example, consider whether we might extend or redefine the notion of social rights (Marshall 1963), extending them from the civil and political to include the quality of life. Here, we might, for example, have to adjust either the legal or constitutional standing of non-humans or perhaps see their welfare in the same "entitlement" framework as we view our own. The possibilities for using such a system to change or promote consciousness is quite formidable. This sense of rights is distinct from Saward's use of rights in his contribution to this volume because here it is the rights of non-humans which are being considered independently of their benefit to humans. Saward is more interested in extending human or civil *rights* to include environmental considerations.

Quality of Democracy

In existing (anthropocentric) political systems the ability of subjects to participate between elections is, in part at least, a measure of democracy. Equally, analysis which considers the role of the institutional arrangements of the state has increasingly suggested that how the state is organised can, and does, affect political outcomes (see, for example, Evans 1984; Hall 1986). For instance, as far as green politics is concerned, Kitschelt (1986) argued that the structures provided for political dissent did make a difference to the nature of participation in anti-nuclear movements.

The same judgements will be applicable to a green democracy. In other words we will have to consider whether such political systems represent a real expansion of the moral community. Precisely what is represented will, of course, depend upon the extent to which we expand our moral community. Nevertheless, it is possible to imagine that consultation exercises might be required to provide opportunities for human representatives of non-human interests to give an opinion; that licensing authorities for industrial plants may have similar constraints placed upon them; that the state may have a statutory responsibility to promote, fund and consult representatives of non-humans in the same way that some political parties are state funded; that regulatory agencies may be required to promote the interests of non-humans' entities and that the representative basis of these agencies should reflect this, and so on.

Decision Making

I mentioned earlier that the British Green Party had promised to change the system of national accounting to accommodate the effects of economic policies on the environment and it is feasible to do this on the basis of discount rates and building in externalities which were not included before. It is, as Ophuls (1977) pointed out, still possible to try to maximise output within more ecologically tolerable limits, but equally, it is also possible to work well within our capacity and not push out to the limits of endurance. Given that we do have to operationalise our philosophical base then Goodin (1983) provides a very good starting place. Working from the assumption that we must now consider (if not apply moral significance to) non-humans then Goodin argues that rather than pursue a utility-maximising strategy, we should opt for decision-making criteria which are biased against irreversible decisions; in favour of protecting the vulnerable; in favour of sustainability; and against causing harm (1983: 16). We do not necessarily have to accept all Goodin's conclusions to see that such a formula could go a long way to satisfying the need we have to accommodate parts of the non-human world into our political community. In particular, the principle of "avoiding harm" would now seem to be a central one given our commitment to some notion of ecocentrism and the moral dispersion of value. This would also tend to accord with an orientation toward ecosystems which is more humble and one which was as concerned with doing the "right" thing, as achieving a "good" result. I do not say this is the only basis upon which decisions may be made within a green political theory, but it is one which, for all the problems it may cause, does illustrate the possibilities.[9]

Political Representation

It has been my intention all along to use the issue of political representation as a means of circumventing arguments about green democracy. To be process oriented *and* to change the nature of political representation (in accordance with an expanded moral community) would mean we could expect green(ish) outcomes, without prescribing the ends of the policy process. We would simply be making a morally defensible case for changing the political process itself. Precisely how this may work out in practice is, I concede, a very difficult question. It is rather unlikely, for example, that individual species would, say, be represented in parliament—since this does seem to lead us into some peculiar possibilities.

Having said this, I think we could make a more plausible case for multi-member constituencies in which some of the representatives were expected to represent the interests of their non-human constituency members. Certainly there could be great benefits from a system like this at national or regional level where representatives are not always (or perhaps often) confronted with the ecological consequences of their actions. This would also fit with the idea that we should be concerned with systems (seen as areas or regions in this case) rather than individual species that might be threatened. It may eventually lead to political boundaries being drawn on ecological lines if, for example, ecological problems are seen as more pressing.

In fact, Kavka and Warren (1983) rehearse many of the arguments as far as the representation of future generations are concerned. They believe it is meaningful to argue that a being is representable if it has interests and if the representative takes instruction from the being, or, has a better than random judgement of their interests (1983: 25). Under these circumstances it should, on Kavka and Warren's formulation, be possible to represent a very diverse range of non-human interests within a green political system because their arguments appear to apply as much to non-humans as they do to future generations. Most arguments in favour of intrinsic value (or other holistic, ecocentric theories) do suppose some idea of the interests of those under consideration, so this is not in itself a problem. Once we have established that something has value or interests then it is representable. I do not propose to go into the questions of how an equitable representation might be achieved, but would say that if we are to make the moral and the political community similar then such an approach is valuable and we cannot dismiss this as an idea, particularly given that it would have a profound and positive effect on the nature of green democracy.

Kavka and Warren also consider, but then dismiss, the possibility that "foreigners" might deserve representation. They argue against this on three grounds: that they are already represented better than other interests (e.g. through the UN); that it would affect sovereignty; and because it may raise moral questions about its effects on diplomacy and self-determination (1983: 33). I am less inclined to dismiss the possibility that those we do damage to might be entitled to a say in the decisions which affect them. In many ways this seems a much more plausible possibility than many others and is quite in line with green thinking on, for example, the transnational nature of pollution and the exploitative economic relations between the North and South. We could, for example, begin by inviting representatives from "foreign" countries to act as observers at debates in national assemblies, or to participate without voting rights in the first instance.

This type of representational diversity is essential to green political theory and to the programmes of green groups and parties. It targets the democratic process as the mechanism which is, at present, failing to protect the vulnerable and the unrepresented. Process, then, is important. It is possible to build into political processes ethical criteria which are logically consistent with the type of moral community we might aspire to under green political theory.

The philosophical basis of such theory lends weight not only to the idea that we have obligations to species and the holistic nature of the ecosystem, but also to the idea that we have rightly to consider individuals and the propriety of the political processes we conduct. If we restore these latter two points within our theoretical schema (without ditching holism or ecocentrism) then democracy has a symbiotic relationship with ecocentrism. In other words, ecocentrism ensures that we do indeed establish a real diversity of political representation and it is this diversity which, in turn, ensures our democracy.

CONCLUSION

In many respects deep/transpersonal ecologists, ecofeminists and those who take a more spiritual line can fill in many of the gaps that are so apparent in a formulation like the one I have just given. It clearly is the case that certain aspects of moral considerability (for example, courtesy, magnanimity, self-sacrifice, humility, compassion) cannot be legislated—they are personal, cultural and spiritual aspects of ourselves (something with which Leopold would agree: Moline 1986: 106–7). I would not for a moment suggest that legal-formalism provides all, or even many, of the answers to the ecological, or the spiritual, crisis. Having said this, it must play some part and such a position is defensible on the basis of green philosophy and green political theory.

My argument has been that if we are only concerned with political outcomes, then green democracy may well be in trouble because such a concern undermines many of the necessary conditions for democracy. However, if we agree that the way we make decisions is a central part of green philosophy, then we begin to set rules by which we must conduct ourselves which, as long as they are not incompatible with democracy (and the important point is that they need not be), provide a solid basis for green decision making. I have also shown, in later sections, the types of theory which may be taken into account (ecocentrism, holism) and the opportunities which polities offer for the incorporation of such ideas (e.g. quality of democracy issues, legal standing, decision making, representation).

Does this mean that green political theory (or a future green government) is unconcerned with political outcomes? Clearly this cannot be the case because the ending of the ecological crisis and the achievement of a sustainable society is a green goal and a green good. We can get around this problem in two ways. First, by arguing for an indirect holism that emphasises tendencies within our orientation towards the "biotic community." Second, we can establish the significance of goals by arguing that political outcomes are important to green political theory to the extent that those outcomes may be adverse or unintended consequences of policy. This does not mean that certain ends have to be prescribed by green political theory, but rather, that it recognises the need to evaluate the decisions which are made to ensure there is some correspondence between what the decision makers intended and what actually happened.

Such a theory cannot ensure green outcomes, but it can ensure a green political process. Indeed, one of the anomalies of green political theory is that it need not guarantee green political outcomes—only a reconstruction of the political process. We can further these arguments by saying that this does provide some guidance to greens on how to behave politically, and how to decide what is politically beneficial to them. In general, anything which works to diversify the political community, to expand its moral constituency, to open up the number of political opportunity structures for that constituency's interests and which encourages tolerance and compassion in decision making will, to a greater or lesser extent, promote some aspect of green democracy. To paraphrase Leopold, anything which tends otherwise, is wrong.

Notes

1. In fact I am not particularly interested in individual theories of intrinsic value, as my argument does not require me to choose between them, only to establish that arguments of this kind are central and, if accepted, affect our views on green democracy.
2. By "the expansion of the moral community" I simply mean the increase in the number of individuals, species or systems which become morally considerable. I do not expect, as some do (Moline 1986) that "extending" the community means that the same ethic applies. I would say that the expansion of the community presupposes a change in the ethic.
3. Personally, I would not have much objection to surrendering some, or perhaps most, of my democratic rights to a loving and trustworthy green council which would pursue ends with which I agreed and which would benefit me and my family. It is quite on the cards that such sacrifices may be necessary at a

practical level and we will have to risk the abuse of the democratic process. This does not mean, though, that, theoretically, we have to be happy at the prospect. Nor does it mean that a little more attention to the anthropocentric nature of the political process would go amiss.

4. I believe that the problems reside principally in translating the spiritual into the political with all the associated problems of accommodating things such as "faith" and "intuition" within a political framework. I have declined to do this, but believe that it is a job well worth trying nevertheless.

5. Holism is more complicated than it appears. Attfield is right to suggest it means more than simple interdependence (1983: 203). Moline (1986: 104–5) recounts that holism is a form of teleology in which what is best is judged in terms of its effects on the whole. He argues for a distinction between direct and indirect holism, in which the latter denotes a broad teleological concern with "tendencies, practices, tastes, predilections, or rules" (105) but they apply distinctive principles or criterion in practice. In this way, we can defend Leopold's holism against the charge that he would be unconcerned about individuals.

6. By process oriented I mean a concern with the way the political system operates, the nature of political representation, the values which are embodied in the system, the opportunities there are for political participation and the constraints there are on political action (e.g. what the state may legitimately do).

7. These are all examples of real complaints that were filed in the USA.

8. Gandhi cautioned the west against thinking we could construct social systems which were so perfect that people no longer had to be good. Perhaps the same applies to green political theory?

9. Of course, Goodin was not arguing for an exclusively process-oriented solution and within his model there is guidance on both ends and means. My position is not that ends should not count, but rather, that axioms like avoiding harm can guide the minute to minute actions of decision makers and can produce desirable (green) results even if they do not prescribe a policy-specific end result.

References

Attfield, R. (1983) "Methods of Ecological Ethics," *Metaphilosophy* 14, 3 and 4: 195–208.

Callicott, J. Baird (1983) "Non-Anthropocentric Value Theory and Environmental Ethics," *American Philosophical Quarterly* 21, 4: 299–309.

Dobson, A. (1989) "Deep Ecology," *Cogito* spring.

——. (1990) *Green Political Thought,* London: HarperCollins.

——. (1993) "Ecologism," in R. Eatwell and H. Wright (eds) *Contemporary Political Ideologies,* London: Pinter.

Eckersley, R. (1992) *Environmentalism and Political Theory: Toward an Ecocentric Approach,* London: UCL Press.

Evans, P. (ed.) (1984) *Bringing the State Back In,* Cambridge: Cambridge University Press.

Goodin, R. (1983) "Ethical Principles for Environmental Protection," in R. Elliot and A. Gare (eds) *Environmental Philosophy,* Milton Keynes: Open University Press.

——. (1992) *Green Political Theory,* Oxford: Polity.

Goodpaster, K. (1983) "On Being Morally Considerable," in D. Scherer and T. Attig (eds) *Ethics and the Environment,* Englewood Cliffs, NJ: Prentice Hall.

Green Party (1992) *New Directions: The Path to a Green Britain Now,* General Election Campaign Manifesto, London: The Green Party.

Hall, P. (1986) *Governing the Economy,* Oxford: Polity.

Hardin, G. (1968) "The Tragedy of the Commons," *Science* 162: 1,243–8.

Johnson, E. (1984) "Treating the Dirt: Environmental Ethics and Moral Theory," in T. Regan (ed.) *Earthbound,* Prospect Heights, NJ: Waveland Press.

Kavka, G. G. and Warren, V. (1983) "Political Representation for Future Generations," in R. Elliot and A. Gare (eds) *Environmental Philosophy,* Milton Keynes: Open University Press.

Kemp, P. and Wall, D. (1990) *A Green Manifesto for the 1990s,* Harmondsworth: Penguin.

Kitschelt, H. (1986) "Political Opportunity Structures and Political Protest: Anti-Nuclear Movements in Four Democracies," *British Journal of Political Science* 16. 1: 56–72.

McClade, J. M. (1991) "The Seas and the Shoreline as Part of the European Biosphere," in S. Parkin (ed.) *Green Light on Europe,* London: Heretic.

Marshall, T. H. (1963) *Sociology at the Crossroads.* London: Heinemann.

Mathews, F. (1988) "Conservation and Self-Realization: A Deep Ecology Perspective," *Environmental Ethics* 10, 4: 347–57.

Moline, J. N. (1986) "Aldo Leopold and the Moral Community," *Environmental Ethics* 8, 2: 99–120.

Ophuls, W. (1977) *Ecology and the Politics of Scarcity,* New York: Freeman.

Ophuls, W. and Boyan, A. S. (1992) *Ecology and the Politics of Scarcity Revisited,* New York: Freeman.

Paehlke, R. (1988) "Democracy, Bureaucracy, and Environmentalism," *Environmental Ethics* 10, 4: 291–309.

Parkin, S. (1989) *Green Parties: An International Guide,* London: Heretic.

Regan, T. (1984) *The Case for Animal Rights,* London: Routledge.

Rolston III, Holmes (1975) "Is There an Ecological Ethic?" *Ethics* 85, 2: 93–109.

——. (1987) *Environmental Ethics,* Philadelphia, Pa: Temple University Press.

Sagoff, M. (1988) *The Economy of the Earth,* Cambridge: Cambridge University Press.

Saward, M. (1993) "Green Democracy," in A. Dobson and P. Lucardie (eds) *The Politics of Nature,* London: Routledge.

Singer, P. (1975) *Animal Liberation,* New York: Random House.

——. (1993) *Practical Ethics,* Cambridge: Cambridge University Press.

Spretnak, C. and Capra, F. (1985) *Green Politics,* London: Paladin.

Sylvan, R. (1984) "A Critique of Deep Ecology, Parts I and II," *Radical Philosophy* 40 and 41.

For Further Discussion

1. What does Mills mean by "ecoauthoritarianism" and "ecoradicalism"?

2. Why might it be difficult for environmentalists to support democracy? In answering this question, explain what Mills means when he speaks of process and outcomes.

3. What is your understanding of "green" political theory? How does Mills use the word *liberal?* Does this correspond to your own understanding of liberalism?

4. Do you agree with the guidance to green political activists that Mills offers at the end of his essay? Are his proposals reasonable? Realistic?

DISCUSSION AND STUDY QUESTIONS FOR CHAPTER 16

1. Are environmental issues political issues? Are certain political parties more environmental than others? To what do you attribute the perception that Democrats are environmentalists and Republicans are not?

2. Conservatives believe that tradition is a good guide—better than public opinion—for deciding public policy matters. In your experience, what traditional values support environmental protection? Which traditional values do not?

3. Do you believe that, with full information, citizens will always choose to support environmental policies? Why or why not? If not, do democratic institutions put the environment at risk?

4. Can you think of a situation in which you would abandon democracy in favor of environmental protection? Do you believe that there is a tension between democratic principles and environmental goals?

CHAPTER 17

Multicultural Perspectives on the Environment

This book has mentioned several times the importance of stepping back from one's own perspective to reflect on it. The very essence of philosophical thinking involves abstracting oneself from one's own framework to critically assess one's understanding. This chapter considers diverse cultural perspectives on environmental issues as another method for this process of critical self-examination.

There are a variety of reasons to consider multicultural perspectives in an environmental ethics textbook. First, it would be foolish to assume that one's own culture provides the last word on any set of concerns as complex as those associated with environmentalism. To assume that one's own beliefs, attitudes, and values are sufficient for living the good life, without considering perspectives from other cultures, is to prejudge the issue. Such prejudice is at the core of ignorance, because it overlooks much relevant information. For the simple reason of making informed judgments, one should consider alternative perspectives.

It is not uncommon to find reference to diverse cultural norms and concepts in environmental writings. For example, deep ecologists often cite Buddhist, Taoist, and Native American views as sources for their philosophy. Familiarity with diverse cultural views can only help to strengthen one's own understanding.

A second reason for considering diverse perspectives lies in the fact that environmental issues seldom respect political and cultural borders. Every culture is affected by global warming, nuclear waste, resource depletion, pollution, overconsumption, and overpopulation. Addressing these issues systematically requires the cooperation of many diverse societies and cultures. International policies that neglect to understand the cultures they affect are bound to fail.

This concern is particularly relevant from an ethical perspective. In many cases, the people most affected by policies favored by Western environmentalists are people of non-Western cultures, and often the most marginalized members of these communities. Such issues as population control, rain forest preservation, protection of endangered species, and sustainable development often disproportionately burden non-Western cultures. Justice, as well as common decency, require that we give due consideration to the points of view of such people. In turn, this entails at least a minimal understanding of diverse perspectives.

Environmental Ethics: A Buddhist Perspective

Padmasiri de Silva

Buddhism is a complex religious philosophy with roots in diverse cultures—India, Tibet, China, Japan, and Thailand. Buddhism often is seen as emphasizing detachment and aloofness from the world through a commitment to asceticism and meditation. But Buddhist ethics accentuates nonviolence and meditation as a means of empathizing with the suffering of others. The ethical goal is self-mastery, not mastery of others. Buddhist ethics rejects materialism, self-centered egoism, and self-aggrandizement.

In this selection, Padmasiri de Silva suggests that Buddhism has much to contribute to ecophilosophy. De Silva believes that a strict dichotomy between science and reason on the one hand and mysticism and intuition on the other characterizes many contemporary environmental debates. He believes that this is a false and misleading dichotomy that Buddhism may help us overcome. Although the Buddha offered no explicit environmental teachings, there are many resources within Buddhism that may contribute to an adequate environmental ethic.

Religious and ethical values play an important role in our attempts to deal with the new challenges generated by expanding technology. Yet, unfortunately, the debate regarding the place of religious values in environmental issues often has been considered a concern of "mysticism and ecology." Apart from its mystical dimensions, Buddhism contains a social ethics with a viable perspective on economic development and an appropriate lifestyle. Environmental ethics is a relatively new concern in moral philosophy. In the light of its current importance, there is a need to develop the environmental values implicit in the doctrine of the Buddha and to articulate the Buddhist contribution to an "eco-philosophy" that can attain global acceptance.

MYSTICISM AND ECOLOGY

Theodor Roszak sees a wide gulf between traditional science and modern ecology, and has been an exponent of a counter-culture standing in opposition to the culture of contemporary science and technology. Ecology, he contends, rejects mathematical quantification and scientific reductionism. This new science deals with the interrelationships among organisms and their environment, and aims at an "intuitive grasp of wholes" rather than an analysis of wholes into their constituent parts. This emphasis on intuition and contemplation, Roszak argues, brings ecology close to art and religion. It constitutes a way of knowing that is different from scientific rationalism: "[I]n the modern period, most of our keenest minds had come passionately to believe, like Dickens' Mr. Gradgrind, that 'in this life, we want nothing but facts, sir: nothing but facts.' What else could follow from this but a culture whose realities are restricted to flat, functional prose, unambiguous quantities, and Baconian inductions. As a result, the one-dimensional language of the logician, scholar, and critic—and eventually of the technician and scientist—has been promoted to a position of omnipotence among us."[1] Roszak also states that traditional science has been the slave of the philosophy of power and has been insensitive to the diversity of human values. He distinguishes between two ways of knowing: the "literal knowledge" of traditional science versus the "resonant knowledge" of myth and mysticism expressed in art and religion. The synthesis of literal knowledge and resonant knowledge is the "rhapsodic intellect" that finds its expression in the science of ecology.

John Passmore opposes Roszak, arguing that mysticism is a greater menace to civilization and our future well-being than science or technology. Passmore contends that ecological problems, if they can be solved at all, can only be solved by "thoughtful action and not by transforming Nature into a mystery." A "science that does not solve problems is not really a science." Passmore also asserts that science is not intrinsically atomistic and that while science understands things by breaking them into parts, it also understands things in terms of larger wholes: "The conception of Nature as sacred, as not to be tampered with or changed but only contemplated, sounds self-defeating coming from professed ecologists because if our ecological problems are to be solved we will have to tamper with or manipulate and control Nature."[2] Passmore bluntly states: "Mystical contemplation will not clean our streams or feed our people," and these problems can be solved only by the old-fashioned method of thoughtful action. The philosopher thus should be concerned with the ethical issues pertaining to environment.

Roszak and Passmore are both right and both wrong. Each has something important to say, but each of their views is also misplaced. Science and technology have generated many ecological issues, yet the problem is not with science, but the uncritical use of technology. Religion and art add diversity to our ways of knowing, an important contribution to the breadth and depth of our ways of knowing. To convert this epistemological strand to a thesis about "environmental knowledge" is a hypothesis that has yet to be scientifically developed rather than left to the rhapsodic intellect. To blame mysticism for the ills of civilization, as Passmore does, again is a misplaced criticism. Mysticism may not help us to clean up our rivers, but religious literature in general, and the discourses of the Buddha in particular, offer useful guidelines for developing a pro-conservation lifestyle. Above all, Buddhism does not see any conflict in contemplating Nature, appreciating it aesthetically, and harnessing it for human use. Despite its strong mystical dimension, Buddhism has a very viable and humanistic framework for ethical concerns related to our environment.

Thus, the Roszak-Passmore debate, while useful for stimulating new thinking, can lead to a stalemate if the discussion continues to focus on dichotomies. In today's world we cannot afford to be divided on different fronts. It is especially necessary to look at the ethical contributions of the major religions to the ecological issue, which is virtually an "ecological crisis." It is to further this objective that a modest attempt is made here to outline a basis for developing a Buddhist environmental ethics.

THE EPISTEMOLOGICAL OUTLOOK OF BUDDHISM

Buddhism has a very balanced epistemological outlook that makes it possible to offer a middle ground between a narrow scientific and empirical perspective on one side and an exclusionary mysticism on the other hand. Thus, it provides a congenial base for mediating the debate between Roszak and Passmore.

The Buddha used logic and rational argument, demonstrations from everyday life, empirical evidence, paradoxes, fables, metaphors, and stories, along with meditational techniques for developing penetrative insights. Thus, he combined and blended the different strands of *rationalism, empiricism,* and *experientialism.* Buddhism may be considered a rational engagement in argument and debate, as well as analysis, which simply points to the facts of existence, an existential encounter with life crises and the changing world, a personal search for the meaning of life, as well as an experiential grasp of life through meditation and self-analysis. The blending of such components in one system makes it unique.[3]

Apart from having an appropriate epistemological outlook to bring values to bear on ecological issues, for an exercise in environmental education, Buddhism can combine all aspects of human potentialities, intellect, moral sensitivity, the ability to make relevant ethical distinctions, a sense of empirical and scientific evidence, intuition, practical experience, and action-oriented programs. The development of environmental ethics has to be understood in terms of human faculties and the possibilities of environmental education. If not, it will remain an abstract and isolated academic study. Finally, Buddhism has a sense of *pragmatism,* so that a Buddhist may view a situation realistically and not be excited by utopian solutions.[4]

THE BUDDHIST CONCEPT OF ENVIRONMENT AND HUMANITY-NATURE ORIENTATION

A very striking feature of the Buddhist orientation to Nature is the attempt to see humanity and the universe as a network of inter-connected relationships, rather than as a static collection of objects. Today, environmentalists emphasize that the natural and the social environments have become integrated into *one universe.* Today more than ever, the Earth has become a delicately balanced system of interdependent parts, an *ecosystem.* Secondly, in Buddhist terms the environment has become a medium for human moral concerns. Thirdly, what may be called the "ecology of the mind," in the words of Gregory Bateson,[5] and the ecology of the physical world interpenetrate in the most complex way.

In this context, the Buddha would say that through human intervention the environment can be polluted or preserved. While understanding the laws of Nature, the Buddhist concept of free will makes it possible for humanity to harness Nature for our use in a sensible and non-destructive way. This makes humans responsible for understanding the laws of Nature and the related scientific information in order to intervene wisely when dealing with the eco-crisis. In this way, we combine the *knowledge* we get from science and the *wisdom* of a deeper eco-philosophy, which is not merely concerned with the *means* for human development but also its *ends* and goals.

The Buddha does talk, in certain situations, of objects like tables and chairs or mountains and forests. Often, he would talk in more analytical terms about solidity (*pathavi*), liquidity (*āpo*), heat (*tejo*), and mobility (*vāyo*). At a deeper level, the Buddha describes the nature of the world as a network of many-sided relationships and causal patterns. Thus, there are laws covering different realms: physical laws (*utu-niyāma*), biological laws (*bija-niyāma*), moral laws (*kamma-niyāma*), psychological laws (*citta-niyāma*) and laws of liberation (*dhamma-niyāma*).[6]

The Buddha rejected a substance-oriented concept of reality and in its place upheld a system of causal interaction. He was very quick to point out that, although these causal patterns help us to understand the natural world as well as the psychological and moral universe, they *condition* our existence without *determining* our actions. Thus, he rejected the naturalism (*svabhāva-vāda*) that held that things possess an absolute essence, and also the theory of determinism concerning human action. By emphasizing the process of reality and free will, and rejecting determinism, the Buddha offered a concept of reality that is ecologically more attractive and useful. The Buddha also opposed other types of determinism, such as theistic and karmic determinism. Thus, in the light of the Buddhist concept of the physical, the psychological, and the moral universe, it is not necessary for humanity to undergo passively the hazards of Nature or aggressively dominate Nature, but rather to meaningfully harness it. Buddhism also advocates harmony with Nature.

Buddhism is critical of the aim to dominate or exploit Nature. The Buddhist perspective on Nature has been well presented by Hajime Nakamura in a study of environment and humanity in Eastern thought: "According to Buddhism, humans constitute just one class of living beings and have no right of unlimited exploitation of animal and plant life, which form a part of Nature."[7] Nakamura also sees in Buddhism the idea of harmony with Nature: "One of the most prominent

features of Buddhist thought is its zest for harmony with Nature. The joy of enjoying natural beauty and of living comfortably in natural surroundings was expressed by the monks and the nuns. . . ."[8] The notion of living in harmony with Nature, the aesthetic appreciation of Nature, as well as the Buddhist contemplative attitude, all support a move toward conservation. This approach does not transform Nature into a mystery, but rather emphasizes that these attitudes have a strong potential for developing a framework for environmental education. Just as scientific information forms an integral part of schemes for environmental education, the understanding of the ethical premises involved in the preservation of the environment—the economic aspects of a viable lifestyle etc., and psychological and aesthetic perspectives on Nature, especially emerging from creative cultural points of view—all can play a useful part in the development of environmental education.

ENVIRONMENTAL ETHICS IN BUDDHISM

Traditionally, Western systems of ethics have reflected concern with the ideals of a good life and the principles and codes of conduct linked to the realization of such an ideal. Apart from such normative ethics, there has been a more epistemologically oriented study of the rationality of our beliefs pertaining to ethical matters and an examination of the meaning of basic ethical terms such as "the good" and "right."

Recently, an interest has developed in what may be called "practical ethics," as evidenced in new areas of study, such as bio-medical ethics, business ethics, and environmental ethics. These are attempts to deal with new ethical issues attendant upon technological development. Environmental ethics is basically a search for normative guidelines regarding human intervention in the physical environment, involving a variety of problems. It raises questions of right and wrong pertaining to plants, animals, technology, population control, questions of health and disease, ideals of economic growth, etc. Many of these questions were non-existent during the Buddha's time. But today Buddhists have to define and articulate attitudes and policies to address these emerging issues. In forming a Buddhist response to these issues, one can find some insights explicit in the doctrines, while others are implicit.

The basic ethical premises found in the discourses of the Buddha that provide significant values for developing a Buddhist ethic of Nature are as follows:

1. The most central concept is the value of life. Ecologists consider the individual to be a part of a whole ecosystem and this concept expands beyond the human community to include animals, plants, the soil, and water resources. Although the Buddha preached the doctrine of the concern for human and animal life, the concern for the physical world and trees is somewhat different. The Buddha did not preach a biotic egalitarianism in which plants have an equal status with animals. Nonetheless, there is a pro-conservation attitude toward the natural environment evident in his teachings. From a modern point of view the Buddhist attitude toward animals is a very significant strand in its environmental ethics.

2. An ecological ethics requires a critical thoughtfulness about the consequences of our actions and our lifestyles. People who care about the environment must live lives that demonstrate their concern. This is an extension of the Buddhist precept of truth (not uttering falsehood) to "truthfulness." If people truly respect environmental values, their care and concern will be manifest in their daily lives, so that the environment can be preserved.

3. While the value of life may be voiced as a concern for life, and may be given legal status as a right to life, or a right for survival, a third important ethical value is the dignity of the human person. This implies reciprocity, with the animals and the natural environment being viewed from a human point of view.

4. The next important value is the right to property for oneself and others. Here we have the Buddhist injunction not to take what does not belong to one. This can be interpreted broadly in terms of our rights and duties with respect to other human beings. Each person has a right to property and other resources necessary for a full life, but should respect the rights of others, or what may be called the common good. When this rule is extended to ecological realities, it becomes necessary to have rules for land and water use, air pollution, and, for those engaged in a profession such as agriculture or industry, professional rules for the protection of natural resources.

These premises, the value of life, truthfulness, and the right to property, are directly or indirectly related to the Five Precepts of Buddhism. The dignity of the human person underlies all the precepts.

5. Living a sane and simple life, though not actually one of the precepts, is a very important Buddhist idea and is relevant to any major restructuring of the ecology. A life of wanton sensuality, conspicuous consumption, unrealistic enjoyment of luxuries beyond one's means, and so forth, is rejected by a devout Buddhist.

6. The rights of the future generations are also valued by Buddhists.[9]

These ethical premises can be more firmly grounded by relating them to special virtues that Buddhists are expected to develop: *self-restraint,* which can help one to develop a sane life style; *benevolence* and compassion, to support our life and the natural environment; truthfulness and *conscientiousness* in implementing an environmental ethics. Looking directly at the Five Precepts—abstention from killing, stealing, sexual misconduct, lying, and intoxicants—they all reflect a pro-conservation code of ethics.

THE VALUE OF LIFE

It is not possible to discuss in detail all these ethical premises and moral virtues, but it would be useful to focus attention on the first of these premises, the value of life, as it is a central concern. The Buddhist premise concerning abstention from killing living creatures focuses attention on the ethical premise concerning the value of life. This applies to both human beings and animals. The Buddha condemned the infliction of pain and suffering on living creatures. He was strongly critical of the practice of animal sacrifice, as well as the hunting enjoyed by the royalty. He discouraged war as a method of settling disputes and demonstrated its utter futility. This sensitivity was extended to the minutest creatures. The rules for monks that prohibit cutting down trees,[10] destroying plants, digging in the soil, and so forth may be interpreted as a warning that the minute forms of life may be destroyed by these actions. According to the Indian ways of thinking, as expressed in the Pāli canon, a certain form of life called "one-facultied" (*ekindriya jiva*), inhabits trees, plants and the soil, and even water may have creatures or "breathers" (*sappanaka udaka*) in it.[11]

As mentioned in the *Cakkāvattaisihanada Sutta,* the ideal Buddhist ruler should provide protection not only to human beings, but also to the beasts of the forests and the birds of the air (*Miga-pakkhisu*).[12] The principle of non-injury (*ahimsā*) to life was found in Jainism and other Indian sects.

Although the Buddha was quite effective in his critique of animal sacrifice as both a religious and social practice, in the case of agriculture and meat-eating, he concentrated on reform within the monastic order. The monks were forbidden to till the soil, to avoid injuring living beings. They were not completely prohibited from eating meat, provided that they were fully convinced that it was not killed on their behalf. The social context in which the monks begged for alms through the streets made it necessary for them not to ask for any particular food, but to accept whatever was offered. Thus, although the Buddha rejects professions like selling armaments and the killing and selling of animals, he did not restrict the monks' food but let them accept what was given (unless, of course, it was poisonous).

The Buddha did not want to encourage food fetishism or a belief that spiritual transformation could be assured by a particular diet. Rather, he adopted a realistic and pragmatic standpoint. If one practices compassion, vegetarianism is the most natural lifestyle. A more modern argument for vegetarianism might mention the ideals of good health and good conservation strategies. The Buddha's concern about the value of life emerges from compassion, which is why he was critical of capital punishment, warfare, hunting, animal sacrifices, suicide, and callousness of a physical or psychological nature toward living creatures. At a social level, respect for life, specially in relation to humans, is a concern common to all religions, to the law, and to the United Nations charter of human rights. It is necessary for the survival of society. At a more psychological level, there are other significant arguments against the destruction of life. For example, those who kill, engage in crime, and inflict pain on others, harden their characters and foster destructive tendencies, leading to defective personalities. In so doing, one cannot live a full and satisfying life or develop the personality needed for the attainment of Nibbāna and extricating oneself from the cycle of Samsāra.

Apart from ethical and psychological considerations, at a more general level, the Buddhist worldview assumes that humans can be born as animals, and animals may be born as humans. This breaks through any anthropocentric conception of life and the environment.

The ethical value of emphasizing the well-being of animals has been at times mentioned in Western ethics, but only recently has it engendered a great deal of discussion in the field of environmental ethics. Be-

fore environmental ethics emerged as a specific discipline, a philosopher like Rashdall had asserted the ethical importance of non-human beings: "The well being of animals . . . seems to me quite distinctively to possess some value, and therefore to form part of that good which constitutes the ethical end."[13] Some moral philosophers have difficulty seeing animals on an equal footing, and deserving of human kindness.[14]

Peter Singer calls for an expansion of our horizons on this point:

> The expansion of the moral circle to non-human animals is only just getting under way. It has still to gain verbal and intellectual acceptance, let alone general practice. Yet the ecology movement has emphasized that we are not the only species in the planet, and should not value everything by its usefulness to human beings; and defenders of rights for animals are gradually replacing the old-fashioned animal welfare organizations that cared a lot for domestic pets but little for animals with less emotional appeal to us.[15]

The Buddhist *Jātaka* stories present an anthropomorphic view of animals, showing their "truly human qualities" of both good and bad, heroic and evil qualities. The most famous is the *Sasa-Jātaka,*[16] about the hare who lived in the woods with a monkey, a jackal, and an otter. The story concerns their decision to observe the holy days and the moral law by giving alms. Recognizing the full moon they decided to consider the next day as a fast day and feed any beggars. While the monkey, the jackal, and the otter collected food to be given to anyone in need of it, the hare was unable to collect any food and offered his own flesh. The hare was rewarded by having its form supernaturally imposed on the face of the moon. The animal hero here is considered as having been a bodhisattva in a previous life. The story offers a very humane picture of its animal characters.

The *Nandiyamaga-Jātaka*[17] is the story of a deer who fearlessly faced a king who was hunting; by his steadfast gaze, he changed the mind of the king and saved the other animals. In the *Dhammapada,* we find the story of Dhanapalaka, an elephant who suffered from homesickness after being separated from his mother. The captive elephant refused food.[18] In the *Mahākapi Jātaka,* a monkey saves his tribe by using his body as part of a bridge for them to cross the Ganges.

While some of the *Jātakas* depict superhuman qualities expressing the life of the bodhisattva, they also reflect a capacity for affection, which is as important as the heroic qualities of courage and sacrifice. Although we may not find a structured moral code among animals, they seem to express certain deeply valued virtues. It has been observed that animals "are devoted to their offspring, sympathetic to their kindred, affectionate to their mates, self-subordinating in their community, courageous beyond praise."[19]

ATTITUDES TOWARD PLANTS AND THE PHYSICAL ENVIRONMENT

Although in a sense plants have life, they are different from animals, and the ethical principles related to plants occupy a slightly different position in the Buddhist context. Peter Singer raises the question, "Will this new stage also be the final stage in the expansion of Ethics? Or will we eventually go beyond animals too, and embrace plants, or perhaps even mountains, rocks, and streams?"[20] In Albert Schweitzer, we find the principle of the reverence for life extended from animals to plants, and Aldo Leopold extends it further to posit a kind of land ethic. Singer continues, "I believe that the boundaries of sentience—by which I mean the ability to feel, to suffer from anything or to enjoy anything—is not a morally arbitrary boundary in the way that the boundaries of race or species are arbitrary. There is a genuine difficulty in understanding how chopping down a tree can matter to *the tree* if the tree can feel nothing. The same is true of quarrying a mountain." Thus, he concludes, "if ethics grows to take into account the interests of all sentient creatures, the expansion of our moral horizons will at least have completed its long and erratic course."[21] While the Buddha attached great importance to the planting of trees, the construction of parks, reforestation, preservation of water, etc., he does not espouse a "biotic egalitarianism," in which plants and animals are considered ethically on an equal footing with humans.

We know that from a scientific point of view plants provide us with a healthy environment and that destruction of trees can be ecologically disastrous. They provide us with shade and food, help preserve our water resources, prevent erosion, etc. In addition they add to the aesthetic dimension of our living space and bring humanity closer to Nature. On ethical grounds this leads to a critique of an aggressive attitude toward the natural environment, and denunciation of an ecologically unsound exploitation of the environment. This value differs somewhat from the value of life.

In the monastic code of discipline monks are prohibited from cutting down the branches of a tree. They also were prohibited from engaging in agriculture, since in the process of ploughing and digging small animals may be destroyed.

There is an interesting passage in the *suttas,* where the notion of gratitude to a tree that gives shade and provides food is cited. There was a huge banyan tree that gave shade to people. One day, a man ate the fruits, broke the branches and went away, callous in the way that only a human being can be callous. The indwelling tree-spirit thought the man was cruel and in response told the tree to refuse to bear fruit, which it did.[22] In another context the banyan tree is referred to as a haven for birds and metaphorically is compared to an *arhat,* who gives shade to those who come in search of the Dhamma.

This is an interesting attitude toward trees. The same attitude also is supported by Buddhist Nature poetry. There are numerous instances where men and women who attained great spiritual heights uttered inspiring verses describing Nature.[23] The monks also asked to go to the roots of trees and the quiet jungle and forest for meditation. All these present a pro-conservationist attitude to Nature, but this does not yet go as far as biotic egalitarianism.

CONCLUDING THOUGHTS

We have examined the philosophical and ethical perspectives of Buddhism and pointed out that implicit in the teachings of the Buddha we find a framework for developing environmental ethics. We have also abstracted some of the ethical values relevant for dealing with environmental ethics and provided a more focused discussion of the central value, the value of life. The examination of the ethics of Buddhism demonstrates the ways in which religion can advance ecological concerns.

An ethic is not complete in this context without an eco-philosophy, and in the search for an eco-philosophy with global relevance, Buddhism has an important contribution to make, due to its attitude toward economic development, lifestyles, and the use of a kind of wisdom for deciding on the foremost ends for the well-being of humanity.[24] One of my objectives has been to point out that Buddhist insights may be brought to bear on the current discussion regarding ecological values, without recourse to mysticism.

Notes

1. Theodor Roszak, *Where the Wasteland Ends* (New York: Doubleday, 1972).
2. John Passmore, *Man's Responsibility to Nature* (London: Gerald Duckworth, 1974).
3. For a development of this theme, see Padmasiri de Silva, *Environmental Ethics in Buddhism* (Paris: UNESCO, forthcoming).
4. de Silva.
5. See Morris Berman, *The Reenchantment of the World* (Cornell: Bantam Books, 1984).
6. *Atthasālini*, 854.
7. Hajime Nakamura in *Encyclopedia of Bioethics*, Vol. I (New York: Macmillan/Free Press, 1978), p. 371.
8. Nakamura, p. 371.
9. See George Kiefer, *Bioethics: A Textbook of Issues* (London: Addison-Wesley, 1979).
10. *Vinaya Piṭaka*, Vol. III, p. 126.
11. *Vinaya Piṭaka*, IV, 49, p. 125.
12. *Further Dialogues*, Vol. III, p. 126.
13. Hastings Rashdall, *Theory of Good and Evil*, Vol. I (London: Oxford University Press, 1907), p. 214.
14. E. W. Hopkins, *Ethics of India*, cited in O. H. de A. Wijesekera, *Buddhism and Society* (Colombo, Sri Lanka: The Baudha Sahitya Sabha), p. 14.
15. Peter Singer, *The Expanding Circle* (Oxford, England: Clarendon Press, 1981), p. 121.
16. See E. B. Cowell, ed. *The Jātaka*, Vol. III (London: Pāli Text Society, 1957), pp. 34–38.
17. Cowell, Vol. III, pp. 171-74.
18. For a discussion of this theme and related material on Buddhist attitude to animals in a small book with great insight, see Francis Story, *The Place of Animals in Buddhism* (Kandy: Bodhi Leaves, Buddhist Publication Society, 1964.
19. Story, p. 3, cites this statement as an observation made by Professor John Arthur Thompson of Aberdeen University.
20. Singer.
21. Singer.
22. *Gradual Sayings*, Vol. III, 368.
23. For a discussion of the Buddhist appreciation of Nature, see de Silva.
24. See de Silva.

For Further Discussion

1. De Silva explains that the Buddha rejected a substance-oriented concept of reality in favor of a more process-oriented concept. What does he mean by "substance-oriented"? Can you think of any other authors in this textbook who would accept a similar view of reality?

2. How, if at all, does the Buddhist understanding of the value of life differ from your own?

3. How does the Buddhist attitude toward trees differ from Christopher Stone's argument (Chapter 6) supporting legal rights for trees?

4. Would de Silva describe Buddhism as "mystical" religion? How do you understand "mysticism" and what role, if any, would you see for mysticism in environmentalism?

5. On what grounds does Buddhism reject the goal of dominating or exploiting nature?

How Much of the Earth Is Sacred Space?

J. Donald Hughes
Jim Swam

Native American beliefs and values are often cited as models of ecological thinking. Some environmentalists adopt a patronizing attitude toward native cultures, portraying native peoples as simple and innocent hunters and gatherers. The romanticism characteristic of books such as James Fenimore Cooper's The Deerslayer *and* The Last of the Mohicans *surely does a disservice to the complexity and variety of native peoples. Yet, at least compared to the Eu-ropean settlers who displaced them, Native Americans did live in closer harmony with the land and in more sustainable ways.*

Among the beliefs of many native peoples is an understanding of certain places as sacred space. In this selection, J. Donald Hughes and Jim Swan describe this understanding of sacred space and contrast it to European and Christian understandings. They believe that contemporary environmental controversies would benefit from a recovery of this ancient and native wisdom. Before reading this essay, it would be beneficial to review earlier readings by Lynn White and Eugene Hargrove.

One of the happiest events of recent years was the return of Blue Lake to the Taos Pueblo. This locality is a holy place for the Taos people, one of whom said, "We go there and talk to our Great Spirit in our own language, and talk to Nature and what is going to grow."[1] In giving back the lake and the forest surrounding it, Congress acknowledged, as it later did more explicitly in the American Indian Religious Freedom Act,[2] that Native American Indian tribes recognize certain places as sacred space, an attitude which is found in all tribes. The Lakota and others have a spiritual relationship to Mato Tipi (Bear Butte) in the Black Hills, and both the Navajo and Hopi regard the San Francisco Peaks near Flagstaff as sacred, although the courts have been remiss in protecting them from desecration.

Sacred space is a place where human beings find a manifestation of divine power, where they experience a sense of connectedness to the universe. There, in some special way, spirit is present to them. People in many parts of the world and in all times have come to designate some places as sacred: in Japan, Mount Fuji is a *kami* or shrine; an island in Lake Titicaca is for the Aymara an altar to the Sun God, Inti; and the Bimin-Kukusmin of Papua New Guinea, revere the area around a spring of ritual oil.[3] Such examples could be multiplied.

But when one asks a traditional Indian, "How much of the earth is sacred space?" the answer is unhesitating: "All." As Chief Seattle, a Suquamish of the Puget Sound area, told the governor of Washington, "Every part of this soil is sacred in the estimation of my people."[4] When tribal elders speak of Mother Earth, they are not using a metaphor. They perceive that earth is a living being, sacred in all her parts. Black Elk, a Lakota holy man, addressed her in these words: "Every step we take upon You should be done in a sacred manner; each step should be as a prayer."[5] For this reason Smohalla, a Wanapum shaman of the upper Columbia River country, required his followers to use the gentle digging stick instead of the plow, which tears Mother Earth's bosom like a knife. He also forbade them to mine, because that would be to dig under her skin for her bones.[6] Those venerable teachers knew that one could experience a sense of connected-ness to the universe virtually anywhere, so there were no boundaries or places that were not sacred.

The last sentence does not contradict what was said before about sacred spaces. In the traditional Indian view, all of nature is sacred, but in certain spots the spirit power manifests itself more clearly, more readily. It is to those places that a person seeking a vision would make a quest. They were localities where the great events of tribal history and the era of creation took place. They were associated with particular beings, whom one ought not even to name unless one were prepared to encounter the energy they wielded, which could either strengthen or destroy. So the Indian view of the universe is that of a sacred continuum that contains foci of power.

Again, this conception of the earth is widespread among traditional peoples around the world and through history. The ancient Chinese practice of *Feng Shui* (geomancy) treats the landscape as a network of potent spots connected by lines of energy.[7] One would be foolish, its practitioners believed, to ignore this sacred geography when locating a house, road, or temple. The Greek philosopher Plato affirmed that the earth is a living organism, alive in every part, and also that there are particular locations where spiritual powers operate positively or negatively. In the *Laws*, he advised founders of cities to take careful account of these influences.[8] . . .

But this is not the only approach that has been taken to sacred space. The Old World produced in ancient times another view that contrasts strongly with the North American Indian version, and which has had a pervasive influence in the history of western thought. That is the idea of sanctuary, an area marked off so as to be separated from the space around it, usually by a wall. The Greek word for such a place, *temenos*, is instructive because it derives from the verb *temnō*, meaning "I cut off." The Latin word *templum*, the root of the word "temple," also means "a part cut off" or "a space marked out." Once dedicated by the proper authorities, such a precinct was protected by all the sanctions of religious custom and local law.

The places chosen were almost always distinguished by some natural feature: an impressive grove of large old trees, a spring, a lake, a fissure in the earth, or a mountain peak. These were often landscapes of great natural beauty. In locating and marking a *temenos*, the seers took account of the lay of the land and the mountain forms visible from it. Within the boundary, all human use other than religious worship was forbidden. There was to be no cutting or removal of wood, no hunting, grazing, or cultivation. The only building permitted was a shelter for the statue of god or goddess. What we call a temple is such a structure, but for the ancients the enclosure itself, and everything within it, served as the temple. There the god lived and became manifest. And there a fugitive could seek sanctuary, a sick person could ask for healing, and anyone seeking wisdom—that is, to know the will of the gods—could sleep overnight in expectation of a meaningful dream. There were hundreds of those places in the ancient world.

Another type of sacred space that deserves mention is a tract of agricultural land dedicated to a god or goddess, the produce of which served as an offering. The Linear B tablets record such land use in ancient Crete. In Athens, the Council of the Areopagus had jurisdiction over groves of sacred olive trees, whose oil was reserved for sacrifices, prize for the winners of the Panathenaic Games, and other purposes sacred to the goddess Athena. The institution of "God's Acre" had a long subsequent history in Europe.

Even within the great ancient cities, sacred spaces retained something of the natural. Babylonian ziggurats were crowned by groves of trees, Egyptian temples were graced with sacred lakes and gardens, and the Acropolis of Athens had its sacred caves, spring, and cypress trees. Those places had to be walled both for protection and to distinguish the sacred space within from the congested streets outside.

But the practice of setting physical boundaries for sacred spaces gives another answer, contrasting with that of the American Indians, to the question, "How much of the world is sacred?" That answer is: "As much as has been consecrated to the gods." Outside the limits, the gods no longer protected the earth, and people were free to use it as they saw fit. Inside the *temenos,* there might be glimpsed a holy light, but outside shone only the ordinary light of day. Thus an enormous step had been taken toward desacralizing nature, but it is also true that the boundaries themselves had been endowed with a numinous quality. . . .

In order to understand how the concept of sacred space entered the medieval and modern mind, one must consider how the Hebrews transformed it. The psalmist proclaimed that all the earth, in a certain sense, is sacred: "The earth is the Lord's and the fullness thereof, the world and those who dwell therein."[9] The early Hebrews had their sacred places: Sinai, where God gave the commandments to Moses, and Bethel, where Jacob had wrestled with the angel. But the dominant view of Judaism held that God the Creator is not

to be identified with His creation, even though it might serve as a marvelous sign of His power and benevolence. Since God is transcendent, He cannot be said to dwell in any spot on earth in an ontological sense. The Hebrews experienced a long struggle with the religions of the surrounding peoples, who worshipped in sacred groves and high places. God had commanded the destruction of those sanctuaries.

The designation of sacred spaces like those of the Canaanites might have suggested that God is present in the natural world in a more intimate way than Judaism was ready to affirm. Since all the world belongs to God, the designation of a particular locale as sacred space could be arbitrary. To avoid the confusion caused by having many sanctuaries, which might have implied to the common people that many gods were being worshipped, the religious authorities under King Josiah centralized the sacrificial worship of the one God in a single space: the Temple of Jerusalem. Even though Mount Zion was undoubtedly a sacred place before it was reconsecrated for the Temple, the Judaic belief was that it was holy because it had been sanctified by God's people at God's command, not because of any special sacredness inherent in the spot. So while the Greeks may be said to have *recognized* sacred space in the landscape, the Hebrews *declared* the Temple space to be sacred.

Christianity took a further step. The early missionaries were anxious that their converts from paganism should not confuse the creation with the creator. Paul the Apostle taught that the natural world had fallen along with mankind and needed to be redeemed through the work of Christ. John urged the Christians not "to love the world or the things in the world."[10] By "world," John doubtlessly meant "non-Christian society," but the Church has insisted on taking the word to mean "the creation." "God who made the world . . . does not live in shrines made by mankind."[11]

It is true that the New Testament does not teach that nature is evil, but that even in its fallen state it exhibits the eternal power and deity of God. Within the first few centuries, however, many Christians were convinced that the natural world was the province of the devil, the adversary of God. Although that idea is not really orthodox, because the sacraments show natural creation as a vehicle of the grace of God, the conception of the world that has fallen into the power of darkness was an image that shaped the imagination of medieval European Christianity. Basil said Satan's "dominion extends over all the earth,"[12] and Synesius of Cyrene prayed to be released from "the demon of the earth, the demon of matter, . . . who stands athwart the ascending path."[13]

The Christians of that time, therefore, were not encouraged to look for sacred space within the world of nature. For them, the churches and monasteries were sacred space, with the enclosed cloisters and churchyards that adjoined them, filled with trees that sheltered the burial places of the sainted dead. They were oases of sanctity in a desert of evil. As outposts of heaven on a fallen earth, they could be established anywhere, though it was considered an act of merit to locate them on the former sites of pagan temples as signs of the victory of Christianity over the demonic gods.

But older attitudes of the pagan converts often surfaced, and some earlier practices continued in the new sacred spaces. From a distance, the appearance of the new sanctuaries was not unlike that of the old, with a sacred building standing within a grove of trees, surrounded by a wall. They still look so today. And within, the right of sanctuary was often given, and those suffering from various maladies were allowed to sleep there in the expectation that God would send them dreams as a means of healing. Once consecrated, they required a different behavior inside the walls, including cloister or churchyard, from that allowed outside. The threshold of the church divided two quite different kinds of space, and boundaries retained a religious sanction.

By the time the Europeans were ready to invade the homeland of the American Indians, the idea of sacred space as a distinct area consecrated by ecclesiastical authority was firmly established. In addition, the concept of a boundary as a sacrosanct limit, whether marked by a wall or an imaginary line, had the force of millennia of tradition. The meeting of the two peoples was foredoomed to tragedy, since the Indians had no way to grasp the alien concept; and because the program of the Europeans amounted to cultural genocide, they had little interest in Indian ideas of the sacred.

A further development, however, was taking place in Europe in regard to sacred space, and that was the final step in desacralizing nature. Nationalism placed the claims of the State above those of the Church, effectively denying that even Church land was sacred space. And rising capitalism defined land as a commodity, subject to division and sale, no more sacred than any other economic resource. The church perforce acquiesced in both of those developments. But the old sense of the inviolability of boundaries persisted; now they were boundaries of nations or of private land rather than religious sanctuaries, but the new order believed in them as firmly as the old. Trespass, the violation of boundaries, was still as heinous a sin.

Indians encountered the strange desire of the Europeans to buy land almost as soon as the foreigners appeared on their coasts. It happened again and again, whether in the "sale" of Manhattan Island or in William Penn's "treaty" in Pennsylvania. The Indians seem to have regarded such arrangements as permission for specific uses of the land, not as "conveyance in fee simple." Indians were incredulous at the idea that the earth could be divided by a line drawn on a map. How could Mother Earth be cut up in that way? In the opening years of the nineteenth century, Tecumseh said, "Sell a country? Why not sell the air, the clouds, and the great sea as well as the earth?"[14] He was protesting not against commercialism, but sacrilege.

For their part, Europeans were shocked at the failure of Indians to respect their boundaries. To Europeans, treaty lines embodied the integrity of the nation state and therefore were inviolable, even if they crossed territory that a European had not seen. Similarly, property lines demanded the same respect as the principle of private ownership itself, and the fact that Indians would return to the hunt on their traditional tribal lands after they had become royal, public, or private property was to Europeans an inexcusable trespass of the limits. For the Europeans to violate the boundaries in the other direction was not a similar trespass in their own minds, because they had convinced themselves that the Indians did not really "occupy" or "use" the land. Indian sacred space was not respected because Indian religions were regarded not as "real" religions, but "superstitions," and Indians were expected to accept the new order, either by adopting European-American ways or by withdrawing beyond the frontiers.

The Indian reservation, as an area set aside within recognized boundaries, represents something of an anomaly within the context of the European-American view. Once the limits had been set, and the Indians recognized as the proper occupants of the enclosed land, all the forces of legality and centuries of customary attitudes should have caused Americans to respect the reservations. That they did not, shows that the economic culture of the late nineteenth-century and early twentieth-century America placed a higher value on the acquisition and exploitation of resources located on Indian lands than on the ideals of sacred property boundaries and whatever relict religious feelings might still have attached to them.

An engine for the destruction of reservations, the General Allotment Act of 1887, was a pious fraud imposing American law on Indian tradition. The congressional advocates of the measure claimed that it would acculturate the Indians by giving them property, but after its enactment, the administrators of allotment managed to alienate two-thirds of all Indian land within fifty years. That process showed no concern for the preservation of Indian sacred sites. Indeed, Indian religions were specifically denied the protection of the First Amendment to the United States Constitution during the same period, and an effort was made to stamp them out through proscription of the religions themselves and the reeducation of Indian children.

At the same time, a similar desire to exploit natural resources was altering the American landscape and destroying the character of the Indians' sacred places. "Wilderness" is a western idea, but it is clear that Indian holy places tended to be unspoiled areas, so that the exploitation of wilder country infringed upon many if not most of them. But also during the late nineteenth and early twentieth centuries, what might be regarded as a resurgence of the idea of demarcating sacred spaces appeared in America. That was the movement to preserve natural areas as natural parks, forests, and wilderness areas. While it might not seem at first glance that those reservations are sacred in the sense used here, many of the most vocal exponents of the new conservation were motivated in large part by a concern for the sacred. True, that concern was not an expression of ecclesiastical religion, but of what has been called civil religion because it involved secular governmental action for conservation. But "civil religion" as a term does not quite capture the way in which they had recovered the perception of sacred space. Deeply religious, the conservationists were highly orthodox. They found their temples in the wilderness, not in churches. And they did regard wild nature as sacred.

John Muir, whose role in the creation and protection of national parks was enormous, believed that what he was doing was saving sacred spaces. He spoke of mountains and meadows as places of healing, renewal, and worship. What better statement, if the sacred is a feeling of connectedness with the all, than Muir's words, "The clearest way into the Universe is through a forest wilderness."[15] Mircea Eliade connects the sacred with the times and places of creation; Muir found Eden in the wild places, saying, "I have discovered that I also live in 'creation's dawn.'"[16] "In God's wildness," he added, "lies the hope of the world."[17] And when Hetch Hetchy—a miniature valley like the more famous Yosemite, located in the same national park and one of the places he honored most—was threatened with flooding for a reservoir to supply the city of San Francisco, he stated the principle of its sacredness unequivocally:

> These temple destroyers, devotees of ravaging commercialism, seem to have a perfect contempt for Nature, and, instead of lifting their eyes to the God of the mountains, lift them to the Almighty Dollar. Dam Hetch Hetchy! As well dam for water-tanks the people's cathedrals and churches, for no holier temple has ever been consecrated by the heart of man.[18]

Why did the conservation movement in America go so far in its perception of the sacred beyond its precedents in the Old World? George Catlin, Henry Thoreau, Muir, John Wesley Powell, and others among its leaders knew American Indians well and reflected upon Indian ideas in their writings. Catlin, the artist whose work did so much to rescue the culture of Indians in the 1830s from oblivion, was the person who first suggested that a national park be set aside to preserve not just the landscape and wildlife, but also the way of life of the Plains Indians.[19] Thoreau spent months in wild country with the Algonquian Indian guide, Joe Polis. Muir stayed for a while among the Tlingits of Alaska, found their ideas of nature and wild animals very much like his own, and was adopted into the tribe.[20] Powell was fluent in Paiute and published perceptive translations of Indian poetry.[21] Those lovers of nature were putting down spiritual roots in the land, and encountering the fact that Indian tribes had already established a relationship with the earth through thousands of years of tradition.

But when national parks and forests were created, unfortunately little provision was made for the Indian people who lived in them. In actual practice, the Park and Forest Services worked out a method of issuing special use permits for Indians who lived in or used the areas, although there were a few cases of attempted eviction like the repeated endeavors to remove the Havasupai settlement from Grand Canyon Village. Especially important is the fact that the outstanding natural features that caused the parks to be created were usually themselves sacred places in tribal traditions.

Indians generally understood that the parks and forests had been established to protect the land, animals, and plants within them. From the Indian standpoint, it was good to have the areas protected because their integrity as sacred space required that forest and park lands be maintained in the natural state. But the natives were frustrated by the way the laws were administered to interfere with their religions, as well as their traditional hunting and fishing rights. As a member of the Crow tribe stated:

> The Laws that protect birds, animals, plants and our Mother Earth from people who have no respect for these things serve to inhibit the free exercise of religion . . . and free access to religious sites when these American Indians pose no threat to them.[22]

For most of this century, no consistent policy granted Indians access to their sacred spaces, and developments such as roads, spraying and removal of trees, ski areas, other recreational facilities, and river channelization and dams, often committed desecrations in Indian eyes.

In recent years, a more considerate attitude has been reflected in congressional action, although it still remains to be put into practice fully. The American Indian Religious Freedom Act of 1978 (AIRFA) guarantees the right to "believe, express, and exercise the traditional religions of the American Indian, Eskimo, Aleut, and Native Hawaiian."[23] The law charges all United States governmental agencies to consult traditional religious leaders in order to preserve Native American religious rights and practices, and to inventory all sacred places on federal lands and come up with proposed policies of management that will preserve the traditional religious values and practices associated with them. The law, it seemed, came none too soon, because the surviving American Indian sacred spaces have never been so threatened with desecration or outright destruction as they now are. The law recognizes that all communities are equally entitled to protection of their freedom of religion, and that in applying this principle to Native Americans, the government must acknowledge the special role of sacred spaces.

But court cases brought under AIRFA to protect sacred sites have failed, with one or two notable exceptions. Among the unsuccessful cases were attempts to protect Navajo rights of worship at Rainbow Bridge,

Utah; to save Cherokee sites threatened with flooding by the Tellico Dam in Tennessee; to prevent intrusion of a ski resort into San Francisco Peaks, Arizona, an area sacred to both Hopis and Navajos; and to allow undisturbed ceremonies by Lakota and Cheyenne people at Bear Butte, South Dakota.[24] In these cases, use by the general public in the form of reservoirs or recreational facilities took precedence over Native American religious rights. Even in the case that succeeded, where the Yurok, Karok, and Tolowa tribes of northern California managed to block construction of a road near their traditional mountain prayer sites, their First Amendment rights, not AIRFA, were held to be decisive.[25] One case where AIRFA appears to have helped was in the denial of a license for Northern Lights, Inc., to build a water project that would have destroyed Kootenai Falls, Idaho, a sacred place for the Kootenai tribe.[26]

The weaknesses of AIRFA are that it is only an advisory resolution of Congress directed at federal agencies, which means it can be ignored with impunity; that it has no enforcement provisions; and that it offers no help in weighing American Indian religious rights against other rights and interests. A federal agency may comply with AIRFA by considering Indian rights; it does not have to decide that they take precedence over skiing, dirt-biking, or electrical power generation. American Indian religious freedom in regard to sacred places can only be guaranteed by a new, stronger, and more carefully drafted law.

But underlying the ineffectiveness of AIRFA is a failure to recognize the difference between the American Indian and European concepts of sacred space. In the successful Yurok-Karok-Tolowa case, usually called the "G-O Road Case," the court defined sacred space in terms of the federal land survey, in a decision quoted here in part:

> It is hereby ordered that the defendants are permanently enjoined from constructing the Chimney Rock Section of the G-O Road and/or any alternative route . . . which would traverse the high country, which constitutes [specified] sections in Six Rivers Forest. . . . It is further hereby ordered that the defendants are permanently enjoined from engaging in commercial timber harvesting.[27]

The decision also requires preparation of Environmental Impact Statements for other future plans in the general area. The provisions resemble those that have been used in demarcating sacred land in the European tradition. The modern officials, without realizing it, are acting like ancient Greeks delimiting a *temenos*. Unlike the Greeks and in accord with modern secular thought, they are doing so without really believing that there is anything inherently sacred inside the lines they are drawing.

Traditional Indians, on the other hand, are faced with difficult alternatives. They can accept the decisions of land managers and/or bring cases to court like the G-O Road Case, thus contenting themselves with saving a few shattered fragments of their heritage and leaving unchallenged the non-Indian idea that sacred space is as much as has been set aside, and no more. This course has the advantage of using federal law to achieve a measure of justice. But another alternative would be to insist on the ancient Indian conception that all the earth is sacred. Taken seriously, this second course could open a dialogue and raise the national consciousness of Indian values. Some tribal elders have already begun to do the latter, as Robert S. Michaelson indicates in an important recent article:

> Some traditionalists have claimed that in keeping with their religion, *all* land on which the tribe has lived, celebrated, and worshipped in the past is sacred and hence essential to tribal free exercise of religion. Such a claim was made before the Federal agencies Task Force by the combined nineteen Pueblo representatives and in Senate testimony on the [AIRFA] resolution by Yakima representatives. It has also been made in court by the Sioux, the Hopi, and the Navajo.[28]

It has been reported that Hopi spokesmen lay claim, in spirit, to the entire North American continent. In the Ghost Dance, Indians of many tribes prayed for the renewal of the whole land, and the spirit of that prayer did not die at the first Wounded Knee.

The federal law, of course, lacks the same vision. Without realizing it, Congress promised far more than can be delivered. It has recognized that the Indian consensus—that certain lands are sacred—can be respected and protected as long as it does not seriously interfere with the rights and interests of others. But Congress and the courts have not even begun to deal with the basic traditional tribal principle that Mother Earth, as a holistic entity, is sacred.

Another fact has to be considered. There are a large, vocal, and increasingly influential number of people in America today who are recovering the ancient idea the sacredness of the earth. They hold, not as a fad or pose, but as a deep conviction, that wilderness is sa-

cred ground, and that visits to places of power enhance wisdom and health. They advocate that we should learn to know the earth, and the plants and animals that inhabit it, in the places where we live. They feel, as Indians long have felt, that human beings are not the lords of creation but fellow creatures with the bears, the ravens, and the running streams.

Who are these people? They have as yet no name and no church, and perhaps they never will. Gary Snyder is one of them, and Paul Winter is another. It is interesting that the first who come to mind are a poet and a musician. Another, the writer Wendell Berry, speaks characteristically of agricultural land and inhabited space. The idea that holiness inheres in the place where one lives is alien to the European tradition, for in that tradition sacred space is sundered, set aside, a place one goes only to worship. But to live in what one regards as sacred space is the most forceful affirmation of the sacredness of the whole earth. Snyder has made a deep and sympathetic study of Indian traditions, and both Winter and Berry acknowledge the closeness of their ideas to Indian insights.[29]

We are now at the point where those people can talk with traditional Native American elders. The Indian sacred places that remain, the places of power within the sacred continuum, must be preserved. A conversation on how to amend and strengthen the AIRFA offers at least a place to start. A better law would give leverage to people who believe in the sacredness of those places. It is encouraging that, as in the G-O Road Case, environmental groups and Indian tribes have joined together. That cooperation has happened because both Indians and environmentalists wish to keep the sacred space in its natural condition. And their motives for wanting to do so, while not identical, are not incompatible either.

The situation should encourage the two groups to engage in a wider dialogue, one in which non-Indians may learn something about the Indian conception of sacred space, while Indians can hear other Americans who feel that many of the same spaces are sacred. Some of the sites, by their nature, must be kept secret and closed to outsiders, a need envisioned by AIRFA and honored in some court cases. In the Kootenai case, for example, tribal elders were allowed to give "limited distribution" testimony that presumably located sites and documented their sacred character; the testimony did not become part of the published court record, however, and its secrecy was preserved. But many of the places that need protection are great shrines that should be held open for most of the year as places where all people may seek wisdom and health. Of course, this must be done in a way that would prevent gross intrusions, vandalism, and theft of holy objects by visitors who did not honor sacred space.

The dialogue has begun. All those who participate in it should be willing to learn. Recent developments in science have shown some support for the Indian view of sacred space. By understanding ecology, we learn the intimate way that all parts of the biosphere are interconnected. Life on earth, as ecologists see it, forms a net in which there are important foci of energy. We cannot allow the net to be broken in too many places without destroying the most basic processes that sustain us. The atmospheric chemist James Lovelock and the biologist Lynn Margulis have advanced a theory called the Gaia hypothesis.[30] This postulates that the living systems of the earth—the animals and plants collectively called the biosphere—regulate physical systems such as temperature and the balance of gases and acidity in the atmosphere so as to protect and support life.

If we grant the truth of the Gaia hypothesis for a moment, it seems that we humans are likely to act to keep the planetary system functioning only if we recognize that every part of it is sacred in the sense of being connected to the whole. Whether we are Indians or others, we can agree with the words of John Muir, "we all dwell in a house of one room—the world with the firmament for its roof—and we are sailing the celestial spaces without leaving any track."[31] We will know that all decisions affecting any part of the natural environment are decisions about sacred space. Thus the visions of the tribal elders can combine with the most daring new conceptions of ecological scientists to show us how to see wholeness—the holiness—of the earth, and how we must act while we live together here.

Notes

1. John Collier, *On the Gleaming Way* (Denver, 1962), 124.
2. Public Law 95-341, Senate Joint Resolution 102, 42 U.S.C. par. 1996, August 11, 1978.
3. Fitz John Porter Poole, "Erosion of a Sacred Landscape," in Michael Tobias, ed., *Mountain Peoples: Profiles of Twentieth Century Adaptation* (Norman, forthcoming from University of Oklahoma Press).
4. W. C. Vanderwerth, ed., *Indian Oratory: Famous Speeches by Noted Indian Chieftains* (Norman, 1971), 120–21.
5. Black Elk, as quoted in Joseph Epes Brown, *The Sacred Pipe* (Norman, 1953), 12–13.
6. James Mooney, "The Ghost-dance Religion," *Fourteenth Annual Report of the Bureau of Ethnology* (Washington, D.C., 1896), 724.
7. Ernest John Eitel, *Feng-Shui: or, the Rudiments of Natural Science in China* (London, 1873).
8. Plato, *The Laws*, 5.747 D-E.
9. Psalm 24:1.
10. I John 2:15.
11. Acts 17:24.
12. J. P. Migne, ed., *Patrologiae Cursus Completus*, Greek Series, vol. 31 (Rome, 1800), 352A.
13. Synesius of Cyrene, *The Essays and Hymns of Synesius of Cyrene*, trans. A. Fitzgerald, vol. 1 (Oxford, 1930), Hymn IV, pp. 240 ff.
14. Glenn Tucker, *Tecumseh: Vision of Glory* (Indianapolis, 1956), 163.
15. Edwin Way Teale, ed., *The Wilderness World of John Muir* (Boston, 1954), 312.
16. Ibid., 311.
17. Ibid., 315.
18. Ibid., 320.
19. George Catlin, *Letters and Notes on the Manners, Customs, and Condition of the North American Indians*, vol. 1 (1841; reprint, Minneapolis, 1965), 261–62.
20. John Muir, *The Writings of John Muir*, manuscript ed., vol. 3 (Boston, 1916), 208–11.
21. John Wesley Powell, *First Annual Report of the Bureau of American Ethnology* (Washington, DC, 1881), 23.
22. Hearings, Senate Committee on Indian Affairs, Joint Resolution 102, February 24, 1978, quoted in *American Indian Religious Freedom Act Report*, P.L. 95-341, Federal Agencies Task Force, Chairman, Cecil D. Andrus, Secretary of the Interior (Washington, DC, 1979), Appendix A. 1.
23. See note 2.
24. *Bandoni v. Higginson*, 638 F.2d 172 (10th Cir. 1980), cert. denied 452 U.S. 954 (1981), regarding Navajo rights at Rainbow Bridge, Utah; *Sequoyah v. Tennessee Valley Authority*, 620 F.2d 1159 (6th Cir. 1980), cert. denied, 449 U.S. 953 (1980), regarding Cherokee rights in the Little Tennessee Valley; *Wilson v. Block*, 708 F.2d 735 (D.C. Cir. 1983), regarding Hopi and Navajo rights in the San Francisco Peaks, Arizona; and *Frank Fools Crow v. Gullet*, 706 F.2d 856 (8th Cir. 1983), regarding Lakota Sioux and Cheyenne rights at Bear Butte, South Dakota.
25. *Northwest Indian Cemetery Protective Association v. Peterson*, 565 F. Supp. 586 (N.D. California 1983).
26. Northern Lights, Inc., Project No. 2752-000, 27 Federal Energy Regulatory Commission, par. 63,024, April 23, 1984.
27. See note 25.
28. Robert S. Michaelson, "The Significance of the American Indian Religious Freedom Act of 1978," *Journal*

of the American Academy of Religion 52 (no. 1, 1984), 93–115, quotation pp. 108–109.

29. Gary Snyder, "Good, Wild, Sacred," *The CoEvolution Quarterly* (Fall 1983), 8–17; Paul Winter, *Missa Gaia: Earth Mass* (Litchfield, CT: Living Music Records, 1982), see notes in 33-rpm disk version; Wendell Berry, *The Gift of Good Land* (San Francisco, 1981), 267–81, and Berry, "The Body and the Earth," in *The Unsettling of America: Culture and Agriculture* (New York, 1978), 97–140.

30. James E. Lovelock, *Gaia: A New Look at Life on Earth* (Oxford, 1979).

31. Teale, *Wilderness World of John Muir*, 310.

For Further Discussion

1. Can you think of any examples, in literature or in your own experience, where a sense of place has proved to be sacred or spiritually important?

2. How is "sacred space" defined by Hughes and Swan? How does this understanding of land and location differ from your own? Are there any similar sacred spaces in your own religion?

3. Describe as clearly as you can the differing understandings of land that were held by the early European settlers of the Americas and the native peoples who lived there. How would you compare these to the understandings of land and property found in Richard Stroup and John Baden (Chapter 4) and Eugene Hargrove (Chapter 10)?

4. What difference, if any, do you see between views that attribute moral value to land and those that attribute religious value to land?

Radical American Environmentalism and Wilderness Preservation: A Third World Critique

Ramachandra Guha

A common error in interactions between cultures occurs when the beliefs and values of one culture are applied to another culture. Such errors range from the trivial, as when tourists expect that their own language will be spoken in a foreign country, to the catastrophic, as when European settlers murdered what they called uncivilized infidels who wouldn't acknowledge their Christian God.

Because so many environmental problems are international, we need to be careful when extrapolating U.S. and Western values to non-Western cultures. In this selection, sociologist Ramachandra Guha claims that the deep ecology movement fails as a truly radical environmental philosophy. Deep ecologists have failed to consider fully how their environmental agenda would affect Third World cultures. In general, their emphasis on biocentric ethics, on wilderness preservation, and on the evils of consumerism and overconsumption make deep ecology an untrustworthy environmental philosophy for the Third World. In reading Guha, you might suspect that the same criticisms could be applied to many environmental philosophies besides deep ecology.

I INTRODUCTION

The respected radical journalist Kirkpatrick Sale recently celebrated "the passion of a new and growing movement that has become disenchanted with the environmental establishment and has in recent years mounted a serious and sweeping attack on it—style, substance, systems, sensibilities and all."[1] The vision of those whom Sale calls the "New Ecologists"—and what I refer to in this article as deep ecology—is a compelling one. Decrying the narrowly economic goals of mainstream environmentalism, this new movement aims at nothing less than a philosophical and cultural revolution in human attitudes toward nature. In contrast to the conventional lobbying efforts of environmental professionals based in Washington, it proposes a militant defence of "Mother Earth," an unflinching opposition to human attacks on undisturbed wilderness. With their goals ranging from the spiritual to the political, the adherents of deep ecology span a wide spectrum of the American environmental movement.

As Sale correctly notes, this emerging strand has in a matter of a few years made its presence felt in a number of fields: from academic philosophy (as in the journal *Environmental Ethics*) to popular environmentalism (for example, the group Earth First!).

In this article I develop a critique of deep ecology from the perspective of a sympathetic outsider. . . . I speak admittedly as a partisan, but of the environmental movement in India, a country with an ecological diversity comparable to the U.S., but with a radically dissimilar cultural and social history.

My treatment of deep ecology is primarily historical and sociological, rather than philosophical, in nature. Specifically, I examine the cultural rootedness of a philosophy that likes to present itself in universalistic terms. I make two main arguments: first, that deep ecology is uniquely American, and despite superficial similarities in rhetorical style, the social and political goals of radical environmentalism in other cultural contexts (e.g., West Germany and India) are quite different; second, that the social consequences of putting deep ecology into practice on a worldwide basis (what its practitioners are aiming for) are very grave indeed.

II THE TENETS OF DEEP ECOLOGY

While I am aware that the term *deep ecology* was coined by the Norwegian philosopher Arne Naess, this article refers specifically to the American variant.[2] Adherents of the deep ecological perspective in this country, while arguing intensely among themselves over its political and philosophical implications, share some fundamental premises about human-nature interactions. As I see it, [the following are three of] the defining characteristics of deep ecology:

First, deep ecology argues, that the environmental movement must shift from an "anthropocentric" to a "biocentric" perspective. In many respects, an acceptance of the primacy of this distinction constitutes the litmus test of deep ecology. A considerable effort is expended by deep ecologists in showing that the dominant motif in Western philosophy has been anthropocentric—i.e., the belief that man and his works are the center of the universe—and conversely, in identifying those lonely thinkers (Leopold, Thoreau, Muir, Aldous Huxley, Santayana, etc.) who, in assigning a man a more humble place in the natural order, anticipated deep ecological thinking. In the political realm, meanwhile, establishment environmentalism (shallow ecology) is chided for casting its arguments in human-centered terms. Preserving nature, the deep ecologists say, has an intrinsic worth quite apart from any benefits preservation may convey to future human generations. The anthropocentric-biocentric distinction is accepted as axiomatic by deep ecologists, it structures their discourse, and much of the present discussion remains mired within it.

The second characteristic of deep ecology is its focus on the preservation of unspoilt wilderness—and the restoration of degraded areas to a more pristine condition—to the relative (and sometimes absolute) neglect of other issues on the environmental agenda. I later identify the cultural roots and portentous consequences of this obsession with wilderness. For the moment, let me indicate three distinct sources from which it springs. Historically, it represents a playing out of the preservationist (read *radical*) and utilitarian (read *reformist*) dichotomy that has plagued American environmentalism since the turn of the century. Morally, it is an imperative that follows from the biocentric perspective: other species of plants and animals, and nature itself, have an intrinsic right to exist. And finally, the preservation of wilderness also turns on a scientific argument—viz., the value of biological diversity in stabilizing ecological regimes and in retaining a gene pool for future generations. Truly radical policy proposals have been put forward by deep ecologists on the basis of these arguments. The influential poet Gary Snyder, for example, would like to see a 90 percent reduction in human populations to allow a restoration of pristine environments, while others have argued forcefully that a large portion of the globe must be immediately cordoned off from human beings.[3] . . .

Third, deep ecologists, whatever their internal differences, share the belief that they are the "leading edge" of the environmental movement. As the polarity of the shallow/deep and anthropocentric/biocentric distinctions makes clear, they see themselves as the spiritual, philosophical, and political vanguard of American and world environmentalism.

III TOWARD A CRITIQUE

Although I analyze each of these tenets independently, it is important to recognize, as deep ecologists are fond of remarking in reference to nature, the interconnectedness and unity of these individual themes.

1. Insofar as it has begun to act as a check on man's arrogance and ecological hubris, the transition from an anthropocentric (human-centered) to a biocentric (humans as only one element in the ecosystem) view in both religious and scientific traditions is only to be welcomed. What is unacceptable are the radical conclusions drawn by deep ecology, in particular, that intervention in nature should be guided primarily by the need to preserve biotic integrity rather than by the needs of humans. The latter for deep ecologists is anthropocentric, the former biocentric. This dichotomy is, however, of very little use in understanding the dynamics of environmental degradation. The two fundamental ecological problems facing the globe are (i) overconsumption by the industrialized world and by urban elites in the Third World and (ii) growing militarization, both in a short-term sense (i.e., ongoing regional wars) and in a long-term sense (i.e., the arms race and the prospect of nuclear annihilation). Neither of these problems has any tangible connection to the anthropocentric-biocentric distinction. Indeed, the agents of these processes would barely comprehend this philosophical dichotomy. The proximate causes of the ecologically wasteful characteristics of industrial society and of militarization are far more mundane: at an aggregate level, the dialectic of economic and political structures, and at a micro-level, the life style choices of individuals. These causes cannot be reduced, whatever the level of analysis, to a deeper anthropocentric attitude toward nature; on the contrary, by constituting a grave threat to human survival, the ecological degradation they cause does not even serve the best interests of human beings! If my identification of the major dangers to the integrity of the natural world is correct, invoking the bogy of anthropocentrism is at best irrelevant and at worst a dangerous obfuscation.

2. If the above dichotomy is irrelevant, the emphasis on wilderness is positively harmful when applied to the Third World. If in the U.S. the preservationist/utilitarian division is seen as mirroring the conflict between "people" and "interests," in countries such as India the situation is very nearly the reverse. Because India is a long settled and densely populated country in which agrarian populations have a finely balanced relationship with nature, the setting aside of wilderness areas has resulted in a direct transfer of resources from the poor to the rich. Thus, Project Tiger, a network of parks hailed by the international conservation community as an outstanding success, sharply posits

the interests of the tiger against those of poor peasants living in and around the reserve. The designation of tiger reserves was made possible only by the physical displacement of existing villages and their inhabitants: their management requires the continuing exclusion of peasants and livestock. The initial impetus for setting up parks for the tiger and other large mammals such as the rhinoceros and elephant came from two social groups, first, a class of ex-hunters turned conservationists belonging mostly to the declining Indian feudal elite and second, representatives of international agencies, such as the World Wildlife Fund (WWF) and the International Union for the Conservation of Nature and Natural Resources (IUCN), seeking to transplant the American system of national parks onto Indian soil. In no case have the needs of the local population been taken into account, and as in many parts of Africa, the designated wildlands are managed primarily for the benefit of rich tourists. Until very recently, wildlands preservation has been identified with environmentalism by the state and the conservation elite; in consequence, environmental problems that impinge far more directly on the lives of the poor— e.g., fuel, fodder, water shortages, soil erosion, and air and water pollution—have not been adequately addressed.[4]

Deep ecology provides, perhaps unwittingly, a justification for the continuation of such narrow and inequitable conservation practices under a newly acquired radical guise. Increasingly, the international conservation elite is using the philosophical, moral, and scientific arguments used by deep ecologists in advancing their wilderness crusade. A striking but by no means atypical example is the recent plea by a prominent American biologist for the takeover of large portions of the globe by the author and his scientific colleagues. Writing in a prestigious scientific forum, the *Annual Review of Ecology and Systematics,* Daniel Janzen argues that only biologists have the competence to decide how the tropical landscape should be used. As "the representatives of the natural world," biologists are "in charge of the future of tropical ecology," and only they have the expertise and mandate to "determine whether the tropical agroscape is to be populated only by humans, their mutualists, commensals, and parasites, or whether it will also contain some islands of the greater nature—the nature that spawned humans, yet has been vanquished by them." Janzen exhorts his colleagues to advance their territorial claims on the tropical world more forcefully, warning that the very existence of these areas is at stake: "if biologists want a tropics in which to biologize, they are going to have to buy it with care, energy, effort, strategy, tactics, time, and cash."[5]

This frankly imperialist manifesto highlights the multiple dangers of the preoccupation with wilderness preservation that is characteristic of deep ecology. As I have suggested, it seriously compounds the neglect by the American movement of far more pressing environmental problems within the Third World. But perhaps more importantly, and in a more insidious fashion, it also provides an impetus to the imperialist yearning of Western biologists and their financial sponsors, organizations such as the WWF and IUCN. The wholesale transfer of a movement culturally rooted in American conservation history can only result in the social uprooting of human populations in other parts of the globe. . . .

3. How radical, finally, are the deep ecologists? Notwithstanding their self-image and strident rhetoric (in which the label "shallow ecology" has an opprobrium similar to that reserved for "social democratic" by Marxist-Leninists), even within the American context their radicalism is limited and it manifests itself quite differently elsewhere.

To my mind, deep ecology is best viewed as a radical trend within the wilderness preservation movement. Although advancing philosophical rather than aesthetic arguments and encouraging political militancy rather than negotiation, its practical emphasis—viz., preservation of unspoilt nature—is virtually identical. For the mainstream movement, the function of wilderness is to provide a temporary antidote to modern civilization. As a special institution within an industrialized society, the national park "provides an opportunity for respite, contrast, contemplation, and affirmation of values for those who live most of their lives in the workaday world."[6] Indeed, the rapid increase in visitations to the national parks in postwar America is a direct consequence of economic expansion. The emergence of a popular interest in wilderness sites, the historian Samuel Hays points out, was "not a throwback to the primitive, but an integral part of the modern standard of living as people sought to add new 'amenity' and 'aesthetic' goals and desires to their earlier preoccupation with necessities and conveniences."[7]

Here, the enjoyment of nature is an integral part of the consumer society. The private automobile (and the life style it has spawned) is in many respects the ultimate ecological villain, and an untouched wilderness the prototype of ecological harmony; yet, for most Americans it is perfectly consistent to drive a thousand miles to spend a holiday in a national park. They possess a vast, beautiful, and sparsely populated continent and are also able to draw upon the natural resources of large portions of the globe by virtue of their economic and political dominance. In consequence, America can simultaneously enjoy the material benefits of an expanding economy and the aesthetic benefits of unspoilt nature. The two poles of "wilderness" and "civilization" mutually coexist in an internally coherent whole, and philosophers of both poles are assigned a prominent place in this culture. Paradoxically as it may seem, it is no accident that Star Wars technology and deep ecology both find their fullest expression in that leading sector of Western civilization, California.

Deep ecology runs parallel to the consumer society without seriously questioning its ecological and sociopolitical basis. In its celebration of American wilderness, it also displays an uncomfortable convergence with the prevailing climate of nationalism in the American wilderness movement. For spokesmen such as the historian Roderick Nash, the national park system is America's distinctive cultural contribution to the world, reflective not merely of its economic but of its philosophical and ecological maturity as well. In what Walter Lippman called the American century, the "American invention of national parks" must be exported worldwide. Betraying an economic determinism that would make even a Marxist shudder, Nash believes that environmental preservation is a "full stomach" phenomenon that is confined to the rich, urban, and sophisticated. Nonetheless, he hopes that "the less developed nations may eventually evolve economically and intellectually to the point where nature preservation is more than a business."[8]

The error which Nash makes (and which deep ecology in some respects encourages) is to equate environmental protection with the protection of wilderness. This is a distinctively American notion, born out of a unique social and environmental history. The archetypal concerns of radical environmentalists in other cultural contexts are in fact quite different. The German Greens, for example, have elaborated a devastating critique of industrial society which turns on the acceptance of environmental limits to growth. Pointing to the intimate links between industrialization, militarization, and conquest, the Greens argue that economic growth in the West has historically rested on the economic and ecological exploitation of the Third World. Rudolph Bahro is characteristically blunt:

The working class here [in the West] is the richest lower class in the world. And if I look at the problem from the point of view of the whole of humanity, not just from that of Europe, then I must say that the metropolitan working class is the worst exploiting class in history. . . . What made poverty bearable in eighteenth or nineteenth-century Europe was the prospect of escaping it through exploitation of the periphery. But this is no longer a possibility, and continued industrialism in the Third World will mean poverty for whole generations and hunger for millions.[9]

Here the roots of global ecological problems lie in the disproportionate share of resources consumed by the industrialized countries as a whole *and* the urban elite within the Third World. Since it is impossible to reproduce an industrial monoculture worldwide, the ecological movement in the West must begin by cleaning up its own act. The Greens advocate the creation of a "no growth" economy, to be achieved by scaling down current (and clearly unsustainable) consumption levels. This radical shift in consumption and production patterns requires the creation of alternate economic and political structures—smaller in scale and more amenable to social participation—but it rests equally on a shift in cultural values. The expansionist character of modern Western man will have to give way to an elite of renunciation and self-limitation, in which spiritual and communal values play an increasing role in sustaining social life. This revolution in cultural values, however, has as its point of departure an understanding of environmental processes quite different from deep ecology.

Many elements of the Green program find a strong resonance in countries such as India, where a history of Western colonialism and industrial development has benefited only a tiny elite while exacting tremendous social and environmental costs. The ecological battles presently being fought in Indian have as their epicenter the conflict over nature between the subsistence and largely rural sector and the vastly more powerful commercial-industrial sector. Perhaps the most celebrated of these battles concerns the Chipko (Hug the Tree) movement, a peasant movement against deforestation in the Himalayan foothills. Chipko is only one of several movements that have sharply questioned the nonsustainable demand being placed on the land and vegetative base by urban centers and industry. These include opposition to large dams by displaced peasants, the conflict between small artisan fishing and large-scale trawler fishing for export, the countrywide movements against commercial forest operations, and opposition to industrial pollution among downstream agricultural and fishing communities.[10]

Two features distinguish these environmental movements from their Western counterparts. First, for the sections of society most critically affected by environmental degradation—poor and landless peasants, women, and tribals—it is a question of sheer survival, not of enhancing the quality of life. Second, and as a consequence, the environmental solutions they articulate deeply involve questions of equity as well as economic and political redistribution. Highlighting these differences, a leading Indian environmentalist stresses that "environmental protection per se is of least concern to most of these groups. Their main concern is about the use of the environment and who should benefit from it."[11] They seek to wrest control of nature away from the state and the industrial sector and place it in the hands of rural communities who live within that environment but are increasingly denied access to it. These communities have far more basic needs, their demands on the environment are far less intense, and they can draw upon a reservoir of cooperative social institutions and local ecological knowledge in managing the "commons"—forests, grasslands, and the waters—on a sustainable basis. If colonial and capitalist expansion has both accentuated social inequalities and signaled a precipitous fall in ecological wisdom, an alternative ecology must rest on an alternate society and polity as well.

This brief overview of German and Indian environmentalism has some major implications for deep ecology. Both German and Indian environmental traditions allow for a greater integration of ecological concerns with livelihood and work. They also place a greater emphasis on equity and social justice (both within individual countries and on a global scale) on the grounds that in the absence of social regeneration environmental regeneration has very little chance of succeeding. Finally, and perhaps most significantly, they have escaped the preoccupation with wilderness preservation so characteristic of American cultural and environmental history.

IV A HOMILY

In 1958, the economist J. K. Galbraith referred to overconsumption as the unasked question of the American conservation movement. There is a marked selectivity, he wrote, "in the conservationist's approach to materials consumption. If we are concerned about our great appetite for materials, it is plausible to seek to increase the supply, to decrease waste, to make better use of the stocks available, and to develop substitutes. But what of the appetite itself? Surely this is the ultimate source of the problem. If it continues its geometric course, will it not one day have to be restrained? Yet in the literature of the resource problem this is the forbidden question. Over it hangs a nearly total silence."[12]

The consumer economy and society have expanded tremendously in the three decades since Galbraith penned these words; yet his criticisms are nearly as valid today. I have said "nearly," for there are some hopeful signs. Within the environmental movement several dispersed groups are working to develop ecologically benign technologies and to encourage less wasteful life styles. Moreover, outside the self-defined boundaries of American environmentalism, opposition to the permanent war economy is being carried on by a peace movement that has a distinguished history and impeccable moral and political credentials.

It is precisely these (to my mind, most hopeful) components of the American social scene that are missing from deep ecology. In their widely noticed book, Bill Devall and George Sessions make no mention of militarization or the movements for peace, while activists whose practical focus is on developing ecologically responsible life styles (e.g., Wendell Berry) are derided as "falling short of deep ecological awareness."[13] A truly radical ecology in the American context ought to work toward a synthesis of the appropriate technology, alternate life style, and peace movements. By making the (largely spurious) anthropocentric-biocentric distinction central to the debate, deep ecologists may have appropriated the moral high ground, but they are at the same time doing a serious disservice to American and global environmentalism.

Notes

1. Kirkpatrick Sale, "The Forest for the Trees: Can Today's Environmentalists Tell the Difference," *Mother Jones* 11, no. 8 (November 1986): 26.
2. One of the major criticisms I make in this essay concerns deep ecology's lack of concern with inequalities *within* human society. In the article in which he coined the term *deep ecology*, Naess himself expresses concerns about inequalities between and within nations. However, his concern with social cleavages and their impact on resource utilization patterns and ecological destruction is not very visible in the later writings of deep ecologists. See Arne Naess, "The Shallow and the Deep, Long-Range Ecology Movement: A Summary," *Inquiry* 16 (1973): 96 (I am grateful to Tom Birch for this reference).
3. Gary Snyder, quoted in Sale, "The Forest for the Trees," p. 32. See also Dave Foreman, "A Modest Proposal for a Wilderness System," *Whole Earth Review*, no. 53 (Winter 1986-87): 42–45.
4. See Centre for Science and Environment, *India: The State of the Environment 1982: A Citizens Report* (New Delhi: Centre for Science and Environment, 1982); R. Sukumar, "Elephant-Man Conflict in Karnataka," in Cecil Saldanha, ed., *The State of Karnataka's Environment* (Bangalore: Centre for Taxonomic Studies, 1985). For Africa, see the brilliant analysis by Helge Kjekshus, *Ecology Control and Economic Development in East African History* (Berkeley: University of California Press, 1977).
5. Daniel Janzen, "The Future of Tropical Ecology," *Annual Review of Ecology and Systematics* 17 (1986): 305–06; emphasis added.
6. Joseph Sax, *Mountains Without Handrails: Reflections on the National Parks* (Ann Arbor: University of Michigan Press, 1980), p. 42. Cf. also Peter Schmitt, *Back to Nature: The Arcadian Myth in Urban America* (New York: Oxford University Press, 1969), and Alfred Runte, *National Parks: The American Experience* (Lincoln: University of Nebraska Press, 1979).
7. Samuel Hays, "From Conservation to Environment: Environmental Politics in the United States since World War Two," *Environmental Review* 6 (1982): 21. See also the same author's book entitled *Beauty, Health and Permanence: Environmental Politics in the United States, 1955–85* (New York: Cambridge University Press, 1987).
8. Roderick Nash, *Wilderness and the American Mind*, 3rd ed. (New Haven: Yale University Press, 1982).
9. Rudolf Bahro, *From Red to Green* (London: Verso Books, 1984).
10. For an excellent review, see Anil Agarwal and Sunita Narain, eds., *India: The State of the Environment 1984–85: A Citizens Report* (New Delhi: Centre for Science and Environment, 1985). Cf. also Ramachandra Guha, *The Unquiet Woods: Ecological Change and Peasant Resistance in the Indian Himalaya* (Berkeley: University of California Press, 1990).
11. Anil Agarwal, "Human-Nature Interactions in a Third World Country," *The Environmentalist* 6, no. 3 (1986): 167.
12. John Kenneth Galbraith, "How Much Should a Country Consume?" in Henry Jarrett, ed., *Perspectives on Conservation* (Baltimore: Johns Hopkins Press, 1958), pp. 91–92.
13. Devall and Sessions, *Deep Ecology*, p. 122. For Wendell Berry's own assessment of deep ecology, see his

"Amplications: Preserving Wildness," *Wilderness* 50 (Spring 1987): 39–40, 50–54.

For Further Discussion

1. Guha describes his critique of deep ecology as coming from a sympathetic outsider. On what issues might he be in sympathy with deep ecology? On what issues might they disagree?

2. Guha describes the environmentalist perspective offered by Daniel Janzen as an "imperialist manifesto." How do you understand imperialism, and do you agree with Guha's judgment?

3. How do you compare Guha's comments on the value of wilderness to the analysis of the idea of wilderness offered by William Cronon (Chapter 11)?

4. Deep ecologists claim to offer a radically new worldview as an alternative to the dominant Western worldview. Guha suggests that deep ecology does not represent a truly radical perspective. Instead, he sees deep ecology as running parallel to consumer society without seriously questioning its basis. Why does he believe this? Do you agree with his analysis?

5. Which other authors share the most views with Guha? What similarities can you find?

DISCUSSION AND STUDY QUESTIONS FOR CHAPTER 17

1. According to de Silva, Buddhism views nature as a "network of inter-connected relationships rather than as a static collection of objects." Compare this to the view of nature found in ecological science and to the conclusions of ecocentric ethics.

2. Do you think concern with wilderness preservation is a peculiarly American issue? Could the value of wilderness have arisen in a widely populated country?

3. Should policies aimed at protecting rain forests and biological diversity be imposed on indigenous peoples living within rain forests? If not, can these values be imposed on Western companies doing business within other countries?

4. Is there a particular location that holds special value to you? Would you describe this place as sacred? Have new acquaintances ever asked you where you are from? How important is that place to your identity?

CHAPTER 18

Environmental Pragmatism

In Chapter 1 and again in the introduction to Part IV, we discussed the philosophical quest for a unified and systematic ethical theory. Lacking a single theoretical basis for environmental ethics, we would seem to have no rational way to resolve conflicts. Given the many environmental controversies that we face, this is not a desirable conclusion.

The philosophical position of ethical relativism holds that all ethical values are relative to particular circumstances of time, place, culture, person. Because values are relative, so the argument goes, there is no rational way to establish the legitimacy of any ethical belief. Relativism denies that there is any right or wrong, good or bad, in general; there is only right or wrong, good or bad, relative to particular situations. In essence, relativism admits defeat for the entire project of philosophical ethics.

The many, deep disagreements surrounding environmental issues might lead one to relativist conclusions. We seem only to have found disagreements, controversies, and conflicts. From a social point of view, the lack of a unifying theory leaves us with little more than chaos. From an individual point of view, we are left with a life without integrity.

But, are there really only two options: a single, unified ethical theory or relativism? Must we be committed to unifying the diverse values and norms that have emerged throughout our readings? According to some observers, there is a middle ground between a single theory (sometimes called "moral monism") and relativism. The selections in this final chapter examine the possibility and consequences of a more pragmatic environmental ethics.

Recall from Chapter 1 that policy disagreements can exist even among people who agree on the goals of the policy. For example, there might be widespread community agreement to preserve wetlands from development. Some defend this policy on grounds that wetlands provide important services such as flood control, water purification, and groundwater renewal. Others view this land as valuable wildlife habitat, although hunters and bird watchers might strongly disagree on the meaning of that value. Still others might value the wetlands as home to native or rare plant species. Others value it as a scarce remnant of presettlement era landscape. The rationales for this policy can be as varied as the people who offer them. Further, even among those who agree on the policy and its underlying rationale, there may still be disagreement concerning the best means for protecting the land: Some argue for government purchase, some for regulation, some for market mechanisms.

Environmental pragmatists focus more on the agreement concerning policy than on the disagreements concerning rationales. The philosophical tradition of pragmatism judges the truth or reasonableness of a belief in terms of the practical difference it makes in life. From this perspective, various theories can offer equally true accounts of the world, even if the theories are themselves incompatible. Two theories can be equally rational if adopting one rather than another makes little practical difference in living one's life. With this conclusion, disagreements among theories do not pose the relativistic threat that some would claim.

Given the wide variety of perspectives that we have examined in this textbook, pragmatism is not a bad point at which to end. Perhaps we should resist the temptation to theorize too quickly and focus instead on agreements in practice. We can interpret environmental pragmatism as advising us to answer first the question, How should we live? Only later should we worry about, Why?

Before Environmental Ethics

Anthony Weston

Philosopher Anthony Weston suggests that it is too early in the history of environmentalism to be looking for a theory of environmental ethics. Employing an ecological meta-phor, Weston suggests that ideas evolve within a complex cultural and social context. Attempts to articulate a new environmental ethics that transcends the contemporary anthropocentric context are bound to fail. At this stage of environmental history, "we can have only the barest sense of what ethics for a culture truly beyond anthropocentrism would actually look like."

Because ideas and practices co-evolve, environmental-ists ought to be more focused on creating conditions that will produce new and stronger environmental values. This project, what he calls enabling environmental practice, would create the social, psychological, and phenomenologi-cal conditions that will allow the emergence of an environ-mental ethics. In the meantime, environmentalists ought to resist attempts to systematize environmental theory.

I INTRODUCTION

To think "ecologically," in a broad sense, is to think in terms of the evolution of an interlinked system over time rather than in terms of separate and one-way causal interactions. It is a general habit of mind. Ideas, for example, not just ecosystems, can be viewed in this way. Ethical ideas, in particular, are deeply interwoven with and dependent upon multiple contexts: other prevailing ideas and values, cultural institutions and practices, a vast range of experiences, and natural set-tings as well. An enormous body of work, stretching from history through the "sociology of knowledge" and back into philosophy, now supports this point.[1]

It is curious that environmental ethics has not yet viewed itself in this way. Or perhaps not so curious, for the results are unsettling. Some theories, in par-ticular, claim to have transcended anthropocentrism in thought. Yet these theories arise within a world that is profoundly and beguilingly anthropocentrized.[2] From an "ecological" point of view, transcending this

context so easily seems improbable. In part two . . . , I argue that even the best non-anthropocentric theories in contemporary environmental ethics are still pro-foundly shaped by and indebted to the anthropocen-trism that they officially oppose.

I do not mean that anthropocentrism is inevitable, or even that non-anthropocentric speculation has no place in current thinking. Rather, as a I argue in part three, the aim of my critique is to bring into focus the slow process of culturally constituting and consolidat-ing values that underlies philosophical ethics as we know it. My purpose is to broaden our conception of the nature and tasks of ethics, so that we can begin to recognize the "ecology," so to speak, of environ-mental ethics itself, and thus begin to recognize the true conditions under which anthropocentrism might be overcome.

One implication is that we must rethink the prac-tice of environmental ethics. In part four, I ask how ethics should comport itself at early stages of the pro-cess of constituting and consolidating new values. I then apply the conclusions directly to environmental ethics. In particular, the co-evolution of values with cultural institutions, practices and experience emerges as an appropriately "ecological" alternative to the proj-ect of somehow trying to leapfrog the entire culture in thought. In part five, finally, I offer one model of a coevolutionary approach to environmental philoso-phy: what I call "enabling environmental practice."

II CONTEMPORARY NON-ANTHROPOCENTRISM

I begin by arguing that contemporary non-anthropo-centric environmental ethics remains deeply depen-dent upon the thoroughly anthropocentrized setting in which it arises. Elsewhere I develop this argument in detail.[3] Here there is only room to sketch some highlights.

For a first example, consider the very phrasing of the question that most contemporary environmental philosophers take as basic: whether "we" should open the gates of moral considerability to "other" animals (sometimes just: "animals"), and/or to such things as rivers and mountains. The opening line of Paul Taylor's *Respect for Nature,* for example, invokes such a model. Environmental ethics, Taylor writes, "is concerned with the moral relations that hold between humans and the natural world."[4]

Taylor's phrasing of "the" question may seem neutral and unexceptionable. Actually, however, it is not neutral at all. The called-for arguments address humans universally and exclusively on behalf of "the natural world." Environmental ethics, therefore, is invited to begin by *positing,* not by questioning, a sharp divide that "we" must somehow cross, taking that "we" unproblematically to denote all humans. To invoke such a divide, however, is already to take one ethical position among others. For one thing, it is largely peculiar to modern Western cultures. Historically, when humans said "we," they hardly ever meant to include all other humans. Moreover, they often meant to include some individuals of other species. Mary Midgley emphasizes that almost all of the ancient life patterns were "mixed communities," involving humans and an enormous variety of other creatures, from dogs (with whom, she says, we have a "symbiotic" relationship) to reindeer, weasels, elephants, shags, horses and pigs.[5] One's identifications and loyalties lay not with the extended human species, but with a local and concretely realized network of relationships involving many different species.

Taylor might respond that his question is at least *our* question: the urbanized, modern, Westerner's question. So it is. But it is precisely this recognition of cultural relativity that is crucial. "The" very question that frames contemporary environmental ethics appears to presuppose a particular cultural and historical situation—which is not the only human possibility, and which may itself be the problem. Cross-species identifications, or a more variegated sense of "the natural world," fit in awkwardly, or not at all.

Consider a second example. A defining feature of almost all recent non-anthropocentrism is some appeal to "intrinsic values" in nature. Once again, however, this kind of appeal is actually no more neutral or timelessly relevant than an appeal to all and only humans on behalf of the rest of the world. Intrinsic values in nature are so urgently sought at precisely the moment that the *instrumentalization* of the world—at least according to a certain sociological tradition[6]—has reached a fever pitch. It is because we now perceive nature as thoroughly reduced to a set of "means" to human ends that an insistence on nature as an "end in itself" seems the only possible response. We may even be right. Still, under other cultural conditions, unthreatened by such a relentless reduction of everything to "mere means," it at least might not seem so *obvious* that we must aspire to a kind of healing that salvages a few non-traditional sorts of ends while consigning everything else to mere resourcehood. Instead, we might challenge the underlying means–ends divide itself, turning toward a more pragmatic sense of the interconnectedness of all of our values.[7]

Also, unthreatened in this way, we might not be tempted to metaphysical turns in defense of the values we cherish. Jim Cheney has suggested that the turn to metaphysics in some varieties of contemporary environmental ethics represents, like the ancient Stoics' turn to metaphysics, a desperate self-defense rather than a revelation of a genuine non-anthropocentrism. Cheney charges in particular that a certain kind of radical environmentalism, which he dubs "Ecosophy S," has been tempted into a "neo-Stoic" philosophy— an identification with nature on the level of the universe as a whole—because neo-Stoicism offers a way to identify with nature without actually giving up control. In this way, abstract arguments become a kind of philosophical substitute for "real encounter" with nature.[8]

Cheney argues that Ecosophy S reflects a profoundly contemporary psychological dynamic. I want to suggest that it also reflects the diminished character of the world in which we live. The experiences for which Ecosophy S is trying to speak are inevitably marginalized in a thoroughly anthropocentrized culture. They are simply not accessible to most people or even understandable to many. Although wild experience may actually *be* the starting point for Ecosophy S, there are only a few, ritualized, and hackneyed ways to actually speak for it in a culture that does not share it. Thus—again, under present circumstances— environmental ethics may be literally driven to abstraction.

Once again it may even be true that abstraction is our only option. Nonetheless, in a different world, truly beyond anthropocentrism, we might hope for a much less abstract way of speaking of and for wild experience—for enough sharing of at least the glimmers of wild experience that we can speak of it directly, even perhaps invoking a kind of love. But such a change, once again, would leave contemporary non-anthropocentrism environmental ethics—whether neo-Stoic or just theoretical—far behind.

As a third and final example, consider the apparently simple matter of what sorts of criticism are generally regarded as "responsible" and what sorts of alternatives are generally regarded as "realistic." The contemporary anthropocentrized world, which is, in fact, the product of an immense project of world reconstruction that has reached a frenzy in the modern age, has become simply the taken-for-granted reference point for what is "real," for what must be accepted by any responsible criticism. The absolute pervasiveness of internal combustion engines, for example, is utterly new, confined to the last century and mostly to the last generation. By now very few Westerners ever get out of earshot of internal combustion engines for more than a few hours at a time. The environmental consequences are staggering, the long-term effects of constant noise on "mental health" are clearly worrisome, and so on. Yet this technology has so thoroughly embedded itself in our lives that even mild proposals to restrict internal combustion engines seem impossibly radical. This suddenly transmuted world, the stuff of science fiction only fifty years ago, now just as suddenly defines the very limits of imagination. When we think of "alternatives" all we can imagine are car pools and buses.

Something similar occurs in philosophical contexts. Many of our philosophical colleagues have developed a careful, neutral, critical style as a point of pride. But in actual practice this style is only careful, neutral, critical in certain directions. It is not possible to suggest anything *different,* for the project of going beyond anthropocentrism still looks wild, incautious, intellectually overexcited. Anthropocentrism itself, however, is almost never scrutinized in the same way. Apparently, it just forms part of the "neutral" background: it seems to be no more than what the careful, critical thinker can *presuppose.* Thus it is the slow excavation and the logical "refutation" of anthropocentrism that, perforce, occupy our time—rather than, for one example, a much less encumbered, more imaginative exploration of other possibilities, less fearful of the disapproval of the guardians of Reason or, for another example, a psychological exploration of anthropocentrism itself, taking it to be more like a kind of lovelessness or blindness than a serious philosophical position. Anthropocentrism still fills the screen, still dominates our

energies. It delimits what is "realistic" because in many ways it determines what "reality" itself is.

III ETHICS IN SOCIAL CONTEXT

The conclusion of the argument so far might only seem to be that we need better non-anthropocentrisms: theories that rethink Taylor's basic question, theories that are not so easily seduced by intrinsic values, and so on. Although such theories would be useful changes, the argument just offered also points towards a much more fundamental conclusion, one upon which very large questions of method depend. If the most rigorous and sustained attempts to transcend anthropocentrism still end up in its orbit, profoundly shaped by the thought and practices of the anthropocentrized culture within which they arise, then we may begin to wonder whether the project of transcending culture in ethical thought is, in fact, workable *at all.* Perhaps ethics requires a very different self-conception.

Here, moreover, is a surprising fact: ethics generally *has* a very different self-conception. Most "mainstream" ethical philosophers now readily acknowledge that the values they attempt to systematize are indeed deeply embedded in and co-evolved with social institutions and practices. John Rawls, for example, who at earlier moments appeared to be the very incarnation of the philosophical drive towards what he himself called an "Archimedean point" beyond culture, now explicitly justifies his theory only by reference to its "congruence with our deeper understanding of ourselves and our aspirations, and our realization that, given our history and the traditions embedded in our public life, it is the most reasonable doctrine for us." For *us,* culture answers "our" questions. "We are not," he says, "trying to find a conception of justice suitable for all societies regardless of their social or historical circumstances." Instead, the theory "is intended simply as a useful basis of agreement in our society."[9] The same conclusion is also the burden, of course, of an enormous body of criticism supposing Rawls to be making a less culturally dependent claim. Rawls, thus, does not transcend his social context at all. His theory is, rather, in a Nietzschean phrase, a particularly scholarly way of *expressing* an already established set of values. That contemporary non-anthropocentric environmental ethics does not transcend *its* social context, therefore, becomes much less surprising. At least it is in good company.

Similarly, John Arras, in an article surveying Jonsen and Toulmin's revival of casuistry, as well as the Rawls–Walzer debate, remarks almost in passing that all of these philosophers agree that "there is no escape from the task of interpreting the meanings embedded in our social practices, institution, and history."[10] Michael Walzer argues for a plurality of justice values rooted in the varied "cultural meanings" of different goods.[11] Alasdair MacIntyre makes the rootedness of values in "traditions" and "practice" central to his reconception of ethics.[12] Charles Taylor localizes the appeal to rights within philosophical, theological and even aesthetic movements in the modern West.[13] Sabina Lovibond updates Wittgensteinian "form of life" ethics along sociologically informed "expressivist" lines.[14]

It may seem shocking that the "Archimedean" aspirations for ethics have been abandoned with so little fanfare. From the point of view of what we might call the "theology of ethics," it probably is. Day to day, however, and within the familiar ethics of persons, justice and rights practiced by most of the philosophers just cited, it is less surprising. Operating within a culture in which certain basic values are acknowledged, at least verbally, by nearly everyone, there is little practical need to raise the question of the ultimate origins or warrants of values. Because the issue remains metaphilosophical and marginal to what are supposed to be the more systematic tasks of ethics, we can acquiesce in a convenient division of labor with the social sciences, ceding to them most of the historical and cultural questions about the evolution of values, while keeping the project of systematizing and applying values for our own. "Scholarly forms of expression" of those values—or at least systematic forms of expression, "rules to live by"—are then precisely what we want.

It now seems entirely natural, for example, to view persons as "centers of autonomous choice and valuation," in Taylor's words, "giving direction to their lives on the basis of their own values," having a sense of identity over time, and so on. It also seems natural to point to this "belief system" to ground respect for persons, as Taylor also points out. He does *not* ask how such a belief system came into being and managed to rearrange human lives around itself. He does not *need* to ask. But we need at least to remember that these are real and complex questions. It is only such processes, finally running their courses, that make possible the consensus behind the contemporary values in the first place. Weber traces our belief system about persons, in part, to Calvinist notions about the inscrutability of fate, paradoxically leading to an outwardly calculating possessiveness coupled with rigid "inner asceticism," both self-preoccupied in a fundamentally new way. In addition, he traces it to the development of a system of increasingly impersonal commercial transactions that disabled and disconnected older, more communal ties between people.[15] The cultural relativity of the notion of persons is highlighted meanwhile, by its derivation from the Greek dramatic "personae," perhaps the first emergence of the idea of a unique and irreplaceable individual. A tribal African or Native American would never think of him or herself in this way.[16]

It may be objected that to stress the interdependence of ethical ideas with cultural institutions, practices and experience simply reduces ethical ideas to epiphenomena of such factors. However, the actual result is quite different. The flaw lies with the objection's crude (indeed, truly "vulgar," as in "vulgar Marxist") model of causation. Simple, mechanical, one-way linkages between clearly demarcated "causes" and "effects" do not characterize cultural phenomena (or, for that matter, *any* phenomena). Thus the question is emphatically *not* whether ethical ideas are "cause" or "effect" in cultural systems, as if the only alternative to being purely a cause is to be purely an effect. Causation in complex, interdependent, and evolving systems with multiple feedback loops—that is, an "ecological" conception of causation—is a far better model.[17]

One implication of such a model, moreover, is that fundamental change (at least constructive, noncatastrophic change) is likely to be slow. Practices, habits, institutions, arts and ideas all must evolve in some coordinated way. Even the physical structure of the world changes. Individualism and its associated idea of privacy, for example, developed alongside a revolution in home and furniture design.[18] Thus it may not even be that visionary ethical ideas (or anything else visionary, e.g., revolutionary architecture) are impossible at any given cultural stage, but rather that such ideas simply cannot be recognized or understood, given all of the practices, experiences, etc. alongside of which they have to be placed, and given the fact that they cannot be immediately applied in ways that will contribute to their development and improvement.[19] To use a Darwinian metaphor, all manner of "mutations" may be produced at any evolutionary stage, but conditions will be favorable for only a few of them to be "selected" and passed on.[20]

It may also be objected that any such view is hopelessly "relativistic." Although the term *relativism* now seems to be confused and ambiguous, there is at least one genuine concern here: if values are thoroughly relativized to culture, rational criticism of values may become impossible. In fact, however, rational criticism remains entirely possible—only its "standpoint" is internal to the culture it challenges, rather than (as in the Archimedean image) external to it. Much of what we tend to regard as radical social criticism reinvokes old, even central, values of a culture rather than requiring us to transcend somehow the culture in thought. Weber, for example, reread Luther's conception of the individual's relation to God as an extension of the already old and even revered monastic ideal to society at large. Likewise, the challenges of the 1960s in the US arguably appealed not to new values but to some of the oldest and most deeply embedded values of our culture. The Students for Democratic Society's "Port Huron Statement" persistently speaks in biblical language; the Black Panthers invoked the Declaration of Independence; the Civil Rights Movement was firmly grounded in Christianity. In his 1981 encyclical "Laborem Exercens," Pope John Paul II appealed to Genesis to ground a stunning critique of work in industrial societies reminiscent of the early Marx.[21]

In general, those who worry about the implications of social-scientific "relativism" for the rationality of ethics should be reassured by Richard Bernstein's delineation of a kind of rationality "beyond objectivism and relativism," a much more pragmatic and processual model of reason built upon the historical and social embeddedness and evolution of ideas.[22] Those who worry that "relative" values will be less serious than values that can claim absolute allegiance might be reassured by the argument that it is precisely the profound embeddedness of our ethical ideas within their cultural contexts that marks their seriousness. For *us*, of course. Nevertheless, that is whom we speak of and to.

Although these last remarks are very sketchy, they at least serve to suggest that a sociological or "evolutionary" view of values is not somehow the death knell of ethics. Instead, such a view seems to be almost an enabling condition of modern philosophical ethics. At the same time, however, "mainstream" ethics does not need to be, and certainly *has* not been, explicit on this point. The actual origins of values are seldom mentioned at all, and the usual labels—for example,

Lovibond's "expressivism" and even MacIntyre's "traditions"—only indirectly suggest any social-scientific provenance. But it is time to be more explicit. As I argue below, large issues outside the "mainstream" may depend upon it.

IV THE PRACTICE OF ETHICS AT ORIGINARY STAGES

In order to begin to draw some of the necessary conclusions from this "evolutionary" view of values, let us turn our attention to the appropriate comportment for ethics at what we might call the "orginary stages" of the development of values: stages at which new values are only beginning to be constituted and consolidated. In the case of the ethics of persons, for example, we must try to place ourselves back in the time when respect for persons, and persons themselves, were far less secure—not fixed, secure or "natural" as they now seem, but rather strange, forced, truncated, the way they must have seemed to, say, Calvin's contemporaries. How then should—how *could*—a photo-ethics of persons proceed in such a situation?

First, such early stages in the development of a new set of values require a great deal of exploration and metaphor. Only later do the new ethical notions harden into analytic categories. For example, although the concept of the "rights" of persons now may be invoked with a fair degree of rigor, throughout most of its history it played a much more open-ended role, encouraging the treatment of whole new classes of people as rights holders—slaves, foreigners, propertyless persons, women—in ways previously unheard of, and in ways that literally speaking, were misuses of the concept. (Consider "barbarian rights." The very concept of *barbarian* seems to preclude one of them being one of "us," i.e. Greeks, i.e. rights holders.) This malleable rhetoric of rights also in part *created* "rights holders." Persuading someone that he or she has a right to something, for example, or persuading a whole class or group that their rights have been violated, dramatically changes his, her, or their behavior, and ultimately reconstructs his, her or their belief systems and experiences. Even now the creative and rhetorical possibilities of the concept of rights have not been exhausted. It is possible to read the sweeping and inclusive notion of rights in the United Nations Declaration of Human Rights in this light, for instance, rather than dismissing it as conceptually confused, as do legalistic thinkers.[23]

Moreover, the process of co-evolving values and practices at originary stages is seldom a smooth process of progressively filling in and instantiating earlier outlines. Instead, we see a variety of fairly incompatible outlines coupled with a wide range of proto-practices, even social experiments of various sorts, all contributing to a kind of cultural working-through of a new set of possibilities. The process *seems* smooth in retrospect only because the values and practices that ultimately win out rewrite the history of the others so that the less successful practices and experiments are obscured—much as successful scientific paradigms, according to Kuhn, rewrite their own pasts so that in retrospect their evolution seems much smoother, more necessary and more univocal than they actually were. Great moments in the canonical history of rights, for example, include the Declaration of Independence and the Declaration of the Rights of Man, capitalism's institutionalization of rights to property and wealth, and now the persistent defense of a non-positivistic notion of rights for international export. *Not* included are the utopian socialists' many experimental communities, which often explicitly embraced (what *became*) non-standard, even anti-capitalistic notions of rights, such sustained and massive struggles as the labor movement's organization around working persons' rights, and the various modern attempts by most social democracies to institutionalize rights to health care.

A long period of experimentation and uncertainty, then, ought to be expected and even welcomed in the originary stages of any new ethics. Again, as I suggested above, even the most familiar aspects of personhood co-evolved with a particular, complex and even wildly improbable set of ideas and practices. Protestantism contributed not just a theology, and not just Calvin's peculiar and (if Weber is right) peculiarly world-historical "inner-world asceticism," but also such seemingly simple projects as an accessible Bible in the vernacular. Imagine the extraordinary impact of being able to read the holy text oneself after centuries of only the most mediated access. Imagine the extraordinary self-preoccupation created by having to choose for the first time between rival versions of the same revelation, with not only one's eternal soul in the balance, but often one's earthly life as well. Only against such a background of practice did it become possible to begin to experience oneself as an individual, separate from others, beholden to inner voices and "one's own values," "giving direction to one's life"

oneself, as Taylor puts it, and bearing the responsibility for one's choices.

Since we now look at the evolution of the values of persons mostly from the far side, it is easy to miss the fundamental contingency of those values and their dependence upon practices, institutions and experiences that were for their time genuinely uncertain and exploratory. Today we are too used to that easy division of labor that leaves ethics only the systematic tasks of "expressing" a set of values that is already established, and abandons the originary questions to the social sciences. As a result, ethics is incapacitated when it comes to dealing with values that are *now* entering the originary stage. Even when it is out of its depth, we continue to imagine that systematic ethics, such as the ethics of the person, is the only kind of ethics there is. We continue to regard the contingency, open-endedness and uncertainty of "new" values as an objection to them, ruling them out of ethical court entirely, or else as a kind of embarrassment to be quickly papered over with an ethical theory.

This discussion has direct application to environmental ethics. First and fundamentally, if environmental ethics is indeed at an originary stage, we can have only the barest sense of what ethics for a culture truly beyond anthropocentrism would actually look like. The Renaissance and the Reformation did not simply actualize some pre-existing or easily anticipated notion of persons, but rather played a part in the larger *co-evolution* of respect for persons. What would emerge could only be imagined in advance in the dimmest of ways, or not imagined at all. Similarly, we are only now embarking on an attempt to move beyond anthropocentrism, and we simply cannot predict in advance where even another century of moral change will take us.

Indeed, when anthropocentrism is finally cut down to size, there is no reason to think that what we will have or need in its place will be something called *non-anthropocentrism* at all—as if that characterization would even begin to be useful in a culture in which anthropocentrism had actually been transcended. Indeed it may not even be any kind of "centrism" whatsoever, i.e. some form of hierarchically structured ethics. It is already clear that hierarchy is by no means the only option.[24]

Second and correlatively, at this stage, exploration and metaphor are crucial to environmental ethics. Only later can we harden originary notions into precise analytic categories. Any attempt to appropriate the moral force of rights language for (much of) the transhuman world, for example, ought to be expected from the start to be *imprecise*, literally confused. (Consider "animal rights." The very concept of *animal* seems to preclude one of them being one of "us," i.e. persons, i.e. rights holders.) It need not be meant as a description of prevailing practice; rather, it should be read as an attempt to *change* the prevailing practice. Christopher Stone's book *Should Trees Have Standing? Toward Legal Rights for Natural Objects,* for example, makes a revisionist proposal about legal arrangements; it does not offer an analysis of the existing concept of rights.[25]

Something similar should be understood when we are invited to conceive not only animals or trees as rights holders, but also the land as a community and the planet as a person. All such arguments should be understood to be rhetorical, in a non-pejorative, pragmatic sense: they are suggestive and open-ended sorts of challenges, even proposals for Deweyan kinds of social reconstruction, rather than attempts to demonstrate particular conclusions on the basis of premises that are supposed to already be accepted.[26] The force of these arguments lies in the way they open up the possibility of new connections, not in the way they settle or "close" any questions. Their work is more creative than summative, more prospective than retrospective. Their chief function is to provoke, to loosen up the language, and correspondingly our thinking, to fire the imagination: to *open* questions, not to settle them.

The founders of environmental ethics were explorers along these lines. Here I want, in particular, to reclaim Aldo Leopold from the theorists. Bryan Norton reminds us, for example, that Leopold's widely cited appeal to the "integrity, stability, and beauty of the biotic community" occurs in the midst of a discussion of purely economic constructions of the land. It is best read, Norton says, as a kind of counterbalance and challenge to the excesses of pure commercialism, rather than as a criterion for moral action all by itself. Similarly, John Rodman has argued that Leopold's work should be read as an environmental ethic *in process,* complicating the anthropocentric picture more or less from within, rather than as a kind of proto-system, simplifying and unifying an entirely new picture, that can be progressively refined in the way that utilitarian and deontological theories have been refined over the last century.[27] Leopold insists, after all, that

> the land ethic [is] a *product of social evolution.* . . .
> Only the most superficial student of history supposes that Moses "wrote" the Decalogue; it evolved

in the mind [and surely also in the practices!] of the thinking community, and Moses wrote a tentative summary of it. . . . I say "tentative" because evolution never stops.[28]

It might be better to regard Leopold not as purveying a general ethical theory at all, but rather as simply *opening* some questions, unsettling some assumptions, and prying the window open just far enough to lead, in time, to much wilder and certainly more diverse suggestions or "criteria."

Third and more generally, as I put it above, the process of evolving values and practices at originary stages is seldom a smooth process of progressively filling in and instantiating earlier outlines. At the originary stage we should instead expect a variety of fairly incompatible outlines coupled with a wide range of proto-practices, even social experiments of various sorts, all contributing to a kind of cultural working-through of a new set of possibilities. In environmental ethics, we arrive at exactly the opposite view from that of J. Baird Callicott, for example, who insists that we attempt to formulate, right now, a complete, unified, even "closed" (his term) theory of environmental ethics. Callicott even argues that contemporary environmental ethics should not tolerate more than one basic type of value, insisting on a "univocal" environmental ethic.[29] In fact, however, as I argued above, originary stages are the worst possible times at which to demand that we all speak with one voice. Once a set of values is culturally consolidated, it may well be possible, perhaps even necessary, to reduce them to some kind of consistency. But environmental values are unlikely to be in such a position for a very long time. The necessary period of ferment, cultural experimentation, and thus, *multi*-vocality is only *beginning.* Although Callicott is right, we might say, about the demands of systematic ethical theory at later cultural stages, he is wrong—indeed wildly wrong—about what stage environmental values have actually reached.

V ENABLING ENVIRONMENTAL PRACTICE

Space for some analogues to the familiar theories does remain in the alternative environmental ethics envisioned here. I have argued that although they are unreliable guides to the ethical future, they might well be viewed as another kind of ethical experiment or proposal rather like, for example, the work of the utopian socialists. However unrealistic, they may, nonetheless, play a historical and transitional role, highlighting new possibilities, inspiring reconstructive experiments, even perhaps eventually provoking environmental ethics' equivalent of a Marx.

It should be clear, though, that the kind of constructive activity suggested by the argument offered here goes far beyond the familiar theories as well. Rather than systematizing environmental values, the overall project at this stage should be to begin *co-evolving* those values with practices and institutions that make them even *un*systematically possible. It is this point that I now want to develop by offering one specific example of such a co-evolutionary practice. It is by no means the only example. Indeed, the best thing that could be hoped, in my view, is the emergence of many others. But it is *one* example, and it may be a good example to help clarify how such approaches might look, and thus to clear the way for more.

A central part of the challenge is to create the social, psychological and phenomenological preconditions—the conceptual, experiential or even quite literal "space"—for new or stronger environmental values to evolve. Because such creation will "enable" these values, I call such a practical project *enabling environmental practice.*

Consider the attempt to create actual, physical spaces for the emergence of transhuman experience, *places* within which some return to the experience of and immersion in natural settings is possible. Suppose that certain places are set aside as quiet zones, places where automobile engines, lawnmowers and low-flying aeroplanes are not allowed, and yet places where people will live. On one level, the aim is modest: simply to make it possible to hear the birds, the winds and the silence once again. If bright outside lights were also banned, one could see the stars at night and feel the slow pulsations of the light over the seasons. A little creative zoning, in short, could make space for increasingly divergent styles of living on the land—for example, experiments in recycling and energy self-sufficiency, Midgleyan mixed communities of humans and other species, serious "reinhabitation" (though perhaps with more emphasis on place and community than upon the individual reinhabiters), the "ecosteries" that have been proposed on the model of monasteries, and other possibilities not yet even imagined.[30]

Such a project is not utopian. If we unplugged a few outdoor lights and rerouted some roads, we could easily have a first approximation in some parts of the country right now. In gardening, for example, we already experience some semblance of mixed communities. Such practices as beekeeping, moreover, already provide a model for a symbiotic relation with the "biotic community." It is not hard to work out policies to protect and extend such practices.

Enabling environmental practice is, of course, a *practice.* Being a practice, however, does not mean that it is not also philosophical. Theory and practice interpenetrate here. In the abstract, for example, the concept of "natural settings," just invoked, has been acrimoniously debated, and the best-known positions are unfortunately more or less the extremes. Social ecologists insist that no environment is ever purely natural, that human beings have already remade the entire world, and that the challenge is really to get the process under socially progressive and politically inclusive control. Some deep ecologists, by contrast, argue that only wilderness is the "real world."[31] Both views have something to offer. Nevertheless, it may be that only from within the context of a new practice, even so simple a practice as the attempt to create "quiet places," will we finally achieve the necessary distance to take what we can from the purely philosophical debate, and also to go beyond it towards a better set of questions and answers.

Both views, for example, unjustly discount "encounter." On the one hand, non-anthropocentrism should not become anti-anthropocentrism: the aim should not be to push humans out of the picture entirely, but rather to open up the possibility of reciprocity *between* humans and the rest of nature. Nevertheless, reciprocity does require a space that is not wholly permeated by humans either. What we need to explore are possible realms of *interaction.* Neither the wilderness nor the city (as we know it) are "the real world," if we must talk in such terms. We might take as the most "real" places the places where humans and other creatures, honored in their wildness and potential reciprocity, can come together, perhaps warily, but at least openly.

The work of Wendell Berry is paradigmatic of this kind of philosophical engagement. Berry writes, for example, of "the phenomenon of edge or margin, that we know to be one of the powerful attractions of a diversified landscape, both to wildlife and to humans." These margins are places where domesticity and wildness meet. Mowing his small hayfield with a team of horses, Berry encounters a hawk who lands close to him, watching carefully but without fear. The hawk comes, he writes, because of the conjunction of the small pasture and its wooded borders, of open hunting ground and the security of trees. . . . The human eye itself seems drawn to such margins, hungering for the difference made in the countryside by a hedgy fencerow, a stream, or a grove of trees. These margins are biologically rich, the meeting of two kinds of habitat.[32]

The hawk would not have come, he says, if the field had been larger, or if there had been no trees, or if he had been plowing with a tractor. Interaction is a fragile thing, and we need to pay careful attention to its preconditions. As Berry shows, attending to interaction is a deeply philosophical and phenomenological project as well as a practical one—but, nonetheless, it always revolves around and refers back to practice. Without actually maintaining a farm, he would know very little of what he knows, and the hawk would not—*could* not—have come to him.

Margins are, of course, only one example. They can't be the whole story. Many creatures avoid them. It is for this reason that the spotted owl's survival depends on large tracts of old-growth forest. Nonetheless, they are still part of the story—a part given particularly short shrift, it seems, by all sides in the current debate.

It is not possible in a short article to develop the kind of philosophy of "practice" that would be necessary to work out these points fully. However, I can at least note two opposite pitfalls in speaking of practice. First, it is not as if we come to this practice already knowing what values we will find or exemplify there. Too often the notion of practice in contemporary philosophy has degenerated into "application," i.e. of prior principles or theories. At best, it might provide an opportunity for feedback from practice to principle or theory. I mean something more radical here. Practice is the opening of the "space" for interaction, for the reemergence of a larger world. It is a kind of exploration. We do not know in advance what we will find. Berry had to *learn,* for example, about margins. Gary Snyder and others proposed Buddhist terms to describe the necessary attitude, a kind of mindfulness, attentiveness. Tom Birch calls it the "primary sense" of the notion of "consideration."[33]

On the other hand, this sort of open-ended practice does not mean reducing our own activity to zero, as in some form of quietism. I do not mean that we simply "open, and it will come." There is not likely to be any single and simple set of values that somehow emerges once we merely get out of the way. Berry's view is that a more open-ended and respectful relation to nature requires constant and creative *activity*—in his case, constant presence in nature, constant interaction with his own animals, maintenance of a place that maximizes margins. Others will, of course, choose other ways. The crucial thing is that humans must neither monopolize the picture entirely nor absent ourselves from it completely, but rather try to live in interaction, to create a space for genuine encounter as part of our ongoing reconstruction of our own lives and practices. What will come of such encounters, what will emerge from such sustained interactions, we cannot yet say.

No doubt it will be argued that Berry is necessarily an exception, that small unmechanized farms are utterly anachronistic, and that any real maintenance of margins or space for encounter is unrealistic in mass society. Perhaps. But these automatically accepted commonplaces are also open to argumentation and experiment. Christopher Alexander and his colleagues, in *A Pattern Language* and elsewhere, for example, make clear how profoundly even the simplest architectural features of houses, streets, and cities structure our experience of nature—and that they can be consciously redesigned to change those experiences. Windows on two sides of a room make it possible for natural light to suffice for daytime illumination. If buildings are built on those parts of the land that are in the worst condition, not the best, we thereby leave the most healthy and beautiful parts alone, while improving the worst parts. On a variety of grounds, Alexander and his colleagues argue for the presence of both still and moving water throughout the city, for extensive common land—"accessible green," sacred sites, and burial grounds within the city—and so on. If we build mindfully, they argue, maintaining and even expanding margins is not only possible, but easy, even with high human population densities.[34]

VI CONCLUSION

In the last section, I offered only the barest sketch of enabling environmental practice: a few examples, not even a general typology. To attempt a more systematic typology of its possible forms at this point seems to me premature, partly because ethics has hitherto paid so little attention to the cultural constitution of values that we have no such typology, and partly because the originary stage of environmental values is barely underway.

Moreover, enabling environmental practice is itself only one example of the broader range of philosophical activities invited by what I call the co-evolutionary view of values. I have not denied that even theories of rights, for instance, have a place in environmental ethics. However, it is not the only "place" there is, and rights themselves, at least when invoked beyond the sphere of persons, must be understood (so I argue) in a much more metaphorical and exploratory sense than usual. This point has also been made by many others, of course, but usually with the intention of ruling rights talk out of environmental ethics altogether. A pluralistic project is far more tolerant and inclusive. Indeed, it is surely an advantage of the sort of umbrella conception of environmental ethics I am suggesting here that nearly all of the current approaches may find a place in it.

Because enabling environmental practice is closest to my own heart, I have to struggle with my own temptation to make it the whole story. It is not. Given the prevailing attitudes, however, we need to continue to insist that it is *part* of the story. Of course, we might still have to argue at length about whether and to what degree enabling environmental practice is "philosophical" or "ethical." My own view, along pragmatic lines, is that it is both, deeply and essentially. Indeed, for Dewey the sustained practice of social reconstruction—experimental, improvisatory, and pluralistic—is the most central ethical practice of all. But that is an argument for another time. It is, nevertheless, one of the most central tasks that now calls to us.

Notes

1. Some landmarks of this body of work come into view in the later discussion. For a general overview of work on ethical ideas in particular from this perspective, see Maria Ossowska, *Social Determinants of Moral Ideas* (Philadelphia: University of Pennsylvania, 1970).

2. I distinguish *anthropocentrism* as a philosophical position, issuing in an ethic, from the practices and institutions in which that ethic is embodied, which I call "anthropocen*trized*."

3. See Anthony Weston, "Non-Anthropocentrism in a Thoroughly Anthropocentrized World," *The Trumpeter* 8, No. 3 (1991): 108–112.

4. Paul Taylor, *Respect for Nature* (Princeton: Princeton University Press, 1986), p. 3.

5. Mary Midgley, *Animals and Why They Matter* (Athens: University of Georgia Press, 1983), p. 118. See also Arne Naess, "Self-Realization in Mixed Communities of Humans, Bears, Sheep and Wolves," *Inquiry* 22 (1979): 231–241.

6. A tradition beginning with Max Weber, *The Protestant Ethic and the Spirit of Capitalism*, trans. Talcott Parsons (New York: Scribner's, 1958) and *Economy and Society: An Outline of Interpretive Sociology*, ed. G. Roth and C. Wittich (Berkeley: University of California Press, 1978), and carried into the present in different ways by, e.g., Morris Berman, *The Reenchantment of the World* (Ithaca, NY: Cornell University Press, 1981) and Albert Borgmann, *Technology and the Character of Contemporary Life* (Chicago: University of Chicago Press, 1984).

7. For an argument in defense of this point, see Anthony Weston "Beyond Intrinsic Value: Pragmatism in Environmental Ethics," *Environmental Ethics* 7 (1985): 321–389.

8. Jim Cheney, "The Neo-Stoicism of Radical Environmentalism," *Environmental Ethics* 11 (1989): 293–325.

9. John Rawls, "Kantian Constructivism in Moral Theory," *Journal of Philosophy* 77 (1980): 318; and "Justice as Fairness: Political, not Metaphysical," *Philosophy and Public Affairs* 14 (1985): 228.

10. John Arras, "The Revival of Casuistry in Bioethics," *Journal of Medicine and Philosophy* 16 (1991): 44.

11. Michael Walzer, *Spheres of Justice* (New York: Basic Books, 1983).

12. Alasdair MacIntyre, *After Virtue* (Notre Dame, IN: University of Notre Dame Press, 1981).

13. Charles Taylor, *Sources of the Self* (Cambridge, MA: Harvard University Press, 1989).

14. Sabina Lovibond, *Realism and Imagination in Ethics* (Minneapolis: University of Minnesota Press, 1983).

15. Weber, *The Protestant Ethic and the Spirit of Capitalism* and *Economy and Society*.

16. For classic examples of selves in other keys, see Louis Dumont, *Homo Hierarchichus* (Chicago: University of Chicago Press, 1980) and Colin Turnbull, *The Forest People* (New York: Simon and Schuster, 1961).

17. Unavoidable here is the Kantian objection that ethical values actually offer "reasons" rather than anything in the merely "causal" universe. My dogmatic response is that, despite its patina of logical necessity, this insistence on seceding from the phenomenal world actually derives from the same misconception of "causal" stories criticized in the text. Let me add, however, that, in my view, the idea that one can somehow understand and systematize ethical values in ignorance of their origins and social dynamics also partakes of the spectacular overconfidence in philosophical reason implicitly criticized in this paper as

a whole. For some support on this point, see Kai Nielsen, "On Transforming the Teaching of Moral Philosophy," *APA Newsletter on Teaching Philosophy,* November 1987, pp. 3-7.

18. Witold Rybezynski, *Home: A Short History of an Idea* (New York: Viking, 1986).

19. I don't mean to deny that rapid change (both cultural and biological) occasionally does occur, perhaps precipitated by unpredictable but radical events. Drastic global warming or a Chernobyl-type accident outside of Washington DC might well precipitate a drastic change in our environmental practices. Still, even in moments of crisis we can only respond using the tools that we then have. From deep within our anthropocentrized world it remains hard to see how we can respond without resorting either to some kind of "enlightened" anthropocentrism or to a reflex rejection of it, still on anthropocentrism's own terms. Thus, when I speak of "fundamental" change, I mean change in the entire system of values, beliefs, practices and social institutions—not just in immediate practices forced upon us by various emergencies.

20. For this way of putting the matter, I am indebted to Rom Harré.

21. In general, the possibility of invoking dissonant strands in a complex culture is part of the reason that radical social criticism is possible in the first place. Cf. Lovibond, *Realism and Imagination in Ethics;* Walzer, *Interpenetration and Social Criticism* (Cambridge, MA: Harvard University Press, 1987); and Anthony Weston, *Toward Better Problems: New Perspectives on Abortion, Animal Rights, the Environment, and Justice* (Philadelphia, PA: Temple University Press, 1992), pp. 167–174.

22. Richard Bernstein, *Beyond Objectivism and Relativism* (Philadelphia, PA: University of Pennsylvania Press, 1983).

23. While Hugo Bedau (in "International Human Rights," in Tom Regan and Donald VanDeveer, eds, *And Justice Toward All: New Essays in Philosophy and Public Policy* [Totowa, NJ: Rowman and Littlefield, 1982]) calls the declaration "the triumphant product of several centuries of political, legal, and moral inquiry into . . . 'the dignity and worth of the human person'" (p. 298), he goes on to assert that "It is . . . doubtful whether the General Assembly that proclaimed the UN Declaration understood what a human right is," since in the document rights are often stated loosely and in many different modalities. Ideals, purposes, and aspirations are run together with rights. At the same time, moreover, the declaration allows considerations of general welfare to limit rights, which seems to undercut their function as protectors of individuals against such rationales (p. 302n). In opposition to Bedau's position, however, I am suggesting that the General Assembly understood what rights are very well. Rights language is a broad-based moral language with multiple purposes and constituencies: in some contexts a counterweight to the typically self-serving utilitarian rhetoric of the powers that be; in others, a provocation to think seriously about even such often-mocked ideas as a right to a paid vacation, etc.

24. See, for example, Bernard Williams, *Ethics and the Limits of Philosophy* (Cambridge, MA: Harvard University Press, 1985); Walzer, *Spheres of Justice,* and Karen Warren, "The Power and Promise of Ecofeminism," *Environmental Ethics* 12 (1990): 125–146.

25. Christopher Stone, *Should Trees Have Standing? Toward Legal Rights for Natural Objects* (Los Altos, CA: William Kaufmann, 1974). G. E. Varner, in "Do Species Have Standing?" *Environmental Ethics* 9 (1987): 57–72, points out that the creation of new legal rights—as, for example, in the Endangered Species Act—helps expand what W. D. Lamont calls our "stock of ethical ideas—the mental capital, so to speak, with which [one] begins the business of living." There is no reason that the law must merely reflect "growth" that has already occurred, as opposed to motivating some growth itself.

26. See Chaim Perelman, *The Realm of Rhetoric* (Notre Dame, IN: University of Notre Dame Press, 1982) and C. Perelman and L. Olbrechts-Tyteca, *The New Rhetoric* (Notre Dame, IN: University of Notre Dame Press, 1969) for an account of rhetoric that resists the usual Platonic disparagement.

27. Bryan G. Norton, "Conservation and Preservation: A Conceptual Rehabilitation," *Environmental Ethics* 8 (1986): 195–220; John Rodman: "Four Forms of Ecological Consciousness Reconsidered," in Donald Scherer and Thomas Attig, eds, *Ethics and Environment* (Englewood Cliffs, NJ: Prentice-Hall, 1983): 89–92. Remember also that Leopold insists that ethics are "products of social evolution" and that "nothing so important as an ethic is ever 'written'"— which again suggests that we ought to rethink the usual reading of Leopold as an environmental-ethical theorist with a grand criterion for ethical action.

28. Aldo Leopold, *A Sand County Almanac* (New York: Oxford University Press, 1949), p. 225.

29. J. Baird Callicott, "The Case against Moral Pluralism," *Environmental Ethics* 12 (1990): 99–124.

30. On "ecosteries," see Alan Drengson, "The Ecostery Foundation of North America: Statement of Philosophy," *The Trumpeter* 7, No. 1 (1990): 12–16. On "reinhabitation," a good starting point is Peter Berg, "What is Bioregionalism?" *The Trumpeter* 8, No. 1 (1991): 6–12.

31. See, for instance, Dave Foreman, "Reinhabitation, Biocentrism, and Self-Defense," *Earth First!,* 1 August 1987; Murray Bookchin, "Which Way for the

US Greens?" *New Politics* 2 (Winter 1989): 71–83; and Bill Devall, "Deep Ecology and its Critics," *Earth First!,* 22 December 1987.

32. Wendell Berry, "Getting Along with Nature," in *Home Economics* (San Francisco: North Point Press, 1987), p. 13.

33. Gary Snyder, "Good, Wild, Sacred," in *The Practice of the Wild* (San Francisco: North Point Press, 1990); Tom Birch, "Universal Consideration," paper presented at the International Society for Environmental Ethics, American Philosophical Association, 27 December 1990; Jim Cheney, "Eco-Feminism and Deep Ecology," *Environmental Ethics* 9 (1987): 115–145. Snyder also speaks of "grace" as the primary "practice of the wild"; Doug Peacock, *The Grizzly Years* (New York: Holt, Henry and Co., 1990), insists upon "interspecific tact"; Berry writes of an "etiquette" of nature; and Birch of "generosity of spirit" and "considerateness." All of these terms have their home in a discourse of manners and personal bearing, rather than moral discourse as usually conceived by ethical philosophers. We are not speaking of some universal categorical obligation, but rather of something much closer to us, bound up with who we are and how we immediately bear ourselves in the world—though not necessarily any more "optional" for all that.

34. Christopher Alexander, et al., *A Pattern Language* (New York: Oxford University Press, 1977). On windows, see secs. 239, 159 and 107; on "site repair," sec. 104; on water in the city, secs. 25, 64 and 71; on "accessible green," secs. 51 and 60; and on "holy ground," secs 24, 66 and 70.

For Further Discussion

1. What does Weston mean by suggesting that ideas can be viewed ecologically?

2. On what grounds does Weston claim that contemporary nonanthropocentric ethics remains deeply dependent upon the thoroughly anthropocentric setting in which it arises?

3. Weston claims that most ethical philosophers acknowledge that ethical values are deeply embedded in and co-evolved from social institutions and practices. Do you think that this makes Weston an ethical relativist? What does Weston say to this?

4. How does Weston interpret Aldo Leopold? How, if at all, does this interpretation differ from J. Baird Callicott's essay in Chapter 7?

5. Explain as fully as you can Weston's concept of "enabling environmental practice."

Diverging Worldviews, Converging Policies

Bryan Norton

Bryan Norton's book, Towards Unity Among Environmentalists, *from which this selection is taken, argues that there is a widespread practical unity among environmentalists. Norton proposes that the major disagreements among environmentalists—disagreements between conservationists and preservationists, for example—are disagreements between values and explanations rather than disagreements concerning actions. Indeed, although there may be a divergence among worldviews, there is a growing convergence concerning policy. Norton believes that there is a rational basis for the emerging consensus and that this basis relies significantly on what he calls scientific environmentalism. Toward the end of this selection, Norton suggests that it might be possible to integrate the* plurality of environmental worldviews under more general principles that acknowledge the particularities of temporal and spatial scale.

THE EMERGING CONSENSUS

This book began with an anecdote, my encounter with an eight-year-old with hundreds of living sand dollars. While I knew what I wanted the little girl to do—I wanted her to put most of the living sand dollars back in the lagoon—I felt in a quandary when I tried to explain *why* she should do so. I had no objection if the little girl took a couple home, to watch them in her aquarium or even to dissect them to learn their structure. But the family's actions showed no respect for life or living systems. I wanted to make a moral point not expressible in the language of economics. I hesitated to introduce, however, without serious qualifications, the moral language of rights. Rights have an individualistic ring about them; if sand dollars have rights, then surely the family should put them *all* back. One language said too little, the other said too much.

This original intuition, that the environmentalists' dilemma is mainly a dilemma of values and explanations, more than preferred actions, has been borne out by the considerations of the second part of this book. An examination of major areas of environmental policy has reinforced the hypothesis that a consensus on the broad outlines of an intelligent policy is emerging among environmentalists, even though there remain significant value differences that affect the explanations and justifications they offer for basically equivalent policies.

Environmentalists of different stripes, as far back as the days of Pinchot and Muir, have often set aside their differences to work for common goals. . . . But those traditional cooperations were, it seemed, almost accidental collaborations originating in temporary political expediency. My hypothesis about the current environmental scene asserts a more than accidental growth in cooperation: In spite of occasional rancorous disputes, the original factions of environmentalism are being forced together, regardless of their value commitments.

For example, a growing sense of urgency led soil conservationists and preservationist groups to work together to pass the 1985 Farm Bill, even though they suffered some ill feelings along the way. Similarly, the National Wildlife Federation, a collection of sportsmen's organizations, and Defenders of Wildlife advocate similar wetlands protection policies. While the value they place on wildlife is very different, the policy of protecting wildlife habitat represents a common-denominator objective, and the National Wildlife Federation is an effective lobbier for legislation to protect nongame endangered species as well as game species and their habitats. Given our present scientific knowledge about wildlife populations, hunters and animal protectionists alike conclude that we must aggressively protect the remaining habitats for migrating waterfowl. Both groups would also agree on the importance of careful management to protect the reserves from the effects of human activity. This consensus signals the end of both the atomistic style of single-species

game management characteristic of early conservationists and of isolationist preservationism, which in its extreme form repudiated management altogether.

Several gradual changes undermined the two extremes of management style. Not the least of these causes was the progressive development of the nation, which increased the likelihood of spillover effects of one activity on another. Another cause was the rapidly increasing demand for outdoor recreation that began in the forties and fifties and has developed steadily since. In general, as population expanded and diversified, more demands were put on more lands, and decisions to use land for productive purposes led to more and more direct conflicts. Leopold's land ethic, which recognizes that the land community is a larger system in which human activities must be integrated, has led environmentalists beyond both atomism and isolationism.

What this means, in more concrete terms, is that all environmentalists, regardless of their allegiance to diverging traditions, must seek to manage the entire mosaic that is the American landscape. If we are to maintain the productivity of American agriculture *and* protect biological diversity, if we are to maintain adequate water supplies for homes and industry *and* preserve some wild and scenic rivers, if we are to provide sufficient opportunities for outdoor activities *and* preserve the pristine nature of wilderness areas, we must make large-scale land use decisions with an eye to their larger context. A landscape that can accommodate all of the varied aspirations of Americans will have to be a patchy landscape, in which urban elements, productive elements, and pristine elements are arranged intelligently. Each of the patches must be managed according to the methods appropriate to goals that define its use, but those methods must also be designed to enhance, or at least not destroy, the values sought elsewhere in the mosaic. Further, the principles of this holistic management must be aesthetic as well as economic, and historically informed as well as forward-looking. But they must be applied to the entire context of human activities, not to specific activities viewed either atomistically or in isolation from other activities.

The forced abandonment of the two extreme styles of management associated with the old split between conservationists and preservationists provides a useful first stab at characterizing the emerging consensus. . . . We see a pattern of emerging policies that unite environmentalists in opposition to the policies usually favored by production-oriented developers.

In each area—resource use, pollution control, protection of biological diversity, and land use policy—environmentalists advocate limits that guide the search for acceptable policy options, insisting against the simple economic Aggregators that there are constraints governing human exploitative activities. These constraints are usually stated in terms of "sustainability," but Leopold and modern environmentalists have gone beyond demand-oriented conceptions of sustainability to recognize limits inherent in the complexity and organizational integrity of larger ecological systems. Similarly, in pollution policy, environmentalists have recognized constraints on activities that pollute the environment based on rights of other individuals to a healthy environment.

The common denominator of these obligations of resource users to limit their activities in these diverse cases cannot be understood as a commitment to any particular moral principle such as the moral equality of all species or of interpersonal equity. The common element is structural: in each case, individually motivated behaviors, which can be understood as activities of economic man, are constrained because of the impacts those behaviors impose on their larger context. Environmentalists emphasize total diversity and biological complexity because the complex processes that constitute biological systems *are* the larger context of all life, human and nonhuman. Rapid alteration of those larger systems will cause serious disruption of both human and nonhuman activities. Land must therefore be used according to patterns that protect the complex processes of nature, so as to avoid destabilizing changes, changes in environing systems that are too rapid to allow human activities and nonhuman processes to respond and adapt.

This consensus represents, in one sense, a victory of Moralists over Aggregators. The essence of the simple aggregationist approach was to reduce all questions in environmental policy to calculations regarding economic efficiency, to judge all questions on a single scale. Environmentalists have rejected simple aggregationism and insist on moral constraints ranging from individual rights to clean air to imperatives to protect the integrity of ecological systems. Pinchot, however, has left his mark as well. In the spirit of the great compromiser, environmentalists play pragmatic politics—they seek their policies from among the politically viable options. In political practice, both moral and economic imperatives exercise a veto power (see Figure 1).

In the context of political debate, individual rights, moral obligations to protect species, and scientifically

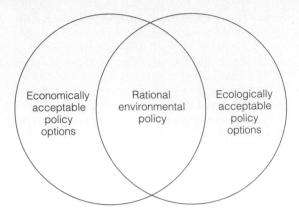

Figure 1. The politically desirable solution, according to contextualism, is chosen from the intersection of ecologically and economically acceptable policy options.

articulated thresholds or constraints inherent in fragile ecological systems must all be factored into a process that sets goals, objectives, and standards for environmental programs. Economics are, of course, not irrelevant. Once goals and standards are set politically, it makes sense to use economic analysis to rank those alternative actions and policies regarding their efficiency in achieving the politically determined goal. Environmentalists know that economic interests will block environmental legislation if it is too costly. It behooves environmentalists, therefore, to propose the least costly option that will fulfill their goals.

Once the crucial distinction between setting goals and choosing means to achieve those goals is recognized, contextualism need not emphasize prohibitions and regulations in order to encourage actions that respect constraints. With an adequate conception of health for the overall ecological context in place, efforts can be shifted from prohibition and regulation to the creation of incentives that encourage individual actors to choose less polluting activities or to choose land uses that will help, rather than harm, efforts to protect biological diversity.

So, for all its emphasis on constraints, the emerging consensus among environmentalists need not operate, on a day-to-day basis, by constraining individual activities. To return to our example of the farmer who clears his woodlot to plant wheat, the emerging consensus need not, when trends in farmers' behavior begin to press against thresholds inherent in the land community, regulate the farmer. It may, instead, choose to institute incentives that will encourage some farmers, either the one in question or others, to let a wheat field go fallow. The beauty of contextualism is that, once ecologically informed constraints are formulated, the society can undertake positive steps to encourage individuals to act in ways that counter dangerous trends. By combining a positive, biological conception of a healthy ecosystem with a program of incentives, the emerging consensus offers an alternative to simple reductionistic economics on the one side and onerous restrictions on the other. Following David Brower, this alternative can be called "restorationism." A positive definition of ecosystem health, one that incorporates human activities as long as they do not threaten thresholds inherent in ecological systems, opens the possibility of a truly positive ideal of humans living, creatively and freely, but harmoniously, within a larger, ecological context.

According to the contextualist synthesis, resource management, in the tradition of Pinchot's wise-use conservationism, is a respected activity, but one that must be understood as limited by the broader, contextual constraints of environmental management. These constraints express Muir's insight that all elements of nature must be seen as ultimately interrelated. As we shall see in the next section, the utilitarian value system so useful to Pinchot and the more holistic ideas of Muir are still strongly represented among environmentalists. What has changed, as the movement has matured, is that these metaphysical and moral worldviews have found distinctive habitats. Wise-use conservation guides resource use policy and largely governs land use decisions on lands already devoted to intense human use. Contextualist thinking, which insists on seeing resource use activities as a part of nature's larger whole, concentrates on problems caused in larger systems in which resource-producing units, such as field or forest, are embedded.

What once appeared as a war between two factions with opposed worldviews now appears as two protective strategies that are applicable in differing situations. As long as the larger, environmental context remains healthy, resource use strategies will be guided by maximization criteria. When exploitative activities threaten to exceed ecological parameters—the fragilities inherent in particular situations—*environmental* managers urge the enforcement of constraints that will limit the resource-producing activities that threaten their environing systems. Because of scientific uncertainty and because the second-order criteria for deciding which lands should be protected and which

should be used wisely are imprecise, environmentalists often disagree about how particular parcels should be used.

When environmentalists differ regarding policies and strategies today, they are usually disagreeing regarding the appropriateness of a given strategy to a particular situation, as, for example, when they disagree about which areas and how much National Forest land should be devoted to protection as opposed to production. These disagreements are often based on very different assessments of the magnitude of certain spillover effects of exploitation—such as stream degradation and loss of diversity—on their larger context and on the values humans derive from that context. Needless to say, self-interest also plays a role in advocacy of land-use decisions, as evidenced by the NIMBY (not-in-my-backyard) syndrome, in which local groups gain their power from quality-of-life arguments, but arguments that emphasize quality of life in very small contexts, such as a suburban municipality.

As these local disagreements continue, a further reach of consensus is emerging: even the American environment, taken as a huge unit, is not self-sufficient. As I interviewed leaders of the environmental movement, one point of consensus stood out, mentioned by almost all interviewees—the American environmental movement must, and will, become more international in the scope of its programs. Environmental groups that have been focused mainly on problems in the United States have instituted international programs, especially in the areas of biological diversity, but increasingly in other areas as well. Partly this change is a matter of the changing nature of environmental problems: Acid rain, the greenhouse effect, pollution of the oceans, and depletion of the marine fisheries do not respect national boundaries. But the urge toward internationalism in environmental affairs is also another expression of the emerging consensus—contextual management is management as part of a system; the system in which the dynamic American mosaic will evolve is the entire world.

SCIENTIFIC ENVIRONMENTALISM

A broadly contextualist approach to environmental policy characterizes the emerging consensus among environmentalists, but the policy consensus is not accidental—it is forged on the basis of a shared view of science. This agreement is evidenced in tentative agreement on some basic scientific "axioms" to guide environmental management and, more profoundly, in a shared acceptance of science as a value-laden enterprise. The underlying scientific consensus can be summarized by listing and drawing out the consequences of five axioms and an associated definition of ecosystem health.

1. *The Axiom of Dynamism.* Nature is more profoundly a set of processes than a collection of objects; all is in flux.
2. *The Axiom of Relatedness.* All processes are related to all other processes.
3. *The Axiom of Systematicity.* Processes are not related equally, but unfold in systems within systems, which differ mainly regarding the temporal and spatial scale on which they are organized.
4. *The Axiom of Creativity.* The autonomous processes of nature are creative and represent the basis for all biologically based productivity.
5. *The Axiom of Differential Fragility.* Ecological systems, which form the context of all human activities, vary in the extent to which they can absorb and equilibrate human-caused disruptions in their autonomous processes.

These five axioms are basic elements of a worldview in the sense that they shape and give context to environmental science. While the most explicitly evaluative aspects of environmentalists' worldviews have remained divergent, these broadly scientific axioms function as a shared scientific component of an emerging worldview. These axioms, I submit, provide the scientific portion of a new and emerging worldview and shape environmentalists' conception of the problems of environmental management.

In practice, these axioms are applied in conjunction with a highly normative definition of ecosystem health: An ecological system is healthy only when its creative processes, represented by the free flow of energy and active competition to utilize it, remain intact. Unhealthy ecological systems will be characterized by a tendency to undergo rapid change, change such as the rapid disintegration of complexity and integrity that Leopold observes as grazing caused deterioration of fragile vegetative communities in the arid Southwest. This shared understanding of the problems of environmental management encourages an analogy between medicine and environmental management. Both are empirical arts; a physician would never administer a heart drug without concern for its effects on the kidneys or on the overall health of the patient. Similarly, resource and environmental managers should practice preventative medicine, moderating dangerous trends and avoiding disruptive "illnesses" that they do not know how to cure.

Environmental managers differ from physicians as well. Since the study of whole ecosystem management is in its infancy compared with medicine, the dangers of iatrogeny (illnesses caused by medical treatment) are great. The treatments prescribed by resource and environmental managers must be carefully considered for their effects on the whole.

The creative processes of nature, as perceived within this worldview, are understood essentially as energy flows up the biotic pyramid. The diversity and complexity of nature result from varied adaptations of systems that exploit that energy. The sun supplies energy and local variation provides the context for ecological and evolutionary processes. Freedom, in an ecologically organized world, represents the creative ability to react to environmental variations, to adapt to an opportunity in the pattern of flows in the energy pyramid. Each species carves a niche in an ecological system either by coevolving with other species or by invading or colonizing a new habitat. The adaptive "choices" species make, the apparent freedom of species to react differentially to varied ecological opportunities and challenges, represents the "freedom" inherent in natural systems. This metaphysical concept of freedom provides the basis for the more inclusive concepts of "integrity" and "autonomy." A system maintains its integrity and autonomy when its future states are a result of ecological processes, interactions among species in ecological time. Human activities that use powerful technologies to alter systems rapidly tend to disrupt these normally slow-changing interactions.

Central to the emerging ecological consensus is the Axiom of Dynamism and its complex interplay with the other four axioms. Because, in the environmentalists' worldview, all is in flux and, according to the Axiom of Differential Fragility, treatment of the environment must be tailored to local conditions, it follows that values (whatever their ultimate formulation) will be given operational meaning in changing local contexts. Values cannot therefore be conceived as static and unchanging—they are given meaning at the sharp cutting edge of the search for a culturally, politically, and environmentally viable conception of the good life.

Because human values are a part of the ever-changing dynamic system, values cannot be conceived as static and unchanging preferences, as simple Aggregators interpret them. The static conception of human values as unquestioned preferences cannot express Muir's insight that observation and scientific understanding can have an ecstatic aspect, causing a shift in worldview toward less consumptive and more contemplative pursuits, for example. Environmentalists conceive environmental management to have public education as one of its central goals. Environmental managers do not have a ready-made, a priori criterion to guide them in the search for a healthy environment. They must engage the public in a dialogue, a political process, of defining such terms as "ecosystem health," "ecological restoration," and "ecosystem integrity." As scientific understanding of the larger, ecological context improves throughout the population, environmentalists believe, citizens will begin to see the world in a new, more holistic light, and the result will be a change in values. This is the transformational element in the search for an adequate ethic for dealing with human impacts on the natural world.

The Axiom of Systematicity, which interprets nature as systems embedded within systems, and as changing according to multiple frames of time, rules out any simple form of aggregationism as the basis for determining environmental policy. Some events will be given meaning, on the contextualist worldview, in smaller systems, others will be interpreted as events in larger systems, and still others will be given meaning in more than one system. One cannot meaningfully calculate the value of events that have different meanings in these different systems within a single system of value such as that of economics. Ecologists, by providing information about what is possible in management and what is not, participate as favored advisors in the decision process. It is from their point of view that holistic constraints, based on projections of change in contextual systems, are formulated. Here, ecologists and conservation biologists adopt tasks analogous to those of physiologists and physicians when they enter public debate about the meaning of, and criteria for, human health.

As citizens begin to see their behavior in an ecological context, there will be a shift toward the ecological worldview, and management problems will be seen as scientific problems in the sense that science, especially ecology, determines the context in which resource use questions will be addressed. Goals will be formulated in ecological language; this means that some objectives will be justified simply as protecting functioning systems and their organization. In this sense, an environmental ethic is "holistic." Public discussion of the goals of environmental management will concern the search for the good life in a good environment, and a good environment presupposes stable, slow-changing environing systems.

The language, environmentalists agree, in which this discussion should be articulated is the language of ecology supplemented with ethics, not the language of economics. Environmentalists are united by their belief that ecological science is capable of transforming worldviews and ultimate values; . . . both Pinchot and Muir thought of themselves as scientists, but that they practiced science in two distinct traditions. Pinchot saw scientific resource management as pursuing socially determined values; he avoided, to the extent possible, value judgments in his work. Muir, however, doubted that science can ever be a value-free enterprise, ridiculed what he saw as the futile and self-defeating attempts to observe the world "objectively," and recognized that the basic assumptions one makes about the order and structure of nature are value-laden. Because the sensibilities of environmentalists evolved within the naturalist tradition, in which the works of scientists, artists, and poets were not sharply separated, naturalists are comfortable with a vocabulary sprinkled with openly evaluative terms, including "healthy ecosystem," "ecosystem integrity," and so forth.

Pinchot, on the other hand, embraced positivistic science on its upswing. For the first half of the twentieth century, objectivity and value-neutrality were the watchwords in the scientific community. Muir's ecstatic science was eclipsed. In the aftermath of the loss of the Hetch Hetchy controversy, Muir's death, and the inexorable growth of government bureaucracies to encourage wise use of resources, the naturalist tradition was unceremoniously dismissed as unscientific.

Only recently has positivism lost its grip on professional philosophy, but it remains strong in many of the special sciences. Popular among philosophers in the 1930s, 1940s, and 1950s, the positivist view is now seen as seriously oversimplified. Philosophers still follow positivists in recognizing the importance of falsifiability in science, but they now emphasize that "falsification" always takes place against a backdrop of theoretical assumptions. Theoretical assumptions are often affected by the values that we hold. It follows that scientists work, even when testing hypotheses, in a world permeated by values. Environmentalists have, in this respect, chosen Muir's worldview over Pinchot's. In the words of Leopold, they have adopted ecology as "the new fusion point of the sciences."

It now becomes clear why it is so difficult to separate scientific, descriptive elements of a worldview from its evaluative components. The very conception of the world as a set of systematic processes that evolved to manage and allocate energy implies that, within the ecological system so described, there is value in maintaining those organizations. Neither economic productivity nor aesthetic and intergenerational moral values can be protected without protecting the complex, organized system that provides the ecological context on which *all* values depend. To assume that all values are economic values is to ignore this implicit, background value in the ecological processes that support economic, and all other, activity.

A contextualist science of management is value-laden in yet another respect. The choice of a scientific management unit—boundaries for a system to be given a "management plan"—is inseparable from moral, aesthetic, and cultural choices regarding goals of management. One cannot, for example, decide on the appropriate management unit for the Chesapeake Bay without considering how important protecting bay water quality is in the management plan; similarly, the choice to manage the GYE for wildness indicates a management unit sufficiently large to contain major migrations of wildlife populations. Choices of the scale on which to conceive management problems are inseparable from the adoption of appropriate management goals.

Environmentalists who endorse a variety of values will nevertheless tend to see environmental problems in a similar way and will pursue common-denominator goals, even though they might explain and justify those goals in quite different evaluative language, because the larger context of environmental management—the context in which all values are pursued—will be understood as the larger ecological context. Environmentalists have been able to fashion a working consensus for addressing environmental problems on an ecological basis precisely because they understand the world as the context of multiple values. This understanding unites them behind goals essential to protect a wide variety of values, however expressed, but the impetus toward the consensus is scientific. Environmentalists are being driven together by their commitment to ecological contextualism, which implies that all human values depend upon a healthy context.

THE WORLDVIEWS OF ENVIRONMENTALISTS REVISITED

This growing scientific consensus has not been accompanied by a corresponding narrowing of the value positions that environmentalists employ. We can separate out at least seven worldviews or fragments of worldviews, all of which are sufficient to trigger, depending on the situation, a constraint on exploitation and development. While each can stand independently against the reductionist approach, environmentalists often appeal to a shifting mixture of all seven worldviews. These are

1. *Judaeo-Christian stewardship.* While some environmentalists are critical of the human-centered emphasis of Genesis I, many Christians and Jews accept a form of benign dominion. This view emphasizes the admonition of Genesis II, that humans were put in the garden to "dress and keep it."[1] Environmentally aware Christians and Jews can, from this biblical admonition, justifiably urge that each generation pass the ecological context on to the next, intact in its creative complexity. This tradition of Christian stewardship recognizes obligations to the Creator not to destroy his handiwork. It recognizes, in other words, theologically based obligations not to destroy ecological systems or the species that compose them.

2. *Deep ecology and related value systems.* These systems are characterized by attributions of *rights* to nonhuman nature or by the related belief that nonhuman elements of nature have *intrinsic value* and that this intrinsic value places obligations and limits on the extent of human prerogatives to use and alter nature.

3. *Transformationalist/transcendentalism.* Within this framework, wild nature has spiritual value because experience of nature can transform human perception and value. Nature becomes, in this worldview, the cure for human alienation caused by modern society, a sacred space that humans enter to reconsider and reform their worldviews and value systems. Although it is possible to link a transformationalist approach with a belief that nature has rights—the transformation may be a recognition that nature is intrinsically valuable—this linkage is not necessary. If shifts in worldview are valued as improving human values and satisfactions—stamping out "quiet desperation" in the immortal words of Thoreau—then this moralistic position is anthropocentric and spiritually instrumentalist.[2]

4. *Constrained economics.* Many environmentalists continue to perceive the problem of resource use as essentially a problem in human economics, but for a variety of reasons they reject Aggregationism. A good example of this approach . . . is the Safe Minimum Standard (SMS) approach to protecting species and natural systems. On this approach, actions to avoid irreversible damage to the environment are assumed to have great, though not quantifiable benefits; these actions should be taken as long as the costs of doing so are not unacceptable. These constraints on economic activity might be justified by appeal to other value systems, such as (1) and (2), but need not be. They might, instead, be based on common sense and a relatively low threshold for risk-taking.

5. *Scientific naturalism.* Scientific naturalists are characterized by their broadly Darwinian view of life's processes and by an associated emphasis on dynamism and contextualism. Naturalists tend to see problems, including resource use problems, scientifically, but they do not accept the economists' suggestion that policy be determined by economic calculation because they believe that a broadly ecological/evolutionary worldview implies contextual limits, limits on population growth and violence to the land. These constraints follow, they infer, from the ecological-evolutionary conception of a species and its niche.

6. *Ecofeminism.* A significant group of feminist scholars argue that the domination of nature is symbolic of gender domination more generally.[3] They oppose the positivist worldview as an ideology of domination and see simply aggregative economics as disguised ideology. They argue that rejection of the ideology of domination will encourage solutions to both environmental and social problems.

7. *Pluralism/pragmatism.* Activists who are not philosophically inclined, and philosophers such as pragmatists who doubt the efficacy of general, overarching moral principles and theories, gravitate toward a pluralistic approach to environmental values. On this approach, practical problem-solving and ethical ideas and principles are enlisted, not so much as a priori principles that enforce consensus, but as useful means to recognize similar features of varied cases—as useful tools, in other words, to aid in the development of a solution to moral quandaries.

Can, or should, these various worldviews of environmentalism be unified? Moral monists, as they are described by Christopher Stone, share two tenets:

(1) that the goal of moral reasoning is "to produce, and to defend against all rivals, a single coherent and complete set of principles capable of governing all moral quandaries"; and (2) that this is a determinate goal in the sense that the favored framework "is to yield for each quandary one right answer."[4] I have argued that, empirically speaking, environmentalists have been pluralists, described by Stone as exhibiting a willingness "to develop a conception of the moral realm as consisting in several different schemata, side by side."[5]

Many commentators imply that environmentalists must, in some ultimate sense, be monists. When Roderick Nash, for example, implies that the history of environmental ethics is the history of the single idea that nonhuman elements of nature have rights, he implicitly assumes moral monism, implying that all goals environmentalists pursue must be supportable on a biocentric principle.[6] This view is no less reductionistic than the monism of economic Aggregators. The moral monist's method is to sort through moral maxims and hold them up against a single measure. If they can be derived from the central principle, they are subsumed under it as corollaries. If not, they are rejected. By this method, the monist would reduce the seven worldviews and fragments to one—the "correct" environmentalists' worldview. Monists, in other words, place a high value on connectedness and pursue the goal of stating a single, rock-bottom principle to guide all moral action.

The strategic question of whether it would be politically useful for environmentalists to have a single worldview must of course be kept separate from the epistemological question of whether there is a correct worldview. . . . We may ask whether environmentalists are dishonest, or otherwise in error, when they use a variety of worldviews to justify their policies. Nash clearly implies that environmentalists' pluralist tendencies are, at the least, devious. Nash describes Muir as fully committed to the rights of all nature by 1867,[7] but "Muir knew very well that to go before Congress and the public arguing for national parks as places where snakes, redwood trees, beavers, and rocks could exercise their natural rights to life and liberty would be to invite instant ridicule and weaken the cause he wished to advance." Nash concludes that Muir therefore "camouflaged his radical egalitarianism in more acceptable rhetoric centered on the benefits of nature for people."[8]

Nash's attitude is further revealed when he compares Joseph Wood Krutch's philosophy of environmentalism with Leopold's: "Like Leopold, Krutch left some ambiguity as to whether his extended ethic was *pure* [my emphasis], in the sense of respecting the rights of other parts of existence, or instrumentally to the successful continuation of human existence. Pragmatically, he knew as did Leopold, that the latter had a better chance to win public favor in the 1950s." This concern for the "purity" of biocentrism is revealing. Even while recognizing that "the main thrust of his [Krutch's] philosophy, as of Leopold's, centered on the idea that an ethical attitude toward nature was 'better in the long run for [humans] also,'" Nash insists that any public reference to human interests in preservation is cheating; they are based on political, rather than moral, goals and are therefore not "pure."[9]

In fact, we have seen, the environmental movement has been pluralistic in its value commitments, given to political compromise, not ideological exclusivity; neither simple aggregationism nor monistic biocentrism could, by itself, express both the wise-use concerns of Pinchot's followers and the systematic constraints characteristic of contextual, ecological thinking. It is moral monism, of course, that gives teeth to the environmentalists' dilemma by encouraging a perception of human utilitarian and nonanthropocentric values as exclusive. The pluralist, on the other hand, can search for value in nature that is reducible neither to the dollars of economists nor to the rights of wild species. The moral pluralist can look for common ground from which to construct a new, philosophically, culturally, and politically viable worldview that sees humans as integrated into larger systems and that values objects as part of their human, cultural, biotic, and abiotic contexts.

But there are also difficulties in accepting pluralism.[10] To simply say that different principles of value apply in different contexts introduces moral chaos unless something is said about which particular principles apply when. Most of the important disagreements among environmentalists—the Hetch Hetchy controversy, for example—have revolved around which criterion should govern particular policies and land uses. Pluralism can provide guidance in environmental policy only if it includes second-order principles that help to determine which of its diverse first-order moral criteria apply in given situations.

A pluralistic system with such second-order principles could be called an *integrated worldview*. It would be integrated in the sense that each of the principles would be given an appropriate domain of application, according to second-order rules based on a determination of the context of the managerial problem faced.

It would also be integrated in a deeper sense, provided those rules are interpreted systematically. A truly integrated system of thought, an adequate environmental worldview, would state rules of application according to the systematic context of the management problem faced. The criterion, according to a contextual approach such as Leopold's land ethic, should be based on the temporal and spatial scale appropriate to the problem at hand.

Notes

1. Wendell Berry, "Amplifications: Preserving Wilderness," *Wilderness* 50 (1987): 39–40, 50–54.
2. Bryan G. Norton, "Thoreau's Insect Analogies: Or, Why Environmentalists Hate Mainstream Economists," *Environmental Ethics* 13/3 (1991).
3. See Karen J. Warren, "Feminism and Ecology: Making Connections," *Environmental Ethics* 9 (Spring 1987).
4. Christopher Stone, *Earth and Other Ethics* (New York: Harper & Row, 1988), p. 116.
5. Ibid., p. 132.
6. Roderick Nash, *The Rights of Nature* (Madison: University of Wisconsin Press, 1989).
7. Nash, *The Rights of Nature,* pp. 38–40.
8. Ibid., p. 41. Nash also cites Bill Devall's criticism of Muir on similar grounds. "John Muir as Deep Ecologist," *Environmental Review* 6 (1982): 63–86. Nash notes that Muir, in this respect, began a trend, and he cites examples of Leopold, the Sierra Club, and species preservationists as having engaged in the same sort of political dishonesty.
9. Ibid., p. 76. Brief quotation from Krutch is from "Conservation Is Not Enough," *American Scholar* 24 (1954): 297.
10. See J. Baird Callicott, "The Case Against Moral Pluralism," *Environmental Ethics* 12 (1990): 99–124.

For Further Discussion

1. What does Norton mean by claiming that the environmentalist's dilemma is mainly a dilemma of values and explanations rather than of preferred actions?

2. How would you characterize the consensus that Norton suggests is emerging among contemporary environmentalists?

3. Compare Norton's discussion of contextualism with Weston's discussion of ethical values being embedded in social institutions.

4. Explain the five basic scientific "axioms." Why does Norton call these views axioms? Who might disagree with these?

5. What does Norton mean by distinguishing between strategic and epistemological reasons for adopting a single environmental worldview?

6. Is pluralism simply a version of ethical relativism? Why or why not?

DISCUSSION AND STUDY QUESTIONS FOR CHAPTER 18

1. There is an old saying that politics makes strange bedfellows. This suggests that there are times when people from widely diverse political perspectives agree on specific policies. Is agreement on policy more or less important than agreement on ideology? What reasons could be given for either answer?

2. Hunters and animal rights groups often agree on policies that protect habitats from development; they disagree radically about how the animals living in such habitats are to be treated. What do you make of this? When people agree on policy, does it matter if they disagree on principle?

3. Are there common principles that underlie and integrate your own views on such issues as animal rights and wilderness protection? Do you have a consistent environmental philosophy?

4. Could a pragmatic approach that emphasizes consensus and agreement on policy ever lead to radical change? Is this a good thing or a bad thing?

Credits

"Genesis" and "Job" from *New English Bible* © 1961, 1970 Oxford University Press and Cambridge University Press. Reprinted with permission.

St. Francis of Assisi, "Canticle of Brother Sun" from *Francis and Clare,* Regis J. Armstrong, O.F.M.CAP. and Ignatius C. Brady, O.F.M. © 1982 by the Missionary Society of St. Paul the Apostle in the State of New York. Used by permission of Paulist Press.

St. Thomas Aquinas, *Summa Contra Gentiles,* Dominican Fathers translation. Benziger Brothers 1923. Reprinted with permission.

René Descartes, "Discourse on the Method," *Philosophical Works of Descartes,* Elizabeth S. Haldane and G. R. T. Ross, translators. Reprinted by permission of Cambridge University Press.

From *Wilderness and the American Mind,* by Roderick Nash. Copyright Yale University Press, reprinted with permission.

Julian Simon, *The Ultimate Resource* © 1981 by Princeton University Press. Reprinted by permission of Princeton University Press.

"Preface," "The Ecorealist Manifesto," from *A Moment on the Earth* by Gregg Easterbrook. Copyright © 1995 by Gregg Easterbrook. Used by permission of Viking Penguin, a division of Penguin Books USA Inc.

Joseph Bast, Peter Hill, Richard Rue, "Eco-Sanity" reprinted with permission from *Eco-Sanity: A Common-Sense Guide to Environmentalism* © 1994 The Heartland Institute.

Reprinted from *Beyond the Limits* by Donella Meadows, Dennis Meadows, Jorgen Randers. Copyright © 1992 by Meadows, Meadows, and Randers. With permission from Chelsea Green Publishing Co., White River Junction, Vermont.

"Property Rights" from *Natural Resources: Bureaucratic Myths and Environmental Management* by Richard L. Stroup and John A. Baden. Copyright © 1983 by Pacific Research Institute for Public Policy. Reprinted by permission of Pacific Research Institute for Public Policy, San Francisco, CA.

A. Myrick Freeman, "The Ethical Basis of the Economic View of the Environment." Reprinted by permission of the Center for Values and Social Policy, University of Colorado at Boulder and the author.

Mark Sagoff, "Free Market vs. Libertarian Environmentalism" from *Critical Review,* vol. 6, nos.2–3. Reprinted by permission of the author.

Herman Daly, "Moving to a Steady-State Economy" from *Beyond Growth* by Herman Daly, Beacon Press. Copyright © 1996 by Herman Daly. Reprinted with permission of the author.

Lynn White, "The Historical Roots of Our Ecological Crisis." Copyright © 1967 American Association for the Advancement of Science. Reprinted with permission.

Martin H. Krieger, "What's Wrong with Plastic Trees?" Reprinted with permission of *Science* vol. 179 (Feb. 1973). Copyright © 1973 American Association for the Advancement of Science.

Reprinted from Robert Elliot "Faking Nature," *Inquiry,* vol. 25 (1982) pp. 81–93, by permission of Scandinavian University Press, Oslo, Norway and the author.

Holmes Rolston III, "Does Aesthetic Appreciation of Landscapes Need to be Science-Based?" *The British Journal of Aesthetics,* vol. 35, no. 4, Oct. 1995, pp. 374–385. Reprinted by permission of Oxford University Press.

Rosemary Radford Ruether, "The Biblical Vision of the Ecological Crisis." Copyright © 1978 Christian Century Foundation. Reprinted by permission from November 22, 1978 issue of the Christian Century.

Joel Feinberg, "The Rights of Animals and Unborn Generations" from *Philosophy and Environmental Crisis,* ed. by William Blackstone, University of Georgia Press, 1974. Reprinted by permission.

Peter Singer, "All Animals Are Equal" first published in *Philosophical Exchange* I (1974) pp. 103–116; reprinted by permission of the author. Copyright © 1974 Peter Singer.

"The Case for Animal Rights" by Tom Regan from *In Defense of Animals,* ed. by Peter Singer. Reprinted by permission of Blackwell Publishers.

Should Trees Have Standing? Toward Legal Rights for Natural Objects by Christopher Stone; Tioga Publishing, 1974. Reprinted by permission of Oceana Publications and the author.

From *A Sand County Almanac: With Other Essays on Conservation from Round River* by Aldo Leopold. Copyright © 1949, 1953, 1966, renewed 1977, 1981 by Oxford University Press, Inc. Used by permission of Oxford University Press, Inc.

From J. Baird Callicott, *A Companion to Sand County Almanac: Interpretive and Critical Essays* Copyright © 1987. Reprinted by permission of the University of Wisconsin Press.

Don E. Marietta, Jr., "Environmental Holism and Individuals" from *Environmental Ethics,* vol. 10 (3) pp. 251–258. Reprinted by permission of the author.

Donald Worster, "The Ecology of Order and Chaos" from *The Environmental Review* 14, (1–2) pp. 1–18. Reprinted by permission of the author.

"Assessing Environmental Health Risks" by Ann Misch from *State of the World 1994: A Worldwatch Report on the Progress Toward a Sustainable Society,* by Lester R. Brown, et al., eds. Copyright © 1994 by the Worldwatch Institute. Reprinted by permission of W. W. Norton & Company, Inc.